CT/Woolstons/ £12.50

PERGAMON INTERNATIONAL LIBRARY
of Science, Technology, Engineering and Social Studies

The 1000-volume original paperback library in aid of education,
industrial training and the enjoyment of leisure

Publisher: Robert Maxwell, M.C.

HIGH EXPLOSIVES
AND PROPELLANTS

THE PERGAMON TEXTBOOK
INSPECTION COPY SERVICE

An inspection copy of any book published in the Pergamon International Library will gladly be sent to academic staff without obligation for their consideration for course adoption or recommendation. Copies may be retained for a period of 60 days from receipt and returned if not suitable. When a particular title is adopted or recommended for adoption for class use and the recommendation results in a sale of 12 or more copies, the inspection copy may be retained with our compliments. The Publishers will be pleased to receive suggestions for revised editions and new titles to be published in this important International Library.

HIGH EXPLOSIVES AND PROPELLANTS

S. FORDHAM
Formerly of Nobel's Explosive Co. Ltd.

SECOND EDITION

PERGAMON PRESS
OXFORD · NEW YORK · TORONTO · SYDNEY · PARIS · FRANKFURT

UK	Pergamon Press Ltd., Headington Hill Hall, Oxford OX3 0BW, England
USA	Pergamon Press Inc., Maxwell House, Fairview Park, Elmsford, New York 10523, USA
CANADA	Pergamon of Canada, Suite 104, 150 Consumers Road, Willowdale, Ontario M2J 1P9, Canada
AUSTRALIA	Pergamon Press (Aust.) Pty. Ltd., P.O. Box 544, Potts Point, NSW 2011, Australia
FRANCE	Pergamon Press SARL, 24 rue des Ecoles, 75240 Paris, Cedex 05, France
FEDERAL REPUBLIC OF GERMANY	Pergamon Press GmbH, 6242 Kronberg/Taunus, Pferdstrasse 1, Federal Republic of Germany

First edition 1966

Second edition 1980

British Library Cataloguing in Publication Data

Fordham, Stanley
High explosives and propellants. — 2nd ed.
— (Pergamon international library).
1. Explosives
I. Title
662'.2'0941   TP270   79-40714

ISBN 0-08-023834-3 (Hardcover)
ISBN 0-08-023833-5 (Flexicover)

Printed and bound in Great Britain by
William Clowes (Beccles) Limited, Beccles and London

Contents

Contents

Part III   Application of High Explosives

Part IV   Deflagrating and Propellent Explosives

Preface to the First Edition

THE writer of a book on explosives is immediately faced with a difficult task of selection. He must decide what relative importances to place on military compared with commercial explosives, and on theoretical against practical aspects of the technology.

The quantity of military explosives made in the Second World War exceeded the total ever made for peaceful use by mankind. On this count it could be argued that military explosives should occupy a major portion of this book. On the other hand, wars are fortunately relatively shorter in duration than peace, so that at any particular time a reader is likely to be interested more in commercial explosives than in military.

More important, however, is to consider the contributions—good and bad—which explosives have made to the history of mankind. Any real assessment of this must show that the benefits which explosives have produced far outweigh their misuse in military pursuits. The explosives technologist, who has usually seen and perhaps even experienced the effects of explosives, is the last to want war or to want his products to be used for warlike purposes. It is no accident that Nobel, who founded the modern explosives industry, also founded the Peace Prize associated with his name. In this book the writer has followed his instincts and given pride of place to commercial, beneficial applications of explosives.

The writer has also dealt in greater detail than many authors would on the more fundamental aspects of his subject. He believes that the reader will be more interested in understanding the bases of the design and performance of explosives than in learning details of individual compounds or devices. If readers consider that the balance is wrong, the writer can only plead that there are no comparable books on the subject with which comparison could be made.

In deference to the wishes of the publishers, references in the text have been kept to a minimum and where possible to books thought to be freely available. The writer has, however, included a short bibliography to Chapter 2, because he is not aware of any general reference of recent date which covers the whole ground adequately.

This book is concerned with the British explosives industry. Practice in other countries has been discussed only when the comparison is thought to be of value. That is the intention of this series of books and is not in any way intended to decry products which satisfy well requirements in other, often widely different conditions.

Acknowledgements are gratefully made to the following for permission to reproduce items from other books: Oliver & Boyd—Fig. 4.1, Newnes—Fig. 7.2; I.C.I. Ltd.—Figs. 14.8 and 14.9; Elsevier—Fig. 19.5; Interscience Publications—Table 17.1; Temple Press—Table 19.3.

Thanks are also due to the writer's many colleagues in the Nobel Division of Imperial Chemical Industries Ltd. for helping, knowingly and unknowingly, in the preparation of this book.

S.F.

Preface to the Second Edition

IN PREPARING the second edition of this book the opportunity has been taken to add sections on slurry explosives and a short account of "Nonel" fuse. Parts which were badly out of date have been modernised, and the opportunity taken to correct some errors and ambiguities. Thanks are again due to my former colleagues in NEC.

S.F.

Please Read This

Do not experiment with explosives or pyrotechnics.

In this volume a considerable amount of information is given on methods of making explosives and pyrotechnics. The book, however, does not attempt to say how these manufactures can be carried out with safety. The writer and publisher would be most distressed if this text led to a single accident by causing any reader to do experiments on his own.

In this country all preparation of explosives, fireworks, rockets and similar devices is illegal unless carried out in a duly authorised establishment.

Throughout the world explosives manufacturers have amassed many years of experience and have spent many millions of pounds to ensure as far as possible the safety of those working for them with explosives. Even so, accidents still occur with distressing injury and loss of life. Where these companies cannot succeed the amateur would be foolish to try.

Do not experiment with explosives—the odds are too much against you.

CHAPTER 1

Introduction

AN EXPLOSION occurs when energy previously confined is suddenly released to affect the surroundings. Small explosions, like the bursting of a toy balloon, are familiar and innocuous, but large-scale explosions, like an atomic bomb, are rare and usually disastrous. Between these two extremes lie the commercial and conventional military fields where explosions are produced on a limited scale to cause specific effects. It is with explosions of this intermediate scale that this book is concerned.

It is unfortunately true to say that the views which most people hold on explosives stem either from first-hand experience of the effects of explosives used during times of war, or from reports of these effects. For military purposes explosives are required to cause destruction and are used in quantities so large or in such a fashion that destruction is inevitable. As a result, the impression is given of an overwhelming force causing uncontrolled devastation.

Yet, in truth, explosives can be used as a controlled and rather precise means of applying energy to a particular system. Many tons of explosive may be used in a single blast at a quarry face, yet the only visible effect will be for the face of that quarry to slump to the ground. It would be bad practice indeed if rocks from that quarry face were thrown any distance across the floor of the quarry or the neighbouring countryside. When an explosive is fired on the surface of a piece of steel, it will harden that steel to a predetermined depth without either breaking the steel or causing any noticeable deformation. The design and application of explosives is a science and explosives are as capable of control as are other products of industry.

Although all explosions are sudden releases of energy, the converse is not true and not all sudden releases of energy are explosions. By common

1

consent, the term explosive is defined to exclude such items as bottles of compressed gas, even though these are capable of exploding on rupture. For the purposes of this book an explosive will be taken to mean a substance or mixture of substances which is in itself capable (1) of producing a quantity of gas under high pressure and (2) of being able to produce this gas so rapidly under certain conditions (not necessarily those of practical use) that the surroundings are subjected to a strong dynamic stress.

The burning of oil in a lamp is a slow process, the rate being determined by the need for the oil to evaporate and for the vapour to mix with the surrounding air to form a combustible mixture. To speed up the rate of combustion, it is necessary to disperse the oil in air before ignition. In a motor car, petrol in the carburettor is mixed with the right amount of air so that when ignited in the cylinder it explodes. Such gaseous mixtures are effective explosives but suffer from many disadvantages, of which the most important is the small amount of power available from any given volume. To obtain a better power ratio it is necessary to use solids or liquids.

The first step in producing more rapid combustion in a condensed phase is to provide a solid which will replace oxygen from the air in supporting combustion. The use of nitrates for this purpose has a long history and it is probable that the stories of old Chinese explosives and of Greek fire relate to combustible mixtures to which nitrates had been added to make their reaction more intense. The first real record of an explosive, however, is the discovery of gunpowder, usually ascribed to Roger Bacon. Realising the possible uses to which his discovery could be put, Bacon concealed it in cypher, and gunpowder was rediscovered in Germany by Schwarz. Gunpowder, or blackpowder as it is now called, consists of a mixture of potassium nitrate, charcoal and sulphur very intimately ground together. It is readily ignited, even in complete absence of air, and then burns very rapidly. Moreover, if it is burned in a confined space, as in a borehole or a military shell, then as the pressure increases the rate of burning also increases to a high value. If a charge in a borehole is ignited at one end, the flame can propagate at a rate of several hundred metres a second.

Even more rapid reaction can be produced if oxygen and fuel are provided in a single chemical molecule. The discovery of nitroglycerine by Sobrero led to the first product of this type to achieve commercial importance. Nitroglycerine contains enough oxygen to burn all its own carbon and hydrogen. It is, therefore, capable of an extremely rapid combustion. In practice, however, combustion of this sort is unstable and readily turns into a form of reaction known as detonation. This process of detonation of nitroglycerine can best be regarded, qualitatively, as the passage through the material of a sudden wave of high pressure and temperature which causes the molecules to break down into fragments which later recombine to give the ultimate explosion products. That the process is more vigorous than combustion is shown by the high speed of 8000 m s^{-1} at which it propagates; this speed is also independent of the pressure of surrounding gas. Explosives which detonate like nitroglycerine are known as high explosives.

Nitroglycerine and other high explosives of this type are difficult to initiate into detonation simply by the use of a flame. Mercury fulminate, discovered by Howard, is an explosive of relatively low power which can, however, always be relied on to detonate when ignited by a flame. Explosives like mercury fulminate are known as initiating explosives.

Although many more modern explosives have been added to the few mentioned above, they all belong to one of the three types described, namely,

1. Deflagrating (or propellent) explosives.
2. High explosives (sometimes called secondary explosives).
3. Initiating explosives (sometimes called primary explosives).

In Great Britain, manufacture of gunpowder probably started in the 14th century. By the 16th century there was certainly manufacture at a number of sites, both privately and by the Government. The Royal Gunpowder Factory at Waltham Abbey dates from this period. Gunpowder factories were best placed near forests, to provide charcoal, and near water power, to drive the mills. Kent and the Lake District became important centres. The invention of the safety fuse by Bickford in 1831 led to its manufacture at Tuckingmill in Cornwall.

Guncotton was made at Faversham in 1847, but manufacture ceased after a serious explosion. It was nearly twenty years before manufacture was recommenced, privately at Stowmarket and also at Waltham Abbey. The initial uses were military.

The starting point of the present British commercial explosives industry was the formation of the British Dynamite Company in 1871 by Alfred Nobel and a group of Glasgow business men. Ardeer factory in Ayrshire was built, and commenced operation in January 1873. Detonators were manufactured from 1877 in a factory at Westquarter in Stirlingshire. TNT was manufactured at Ardeer from 1907. The later history of the company is too complex to be given in detail here. It is sufficient to say that by 1926, as Nobel Industries Ltd., it had acquired many interests other than explosives. In 1926 Nobel Industries Ltd., with Brunner Mond Co., the United Alkali Co., and the British Dyestuffs Corporation, merged to form Imperial Chemical Industries Ltd. At the present time Nobel's Explosives Co. Ltd., a wholly owned subsidiary of I.C.I., has responsibility for the manufacture, distribution and sale of explosives and accessories.

N.E.C. is the major manufacturer with a complete range of explosives and accessories, with factories in Scotland, Wales and England. Explosives and Chemical Products Ltd., with factories in England, is the other manufacturer of explosives for sale. The major commercial manufacturer of ammunition is Imperial Metal Industries (Kynoch) Ltd. at Witton near Birmingham. The British Government has of course a number of Royal Ordnance Factories and establishments to cover all aspects of military explosives.

Explosives of all types are made for commercial and military purposes in many countries throughout the world. It is, however, difficult to obtain any figures which give a worthwhile idea of the magnitude of the explosives industry. Military explosives are usually made under conditions of secrecy and no figures of output are published. Even for commercial explosives published figures are scanty and vary considerably from country to country. Data which are available are given in Table 1.1.

TABLE 1.1   *Annual Production of Commercial Explosives, 1977*

Country	Tonnes per annum
U.S.A.	1 680 000[a]
German Federal Republic	59 000
France	56 000
Spain	43 000
Sweden (1975)	31 000
Greece (1973)	21 000
Italy	16 000
Portugal	5 000

[a] Consumption.

The largest commercial explosives factory in the world is at Modderfontein in South Africa.

The most complete set of statistics is that published in the United States of America, which showed a total consumption of industrial explosives in 1977 of 1 680 000 tonnes. Details of the types of explosives consumed and the industries using the products are given in Tables 1.2 and 1.3 respectively.

TABLE 1.2   *U.S.A. Consumption of Industrial Explosives, 1977*

Type of explosive	Tonnes
Permissibles [a]	22 000
Other high explosives	85 000
Water gels and slurries	144 000
Blasting agents and ANFO	1 432 000
Total	1 683 000

[a] U.S.A. equivalent of British "Permitted"

TABLE 1.3   *U.S.A. Explosives Consumption by Industry, 1977*

Industry	Tonnes
Coal mining	950 000
Metal mining	202 000
Quarrying and non-metal mining	237 000
Construction work	159 000
All other purposes	135 000
Total	1 683 000

No comparable figures are available for Great Britain. Deep coal mines currently record firing 13 million shots per annum using about 8 000 tonnes, but quarrying is certainly a larger user of explosives. A rough guess of the British market for explosives and pyrotechnics could be given as £50 million. There is an appreciable export market; the U.K. Chemical Industry Statistics Handbook gives exports of explosives and pyrotechnics for 1977 as £34·7 million.

The following figures give an indication of British prices. A user of nitroglycerine explosives will pay 55–110p per kg according to type. To fire a charge of explosive, whatever its size, he will require a plain detonator at 3p (with perhaps 10p worth of safety fuse), an electric detonator at 16p, or a delay detonator at 25p. These are just a few possible figures from a total range of several thousand products!

Although these costs are not high, they do represent a greater expenditure per unit of energy than more conventional means. For example, the following table gives approximate relative costs of energy from explosives and from other well-known sources (neglecting efficiency of use for mechanical purposes).

TABLE 1.4   *Relative Costs of Energy from Various Sources*

Source of energy	Relative cost
Nitroglycerine gelatine explosive	50
Electricity	4
Fuel oil	1

The particular advantage of explosives is the rapid generation of energy. Thus a single cartridge of blasting gelatine 3 cm in diameter produces on detonation about 60 000 MW—appreciably more than the total electric power station capacity of the United Kingdom.

Explosives are used mainly for doing mechanical work and particularly for breaking rock and coal. The advantages in cost of electrical and similar forms of energy mean that there is a continued incentive for users to replace explosives by mechanical methods of working. This is particularly noticeable in the coal industry, where mechanical operation offers other advantages at the same time. The amount of explosive used per ton of coal is therefore diminishing, and with production of coal in Great Britain remaining static, coal mining explosives are in smaller demand now than in previous years. Throughout the world, also, the conventional explosives industry is suffering from increasing competition from cheap mixtures of ammonium nitrate and oil which can be made to detonate and which in some countries may be mixed by the user on the site of operation. Such mixtures have replaced conventional explosives on many sites, particularly in North America. A qualitative picture of the trend of the explosives industry is given in Fig. 1.1, from which it will be seen that the total usage has increased (although the rate has been less than the general expansion of the world economy). Conventional explosives, on the other hand, have passed through a peak and the industry is operating at a lower level than previously. There are signs that this level is now being kept steady. World usage of detonators and other accessories is thought to have shown a general increase over the years.

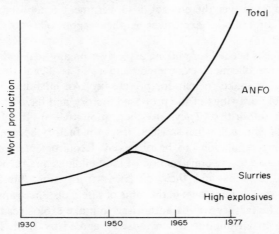

FIG. 1.1. Trend of world production of commercial explosives.

In all countries the manufacture, transport and sale of explosives are strictly controlled by law. The nature of the regulations does, however, vary considerably throughout the world. In Communist countries, commercial and military explosives are made in the same factories and under precisely the same conditions by the State. In the rest of the world, manufacture of commercial explosives is mainly by private firms, as is indeed the case in Britain. The operations of these firms is however closely controlled by the Government, whilst conditions of transport are increasingly becoming matters of international concern. In the U.S.A., manufacture is again by private companies, but the conditions are laid down by the individual states and only inter-state transport is regulated by the Federal Government. In almost all countries, governments maintain their own research and testing facilities, and in most countries private industry also carries out research and development work on explosives.

In Great Britain manufacture, storage and transport of industrial explosives are governed not only by the general provisions of the Health and Safety at Work Act but more specifically by the Explosives Acts of 1865 and 1923. There are also numerous Statutory Instruments made under the 1865 Act. Administration of these regulations is by H.M. Inspectors of Explosives who form part of the Health and Safety Executive. However the control of Government Establishments and the transport of military explosives is the responsibility of the Ministry concerned.

A licence is needed for making explosives on any scale other than small amounts for chemical experiment and so it is illegal to manufacture explosives or make rockets for private use. An intending manufacturer must supply drawings of the proposed factory and have these agreed by the local authorities. They are then submitted to the Inspectors of Explosives who will, when satisfied, issue the factory licence and indicate the working regulations to be observed. Requirements which must be satisfied relate to the construction of the buildings and their surrounding protective mounds and their distances from other buildings, public highways etc., as laid down in the table of safety distances approved under the Act. The factory will be authorised to make explosives in a Schedule according to Definitions which must be agreed by H.M. Inspectors of Explosives. Should a new explosive be considered for manufacture, then an Authorised Definition must be approved beforehand. Such approvals are based on tests essentially of stability and safety in handling.

Special authorisations are available for manufacture at the site of use of ammonium nitrate/fuel oil and certain slurry explosives provided that the equipment used and its position on the site are suitable.

The Explosives Acts and Orders made under them also stipulate how explosives may be packed for storage and transport. Blasting explosives are packed in an inner wrapper which prevents spillage and gives protection against moisture and then in an outer wrapper which provides strength in handling. In general the outer case is nowadays of fibreboard, but wooden cases are still used in some parts of the world.

The transport of explosives is covered strictly by regulation and this is becoming increasingly international in character. A committee of the United Nations Organisation has produced an improved classification of explosives and specified agreed methods of packing suited to each item. These proposals now form the basis of the conditions of transport by air laid down by IATA (International Air Transport Association) for the limited range of explosives which may be carried on aircraft. The same proposals are used for transport by sea as laid down by IMCO (Intergovernmental Maritime Consultative Organisation) and the British "Blue Book" has been adapted to correspond. It is hoped that these leads will be followed in the near future by the international bodies concerned with transport by road, rail and inland waters. Meantime, in Britain, road transport of more than 50 kg of blasting explosive requires specially designed vehicles which may carry up to a maximum of 4 tonnes. In rail transport suitable wagons are required and the load must not exceed 20 tonnes.

Regulations concerning the manufacture and use of explosives appear complicated, but their necessity is obvious. Experience and goodwill are however always essential to ensure the smooth running of the system and thus public safety.

References

Guide to the Explosives Act. H.M.S.O., London, 4th ed. 1941.
PARTINGTON, J. R., *A History of Greek Fire and Gunpowder.* Heffer, Cambridge, 1960.
WATTS, H. E., *The Law Relating to Explosives.* Griffin, London, 1954.

Part I.   High Explosives

CHAPTER 2

General Principles

History

The first explosive used was not a high explosive but gunpowder, discovered in the 13th century and rediscovered in the 14th century. It was used for military purposes from the 14th century onwards, and first introduced into blasting practice in Hungary in the 17th century. It soon spread to Germany and to Britain. Gunpowder, as noted above, is a deflagrating explosive and has long been overshadowed in importance by high explosives.

The first high explosive discovered was probably nitrocellulose, in the period 1833 to 1846, but its development was long delayed by difficulties in obtaining a stable product. The two major discoveries in this field were of nitroglycerine by Sobrero in 1847 and TNT by Wilbrand in 1863. Of these, the first to attain commercial importance was nitroglycerine.

Sobrero early recognised the dangers of handling nitroglycerine and it was only the tenacity of Alfred Nobel which finally succeeded in making nitroglycerine a commercially useful material. Alfred Nobel followed his father's interest in explosives and in spite of a number of explosions and accidents, including one which killed his younger brother and indirectly his father, he devised in a laboratory on a barge a safe method of producing nitroglycerine. Equally important, he realised that nitroglycerine, unlike gunpowder, could not be set off by flame, but needed a shock to cause effective initiation. He therefore invented first his patent detonator incorporating gunpowder, and then finally the modern detonator containing mercury fulminate. This he introduced in 1865. He realised, also, the hazards involved in handling liquid nitroglycerine and invented first guhr dynamite using kieselguhr (diatomaceous earth), and second blasting gelatine, both of which are safe and highly powerful detonating explosives. He early realised, and acquired patents for, the inclusion in explosives of oxidising salts such as nitrates together with combustibles. By 1875 Nobel had completed the invention of ordinary blasting explosives which were to dominate the field well into the 20th century. Nobel was a prolific inventor and a man whose life story makes fascinating reading.

From the days of Nobel to about 1950 the scientific basis of commercial explosives remained relatively unchanged, although continuous and numerous improvements in manufacturing methods occurred throughout the world. There were, however, many advances in military explosives, note of which will be made later. These advances were, of course, largely due to the two world wars, which occurred since the death of Alfred Nobel. There were also many advances in the development of permitted explosives designed for use in gassy coal mines.

In the 1950's a sudden and dramatic change affected the explosives industry in many parts of the world. This was the introduction in the U.S.A., Sweden and Canada of ammonium nitrate sensitised with fuel oil as a major blasting explosive. A slower but also important change started in the 1960's with the development of slurry explosives in the U.S.A., Canada and other countries.

The Nature of Detonation

The process of detonation in an explosive, the blasting effects of such an explosive in coal or rock, and the destructive effects of military high explosives all depend on the operation of shock waves. It is, therefore, intended to give here an elementary account of shock waves in inert media and in explosives before discussing individual high explosives.

A compression wave of low intensity is well known in ordinary sound waves in the air, or in other media. Sound is propagated with a velocity determined by the following equation:

$$c^2 = \frac{\partial p}{\partial \rho}\bigg|_S \qquad (1)$$

where c is the velocity of the sound wave, p is the pressure, ρ is the density and S is the entropy (which remains constant). Qualitatively it is convenient to remember that the velocity of sound increases as the compressibility of the medium decreases.

In the case of sound waves, which are of very low intensity, the pressure and density of the medium remain effectively constant throughout the process. Therefore, all parts of a sound wave are transmitted at the same velocity, so that a sinusoidal (sine) wave, for example, remains sinusoidal indefinitely during propagation.

With shock waves it can no longer be assumed that pressure and density remain constant. Indeed, at the peak of a strong shock wave the pressure can be many thousands of atmospheres and the density appreciably increased. Under such conditions the velocity no longer is that of a sound wave. In practice, as pressure or density is increased, the compressibility decreases so that the velocity of propagation of the disturbance increases. If we can imagine an intense sinusoidal half-wave generated in a medium, then the velocity of propagation of the peak of the wave where the material is at high density would be greater than the velocity of the front of the wave where the material is almost at its original condition. Therefore, the peak would overtake the front and the shape of the wave would alter as shown in Fig. 2.1, until ultimately the wave form becomes a sudden and discontinuous jump to a high pressure followed by a gradual fall. This is the typical and inevitable profile of a shock wave in an inert medium. As transmission continues, however, losses gradually reduce the peak pressure until ultimately the shock wave degenerates into an ordinary sound wave.

FIG. 2.1. Development of a shock wave.

Shock waves, like all other waves, undergo the normal processes of reflection and refraction in passing from one medium to another. The case of reflection is of particular importance. Three possibilities are shown in Figs. 2.2(a), 2.2(b) and 2.2(c). Figure 2.2(a) shows the position where a shock wave in a non-compressible (dense) medium meets a boundary with a highly compressible (light) medium. The shock wave is reflected at the

High Explosives

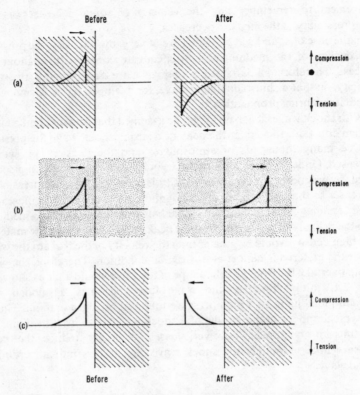

FIG. 2.2. Reflection of shock waves. (a) Shock wave in denser medium.
(b) Matched media. (c) Shock wave in lighter medium.

interface as a rarefaction wave, provided that the tensile strength of the medium makes this possible. Figure 2.2(c) shows the extreme contrast to this where a wave in a compressible medium meets a medium of low compressibility. In this case again the wave is reflected, but as a compression wave rather than a rarefaction wave. Case 2.2(b) is the intermediate case where the media have similar compressibility and in this case, and this case alone, is the shock wave transmitted across the boundary without alteration.

A shock wave in an inert medium is not propagated indefinitely without change because rarefaction waves can always overtake the pressure wave and reduce the peak pressure until the conditions of a sound wave are reached. For a stable shock wave to be maintained, a source of energy must be available which will enable the wave to be propagated without rarefaction waves occurring. This is what happens when detonation occurs in an explosive.

When a detonation wave passes through an explosive, the first effect is compression of the explosive to a high density. This is the shock wave itself. Then reaction occurs and the explosive is changed into gaseous products at high temperature. These reaction products act as a continuously generated piston which enables the shock wave to be propagated at a constant velocity. The probable structure of the detonation zone is illustrated in Fig. 2.3.

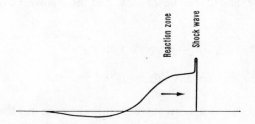

FIG. 2.3. Structure of detonation wave.

Mathematically, the following three equations can be written down representing respectively the application of the laws of conservation of mass, momentum and energy.

Mass:
$$\frac{D}{v_1} = \frac{D - W_2}{v_2} \tag{2}$$

Momentum:
$$\frac{D^2}{v_1} + p_1 = \frac{(D - W_2)^2}{v_2} + p_2 \tag{3}$$

Energy: $\quad E_1 + D^2 + p_1 v_1 = E_2 + \tfrac{1}{2}(D - W_2)^2 + p_2 v_2 \tag{4}$

where D = velocity of detonation,

W_2 = velocity of material behind the wave, relative to that in front,

v = specific volume,

p = pressure,

E = specific internal energy,

and subscripts 1 and 2 relate to the initial and final states of the explosive respectively.

It will be noted that E_1 is the specific internal energy of the unreacted explosive, whereas E_2 is the specific internal energy of the explosion products at pressure p_2 and specific volume v_2. These equations are deduced from physical laws only and are independent of the nature or course of the chemical reaction involved.

Equations (2) and (3) can be solved to give the following equations for D and for W_2:

$$D = v_1 \sqrt{[(p_2 - p_1)/(v_1 - v_2)]} \tag{5}$$

$$W_2 = (v_1 - v_2) \sqrt{[(p_2 - p_1)/(v_1 - v_2)]} \tag{6}$$

It will be noted that as p_2 is greater than p_1, v_2 must be less than v_1, and W_2 (known as the streaming velocity) is positive, meaning that the explosion products travel in the same direction as the detonation wave. This positive streaming velocity is a characteristic and identifying property of a detonation wave.

These are the basic equations of the hydrodynamic theory of detonation. If p_2 and v_2 can be determined, they enable the remaining features of the detonation wave to be calculated. Unfortunately p_2 and v_2 relate to conditions in the detonation wave and not to the lower pressure conditions which the explosion products would reach at equilibrium in, for example, a closed vessel. Therefore, further calculations are needed to determine p_2 and v_2.

In studying detonation, it is convenient to use diagrams of the type illustrated in Fig. 2.4. In these, the pressure is plotted over the specific volume. The original explosive corresponds to point A. The reaction products are represented by the curve BC, known as the Rankine–

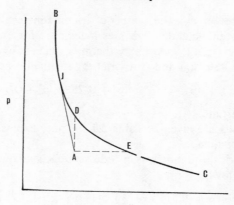

FIG. 2.4. Rankine–Hugoniot diagram.

Hugoniot curve, which represents all possible points in the area which satisfy the chemical equilibria and entropy conditions of the explosive reaction products. It is a locus which is defined by, and can be calculated from, the energy equation — equation (4). (p_2, v_2) must therefore be a point on this line. As noted above, v_2 must be less than v_1, so the point must lie on the part of the curve BD. To determine exactly where the point (p_2, v_2) lies in the curve BD requires difficult arguments which are not reproduced here, but which show that the point can be obtained by drawing a tangent from A to the curve; this is known as the Chapman–Jouguet condition, and the tangent is the Raleigh line. Thus the point J shown in Fig. 2.4 gives p_2 and v_2. Substituting these values in equation (5) gives the calculated detonation velocity.

The major difficulty in applying this hydrodynamic theory of detonation to practical cases lies in the calculation of E_2, the specific internal energy of the explosion products immediately behind the detonation front, without which the Rankine–Hugoniot curve cannot be drawn. The calculations require a knowledge of the equation of state of the detonation products and also a full knowledge of the chemical equilibria involved, both at very high temperatures and pressures. The first equation of state used was the Abel equation

$$p_2(v_2 - a) = n_2 R T_2 \qquad (7)$$

where a is a constant, n_2 is the number of gramme-molecules of gas, and R is the gas constant, but this becomes inaccurate for explosives with a density above $0 \cdot 1$ g ml^{-1}. Probably the most successful equation so far used is that by Paterson and is a virial type equation as follows

$$p_2 = n_2 RT(1 + b\rho + 0\cdot625b^2 \rho^2 + 0\cdot287b^3 \rho^3 + 0\cdot193b^4 \rho^4) \qquad (8)$$

where b is a constant.

Typical results are given in Table 2.1 for pure explosives which give only gaseous products.

A semi-empirical equation was introduced by Kistiakowsky and Wilson and took the following form:

$$p_2 v_2 = n_2 RT(1-xe^{\beta x}) \quad \text{where} \quad x = \rho kT^{-\alpha}$$

TABLE 2.1   *Calculated Properties of Explosive Compounds*

Compound	Density (g ml^{-1})	Energy (J g^{-1})	Streaming velocity (m s^{-1})	Velocity of detonation (m s^{-1})	
				Calculated	Observed
Nitroglycerine	1·60	6283	1550	8060	8000
PETN	1·50	5881	1550	8150	7600
Tetryl	1·50	5810	1320	7550	7300
TNT	1·50	5413	1140	6480	6700
Nitroguanidine	0·60	2658	1027	4040	3850
Ammonium nitrate	1·00	1580	832	3460	—

The constants α and β were first chosen to give the best fit to experimental detonation velocity measurements for a wide variety of materials. They have more recently been revised by Cowan and Fickett to give better agreement with experimentally measured detonation pressures. For numerous other approaches to the problem of the equation of state under detonation conditions, readers are referred to the book by Cook and a paper by Jacobs.

Commercial explosives frequently contain salts, or give other solid residues. In calculations these cause difficulties, as it is not certain whether solid ingredients reach equilibrium with the explosion products. In the calculations it is possible either to assume thermal equilibrium, or to assume no heat transfer to the solids. In Table 2.2 are given the results by Paterson of calculations on a number of commercial explosives, when necessary according to each method.

TABLE 2.2 *Calculated Properties of Commercial Explosives*

Explosive type	Energy	Streaming Velocity	Velocity of detonation (m s⁻¹)	
	$(J g^{-1})$	$(m s^{-1})$	Calculated	Maximum observed
Blasting gelatine	6584	1540	7900	7800
Gelatine	5559	1290	6520	6600
Ammon gelignite	4218	1220	6310	6000
Permitted gelignite (P1)	3277	1065[a]	5850[a]	5800
		900[b]	5040[b]	
Nitroglycerine powder	3737	1090	4270	—
Permitted powder (P3)	2132	830[a]	3490[a]	3000
		730[b]	3060[b]	
TNT powder	3984	1172	5060	5250

[a] Salt remaining cold.
[b] Salt in thermal equilibrium.

Tables 2.1 and 2.2 show that theory enables detonation velocities to be calculated in close agreement with those observed experimentally. This, unfortunately, is not a critical test of the theory as velocities when calculated are rather insensitive to the nature of the equation of state used. A better test would be to calculate the peak pressures, densities and temperatures encountered in detonation, and compare these with experimental results. The major difficulties here are experimental. Attempts to measure temperatures in the detonation zone have not been very successful, but better results have been obtained in the measurement of densities and pressures. Schall introduced density measurement by very short X-ray flash radiography and showed that TNT at an initial density of 1·50 increased 22% in density in the detonation wave. More recently detonation pressures have been measured by Duff and Houston using a method (introduced by Goranson) in which the pressure is deduced from the velocity imparted to a metal plate placed at the end of the column of explosive. Using this method, for example, Deal obtains the detonation pressures for some military explosives recorded in Table 2.3. More

TABLE 2.3   *Detonation Pressures of Military Explosives*

Explosive	Density (g ml^{-1})	Detonation velocity (m s^{-1})	Streaming velocity (m s^{-1})	Detonation pressure (10^8 Pa)
RDX	1·767	8639	2213	337·9
TNT	1·637	6942	1664	189·1
RDX/TNT (77/23)	1·743	8252	2173	312·5

recently Paterson has measured detonation pressures of some commercial explosives using a similar method with the results given in Table 2.4. Comparison of these measured pressures with those calculated by the hydrodynamic theory show that so far only equations of state containing empirical constants give satisfactory agreement.

TABLE 2.4   *Detonation Pressures of Commercial Explosives (3·18 cm diam.)*

Name	Type	Detonation pressure (10^8 Pa)
Polar blasting gelatine	Gelatine	160 (high velocity) 19 (low velocity)
Belex	Semi-gelatine	31
Polar Ajax	P1 gelatine	13
Unigel	P3 gelatine	9
Polar Viking	P1 powder	12
Unifrax	P3 powder	6
Carribel	P4 powder	5
ANFO (2% oil, 5·08 cm diam.)		11

The simple theory outlined here can apply only to charges which are infinite in size, or which are so large that lateral losses in the reaction zone are negligible. In many practical cases the cartridge diameter is no longer sufficiently large in comparison with the thickness of the reaction zone for this assumption to be true and lateral losses lead to observation of reduced detonation velocities. With many explosives, if the diameter is decreased below a certain critical diameter, losses are so great that detonation no longer can occur. Typical examples of the variation of observed detonation velocity with cartridge diameter are given in Table 2.5.

TABLE 2.5   *Variation of Velocity of Detonation with Cartridge Diameter*

Diameter (cm)	Velocity of detonation (m s⁻¹)	
	TNT powder	Nitroglycerine powder
1·9	3190	1830
3·2	3680	2250
5·1	4060	2610
6·4	4030	—
7·6	4100	3150
10·2	4560	3290
12·7	—	3440
15·2	4815	—
21·6	—	3920

Extension of the hydrodynamic theory to explain the variation of detonation velocity with cartridge diameter takes place in two stages. First, the structure of the reaction zone is studied to allow for the fact that the chemical reaction takes place in a finite time: secondly, the effect of lateral losses on these reactions is studied. A simplified case neglecting the effects of heat conduction or diffusion and of viscosity is shown in Fig. 2.5. The Rankine–Hugoniot curves for the unreacted explosive and for the detonation products are shown, together with the Raleigh line. In the reaction zone the explosive is suddenly compressed from its initial state at

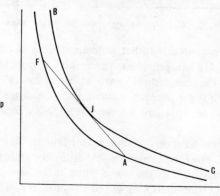

FIG. 2.5. Rankine–Hugoniot diagram for detonation.

point A to a state at point F. Reaction then occurs with the pressure falling along the line FJ until at the end of the reaction zone a point J is reached and stable detonation conditions arrived at. The actual shape of the pressure wave is shown in Fig. 2.6 where the so-called Neumann peak of twice the detonation pressure is shown. A more complete solution by Hirschfelder and Curtiss shows that the effects of viscosity and thermal transfer by diffusion and conduction are to reduce the magnitude of the Neumann peak and the steepness of the initial increase in pressure. One such solution is sketched in Fig. 2.6 where a first order chemical reaction has been assumed. The shape will depend, however, on the magnitude of the heat losses and in some cases the initial peak need not occur at all. Some evidence for the existence of the Neumann peak has been obtained as, for example, by Duff and Houston. In their measurements the peak pressure reached was 42% above p_2.

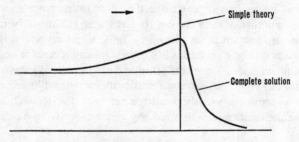

FIG. 2.6. Calculated shape of detonation wave head.

Cook has propounded a rather different theory of the nature of the reaction zone. He emphasises that the demonstrable electrical conductivity of the detonation wave is evidence of a high thermal conductivity. Both these effects are ascribed to ionisation of the explosion products. In terms of the reaction zone, this implies a steady pressure with no peaks.

Theoretical considerations of charges of limited diameter have taken one of two forms. The former assumes that the effects are best described as a result of the curvature of the wave front in the explosive (this can be demonstrated experimentally), or of reduction of the driving pressure by lateral expansion. Solutions of this type have been given by Eyring and co-workers and by Jones. Alternatively, the variation in velocity of detonation can be explained as a result of incomplete reaction during the actual detonation wave. Explanations of this type have been put forward by Cook and also by Hino. The following equations illustrate the dependence of detonation velocity on charge diameter according to the various theories.

Eyring, Powell, Duffey and Parlin:

$$D/D_0 = 1 - z/d \quad \text{for } z/d < 0.25$$

Jones (cf. Jacobs):

$$(D/D_0)^2 = 1 - 3.2 \, (z/d)^2$$

Cook:

$$(D/D_0)^2 = 1 - (1 - \frac{4ad'}{3tD})^3$$

where d = diameter of charge,
$\quad z$ = reaction zone thickness,
$\quad a$ = constant,
$\quad d' = d - 0.6$ cm,
$\quad t$ = reaction time.

The thickness of the reaction zone in high explosives is usually in the range 1–10 mm.

Liquid nitroglycerine and gelatinous explosives made from it can exhibit two stable velocities of detonation, of approximately 2000 and 8000 m s^{-1} respectively. The phenomenon is complicated by the occurrence of air bubbles in such explosives and has not yet been completely explained.

The theory of detonation has also been extended to study the process of initiation of reaction by the commonest means used in practice, namely, by the shock wave arising from another high explosive. Campbell, Davis and Travis have studied the initiation by plane shock waves of homogeneous explosives, particularly nitromethane. Initiation occurs at the boundary of the explosive after an induction period which is of the order of a microsecond and which depends markedly on initial temperature. During the induction period the shock wave has proceeded through the explosive and compressed it. The detonation initially in compressed explosive has a velocity some 10% above normal, but the detonation soon overtakes the shock front and the detonation velocity falls abruptly to the normal value. The process is shown in Fig. 2.7(a). These results can be explained adequately on the basis of thermal explosion theory as developed, for example, by Hubbard and Johnson.

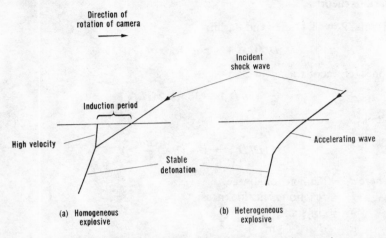

FIG. 2.7. Initiation of detonation in explosives, as shown in a rotating mirror camera.

Even slight lack of homogeneity in explosives or shock wave form leads to an alternative mode of initiation characteristic of ordinary military or commercial explosives in solid or gelatinous form. This process has been studied by Campbell, Davis, Ramsay and Travis and is depicted in Fig. 2.7(b). The shock wave first proceeds through the explosive with slow acceleration. After a few centimetres there is abrupt transition to detonation at normal velocity. The distance at which transition occurs is independent of temperature but can be reduced by (i) increasing the pressure of initiating shock, (ii) increasing the fineness of the active ingredients, or (iii) for powder explosives reducing the loading density. Deliberate introduction of centres of heterogeneity, such as air bubbles or barium sulphate, is well known to reduce the distance. These results cannot be explained by general thermal explosion as there is insufficient energy to give the required temperature rise. They can be explained on the basis of reaction at local centres or "hot spots".

A similar explanation had already been given for the initiation of explosives by impact and friction. These phenomena have been extensively studied, particularly by Bowden and his school. Their work demonstrated two particularly important modes of initiation:

1. By adiabatic compression of gas. This is particularly noticeable in liquid explosives such as nitroglycerine, where even the moderate compression of small gas bubbles can readily lead to initiation of the explosive.

2. By the development of hot spots by friction. This is shown particularly by the effect of added materials of a gritty nature. For initiation to occur, the melting point of the grit must be above a limiting temperature dependent on the explosive. Initiation is favoured by a low thermal conductivity and also by a high hardness value.

Burning of high explosives can on occasion lead to detonation, particularly if large quantities are involved. In close bomb tests Wachtell, McKnight and Shulman find a reproducible pressure above which burning becomes progressively faster than would be expected from strand burner tests (see p. 179). For TNT this pressure is about 45 MPa, but for propellants much higher. They regard this characteristic pressure as a measure of the liability to detonation.

The external effects of a detonating explosive are of two types, due in the first place to the shock wave and in the second place to the expansion of the detonation products. The detonation wave reaching the end of a cartridge is propagated into the further medium, whether this be air, rock, or water. Because of the positive streaming velocity of the detonation wave, the effects are particularly strong at the terminal end of the cartridge. Although a cartridge detonated in air can produce a shock wave with a velocity even higher than that of the explosive itself, the energy in this wave is relatively small. If the explosive completely fills a metal case, or a borehole in rock, much greater amounts of energy can penetrate into the surrounding media. In either case, however, by far the greater proportion of the energy of the explosive is liberated during the expansion of the gaseous products and it is in general this work of expansion which causes the explosive to have its desired effect. The amount of energy available from an explosive for this purpose can be calculated by integrating the mechanical work done during the expansion of the products to atmospheric pressure. More often, however, it is measured by some practical test (cf. Chapter 6).

A special simple case of work done by an explosive is observed when charges are fired under water. The observed effects are first a shock wave which is transmitted through the water, secondly an expanding bubble of gas. This bubble expands to a maximum size and then collapses, to expand again and vibrate until the energy has been dissipated. Rather more than half the energy remaining available in the gas is transmitted through the water during each expansion of the bubble. These phenomena are described in detail by Cole.

References

Historical

BERGENGREN, E., *Alfred Nobel*. Nelson, London, 1960.
MACDONALD, G. W., *Historical Papers on Modern Explosives*. Whittaker, London, 1912.
SCHÜCK, H. and SOHLMAN, R., *The Life of Alfred Nobel*. Heinemann, London, 1929.

General

BOWDEN, F. P. and YOFFE, A. D., *The Initiation and Growth of Explosion in Liquids and Solids*. University Press, Cambridge, 1952.
COLE, R. H., *Underwater Explosion*. Princeton University Press, New Jersey, 1948.
COOK, M. A., *The Science of High Explosives*. Reinhold, New York, 1958.
JACOBS, S. J., *Am. Rocket Soc. J.* **30,** 151 (1960).
TAYLOR, J., *Detonation in Condensed Explosives*. Clarendon Press, Oxford, 1952.
ZELDOVICH, J. B. and KOMPANEETS, A. S., *Theory of Detonation*. Academic Press, New York, 1960.

Specific

CAMPBELL, A. W., DAVIS, W. C., RAMSAY, J. B. and TRAVIS, J. R., *Physics of Fluids*, **4,** 511 (1961).
CAMPBELL, A. W., DAVIS, W. C. and TRAVIS, J. R., *Physics of Fluids*, **4,** 498 (1961).
COWAN, R. O. and FICKETT, W., *J. Chem. Phys.* **24,** 932 (1956).
DEAL, W. E., *J. Chem. Phys.* **27,** 796 (1957).
DUFF, R. E. and HOUSTON, E. *J. Chem. Phys.* **23,** 1268 (1955).
EYRING, H., POWELL, R. E., DUFFEY, C. H. and PARLIN, R. B., *Chem. Rev.* **45,** 69 (1949).
HINO, R., *J. Ind. Expl. Soc., Japan*, **19,** 169 (1958).
HUBBARD, H. W. and JOHNSON, M. H., *J. Appl. Phys.* **30,** 765 (1959).
JONES, H., *Proc. Roy. Soc.* A **189,** 415 (1947).
KISTIAKOWSKY, G. B., and WILSON, E. B., *OSRD* No. 114 (1941).
PATERSON, S., *Research* **1,** 221 (1948).
SCHALL, R., *Nobel Hefte*, **21,** 1 (1955).
WACHTELL, S., MCKNIGHT, C. E. and SHULMAN, L., *Picatinny Arsenal Technical Rep. DB–TR: 3–61* (1961).

CHAPTER 3

Military High Explosives

FOR military purposes high explosives are used as filling for shell, bombs and warheads of rockets. The basic requirements for such explosives are the following:

1. Maximum power per unit volume.

2. Minimum weight per unit of power.

3. High velocity of detonation.

4. Long-term stability under adverse storage conditions.

5. Insensitivity to shock on firing and impact.

Requirements 1 and 3 follow immediately from the considerations of the theory of detonation when it is remembered that the purpose of the charge is to obtain maximum effect, both from the shock wave of the explosive and also from the destructive effect of expansion of the explosion products. Requirements 1 and 2 follow from the consideration that any reduction in size and weight of the warhead of a missile, or in a shell, immediately makes it possible to increase the range and therefore the usefulness of the weapon. Requirement 5 relates not only to safety, but also the desirability, particularly for armour-piercing ammunition, for the time of detonation to be determined solely by the functioning of an appropriate fuze.

In this chapter the explosives employed are discussed; their actual application is described in Part III. The most important properties of the commonest military explosives are listed in Table 3.1.

High Explosives

TABLE 3.1 *Properties of Military Explosives*

Explosive	m.p. (°C)	Density (g ml^{-1})	Weight strength % Blasting gelatine[a]	Maximum detonation velocity (m s^{-1})
TNT	80·7	1·63	67	6950
PETN	141·3	1·77	97	8300
RDX	204	1·73	100	8500
Tetryl	129	1·6	84	7500

[a] See p. 62.

TRINITROTOLUENE

Trinitrotoluene is the 2, 4, 6-isomer of the following constitution:

The starting material is pure toluene, specially free from unsaturated aliphatic hydrocarbons. This is nitrated in several stages to avoid oxidation side reactions which occur when toluene itself is mixed with strong nitrating acids. The traditional process employed three nitrating stages; as an example the nitrating acids used in France are given in Table 3.2.

TABLE 3.2 *French Nitrating Acids for TNT Manufacture*

	1st stage	2nd stage	3rd stage
Nitric acid wt. %	28	32	49
Sulphuric acid wt. %	56	61	49
Water wt. %	16	7	2

During and since the Second World War, the three-stage process has been replaced by continuous methods of nitration employing a larger number of stages. In these, the chemical engineering can differ widely, but in principle toluene enters the process at one end and trinitrotoluene is produced at the other. The nitrating acid flows in the opposite direction, being fortified as required at various points. One of the most important factors is stirring in the nitrators, as this markedly affects the speed and completeness of nitration, particularly in the later stages. For a discussion of these methods the reader is referred to a book by Urbanski.

The crude product contains isomers other than that required and also nitrated phenolic compounds resulting from side reactions. The usual method of purification is to treat the crude product with sodium sulphite, which converts asymmetric trinitro compounds to sulphonic acid derivatives, and to wash out the resulting soluble products with alkaline water. The purity of the product is determined by the melting point, the minimum value for Grade I TNT commonly being 80·2°C. Unless adequate purity is achieved, slow exudation of impurities can occur during storage and the TNT then becomes insensitive.

TNT is relatively safe to handle and of low toxicity. It is, therefore, preferred to picric acid and ammonium picrate which give sensitive compounds with a variety of metals, and to trinitrobenzene or hexanitrodiphenylamine which are highly toxic.

TNT contains insufficient oxygen to give complete combustion of the carbon on detonation. It can, therefore, usefully be mixed with ammonium nitrate, which has an excess of oxygen. The resulting explosives, known as amatols, are more powerful and cheaper than TNT itself, but in general have a lower velocity of detonation. A proportion of 60% ammonium nitrate is perhaps the commonest of these compositions.

PENTAERYTHRITOL TETRANITRATE (PETN, PENTHRITE)

This material has the following formula:

$$\text{O}_2\text{NOH}_2\text{C} \diagdown \quad \diagup \text{CH}_2\text{ONO}_2$$
$$\text{C}$$
$$\text{O}_2\text{NOH}_2\text{C} \diagup \quad \diagdown \text{CH}_2\text{ONO}_2$$

Pentaerythritol is made commercially by the reaction of formaldehyde and acetaldehyde in the presence of alkali. It can be nitrated by adding it to strong nitric acid at temperatures below about 30°C. An excess of nitric

High Explosives

acid is used so that the refuse acid from the reaction contains at least 75% nitric acid, as refuse acids of lower strength can be unstable. The washed product is reprecipitated from acetone to give a suitable crystal size and adequate stability on storage. PETN is always transported wet with water, and dried only as required.

PETN, discovered in 1895, is a solid melting at 141°C and is a very powerful explosive. It is very stable both chemically and thermally.

Pure PETN is too sensitive to friction and impact for direct application for military purposes. It can usefully be mixed with plasticised nitrocellulose, or with synthetic rubbers to obtain plastic or mouldable explosives. The commonest application, however, is in conjunction with TNT in the form of pentolites. Pentolites are usually obtained by incorporating PETN into molten TNT. A small amount of the PETN goes into solution, but the bulk remains suspended in the liquid and the whole mix can suitably be used in preparing cast charges. Pentolites containing 20–50% PETN are the commonest in practice.

RDX (CYCLOTRIMETHYLENE-TRINITRAMINE, CTMTN, CYCLONITE, HEXOGEN)

RDX has the following formula:

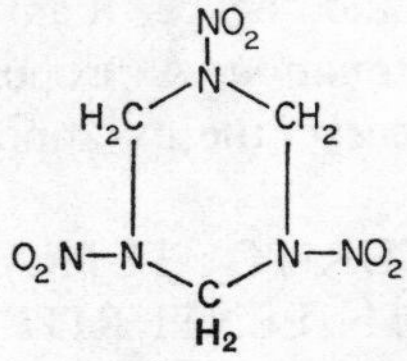

It is made by the nitration of hexamine (hexamethylenetetramine), itself prepared from formaldehyde and ammonia. Hexamine was originally nitrated with a large excess of concentrated nitric acid at temperatures below 30°C and the product recovered by adding the reaction liquor to an excess of chilled water. Later the yield was improved by adding ammonium nitrate to the reaction as this reacts with the liberated formaldehyde. A much-used process converts the hexamine first to its dinitrate, which is then reacted with ammonium nitrate, nitric acid and acetic anhydride (the last reagent being re-formed from the product by use of ketene). The RDX can be isolated by evaporation of the mother liquor, and then purified by washing.

RDX is a white solid melting at 204°C. Originally discovered by Henning in 1899, it attained military importance during the Second World War owing to its lower sensitiveness than PETN. It is very stable, both chemically and thermally.

RDX may be used alone in pressed charges, although for this purpose tetryl is a more general choice. For shell and bomb fillings it is too sensitive alone to initiation by impact and friction and is either desensitised with wax, or else used like PETN in admixture with TNT. RDX may also be compounded with mineral jelly and similar materials to give a useful plastic explosive.

Nitration of hexamine dinitrate in the presence of acetic anhydride can also give another explosive of high power and high stability called HMX (tetramethylenetetranitramine). This has the following structure:

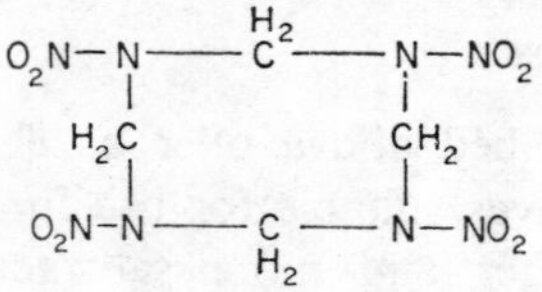

$$O_2N-N \underset{\underset{O_2N-N}{\overset{|}{\underset{}{H_2C}}}{\overset{|}{}}}{} \overset{\overset{H_2}{C}}{\underset{\underset{}{\overset{|}{\underset{}{C}}}{H_2}}{}} N-NO_2$$

TETRYL (CE, 2, 4, 6-TRINITROPHENYLMETHYLNITRAMINE)

Tetryl has the following formula:

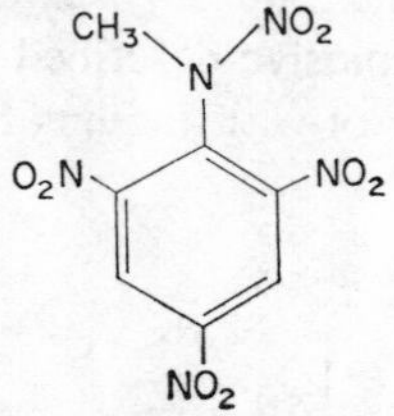

It was discovered by Michler and Meyer in 1879 and was made by the nitration of dimethylaniline. One methyl group is oxidised and at the same time the benzene nucleus is nitrated in the 2, 4, 6-positions. The usual method of preparation from dimethylaniline is to dissolve 1 part of dimethylaniline in 14 to 15 parts of sulphuric acid; to this solution about 9 parts are added of a mixed acid containing 67% of nitric acid and 16% of sulphuric acid. The nitration is carried out at about 70°C. The water content of the mixture must be kept reasonably low or benzene insoluble impurities of benzidine derivatives are produced.

More recently methylamine is treated with 2, 4- or 2, 6-dinitrochlorobenzene (usually in the form of an unseparated mixture of isomers) to give dinitrophenylmethylamine. This without separation is then nitrated to tetryl.

In either case purification is carried out by washing in cold water and boiling water, the latter hydrolysing tetra-nitro compounds. Finally, the tetryl is recrystallised by solution in acetone and precipitation by water, or recrystallised from benzene.

Tetryl is a pale yellow solid, melting at 129°C. It is moderately sensitive to initiation by friction or percussion. Tetryl is most used in the form of pressed pellets as primers for other less easily initiated explosives.

Other Military Explosives

The explosives described above offer a selection of powerful and thermally stable explosives suitable for use for military purposes. The search for even stronger and more satisfactory explosives always continues. Objects of present research are in general to give compounds which are more stable at high temperatures and which, if possible, have higher strength. Many such compounds have been described, but apparently none have come into any extensive use. Mention may, however, be made of an explosive described by DuPont which has the remarkable thermal stability of withstanding heating to 350°C. This has the following structure:

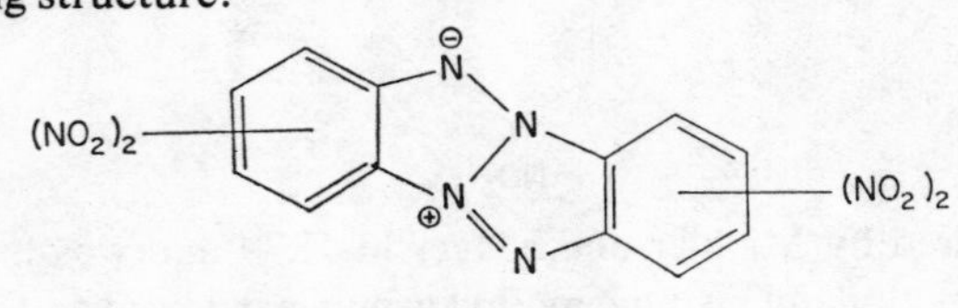

References

BARLOW, E., BARTH, R. H. and SNOW, J. E., *The Pentaerythritols*. Reinhold, New York, 1958.
URBANSKI, T., *Chemistry and Technology of Explosives*. Vol. 1, Pergamon Press, London, 1964.

CHAPTER 4

Manufacture of Commercial Explosives

Ammonium Nitrate

Ammonium nitrate is the cheapest source of oxygen available for commercial explosives at the present time. It is used by itself in conjunction with fuels, or to give more sensitive explosives in admixture with solid fuels and sensitisers such as nitroglycerine and TNT. It is, therefore, a compound of particular importance for the explosives industry.

Ammonium nitrate is made by the neutralisation of nitric acid with ammonia. The details of these processes are given in other volumes of this series. For particular application in explosives, ammonium nitrate is required in specialised forms, of which the following two are the most important.

For use in explosives sensitised by high explosive ingredients, ammonium nitrate should be of a dense and non-absorbent character. Whilst absorbent ammonium nitrate can be employed, it tends to be of lower density and therefore gives lower bulk strengths to the final explosive; it also absorbs a larger amount of nitroglycerine and requires more of this expensive ingredient to give a suitable gelatinous consistency. Dense ammonium nitrate is made either by crystallisation from solution, followed, if necessary, by grinding and screening, or more generally by spraying a melt containing at least 99·6% of ammonium nitrate down a short tower. The spray process produces spherical particles which can be taken from the bottom of the tower and cooled with the minimum of drying. For many years setting of ammonium nitrate, due to absorption of moisture and subsequent temperature change or drying out, led to caking of the salt and corresponding difficulties in handling. Nowadays this is overcome by adding either crystal habit modifiers which cause recrystallised ammonium nitrate to have low physical strength, or else by other additives which appear to lubricate the surface of the crystals. In this way, provided storage conditions are reasonable, ammonium nitrate can be kept in a condition suitable for easy mechanical handling.

For use in conjunction with fuel oil an absorbent form of ammonium nitrate is required. This is produced by spraying a hot 95% solution down a high tower, so that some drying occurs before the spherical droplets reach the bottom. The resultant spheres must be carefully dried and cooled to prevent breakdown in handling. They are then usually coated with a mixture of diatomaceous earth and a wetting agent in approximate proportions of 0·5% and 0·05% respectively. The bulk density of the product is about 0·7 to 0·8 compared with about 1 for the dense material and it will absorb 7–8% of light oil without appearing unduly wet. Ammonium nitrate "prills" of this type were originally made in Canada, but have since become popular in many parts of the world.

Ammonium nitrate undergoes phase changes at 32° and 83°C and melts at 170°. It is not normally considered an explosive when pure, although under suitable conditions it can be made to detonate. When mixed with small amounts of organic matter it becomes much more sensitive and several serious explosions have occurred with such mixtures. The limit of organic matter so far allowed in the U.K. is 0·05%, but in some countries 0·1% is accepted.

Nitroglycerine

Nitroglycerine, or glycerine trinitrate, has the following formula:

$$H_2CONO_2$$
$$|$$
$$HCONO_2$$
$$|$$
$$H_2CONO_2$$

It was discovered by Sobrero in 1847, but was developed to a commercial scale by Nobel. It has for a long time been, and still is, the most important sensitiser for commercial explosives.

Nitroglycerine is made by reacting purified glycerine with a mixed acid containing nitric acid, sulphuric acid and water. The temperature must be carefully controlled and the product when separated from the refuse acid has to be washed free from surplus acid before it becomes stable. It is a sensitive explosive, easily initiated by certain forms of friction and impact. For this reason and because of the importance of economic production of the compound, considerable study has gone into the design and operation of nitroglycerine plants.

Originally nitroglycerine was made by batch processes in which glycerine was added slowly to mixed acid in large vessels containing cooling coils. The acid contained 40–50% nitric acid with the remainder sulphuric acid. The worker controlled the flow of glycerine so that the maximum temperature allowed, usually 18°C, was not exceeded. Nitroglycerine, being less dense than the refuse acid, separated to the top and could be skimmed off. It was then washed with water and dilute sodium carbonate solution in air-stirred vessels before being allowed to stand and was then weighed for use. The nitration, weighing and washing were usually carried out in separate houses. Should any untoward incident occur during nitration, the whole mixture of acid and nitroglycerine could be discharged rapidly into a large volume of water in a drowning tank. As large quantities of nitroglycerine were involved, accidents when they happened were usually severe, and for this reason continuous processes involving smaller amounts of nitroglycerine in process at any time were evolved.

The first continuous process was that of Schmid in which glycerine and mixed acid were fed continuously into a specially designed, stirred and cooled nitrator; cooling was by chilled brine. The mixture from the nitrator went into a separator of special design from which the nitroglycerine overflowed from the top and refuse acid was removed from the bottom. The crude nitroglycerine was then washed in a series of columns and flowed by gravity to the weighing house.

A more recent process was developed by Biazzi and is somewhat similar in general principle to the Schmid process. It uses, however, improved chemical engineering designs and in this way is suited to operation by remote control.

The most recent process was introduced by Nitroglycerine AB (NAB) in Sweden and has a radically different system of nitration. An injector is used for mixing glycerine and nitrating acid and the nitration is carried out in a tube. The mixture of acid and nitroglycerine passes through a tubular cooler and is then separated in a centrifugal separator. The short residence time makes possible the use of high nitration temperatures and the throughput of the plant is high. Nevertheless, only very limited amounts of nitroglycerine are in process at any time.

Nitroglycerine can detonate in pipes of diameter down to approximately 5 mm. In nitroglycerine manufacture there is, therefore, an inherent danger of transmission of detonation from one manufacturing house to another in the series. Even a pipe which has been emptied of nitroglycerine can have on it a skin of the product sufficient to enable transmission of detonation from one end of the pipe to the other. To prevent the spread of an accident it is now usual to transfer nitroglycerine as a non-explosive emulsion in an excess of water. Such emulsion transfer is particularly convenient with the NAB process, as the emulsion transfer lines can also carry out the necessary process of washing and purification.

Nitroglycerine is a viscous yellow liquid which freezes at 13·2°C to a sensitive solid explosive. Because of the danger of freezing, pure nitroglycerine is now only rarely used in making explosives. The common practice is to mix ethylene glycol with the glycerine and nitrate the mixture so as to give a product which contains from 20 to 80% of ethylene glycol dinitrate. For most climates any mixtures in this range give satisfactory results, although under the very coldest conditions the extremes should be avoided.

Nitroglycerine when heated rapidly explodes somewhat above 200°C, but on storage it proves unstable at temperatures exceeding 70–80°C. The thermal decomposition products are very complex. In large quantities of water it is hydrolysed to nitric acid and glycerine, but this reaction is very slow at ordinary temperatures. Nitroglycerine has a marked physiological effect producing dilation of the arteries and severe headaches. Ethylene glycol dinitrate, or nitroglycol, has even more severe effects and with a higher vapour pressure is more prone to cause unpleasant reactions. It is also more toxic than nitroglycerine. These effects call for precautions during manufacture, but are not severe enough to affect the user.

Nitrocellulose

Nitrocellulose is used in commercial high explosives mainly to thicken the nitroglycerine in the preparation of gelatine and semi-gelatine compositions. The raw material is cotton. Nitration is carried out with mixed acid containing nitric and sulphuric acids and a proportion of water adjusted so that the nitrogen content of the nitrocellulose produced is about 12·2%. The relationship between the nitrogen content of the product and the acid left after nitration is shown in the ternary diagram (Fig. 4.1). Nitrators

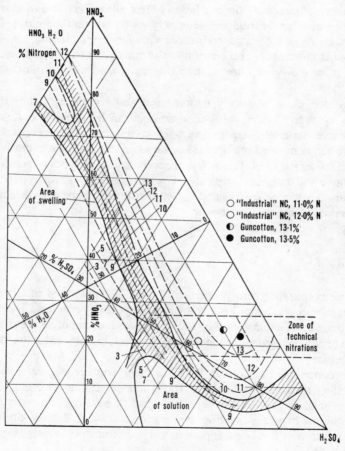

FIG. 4.1. Cellulose nitration diagram.

may be stirred or unstirred, depending on the nitration method used and the particular type of nitrocellulose required. After nitration, excess acid is removed in a centrifuge and the acid-wet nitrocellulose drowned in a stream of water. The nitrocellulose is stabilised by treating it with hot acidic water followed by hot dilute sodium carbonate solution. It is then pulped to a fine form so that it will dissolve rapidly in nitroglycerine.

The nature of the nitrocellulose used is of particular importance in explosives if freedom from exudation of free nitroglycerine during storage is to be avoided. Nitroglycerine is only a poor solvent for nitrocellulose and stability of the gel depends on continuous formation and breakdown of gelled structures. The distribution of nitrogen content and viscosity, even in the individual fibres of the nitrocellulose, is therefore of paramount importance.

Nitrocellulose is usually handled wet and containing approximately 30% of water. Under these conditions it can be considered as a non-explosive material when the nitrogen content does not exceed 12·6%. More highly nitrated cellulose is known as guncotton and is explosive even when moderately wet. When dry, nitrocellulose of all types is an extremely sensitive and dangerous explosive. Dry nitrocellulose is required for use in certain types of explosives and is then prepared by slow drying of the wet material in a current of warm air.

TNT

The preparation and properties of TNT are described in Chapter 3. Next to nitroglycerine, TNT is the most important sensitising constituent of commercial explosives. For such purposes it does not need to have the high purity demanded for the military product, but otherwise the material is identical.

In commercial explosives TNT has the advantage of greater safety in handling than nitroglycerine and also less physiological effect. On the other hand, it gives explosives which tend to be unreliable under certain conditions. Its use is therefore tending to diminish in Britain, although in some countries it is being applied in slurried explosives for large boreholes (see p. 56).

Powder Explosives

The preparation of powder explosives is in essence simple. In the case of mixtures of ammonium nitrate and fuel oil in particular, the only requirement is a method of mixing which does not cause undue breakdown of the absorbent grains of ammonium nitrate. Hand mixing is employed for small quantities, otherwise some form of rotating container or gently stirred vessel.

Powder explosives containing nitroglycerine are naturally more difficult to manufacture. Full precautions against explosion must be taken at all stages. The mixing equipment is generally a vessel containing a stirrer on a horizontal axis, or with two parallel stirrers as in the familiar Werner Pfleiderer type. In modern installations addition of ingredients and mixing and emptying operations are frequently carried out by remote control. The process consists simply of adding the ingredients to the equipment and stirring until adequately mixed, usually for a period of about 15 min. The mixed explosive is then cartridged, and for this purpose a variety of machines is available. In the commonest types, pre-formed waxed paper shells are placed under a row of nozzles and the powder explosive is then tamped or fed by worms into these shells. When the shells are sufficiently full, the spare paper at the upper end is closed over in a crimping operation. Cartridges can afterwards, if necessary, be further dipped in wax as an extra protection against ingress of moisture.

Powder explosives based on TNT are manufactured in a different manner. The commonest is to mill the TNT in large mills with suspended steel wheels which grind the explosive to a powder. Other ingredients are added and milling continued until the mixture is sufficiently fine and well mixed to have the required sensitiveness. Alternatively, the TNT and other ingredients can be mixed by stirring them together at a temperature above the melting point of TNT. Explosives so mixed are usually less sensitive but have improved resistance to moisture. Cartridging of TNT explosives is usually carried out with screws of the auger type with large open flutes. For blasting operations the filling is into pre-formed paper shells, which are afterwards dipped in wax. TNT explosives are also often used for larger charges employed in seismic prospecting and in this case are usually filled into tins to give complete protection against water even under hydrostatic pressure.

Semi-gelatine explosives (see p. 49) are manufactured and cartridged as powder explosives, although the presence of a thickened nitroglycerine base gives them properties which can approach those of gelatines.

Gelatine Explosives

Gelatine explosives contain a sufficient quantity of nitroglycerine thickened with nitrocellulose to give the mixture a plastic or gelatinous consistency. Advantage of this is taken in the manufacturing operations. To obtain the best results, it is desirable to ensure that at least a high proportion of the nitrocellulose is dissolved in the nitroglycerine before the other ingredients are added.

The original practice, still often employed, consists of weighing out the required amount of nitroglycerine on to nitrocellulose in a rubber-lined box; the two are stirred by hand and the mixture allowed to stand, sometimes for several hours. The resulting jelly is placed in a mixer which frequently takes the form of a bowl of figure eight section with two sets of revolving blades on vertical axes; the remaining ingredients are added and the whole mixed until uniform. Frequently hot water is circulated through a jacket round the mixer so as to speed the final gelation of the nitrocellulose.

More recently advances in the manufacture of nitrocellulose and in mixer design have enabled a much shorter process to be adopted. Bowl mixers with twin sets of blades on horizontal axes are employed most frequently, although various other designs have been found satisfactory. Practice is to add the nitroglycerine and nitrocotton and stir for a few minutes to enable gelation to occur. The remaining ingredients are then added and mixing continued for a further period. Mixers are usually arranged to tilt so that the mixed explosive is discharged into bogies or carrying boxes.

There are many ways of preparing cartridges of gelatinised explosives. The original method employed the screw extrusion of the plastic in machines very similar to those used for making sausages. The extruded cord was cut by hand and wrapped in paper. Many machines have been designed for carrying out such filling processes automatically, usually by extruding the explosive directly into pre-formed paper shells. One such machine, developed by the DuPont Company, extrudes several cords downwards into shells simultaneously, and stops the extrusion when the required depth of filling has been achieved.

A recent cartridging machine is the Rollex, designed by the firm of Niepmann & Sons. This operates on an entirely different principle in that the explosive is first rolled into a sheet and then portions of this sheet are cut off and automatically wrapped in paper. It can also be used for cartridging semi-gelatine explosives.

Slurry Explosives

Plants for the manufacture of slurry explosives vary considerably from the simplest to fully automated. A simple plant for example can consist of a jacketed ribbon mixer; a hot nitrate solution is first made and the other solid ingredients and gelling agent are mixed in. Finally the cross-linking agent and any sensitisers are added. The mixture is immediately transferred to plastic bags or cartridges and it is in these that the actual cross-linking and cooling takes place.

The more sophisticated plants may be typified by the mix-trucks used for mixing slurries on site and pumping them direct into boreholes. Such a truck will have a number of hoppers and tanks containing pre-prepared hot nitrate solution; solid ingredients such as ammonium nitrate, aluminium powder and guar gum; and solutions of cross-linkers and gasifying agents. Controlled feeds from these are led to suitable points along a continuous mixing system and the product pumped direct to a borehole before cooling or cross-linking has occurred. The diameters of all hoses should be less than the critical diameter of detonation of the explosive so that the plant is intrinsically protected from detonation.

Packaging

The packing of explosives is a matter of importance if they are to reach the user in satisfactory condition. The original wooden boxes have largely been replaced by more efficient fibreboard cases, although certain authorities still insist on the former. In either case, provision of an adequate waterproof barrier is important to prevent moisture causing hardening and desensitisation of the explosive. The cartridges are sometimes placed in cartons, containing 2·5 to 5 kg, wrapped in waxed paper and sealed with wax. More recently the tendency has been to obtain waterproofness by using a polythene layer inside the packing case and doing away with the cartons. The final case usually contains 25 kg of explosives.

Factory Construction and Operation

In the manufacture of high explosives, the possibility of accidental detonation must always be borne in mind and buildings are constructed and arranged so as to minimise the possible effects of such an explosion. The buildings are placed sufficiently far apart so that an explosion in one building will not cause sympathetic detonation in the other. Practical distances do not make it completely impossible for detonation in one house to leave all other houses unaffected, for two reasons:

1. Debris from the first building may be projected for long distances and can fall in other buildings, causing the explosive therein to explode.
2. Explosives must be transported through the factory between the buildings and can assist the propagation of detonation from one building to another.

To assist in safety, buildings are specially constructed according to one of two schemes:

(a) The buildings can be of light construction with the machinery placed as far as possible below the level of the explosive and therefore unlikely to be projected into the air. Such buildings are usually surrounded by a mound or barricade consisting of concrete or earth, and at least as high as the eaves of the building.

(b) The buildings can be completely enclosed by concrete and layers of earth. If the amounts of explosives involved are small, such construction can completely contain any accidental explosion. More generally, however, the purpose of the overmounding is to prevent shrapnel from an explosion in one building penetrating to the explosive in adjoining buildings.

The safety of operators for the processes of mixing and cartridging can be much assisted by remote control. In this case a concrete block house is built outside the explosive building and the controls for the machine are placed in the block house. The process itself is viewed either by a system of periscopes, or by closed-circuit television. Such a method of working leads readily to semi-automatic or automatic processes.

Originally, equipment was made largely from wood or gun-metal and often rubber lined. These materials give the lowest hazards from friction with explosives. Nowadays, improved standards of engineering and of design have made it possible to employ stainless steel and plastics in the construction of explosive machinery with considerable increase in mechanical efficiency. In this way not only can processes be carried out more rapidly, but the quantity of explosive present at any time is reduced, with consequent increase in overall safety.

It is essential in explosive operations to avoid the presence of grit or extraneous materials. In Britain the general practice is to provide buildings with "clean" floors, which can only be approached by the donning of special shoes or overshoes. The carrying in of unnecessary objects and particularly ferrous tools is strictly forbidden, and, of course, the presence of matches or smoking materials is illegal.

References

MILES, F. D., *Cellulose Nitrate*. Oliver & Boyd, London, 1955.
NAUCKHOFF, S. and BERGSTRÖM, O., *Nitroglycerine and Dynamite*. Nitroglycerin AB, Sweden, 1959.
URBANSKI, T., *Chemistry and Technology of Explosives*. Vol. 2, Pergamon Press, London, 1965.

CHAPTER 5

Design of Commercial Explosives

THE design of explosives is a matter of major importance to the manufacturer. The explosives industries of the world are extremely competitive and it is essential that products should be as cheap as is possible. A range of products must be provided such that one of them at least will be suitable for every type of work for which explosives may be required. A typical range, apart from special sizes, would be about thirty.

Explosives contain oxidising and combustible ingredients, whether united in a single molecule or present in different chemicals. A proper balance is essential, particularly if the explosive is to be used in confined places, especially underground. The "oxygen balance" of an explosive is the percentage excess of oxygen in the composition. It is calculated, usually, on the unwrapped explosive, as, in practice, only a part of the wrapper (usually considered to be about a half) takes part in the chemical reaction. Too high an oxygen balance leads to the production of nitric oxide and nitrogen dioxide in the explosion fumes which, therefore, become toxic. Too low an oxygen balance, whilst having advantages in generally giving the explosive a higher power, leads to the production of excessive amounts of carbon monoxide which is also poisonous. The production of poisonous gases by explosives cannot be entirely eliminated, but by good design can be reduced to negligible proportions.

Complete reaction of explosives is important in obtaining the best results for both power development and fume avoidance. It is aided by the use of fine materials, but if this is carried too far other disadvantages of hardness and low density are introduced.

Some typical compositions of commercial explosives are given in Table 5.1.

TABLE 5.1 Compositions of Non-permitted Commercial Explosives

| Explosive | Explosive sensitiser | | Oxidiser | | Fuel | | | Miscellaneous | |
	Nitroglycerine /nitroglycol	TNT	Ammonium nitrate	Sodium nitrate	Cellulosic	Aluminium	Oil	Nitrocellulose	Other
ANFO	—	—	94	—	—	—	6	—	—
NG powder	10	—	80	—	10	—	—	—	—
NG semi-gelatine	15	—	77·1	—	6	—	—	0·4	1·5
Ammon gelignite	32	—	60·5	—	6	—	—	1·0	0·5
NS gelignite	60	—	—	29	7	—	—	3·5	0·5
High velocity gelatine	50	—	—	29	9	—	—	2	10
Blasting gelatine	91·4	—	—	—	—	—	—	8	0·6
TNT powder (strong)	—	10	85	—	—	5	—	—	—
TNT powder (normal)	—	10	86	—	4	—	—	—	—

Ingredient %

ANFO

The simplest explosive to design is ANFO, or ammonium nitrate mixed with fuel oil. The choices to be made here are the grades of ammonium nitrate and oil and the proportion in which they are mixed.

As noted above (p. 36), the ammonium nitrate used for these explosives should be porous and should retain 7% of the oil without tendency to segregate on standing. A further requirement in modern practice is that the mixture, when blown into boreholes by compressed air, should break down to give a proportion of fine powder which increases both the density of loading and the sensitiveness of the mixture (see p. 141).

The oil used should be of a volatile type, as the more volatile oils gives the greatest sensitivity. However, the use of petroleum fractions with too low a flash point is hazardous. The choice is, therefore, usually made of a fuel oil similar to those used for diesel engines. It is common practice to add a proportion of dyestuff to the oil, partly to make adequate mixing of the explosive immediately visible, and partly to assist the user in seeing proper loading of boreholes, particularly in salt and other white materials.

A choice of composition is generally determined by the necessity for oxygen balance. A range of 5·5–6% of fuel oil gives a balanced mixture which produces negligible poisonous fume and maximum power on detonation. For certain uses, as in quarries, lower proportions of oil are sometimes used, as 2–3% of oil gives maximum sensitiveness to initiation; such compositions are, however, unsuitable for underground use as they produce nitrogen dioxide on detonation.

Nitroglycerine Powder Explosives

Another type of explosive which is of relatively simple design is the nitroglycerine powder explosive. This type of explosive is made from ammonium nitrate, sometimes with sodium nitrate, a combustible material, and nitroglycerine as a sensitiser. As this is essentially a cheap type of explosive, the relatively expensive nitroglycerine component must be kept to the minimum possible value. Figures as low as 6% have been used, but general practice for smaller sizes of cartridges is 8 or 10% so as to ensure adequate sensitiveness and reliability of propagation. If more than 10% of nitroglycerine is included, care is needed to ensure that the liquid does not tend to leak out of the explosive on standing.

As nitroglycerine itself is nearly oxygen balanced, the mixture of ammonium nitrate and fuel used in a powder explosive should also be approximately oxygen balanced. The actual proportions depend on the nature of the combustible chosen and this choice is in itself dependent on the explosive properties required. In general, the aim is to obtain the highest strength per unit volume and this is achieved by attaining the maximum density possible in the explosive. To this end, the ammonium nitrate must be used in a dense form. The choice of combustible is more restricted because it must not be bulky but must still be absorbent enough to retain the nitroglycerine in the explosive. For economic reasons, one of the commonest choices of combustible is a fine form of sawdust, known as woodmeal, made from selected woods, as this gives a suitable combination of density and absorption.

There is always a search for higher power in explosives, and one common way of achieving this end is to add materials which liberate the maximum energy on combustion. Of such materials, the only one which has achieved common usage is aluminium, but such others as silicon and ferro-silicon have been considered. Magnesium is in general too reactive chemically to be safe on storage. Aluminium, when added, is used in a relatively fine, sometimes flake, form as in this condition it increases the sensitiveness of the explosive as well as the power. Aluminium, indeed, is frequently added to ANFO for similar reasons, but the greater sensitiveness of the resulting explosive makes it suited only to manufacture in specially designed premises, and in most countries it is not considered safe for such an explosive to be made with simple equipment on the site where it is used.

Semi-Gelatine Explosives

The strength of nitroglycerine powder explosives is limited to about 80% blasting gelatine and the density to approximately 1. To achieve greater concentrations of energy it is necessary to increase the nitroglycerine to an extent such that it can no longer be absorbed by the other ingredients of the powder explosives. When this stage is reached, nitrocellulose must be added to the nitroglycerine to prevent exudation from the explosive cartridges. This leads to the semi-gelatine type of explosive which forms a rather indistinct class between the powders and the gelatines. The true powder explosives contain no nitrocellulose and, therefore, can only be made at relatively low densities and are also susceptible to the action of water. The true gelatine explosives, on the other hand, have a continuous phase of gelled nitroglycerine and, therefore, have a high density and are relatively unaffected by water for an appreciable length of time. Semi-gelatine explosives can be made with proportions of nitroglycerine from 10 to 30% and with properties which range over the extreme limits between powders and gelatines.

The choice of composition of a semi-gelatine depends ultimately on two requirements, namely the strength required and the resistance to water needed for the particular application. For economic reasons the lowest nitroglycerine content which satisfies both these requirements is always chosen.

With mixtures of ammonium nitrate and ordinary combustibles, the highest weight strength (see p. 61) which can readily be achieved is about 85% of blasting gelatine. To obtain high bulk strength it is, therefore, necessary to increase the density to the maximum possible. In practice, the maximum density usefully achieved is about 1·25 and this with relatively dense combustibles means a nitroglycerine content of 15–20%. If either the ammonium nitrate or the combustible available is not of high density, increased quantities of nitroglycerine may be necessary. As with powder explosives, alluminium may be added to give extra power, although this is not necessarily economic.

An explosive designed as just stated would give very satisfactory results for many purposes when used in dry or almost dry conditions. Underground, however, boreholes are frequently very wet and in work above ground running water is often encountered. Semi-gelatine explosives, aluminium may be added to give extra power, although this is not necessarily economic.
consist in general of sodium carboxymethylcellulose, starches, or natural gums, prepared in a way to swell rapidly in contact with water. A small percentage of such an additive has little effect on explosive properties, but should the explosive come into contact with water, the gum on the outside layers immediately swells to form a gelatinous layer which impedes further

ingress of the water. Such protection cannot be permanent, but with suitable design it can last for a number of hours sufficient to enable boreholes to be loaded and fired.

Gelatine Explosives

Gelatine explosives are more costly in raw materials than the powder or semi-gelatine types, although this can to some extent be offset by greater ease, and therefore less expense, in manufacture. Their popularity throughout the world rests on a number of important advantages over other available explosives:

1. They provide high bulk strength.
2. They are very resistant to the effects of water.
3. Under conditions of use they propagate extremely well from one cartridge to another, so that failures are unlikely even under bad conditions.

In designing a gelatine explosive, the same questions must first be answered as in the case of all other explosives, namely, the bulk strength required and the degree of resistance to water. Again the practical requirement is to provide these properties with the minimum proportion of nitroglycerine.

Blasting gelatine, containing 92% of nitroglycerine and 8% of nitrocellulose (usually also small amounts of chalk) is the strongest and most water-resistant of these explosives. Cartridges have in fact been fired successfully after being left accidentally for many years immersed in water. Although for economic reasons blasting gelatine is little used nowadays, its properties are of importance in understanding those of other gelatine explosives and the principles of their design.

If the most uniform obtainable nitrocellulose of nitrogen content about 12·2% is dissolved in nitroglycerine (containing nitroglycol) and the solution made as complete as possible, by heating and prolonged standing (and particularly with the assistance of gelatinising accelerators such as dimethylformamide, or certain ketones and esters), a clear yellow gel can be obtained. Vacuum mixing is necessary to avoid the occlusion of air. Such a gel has two unexpected properties. In the first place it is extremely difficult to initiate this explosive, a primer as well as a detonator being required. Secondly, the explosive on standing tends to exude liquid nitroglycerine. These tendencies are present in all gelatine explosives and must be avoided by careful design.

Control of exudation depends mainly on the suitable choice of the nitrocellulose used. Some lack of uniformity in this product is certainly desirable. This offers no serious difficulty, although it is necessary to ensure a constant watch on manufacturing processes to see that quality is maintained. In other gelatine explosives, particularly those containing ammonium nitrate, exudation can be induced by slow chemical reaction. The addition of alkalis, for example, can liberate ammonia which in turn can react with nitrocellulose and cause it to lose its power of binding nitroglycerine. Such effects are accelerated at high temperatures and under wet conditions and it is usual practice to test all explosives under such adverse conditions before they are put on the market.

Blasting gelatine as normally manufactured is easily initiated by commercial detonators. This has been shown to be due to the trapping of small bubbles of air in the explosive during the operations of mixing and cartridging. Such an explosive shows two velocities of detonation, one at about 2000 m s^{-1} with low strength initiators, called the low velocity of detonation, and the other approaching 7000 m s^{-1} with high strength initiation, called the high velocity of detonation. This high velocity of detonation tends to fall as the degree of aeration is increased, but at the same time the minimum detonator strength needed to produce it is also decreased. These air bubbles are effective only within certain limits of size; thus during storage the explosive loses sensitiveness because the bubbles gradually coalesce into larger and ineffective bubbles, and indeed the density of the explosive increases due to loss of air. The rate of these changes depends on the type of nitrocellulose used and special grades are necessary if explosives of this type are to remain easily initiated after long storage under adverse conditions.

In other explosives of a gelatine type some form of combustible is normally present. Most combustibles contain pores which hold air and they also assist mechanically in the trapping of air bubbles in the explosive. As a result, the lower the nitroglycerine content of a gelatine explosive, the easier it is to obtain adequate sensitiveness to commercial detonators. In the particularly difficult cases such as the higher strength gelignite explosives, it is advantageous to add a proportion of a low density combustible, such as dried sugar cane pith.

Gelatine explosives, initiated by commercial detonators, will normally fire at the low velocity of detonation initially, although this may well build up quite quickly into the high velocity. For some applications a high velocity of detonation is essential. This can be ensured by the addition of barium sulphate, or other material with density exceeding 2·8, in a fine form. Such additives have the property of ensuring rapid transition to the high velocity of detonation. This is, for example, of particular importance when the explosive is to be fired under a hydrostatic head, as in submarine work.

The other important factor in the design of gelatine explosives is the consistency of the product, which must be suited to the manufacturing facilities available, and which must remain usable throughout the life of the explosive. This entails that there should be a suitable balance between the solid and gelatinous phases of the explosive, the former not being too bulky and the latter not being too thin or too highly gelled. The explosives of this class, which are the most popular because they are the cheapest, usually contain the minimum proportions of nitroglycerine. Commonly, they can include also nitrobodies such as dinitrotoluene which are soluble in nitroglycerine and therefore assist in increasing the proportion of the liquid phase. When allowance is made for absorption by ammonium nitrate and by solid combustibles, nitrocellulose is added to an extent necessary to bind the nitroglycerine into the explosive. If the proportion falls, however, much below 1% on the total composition, the nitroglycerine phase becomes too fluid and the explosive tends to have reduced cohesion. At the same time, the large relative volumes of the solid phase make the explosive more sensitive to hardening under adverse conditions of temperature and humidity. These are the factors which effectively limit the extent to which the nitroglycerine content of such explosives can be reduced in efforts to attain economy.

The most important explosives of this class are the ammon gelignites, so called because they are based essentially on ammonium nitrate which is the cheapest and most powerful source of oxygen. Sodium nitrate is sometimes added as well in order to improve the oxygen balance for certain types of these explosives. The ammon gelignites are explosives with density about 1·45 which can be made in all powers up to nearly 100% of blasting gelatine strength. They have good resistance to water and can be used freely under most wet conditions. In time, however, the ammonium nitrate, being extremely water soluble, is leached out of the explosive, which then becomes insensitive.

For the wettest conditions where the high strength of blasting gelatine is not required, the explosives used are the straight gelignites based on sodium nitrate instead of ammonium nitrate. Because of the low explosive strength obtainable when sodium nitrate is used, the proportion of nitroglycerine in these explosives must be high. On the other hand, when properly made to ensure adequate sensitiveness, these explosives have exceptionally good storage properties and can be used under even the most adverse conditions of wet working.

It will be clear from the above that the optimum types of oxidising materials are those of highest density and dense forms of ammonium nitrate are always used. The combustibles can be dense also, although it is sometimes necessary to add at least a proportion of the combustible in an absorbent form to ensure adequate sensitiveness. Wheat flour may be regarded as typical of a dense combustible; woodmeal is a useful and cheap combustible of intermediate properties.

TNT Explosives

The oxygen balanced mixture of ammonium nitrate and TNT contains 79% of ammonium nitrate and 21% of TNT. It has a power of about 85% of blasting gelatine. As a commercial explosive it would normally be considered too expensive because of the high proportion of the expensive TNT ingredient. Practical explosives of this type contain less TNT, usually in the range of 10–15% for explosives to be used in small diameters (below 5 cm) and 5–10% for explosives to be used in larger sizes. Oxygen balance is then achieved by the addition of any cheap finely divided combustible.

If the explosive is to be mixed hot, a fairly fine grade of ammonium nitrate is advantageous, as its high surface area gives a high reaction rate and therefore maximum sensitiveness in the explosive. Compositions intended for mixing in edge-runner mills are ground together in the process so that the type of ammonium nitrate or TNT used becomes of much less importance. Indeed, a friable form of ammonium nitrate is often useful in reducing the time of mixing.

Simple mixtures of this type have little resistance to water and are suitable for use only under the driest conditions. Many methods have been used for giving water resistance to these explosives, particularly to the milled varieties which tend to be less satisfactory in this respect than the hot mixed compositions. A small proportion of wax may be used, particularly if mixed in hot, but this tends to desensitise the explosive. Greater success is achieved by the addition of calcium soaps such as calcium stearate, in finely divided form. These also cause some desensitisation, but to an extent which is less marked in proportion to the degree of waterproofness achieved.

A proportion of finely divided aluminium is often added to TNT explosives in order to increase the power. As aluminium has also a sensitising effect, it is particularly useful in waterproofed compositions. Another power producing additive which is sometimes employed in large diameter charges where its slow reaction is of less disadvantage is calcium silicide. Care must be taken with this material, however, to ensure that it does not lead to sensitiveness to friction and impact.

Slurry Explosives

Slurry explosives, also known as water-gel explosives or dense blasting agents, resulted from the work of M. A. Cook and others in North America and have now found world-wide application particularly for large scale operations. Whilst slurries are made in many forms to suit almost all types of use, they may be divided into two essentially distinct types:

1. Dense slurries which are not aerated and therefore have a density of about 1·4. The required sensitivity is achieved by the addition of substances such as TNT.
2. Aerated slurries, where sensitivity is attained by introducing air or other gaseous bubbles to give densities ranging from 1·05 to 1·3 according to the proposed use. The practical lower limit to the density is usually set by the requirement that the explosive should not float to the surface in a water-filled borehole.

In either case the base of the explosive is an aqueous solution of ammonium nitrate with another nitrate such as that of sodium or calcium. Compared with ammonium nitrate alone, such mixed nitrate solutions contain less water and therefore give increased explosive power; in addition they are more easily sensitised and retain their properties better at low working temperatures. The nitrate solution is thickened by addition of guar gum which is then cross-linked, typically by adding a chromate. The base of the explosive is therefore a more or less rigid gel, the term slurry being in fact rather inappropriate.

The process of gelling is of importance as it imparts water-resistance which makes it possible to use the final explosive under wet conditions. Particularly for explosives which have to be stored before use, careful control of gelling is necessary to ensure retention of sensitivity and absence of physical breakdown. Suitable grades of gum and cross-linker must be used and an appropriate pH of the solution maintained.

This base is used to the minimum possible extent in the final explosive as the water it contains does not contribute to the power and indeed requires energy for its evaporation. All slurry explosives therefore contain further ammonium nitrate in solid form and also a fuel for combustion. The ammonium nitrate is usually in dense form similar to that used in nitroglycerine explosives as this gives the best physical properties. However, it is common practice to mix the explosive hot so that much or all of the solid ammonium nitrate results from crystallisation during cooling.

The gums and other ingredients can often provide much of the necessary fuel but the addition of aluminium has special advantages. Aluminium gives a highly energetic reaction in these explosives and helps to remedy their otherwise low power. The metal must be finely divided if it is to react completely in the borehole. It must also be relatively pure, otherwise even in the presence of chromates it will react with the aqueous base at room temperature to give gassing and instability. Other energetic fuels which have been used with some success are silicon and ferro-silicon, but these tend to be slow in reacting during detonation.

The first and still successful dense slurry was sensitised with TNT and therefore consisted of a suspension of TNT and solid ammonium nitrate in a solution of ammonium and sodium nitrates gelled with cross-linked guar gum. The TNT is preferably in the form of small pellets. No further fuel than the TNT is essential but aluminium can be added for increased strength. The degree of sensitivity which can be achieved with TNT is limited and it is usual to fill the slurry in plastic cylinders 10 cm or more in diameter. If required the plastic can be slit immediately before use to allow the explosive to slump and completely fill the borehole. Initiation is by booster, such as cast pentolite.

More recently isopropyl nitrate has been used in place of TNT for sensitising dense slurries. Although not itself explosive, this liquid gives a sensitivity at least equal to that obtained with TNT, whilst at the same time reducing the proportion of aqueous base and therefore water needed in the composition. Dense slurries have also been made with such ingredients as pentolite and smokeless powder as sensitisers but these have no special advantages and are usually uneconomic.

The more widespread current use of slurry explosives is undoubtedly due to the development of aerated slurries. These have much greater versatility than the dense variety and can be made to a wide range of sensitivity, density and power. Moreover they are particularly suited to manufacture on site for pumping direct into a borehole without intermediate packing and transport.

The sensitivity depends on the presence of gas bubbles, but to be effective these bubbles must be of suitable size. The following are the most important ways of achieving the required result:

1. Mechanical entrapment of air.
2. Introduction of microballoons—i.e. hollow spheres of plastic.
3. Production of gas in the explosive by chemical reaction.

Mechanical aeration by prolonged mixing alone is not an effective means of sensitisation as the bubbles produced are predominately too large. The addition of some material which facilitates the occlusion of small air bubbles is therefore necessary. A particularly useful material of this type is paint fine aluminium. This material, used in the manufacture of paints and lacquers, consists of beaten flakes of aluminium made water-repellent with a coating such as a stearate. When stirred into a slurry explosive the flakes are only imperfectly wetted and retain adhering to them small bubbles of air which are very effective in sensitising the slurry. For best results it is necessary to carry out the addition of the aluminium carefully and as the final stage of the process of manufacture; moreover the base of the slurry must be of optimum consistency. Used in this way only a few per cent of paint fine aluminium can give a slurry explosive which is sensitive to a commercial detonator. Although the gap sensitiveness is likely to be low compared with explosives based on nitroglycerine, these slurries can be used with care in small diameters and by variations in other ingredients can be designed for all general purposes including permitted explosives of the P1 and P3 types. Disadvantages of this type of slurry however are that paint fine aluminium presents special handling difficulties in explosives manufacture and that it is expensive even in the small quantities required.

Air can also be introduced with the help of the thickeners, either guar gum itself or alternatives such as other natural gums or starches. Part of the total gum is first dissolved in the aqueous base, without cross-linking. Addition of the remaining gum to this pre-thickened solution automatically occludes air bubbles which can be stabilised by cross-linking. This process is particularly suited to slurries mixed on site; in this case the process of cross-linking is critical and often controlled by activating the chromate by adding a small proportion of a reducing agent.

Microballoons have the advantage of giving a well controlled addition of sensitiser and also improved stability on storage. They are however less effective than direct gas occlusions when measured at constant explosive density and therefore tend to have most use in conjunction with the soluble additives described below.

Chemical production of gas, for example by the decomposition of dissolved hydrogen peroxide, is again a very controlled method of sensitisation. The same method can be used for adjustment of the final density of the explosive. Thus a small addition of a peroxide gives an explosive of adequate sensitivity for large boreholes together with maximum density and bulk strength of explosive. To attain the sensitivity needed for smaller boreholes it is only necessary to increase the addition of peroxide, although there will of course be a reduction in density and bulk strength.

It has been pointed out that the water present in a slurry explosive must be vaporised during detonation and thus absorbs energy. Although the water vapour increases the total volume of gas produced the nett effect is still a reduction in strength. Many attempts have therefore been made to find a soluble additive which will replace part of the water to give both increased power and sensitivity. One such substance which has found success is ethylene glycol mononitrate. This when pure is explosive, but for use in slurries it can be handled throughout in safe solution form. Another such substance is mono-methylamine nitrate which again is handled in solution. In both cases there is an increase in power and sensitivity but in most practical applications further power is attained by the addition of aluminium and the sensitivity is increased by the addition of microballoons or by gassing.

In many ways slurries may be considered as intermediate between ANFO and nitroglycerine explosives. They are more expensive than ANFO but can be used in wet conditions; they are often cheaper and safer than nitroglycerine explosives but are more critical of conditions of use if misfires are to be avoided.

Other Explosives

For reasons of safety it is always desirable to use the least sensitive explosive which is adequate for any given operation. This has led to development of a class of explosives which in the United States are called nitrocarbonitrates. These explosives contain no self-explosive ingredient and are themselves insensitive to initiation by a single commercial detonator of ordinary strength. More important, when they are properly designed they are insensitive to impact and friction and unlikely to detonate when involved in a fire. Such explosives are usually based on ammonium nitrate, sensitised with nitrobodies, typically dinitroluene. Ordinary combustibles may well be added to give oxygen balance, and waterproofing ingredients similar to those used in TNT explosives may also be considered desirable. The addition of most power producing ingredients such as aluminium, however, is unwise in view of the reduction in safety to which they lead. Explosives of this type must be used with adequate care, either in large diameter holes, or else sealed in tins for oil prospecting. Although relatively safe, they are still explosives and should be handled as such.

A type of explosive, not based on nitroglycerine or on TNT, which achieved popularity for a time and is still used in some countries, is the liquid oxygen explosive or LOX. This is made on the site of the blasting operations by immersing in liquid oxygen a pre-formed cartridge of absorbent charcoal. Charcoal saturated with a suitable proportion of liquid oxygen is readily detonated by a commercial detonator. The cartridge thus prepared is loaded and fired before the oxygen evaporates. For large-scale operations this represents a very cheap way of preparing explosive. Unfortunately, the process is dangerous for reasons not fully understood and the history of LOX explosives is marred by a series of fatal accidents. For this reason, these explosives are little used today in most countries.

Before nitrates and particularly ammonium nitrate were readily available commercially, explosives were developed based on chlorates and perchlorates. These also are still used in some countries. In general perchlorates are considered less dangerous than chlorates and therefore preferred. They are easily sensitised, so that in addition to explosives of this type based on nitroglycerine, others have been based on various organic liquids, particularly nitrobodies. History shows that chlorates and perchlorates must be regarded as temperamental substances, liable in bulk to lead to inexplicable accidents. Particularly when mixtures of chlorates and oxidising materials are allowed to become wet and then dry out, conditions can arise in which there is an appreciable sensitiveness to friction and impact. Explosives of this type have an unfortunate record of accidents. They are used, therefore, to a limited extent only, now that safer compositions are available.

Reference

TAYLOR, J., *Detonation in Condensed Explosives*. Clarendon Press, Oxford, 1952.

CHAPTER 6

Assessment of Explosives

MUCH information concerning any given explosive can be obtained by calculating its properties on theoretical grounds. This is particularly valuable as the calculations indicate the performance which the explosive may have under ideal conditions, namely, infinite charges. In practice, such perfection is never achieved and it is a matter of practical importance to assess the properties of an explosive under conditions more appropriate to its use. Such an assessment is usually made by a series of tests chosen to measure the performances of the explosive under various conditions. Whilst these tests are of considerable value, particularly for comparative purposes, it must be remembered that no laboratory test or series of tests can predict precisely the performances of explosives which may themselves be used for widely varying purposes. It is obvious that the assessment of an explosive for use under water should follow different lines from the assessment of an explosive for use in a mine. It is less clear, but equally important, that the assessment of mining explosives should depend on the nature of the rock in which the mining is to be carried out. The ultimate test for all explosives is use in the field.

Power

The power, or strength of an explosive, is one of the most important properties of interest to the user. It is usually expressed in terms of power per unit weight, which is appropriate for comparing explosives used in charges measured by weight. It can alternatively be expressed as power per unit volume, which is appropriate for explosives which are used to fill boreholes of a given size. The relation between the two depends solely on the density, so that the one is readily calculated from the other.

High Explosives

The ultimate test used by NEC for commercial explosives consists in carrying out small-scale blasting operations in a quarry. The maximum weight of rock adequately broken per 0·5 kg of the explosive is calculated and used to indicate the power of the explosive. The rock involved is of fairly average nature for the district and the results therefore form a useful general comparison of strengths. Such testing is tedious and expensive and carried out only as a final assessment.

The maximum potential power of an explosive can be calculated, or it can be measured by techniques such as those developed by Cook. A typical method consists of firing the explosive under water and measuring the energy liberated in the various forms, such as shock wave in the water, the work of expansion of the gas bubble, etc. These figures have limited practical value as the methods of application of explosives are of low and variable efficiency. A more practical measurement of strength can be obtained by the measurement of cratering efficiency. This, again, demands considerable expense and also requires the availability of uniform rock.

Most measurements of strength are done by laboratory methods and of these the most satisfactory is the ballistic mortar. Various forms of mortar have been designed, but the method accepted as an international standard by the International Committee on the Standardisation of Test on Explosives at Sterrebeck, Belgium, in 1962 rests on a design by the DuPont Company of the U.S.A. and is illustrated in Fig. 6.1. This consists of a pendulum 3 m long, at the bottom of which is a bob weighing 333 kg. A shot is used of 12·37 cm diameter, weighing 16·6 kg, and fitting with a clearance of approximately 0·08 mm. Ten grammes of the explosive under test are wrapped in tin foil and fired with a standard copper detonator. The explosive ejects the shot on to rubber matting or into a mound of suitably non-abrasive mixture. The corresponding recoil of the bob is shown on a scale and is a measure of the energy imparted by the explosive to the system. The mechanical efficiency of the ballistic mortar has been shown to be rather low, but it is constant and results are reproducible. Gradual wear causes the efficiency to change during use, so that it is general practice to use standard explosive as a reference material. In Great Britain the standard explosive is blasting gelatine and results are expressed as a percentage of the strength of blasting gelatine. In the U.S.A. the standard is usually TNT and the results are expressed relative to that explosive.

In the ballistic mortar test the explosive is well confined and develops almost maximum power. To give results more comparable with practice for explosives which are slow in reacting, tests have been devised in which the explosive is fired in a less confined condition. Such tests are, of course, arbitrary and the results must be compared directly with practical results.

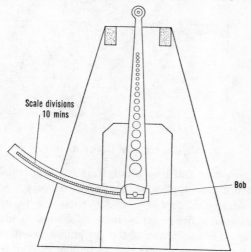

Scale divisions
10 mins

Bob

FIG. 6.1. Ballistic mortar.

An older form of measuring power was the Trautzl lead block test. This is illustrated in Fig. 6.2. The explosive under test is placed in a cylindrical hole in a block of specially cast lead and the remainder of the hole filled with sand. When the explosive is fired it causes an expansion of the hole in the lead block and this expansion is measured by filling with water. After subtracting the expansion caused by the detonator the result gives an indication of the strength of the explosive. This test has been particularly developed in Germany and the method adopted there has been accepted as an international standard. For most explosives there is an adequate correlation between lead block results and ballistic mortar measurement, as shown in Fig. 6.3. The lead block is less accurate for practical explosives, but is of value in studying weak explosives which are marginal in properties and therefore unsuitable for measurement in the ballistic mortar.

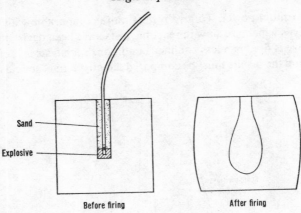

Sand ————

Explosive ————

Before firing   After firing

F IG . 6.2. Lead block test.

The ballistic mortar and lead block tests use only small amounts of explosive and are not applicable to slurry explosives which are too insensitive to detonate properly under such conditions. For these explosives it is useful to fire larger amounts of several kg under water and measure the period of oscillation of the gas bubble produced. The longer the period the greater the energy of the gas bubble and this part of the total energy of the explosive has been found to correlate well with the blasting effect of the explosive.

Velocity of Detonation

A particularly valuable method of measuring velocity of detonation is by high-speed camera, usually of a rotating mirror type. The layout is illustrated in Fig. 6.4. A slit is placed in front of the explosive, or of an image of the explosive in the optical system. This slit is in turn photographed with the rotating mirror camera, the velocity of which is known from stroboscopic measurements. As the detonation front is luminous, the illuminated point travels along the slit and gives a photograph consisting of a line inclined to the axis of the camera at an angle which depends on the velocity of detonation and on the speed of rotation of the mirror. By this method small charges can be studied and constancy of velocity can be observed or any changes in velocity calculated.

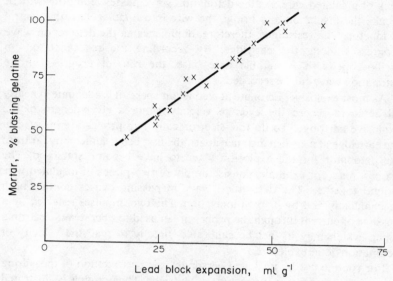

FIG. 6.3. Relationship between ballistic mortar and lead block tests.

Velocities can also be measured by electronic means and methods have been devised to give either continuous or intermittent readings. To obtain a

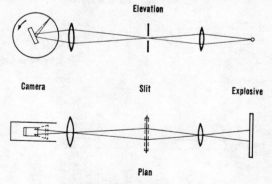

FIG. 6.4. Rotating mirror camera.

High Explosives

continuous record, a probe is placed along the cartridge, usually along the centre. This probe consists of a conducting rod around which is wound a helix of insulated wire. As the detonation wave passes along this wire, it breaks the insulation and brings the wire into contact with the central conductor. The resistance, therefore, diminishes as the detonation wave progresses along the cartridge. By recording the resistance on an oscilloscope with a suitable time base, the rate of progress of the detonation wave can be recorded.

As most explosives detonate at a constant speed, it is adequate in nearly all cases to record the average velocity along a given length of the explosive cartridge. To do this electronically two probes are inserted in the cartridge at a known distance apart, the first being sufficiently far from the detonator for the explosion wave to have become stabilised. The probes, may, for example, consist simply of two pieces of insulated wire wound together. The detonation wave in passing causes contact both mechanically and by its own ionisation. The two impulses received by a passage of current through the probes are caused to operate an electronic timer, which may in fact be calibrated directly to read the velocity of detonation of the explosive.

For routine use by unskilled personnel, the best method of measuring velocities of detonation is one due to Dautriche. The principle is illustrated in Fig. 6.5. The two ends of a length of detonating fuse are inserted in the explosive under test at a known distance apart. The mid point of the piece

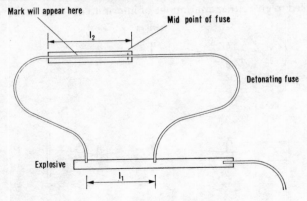

FIG. 6.5. Dautriche test.

of detonating fuse is known and the part of the fuse near this point is placed over a V-shaped groove in a thin lead plate. When the explosive is fired the fuse commences to detonate first at the end nearer the detonator and later at the other end. When the two detonation waves in the fuse meet they reinforce and produce a distinct mark on the lead plate, visible particularly as a split on the back. This point will be removed from the mid-point of the fuse by a distance which depends on the velocity of detonation of the explosive under test according to the following equation:

$$D = dl_1/2l_2$$

where D is the velocity of detonation of the explosive under test,
 d is the velocity of detonation of the detonating fuse,
 l_1 is the length of explosive under test,
 l_2 is the distance between the centre of the fuse and the mark on the plate.

Thus by measuring the distance of the mark on the lead plate from the mid point of the fuse the velocity of detonation of the explosive under test may be calculated. In practice, it is simple to construct a rule by which the velocity can be measured directly from the lead plate.

Sensitiveness

The detonator sensitiveness of an explosive is measured by firing in it detonators of increasing strength. It is convenient to stand the cartridge of explosive vertically on a lead plate with the detonator at the top. The indentation produced on the lead plate then gives a good measure of the detonation of the explosive. The weakest detonators for this test contain only mercury fulminate. Stronger detonators have base charges of PETN, increasing in amount as the strength of the detonator is increased (see p. 102).

An equally important and more severe requirement for commercial explosives is that they should propagate in a train of cartridges. It must be remembered that in practice, when several cartridges are placed in a borehole it cannot be ensured that they are in contact and free from rock or coal dust between them. It is, therefore, necessary for the explosive to be able to fire over a gap, whether this be air or other substance. The Ardeer double cartridge test (ADC test) was the first routine test of this nature. In this test two cartridges, as used in practice, are rolled at a selected distance apart in manilla paper. The assembly is placed on a flattened steel bar and fired from one end. The gaps are measured at which the second cartridge is detonated and also fails to detonate. Differences of distance of 1·25 cm at lower gaps and 2·5 or 5·0 cm at higher gaps are used. In the international test for gap sensitiveness, the paper is omitted and the cartridges are attached to a wooden strap which is suspended above the ground. For special purposes, such as with some modern weak coal mining explosives, a similar test can be carried out with the explosives enclosed in a cement or similar tube to simulate the confinement of a borehole. It is not normally necessary to do tests with materials other than air between the cartridges. In America, however, a test has been developed in which the sensitiveness is measured by placing increasing thicknesses of card or similar material between the cartridges.

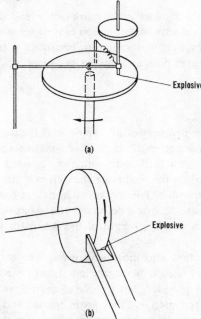

Fig. 6.6. Friction tests. (a) Liquid explosives. (b) Solid explosives.

Whilst an explosive must have adequate sensitiveness to initiation by detonators and to propagation of detonation through a train of cartridges, it must be sufficiently insensitive to friction and impact to be safe to handle. Sensitiveness to friction is commonly measured by some device as that illustrated in Fig. 6.6. The explosive is smeared on the surface of a rotating disc on which rests a rod of similar or dissimilar material, carrying a known weight. The speed of rotation can be varied and also the load employed. The higher the speed and the greater the load before initiation of the explosive occurs, the lower the sensitiveness of the explosive. Some-times an oscillating plate is used instead of a rotating disc.

The sensitiveness of an explosive to impact is measured by determining the minimum height from which a given weight must be dropped in order to initiate detonation. Many forms of "fall hammer" test have been devised, the most important point of the various designs being the means adopted for retaining the explosive. A simple and practical method, used at Ardeer for many years, is shown in Fig. 6.7. In this the explosive is put

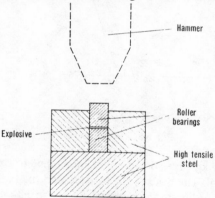

FIG. 6.7. Fall hammer test.

between two roller bearings, themselves placed in a ring of hardened steel and resting on a hardened steel base. The falling weight is arranged to hit the upper of the steel cylinders. The roller bearings must be changed after each ignition, as the condition of their surface markedly affects the results of the test. Similarly, the closeness of fit of the hardened steel ring is of importance. Typical weights employed are from 0·5 to 5 kg and heights of fall may be up to 200 cm.

High Explosives

A common hazard in the handling of explosives is for them to be subjected to the effects of a blow which is to some extent at a glancing angle. This corresponds neither to pure impact, nor to pure friction. A corresponding simple and useful test for the safety of an explosive during handling is the torpedo friction test illustrated in Fig. 6.8. In this, a torpedo

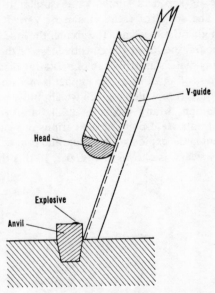

FIG. 6.8. Torpedo friction test.

weighing 0·5 to 5 kg slides down an inclined plane to strike the explosive resting on an anvil. The head of the torpedo and the material of the anvil can be varied according to the materials of construction of an explosives plant, or can be kept as mild steel in each case if an indication of relative hazards of explosives is to be obtained. The angle of fall is commonly 70 or 80°.

Numerous tests for safety have been described in the past. It is probably far less important what test is employed than that some test should be undertaken before any explosive is handled on any but the smallest scale. All tests of the safety of explosives are in fact relative, so that the particular test employed is a matter of convenience. As the presence of grit reduces the safety of explosives, tests are usually repeated after a quantity of fine sand has been added.

Results of sensitiveness tests on typical explosives are given in Table 6.1, for steel to steel surfaces, in the absence of grit.

TABLE 6.1   *Sensitiveness of Explosives*

Explosive	Fall hammer (0·5 kg) (cm)	Torpedo friction (1 kg at 80°) (cm)	Friction wheel (0·5 m s⁻¹) (kg)
TNT	>200	80–120	>50
RDX/TNT	80–100	40–45	
RDX	25–30	10–20	
PETN	60–80	35–40	10
Gelignite	5–10	40–60	4
Ammon gelignite	30–40	40–60	30
NG powder	20–30	>150	>50
TNT powder	160–200	100–120	>50

Stability on Storage

All explosives are kept in magazines before use for periods which may be up to several years. These magazines are normally buildings which have neither special heating nor cooling, so that the temperature can vary widely in different parts of the world. Throughout this storage it is important that the explosive should remain safe and also retain satisfactory properties.

Explosives are exothermic in decomposition and can be considered as being in a metastable condition. To determine whether instability can set in, it is necessary to subject the explosive to conditions which are more severe than those occurring in practice. This in effect means that the explosives must be subjected to a higher temperature than those normally encountered. Unfortunately, as the temperature is increased the type of decomposition reaction changes, so that experiments at the highest temperatures, although rapidly carried out, do not necessarily indicate the stability of the explosive under practical conditions. The ultimate test of stability of an explosive must remain the maintenance of that explosive for several years at a temperature which somewhat exceeds the highest which is to be expected in practice. Common temperatures range from 35° to 50°C. To obtain a quicker indication of the stability of an explosive, tests are carried out at higher temperatures. Storage at 60°C for several months gives useful information on the properties of an explosive, both chemical and physical.

In determining stability by the above tests, it is usual to note the onset of mass deterioration in the explosive. To obtain rapid results in the course of hours or minutes, it is necessary to modify the tests so as to determine the initial stages of decomposition. Many tests have been devised in which the explosive is heated at a relatively high temperature and the extent of initial decomposition determined by measuring the amount of gas evolved or the amount of acidity generated. Probably the best known of these tests is the Abel heat test in which a small amount of the explosive is heated (often at 70°C) and the time determined in which the gases liberated will produce a standard coloration on a starch–iodide paper. In Britain it is a legal requirement that explosives and their explosive ingredients shall satisfy this test for times in the region of 10 min depending on the class of explosive. The heat test, like all similar tests, does not measure the true stability of an explosive, although an unstable explosive cannot give a good result. It can be markedly affected by minor impurities which have no effect on the long-term life of the explosive, and the trend of the results with time is frequently more important than the results themselves. Rapid tests are invaluable as a control of production in comparing one batch of explosive with another; their application to new explosives can, however, be very misleading.

For nitrocellulose, including guncotton, the most suitable stability test is the B and J (Bergmann and Junk) test in which 2 g dried material is heated for 2 h at 132°C, the gases evolved being dissolved in water. The nitric acid in this water is reduced to nitric oxide, which is measured by volume.

For PETN and similar military explosives a valuable test is a vacuum stability test, in which some of the explosive is heated in vacuum and the rate of evolution of gas measured.

Apart from maintaining chemical stability on storage, an explosive must also maintain its physical form and its sensitiveness. Explosives containing nitroglycerine would, for example, be dangerous if they exuded nitroglycerine during storage. To determine these properties there is no real alternative to storing the explosive and examining it afterwards. The length of storage can, however, be reduced by marginally increasing the temperature during the period.

Fume

The fume from an explosive is best determined by firing a round in a part of a mine which can be completely cut off from the circulating air. After the air and fume have been circulated by a fan for a sufficient period, samples of the resulting mixture are taken and analysed.

Results of value can also be obtained by firing the explosive in a steel vessel reinforced with concrete. Care must be taken that by the use of lead tube or similar method the explosive is adequately confined, as unconfined explosives can produce abnormal fumes.

In measuring the nitrogen oxides produced by explosives, it must be remembered that ammonium nitrate can be left unreacted in small amounts from explosives containing this substance. The method of analysis adopted should, therefore, be insensitive to ammonium nitrate, for example the Griess–Ilosvay method. The composition of the gases will change with time, as the oxidation of nitric oxide to nitrogen dioxide at these low concentrations is extremely slow. A common procedure is to determine nitrogen dioxide after a period sufficient to allow oxidation to be complete. Carbon monoxide in the gases can be estimated by reaction with iodine pentoxide.

Whilst carbon monoxide and nitrogen oxides are the toxic products of explosives, other constituents of the fume cause a characteristic smell. As the nitroglycerine content of explosives is reduced, this smell tends to become rather unpleasant. Subjective tests must be used for its estimation.

Miscellaneous

Of other tests which are sometimes applied to high explosives, mention should be made of the tests for brisance. Brisance is an ill-defined word, best described by saying that an explosive of high brisance, when fired unconfined on a steel plate, will bend or shatter that plate more effectively than an explosive of low brisance. The Hess and Kast tests for brisance depend on this property, using the deformation of a metal cylinder by the explosive as a measure of the property. In most countries these tests are now little used.

A qualitative test of some value is to fire the explosive standing on its end on a plate of lead about 2·5 cm thick. A strong high velocity explosive will punch a hole completely through the plate, whereas a weak low velocity explosive will cause merely a minor indentation. This sort of test gives a good empirical indication of the completeness of detonation of the explosive in the unconfined state, provided that it is used along with measurements of power and velocity.

Particular tests are used for nitrocellulose for use in blasting explosives. A typical test is the clearing test, which consists essentially of mixing blasting gelatine under controlled conditions and determining the time at which the explosive is lifted by the stirrers from the bottom of the mixing vessel. This test measures the speed of gelatinisation of the nitrocellulose.

References

COOK, M. A., *The Science of High Explosives*. Reinhold, New York, 1958.
McADAM, R. and WESTWATER, R., *Mining Explosives*. Oliver & Boyd, London, 1958.
MARSHALL, A., *Explosives*. Churchill, London, 2nd ed. 1917.

CHAPTER 7

Permitted Explosives

IN MANY coal mines there is a continual evolution of methane (firedamp) into the air of the workings. The methane is trapped in the coal or rock, often in pockets within the veins, and is sometimes at relatively high pressure. It is usually of fairly high purity, containing only minor amounts of other hydrocarbons and nitrogen. Methane, when mixed with air in proportions between 5 and 14%, forms an explosive mixture. In gassy mines, therefore, there is always the danger that a mixture may be formed which, if ignited, can cause serious damage and loss of life.

In coal mines, also, there is a further danger in that the working of the coal produces coal dust. Coal dust when mixed with air gives a mixture which when suitably ignited can undergo a dust explosion. Indeed, if an explosion of firedamp (or coal dust) occurs, the wave produced can stir the dust lying in the mine into the air, producing a mixture which can lead to further propagation and devastating explosion. In the history of coal mining there have been many examples of both firedamp and coal dust explosions, and the latter in particular have led to serious loss of life.

The firing of explosives in a gassy mine must always be undertaken with suitable consideration of safety. The same is, of course, true of all work, as any electrical fault, or spark from some metals and rock, can give rise to initiation of explosion should firedamp be present in suitable amounts. Most ignitions which occur are local and limited in extent, but there is always the possibility of their spreading to major proportions. The main method of combating this risk is to ensure suitable ventilation so that the firedamp liberated is carried away at a concentration below that at which explosion can occur. Frequent testing of atmospheres, particularly at points where ventilation is less effective than usual, is essential. Stone dusting is used to prevent coal dust explosions. A further and important

75

measure of safety is the provision of explosives which do not ignite methane/air mixtures under the condition of use. In Great Britain, an explosive for use in gassy mines must pass certain tests carried out by the Safety in Mines Research Establishment at Buxton, as a result of which the Ministry of Power issues the necessary authorisation for use and places the explosive on the Permitted List. Arrangements similar in principle are made in nearly all countries. In Britain the safety achieved is such that ignitions associated with explosives are less than 1 per 10 million shots fired, and are only about a quarter of the total ignitions in coal mines.

Initiation of Firedamp and Coal Dust Explosions

The high temperature reached in a detonation wave makes it seem improbable that explosives can be designed which do not cause ignition of an explosive mixture of methane and air. The possibility of doing so depends on the very short time for which the mixture is subjected to these temperatures. Methane/air mixtures have a finite induction period and unless energy is applied for at least a considerable proportion of that period, explosion will not result. Deflagrating explosives, such as black-powder, are more dangerous in coal mining conditions than high explosives, because of the much greater time which they take in burning. The strongest high explosives also readily ignite methane/air mixtures, because of their extremely high reaction temperatures. Weaker explosives, but having high speeds of reaction, are relatively safer. Present-day permitted explosives, therefore, always consist of high explosives of which the reaction temperature has been adjusted to a suitably low figure. At the same time care is taken to avoid constituents in the explosive which can continue to burn after the main reaction has been completed, thus setting up in the methane/air mixture a continuing source of ignition.

The mechanisms by which explosives can cause ignition of methane/air mixtures are the following:

1. By direct action of the shock or expansion wave from the explosive.
2. By indirect action of the shock wave after it has been reflected from solid surfaces in the vicinity of the explosion.
3. By ignition of the methane/air mixture on mixing with the hot gaseous products of the explosion.

4. By hot reacting particles of explosive escaping into the methane/air mixture.

It is probable that all of the above mechanisms can operate with certain explosives under some conditions. Strong high velocity explosives suspended in methane/air mixtures cause ignition by the effect of their shock wave. With some high energy explosives, calculations show that the hot gaseous products, when mixed with suitable proportions of methane/air mixtures, can give temperatures and concentrations which lead to gas explosion. Recent studies indicate that explosives can cause centres of initiation of methane/air mixtures at a distance from the exploding charge, and the fact that in many tests the probability of initiation varies according to the shape and size of the containing vessel suggests that shock wave reflections can play an important part. Finally, many workers have shown that reacting particles can penetrate beyond the expanding reaction products of the explosive into the methane/air mixture and can at least assist in causing initiation.

In the practical case, explosives designed for use in coal mines are such that they are unlikely to initiate methane/air explosions by their shock waves when fired inside boreholes. The addition of cooling salts reduces the temperature of the explosion below that at which the gaseous products can cause initiation by thermal effects alone. Therefore only mechanisms (4) and (2) are likely to be of practical importance. Experience shows that initiation by reacting particles of explosive, whilst always possible, is unlikely to be dominant. It is found rather that, in most practical conditions, there is a good correlation between the power developed by the explosive under the actual conditions of firing and the likelihood of initiation of methane/air mixtures. The chance of an explosion therefore depends not only on the explosive, but also on the geometry of the conditions of use, and the safety of a coal mining explosive depends on the operations in the mine for which it is employed.

Research throughout the world has rightly been aimed for many years at the production of explosives incapable of initiating firedamp explosions under any practical conditions. The attainment of such an ideal depends on reducing the effective power of the explosive, so that there are no geometrical arrangements in the coal mine which can cause that explosive to ignite the gas mixture. Such low power unfortunately can only be achieved by limiting the proportion of nitroglycerine and other reactive ingredients of the explosive to the minimum possible values. As a result the explosives inevitably become less sensitive and less certain in their use. There is an increasing danger that when a number of cartridges are fired in a borehole, particularly when they can be affected by rock movements, some of the cartridges will fail to detonate. These undetonated cartridges can also be liable to deflagrate, that is commence to burn when in the borehole and continue to decompose when the coal or rock is displaced. Such an effect introduces a danger into the coal mine of a different, but still serious, nature. Even if by suitable design of explosive deflagration be avoided, the presence of undetonated cartridges in the product is most undesirable.

It will, therefore, be seen that the best and safest practice is to use in each situation in the coal mine an explosive which is best adapted for the operation in hand. The use of unnecessarily weak explosives, even though apparently safer in some tests, is often a disadvantage and can be dangerous. These considerations have led to the development of a series of types of explosives suited for particular purposes in coal mines and subjected to tests relevant to their individual uses. Thus, in Britain there are now five classes of permitted explosives. In Germany there are three classes. In the U.S.A. geological conditions are different and the hazards are less than in Europe, and it has been possible to work throughout with the strongest types of permissible explosives.

Gallery Testing of Permitted Explosives

It was early recognised that an explosive is more hazardous in a coal mine if it is fired in a borehole from which the stemming is omitted or blown out early by the explosive than if it is fired in a properly stemmed hole and does adequate work in bringing down rock or coal. The tests which led to the original permitted explosives, now called P1 explosives, were therefore designed to test the product under these conditions.

The testing gallery consists of a steel cylinder, 1·5 m in diameter. The first 5·5 m is sealed by a paper or polythene diaphragm and the remaining 9·8 m is left open to the atmosphere. The general arrangmement is shown in Fig. 7.1. At the end opposite the diaphragm there is a hole about 30 cm in diameter against which a heavy cannon is placed. The joint is sealed by

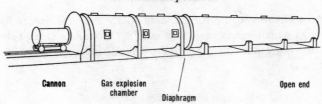

Cannon Gas explosion   Open end
chamber
Diaphragm

FIG. 7.1. Testing gallery for permitted explosives.

a rubber ring. The cannon is 1·5 m long and has in it a borehole 5·5 cm in diameter and 1·2 m long. The explosive is placed in this borehole with or without stemming according to the test being carried out. After the cannon has been placed in position, methane is introduced into the enclosed portion of the gallery and thoroughly mixed with the air. The explosive is then fired and the ignition, or otherwise, of the methane/air mixture observed from a safe distance.

The results obtained in gallery testing depend appreciably on whether the explosive is stemmed or not and also on the method of initiation. In the British test the stemming used is a close-fitting plug of clay 2·5 cm long. In spite of its short length in comparison with practical stemming, it has a marked effect on the likelihood of ignition of the gas mixture in the gallery. In the original tests, the detonator was inserted last into the cannon and this method of initiation is called direct initiation. If the detonator is at the opposite end of the train of cartridges, and is inserted first into the cannon, the initiation is called inverse or indirect and the probability if ignition of the methane/air mixture is appreciably increased. Figures illustrating this effect have been given by Taylor and Gay and are shown in Fig. 7.2. The same authors have given figures showing how the severity of the test increases as the diameter of the gallery is decreased.

In other countries galleries of similar construction are used, but the dimensions are frequently somewhat different. Data are given in Table 7.1.

Gallery testing in the equipment above is employed for studying the safety of explosives for general applications in coal mines. For explosives for particular purposes special tests have been devised, often to simulate to some extent the hazards which may occur in practice. Typical of these tests are the break tests devised in Britain. These tests are for studying explosives intended for use in ripping, that is in increasing the height of roadways in mines after the coal has been extracted. The extraction of the

80 *High Explosives*

coal relieves stresses in the surrounding rocks and is therefore liable to cause breaks which can contain methane/air mixtures. Three tests were devised to indicate the hazard involved in shotfiring in these circumstances.

TABLE 7.1   *Galleries for Testing Coal Mining Explosives*

Country	Gallery		Cannon	
	Diam. (m)	Length[a] (m)	Bore diam. (mm)	Bore length (cm)
Britain	1.5	5.5	55	120
U.S.A.	1.9	6.1	57	55
Germany (Dortmund)	1.8×1.35[b]	5	55	60
Belgium	1.6	5	55	50
France	2	3.5	30	60
Poland	2	3.3	50	70 or 120
U.S.S.R.	1.65	5	55	90

[a] Explosion chamber only.
[b] Elliptical section.

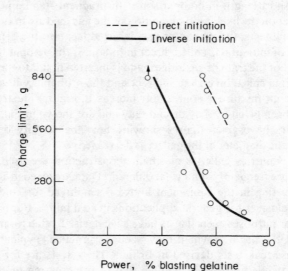

FIG. 7.2. Effect of initiation on gallery testing.

Break 1 test consists of a shothole in which a break occurred across the hole in the middle of the explosive charge. It is simulated by placing a train of explosive across the gap between two steel plates in a test gallery. Break 2 test consists of a break parallel to the shothole and formed in such a way that half the shothole is in one piece of rock and the remainder of the explosive in the space between the rocks. It is simulated by two parallel plates of which the lower one is grooved, the explosive resting in this groove. Break 3 test consists of a break occurring across a shothole, but at the end of the shothole. This is simulated by a cannon fired with the mouth in close proximity to a steel plate. Of these tests, experience has shown that the most severe condition by far is the No. 2 break test and this only will be described in more detail here.

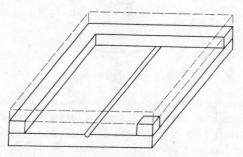

FIG. 7.3. No. 2 break test. (The position of the upper plate is shown by the broken lines.)

The No. 2 break test is shown in Fig. 7.3. Two heavy steel plates 1·8 m square are held either 5 or 15 cm apart. The lower plate has a semi-cylindrical groove of diameter equal to that of a standard borehole. The plates are in a gastight enclosure conveniently formed from steel sides with sheet polythene at the ends and top. The explosive cartridges rest in the groove, as if in a borehole lying in a break. As in gallery testing, the enclosure is filled with an explosive gas mixture. The explosive is fired by a copper detonator and the ignition or otherwise of the gas mixture observed visually.

In the No. 2 break test the explosive is fired almost unconfined. Under these conditions weak coal mining explosives detonate only partially and the probability of ignition of the gas mixture is reduced. Measurement of power carried out under conditions of poor confinement correlate reasonably well with the likelihood of ignition of methane/air mixtures in the test.

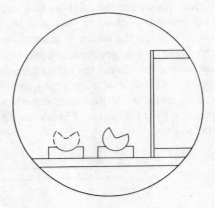

FIG. 7.4. Angle shot mortar.

In other countries other approaches to the study of specially safe coal mining explosives have been employed. Of these, the most popular is the angle shot mortar. This is illustrated in Fig. 7.4. It consists of a heavy cylindrical steel bar with one quadrant removed. The inner angle of this quadrant is rounded to simulate a portion of a shothole. A steel plate is fixed at a predetermined distance and angle from the bar and the whole assembly placed inside a gallery full of methane/air mixture. This test may be considered to simulate a shothole which blows out sideways into a space filled with explosive gas mixture. This test is particularly used in Germany and Belgium.

P1 Explosives

The original permitted explosives are now known as P1 explosives and must pass the following tests:
1. Twenty-six shots are fired of 142 g of explosive with inverse initiation, unstemmed into methane/air mixture. Not more than thirteen ignitions may occur.

2. Five shots are fired of 795 g explosive with direct initiation, stemmed into methane/air mixture. No ignitions may occur.
3. Five shots are fired of 795 g explosive with direct initiation, stemmed into coal dust/air mixture. No ignitions may occur.

In designing P1 explosives, the major consideration is the power. The maximum power of an explosive likely to pass the test depends on the bulk density and Table 7.2 gives approximate figures. Powers exceeding those quoted are liable to cause ignitions in the unstemmed test.

TABLE 7.2   *Maximum Strengths of P1 Explosives*

Type	Density	Max. strength % blasting gelatine
Gelatine	1·5	58
Semi-gelatine	1·25	62
Powder	1·0	66

Ignitions when firing into coal dust suspension are relatively uncommon and appear to depend on the inclusion in the explosive of certain types of ingredients. This effect is, for example, often observed with explosives containing sodium nitrate. For this reason this ingredient is usually avoided and the oxidising agent preferred is ammonium nitrate. To reduce the power to a suitable level, the usual additive is sodium chloride in finely divided form. Typical compositions are quoted in Table 7.3 for Polar Ajax and Polar Viking, gelatine and powder explosives respectively.

P2 Explosives

Lemaire, in Belgium, introduced the concept of enclosing explosives in an inert sheath of cooling material. Most commonly sodium bicarbonate has been used for this purpose. Originally it was packed as a powder round the explosive cartridge and inside an outer paper shell. A further method used in Great Britain was to prepare a "felt" consisting of sodium bicarbonate with a small amount of woodpulp to bind it in usable form; this bicarbonate felt was wrapped round the cartridge of explosive. More recently in Belgium the sheath has been produced as compressed hollow cylinders of sodium bicarbonate into which the explosive is placed.

Sheathed explosives appeared very successful at their introduction, but

TABLE 7.3   *Compositions of Permitted Explosives*[a]

P1 Explosives

Explosive	NG	NC	AN	Salt	Barytes	Combustible	Other
Gelatine (Polar Ajax)	26·5	0·8	42·7	24·6	—	3·8	1·6
Powder (Polar Viking)	10·5	—	70·7	10	—	8·8	—
Pulsed Infusion (Hydrobel)	40	2	20	27	9·5	1	0·5

P3 Explosives

Explosive	NG	NC	AN	Salt	Combustible	Other
Gelatine (Unigel)	29·2	0·8	25	42·5	2	0·5
Semi-gelatine (Unigex)	16	0·4	46·5	25·5	3·5	8·1
Powder (Unipruf)	8	—	53	29	9·5	0·5

P4 Explosives

Explosive	NG	AN	Amm. chloride	Sod. nitrate	Combustible	Other
Powder (Carrifrax)	9	10	28	46·5	6	0·5

[a] For each composition, official permitted limits are in fact ascribed for each ingredient. The compositions quoted are the mid points of these limits, adjusted to total 100%.

later experience was rather disappointing. There is always the danger of breakage of the sheath and therefore loss of safety. In Britain they have been almost entirely superseded by P3 explosives described below. The tests required for P2 explosives are similar to those for the P3 variety.

P3 Explosives

These explosives, when introduced into Britain, were called "equivalent to sheathed" or Eq.S. explosives. Currently, the explosives must pass the following tests:

1. Twenty-six shots are fired, consisting of 397 g with inverse initiation, unstemmed into methane/air mixture. Not more than thirteen ignitions may occur.
2. Five shots are fired consisting of 1020 g with direct initiation, stemmed into methane/air. No ignitions may occur.
3. Five shots are fired consisting of 567 g with inverse initiation, unstemmed into coal dust/air mixture. No ignitions may occur.

In the design of P3 explosives the first consideration is again that of power. Approximate limits, above which ignition of gas is likely to occur in unstemmed tests, are given in Table 7.4.

TABLE 7.4   *Maximum Strengths of P3 Explosives*

Type	Density	Max. strength % blasting gelatine
Gelatine	1·65	42
Semi-gelatine	1·3	47
Powder	1·0	50

When explosives of the power shown in this table are prepared, the power per unit length of the explosive cartridge is similar to that of the power of a sheathed explosive made from a P1 composition. Effectively, therefore, the difference between a P2 and a P3 explosive is that the inert material in the sheath of the former is distributed uniformly through the explosive composition of the latter. For reasons of stability, however, sodium bicarbonate is no longer used; instead an increase in the proportion of sodium chloride gives the required cooling effect.

In this way P3 explosives of powder or semi-gelatine type can be directly designed. The addition, however, of sodium chloride to a gelatine explosive, in the proportions required, is not possible without loss of the gelatine consistency. In designing a P3 gelatine explosive, therefore, the extra cooling salt is substituted mainly for the oxygen-balanced mixture of ammonium nitrate and combustible. The development of P3 explosives has been described by Taylor and Gay and typical compositions are given in Table 7.3.

P4 Explosives

P4 explosives were specifically designed for the operation of ripping with delay detonators (see p. 143) and must satisfy the following tests:

1. Twenty-six shots are fired consisting of 397 g inversely initiated, unstemmed into methane/air mixtures. Not more than three ignitions may occur.
2. Five shots are fired of the maximum permitted charge weight of explosive into methane/air mixture in Break test I. No ignitions may occur.
3. Break Test II uses a gas mixture of 3·60% propane with air and nitrogen, which is more easily ignited than methane/air. Preliminary shots determine the most hazardous charge of explosive not exceeding 227 g. Twenty-six shots are then fired at this weight and not more than thirteen ignitions may occur.
4. Five shots of 30·5 cm length and 3·7 cm diameter are fired in methane/air in Break Test III. No ignitions may occur.

In designing P4 explosives, it must be remembered that the overall strength is limited by the requirement of the gallery test, which is even more severe than that employed for P3 explosives. It is, therefore, necessary to include even greater proportions of cooling salts into the explosive than is the case with the P3 class.

Equally severe in practice is the requirement of the break 2 test. To pass this test the explosive must have a low power when fired in an unconfined condition. The actual power has not been quantitatively measured, but is probably in the region of 15% blasting gelatine. To achieve such a result, it is necessary deliberately to design the explosive in such a way that only partial reaction occurs in the unstemmed condition. Such partial reaction can be achieved by either of two ways.

The first method, popular in Britain, consists of adjusting the nature of the ingredients and particularly their specific surface in such a way that the reaction of the oxidiser and combustible is slow. The ingredients are also such that a flame-quenching solid is released in a form having a high specific surface. For example, in the first explosive of this class, N.E. 1235, much of the ammonium nitrate was introduced in the form of an intimate granular mixture containing calcium carbonate.

In the other method, particularly popular in Germany, the ammonium nitrate is replaced by an equimolar mixture of ammonium chloride and potassium or sodium nitrate. The reaction between the salts, which gives potassium or sodium chloride and ammonium nitrate or its decomposition products, is relatively slow and does not occur to a marked extent when the explosive is fired in an unconfined condition. This method of working is particularly effective in reducing the power of an explosive in the unconfined condition. Used alone it has not proved popular in Britain, because of the low power which tends to be developed under practical firing conditions. Moreover, the finely divided sodium chloride smoke which is produced by the explosive tends to be unpleasant for the miners.

The most recent practice in Britain is to employ a combination of the above methods so as to give the required overall effect on the power of an explosive in an unconfined condition. Compositions and properties of such an explosive are given in Table 7.3.

P5 Explosives

P5 explosives were specifically designed for blasting solid coal with millisecond delay detonators. The nature of this application and its advantages are described in Chapter 14 (see p. 144). In Europe, exchanged ion explosives are used for this purpose. Originally, they gave undetonated cartridges liable to deflagration, but this problem has been overcome by enclosing the cartridges in a plastic sheath and redesigning the explosive to be more powerful. In Britain a consideration of the basic requirements of the explosive has led to an approach in a different direction. It is considered that three special hazards to be overcome relate to ignition of firedamp, desensitisation of the explosive and deflagration.

The particular hazard of firedamp ignition relates to the circumstance when a hole fired early in the round breaks the coal at another hole and exposes the explosive before it detonates. This is simulated by firing gallery tests similar to those described above, but with inverse initiation of a column of cartridges which reach to 5 cm from the mouth of the cannon. Twenty shots of 567 g explosive are fired into methane/air mixture and no ignitions may occur. The explosive must in addition pass the second and third tests applied to P3 explosives.

Another hazard is that cartridges fired early will cause compression in other holes and thus desensitise the explosive. To overcome this a sensitive explosive is required and one of semi-gelatine type is used in Britain.

Deflagration, or smouldering, is more likely to occur with this technique of off-the-solid blasting than in other methods. This is because coal dust makes deflagration more likely and more dangerous; because there is the tendency noted above towards desensitisation; and because of the high pressures which can be developed in boreholes in view of the good stemming made necessary because blasting is not towards a free face (see p. 138). Deflagration is tested by enclosing two cartridges in a steel tube with controlled venting and with a 10 cm gap between the cartridges filled with coal dust. When one cartridge is detonated the other should not deflagrate if the venting is more than minimal. The nature of the fuel used has a major influence on the liability to deflagration: salts of organic acids are often used. The first P5 explosive (Dynagex) used a special combined fuel and coolant.

P4/5 Explosives

The requirement to use both P4 and P5 explosives in a single mine working leads not only to inconvenience in supply but also to a potential hazard should either explosive be used for the wrong purpose. This has led to the development of P4/5 explosives which pass the tests for both P4 and P5 types and which can therefore be used either for ripping or for firing off the solid with delays.

These are essentially P4 explosives which incorporate the fuels developed for P5 types to avoid the danger of deflagration. When fired off the solid the confinement is sufficient to ensure that the explosive develops its full power and can perform the required work.

Other Coal Mining Explosives

The first successful method of firing explosives in solid coal was the pulsed infusion technique (see p. 143). This requires explosive which will fire under a pressure of water of 1·4 MPa and which also passes suitable permitted tests, either P1 or P3. Such explosives are based on the addition of high density materials, such as barium sulphate, to a gelatine composition, so as to ensure that it will detonate at high velocity. When this is done propagation under high water pressure can be achieved. By adding appropriate quantities of cooling salts, either P1 or P3 explosives can be designed. The former propagate better under pulsed infusion conditions, whereas the latter enable the shotfirer to use the same explosive for further operations at the coal face. With the introduction of P5 explosives, pulsed infusion techniques are likely to diminish in importance, but they have the advantage of reducing airborne dust and are therefore particularly valuable in anthracite mines.

Mention should be made of devices which are not, strictly speaking, explosives, but which have an explosive effect and can be used for similar purposes in coal mining. Such devices are Airdox, Cardox and Hydrox. All consist of strong metal tubes containing at one end a bursting disc or device of known venting strength. Inside the tube is a cartridge which is caused to liberate high pressure gas at low temperature. When the pressure reaches the designed value the bursting disc breaks and the gas is liberated violently from nozzles at the end of the tube. This sudden liberation of gas causes breaking of the coal. Because of their inconvenience and other disadvantages in operation, these devices are now of limited importance.

References

TAYLOR, J., and GAY, P. F., *British Coal Mining Explosives*. Newnes, London, 1958.
Thorpe's Dictionary of Applied Chemistry, **4,** 558. Longmans Green, London, 4th ed. 1940.

For details of permitted tests, reference should be made to the Health and Safety Executive. For uniformity in text, figures chosen for the above account have been converted to their nearest equivalents in SI units.

Part II. Blasting Accessories

Introduction

IN EARLY attempts to use nitroglycerine for blasting purposes, the practice was to ignite the nitroglycerine with blackpowder charges. Whilst this caused combustion, it was soon found to be unreliable in producing detonation. Satisfactory results were first achieved by Nobel by the design of the detonator, which employed fulminating substances as ignition compounds. These materials, of a type now called initiating explosives, are characterised by the fact that even in small quantities they detonate on application of flame. Although initiating explosives are sensitive, not only to flame, but also to friction and percussion, they need be made only in such small quantities that special precautions in their handling can be taken. In the course of manufacture, the initiating explosives are enclosed in metal tubes and thereby so protected that the finished product can be handled without danger.

It is always necessary for the shotfirer to be at a distance from the explosive when it is detonated. This can be achieved in either of two ways. The earlier method was to use safety fuse which was ignited at one end and burned slowly towards the detonator, thus giving the shotfirer time to retire to safety. The development of satisfactory and reliable safety fuse was of considerable importance to the explosives industry. But if a large number of shots had to be fired, the operator might not have time to ignite the many ends of safety fuse and still retire to protection before the first explosive detonated. Therefore, a method was sought by which a number of lengths of safety fuse could be ignited by a single operation. This led to the development of igniter cord, which can be joined to the ends of the lengths of safety fuse by special connectors. The shotfirer then only has to ignite one end of igniter cord before retiring.

Although safety fuse is reliable and relatively constant in burning speed, a number of shots fired with this aid will naturally detonate at slightly different times. This can be an advantage, particularly if the times of detonation are varied by altering the length of safety fuse, or by ensuring suitable delays by the use of igniter cord. On other occasions, however, it is desirable to be able to fire a number of shots simultaneously. For this purpose detonating fuse can be employed by which ignition is caused by a detonation wave instead of burning. A detonation wave in the fuse travels at 6500 m s^{-1} so that a larger number of shots can be fired almost simultaneously. The detonating fuse itself must be initiated by a detonator, but can be used without further attachments for initiating most nitroglycerine explosives. For less sensitive explosives such as slurries the detonating fuse can be used to initiate a primer or booster of a cast explosive such as pentolite.

The alternative method of ensuring safety for the shotfirer is to enable him to fire the round of shots from a safe distance by using electrical means. For this purpose electric detonators have been developed. An electric detonator is placed in each shothole, the leads are connected into a single electrical circuit and the whole fired from a central point. The original electric detonators did not allow use to be made of the advantages described in Chapter 14, of firing shotholes in a prearranged order in time. This led to the development of delay detonators, which are electric detonators incorporating delay elements. When an electric current is passed through delay detonators they are actuated immediately, but the base charges do not detonate till after prearranged times. The first delay detonators manufactured had intervals of either 1 or 0·5 s, but more recently delays of the order of 30 to 50 ms have been produced and are in wide use.

The above represent accessories which do not directly perform blasting operations, but which are essential for the efficient and safe use of blasting explosives. For the explosives industry and for the user they are as important as the blasting explosives themselves.

CHAPTER 8

Initiating Explosives

Mercury Fulminate

Mercury fulminate has the formula $Hg(ONC)_2$. It was probably discovered by Howard in 1800 and its constitution was established by Nef. The method of preparation is known as the Chandelon process and is a complex reaction studied by Wieland. Mercury is dissolved in an excess of warm nitric acid and ethyl alcohol added to the resulting solution. Considerable bubbling occurs in the strong reaction, so this is usually carried out in capacious glass "balloons". At the end of the reaction the mercury fulminate remains as a dense precipitate which is filtered and washed several times.

Mercury fulminate is a pale brownish solid, insoluble in cold water, but dissolving slightly in hot water to a solution which does not give the normal mercury reactions. In cold conditions it is stable, but at higher temperatures gradually decomposes and loses strength as an explosive. It has a density of $4 \cdot 45$ g ml^{-1} and a velocity of detonation, when compressed to a practical density of $2 \cdot 5$, of about 3600 m s^{-1}.

When used in detonators, mercury fulminate is frequently mixed with 10 or 20% of potassium chlorate. Such mixtures have a better oxygen balance and therefore give improved and more reliable initiation of other explosives.

Lead Azide

Lead azide, discovered by Curtius in 1891, has the formula $Pb(N_3)_2$. Hydrazoic acid, HN_3, is a liquid boiling at $37°C$ and because of its sensitiveness is an extremely dangerous substance. As a strong acid it

95

gives salts with many metals and these can range from stable compounds such as sodium azide to very sensitive solids such as the copper and silver azides. The starting point in azide manufacture is sodium azide, itself made from sodamide and nitrous oxide. Sodium azide is one of the few soluble azides and therefore the salts of other metals can readily be prepared by precipitation in aqueous solution.

Lead azide is manufactured by reaction of sodium azide with either lead nitrate or lead acetate. It is a white crystalline solid, insoluble in cold water and stable on storage. It is very sensitive to friction and impact and has a velocity of detonation, when pressed to a density of $3 \cdot 8$, of 4500 m s^{-1}.

Two crystallographic forms of lead azide are important, the ordinary alpha form which is orthorhombic and the beta form which is monoclinic. The densities of these forms are $4 \cdot 71$ and $4 \cdot 93$ respectively. It was for many years believed that the beta form is the more sensitive to friction and impact and accounted for detonations which have occurred in the manufacture and handling of the substance. It is now known that the beta form is in fact no more sensitive than the alpha. Even the alpha form, when present as large crystals, is very sensitive and conditions can arise (particularly when the formation of the lead azide is controlled by diffusion effects) where spontaneous detonation occurs. Although with modern knowledge these hazards can be avoided, pure lead azide is nevertheless a dangerous compound and is now made only for military purposes.

Commercially, lead azide is usually manufactured by precipitation in the presence of dextrine, which considerably modifies the crystalline nature of the product. The procedure adopted is to add a solution of dextrine to the reaction vessel, often with a proportion of the lead nitrate or lead acetate required in the reaction. The bulk solutions of lead nitrate and of sodium azide are, for safety reasons, usually in vessels on the opposite sides of a blast barrier. They are run into the reaction vessel at a controlled rate, the whole process being conducted remotely under conditions of safety for the operator. When precipitation is complete, the stirring is stopped and the precipitate allowed to settle; the mother liquor is then decanted. The precipitate is washed several times with water until pure. The product contains about 95% lead azide and consists of rounded granules composed of small lead azide crystals; it is as safe as most initiating explosives and can readily be handled with due care.

Lead azide has virtually supplanted mercury fulminate in the manufacture of commercial detonators, having better storage properties under hot conditions and also greater initiating power for base charges. Lead azide does, however, suffer from two disadvantages. One is its ready reaction under moist conditions with copper, or copper salts. The second is its relative insensitiveness to initiation by flame, such as the spit of a safety fuse.

Many other methods of making lead azide in a safe form have been described, but the only one to have found commercial importance consists of replacing the dextrine by a small proportion of gelatine. When properly made this form of lead azide is as safe to handle as the dextrinated form and has improved sensitiveness to flame. It can therefore be used by itself in electric and delay detonators, but not in plain detonators as it is not ignited with certainty by safety fuse.

Lead Styphnate (Lead 2, 4, 6-Trinitroresorcinate)

This compound has the formula $(NO_2)_3C_6HO_2Pb$, but usually contains also one molecule of water as water of crystallisation. Trinitroresorcinol is manufactured by the nitration of resorcinol and is then usually converted to the magnesium salt, which is reacted with lead nitrate solution under warm conditions and with good stirring. The product is precipitated as a red crystalline material which can be washed by decantation and separated by filtration.

Lead styphnate is a poor initiating explosive which when dry is very sensitive to friction and impact, to electrostatic discharge, and to flame. Its main use is as an additive to lead azide to improve flame sensitiveness (see p. 101). When pressed to a density of $2 \cdot 6$ g ml^{-1} it has a velocity of detonation of 4900 m s^{-1}.

Diazodinitrophenol (DDNP, DINOL)

This substance has the following formula:

Originally prepared in 1858 by Griess, it is made by diazotising picramic acid with sodium nitrite and hydrochloric acid according to the following reaction:

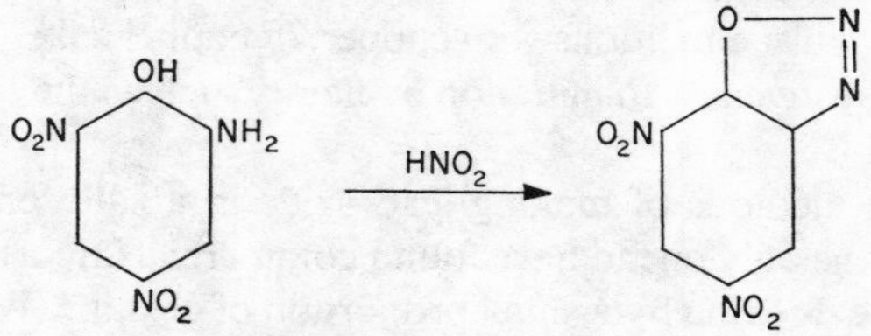

Diazodinitrophenol is a yellow powder, almost insoluble in cold water. It does not detonate when unconfined, but when confined has a velocity of 6900 m s^{-1} and a density of 1·58 g ml^{-1}. For an initiating explosive it is relatively insensitive to friction and impact, but still is powerful when confined. DDNP has good properties of storage and has found application in detonators, particularly in the U.S.A.

Tetrazene

This compound has the following formula:

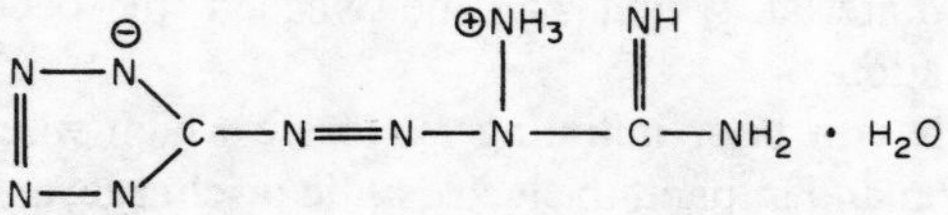

Tetrazene was discovered by Hofmann and Roth in 1910 and the structure determined by Duke. It is made by the action of sodium nitrite on aminoguanidine sulphate or nitrate under slightly acid conditions.

Tetrazene is a light yellow crystalline substance, insoluble in water and most organic solvents. The density is low under normal conditions, but on pressing can reach approximately 1 g ml^{-1}. Tetrazene is weak as an initiating explosive, and is therefore not used alone. It has no advantages to commend it for use in commercial detonators, but does find application in the manufacture of military and other percussion caps. Like diazodinitrophenol, tetrazene does not detonate when ignited in the open, but only when ignited under confinement.

Reference

Thorpe's Dictionary of Applied Chemistry, **4,** 558. Longmans Green, London, 4th ed. 1940.

CHAPTER 9

Plain Detonators

A PLAIN detonator consists of a metal tube, closed at one end and containing an explosive charge of which at least a part is an initiating explosive. The normal use is for firing in conjunction with safety fuse. The main factors governing detonator design are safety, stability on storage, certainty of ignition and initiating power.

Safety

Initiating explosives are sensitive to friction and impact so that the safety of the device must be provided by the detonator tube. For this reason tubes are invariably made of metal and proposals to use plastic have met with no success. The tubes used have a diameter of approximately 6 mm, this being a convenient size for insertion into a cartridge of blasting explosive. With a wall thickness of about 0·3 mm, such tubes are adequately strong for purposes of both manufacture and use. The safety achieved is often remarkable; indeed, occasions have occurred when detonators have been completely flattened by being run over by vehicles without exploding. Such safety cannot, however, be relied upon, and in particular any friction inside a detonator from inserted hard objects is extremely dangerous.

Metal cups are sometimes inserted in detonators to provide extra confinement for the composition. It is often claimed that such cups, by increasing the mechanical strength, also increase the safety of handling of the detonator. In the case of plain detonators this is true to a limited extent, but the difference is not of practical importance. Of much greater importance is the ensurance of the absence of grit or hard particles, the presence of which can cause dangerous sensitiveness in the detonator.

Stability on Storage

Detonators must be capable of storage for long periods under various types of climate. After such storage the detonators should still be in safe condition and also perform normally.

The original initiating explosive used by Nobel and all manufacturers for many years was mercury fulminate. This had the disadvantage of decomposing slowly in hot climates, particularly under moist conditions. For this reason mercury fulminate is no longer widely used. In most countries it has been replaced by a mixture of dextrinated lead azide and lead styphnate. In the U.S.A. some detonators are made containing diazodinitrophenol.

The introduction of lead azide led to a difficulty in the choice of metal for the detonator tube. Under moist conditions, lead azide and copper can react to form cuprous azide on the inner wall of the tube and thus in a particularly dangerous position. Therefore with plain detonators, which cannot be sealed, copper cannot be used when lead azide is employed. Such detonators are usually made from aluminium tubes, or occasionally zinc.

Certainty of Ignition

The initiating explosive used must ignite with certainty from the spit of a safety fuse. It must be remembered that the intensity of the spit can be reduced if the safety fuse is not cut squarely and also that the fuse may in practice not always be fully inserted into the detonator. Lead azide by itself is not sufficiently easily ignited to give a satisfactory plain detonator and it is therefore used in admixture with lead styphnate, which is very readily ignited by flame. The proportions of such mixtures vary from 25 to 50% of lead styphnate. Mercury fulminate and diazodinitrophenol are sufficiently sensitive to flame not to require such additives.

Although the requirement for flame sensitiveness is the main consideration for initiating explosives for plain detonators, others are important in manufacture. The explosive must be capable of compression into a coherent mass and at the same time leave the equipment free from adhesions. Lead azide can be somewhat deficient in cohesion, and to improve this a small proportion of tetryl is sometimes added to the mixture. In Britain, but not in other countries, it is common practice to add a small proportion of fine flake aluminium to the mixture, the purpose being to improve lubrication in the presses and to prevent adhesion to the punches.

Initiating Power

To some extent detonators and blasting explosives are designed together so that one will initiate the other with certainty. Experience with fulminate detonators showed that the No. 6 detonator containing 1 g of a mixture containing 80% mercury fulminate and 20% potassium chlorate could reliably initiate the nitroglycerine and TNT explosives in use. Such a detonator can be considered the normal minimum strength in most countries. Changes in design have, however, led to an overall increase in strength so that the No. 6 detonators now being sold are in fact more powerful than the original.

For both technical and economic reasons, current detonators contain a base charge of high explosive which provides the main initiating power of the device. The most satisfactory high explosives for use as base charges are PETN, tetryl and RDX, and of these the first is by far the commonest, because of its sensitiveness and relatively low cost.

The strength of the detonator depends to a large extent on the weight of base charge employed, but for a given weight of base charge the strength may be increased by the following measures:

1. Increasing the pressing of the base charge to give a higher density. This leads to a higher velocity of detonation and therefore greater initiating power.
2. Use of a narrower diameter tube. This leads to a longer length of base charge which in general is more effective in initiating blasting explosives.
3. Increasing the thickness of the metal of the tube. Within limits this gives greater confinement and therefore more effective detonation of the base charge.

In practice detonators of increased strength are made by increasing the weight of the base charge rather than by other means. The only important exception to this is the German Briska detonator (see p. 103).

Construction of Plain Detonators

The general construction of a British plain detonator is shown in Fig. 9.1; other countries in general adopt similar designs.

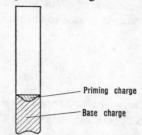

FIG. 9.1. Plain detonator.

The tube is of aluminium, 6·4 mm external diameter and with an overall length of 35 mm; the wall thickness is 0·3 mm. The base is dimpled inwards, as this construction gives a miniature cone charge and therefore a somewhat increased initiating power at the end of the detonator (see p. 158). These tubes are made from aluminium sheet by drawing into cups, annealing and then drawing by several stages to the final size. After drawing, the tubes are trimmed to length to give a neat open end. The tube length is designed to give an unoccupied space of approximately 2·5 cm in the finished detonator. In use the safety fuse is inserted into this portion of the tube, which is then crimped round the fuse to hold the assembly together.

The base charge consists of PETN and the No. 6 detonator contains 0·24 g. The charge is pressed at 28 MPa, this pressure being chosen to give adequate density and velocity of detonation without imposing undue strain on the tube with any liability of causing splitting. By use of a flat, or dimpled, punch, the surface of the base charge can be produced either flat or with a conical depression. The latter construction gives advantages in the initiation of the base charge, but by blocking of punches can give rise to more frequent faults in manufacture.

In Briska-type detonators the base charge is loaded in two portions. The first is pressed at about 140 MPa, the tubes being specially supported. The second addition is pressed to a normal pressure and serves to initiate the more highly compressed explosive. In Britain it is preferred to increase detonator strength by increasing the weight of charge; thus the strongest commercial detonator (No. 8 Star) contains 0·8 g PETN in a tube of slightly increased diameter.

The normal initiating charge in a British detonator is lead azide modified with gelatine. In the case of plain detonators a small proportion of lead styphnate is added to the azide to ensure satisfactory ignition from safety fuse.

Manufacture

The tubes after manufacture are examined for any faults and then loaded, open end uppermost, in holes of a carrying plate, usually 181 at a time. The plate is first placed in a charging machine for PETN. This machine has a charging plate with 181 holes of standard size. It rests on a base plate, also with 181 holes, but initially displaced from these holes by finite horizontal distance. A charging hopper containing PETN is moved backwards and forwards across the charging plate until the holes are full. The plate is then slid until the holes coincide with those of the base plate, so that the PETN in measured volume drops through into the detonator tubes. The charging plate is then removed and placed under a press with 181 punches, where the base charge is consolidated.

The ingredients for the priming composition are dried separately under carefully controlled conditions. They are then weighed and carefully mixed behind blast protection, usually in a simple conical cloth bag. The mixing may be facilitated by adding a number of rubber balls. The mixture is then passed through a coarse sieve, still by remote control, into bags for transport to the loading unit.

The charging of the detonators with priming composition is carried out by a method similar to that used for the base charge, although the quantities involved are smaller and the volumes added similarly small. Again after charging, the charging plate is taken to a press where the priming charge is compressed to a total load of about 15 tonnes. Both charging and pressing of the priming composition must be carried out under full remote control.

The detonators as made may be dirty, that is have free priming composition as a dust on the walls. They are therefore transferred to a drum containing a quantity of dry sawdust. The drum is rotated for a predetermined time and the detonators then removed and shaken free of sawdust. In this way any free priming composition is removed and also any loose charges would be broken down. The detonators are then carefully inspected and finally packed. Most detonators in Britain are now made on fully automatic equipment. The physical principles are the same except that the final removal of loose priming composition is by vacuum cleaning.

Testing of Detonators

The essential property of a detonator is its initiating power, but this is very difficult to measure. The best type of test is carried out by manufacturing a series of explosives of decreasing sensitiveness and attempting to initiate them with the detonator under examination. The first of these tests is known as the Esop test and employs picric acid as a standard explosive, gradually diluted with olive oil as a means of reducing sensitiveness, but the results depend critically on the grist of the picric acid. A rather more reproducible test has been based on milled TNT ammonium nitrate explosives desensitised with salt. A standard range of such explosives is used, and for each composition the sensitiveness can be further varied by pressing the explosive to a varying extent. In tests of this sort it is important to assess the detonation of the high explosive by indentation of a lead plate or by a lead crusher. Tests of the type described here are tedious to carry out and also give results which depend to some extent on the nature of the high explosive used.

For routine testing simpler tests can be employed. One such uses a lead plate 0.5 cm thick and resting on an iron support with a circular hole of radius 18 mm. The detonator is fired when standing vertically on the plate above the centre of the hole and in contact. The dent or hole in the lead plate is compared with a series of standard deformations from detonators of increasing strengths. This test measures only the end blow of the detonators, whereas in initiating explosives the side blow can be at least as important. Another test commonly used is to fire a detonator lying on a standard iron nail and observing the bending produced in the nail. Such a test can give a rough indication of the side blow of the detonator. All tests of this sort should be used only for comparing constancy of manufacture in a single type of detonator. Their use in comparing detonators of different constructions can be grossly misleading.

Reference

MARSHALL, A., *Explosives Vol. 2, Properties and Tests*. Churchill, London, 2nd ed. 1917.

CHAPTER 10

Electric Detonators

ELECTRIC detonators consist of plain detonators to which a device has been added for generating a flash on receipt of an electric impulse. The first devices of this sort were invented by Watson in England and Benjamin Franklin in the U.S.A. and utilised electric sparks for igniting gunpowder charges. The modern method of igniting a flashing composition by passing the electric current through a fine wire was invented by Hair in 1832. Suitable ignition systems were therefore available long before detonators themselves were invented.

It is convenient to distinguish four systems by which the electrical energy can be converted into a flash.

1. A bridge of resistance wire is connected across the ends of two leading wires and surrounded by a loose charge of flashing composition or of initiating explosive.
2. A bridgewire similarly attached to the ends of leading wires is coated with a coherent layer or layers of flashing materials.
3. The leading wires are attached to a fusehead, originally a separate entity and containing a bridgewire surrounded by coatings of flashing composition.
4. The leading wires are again attached to a fusehead, but this contains no resistance wire, having a flashing composition which is itself made sufficiently conducting to ignite when a high voltage is placed across it.

The construction of these types is shown in Fig. 10.1 (p. 108).

Type 4 gives a high tension detonator, so called because it requires at least 36 volts to fire it. The electrical conductivity is achieved by adding graphite to the flashing composition, but control of electrical properties is difficult and the finished product liable to change on mechanical handling or on storage. For this reason detonators of this type are no longer employed.

Detonators of the first three types all employ bridgewires and therefore require low voltages for their initiation. The choice is one of manufacturing convenience, but as most makers prefer type 3, only this will be described in detail.

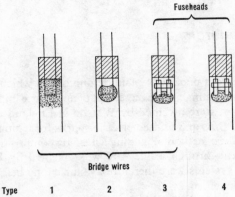

FIG. 10.1. Types of detonator ignition systems.

The flashing device, without leading wires, is known as a fusehead and its construction and properties will be considered first.

Fuseheads

The first successful type of fusehead was invented by Krannichfeldt in Germany. This "sandwich" type of construction is used in many countries, including Great Britain, and is illustrated in Fig. 10.2.

The manufacture of a sandwich fusehead proceeds in the following manner. Brass or other metal foils are fixed on each side of a sheet of pressboard with a suitable adhesive. The pressboard is then stamped into combs of the shape shown in Fig. 10.3 and steps are cut in the tips of the heads. Fine resistance wire is stretched across the heads and soldered to the foil on each side of the pressboard. These operations were originally all carried out by hand; now many are carried out mechanically.

A number of these combs are fixed side by side with spacers in a carrying plate. The tips of the combs are then dipped into a solution of paint-like consistency containing flashing compositions to be described

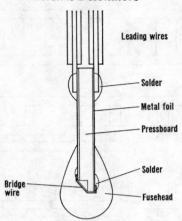

Leading wires

Solder

Metal foil

Pressboard

Solder

Bridge wire

Fusehead

FIG. 10.2. Sandwich fusehead.

later. A series of dips is given with drying between each stage. The combs are then removed, cut into individual fuseheads and at the same time tested for suitability of resistance.

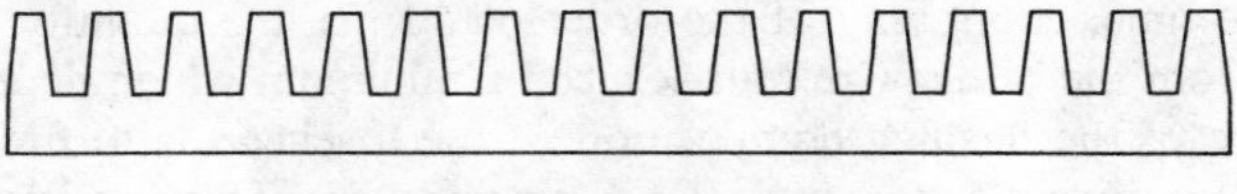

FIG. 10.3. Fusehead comb.

An alternative design of fusehead much used in Europe is known as the Unifoil fusehead, invented by Schaffler. The principle of the production of modern fuseheads of this type is shown in Fig. 10.4. A metal strip, usually of steel, is first stamped as shown at *A*. A band of plastic is then moulded as shown at *B*. Further stamping produces pole pieces as at *C*, the ends of which are first bent upwards and then clinched over resistance wire which is introduced into the angle formed. The bridgewire is then welded to the pole pieces. Excess plastic and wire is removed and the combs thus formed are dipped in the same manner as described for sandwich type fuseheads. After dipping, the individual fuseheads can be removed and tested. In the original Schaffler design the foil was tinned on one side so that the bridgewire could be soldered in place rather than welded. The plastic described was not used, but the pole pieces were instead held in place by insulated metal clips. Both types of fusehead are currently made.

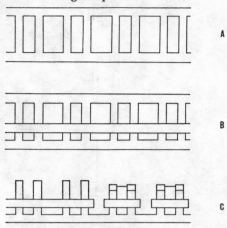

Fig. 10.4. Unifoil fusehead manufacture.

The properties of a fusehead depend very largely on the bridgewire employed. The energy liberated per unit length of wire is proportional to I^2r, where I is the current applied and r the resistance. If ignition is to occur in a finite time, commonly of the order of 50 ms, the quantity of heat evolved from the bridgewire must exceed a minimum which depends on the nature of the flashing composition. If the fusehead is to fire with a suitably low current, for example 0·5 A, a bridgewire of high resistance per unit length must be employed. Suitable high resistances can be achieved by using nickel chromium alloys in gauges of about 48 S.W.G. For fuseheads to be fired with a different current the bridgewire can be made of other materials, or of other diameters.

The first dip given to a fusehead is known as the flashing composition and is of particular importance. Originally copper acetylide was used for this purpose, but it has been superseded by more stable materials. Three common compositions are based on lead picrate, lead mononitroresorcinate and a mixture of charcoal and potassium chlorate respectively. These materials are suspended in a solution of nitrocellulose in amyl acetate and amyl alcohol, known as Zapon. One or more dips, with intermediate drying, give a layer of suitable thickness.

The second dip, or series of dips, is intended to provide a suitable flame which can ignite the initiating explosive in the detonator. These dips are commonly based on potassium chlorate and charcoal, again suspended in

Zapon. The fusehead is finally given a coat of a nitrocellulose solution as a protective layer and this coat may well be coloured with pigment so that the type of fusehead can readily be identified.

Assembly of Electric Detonators

Leading wires are made from tinned iron or copper, of gauges from 23 to 25 S.W.G. They are insulated with plastic, commonly PVC, coloured to enable different types of detonator to be readily distinguished. Two such wires are wound, usually by machine, to form figure eight coils of total lengths ranging from 1·2 to 5·5 m, according to requirements. The coils are held together by a few turns of the two leading wires around the bundle. The four ends of wires are stripped of insulation to a suitable distance.

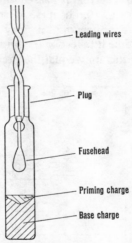

FIG. 10.5. Electric detonator.

A small cylindrical plug is fixed near one end of the leading wires, either by threading on a pre-formed plug, or by moulding plastic around the wires at a suitable point. The bared ends of the wires immediately adjacent to the plug are then soldered to the two conducting foils of the fusehead. In the case of Unifoil fuseheads the wires are soldered or welded to the metal tags. The completed assembly is then inserted to the correct depth inside a plain detonator and the tube crimped round the plug to hold the detonator together and to provide a watertight seal. The structure of the completed detonator is shown in section in Fig. 10.5.

For detonators to be used where electrostatic charges may occur, it is desirable to prevent sparking from the fusehead to the case, should a high voltage be generated on the leading wires. For this purpose, an insulating sheath is inserted, either into the detonator tube or else immediately around the fusehead.

Electric detonators are made from aluminium or copper tubes, the latter being for use in coal mines where aluminium may lead to a possibility of ignition of methane/air mixtures.

Firing Characteristics of Electric Detonators

When an electric current is passed through a fusehead, the sequence of events shown in Fig. 10.6 occurs. After a time known as the excitation time, the fusehead ignites and this may or may not cause rupture of the bridgewire. The time it takes for the bridgewire to be broken is known as the lag time and this may equal the excitation time, or be as long as the

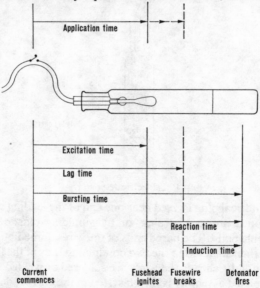

FIG. 10.6. Sequence of events in firing electric detonators.

bursting time, which is the time between the application of current and the explosion of the detonator. The interval between the ignition of the fusehead and the bursting of the detonator is known as the reaction time and the interval between the breaking of the fuse wire and the bursting of the detonator is known as the induction time.

For any particular kind of fusehead these times depend on the magnitude of the electric current applied. If a steady current, less than the minimum firing current of the fusehead, is applied, the bridgewire does not get hot enough to ignite the flashing dip, so that the fusehead remains unaffected. As the magnitude of the current is increased, the fusehead ignites with gradually reduced excitation times and therefore with gradual reduction in overall bursting time. In applications where short bursting times are essential, as in seismographic work, not only are special detonators used, but the firing currents employed are maintained as high as possible.

Although steady currents are sometimes applied to fire detonators, it is equally common to use exploders which have a finite time of current application. This may be either incidental, because the mechanism is the discharge of a charged condenser, or deliberate in the case of coal mining applications where finite time of pulse is desirable for safety reasons. Such finite pulse times, usually of the order of 4 to 5 ms, prevent the possibility of later sparking between leading wires, etc., in the presence of methane/air mixtures. When a pulse of current is used to fire a detonator, the time for which it is applied is known as the application time. If the detonator is to be fired successfully, the magnitude of the current must be such that the application time of the exploder exceeds the corresponding excitation time of the fusehead.

When a large number of shots are to be fired it is common practice to connect the electric detonators in series and fire them all with the single application of an exploder. The current from the exploder ceases when the first fuse wire breaks and therefore at the time of the shortest lag time of the detonators involved. This time will depend on the magnitude of the firing current, as shown in Fig. 10.7. In this figure the curve A represents the minimum lag time shown by any detonator in a large number selected from those made to a single specification. Owing to minor variations in ingredients, etc., during manufacture, the average lag time will be at each point longer than that shown on the curve. Also shown in Fig. 10.7 in curve C is the maximum excitation time characteristic of the fuseheads involved. Again because of random variations during manufacture, the average excitation time will be less than that shown in the curve. For any single fusehead, of course, the excitation time cannot exceed the lag time. If a number of fuseheads in series is to be fired successfully, the shortest lag time of the fuseheads involved, which determines the application time, must exceed the longest excitation time present in the circuit. This is the case in Fig. 10.7 for currents greater than those corresponding to point Y where the two curves cross. Point Y is therefore the minimum series firing current, which is in all cases greater than the ordinary minimum firing current for a single fusehead.

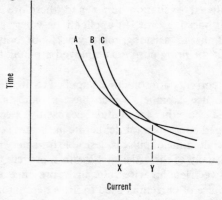

A : Minimum lag time

B : Minimum excitation time

C : Maximum excitation time

X : Minimum firing current for single detonator

Y : Minimum series firing current

FIG. 10.7. Series firing of electric detonators.

The curves of Fig. 10.7 apply only to fuseheads made to the same specification. If fuseheads are made with different bridgewires or different flashing dips, the corresponding curves could be completely different and there may be no point at which the minimum lag time of one type of detonator exceeds the maximum excitation time of the other. Under these conditions series firing would not be practicable and it follows that it is not feasible to use different types of detonators in a single series firing circuit.

CHAPTER 11

Delay Detonators

Introduction

As is discussed in Chapter 14, it is often advantageous if a number of detonators are fired not simultaneously but in a predetermined order. With ordinary electric detonators, such sequential firing can be achieved only by the use of complicated wiring circuits and of special switches attached to the exploder. These disadvantages are overcome by the use of electric delay detonators. In delay detonators a time lag is deliberately introduced between the firing of the fusehead and the explosion of the detonator. If a number of such detonators are fired in series the fuseheads all ignite simultaneously, but the detonators fire at predetermined intervals depending on the construction of the delay detonators. Only a single wiring circuit and single exploder is needed to carry out the actual firing operation.

The earliest delay detonators were introduced in Great Britain in 1910 and had a length of safety fuse between the fusehead and the detonator proper. The construction is shown in Fig. 11.1. The fusehead was sealed in a paper tube by a sulphur plug and this tube in turn sealed into a metal sleeve. The other end of the sleeve was crimped onto safety fuse and the free end of the safety fuse crimped into a detonator. An important part of the construction of such a detonator is the presence of a small hole in the metal sleeve, usually covered initially with adhesive tape. The firing of the fusehead and the safety fuse punctures the adhesive tape and the hole then provides a vent from which the hot gases can escape without an increase in pressure sufficient to cause too rapid burning of the safety fuse. The emission of hot gases which is essential with this structure of detonator is a serious disadvantage in practice as it can lead to premature ignition of the primer of high explosive. For this reason, and because of relative inaccuracy in delay times, this type of delay detonator is now only rarely used.

A real advance in the construction of delay detonators was made by Eschbach, who introduced delay compositions which evolved so little gas that there was no longer need to vent the detonator. This eliminated risk of pre-ignition of high explosive and also made it possible to provide fully waterproofed assemblies which gave delay times much more regular because of the removal of variable venting effects. Delay detonators of this type have now virtually superseded all others.

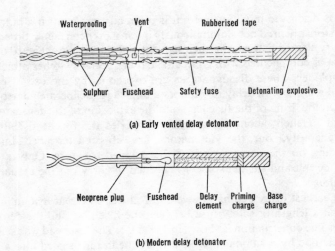

(a) Early vented delay detonator

(b) Modern delay detonator

FIG. 11.1. Construction of delay detonators.

Gasless delay detonators are manufactured to fire after pre-arranged delay times, each manufacturer providing a series of detonators with intervals of usually 1 s or 0.5 s. More recently a further type of delay detonator has been introduced, usually known as the millisecond, or short delay detonator. Again, manufacturers provide a series of fixed delay times, but in this case the interval between each number of the series is much shorter and may be from 25 to 50 ms.

Delay Compositions

The mixtures used in providing delay elements contain an easily oxidised element, often a finely divided metal, together with an oxy-salt or an easily reduced metal oxide. The first of such mixtures to attain wide use was that employed by Eschbach and consisted of a mixture of elemental antimony and potassium permanganate. When such a mixture is ignited, the antimony is oxidised to antimony oxide and the potassium permanganate reduced either to the manganate or to a mixture of potassium and manganese oxides, depending on the proportions present. Whilst a little gas is formed in a side reaction of permanganate decomposition, the amount is small and the pressure developed inside a detonator is not large. The reaction products form a solid slag of indeterminate composition. The proportions commonly used vary from 55 to 70% of permanganate with 45–30% of antimony.

In the U.S.A. a common mixture for delay elements consists of selenium and barium peroxide which reacts to give selenium oxides and barium oxide. Usual proportions are 85% of barium peroxide with 15% of selenium.

For the manufacture of millisecond delay detonators, faster burning compositions are necessary. Many have been suggested, but the two most commonly used are based on silicon mixed with either red lead or lead dioxide. Typical mixtures would contain between 30 and 50% of silicon and the remainder as oxidising material.

The burning of a column of delay composition takes place by the passage of a reaction front along the column. The temperature profile of this reaction front can be measured by the use of suitable thermocouples and recording instruments. By analysing the shape of the front it can be shown that the reaction is a solid/solid reaction initiated by thermal conduction of heat through the unreacted material. It follows that to obtain reproducible reaction rates there must be (1) constant amount of solid to solid contact and (2) constant thermal conductivity.

To attain adequate surface contact between the reacting solids it is necessary to use the ingredients in fine condition and to bring them into contact by pressure. To maintain both solid contact and thermal conductivity correct, the density of the column is controlled at a uniform value. The manufacture of satisfactory delay compositions, therefore, entails the provision of suitably sized ingredients followed by adequate mixing and accurate pressing.

Manufacture of Delay Elements

Delay compositions are not normally pressed direct into detonator shells because of the irregular delay times which result from the relatively low compression thus possible. Instead they are made into delay elements by either of two basic methods. The more general method comprises pressing into brass or aluminium tubes which are a sliding fit into the detonator shell and which are thick enough in the wall to withstand heavy consolidating pressures. The other method consists of filling a lead tube with the delay composition and drawing it down in diameter by conventional means. This process gives a length of lead tube filled with delay composition ready for cutting into the required lengths. The process of drawing itself consolidates the delay composition.

For either method, the delay composition must be made in a free-flowing form. In the case of the antimony/potassium permanganate mixtures, this is done by mixing fine antimony and fine potassium permanganate and then pelleting the mixture in a press. The pellets are then broken down and after sieving to remove fines a suitable free flowing granule is obtained. For other types of compositions this method is not generally applicable because of the difficulty of forming pellets by pressure. Instead the mixture is granulated with a small amount of nitrocellulose solution by working in a suitable mixer.

The actual filling of brass or aluminium delay elements is similar in principle to the filling of plain detonators described above. The loading is, however, carried out in a number of stages so as to ensure uniformity of density of the column throughout its length.

The manufacture of lead elements commences with the filling of a lead tube, probably 1 m long, by tamping in delay composition. This tube is gradually drawn down until it is of the correct external diameter to be a sliding fit into the detonator shell. It is then accurately cut to the lengths required to give the delay times.

Assembly of Delay Detonators

Assembly commences with a plain detonator containing its base charge and initiating explosive. These detonators are held in plates and the processes are insertion of suitable delay elements, followed by pressing in the normal way. They are then checked by X-ray examination to ensure proper assembly. A fusehead, complete with plug and leading wires, is placed in position in each detonator and crimped to give the finished product.

Design of Delay Detonators

Some special points of design may be noted. Thus, some delay compositions, particularly antimony/potassium permanganate mixtures, have relatively poor igniting powers for initiating explosives. This can be controlled by suitable design of the delay elements, or alternatively, an intermediate layer of priming explosive of high sensitiveness, such as lead styphnate, can be introduced between the element and the lead azide charge. Another point to be watched is the possibility of "flash past", particularly in the shorter delay periods where the lengths of the elements used are at a minimum. In the past it has sometimes been the custom to introduce a layer of delay composition above the initiating explosive before placing in the delay element. Nowadays, however, improved overall design and technique has made this unnecessary.

Some delay compositions are difficult to ignite and ordinary fuseheads may not be adequate for this purpose. In such cases special fuseheads are used which commonly contain cerium powder, or some similar additive which burns with the evolution of a large amount of heat.

In the design of the detonator attention must also be paid to the effects of the liberation of gas which, though small, is still sufficient to require attention. As the speed of the delay composition is affected by pressure, it is necessary that the free space in the detonator should be carefully controlled. Also, the plug which seals the leading wires in place must withstand this pressure for more than the delay period of the detonator. This task is made more difficult by conduction, along the metal walls of the detonator shell, of heat liberated by the fusehead and the delay composition. Plastic plugs in particular are liable to soften and be ejected from the detonator with probable failure of burning of the delay column.

Delay detonators for use in coal mines must be constructed so as not to ignite methane/air mixtures even if fired accidentally outside a cartridge of a blasting explosive. This requires a suitable selection of fusehead and the provision of the delay element in a form which will not produce large particles of hot slag on burning. In the British design the delay elements are of the lead type, but with five or six narrow cores instead of the single central column. Such "Carrick" detonators will satisfy even the rigorous British test, which involves firing the detonators in a simulated break, formed by two steel plates 5 to 15 cm apart and filled with methane/air mixture.

CHAPTER 12

Detonating Fuse

Introduction

A detonating fuse is a narrow cord of explosive which is capable of detonating from one end to the other at high velocity and therefore of transmitting detonation almost instantaneously from one cartridge to another some distance away. The thinness of detonating fuse makes it essential to use explosives which are capable of propagating reliably at such small diameters. The earliest fuses of this sort were made with either dry nitrocellulose or more commonly with mercury fulminate phlegmatised with wax. Such fuses were irregular and dangerous in use as they were sensitive to initiation by shock. The first successful detonating fuse was Cordeau, which was made in the following manner. TNT was cast into a lead tube and allowed to solidify. The whole tube was then drawn down to a diameter of 4 mm. This drawing process at the same time broke down the cast TNT and brought it into a sensitive form which would propagate whilst still confined by the lead tube. Cordeau had a uniform velocity of detonation of about 5000 m s^{-1} and was also safe to handle.

From about 1930 onwards a new type of detonating fuse appeared on the market. This consisted of a core of PETN with textile and plastic coatings. Compared with Cordeau, this detonating fuse has several advantages in being more flexible, more easily jointed, lighter in weight and cheaper to manufacture. Under such names as Cordtex and Primacord it has completely replaced Cordeau in normal use.

Detonating fuse is made by either a dry process or a wet process. The former is generally more popular because of its lower costs for ordinary types of fuse. The latter is, however, used particularly in North America.

Manufacture by the Dry Process

The PETN must be sufficiently fine to propagate reliably in small diameter; at the same time it must be relatively free flowing. These two properties are to some extent contradictory, but suitable grades can be manufactured by careful attention to the method of precipitating the PETN during purification.

The PETN is placed in a hopper with a conical base leading to an orifice. This is shown schematically in Fig. 12.1. A centre cord of textile, usually cotton, passes down through the hopper and assists the flow of the PETN.

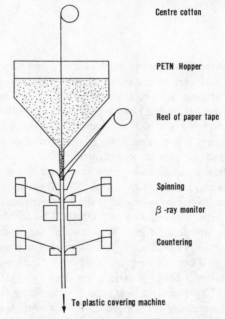

Centre cotton

PETN Hopper

Reel of paper tape

Spinning

β -ray monitor

Countering

To plastic covering machine

FIG. 12.1. Principle of dry process for detonating fuse.

Immediately below the nozzle a tube is formed from a strip of paper which is bent round in a forming die. The paper tube thus formed and containing relatively loose PETN passes through a second die, at which stage it is also spun with textile, often in this case jute to give maximum tensile strength. This second die is smaller than the first and therefore exerts a consolidating effect on the core of the fuse. The fuse is finally countered and covered with plastic by extrusion.

In any process for manufacturing detonating fuse it is important to ascertain that there are no gaps in the PETN core. Such gaps would cause failure of the fuse to propagate detonation. A test is therefore done, either with a roller diameter detector or by beta-ray monitor to determine the presence of the core. The roller diameter detector consists of two spring-loaded rollers which approach each other in the absence of a core and indicate the fault. A beta-ray monitor determines the total weight of matter in the fuse by measuring its beta-ray absorption. Any lack of a core then is indicated in an alarm or other device. The fuse is also examined at various stages of manufacture for flaws in textile or plastic covering.

Manufacture of Detonating Fuse by the Wet Process

The PETN used in the wet process is finer than that required for the dry process. It is no longer required to flow freely in the solid dry condition, but is suspended to a fluid paste in water with a suspending agent. This paste is then braided with textile fibre and dried in a continuous drier. The remaining stages in the process consist of countering with textile, covering with plastic and possibly final braiding. The speed of operation is determined by the rate of drying of the PETN paste and is therefore slower than that of the dry process where the limiting speed is usually the operation of the countering machinery.

Properties of Detonating Fuse

The velocity of detonation of a fuse is about 6500 m s^{-1}. This is attained in the dry process by adequate consolidation of the core during manufacture. During the wet process the necessary density is attained automatically by the method of preparing the core. Fuse usually contains about 10 g PETN per metre length.

A satisfactory fuse must be capable of propagating laterally from one line to another. In this way it is practicable to join lengths of fuse by ordinary knots, such as the reef knot. In test, the usual method is to lay two strands of fuse at a measured distance apart and ascertain that one initiates the other. The gaps necessary for satisfactory use are small and in the region of 6 mm. Alternatively, the test can be for propagation between fuses separated by varied thicknesses of card.

The tensile strength of the fuse is of some importance, as it is common practice to lower charges down a large hole by means of the detonating fuse. A strength of 35 kg is adequate, but higher strengths can be achieved when necessary by the addition of further amounts of textile.

Further properties which a detonating fuse should have are the ability to initiate blasting explosives (tested with suitable relatively insensitive mixtures usually of TNT and ammonium nitrate); resistance to low temperatures without cracking on flexing and to hot storage without desensitisation; and toughness to prevent damage from stones, etc. The fuse must always be waterproof and must often withstand diesel oil, which can separate from ANFO.

Low Energy Detonating Cord (LEDC)

When detonating cord is not required directly to initiate high explosives, but solely to transmit detonation from one place to another, it is sometimes an advantage to use a cord with a very low charge weight. Two types of such cord are at present available in certain countries.

The first type to be introduced consisted of PETN or RDX in a metal sheath drawn down to a small diameter. More recently, cord made by the wet process, covered with plastic instead of metal, has been introduced in North America and is available in charges down to 0·6 g m^{-1}. The methods of manufacture of these cords are believed to be similar to that of Cordeau and the wet detonating fuse process described above.

When used for laying above ground for connecting shots in civil engineering, quarrying, etc., a low energy detonating cord has the advantage of producing much less noise than the normal grade. This avoids any requirement of covering the cord with earth or sand when used in populous areas. Another use for the cord, particularly the plastic variety, is in connection with blasting with ANFO. The cord itself has insufficient initiating power to cause ANFO to detonate and can be attached to a special delay detonator. In this way propagation between holes by detonating fuse can be achieved with the extra advantage of the introduction of suitable delay times between individual holes. The particular advantage of this method with ANFO is the absence of electric fuses and therefore complete safety in dry conditions where static electricity may be generated. Special connectors must be used if LEDC is to be joined to ordinary detonating fuse.

Nonel

As the loading density of a powdered solid explosive is reduced the detonation velocity becomes less. This follows from equation (5) of Chapter 2, as p_2 is smaller the lower the density. Calculation shows that at the lowest densities the detonation velocity tends towards a limiting lower value, typically about 2000 m s^{-1}. This is also the detonation velocity in a dust explosion of these substances.

Nonel fuse, invented by Nitro Nobel AB in Sweden, consists of a thick plastic tube of bore about 1 mm, the inside surface of which is dusted with a small amount of powdered high explosive. If a shock wave is formed at one end of the tube the explosive powder is raised to a dust and a stable detonation at velocity 2000 m s^{-1} proceeds indefinitely along the fuse. The plastic itself is unaffected and the only outside effect is a flash of light seen through the tube walls. This therefore is an extremely safe method of propagating a detonation from one place to another.

In practice lengths of fuse are supplied sealed in the factory to prevent ingress of dust or moisture. The detonation may be started in various ways such as a small detonator or by firing a blank cartridge. At the other end of the fuse the flash is sufficient to ignite the priming explosive of a plain detonator or the delay element of a delay detonator. Plastic mouldings are available so that branch lines may be introduced. It is therefore possible to fire from one point any number of detonators in any predetermined delay sequence. The system is very simple to operate and is immune to stray electric currents in the ground due to electric circuits or lightning discharges.

CHAPTER 13

Safety Fuse

WHEN blackpowder was used for blasting, the original method of initiating the charge was by means of a straw or goose quill filled with loose blackpowder, and ignited with paper or string coated with blackpowder paste. Such initiation was irregular and caused many accidents. In 1831 William Bickford introduced safety fuse which consists of a core of blackpowder enclosed in textile sheaths and suitably waterproofed. Regularity of burning is of extreme importance and in Great Britain the fuse must burn within set limits, 87 to 109 s m^{-1}. It must also not emit any side sparks during burning, nor be capable of igniting similar fuses placed alongside it.

Manufacture

Safety fuse is made from fine grain blackpowder, although certain processes use mill cake, or indeed blackpowder ingredients themselves. The normal process is illustrated further in Fig. 13.1.

Blackpowder (fuse powder) is fed down a tube from a safety loft into a spinning die, usually made of hardened steel or tungsten carbide. At the same time, centre cottons pass down with the blackpowder so as to maintain adequate flow. Also introduced into the spinning die are the spinning yarns of jute, the purpose of which is to enclose the blackpowder core for later processing. This processing consists of consolidation by passing through further dies of smaller diameter and then countering in textile yarns, usually jute.

The semi-fuse thus produced is waterproofed by passing through troughs of molten bitumen or by coating also with plastic.For higher grade safety fuse, a further countering of textile, such as cotton, is followed by a final varnish, which may be coloured for identification purposes.

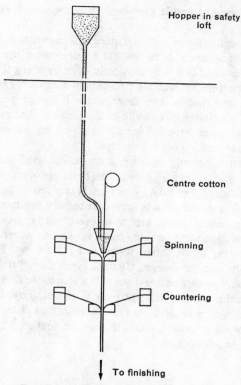

Hopper in safety loft

Centre cotton

Spinning

Countering

↓ To finishing

FIG. 13.1. Principle of safety fuse manufacture.

To obtain the best results, the fuse powder should have a relatively low potassium nitrate content, and a typical formula would be

Potassium nitrate	65%
Sulphur	24%
Charcoal	11%

Such a powder gives less smoke than the 75% nitrate composition and eliminates side sparking.

During manufacture it is important that the fuse should not be allowed to stand in the molten bitumen, or the core may be penetrated. The time for such faults to develop depends on the construction of the fuse, and is shorter the lower the potassium nitrate content of the powder and the thinner the covering of textile provided by the spinning and countering yarns.

To maintain regularity of burning speed it is essential that high quality fuse powder is employed. It is normal to determine the quality from samples before use and this is conveniently done by filling the powder into a lead tube, which is then rolled down to a small diameter. The regularity of burning of the lead fuse is then determined. If a high-grade powder is employed, then regularity of burning of the finished fuse may be secured provided that sufficient attention is paid to giving regular confinement by control of the construction of the envelope.

Safety fuse covered only with bitumen can be used with care in dry conditions but normal practice requires better protection from rough handling and moisture. For many years this was given by gutta percha and textiles but the former material is no longer readily available. Coatings of plastic such as polyethylene are now used. Early products had the disadvantage of speeding up in burning under water and causing irregularity of firing of shotholes under wet conditions. This defect has been removed by careful design. Current fuses therefore have excellent resistance not only to abrasion and water but also to the fuel oil in ANFO.

In the Fritzsche process blackpowder is not employed, but only the mixed ingredients of potassium nitrate, sulphur and charcoal. These are fed into the process by special methods, but the final construction and use of the fuse remain unchanged.

Properties of Safety Fuse

As the blackpowder core of a safety fuse burns, it produces gases which must escape. At the same time the heat of the combustion melts the bitumen and plastic and thus produces side venting through the textile layers. This results in the production of an increased but constant gas pressure, determined by the equilibrium between gas generation and gas lost sideways. As the rate of burning of blackpowder depends markedly on the pressure, it is this process of equilibration which determines the speed of burning of the fuse.

With a properly constructed safety fuse the core is almost impermeable to gases, but should faults occur, either during manufacture or due to mishandling afterwards, cracks may develop which enable the hot gases to penetrate forward. In this case the speed of the burning of the safety fuse can increase and even become violent.

It follows also that the actual rate of burning of safety fuse depends on the ambient pressure. Indeed, if the pressure is reduced to less than about a fifth of an atmosphere the burning ceases altogether. In deep mines the extra pressure can be sufficient to give an increase in burning speed of safety fuse. Compared with the effect of pressure other influences on the burning speed are small. Temperature has little effect and humidity also has little effect unless the fuse is kept for a prolonged period at a humidity sufficient to cause deliquescence of the potassium nitrate in the core.

The gas evolved by safety fuse consists mainly of carbon dioxide and nitrogen with some carbon monoxide and oxides of nitrogen. The amount of gas produced is likely to be 15 to 20 ml per cm of fuse.

Testing of Safety Fuse

Because of the extreme importance of regularity and freedom from failure which must characterise safety fuse, detailed and exhaustive tests must be carried out on the product. Certain controls are obvious, namely, measurement of powder charge and of burning speed, both before and after immersion in water. Other tests usually carried out include resistance to cracking on flexing at low temperatures round a mandrel, "coil" tests in which the fuse is bunched into flat or complex coils and freedom from failure after immersion in water is determined. New types of fuse are usually tested to indicate the amount of smoke produced and also to determine the adequacy of the end spit. The end spit is the projection of particles of burning powder from a cut end and is of importance because it provides the mechanism which enables the fuse to ignite a detonator.

Instantaneous Fuse

For some purposes, particularly in fireworks, instantaneous or almost instantaneous transmission of flame is required without detonation such as is characteristic of detonating fuse. Instantaneous fuse is used for this purpose, deliberate use being made of one of the possible faults in safety fuse, namely, the rapid speeding up which can occur if the core of the safety fuse is porous.

Instantaneous fuse consists of a highly combustible thread, often made from nitrated paper, inside a tube of known diameter. The nitrated paper occupies only a small part of the cross section of the tube, so that the gases can penetrate along it freely. The high speed of burning is caused by the rapid passage of hot gases along the tube, igniting the core well in front of the portion already consumed. The rate of burning is not easily controlled, but the uses to which the fuse is put are such that this is not of great importance. Threads coated with blackpowder may be used in place of a nitrated paper core. Instantaneous fuse is usually made only in small quantities by simple hand methods.

Igniter Cord

If the outer covers of safety fuse are pierced, e.g. by falling rock, water can penetrate and stop the combustion of the fuse. Much work has been done in efforts to obtain a waterproof fuse which will obviate this difficulty. This work, although not successful in its original purpose, has led to the development of igniter cord which has proved of considerable value in secondary blasting (see p. 148). Two types of igniter cord are manu-factured, fast cord with a burning speed of 3 s m^{-1} and slow cord with a speed of 30 s m^{-1}.

In the manufacture of fast cord the first process is to coat paper or textile yarns with a blackpowder/nitrocellulose dope by a dipping process. These yarns are thoroughly dried and a number of them passed through an extruder and given a thick covering of plastic incendiary composition. The cord thus produced is covered with a protective layer of plastic, usually polyethylene. The overall diameter of the cord is about 2·5 mm.

The speed of burning of fast igniter cord depends on the speed of burning of the blackpowder-coated yarns and this depends in turn on the gas channels left in the construction of the fuse. The remainder of the fuse is also combustible, so that there is no pressure build up due to lack of venting. Should a minor pin hole or break in the outside layers be produced and water enter, the incendiary composition will continue to burn so that the fuse will not fail, although it will slow down for the distance of the damage.

Slow igniter cord consists of the same type of plastic incendiary composition extruded not over combustible blackpowder-coated yarns, but over a metallic wire. The function of the wire is to conduct heat from the burning front into unburnt composition and so control and speed up the rate of burning of the igniter cord. This centre is usually of copper, but may be of iron or aluminium. Slow igniter cord is covered with a thin layer of protective plastic.

The basis of igniter cords is the incendiary plastic. Of the possible compositions those which have achieved most importance are based on a nitrocellulose binder with oxidising and combustible additives. The nitrocellulose is normally plasticised with dibutylphthalate and contains also the usual stabilisers. A low viscosity lacquer grade of nitrocellulose is most conveniently employed. The oxidising components consist of a balanced mixture of red lead with either potassium nitrate or potassium perchlorate. The addition of a potassium salt is found to give certainty of pick up of flame and suitable vigour in combustion, but its quantity is restricted by the requirement that the finished product should be safe under conditions of impact such as can occur in use. The combustible additive is usually finely divided silicon. The incendiary composition is thermoplastic and extruded hot. In bulk it can burn vigorously and therefore special safety precautions are required in processing, including automatic guillotines to cut the fuse at suitable points in case of fire.

Igniter cord is used for the purpose of igniting a number of lengths of safety fuse at predetermined intervals. Special connectors are therefore supplied for transmitting the flame from the igniter cord to the safety fuse. The connector consists of an aluminium tube, closed at one end, into which a pellet of incendiary composition is pressed. An elliptical "bean hole" is then cut through the two sides of the tube and the composition. In use, the connector is crimped to the cut end of safety fuse; the igniter cord is then bent double and the doubled portion passed through the bean hole, which is crimped firmly with a special tool. The side flame from the igniter cord sets fire to the incendiary composition in the connector and this in turn ignites the safety fuse. Some designs of connector contain also blackpowder to assist propagation of the flame, but such connectors are generally employed only when supplied from the factory already fastened to lengths of safety fuse so that there is no danger of penetration of moisture to the blackpowder during storage.

Part III.   Application of High Explosives

CHAPTER 14

Commercial Applications

Introduction

The major applications of blasting explosives are in mining and quarrying, where the purpose is to break solid rock into smaller fragments. In the British Isles the most important operation is coal mining, for which permitted explosives are almost entirely employed. Other underground mining is for gypsum, anhydrite, non-ferrous metal ores, iron ore and to a small extent rock salt. Surface uses include opencast (strip) coal mining, quarrying and civil engineering work such as is encountered in hydroelectric schemes and road building. In all these cases the general procedure is to drill a hole into the solid rock or coal, insert cartridges of explosives with a detonator and thereby use the explosives to fracture and bring down the rock. The nature of this process has been studied in some detail, particularly by Livingston, Langefors and Hino.

The first effect of the firing of an explosive in a borehole is the production in the surrounding rock of a shock wave, either directly from the explosive in contact with the rock, or by the impact on the rock of the expanding detonation products. This shock wave produces in the rock both compressive and shearing forces, the former being by far the most important. The compressive forces themselves are unable to break rock, but when they reach a free surface, or a fissure in the structure, they are reflected as rarefaction waves. The stages of this reflection process are shown in Fig. 14.1 for various time intervals. It will be seen that at the free surface and immediately adjacent to it the rarefaction and compression waves compensate each other. At a distance from the free surface the rarefaction wave can exceed the compression wave at a suitable instant of time and the net stress can exceed the tensile strength of the rock. When this happens the rock is fractured and fragments break off. In a perfectly uniform material this can happen only at the outside surface and the effect is to produce a spalling of a surface layer. In heterogeneous materials such as ordinary rocks, reflections can occur at other points and fissures are produced at a wide range of places throughout the mass.

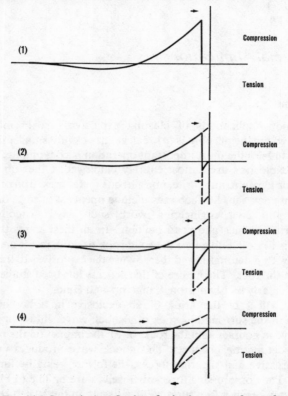

FIG. 14.1. Stages in the reflection of a shock wave at a free surface.

After the shock wave has passed through the rock the borehole still contains hot gases at high pressure. As there are now fissures in the rock, however, the strength is reduced to a negligible amount and the gases can expand and throw the broken rock away from the solid mass.

If the amount of explosive used is too large, the broken rock can be projected for great distances. If the quantity used is too small, the amount of fissuring of the rock can be insufficient to free the explosive gases adequately and these are then most likely to blow the stemming from the borehole, but to leave the mass of the rock intact. In a properly balanced explosive shot the rock is broken into fragments, but the expansion of the gases causes only sufficient movement to move these fragments a short convenient distance. Many factors influence the optimum charging rate for boreholes, and skill and experience of the shotfirer are of considerable importance in attaining optimum effects.

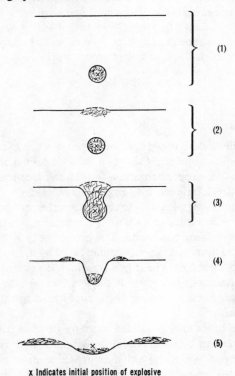

x Indicates initial position of explosive

FIG. 14.2. Effect of depth of charge on cratering.

The efficiency of an explosive in any given rock can be determined by cratering experiments. In these, boreholes are drilled vertically downwards into the rock and loaded with increasing charges of explosive. The effects which are produced are shown in Fig. 14.2. The optimum charge produces a crater extending to the bottom of the borehole, but does not project the broken rock so far that most of it cannot fall back into the crater originally produced. By utilising the results of such trials, it is possible by mathematical formulae and by experience to estimate accurately the charges needed in practical mining and other operations.

It will be seen that the reflection of the shock wave from a free face is of considerable importance in the use of explosives. In practice, the most economic use of an explosive can only be achieved if it is fired at a suitable distance from such a free face. This distance is known as the burden of the shot. If, as in tunnelling, no free face exists naturally, the first shots fired are heavily loaded and arranged so as to produce such a free face for the later charges.

Tunnelling

As an illustration of practical methods of using explosives it is proposed to describe in some detail the way in which a tunnel may be driven through rock. This is a process common to all mining operations and frequently used in hydroelectric and civil engineering work.

The principle employed is to drive a number of boreholes into the rock, load them with explosive and fire the explosive. This is known as drilling, charging and firing the round. The broken rock is cleared away and the length of the tunnel has then been increased by approximately the depth of drilling the boreholes. The length by which the tunnel is increased by each cycle of operations is known as the advance. As the tunnel advances, its inner surface is made smooth and secure by lining with steel and concrete, or other means.

The boreholes may be drilled in many patterns, differing essentially in the arrangement by which the first section of rock is removed and a free face developed. A method which is applicable under most circumstances is known as the wedge cut and this is illustrated in Fig. 14.3 for a medium-sized tunnel. In Fig. 14.3 the boreholes are shown in elevation and section; the figures in brackets refer to the period of delay of the detonator used in each individual hole.

In approximately the centre of the line of the tunnel, but usually somewhat below this, there is a series of six holes, three on each side, arranged in the form of a wedge. These are drilled at an angle so as to be inclined towards each other and almost meet at the back of the holes. These six holes are fired simultaneously with detonators containing no delay elements.

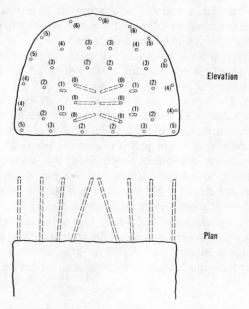

Fig. 14.3. Wedge cut.

Around the wedge are drilled a number of holes known as easers, arranged in approximately concentric rings. The detonators used in these holes have delays which gradually increase in number as the hole is farther from the wedge. This is to ensure that the inner holes nearest the free face always fire first and can produce a fresh free face against which the farther shots can work. Finally, an outer row of holes, known as trimmers, are drilled close to the intended outline of the tunnel. The trimmers have the longest delays, particularly at the top of the tunnel.

The sequence which occurs on firing is that the rock in the wedge is first blown out and then the hole thus produced is expanded outwards as each successive ring of charges detonates. Apart from the first six holes, which in any case are angled suitably, the explosives in the other boreholes all work towards a free face.

In an operation of this sort, the commonest explosive used is an ammonium nitrate gelatine such as Polar Ammon Gelignite. The wedge holes are likely to contain about 1 kg each of explosive and the other holes less than half this amount. The appropriate charge is usually calculated according to the nature of the rock, varying from 1 to 4 kg per m³ of rock between the softest and the hardest strata.

The charges in the boreholes are made up of a number of cartridges, paper wrapped, each containing 110, 170 or 230 g of explosive. One of these cartridges is made into the "primer" cartridge by the insertion of a detonator. The first operation is to insert a soft metal pricker into the end of the cartridge so as to make a hole for the detonator. The detonator is then inserted firmly and completely into this hole. It is important to ensure that the detonator is not pulled out of the cartridge during subsequent operations, and a convenient way of doing this is to tie the leading wire of the detonator round the cartridge with a half hitch. The cartridges of explosives are placed in the borehole and pushed firmly but gently home with the help of a wooden or brass-tipped rod, known as a stemming rod. The primer cartridge can be inserted first or last, but always with the base of the detonator pointing towards the main length of the charge as this is the direction in which initiation is best. When all the explosive is in, the hole is stemmed by the insertion of clay, sand or similar material so as to

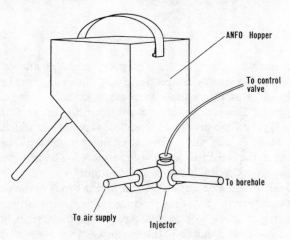

FIG. 14.4. "Anoloder" for ANFO.

provide a resistance against the explosive when the latter is fired. This stemming is packed tightly with the stemming rod.

When all the holes have been charged, the leading wires from each detonator are separated and connected together with the detonators in series. The two free ends of the circuit are then connected to the shotfiring cable, the continuity tested and the round fired when all the personnel are in safe places.

When ANFO is used underground it is loaded into the hole by means of compressed air. The equipment used for this purpose can be of two types illustrated in Figs. 14.4 and 14.5. Figure 14.4 shows an injector loader of which the best known is the "Anoloder". This consists essentially of a hopper at the base of which is an injector similar to a steam injector, but using compressed air which follows the external annulus, allowing the ammonium nitrate/oil mixture to pass through the central portion. The mixture of ANFO and air is carried through a hose into the borehole. The other type, shown in Fig. 14.5, is known as the blow case and in this

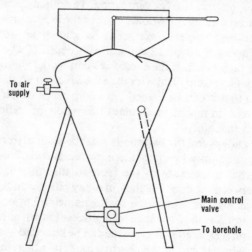

FIG. 14.5. Blow case loader for ANFO.

design the container is sealed and pressurised with compressed air. This blows the ANFO out through the bottom of the container, through a hose and into the borehole. When ANFO is used the primer is usually inserted last to allow time for dispersal of any static electrical charges caused by the loading of the hole.

Coal Mining

In order to describe the uses of explosives in coal mining, it is first necessary to give a brief outline of the method of mining usually adopted in the British Isles. This is known as the long wall system.

Before mining commences, shafts are driven from the surface to the required depth. At least two shafts are required for ventilation to be possible. From these shafts, tunnels called stone drifts are made, roughly horizontally, to the coal seams. The seams may be 1 m or more thick and are usually inclined at an angle to the horizontal.

The actual work of getting coal is at a long straight face of 100 to 200 m in length. At each end of this face is a road which is used for conveying coal and materials and also for ventilation. The mining process consists of extracting the coal along the whole length of the face to give a regular daily advance. The first stage in extraction is known as undercutting and consists of cutting a 10 or 12 cm slot to a depth of about 1·5 m. This creates a free face to which the explosives can work. Shotholes are drilled near to the roof level and filled with the explosive and fired. This breaks the coal down so that it can be loaded onto conveyors which pass along the face to the road. In mechanised mines this whole operation of obtaining coal is done by machine.

As the face advances, the two roads must be similarly brought forward. Indeed, with mechanised mining, there is at each end of the face a "stable hole" somewhat in advance of the face, so that the machinery can be positioned between the cuts. Further, in many pits the thickness of the coal may not be as high as that of the roads, namely, 3 m or so, so that the roads are enlarged as they advance by a process known as ripping.

It must be remembered that methane can be liberated in a coal mine, not only in the coal itself, but in the nearby stone. It is therefore necessary to use permitted explosives, both in and near a seam of coal. Further, before any shot is fired in a coal mine, tests are made for the presence of methane in the air by means of a safety lamp. This safety lamp is the well-known Davy lamp and an experienced operator can judge the presence of methane in the air from the appearance of the flame. Tests are made particularly near the roof where methane can accumulate, being lighter than air. As a precaution against coal dust explosions, the neighbourhood in the mine is sprinkled liberally with limestone dust which has the property of suppressing these explosions.

Explosives are used in coal mines for a wide variety of purposes. For example, in the initial sinking of shafts P1 explosives of the gelatine type are frequently employed. In making the roadways, similar explosives can be used following the principles described above in tunnelling. The other place in coal mines where explosives are fired in stone is in the process of ripping. This in principle is a simple application to bring down stone above an existing opening. The particular situation in the mine is, however, unusually dangerous and high concentrations of methane are liable to occur in this vicinity. At the same time, for speed of operation the use of millisecond delay detonators is desirable. As the bottom of the rock is unsupported, separation of the rock is liable to occur, either before shotfiring or between the succession of shots. These small partings, known as breaks, can contain methane and then give rise to exceptional hazards in firing. To meet this situation, the P4 class of explosives has been introduced into Britain. These are specially tested for safety when fired in the presence of breaks containing methane/air mixtures (see p. 86).

Explosives are fired in coal on the face, in the stable holes and sometimes in the making of roads. The commonest practice, as stated above, is to undercut the coal mechanically. Often a single row of holes with a permitted P3 explosive is then sufficient to bring down the coal. Powder explosives are most commonly used for this purpose, water-proofed when necessary.

The process of undercutting is time consuming and also requires machinery in a congested part of the mine. It is also one which produces a large amount of dust and adds to ventilation difficulties. It is there-fore natural that means have been sought of avoiding undercutting. Historically, the first was the use of pulsed infusion shotfiring, illustrated in Fig. 14.6. A special explosive (see p. 89) which will fire when subjected to a high pressure of water is placed at the back of the borehole. In the front of the borehole is placed a seal to which a water infusion tube is connected. Water is pumped through this tube until it is at the desired pressure and the whole of the borehole full of water. When the explosive is fired it is in fact surrounded by water and this considerably reduces the risk of gas ignition and also the amount of dust and fume set free.

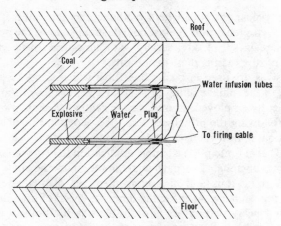

FIG. 14.6. Pulsed infusion round.

An even simpler means of avoiding undercutting is now being pursued, which introduces also the advantage of allowing short delay detonators to be employed. In solid coal there is no danger of breaks as in the situation considered above in ripping, but there is the possibility that the first explosive fired may cause the end of the boreholes of later shots to be broken, so exposing the explosive. A new class of explosive, known as P5, has been designed with the object of avoiding ignitions under these circumstances (see p. 87). This is particularly valuable in stable holes used in mechanical methods of mining. A wedge cut, or similar cut, as used in tunnelling is employed to produce a free face with zero delay detonators. Holes on each side are then fired with longer delay periods and the stable hole can then be cut with a single round of shotfiring. For reasons of safety, the longest delay period allowed is usually 100 or 200 ms so that there can be no time for appreciable release of methane from pockets of gas. Also the number of shots is, in practice, limited to twelve, as this is currently the capacity of the most powerful exploder which can be guaranteed to be safe for use in gassy mines.

The policy of the National Coal Board is to increase mechanisation in the mines for reasons both of safety and economy in personnel. This process has had considerable success on long wall faces and the majority of these are now worked mechanically. On a few faces complete remotely operated mining is possible at present. The process of mechanisation will no doubt extend to other parts of the mine, but here the rewards are less and the convenience and simplicity of explosives ensure that they are likely to continue in use for many years.

Other Mines

In Britain mining is carried out, apart from coal mines, for anhydrite and gypsum, for iron ore, for non-ferrous ores and for salt. A common method used in this type of mining is known as pillar and stall working, illustrated in Fig. 14.7. The ore, or rock, which is usually in thick seams, is extracted over the whole area of the field, but leaving pillars which support the strata above and prevent the roof from falling in. The area occupied by

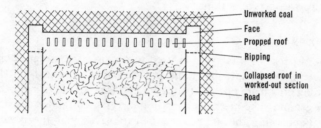

(a) Long-wall working

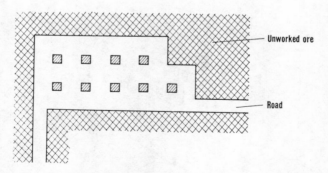

(b) Pillar and stall working

FIG. 14.7. Two methods of mining.

the pillars may be as much as a quarter of the whole area of the mine. When a particular mine is worked out it is possible to extract the pillars and allow the upper strata to subside.

The method of mining is somewhat like driving two sets of tunnels at right angles through the strata. Most mines of this type do not present any hazard from methane and therefore non-permitted explosives can be used. The selection of powder or gelatine type of explosive depends on the nature of the rock and particularly on the wetness of the mine.

As Britain is relatively poor in mineral deposits, this type of mining is less important here than in other countries of the world. The gold mines in South Africa, metal mines in the U.S.A., Canada and Sweden all use considerable amounts of explosive. In such mines the methods of working are often appreciably different from those described above and adapted to very large-scale production.

Quarrying

The process of quarrying is used to obtain road metal, rock for civil engineering purposes, limestone for steel making and is analogous to obtaining coal from opencast sites where the coal is at or near the surface of the ground. The methods used in quarrying can differ considerably in detail concerning the application of explosives, but the commonest is illustrated in Fig. 14.8. A quarry is worked as a series of benches, the

FIG. 14.8. Benching method of quarrying.

height of each being conveniently 15 or 18 m. The width of each bench is sufficient to allow access to loaders and lorries, so that each bench can be used as an entity.

To bring down the rock a series of holes is drilled from the surface down to the depth of the bench, or 0·3 to 0·6 m greater. If the bench is 18 m high, the holes can conveniently be 10 cm in diameter, 3·5 m back from the existing face and 3·5 m apart. They may be drilled either vertically, or at an angle of 70 to 80°. A row of such holes is fired at a time and the effect should be that the rock in front of them should slump into a heap of boulders of a size suitable for loading and crushing.

As the major expense in this operation is the drilling of the holes, it is usual to space them as far apart as possible and this is best achieved by the use of the strongest explosives. Under these conditions 9 tonnes of rock can usually be broken per kg of explosive. The amount of explosive to be charged in the drill holes is calculated from a ratio such as this. In the example quoted, each borehole will contain approximately 70 kg of explosive. If the conditions are at all wet, a gelatinous explosive is used, particularly at the foot of the hole. Under dry conditions, or in the upper part of the hole, where the work to be done by the explosive is less, weaker explosives can be used and ANFO is often suitable. The holes are charged, either by lowering cartridges downwards, or else by pouring in free-flowing explosive, such as ANFO.

The charges can be fired with electric detonators, but the commoner method is to use detonating fuse, usually known in Great Britain as Cordtex. This is capable of initiating gelatine explosives, such as Opencast Gelignite, without the extra use of a detonator. It is also a convenient method of connecting boreholes together. The method of application is shown in Fig. 14.9 A trunk line of Cordtex lies on the bench of the quarry and from this branch lines extend down the holes to the bottom. Frequently with large holes two lines of Cordtex are employed to make certain of initiation under all conditions, because in a large quarry blast a single failure from any cause can be extremely expensive. It is an advantage to arrange for the holes to be fired in succession by a series of short delays and this is done by inserting in the trunk line special delay connectors which give time lags of 15 ms. When the shot is fired, most of the noise comes not from the main charge of explosive itself, but from the Cordtex on the surface of the quarry. This noise can, however, be considerably reduced by covering the Cordtex with sand or stone dust.

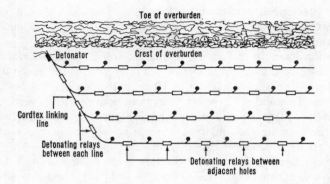

Fɪɢ. 14.9. Quarrying with Cordtex and detonating relays.

Removal of rock is required for many civil engineering purposes, and uses explosives in a manner similar to quarrying. An interesting technique is that known as pre-splitting. In this holes are drilled on the line of the proposed edge of the roadway, or cutting. These are filled with blasting explosives in relatively low charge, but the holes are drilled close together. When the explosive is fired a crack is formed along the line of the holes, but with little or no damage to surrounding rock. Subsequently the rock is removed by normal blasting and breaks away cleanly at the line of holes. This in suitable cases leads to a well-formed face of solid rock which can be finished by the minimum of concrete or other preparation.

No matter how well a blasting operation is carried out, there is liable to be some amount of rock left in the form of boulders too large to be handled by normal equipment. This is an inevitable result of lack of uniformity of rock. These boulders are broken down in size by a process known as secondary blasting. The usual method is to drill holes into the boulders, load them with explosives and fire. In certain cases the explosive can be placed on the surface of the boulder, covered liberally with clay or other material, and then fired. A cheap and effective way of detonating such charges is with safety fuse and detonators, the lengths of the fuse being arranged so that the shotfirer has ample time to get to safety. The Regulations do not allow more than six ends of safety fuse to be lit at one time by a shotfirer. Instead, the ends are connected to igniter cord (see p. 131). The shotfirer then has to light only the end of the igniter cord before retiring to a prepared position.

Seismic Prospecting

Sound waves in the ground are reflected at the boundaries between strata in the same way as sound waves are reflected from walls of buildings, etc. In seismic prospecting use is made of this by generating a wave with a charge of explosive and observing the reflections by means of geophones placed at suitable positions. This enables a geological map of the substrata to be drawn and indicates to the explorer the sites most likely to contain oil. This prospecting can be carried out either on land or at sea.

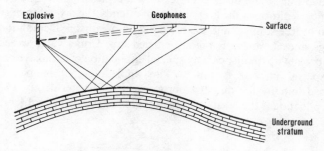

FIG. 14.10. Seismic prospecting by reflection method.

Figure 14.10 illustrates the method of seismic prospecting on land by what is known as reflection shooting. A hole usually 10 to 12 cm in diameter is drilled to a depth of 15 to 30 m. The charge of explosive is likely to be 5 to 12·5 kg and the stemming used is usually water. As the explosive must fire under a depth of water which may exceed 45 m, special varieties of gelatines are employed (see p. 53). Alternatively, a powder explosive can be sealed into pressure-resistant metal containers. Special detonators are also employed, not only to withstand the possible head of water, but also to have a specially short bursting time (see p. 113).

Seismic prospecting is also carried out at sea, the explosive charge being in a metal container and the geophones stretched along a cable from the stern of a ship. Either one or two boats may be used; charges are fired and records taken at regular intervals while the vessels are steaming. The explosive must be fired near the surface of the water, otherwise, as noted in Chapter 2, a bubble is formed which oscillates and passes into the water a series of shocks; the seismographic recordings then obtained are unsatisfactory for analysis. When the explosive is fired within 1·2 or 1·5 m of the surface, the bubble is not formed, as the gases are immediately discharged into the atmosphere. In recent years prospecting for oil in the North Sea has been on a large scale and several thousand tonnes of explosive per annum have been used.

Miscellaneous Uses

There are many miscellaneous applications of explosives which may be mentioned but which do not merit individual description. Such uses are those in agriculture, in the preparation of ditches, the diversion of streams, removal of tree stumps, and the breaking up of subsoil. Demolition of old buildings and chimneys is readily carried out. Underwater wrecks may also be broken up for disposal by special application of explosive charges.

Of special interest is the civilian application of shaped charges developed initially for military purposes (see Chapter 15, p. 158). One such application is the blast furnace tapper illustrated in Fig. 14.11. In this

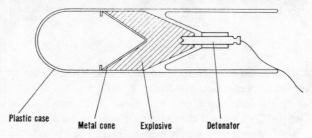

Plastic case   Metal cone   Explosive   Detonator

FIG. 14.11. Shaped charge for tapping blast furnaces.

the conical charge is used to penetrate a clay plug which seals the outlet of a blast furnace, and affords a quicker and safer method than the normal oxygen lance. Wedge-shaped charges have been developed for the cutting of submarine cables for lifting for repair.

In recent years much attention has been paid to the use of explosives for the shaping and working of metals. Figure 14.12 illustrates a simple application of the process for forming a dished end of a vessel. A flat metal blank is placed over a suitable mould and the space between them evacuated. Above the blank is water and in this a suitable explosive charge is fired. The metal takes the form of the mould with little or no spring-back and usually does not require further treatment. The process is particularly

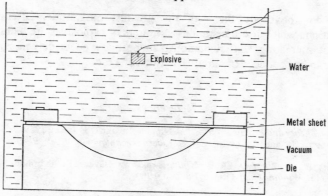

FIG. 14.12. Forming metal with explosive.

suited to the preparation of small numbers of complicated shapes, as no expensive equipment is required. It is therefore of particular interest in the production of prototypes for aircraft, missiles, etc.

An application of increasing importance is in the cladding of metals. In this a layer of explosive is used to project the cladding on to the base metal at such a speed that the two become firmly welded together. Fig. 14.13 shows one technique for small items where the cladding is at an angle to

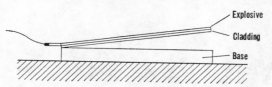

FIG. 14.13. Cladding metal with the use of explosive.

the base, which is firmly supported. The best results are however obtained by supporting the cladding parallel to the base plate and a short distance from it. The explosive used must have a suitable velocity of detonation, somewhat lower than the velocity of sound in the metals. The cladding and the base become welded with a wavy interface; the durability of the bond is such that the clad can readily be rolled or fabricated. Many combinations have been successfully made, of which the most note-worthy is probably titanium clad steel.

Many other applications of explosives could be described, but the ones mentioned above are those of the greatest magnitude. Often those applications which use the smallest amounts of explosives are technically the most interesting, as they demonstrate best how accurately mechanical effects can be produced in minute fractions of a second.

References

HINO, K., *Theory and Practice of Blasting*. Nippon Kayaku Co., Japan, 1959.
LANGEFORS, U., and KIHLSTROM, B., *Modern Techniques of Rock Blasting*. Wiley, London, 1963.
MCADAM, R., and WESTWATER, R., *Mining Explosives*. Oliver & Boyd, London, 1958.
SINCLAIR, J., *Winning Coal*. Pitman, London, 1960.

CHAPTER 15

Military Applications

THE first offensive weapons used by man were probably stones, and
similarly the first objects thrown when mortars were developed were solid,
usually spherical, balls of stone or iron. With the development of
explosives it was soon realised that it would be more effective to use a
hollow missile filled with explosive, designed to burst in the middle of the
enemy. Gunpowder was originally used as filling, but has now been
completely superseded by high explosives.

It is convenient to differentiate between three broad classes of use of
high explosive fillings.
  (a) The explosive is designed to rupture its container into fragments
      which are projected as shrapnel against enemy personnel.
  (b) The explosive is used to produce a blast effect against enemy
      buildings and equipment.
  (c) The explosive is used to penetrate targets such as armoured
      vehicles.
For the first purpose relatively small amounts of explosive are used and the
nature of the explosive is of secondary importance. For the other purposes,
however, larger proportions are necessary and for maximum effect the
most powerful explosives are required.

Grenades

The modern equivalent of a stone thrown by hand is the hand grenade,
of which the best-known form is the Mills bomb illustrated
diagrammatically in Fig. 15.1. The steel shell is thinned along crossing
lines, so that on explosion it is broken into fragments of predetermined
size. When the grenade is thrown the lever is released and allows the

striking pin to fire a percussion cap. This in turn lights a short length of safety fuse which takes 4 s to burn. At the end of this time the fuse fires the detonator which in turn initiates the main explosive charge. The normal modern filling for this type of device is an explosive based on TNT, filled as a powder. Other variants of hand grenades can use pressed charges of TNT or Amatol, or cast explosive. Instead of using the shell of the grenade to provide the shrapnel, it is possible also to use small metal objects, such as nuts and bolts, or coils of wire wound round the explosive charge.

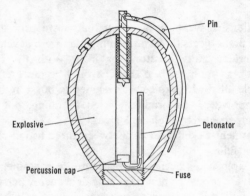

FIG. 15.1. Hand grenade.

Grenades can also be projected from rifles and then have a longer range. Their construction is, however, essentially the same.

Many other devices for use by hand have been invented. Mention may be made of limpet charges which usually employ magnets to make them adhere to the metal sides of tanks or ships. The explosive filling for such charges is a high velocity high power explosive such as cast RDX/TNT.

Shell

The first explosive bodies fired by mortars consisted of hollow spheres containing gunpowder which was ignited by a length of fuse, in turn ignited by the propellent gunpowder. The modern general purpose shell is illustrated in Fig. 15.2. It consists of a hollow steel shell with a cylindrical body and a head of ogive shape. Near the base of the shell is the driving band, made of copper and which takes the form of the rifling of the gun during firing. The base of the shell is covered by a metal plate, the purpose of which is to prevent any hot gases from the propellant penetrating through joints or flaws in the body and igniting the high explosive charge prematurely. In a typical shell the fuze is contained in the nose and may be of several types according to the time at which the shell is required to explode.

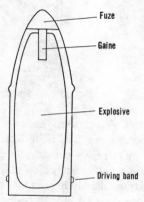

FIG. 15.2. High explosive shell

The shell has a dual function of producing fragments as an anti-personnel weapon and also producing blast against enemy installations. For the latter reason, the explosive charge should be of high density and power. Common fillings are therefore TNT, Amatol, or RDX/TNT mixtures, usually filled into the shell by casting. Because of the low density of the explosive compared with the metal, the actual weight percentage of explosive in the total shell is likely to be about 15 to 20%.

Shell for penetrating armour have heavier steel bodies with at least the nose of specially hardened metal. The proportion of explosive is smaller and it must also be exceptionally resistant to detonation by impact, so that the shell can penetrate the armour before the explosive is detonated by the fuze. Suitable fillings are therefore TNT or desensitised TNT/RDX mixtures. The latest armour-piercing projectiles for anti-tank use contain no explosive, but have high density cores made of tungsten.

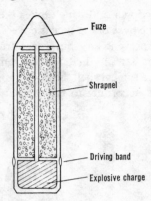

FIG. 15.3. Shrapnel shell.

The well-known shrapnel shell, used for anti-personnel purposes, is of quite different construction and is illustrated in Fig. 15.3. The base and sides of the shell are of heavy construction, but the nose is relatively weak and contains a fuze which is easily ejected by the explosive charge from the body of the shell. Immediately behind the fuze is a filling of metal shrapnel, often bonded relatively weakly with resin. The explosive charge is at the base of the shell behind the shrapnel. After a time of flight determined by the setting of the fuze, the explosive charge detonates and projects the shrapnel forward out of the shell as an expanding cone. In modern warfare the shrapnel shell is becoming of limited importance.

Bombs

Bombs carried by aircraft can be made of lighter construction than shell, because they do not have to resist the acceleration of firing from a gun. The general construction is shown in Fig. 15.4, but differences exist according to the detailed purpose for which the bomb is to be used.

Anti-personnel bombs have a relatively heavy casing containing an explosive such as Amatol, sufficiently strong to break the casing into fragments on impact. Bombs intended to produce a blast effect against buildings have a lighter casing and are usually filled with an explosive containing aluminium to increase the blast effect. Armour-piercing bombs

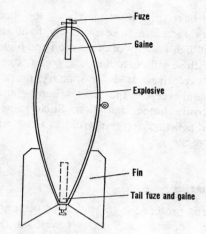

FIG. 15.4. Aircraft bomb.

for use against warships resemble armour-piercing shell in their type of construction and have heavy bodies with smaller high explosive charge. In the proper functioning of all these bombs, design of the fuze is of paramount importance.

Torpedoes

The explosive charge in a torpedo is carried in the nose, the rear compartments containing fuel and motor, together with the control equipment. As the torpedo must penetrate the ship to give the best effects, the nose is of heavy steel construction and the fuze operates with a delay. The high explosive charge must therefore be of maximum density and power. As the torpedo operates under water, the shock wave is considerable and advantage can well be taken of the increased energy given to explosives by the addition of aluminium. A common filling is therefore Torpex which consists of a mixture of RDX, TNT and aluminium and which has a high density, high power and high velocity of detonation. Maximum density is attained by filling the head of the torpedo by a carefully controlled casting process.

Depth charges for use against submarines are similar in principle to bombs, but have their effect mainly by the shock wave produced under water. They therefore have relatively light shells containing a dense explosive of high velocity of detonation and high power. Cast RDX/TNT mixtures, with or without aluminium, are suitable for this purpose.

All the explosives used for military purposes are in general very insensitive and except in the smallest hand grenades a gaine or booster is used to ensure proper initiation. Such gaines are usually made by compressed pellets of tetryl inserted as a column into a metal tube inside the explosive charge.

Shaped Charges

Increased protection to military personnel, particularly in the development of armour-plated vehicles, has led to a rapid growth in importance of weapons using shaped charges of explosive. The effects of shaping the explosive itself were observed independently by Munroe and

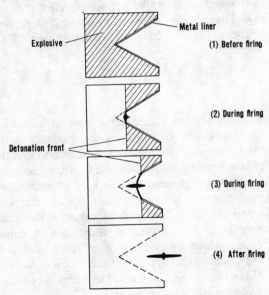

FIG. 15.5. Shaped charge-formation of jet and slug.

Neumann, but the development of practical charges with linings was due to a large number of workers.

If a conical depression is made in the end of a column of explosive, the shock wave is concentrated along the axis of the charge, which then has increased local penetration of metal or stone. A different and much more pronounced effect is achieved if the cone is lined with metal, as this forms a jet followed by a slug; the jet is projected forward at high velocity and can penetrate to great depths, giving a long, almost cylindrical hole. The stages of formation of the jet are illustrated in Fig. 15.5. In a sense, the cone is turned inside-out and collapsed, a process only complete some distance in front of the charge, so that a stand-off distance of one or more charge diameters is needed to ensure maximum penetration. A diagrammatic practical design of a shaped charge is shown in Fig. 14.11 (p. 150), where the explosive is shaped to give maximum economy, and the case provides an automatic stand-off.

The penetrating power of a shaped charge is approximately proportional to the cube of its diameter, but also very dependent on maintenance of exact axial symmetry during construction. It is also proportional to the detonation pressure of the explosive used, so that suitable fillings are cast Pentolite or RDX/TNT. Well-known applications of shaped charges are in the British PIAT and American bazooka.

A recent alternative to shaped charges is known as the squashhead projectile. As the name implies, this contains a plastic explosive which spreads on impact so as to make contact with the largest possible area of the tank before detonating. When the explosive detonates, reflection of the shock wave causes a scab of metal to be displaced from the inside surface of the armour plate (see p. 135). The effect inside the tank can therefore be greater than is the case with a shaped charge which may do little more than penetrate the armour. The amount of scabbing is approximately proportional to the area of contact of the explosive at the moment of detonation.

References

COOK, M. A., *The Science of High Explosives*. Reinhold, New York, 1958.
OGORKIEWICZ, R. M., *Engineering*, 21 July 1961, p. 78.
OHART, Maj. T. C., *Elements of Ammunition*. Chapman & Hall, London, 1946.

Part IV.   Deflagrating and Propellent Explosives

Introduction

GUNPOWDER was first employed as a propellent explosive in guns, although later it was also employed for blasting. In neither case, however, does it detonate in the same way as the blasting explosives described in Part I, or the initiating explosives described in Part II of this book.

As a gun and rifle propellant, gunpowder had serious disadvantages. It was rather unpredictable in use, extremely dirty and caused considerable fouling of gun barrels. It also emitted an excessive amount of smoke and flash which immediately disclosed the position of the gun to the enemy. When Nobel and Abel invented ballistite and cordite respectively, many of these disadvantages were immediately overcome. Gelatinised nitro-cellulose was soon found in France to have similar advantages. Gunpowder is no longer used as a propellent explosive.

Gunpowder was supplied in pellet form as a propellant and the substitutes were similarly manufactured. They were therefore called powders. Subsequent developments have led to the provision of propellent explosives in special and often massive form. Nevertheless, the term powder is still retained for their nomenclature, and the individual unit of the charge, no matter how large, is still called a grain.

Although these explosives do not detonate under conditions of practical use, this does not mean that they are incapable of detonation. The possibility cannot be ignored when large quantities are being handled.

CHAPTER 16

Blackpowder

Manufacture

The raw materials potassium nitrate, charcoal and sulphur are first brought to a fine condition. Potassium nitrate must be ground by itself, but the charcoal and sulphur may be ground together or separately according to individual requirements. This grinding is frequently carried out in steel ball mills. The products are then sieved to remove any oversize or extraneous matter.

Appropriate quantities of the raw materials are added to a mill for intimate mixing and grinding. The mills used are of edge runner type in which a cast iron pan carries the blackpowder and the edge runners are large cylindrical wheels of steel, each weighing about 5·5 tonnes. These wheels are suspended with a clearance of about 6 mm from the bed to reduce hazard, as contact with the wheels is solely with blackpowder and friction between iron and steel is avoided. The charge of a blackpowder mill is about 125 kg. During mixing the wheels are rotated and roll on the blackpowder, whilst ploughs carry the material spread sideways by the wheels back into their path. Throughout the milling operation water is added to maintain a moisture content of 1 to 1·5%; this reduces the danger and also increases efficiency by assisting the incorporation of potassium nitrate into the charcoal. The milling process takes from one to seven hours according to the nature of the product required.

The mill cake taken from the mill is broken by passing between toothed gunmetal rolls to produce powder, which is then pressed between brass plates in a hydraulic press to a pressure of about 7 MPa, depending on the properties finally required. The purpose of the pressing is twofold. On the one hand it increases the intimacy of contact of the ingredients and improves the burning properties of the powder. On the other hand, it increases the physical strength of the powder and reduces breakdown of the finished grain during transport and use.

The press cake is broken by passing it between gunmetal rolls, usually toothed. The coarse product thus produced is subjected to the corning operation. This consists of passing the material through a series of metal rolls which progressively break down the grain into smaller size, and sieving by shaker sieves. The required sieve size is extracted for further processing, coarse material returned for regrinding and fine material for reprocessing.

The sieved grains are glazed or polished by rotating in a drum for a period of about 6 h, frequently with the addition of a small amount of graphite to give a surface polish. This glazing process gives a rounded grain which is free flowing and which has increased resistance to moisture. The product is finally sieved to remove any dust.

To ensure adequacy of manufacture, blackpowder is subjected to a number of tests of which the following are probably the most important. The burning speed of the powder is determined, usually by making up a length of lead fuse and determining its burning speed. The rate of burning under confinement is measured in a "prover" which is a closed vessel with a piezo-electric pressure gauge and cathode ray oscillograph. The trace on the oscillograph gives the maximum pressure reached and also the rate of combustion to this stage.

Most blackpowders contain 75% of potassium nitrate, 15% of charcoal and 10% of sulphur. For safety fuse, however, blackpowders are made with reduced amounts of potassium nitrate. Also manufactured are sulphurless powders, containing approximately 70% of potassium nitrate with 30% of charcoal. These sulphurless powders are used for ignition purposes where sulphur could cause corrosion of metallic components.

Properties of Blackpowder

The density of blackpowder is usually about 1·7, but depends on the compression during the pressing process. There is a wide range of sizes, varying from about 12 mm diameter to material which passes a 150 mesh sieve. Blackpowder is ignited readily at a temperature of about 300°C for the normal product, or 340°C for the sulphurless meal. It is hygroscopic at humidities exceeding 90% and then rapidly ceases to burn.

The properties of blackpowder depend considerably on the charcoal used. Soft woods freed from bark give the best results. Alderwood and dogwood are best, but the cheaper birch and beech are frequently used for ordinary purposes. In any case, it is important that the woods should be carbonised to the correct extent and this depends on the nature of the wood. In the case of alder, the optimum carbon content is 74%, whereas with birch a figure of 82% gives the best results. In general, if the carbon content is too low a readily ignited blackpowder is obtained but it has slow burning properties. It is also difficult to manufacture. If the carbon content is too high the material is easy to grind and milling is rapid, but the final blackpowder may be difficult to ignite and irregular in burning.

The burning speed of the powder in the lead fuse test can also be varied by changing the potassium nitrate content. The maximum burning speed is usually observed at a content of rather less than 70% of potassium nitrate.

It was shown by Vieille that the rate of burning, R, of blackpowder depends exponentially on the pressure, p, by the following equation:

$$R = ap^n$$

a and n being constants.

Mechanism of Combustion of Blackpowder

The mechanism of combustion of blackpowder is extremely complex and only an outline can be given here. It will be convenient to consider first the initial chemical reactions in ignition, then the reactions which occur during the main combustion and finally the nature of the products.

Hofmann and Blackwood and Bowden have studied the chemical changes which occur at temperatures approaching 300°C, which is the temperature of thermal ignition of blackpowder. That the process involves gases is shown by the proof by Blackwood and Bowden that ignition can occur at hot spots at temperatures as low as 130°C, provided that the local pressure is at least 150 atm. An important part in the reaction is played by a constituent of charcoal which can be extracted with acetone and shown to contain carbon, hydrogen and oxygen. Sulphur can react with this substance at temperatures as low as 150°C to produce hydrogen sulphide, which in turn can react with potassium nitrate at about 280°C to give potassium sulphate. Also at about 280°C the organic constituent of charcoal can react with potassium nitrate with the liberation of nitrogen dioxide. The direct reaction of potassium nitrate and sulphur to give potassium sulphate and oxides of nitrogen commences at about 250°C, but becomes rapid only above the melting point of potassium nitrate, 330°C.

At temperatures below 300°C, the evolution of hydrogen sulphide is sufficient to reduce the nitrogen dioxide also produced, with the formation of sulphur and nitric oxide. As the temperature is raised, however, the evolution of nitrogen dioxide increases until there is excess which can react with sulphur to give sulphur dioxide and nitrogen. This sulphur dioxide can then react with potassium nitrate to give potassium sulphate in a strongly exothermic reaction. The heat produced in this reaction leads to further temperature rise and thus to mass ignition of the blackpowder.

The main process of combustion of blackpowder was studied exhaustively by Nobel and Abel and by Berthelot. These experimental results were examined in much greater detail by Debus, who has provided a self-consistent account of the chemical reactions involved. Debus considers that the overall reaction can be divided into two distinct stages: (a) a rapid oxidation process and (b) a slower reduction process.

The oxidation process is responsible for the actual explosion. Whilst it is no doubt complex, it can be simplified to the following overall equation:

$$10KNO_3 + 8C + 3S \rightarrow 2K_2CO_3 + 3K_2SO_4 + 6CO_2 + 5N_2$$

As the initial composition of blackpowder contains, for each 10 molecules of potassium nitrate, 14 molecules of carbon and 4 molecules of sulphur, this equation does not account for 6 molecules of carbon and 1 molecule of sulphur.

The excess carbon and sulphur take part in slower reduction reactions which are as follows:

$$4K_2CO_3 + 7S \rightarrow K_2SO_4 + 3K_2S_2 + 4CO_2$$
$$4K_2SO_4 + 7C \rightarrow 2K_2CO_3 + 2K_2S_2 + 5CO_2$$

As the reduction is a slow process, it is not necessarily complete when the blackpowder has done its work. The reduction reactions are endothermic and lower the total heat evolution. On the other hand they increase the amount of gas evolved. Hofmann considers that these equations represent only the overall reactions and that the actual paths are more complicated, involving the intermediate production of potassium monosulphide and of various gaseous products, including oxides of nitrogen.

As the reduction stage of the reaction does not necessarily go to completion the reaction products depend to some extent on the conditions of firing. In all cases, however, the chief products appear to be potassium carbonate, potassium sulphate, potassium disulphide, carbon dioxide, nitrogen and carbon monoxide, the last named arising either from the oxygen in the charcoal, or from side reactions. Side reactions give the by-products usually observed, namely, hydrogen, hydrogen sulphide, methane, ammonia, water and potassium thiocyanate. In most analyses small amounts of unburnt powder have also been observed.

CHAPTER 17

Manufacture of Propellants

Introduction

Current propellent explosives may be divided into three classes: single base, double base and composite; however double base propellants which contain picrite are often considered a separate class and called triple base.

Single base propellants are basically nitrocellulose which has been made colloidal by the action of solvent.

Double base propellants contain nitroglycerine in addition to nitrocellulose and frequently other additives to give special properties. They can be manufactured by three methods and are then known as solvent type, solventless type and cast double base compositions. In the first method solvent is employed to ensure completeness of gelation of the nitrocellulose by the nitroglycerine. Such powders are used in small grain sizes. The solventless process gives a dimensionally more stable product as gelatinisation is performed without the aid of solvent. The method can therefore be employed for making larger grains for burning in small rockets and in gas-producing devices. Limitations of reasonable press capacity prevent even this method from being used for making the largest rocket charges. For this purpose the casting process is employed and the product is known as cast double base (CDB) composition. This process is generally employed only for the manufacture of rocket charges.

The third type of propellent explosive, the composite type, is a more recent development, the purpose of which is to provide rocket propellants of increased thrust, compared with the ordinary varieties. Composite propellants are based on an oxidising solid, commonly a perchlorate, together with an organic binder which both acts as fuel and gives adequate mechanical strength to the mixture. The search for even more energetic compositions continues, but because of the military importance of the

169

results, little has been published. Table 17.1 from Kirk and Othmer gives typical compositions of single and double base propellants.

Single Base Propellants

As single base powders have been developed particularly in the U.S.A., their method of manufacture may be described as illustrating the principles involved. British practice is similar but often with omission of the "macaroni" pressing stage.

The nitrocellulose is first dehydrated, that is the water present is replaced by alcohol. This is done by compressing the wet nitrocellulose in a hydraulic press and passing alcohol through the press until the strength in the block is about 92%. The resulting block of alchohol wet nitrocellulose is broken down to small pieces with toothed rolls.

The incorporating process is carried out in mixing machines of the Werner Pfleiderer type, that is with two heavy horizontal blades in a specially shaped bowl. Incorporation consists of mixing the alcohol wet nitrocellulose and other ingredients with a solvent, normally ether and alcohol mixture. The amount of solvent used is adjusted so that the final consistency is that of a stiff dough.

This dough is transferred to a "blocking" press and there formed by the help of hydraulic pressure into solid blocks free from air. The blocks are then placed in a "macaroni" press, in which they are extruded through gauzes and dies into a mass of cords similar in appearance to macaroni. These cords are passed to a second block press so that a further block form is produced. These blocks are then placed in the final press, which has a die plate with dies of the correct size to give the physical dimensions required in the finished product. Above this plate is a series of gauzes to strain out any ungelatinised or foreign particles. The strands emerging from this press may be collected in a number of individual containers and then fed into a cutting machine. This cutting machine has rotating angled blades which brush the face of a perforated cutter bar. Strands are fed through the cutter bar by rollers geared to the rotating knives so that a constant length of cut grain is assured.

In an older cutting process still much used, strands are formed into a bunch which is then cut by guillotine.

TABLE 17.1   *Nominal Compositions of Smokeless Powders*

Constituent (%)	1	2	3	4	5
A. Partially colloided single base powders					
Nitrocellulose	80.00	89.00	87.00	84.00	
Barium nitrate	8.00	6.00	6.00	7.50	
Potassium nitrate	8.00	3.00	2.00	7.50	
Starch	2.75	1.00	—	—	
Paraffin oil	—	—	4.00	—	
Diphenylamine	0.75	1.00	1.00	1.00	
Dye (Aurine)	0.25	—	—	—	
B. Colloided single base powders					
Nitrocellulose	99.00	97.70	90.00	85.00	79.00
Dinitrotoluene	—	—	8.00	10.00	—
TNT	—	—	—	—	15.00
Di-n-butyl phthalate	—	—	2.00	5.00	—
Triacetin	—	—	—	—	5.00
Tin	—	0.75	—	—	—
Graphite	—	—	—	—	0.20[a]
Diphenylamine	1.00	0.80	1.00[a]	1.00[a]	1.00
Potassium sulphate	—	0.75	—	—	—
C. Double base powders[b]					
Nitrocellulose	77.45	52.15	51.50	56.50	
Nitroglycerine	19.50	43.00	43.00	28.00	
Diethyl phthalate	—	3.00	3.25	—	
Potassium sulphate	—	1.25	1.25	1.50[a]	
Potassium nitrate	0.75	—	—	—	
Barium nitrate	1.40	—	—	—	
Carbon black	—	—	0.20[a]	—	
Candelilla wax	—	—	0.08[a]	0.08[a]	
Methyl cellulose	—	—	0.10[a]	0.50[a]	
Dinitrotoluene	—	—	—	11.00	
Ethyl centralite	0.60	0.60	1.00	4.50	
Graphite	0.30	—	—	—	

[a] These constituents are added to the basic composition.
[b] Type 1 is the solvent type; types 2, 3 and 4 are the solventless type.

The cut powder is dried in stoves fitted with solvent recovery and operated to a carefully determined time schedule. If drying is carried out too quickly the surface can be hardened and afterwards crack, with serious effects on the ballistics of the resulting powder. This process of stoving does not satisfactorily remove all the solvent and the last traces are removed by steeping in water at carefully controlled temperatures. The steeped powder is then dried by hot air.

Adequate ballistics cannot be achieved in single batches of propellants, even with closest control. It is therefore universal practice to take material from a number of batches of manufacture and blend them until correct and uniform ballistics can be ensured.

A common method of obtaining high rates of burning for shotguns, revolvers and pistols is the incorporation of potassium nitrate in the dough, followed by leaching of this potassium nitrate during the steeping process. In this way a product of controlled porosity and therefore controlled rate of burning is obtained. Such powders are also sometimes prepared in flake form by rolling into sheet and then cutting.

Rifle powders, particularly when porous, can burn so rapidly that the initial rise of pressure can be faster than necessary. In this case, the grains can be surface moderated, or given a surface coating of a nitrocellulose gelatiniser, such as dinitrotoluene, dibutylphthalate, or carbamate. This process is often carried out at the same time as glazing, with a small amount of graphite, which improves the flow properties of the powder as well as increasing its loading density.

Double Base Powders—Paste Mixing

In modern practice the use of dry guncotton is avoided for reasons of safety and the first stage in the process of manufacture, whether for solvent type or solventless products, is the formation of a paste from wet guncotton. The British methods by which this is done are described by Wheeler, Whittaker and Pike. There are in fact two methods which have been used. In one a slurry of finely pulped guncotton is made in water and circulated through a ring main. The concentration of the slurry is determined and then a suitable volume is measured in a tank. This is then passed, either with the original process water, or with previously used water, into a tun dish in which the slurry is sprayed with nitroglycerine. The slurry then passes to a mixing tank in which a stabiliser and chalk are added and the whole is stirred for a period of not less than 30 min.

Alternatively, the guncotton can be obtained in a water-wet condition in which it contains approximately 30% of water. The water content is determined and the appropriate amount of wet guncotton for a batch is weighed out. This is slurried in water in a pre-mix tank and then nitroglycerine is run in with stirring. The product is transferred to a final mixing tank where it is processed as above.

The mixture is then run on to a sheeting table which is made on the same principle as a paper-making machine, but with the wire replaced for safety by a suitable cloth filter. The pulp passes over suction tubes and between rollers in order to remove as much water as possible, so that the paste when stripped from the cloth contains 20–25% of water. It is dried on trays or else in trucks through which air at 50°C is blown.

Solvent Type Double Base Propellants

Dried paste is weighed into incorporators, usually of the Werner Pfleiderer type. To the incorporators are added the appropriate amounts of other ingredients and also solvent which is usually acetone or a mixture of acetone and water. When mixing is finished a stiff dough is obtained. This is rammed into the cylinder of a hydraulic press at one end of which is a die containing holes of diameter depending on the cord size required. For cannon powders, for example, cord diameters are large and only a few are extruded. The cut cords are taken to a solvent recovery stove, where the acetone is removed and recovered; the final stoving to remove the last traces of acetone is sometimes given at a higher temperature. The powder is blended with extreme care so as to give completely regular performance.

Ball Powder

In the ball powder process, developed by the Western Cartridge Company, nitrocellulose is agitated with ethyl acetate solvent in aqueous suspension with emulsifying ingredients. This gives a suspension of spheres, the size of which can be controlled by the speed of stirring. The suspension is heated to distil off the solvent and harden the spheres, which after cooling can be impregnated with nitroglycerine and dried. If necessary the spheres can be rolled to decrease the web thickness and time of burning and they can be graphited according to requirements. The process is particularly applicable to the manufacture of rifle powders, but it is also used in producing grains suitable for the slurry casting process.

Solventless Double Base Propellants

The use of solvent limits the size of grain which can be produced because of the difficulty of removing final traces of solvent and thus ensuring ballistic stability during storage. The solventless process overcomes this difficulty by attaining gelatinisation of the paste by the effect of heat and work. The dried paste is first passed repeatedly through hot rolls to a definite schedule of passes and roll settings. Reworked material can be blended in at this stage.

The sheets thus produced are cut into discs, or else formed into carpet rolls. The former process involves cutting circular or square portions from the sheets so as to fit the extruding press. These are examined for flaws and then compressed to form a "cheese". The latter process involves cutting the sheet into strips which are then rolled into "carpet rolls" of a diameter suitable to fit the cylinder of the extruding press.

The propellant is extruded hydraulically at pressures of 15–35 MPa. The press cylinders are heated and filled with heated discs or carpet rolls. In most modern practice the cylinders can be evacuated to avoid air bubbles in the final propellant. For the larger sizes, the press dies give a single large cord which is cut to length by a guillotine while still hot. After cooling it can be handled in wood-working machinery under suitable conditions to give exact length and for such purposes as slotting or drilling. For smaller sizes of grain, the process is similar to that for the solvent type product, but without the necessity of the solvent recovery stage.

Cast Double Base Propellants

Although the solventless cordite process enables grains of larger diameter to be made than are possible by the solvent process, it is nevertheless restricted to about 10 cm in diameter, owing to difficulties in maintaining dimensional stability and to difficulties in construction of adequate extrusion presses. The problem of producing larger grains of this type of propellant has been overcome by the introduction of the casting process. This utilises a powder consisting of nitrocellulose and nitroglycerine, with any necessary modifiers, and a casting liquid which consists of nitroglycerine with suitable desensitisers. A case is filled with these ingredients to give an air-free heterogeneous mixture. It is then subjected to prolonged heating, so that the nitroglycerine swells the nitrocellulose until finally a homogeneous structure results.

In the simpler version of this method, a double based powder of small size is made by conventional solvent methods and thoroughly dried. The required amount of this powder is then placed in a "beaker" of cellulose acetate or ethyl cellulose and the voids are all filled with desensitised nitroglycerine. The curing process consists of heating to temperatures of the order of 80°C for a prolonged period and on cooling, the mass becomes a gelatinous body similar to cordite or ballistite.

In another version of this process developed in America, the grain and the nitroglycerine are formed into a slurry which is poured as such into the casing before curing. This process has been described by Boynton and Schowengerdt. The nitrocellulose is used in the form of a ball powder in granules which may vary from a few microns to a coarse mesh size. The nitroglycerine and other liquid ingredients are then placed in a simple mixing pot and the solid ingredients, including the nitrocellulose, added. After stirring for a few minutes the slurry is poured into a case and cured by heat.

It is possible in both the above processes to add oxidisers such as ammonium perchlorate so as to give propellants which combine the properties of the composite propellants and the more conventional double base type. It is claimed that the product can have an ultimate tensile strength of 800 kPa with an elongation of roughly 30%. These properties must, however, be sacrificed to some extent if the highest propellent performance is required.

Composite Propellants

Composite propellants consist of an oxidiser together with a plastic which serves the dual purpose of a binder and a fuel. Other ingredients, such as aluminium, may be added to increase the heat of combustion. The commonest oxidiser is ammonium perchlorate and the method of manufacture will be described on this basis. Fuels of many types are used, mainly polymeric and usually of a rubbery consistency. Grains are made by casting, moulding or extrusion.

The first process in all cases is the production of the oxidiser in a suitable fine crystal size. A bimodal particle size distribution, obtained by mixing very fine with slightly coarser particles, often gives the best product. The fuel/binder is frequently prepared as a prepolymer so as to assist mixing and also to reduce the time of the later curing process.

In casting grains, the oxidiser and the prepolymer are mixed in equipment similar to the Werner Pfleiderer mixer. The resulting thick dough is subjected to vacuum to remove air bubbles, and cast either into moulds or directly into motors. With thermoplastic binders the mixing and casting are carried out hot and the charge is then allowed to cool by a carefully controlled process. With thermosetting binders the mixing and casting are carried out cold, but the charge is then cured at an elevated temperature for a time which may be one or more days. It is frequently necessary with charges of this sort for them to be bonded to the case and this is done by applying a case bonding to the case before casting. Careful control of all temperature changes is necessary to avoid shrinkage and to allow stress relaxation, otherwise the charge is likely to become separated from the case and may even crack. A continuous process for mixing and casting composite propellants has been described.

Polymers which give mouldable propellants are mixed with the oxidiser in a similar manner, but the product is usually worked mechanically between rolls and evacuated to remove air bubbles. The powder is then moulded by pressure into the metal casing. Alternatively, such plastics can give a propellant which can be extruded into charges which are afterwards cut and machined to suitable shape. The Rocket Propulsion Establishment at Westcott has used ammonium perchlorate and polyisobutene to produce a propellant of putty-like consistency.

In Britain grains of composite propellant containing 4000 kg have been fired. In the U.S.A. even larger charges are recorded, including a 3 m casing containing 100 000 kg of propellant.

Factory Construction and Operation

The main risk in the manufacture of propellants is that of fire and only in isolated circumstances is there also a detonation risk. A common construction of a building is therefore of reinforced concrete, and frequently one single building contains a number of compartments separated by strong partition walls. Each compartment will, however, have a blow-out panel of large size and flimsy construction. Should a fire occur the panel blows out and prevents any build up of pressure which could cause the deflagration to become more severe.

In certain operations, such as the rolling of solventless cordite, experience shows that fires must sometimes be expected. Under these circumstances it is usual to provide a water drenching system operated by a photoelectric device. Rapid application of a mass of water successfully limits the fire to the material originally enflamed.

It is an unfortunate characteristic of propellants that they invariably burn to detonation if there is more than a critical depth of powder above the point of ignition. This depth depends greatly on the composition and on the grain size. It may vary from about 10 cm to several metres. In processing, the critical depth for the product being made is not exceeded unless full precautions for handling a detonating explosive are taken.

References

General

DAVIS, T. L., *The Chemistry of Powder and Explosives*. Chapman & Hall, London, 1956.
KIRK, R. E., and OTHMER, D. F., *Encyclopedia of Chemical Technology*. Vol. 6, Interscience, New York, 1951.
PENNER, S. S., and DUCARME, J., *The Chemistry of Propellants*. Pergamon Press, London, 1960.
TAYLOR, J., *Solid Propellent and Exothermic Compositions*. Newnes, London, 1959.
WARREN, F. A., *Rocket Propellants*. Reinhold, New York, 1958.

Specific

BOYNTON, D. E., and SCHOWENGERDT, J. W., *Chem. Eng. Progr.* **59**, 81 (1963).
WHEELER, W. H., WHITTAKER, H., and PIKE, H. M., *J. Inst. Fuel*, **20**, 137 (1947).

CHAPTER 18

Properties of Propellants

PROPELLANTS, like all explosives, are intended to do work in a finite time. It is, therefore, with energy and burning speed that this chapter is mainly concerned. The method of approach differs from that with high explosives, because the hydrodynamic theory, so useful in that case, is of little value with propellants. If we refer to Fig. 2.4 of Chapter 2, it was shown that burning consists of a transfer from the initial point A to the lower portion EC of the Rankine–Hugoniot curve. In view of the relative slowness of the process, the pressure of the propellant and the products can be considered equal, so that the transfer is effectively along the line AE. Propellants, however, are always used in closed systems and the initial pressure (and final pressure) is no longer equal to atmospheric. Because the point A is defined only by the application of the propellant, the hydrodynamic theory gives no direct information. It does, however, tell us that the products of combustion stream backwards relative to the burning front and not forwards as in the case of a detonation wave.

The heat given out in the combustion of propellants is readily measured by exploding them in special calorimetric bombs built to withstand the high pressures produced. The result thus obtained, however, does not necessarily give exactly the heat available under practical circumstances, as in the calorimetric bomb the products are cooled and can undergo reactions which have different equilibria at high and low temperatures. It is, therefore, usual for design purposes to obtain the calorimetric value of a propellant also by calculation.

The calculation of the heat liberated in the burning of the propellant can be carried out in a manner similar to that described for high explosives. The pressures encountered are, however, much lower and correspondingly the gas densities also are lower. Simple equations of state are therefore

178

adequate, and indeed for rocket propellants the combustion products are usually considered to behave as ideal gases. For gun propellants where the pressures are higher, the Abel equation of state is usually employed:

$$p(v-a) = RT$$

where a is a constant.

By constructing the Rankine–Hugoniot curve in the lower pressure region in this way, the volume, temperature and composition of the combustion gases can be calculated for various given operating pressures. In general the calculations must be carried out by successive approximation, but in the particular cases of single and double base compositions, simpler methods have been derived.

Calorimetric values of single base powders lie in the range 4000–3000 J g^{-1} and for double base 5200–3000 J g^{-1} for ammunition or down to 1750 J g^{-1} for use in power cartridges.

The other particularly important property of propellent explosives is their rate of burning. This cannot be calculated but must be measured. The measurements can be carried out in a number of ways, but the simplest is by the strand burner particularly used for rocket compositions. This is illustrated in Fig. 18.1. A strand of the propellant composition, commonly 3 mm diameter and 15 to 18 cm long, is coated by a plastic so that it can burn only from one end. This strand is held inside a bomb pressurised with nitrogen and allowed to burn. The rate of burning is measured by timing the interval between the melting of fuse wires resting on the strand. Tests are carried out at a number of pressures and the variation of burning rate with pressure is thus determined.

It is found that the rate of burning increases approximately linearly with the calorimetric value and lies in the range 0·4–1·3 cm s^{-1} at 7 MPa, although higher rates can be achieved.

This technique is much less suited to measuring the burning rates of propellants for use in rifles and guns. In these cases, it is not usually necessary to determine this rate accurately, but instead a measurement is made of the pressure developed in the actual use of the propellant. This is done by connecting a piezo-electric gauge to the chamber of a gun and recording the pressure change, during firing, by an oscilloscope.

The results of measurements made in these ways show that the rate of burning of a propellant depends markedly on the pressure. At the high

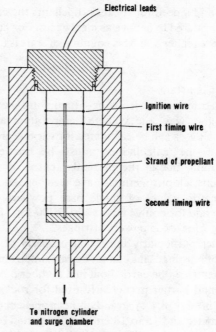

Electrical leads

Ignition wire

First timing wire

Strand of propellant

Second timing wire

To nitrogen cylinder
and surge chamber

FIG. 18.1. Strand burner.

pressures used in guns (perhaps 300 MPa) it is often sufficient to assume that the rate of burning, R, is directly proportional to the pressure, p. Rather more accurate results can be obtained by using a formula of the following type:

$$R = b + ap$$

where b and a are constants.

At lower pressures, such as are encountered in rockets and power cartridges, the relationship preferred is the exponential form:

$$R = ap^n$$

This is known as Vieille's Law. The exponent, n, is known as the pressure index of the propellant and is usually in the range 0.5 to 0.8.

Vieille's Law is obeyed well by the older types of single and double base propellants. Many of the more modern compositions do not obey this law. This is particularly true of low pressure propellants containing a platonising agent, the purpose of which is to make the burning rate practically independent of pressure over a useful working region. Nevertheless, it is convenient to use the Vieille equation even if it can be regarded only as an approximation and used only over a limited range of pressure.

The rate of burning of a propellant is also influenced by the initial temperature of the charge. The effect is much less than the normal effect on chemical reactions, but can still be important when the material is to be used over a wide range of temperatures from arctic to tropical and those encountered in supersonic flight.

A characteristic of propellent burning is that it proceeds by layers with the burning front always parallel to the surface. This is known as Piobert's Law, and it is on this law that the design of propellent grains depends. If, for example, we consider a long solid cylinder of propellant, then as burning proceeds the cylinder remains of the same shape but with gradually reducing radius. It is clear that the surface area of the propellant gradually decreases so that the mass rate of burning of the propellant also decreases. This is obviously undesirable and many means of avoiding it have been designed.

One common method of designing propellent grains is to use a long annulus. This can burn on both the interior and exterior surfaces, so that as burning proceeds the outer surface decreases, but the inner suface increases in such a way that the total surface and therefore mass burning speed remains constant. Similar properties are also shown by thin flakes or discs of propellant which are suitable for use for small arms ammunition.

For some purposes, it is indeed desirable that the rate of burning should increase during the process rather than remain constant. This is true in certain guns. For such purposes grains can be produced perforated by a number of holes so that the burning surface increases as combustion proceeds. For rockets special constructions are used which are discussed in Chapter 19.

Whilst the shape of a propellent grain determines the constancy or otherwise of the burning process, the actual time occupied depends on the grain dimensions. The term "web thickness" is used to denote the shortest distance in a grain through which burning can go to completion; it is measured normal to the burning surface. The number of burning surfaces is ignored. Thus in small arms propellants it is customary to denote by web thickness the diameter of a cylinder or the thickness of a flake, even though burning is from both sides. On the other hand, rocket charges burn from one surface only, but the total thickness is still called the web thickness.

The actual process of burning of single and double base propellants has been studied in some detail and shown to consist of a number of stages, as shown in Fig. 18.2. The succession of stages is as follows.

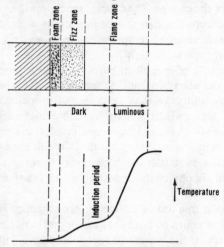

FIG. 18.2. Process of propellent burning.

Solid propellant is first caused by radiation and convection to melt and evolve sufficient gas to give a foamy structure. This is known as the foam zone. The gaseous products from this zone pass through the fizz zone where an initial reaction occurs. These intermediate products enter the flame zone where after a brief induction period they undergo the final reaction to the combustion products. It is only in this final reaction that there is any luminosity, so that the fizz zone and the initial stages of the flame zone are in fact dark. The actual thickness of this dark zone depends on the pressure under which the propellant is burning. Below 1·4 MPa the flame zone does not exist. When the pressure increases the dark zone decreases in thickness until at 7 MPa it can no longer be observed. These changes explain well the marked dependence of burning rate on pressure.

The burning mechanism of composite propellants differs from that described above. There is no exothermic reaction which can lead to a self-sustaining fizz zone. Instead, the first process appears to be the softening and breakdown of the organic binder/fuel which surrounds the ammonium perchlorate particles. Particles of propellant become detached and enter the flame. The binder is pyrolysed and the ammonium perchlorate broken down, initially to ammonia and perchloric acid. The main chemical reaction is thus in the gas phase, between the initial dissociation products.

The stability of propellants from both the chemical and physical points of view is of considerable importance, because they frequently have to be stored for many years under adverse conditions of temperature. Minor changes, such as could well be tolerated with high explosives, cannot be allowed with propellants because they would seriously affect the performance of the gun or other weapon. Accelerated storage tests, usually involving cycling between the extremes of temperatures likely to be encountered, are used to determine the long term stability of products. In the case of large grains, such as those used in rockets, physical strength and stability can be of great importance. Thus, it is usual to determine tensile strength and elongation under both static and dynamic conditions. Even more important is the examination of the grain for tendency to crack under conditions of varied temperature.

Many methods have been proposed and are used to study the thermal stability of propellants and to ensure the absence of possible autocatalysed decompositions during storage. None are sufficiently reliable to merit individual description. In practice, stabilisers are added, the usual being diphenylamine for nitrocellulose powders and symmetrical diethyl diphenyl urea (carbamate or centralite) for double base propellants. Provided a reasonable proportion of stabiliser remains, the propellant can be assumed to be free from the possibility of autocatalytic decomposition. The best test of stability is therefore a chemical determination of the stabiliser present.

References

SIEGEL, B., and SCHIELER, L., *Energetics of Propellant Chemistry.* Wiley, New York, 1964.

TAYLOR, J., *Solid Propellent and Exothermic Compositions.* Newnes, London, 1959.

CHAPTER 19

Design and Application of Propellants

AS PROPELLENT explosives comprise a wide range of products, each designed for a specific application, it is convenient to discuss their detailed design and applications at the same time. It is proposed to proceed in the order roughly of the smallest grains to the largest. In this way progression will be the more natural, even though a considerable degree of overlap must still remain between the individual classes.

Small Arms Ammunition

Small arms may be taken to mean weapons with a bore of less than 2·5 cm, whether the barrel is smooth or rifled. Compared with larger weapons, they have light bullets or shot and have relatively short barrels. Propellants used must therefore be fast burning and have small web thickness. The factors which determine design of powder depend also on the individual characteristics of the weapons.

In the case of shotguns, for example, experience shows that it is best to use shot weighing about 30 to 35 g in a weapon weighing between 3 and 4·5 kg. Heavier loads of shot, or lighter guns, give unpleasant recoil, whereas heavier guns become too unwieldy. Even at this weight of weapon, the barrel must be relatively thin, particularly at the muzzle where extra weight would lead to slowness in aiming. The working pressure which the chamber will stand is therefore relatively low and not likely to exceed 45 MPa. A powder which will burn fast at low pressures is required in small web thickness. The structure of a typical shotgun cartridge is shown in Fig. 19.1.

Pistols and revolvers have very short barrels and therefore the time during which the propellant must burn is very limited. Burning pressures can, however, be higher and are often in the range 75–110 MPa.

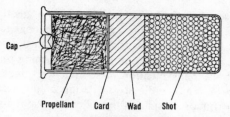

Cap

Propellant   Card   Wad   Shot

FIG. 19.1. Shotgun cartridge.

Compared with shotguns the higher pressure and shorter barrel length tend to cancel each other out and somewhat similar powders may be used.

The smallest common rifle is of 0·22 in (0·56 cm) calibre as used for target shooting. The weapons themselves have barrels which are neither particularly short, nor with any noteworthy restrictions on strength. On the other hand, the ammunition is of a particular type known as rimfire, as illustrated in Fig. 19.2. The name comes from the method of firing which is

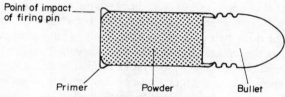

Point of impact
of firing pin

Primer   Powder   Bullet

FIG. 19.2. Rimfire cartridge.

by the use of an eccentric striker which crushes pyrotechnic composition contained in the base of the cartridge between the walls and the rim. Such cases are soft and relatively weak and cannot withstand high pressures. The bullets also are of a relatively soft nature. It is these factors which limit the pressure at which the powder must burn to about 110 MPa. The small size of the cartridge also means that the grain must be small.

The larger rifles such as the familiar military rifle use cartridges of the type shown in Fig. 19.3. The relatively heavy base contains centrally a pyrotechnic percussion igniter (cap) and also a rim on which the ejector mechanism operates. Bullets can be made of various materials, but the detailed construction depends on whether they are to be used for armour piercing, anti-personnel, incendiary or tracer purposes. In operation, the primer is fired by the striking pin of the rifle, the propellant ignites and the gas pressure causes the case to expand and form a gas-tight seal in the breech of the rifle. The bullet is projected forward, engages in the rifling where it forms a seal and also is driven forward in a spiral fashion. A working pressure of 300 MPa is common and burning times are longer than in the other weapons mentioned, so that web thicknesses can be somewhat higher. A major difficulty in design of these cartridges is to ensure that tracer rounds, in spite of the different weight and shape of the bullet, follow the same trajectory as the ball rounds.

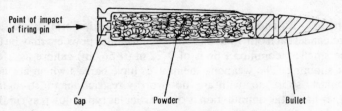

Point of impact
of firing pin →

Cap   Powder   Bullet

FIG. 19.3. Rifle cartridge.

For all these weapons there is a choice between single base and solvent type double base compositions. Much has been said on the relative advantages of these two types of powder. Double base powders are less susceptible to the effects of moisture and varying humidity, but they cannot successfully be surface moderated. Modern practice is to use these powders in such cases as shotgun shells where complete protection against humidity changes is difficult. Double base powders are used in smaller powder charge weights than single base and this largely offsets their tendency to give increased barrel erosion, due to the higher temperature of the combustion products. For the smallest grains, as for shotgun and rimfire cartridges, the usual form is a thin disc. The powder is made porous in order to give sufficiently rapid combustion. For the larger sizes, as in rifle cartridges, the larger web thickness makes possible the use of a tubular powder which has the advantage of improved loading characteristics. Details of some typical small arms propellants are given in Table 19.1.

Ordnance Propellants

The general design of a large round is similar in principle to that of a rifle cartridge, but the shell has the special construction described on p. 154.

TABLE 19.1   *Typical Small Arms Propellants*

Ammunition	Powder			
	Type	Shape	Dimensions	Web thickness
Shotgun	Double base	Porous disc	1·27 mm diam.	0·15 mm
Rimfire	Double base or single base	Porous disc	0·89 mm diam.	0·10 mm
Revolver	Double base	Porous disc	1·02 mm diam.	0·13 mm
Rifle	Single base	Tubular	1·27 mm diam. 0·38 mm diam.	0·46 mm

For weapons of larger size, the design of propellent charge can become a complicated matter of interior ballistics. Because of the long burning time, however, web thicknesses are large and multi-perforated grains can be produced to give suitable pressure–time curves.

A choice has again to be made between single and double base propellants. In the past double base propellants tended to be unpopular, because their original high calorimetric value caused gun erosion. This has now been overcome. Also the presence of nitroglycerine can give physiological effects under bad conditions; these, however, no longer occur in modern equipment. The main advantage of double base propellants is that they can be produced with low quantities of volatile material and also with high stability, so that their ballistic change during storage is very small. On the other hand, single base propellants have some advantages in manufacture and in general can be blended more readily to give required ballistics. Because of the wide variety of weapons, it is not possible to quote typical grain sizes of web thicknesses.

Ordnance propellants are required to give the minimum of muzzle flash, smoke and barrel erosion. All these objectives are assisted by adding to the propellant a proportion of nitroguanidine (picrite), made by treating guanidine nitrate with sulphuric acid and brought to very fine particle size by recrystallisation and disintegration.

For the largest weapons, the propellent charge can be kept separate from the rest of the round and in this way it is possible to adjust the weight of the propellant to allow for wear of the gun during continued use. The storage and handling convenience of a complete sealed round are, however, such that this is the form usually adopted.

Rockets

The design of propellants for solid fuel rockets differs considerably from that for ordnance, because of the lower operating pressures, usually below 15 MPa. To understand the principles involved it is first necessary to give a brief account of rocket propulsion. In this account considerations will be restricted to motors based on solid propellants. Motors based on liquid fuels, such as petroleum fractions and liquid oxygen, depend on combustion processes of non-explosive type.

A sketch of a rocket motor is shown in Fig. 19.4. It consists of a chamber, containing the propellant and an igniter, at one end of which is the nozzle. The nozzle has a restricted portion, or throat, which controls the rate of flow of the gas and a divergent portion which causes the exhaust gas to attain a high and supersonic velocity. The thrust on the rocket consists of two parts, the first due to the backward momentum of the gases and the second due to the difference in static pressure of the exhaust gases at the nozzle exit and the surrounding atmosphere. The latter component is usually designed to be small and is therefore neglected in the following discussion.

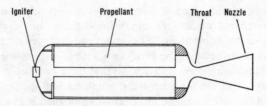

FIG. 19.4 Solid fuel rocket motor.

As there is no appreciable loss of heat from the rocket to the surrounding atmosphere, the internal energy of the propellant which is released on combustion appears in the exhaust gases, partly as kinetic energy, and partly as internal energy of the reaction products. If the internal energy of the original propellant is H_o and the internal energy of the exhaust gases H_e, then the kinetic energy is by difference $H_o - H_e$. If V is the velocity of the exhaust gases and M their mean molecular weight, it follows that

$$H_o - H_e = \tfrac{1}{2}MV^2$$

therefore
$$V = \sqrt{[2(H_o - H_e)/M]}$$

For comparative purposes it is usual to consider the effect when unit weight m of the propellant burns in unit time t. As momentum is equal to impulse in a flow system, then $Ft = mV$ or, as t and m are both equal to 1, $F = V$. Thus for the propellant the specific impulse I_{sp} is given by

$$I_{sp} = F = \sqrt{[2(H_o - H_e)/M]}$$

H_o can be calculated from the propellant composition, but H_e must be obtained by successive approximation, assuming that the final state of the exhaust gases is known. For present purposes, it is sufficient to note that $H_o - H_e$ correlates well with the heat of explosion of the solid explosive. In order to obtain the maximum thrust from a rocket it is therefore necessary to achieve the highest combustion temperature, but also necessary to produce gases with the lowest mean molecular weight.

The composition and properties of an American solventless double base composition have been published and are quoted in Table 19.2. This powder is known as JPN and is processed by solventless extrusion.

TABLE 19.2   *Rocket Propellant JPN*

Composition	%
Nitrocellulose	51·5
Nitroglycerine	43·0
Diethylphthalate	3·25
Carbamate	1·0
Potassium sulphate	1·25

Carbon black (0·2%) and candelilla wax are also added.

Properties	
Burning rate (7 MPa)	16·5 mm s^{-1}
Pressure index	0·7
Molecular weight of products	26·4
Specific impulse	250 s

According to Sutton, cast double base charges are likely to contain 45–55% nitrocellulose, 25–40% nitroglycerine, 12–22% plasticiser and 1–2% other ingredients such as stabilisers. Typical properties are:

Burning rate (7 MPa)	5·6–9·4 mm s^{-1}
Pressure index	0·1–0·8
Molecular weight of products	22–28
Specific impulse	160–220 s

For composite propellants the properties depend on the proportion of binder, and also on whether high energy fuels such as aluminium have been added. Figure 19.5, taken from Barrère, Jaumotte, de Veubeke and Vandenkerckhove, shows how the specific impulse depends on these factors.

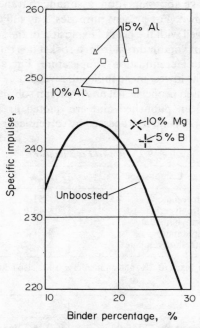

FIG. 19.5. Specific impulse of composite propellants.

The design of the grain is as important as the selection of the propellant. A rocket is required to have a uniform thrust throughout the period of burning. As the grains, like all propellants, obey Piobert's Law, it is necessary to achieve a uniform area of burning surface through the whole process of burning. As the motors are in general cylindrical, the propellent grain must follow this shape in external diameters. In the most important case, the outside of the grain is either bonded to the case, or is otherwise inhibited from burning. Burning, therefore, occurs only on the internal surface of the charge. A simple tubular charge burns with gradually increasing surface and therefore with gradually increasing release rate of gas. In order to achieve uniform release rate more complicated shapes are employed. Two such shapes are shown in Fig. 19.6 and are known as star centre and clover leaf charges. In each of these the initial burning area is equal to the outside area of the charge, so that the burning area remains essentially constant. Sometimes more than one composition may be employed in order to achieve suitable burning rates. Some allowance must also sometimes be made for increased burning speed in the initial stages of combustion due to the erosive effects of the products of combustion passing along the initial narrow channel of the charge.

Clover leaf   Star centre

FIG. 19.6. Rocket charge sections.

A list of missiles using solid fuel rockets is quoted from Daboo in Table 19.3.

In the design of a charge for a rocket the quantities given are the expected total weight of the rocket, the maximum acceleration which the charge may give and the time for which this stage of the rocket should burn. The last two factors will determine the ultimate speed of the rocket. The maximum pressure for which the motor body is to be designed will also be given.

From the time of burning and the known pressure, together with the rate of burning equation for the selected propellant, the web thickness is calculated. This gives the size and cross-sectional area of the grain, assuming that it is radial burning, probably with a star centre. The total thrust required divided by the specific impulse gives the weight of the charge. As the cross-section is known the length of the charge can then be calculated. In actual practice, calculations are much more complicated, as they must allow for inefficiencies in the system and for air resistance and must also be carried out for various ambient temperatures which may

TABLE 19.3   *Missiles with Solid-fuel Rocket Propulsion*

Role	Range (km)	Launch weight (kg)	Designation	Propulsion
Air-to-air	3–6.5	50–200	Fireflash	Wrap-round separating boost. No sustainer.
			Firestreak Sparrow	Internal boost. No sustainer.
			Falcon	Internal boost-sustainer.
Surface-to-air	5–10	50–200	Seacat	Internal boost-sustainer.
	16–24	500–1500	Terrier	Tandem separating boost. Internal sustainer.
			Nike–Ajax	Tandem separating boost. Internal liquid-propellant rocket sustainer.
			Hawk Tartar	Internal boost-sustainer.
	32–48	2000–2500	Seaslug Thunderbird	Wrap-round separating boost. Internal sustainer.
	64–160	2000–5000	Nike–Hercules	Tandem separating boost. Internal sustainer.
			Bloodhound	Wrap-round separating boost. Ramjet sustainer.
			Talos	Tandem separating boost. Ramjet sustainer.

Category			Name	Description
Air-to-surface	5–11	250–500	Bullpup	Internal boost. No sustainer.
	1600–3200	5000–10 000	Nord A.S. 30	Internal boost-sustainer.
Surface-to-surface			Skybolt	Ballistic missile.
	1·5–5	10–100	Vigilant Dart	Internal boost-sustainer.
	16–32	500–1500	Lacrosse	Internal boost. No sustainer.
	80–160	5000–7500	Sergeant	Single-stage ballistic missile.
	800–3200	7500–15000	Matador Regulus	Separating boost. Turbojet sustainer.
			Triton	Separating boost. Ramjet sustainer.
			Polaris Pershing	Two-stage ballistic missile.
	8000+	25 000–50 000	Snark	Separating boost. Turbojet sustainer.
			Minuteman	Three-stage ballistic missile.
High-altitude sounding rocket	(Height up to 160 km)	1000–2000	Skylark	Single-stage vehicle.
Satellite launcher	(Payload up to 100 kg)	15 000–20 000	Scout	Four-stage vehicle with single solid-propellant motor in each stage.

occur during use. Moreover, the burning pressure is probably not originally quoted, but deduced from consideration of its effect on the size and design of the motor itself.

Such a charge must in use perform regularly and give constant thrust. Should, for any reason, a slight increase in pressure in the chamber occur, two opposing factors operate:

1. The gases flow out from the chamber through the nozzle at an increased rate.
2. Gases are produced more rapidly by the propellant because of the increased pressure.

It can be shown that if the pressure index of the propellant exceeds 1 the rate of gas increase by factor 2 exceeds the rate of gas loss by factor 1, so that the pressure builds up in the motor, which finally explodes. Quite apart from such an extreme case, a low pressure index in the propellant is desirable so that irregularities in burning are quickly smoothed out with the least effect on rocket performance. It is for this reason that platonising agents mentioned on p. 181 are important, because they enable a very low pressure index to be achieved at ordinary operating pressures of the order of 14 MPa.

Power Cartridges

Special slow burning cool propellants can be used to generate gas under pressures suitable for the operation of mechanical devices. Perhaps the most important of these applications is in cartridge starting of jet engines for aircraft. The principles involved in this application of propellent explosives are illustrated in Fig. 19.7.

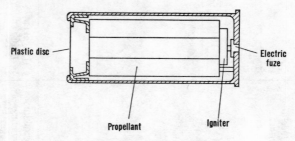

FIG. 19.7. Engine starter cartridge.

The propellant is usually either tubular or cigar burning, depending on the required time of operation. It is enclosed in a pressure chamber from which the gases are led through an orifice which controls the flow and regulates the pressure. A pressure relief valve is included for safety reasons. The hot gases from the orifice are taken to an impulse turbine which is geared to the rotor of the jet engine.

A jet engine has a compressor which provides compressed gas to a combustion chamber where the fuel is burnt. The hot gases from the combustion chamber pass through a turbine, which drives the compressor, and thence go to the jet. For the engine to be started the compressor must be driven above a critical speed, below which the power given by the turbine is insufficient to drive the compressor. The advantage of propellent cartridges for starting jet engines is that they enable the compressor and turbine to be brought to the critical speed in a matter of seconds. When this speed is reached fuel injection is provided into the combustion chamber and the engine is then capable of being run to its idling speed or above.

For civilian aircraft the facility for rapid starting is not important and cartridge operation is not often employed, particularly because it involves storing and handling explosives, even though the hazards of these explosives are those of fire and not of detonation. For military purposes, however, particularly for fighter aircraft which are best scattered on an airfield, a rapid start is of considerable importance. Therefore cartridge operated starters are much used for these aeroplanes. In Britain, development has been essentially with propellants based on ballistite, namely double base propellants of the solventless type, whereas in the United States composite propellants based on ammonium nitrate have proved more popular.

Electrically Actuated Devices

Electrically actuated devices (explosive motors) of small size are finding varied and increasing uses. Typically they contain a fusehead and in the larger sizes also a small charge of a propellent explosive such as blackpowder. This is sealed in a strong metal tube so that the finished device offers no explosive hazard. The firing of the fusehead can for example drive a captive piston a small distance outwards at the end of the tube. This in turn can be used to open or close a switch or valve or to work a guillotine cutter. These actuators have great reliability even after long storage under adverse conditions and give strong rapid action from small electrical inputs. These properties make them of value in applications such as safety and security systems and in fire protection.

References

BARRÈRE, M., JAUMOTTE, A., DE VEUBEKE, B. F., and VANDENKERCKHOVE, J., *Rocket Propulsion*. Elsevier, Amsterdam, 1960.
CORNER, J., *Theory of the Interior Ballistics of Guns*. Chapman & Hall, London, 1950.
DABOO, J. E., *Solid-fuel Rocket Propulsion*. Temple Press, London, 1962.
HUGGET, C., BARTLEY, C. E., and MILLS, M., *Solid Propellant Rockets*. Princeton Univ., Princeton, 1960.
OHART, Maj. T. C., *Elements of Ammunition*. Chapman & Hall, London, 1946.
SUTTON, G. P., *Rocket Propulsion Elements*. Wiley, New York, 2nd ed. 1956.
TAYLOR, J., *Solid Propellent and Exothermic Compositions*. Newnes, London, 1959.
WHELEN, Col. T., *Small Arms Design and Ballistics*. Small Arms Technical Publishing Co., Plantersville, S. Carolina, 1945.

Glossary of Common Terms and Abbreviations

Abel Heat Test. A quick routine test for demonstrating the absence from explosives of impurities causing low thermal stability (see p. 72).

A.D.C. Ardeer Double Cartridge Test. A test measuring the ability of cartridges of explosive to propagate over air gaps (see p. 68).

Advance. In mining or tunnelling, the distance by which the face or tunnel is moved forward by each round of blasting (see p. 138).

Airdox. A blasting device based on compressed air (see p. 89).

Amatol. A mixture of ammonium nitrate and trinitrotoluene.

Ammon gelignite. A nitroglycerine gelatine explosive containing ammonium nitrate as the main oxidising ingredient.

Ammonal. An explosive containing ammonium nitrate, trinitrotoluene and aluminium.

AN. Ammonium nitrate.

ANFO. An ammonium nitrate/fuel oil explosive.

Application time. In firing electric detonators, the time for which the electric current is applied (see p. 112).

B. and J. Bergmann and Junk. Test of stability (see p. 72).

Base charge. In a detonator, the charge of high explosive which makes the major contribution to the power of the device.

Boost. In rocketry, a motor which gives rapid acceleration for a short period.

Braid. To enclose in criss-crossing yarns of textile.

Break. In mining, a separation between strata or along a cleavage plane in coal or rock.

Break gallery. An experimental installation to assess the hazards of firing explosives in coal mines in the presence of breaks (see p.80).

Bridge wire. The fine wire in an electric detonator which is heated by the firing current.

Brisance. The shattering power of an explosive.

Briska detonator. A detonator in which extra power is achieved by extremely heavy pressing of the base charge.

Bulk strength. The strength per unit volume of an explosive, usually expressed as a percentage of the strength per unit volume of blasting gelatine.

Burden. In blasting, the shortest lateral distance between a borehole and a free face.

Bursting time. In firing electric detonators, the time between the application of the electric current and the explosion of the detonator (see p. 112).

Cap. A metal shell with pyrotechnic filling, such as is used in small arms ammunition for causing a mechanical blow on the outside of the shell to ignite the propellent charge inside. In the U.S.A. detonators are known as blasting caps.

Carbamate. Symmetrical diethyldiphenylurea.

Cardox. A metal tube device using carbon dioxide to produce a blasting effect (see p. 89).

C.D.B. Cast Double Base Propellant (see p. 174).

C.E. Composition Exploding. Tetryl, usually in the form of a pellet.

Centralite. Symmetrical diethyldiphenylurea.

Cigar burning. In propellants, the burning of a cylindrical charge from one end only, the other surfaces being inhibited.

C.-J. Chapman Jouget (see p. 19).

Clearing Test. A test of the speed of solution of nitrocellulose in nitroglycerine (see p. 74).

Composite propellant. A propellant based on an oxidising salt and an organic fuel/binder.

Cordtex. British trade name for detonating fuse.

Corning. In blackpowder manufacture, the stage of the process which produces uniform spherical grains.

Counter. In fuse manufacture, to wind on textile yarns in a direction opposite to a previous spinning process.

Crimping. Squeezing a metal tube round a plug, fuse, or similar body to secure the latter firmly.

DDNP. Diazodinitrophenol (see p. 97).

Deflagrating explosive. An explosive which burns rapidly but does not detonate.

Deflagration. In high explosives, a relatively slow decomposition accompanied by fumes but not normally by flame.

Delay detonator. A detonator in which a time lag is introduced between application of the firing current and explosion of the detonator.

Detonating fuse. A fuse with a core of detonating explosive.

Detonation. An explosion process of high speed involving a sustained shock wave.

Detonator. A metal tube containing a primary explosive used for initiating a secondary explosive.

Double base propellant. A propellant based on nitrocellulose and nitroglycerine.

Drift. In coal mining, an underground tunnel through stone.

Electric detonator. A detonator for firing by electric current. The term does not normally include delay detonators.

End spit. The flash of burning material ejected from safety fuse when burning reaches a cut end.

Eq.S. Equivalent to Sheathed. The original name for Group P3 Permitted Explosives.

Excitation time. In firing electric detonators, the interval between the application of the current and the firing of the fusehead (see p. 112).

Explosion. A liberation of energy sufficiently sudden to cause dynamic stress to the surroundings.

Explosive. A chemical, or mixture of chemicals, which can react so rapidly and with such liberation of energy that there can be damage to the surroundings.

Explosive motor. A device in which explosive is completely enclosed and which on operation causes a mechanical movement as of a piston (see p. 195).

Exudation. In nitroglycerine gelatine explosives, a liberation of nitroglycerine following breakdown of the gelatinous base.

Fall hammer test. A test of the sensitiveness of explosives to impact using a weight which falls vertically (see p. 69).

Fizz zone. In the burning of propellants, the zone in which the solid propellant is converted to gaseous intermediates (see p. 182).

Flame zone. In the burning of propellants, the final stage in which gaseous intermediates react with the production of a flame (see p. 182).

Flash past. In assemblies such as delay detonators and military fuzes the possibility of an igniting flash by-passing a delay element.

Foam zone. In the burning of propellants, the initial stage of partial gasification (see p. 182).

Free face. In blasting, a face of rock or coal approximately parallel to the line of boreholes (see p. 138).

Fulminating compound. An early term, applied to mixtures of chemicals usually containing silver and used essentially for pyrotechnic purposes.

Fume. In mining, the gaseous products of an explosion.

Fuse. A cord for transmitting explosion from one site to another.

Fusehead. Ignition element for an electric detonator.

Fuze. Military device for initiating an explosive charge.

Gaine. An intermediate booster charge used between a detonator and an insensitive high explosive.

Gallery. Equipment for firing explosives into incendive mixtures of methane and air.

Gasless delay detonator. Original name for modern delay detonator.

Gassy coal mine. A mine in which methane may be present; also known as a safety lamp mine.

Gelatine. An explosive which is a jelly of nitroglycerine containing nitrocellulose, usually with oxidising salts and solid fuel dispersed in it.

Gelignite. Originally a gelatine explosive containing potassium nitrate as oxidising material. Now often applied to any gelatine.

Grain. A unit of propellent powder.

Guncotton. Nitrocellulose containing more than 12·6% of nitrogen.

Heat Test. A rapid stability test depending on detection of traces of products of decomposition of an explosive at an elevated temperature. In Britain this usually refers to the Abel Heat Test.

Hess Test. A German test for brisance (see p. 73).

High explosive. Literally any explosive which detonates. In practice, the term is usually confined to explosives which do not normally burn to detonation but which require a detonator for use.

High tension detonator. An early form of detonator which required a voltage exceeding 36 volts to fire it (see p. 107).

HMX. Cyclotetramethylenetetranitramine.

Hydrox. A steel tube device using low temperature gas produced from a chemical cartridge for producing a blasting effect (see p. 89).

Igniter cord. A cord for igniting safety fuse.

Induction time. In firing electric detonators, the time between the breaking of the fuse wire and the detonation of the base charge (see p. 112).

Initiating explosive. An explosive which when lit by a flame immediately detonates.

Instantaneous fuse. A fuse which propagates by burning at high velocity.

Kast Test. A German test for brisance (see p. 73).

Lag time. In firing electric detonators, the interval between the application of the current and the breaking of the bridge wire (see p. 112).

L.E.D.C. Low Energy Detonating Cord. A detonating fuse with a core charge too low to enable it to be used reliably for initiating high explosives (see p. 124).

Long wall mining. In coal mining, a method of working in which the coal is won along a face 100 to 200 m long.

LOX. Liquid Oxygen Explosive (see p. 59).

Mill cake. In blackpowder manufacture, a product taken from the edge runner mills.

Millisecond delay detonator. A short delay detonator.

Moderated. In propellants, implies the presence of a surface coating to the grain which slows down the initial rate of burning.

NC. Nitrocellulose.

NG. Nitroglycerine.

NS gelignite. A nitroglycerine gelatine explosive containing sodium nitrate as its main oxidising ingredient.

Opencast mining or strip mining. Obtaining coal or ore which is near the surface by removing the overlying soil and rock to expose the coal or ore for direct recovery.

Ordnance. The larger size military guns and mortars.

P1. The first and strongest class of Permitted Explosive.

P2. Originally called sheathed explosives, consist of P1 explosives enclosed in a sheath to give increased safety.

P3. Class of Permitted Explosives for general use with instantaneous detonators.

P4. Class of Permitted Explosives particularly for ripping with short delay detonators.

P5. Class of Permitted Explosives particularly for firing with short delay detonators in solid coal.

Paste. In double base propellant manufacture, the initial mixture of guncotton and nitroglycerine.

Pentolite. A mixture of TNT and PETN.

Permissible Explosive. The American equivalent to British Permitted Explosive.

Permitted Explosive. An explosive which is authorised for use in gassy coal mines.

PETN. Pentaerythritol tetranitrate.

Phlegmatise. To add a proportion of oil or other ingredient to render an explosive insensitive.

Picrite. Nitroguanidine.

Pillar and stall mining. A method of mining in which pillars are left to support the roof.

Plain detonator. An open detonator with no means of ignition attached.

Platonisation. In propellants, the addition of ingredients to produce a low pressure index over a working range of pressures.

Powder. A generic name for a propellent explosive.

Presplitting. A technique of blasting which gives accurate finished contours (see p. 148).

Press cake. In blackpowder manufacture, a cake taken from the presses.

Pressure index. In propellants, the variation of burning speed following changes in pressure (see p. 180).

Prill. An absorbent spherical form of a product, particularly ammonium nitrate.

Primary explosive. An alternative name for an initiating explosive. A flame causes the explosive to detonate immediately.

Primer. High explosive charge used to initiate other high explosive.

Priming charge. In detonator manufacture, the charge of initiating explosive.

Propellant. An explosive which normally burns and does not detonate.

Propellent. Adjective implying the possession of a propulsive effect.

RDX. Cyclotrimethylenetrinitramine.

Reaction time. In firing electric detonators, the time between the ignition of the fusehead and the explosion of the detonator (see p. 112).

Ripping. In coal mining, the removal of stone after recovery of coal to produce a road of normal size (see p. 143).

Safety fuse. A fuse which propagates by slow burning.

Safety lamp mine. Alternative name for gassy mine, implying possible presence of methane.

Secondary blasting. A process of breaking, with explosives, boulders from an initial blast which are too large for immediate handling.

Secondary explosive. Alternative name for high explosive indicating that the explosive does not burn to detonation but is detonated by suitable devices.

Semi-gelatine. An explosive containing nitroglycerine gelled with nitrocellulose, but in quantity insufficient to fill the voids between the salt and combustible particles and thereby produce a gelatine.

Shaped charge. An explosive charge designed to produce specific effects by the inclusion of a re-entrant conical or V shape usually lined with metal.

Shock wave. A pressure wave of high intensity characterised by a very rapid initial increase in pressure followed by a slow falling off.

Short delay detonator. A delay detonator with time interval between individuals of the series of 25 or 50 ms.

Single base propellant. Propellant based on nitrocellulose without the inclusion of nitroglycerine.

Slurry explosive. An explosive made by sensitising a thickened aqueous slurry of oxidising salts (see p. 55).

Small arm. A gun or rifle of up to about 2·5 cm in calibre.

Solid coal. Implies coal which is being worked without the provision of a free face by undercutting or similar means.

Solvent type propellant. A double base propellant in which solvent is used to assist the gelatinisation of the nitrocellulose.

Solventless double base propellant. A double base propellant in which gelatinisation is effected by mechanical means without the addition of solvent.

Spalling. In explosives technology implies the breaking off of a scab of material from a free face as a result of the reflection of shock waves (see p. 135).

Special gelatine. A gelatine explosive in which the main oxidising ingredient is ammonium nitrate.

Spin. In fuse manufacture, to wind on a spiral of textile yarns.

Stemming. The insertion, into the end of a borehole, of clay or other material which will resist the pressure of the explosive when the latter is fired.

Streaming velocity. The velocity of the products of detonation in the direction of travel of the detonation wave.

Strip mining. *See* Opencast mining.

Sustainer. In rocketry, a slow burning motor to produce a continued thrust.

TNT. Trinitrotoluene.

Torpedo friction test. A test of sensitiveness to impact and friction (see p. 70).

Triple base propellant. A propellant containing nitroglycerine, nitrocellulose and Picrite.

Tun dish. A large unstirred cylindrical vessel used for slow processes of steeping or steaming.

Undercutting. In coal mining, the production of a free face by cutting out mechanically the lowest 10 to 12 cm of a seam (see p. 142).

VOD. Velocity of detonation, usually measured in metres per second.

Web thickness. The distance of travel of a burning surface in a propellent grain to give complete combustion.

Wedge cut. A method of tunnelling, etc., in which a free face is first produced by blowing out a wedge of rock.

Weight strength. The strength per unit weight of an explosive, usually expressed as a percentage of the strength per unit weight of blasting gelatine.

Zapon. A solution of nitrocellulose used in fusehead manufacture.

Index

PATHOLOGY
OF THE LUNG
(Excluding Pulmonary Tuberculosis)

Third Edition

IN TWO VOLUMES

H. SPENCER
M.D.(Lond.), Ph.D., F.R.C.S.Eng., F.R.C.P., F.R.C.Path.

*Professor of Morbid Anatomy in the University of London at St. Thomas's Hospital Medical School
and Honorary Consultant Pathologist to St. Thomas's Hospital, London*

With a Foreword by
AVERILL A. LIEBOW, M.D.

*Formerly Professor of Pathology, Yale University School of Medicine,
and Emeritus Chairman and Professor of Pathology, University of California,
San Diego, La Jolla, California*

Volume 2

PERGAMON PRESS
OXFORD · NEW YORK · TORONTO
SYDNEY · PARIS · FRANKFURT

U.K.	Pergamon Press Ltd., Headington Hill Hall, Oxford OX3 0BW, England
U.S.A.	Pergamon Press Inc., Maxwell House, Fairview Park, Elmsford, New York 10523, U.S.A.
CANADA	Pergamon of Canada Ltd., 75 The East Mall, Toronto, Ontario, Canada
AUSTRALIA	Pergamon Press (Aust.) Pty. Ltd., 19a Boundary Street, Rushcutters Bay, N.S.W. 2011, Australia
FRANCE	Pergamon Press SARL, 24 rue des Ecoles, 75240 Paris, Cedex 05, France
FEDERAL REPUBLIC OF GERMANY	Pergamon Press GmbH, 6242 Kronberg-Taunus, Pferdstrasse 1, Federal Republic of Germany

First edition 1962
Reprinted 1963
Second edition 1968
Reprinted 1969, 1973, 1975
Third edition 1977
Reprinted 1977, 1978

Library of Congress Cataloging in Publication Data
Spencer, Herbert, 1915–
Pathology of the lung. 3rd Edition

Bibliography: p.
Includes index.
1. Lungs—Diseases. I. Title. [DNLM: 1. Lung diseases.
WF600 S745p]
RC756.S65 1976   616.2′4′07   76–11763
ISBN 0–08–021021–X

Exclusive distribution in North and South American continents granted to W. B. Saunders Company, Philadelphia and Toronto

Typeset in Northern Ireland at the Universities Press (Belfast) Ltd.
Printed in England by Western Printing Services Ltd, Bristol

Contents

Volume 1

Volume 2

CONTENTS

Pulmonary Thrombosis, Fibrin Thrombosis, Pulmonary Embolism and Infarction

Pulmonary Thrombosis

The importance and frequency of pulmonary embolism have caused it to over-shadow pulmonary thrombosis which in the past was regarded as of little importance in lung pathology.

Because the conditions under which thrombosis is likely to occur are also likely to favour embolism, it is often impossible to be certain whether intravascular thrombi are autochthonous or embolic in origin. The problem has been extensively reinvestigated during recent years, particularly in connection with certain forms of congenital heart disease, but the findings apply equally to all conditions in which the pulmonary blood-flow is reduced, including mitral stenosis and primary pulmonary hypertension.

The introduction of more exact methods of measuring intracardiac pressure and determining cardiac output, together with the introduction of surgical procedures to mitigate the effects of congenital heart disease and to repair certain of the abnormalities, has resulted in a careful reappraisal of the circulatory states in these various conditions and of the causes of surgical failure.

Thrombosis of arteries will occur when (a) the flow through the vessel is reduced below a certain critical level, (b) when the state of the circulating blood renders it more liable to clot and (c) when the vessel walls are damaged. Aschoff (1909) also regarded turbulence of the blood-stream as a predisposing cause of thrombus formation.

Thrombosis of the *major pulmonary arteries* may result from traumatic injury of the chest (Dimond and Jones, 1954) and may complicate an aneurysm of the pulmonary artery, and very occasionally severe pulmonary arterial atheroma or syphilitic pulmonary arteritis. It may occasionally result from such primary lung diseases as pulmonary tuberculosis, emphysema, pneumoconioses and conditions of chronic cardiac failure. Spreading pulmonary artery thrombosis may also follow pneumonectomy, the thrombotic process originating in the stump of the ligated branch of the pulmonary artery. Clinically, the picture is one of increasing dyspnoea, syncopal attacks, fatigue and evidence of rapidly increasing right heart failure, and closely resembles that seen in any patient with chronic pulmonary hypertension. Owing to the failure of lung perfusion in the presence of continuing adequate ventilation, the end tidal (i.e. alveolar) CO_2 tension falls below the arterial pCO_2. Angiocardiographic studies enable the obstruction to be visualized in the pulmonary arteries.

At post-mortem, the pulmonary arteries on one or both sides are partly filled with a smooth polypoid mass of laminated ante-mortem thrombus which projects into the main divisions (Fig. 15.1), and is attached within the intrapulmonary part of the arteries to the endothelial surface. The thrombus is not coiled upon itself.

Thrombosis of the *smaller pulmonary muscular arteries and veins* occurs when the pulmonary blood-flow is greatly reduced by such congenital abnormalities as Fallot's tetralogy, uncomplicated pulmonary stenosis, congenital stenosis of the tricuspid valve, and may also occur in

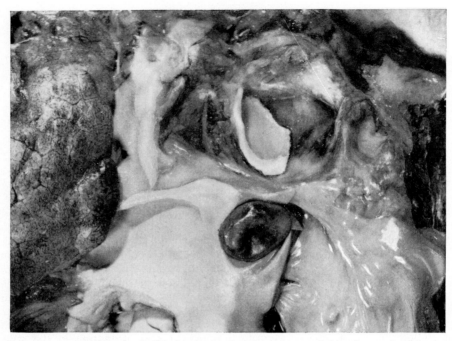

Fig. 15.1. Pulmonary artery thrombosis showing a smooth polypoid mass of thrombus projecting centrally from the left pulmonary artery. The right pulmonary artery is free of thrombus. Approx. three-quarters natural size.

association with neonatal sepsis and sickle-celled anaemia (Haemoglobin-S disease). Heath and Thompson (1969) described a case of sudden death in a young negro man with haemoglobin-S disease in whom many of the elastic branches of the pulmonary arteries were occluded by recent and recanalized thrombi. Newly formed capillary channels traversed the walls of the affected pulmonary arteries and joined branches of the bronchial arteries. The muscular arteries distal to the anastomoses, however, showed no hypertensive changes. Previously Wintrobe (1961) had noticed that arterial blood was frequently unsaturated with oxygen in haemoglobin-S disease and had suggested this might be caused by shunting of pulmonary arterial blood in the lung.

Heath *et al.* (1959a) found that pulmonary thrombosis occurred in congenital heart disease when the pulmonary index fell below 2·5 litres (the pulmonary index = pulmonary blood-flow in litres per minute per square metre of body surface, the normal figure being 2·5–4·4 litres per minute).

Rich (1948) reported that thrombosis occurred in about 90 per cent of untreated cases of Fallot's tetralogy, but Ferencz (1960a) found that it was present in only about 73 per cent of such cases. Thrombosis may extend to the major branches of the pulmonary arteries and is often accompanied by thrombosis of the pulmonary veins and dilatation of the bronchial arteries. Ferencz found that little correlation existed between the degree of polycythaemia and the incidence of arterial thrombosis, but Heath *et al.* (1958a) stated that when the haemoglobin content exceeded 20 g/100 ml pulmonary thrombosis was likely to occur especially in older patients. The thrombosis may occur in the small pulmonary arteries during cyanotic attacks because Hamilton *et al.* (1950) found that during such attacks the pulmonary blood-flow, as judged by oxygen uptake, fell to very low levels. Ferencz (1960b) described the post-operative vascular changes following anastomoses of systemic to pulmonary arteries (Blalock operation) and found that in those cases with an inadequate anastomosis, thrombosis tended to occur in the pulmonary circulation, but thrombus formation was discouraged by the establishment of an adequate circulation. In the later stages of Fallot's tetralogy, pulmonary stenosis, and especially congenital tricuspid stenosis, by-pass operations may no longer afford relief as so many of the smaller intrapulmonary arteries become permanently obliterated by organized intravascular thrombi.

Pulmonary arterial thrombosis due to decreased pulmonary blood-flow may be accompanied, as already stated, by pulmonary venous thrombosis. It has been suggested that the latter change is largely responsible for causing lung infarction in pulmonary embolism. Embolization of infected pulmonary venous thrombus can be responsible for the development of metastatic abscesses in the brain in those forms of congenital heart disease with a decreased pulmonary index. Campbell (1957), however, believed that the cerebral abscesses, which are usually caused by microaerophilic streptococci, resulted from cerebral thrombosis or embolism followed by bacteriaemia and localization of the infection in the devitalized brain tissue.

Pulmonary arteriolar thrombosis may occur as a complication of generalized infections in the neonate (Groniowski, 1963), and both pulmonary thrombosis and embolism occur as terminal complications in patients dying from chronic emphysema (Ryan, 1963). Thrombosis of the small pulmonary muscular arteries may occur in heterozygous sickle-cell anaemia although the majority of the complications in this disease, including haemolytic anaemia, are more commonly found in homozygous cases. The first cases of pulmonary thrombosis in sickle-cell anaemia were described by Yates and Hansmann (1936).

An increased tendency for the red blood-cells to agglutinate (sludge) in the pulmonary capillaries and for thrombosis to occur is found in conditions of profound circulatory collapse such as occurs in severe burns, in haemolytic anaemias and in leukaemia. The capillary changes are usually restricted to the subpleural region and macroscopically the affected region of the lung resembles pressure collapse.

Microscopically, the organized thrombi in the muscular arteries may result in lesions with more than one type of appearance. The resulting fibroelastic tissue which replaces the thrombus may grow either as an eccentrically situated plaque or polypoid mass attached to the wall, or it may organize, leaving the lumen divided into two or more channels by narrow strands of fibro-elastic tissue that bridge the lumen (Fig. 15.2).

Another change, which is less common, is concentric intimal fibroelastosis in the smaller pulmonary arteries. This type of lesion was described by Barnard (1954) who regarded it as a sequel to periodic or sustained vasoconstriction with diminished blood-flow through the artery concerned. The same concentric intimal fibroelastosis is met in pulmonary hypertension, where vasoconstrictive impulses are known to occur and initially play a large part in the development of the arterial changes.

In conditions with low pulmonary blood-pressure such as those mentioned above, the media of the elastic and muscular arteries is hypoplastic and reduced in thickness.

Fibrin Thrombosis of Pulmonary Vessels

This rare condition may follow as a delayed complication of amniotic fluid embolism (Ratnoff and Vosburgh, 1952; Tuller, 1957), and it has also been reported by Schneider (1951) in a pregnant woman who died after developing abruptio placentae; in the latter fibrin blocked many of the pulmonary arterioles. A similar case was also recorded by Johnstone and McCallum (1956). The condition has also been reported in patients suffering from carcinoma of the prostate, carcinoma of the pancreas, leukaemia, and following pneumonectomy.

In some of these cases fibrinolysins have developed and have dissolved the intravascular

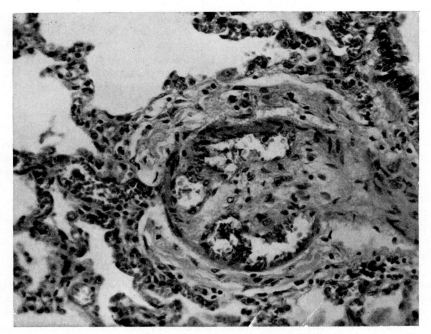

FIG. 15.2. Organizing thrombus in a pulmonary artery from a case of congenital pulmonary stenosis. × 280 H and E.

fibrin thrombi. In such cases afibrinogenae-mia was present but was likely to remain undetected as signs and symptoms of disease due to the presence of thrombi were completely absent.

Various components necessary for the normal clotting mechanism have been found to be absent in this rare condition, including blood platelets, Factor V, Factor VII and anti-haemophilic globulin. Placenta, decidua, and presumably the amniotic fluid which gains entrance to the circulation through the placental site are unusually rich in thromboplastic sub-stances. Schneider (1952) discussed the routes whereby thromboplastic materials could reach the maternal circulation from the site of placental detachment.

Boyd (1958) claimed that fibrin thrombosis occurred in lungs of infants borne by mothers who subsequently showed an abnormal tendency to bleeding.

Apart from the general tendency for hae-morrhages to occur in various sites and the production of sticky mucoid blood-stained sputum, there are no characteristic changes whereby the condition may be recognized clinically or naked-eye at post-mortem. Mi-croscopically, numerous intravascular fibrin thrombi are found filling the alveolar capillaries and are stained dark blue with phosphotungstic-haematoxylin (Fig. 15.2A).

The thrombi in small arteries display a laminated appearance suggesting their formation and deposition from a fast-moving blood-stream.

Later Boyd (1960) claimed to have shown the presence of fibrin thrombo-embolism in the vessels of stillborn children and others dying within 3 days of birth and considered that it was responsible for death. In such cases laminated fibrin thrombi were found in the pulmonary vessels together with similar vege-tations on the heart valves. Children surviving for a few days developed haemorrhages. The route by which the coagulating factor entered the circulation remained uncertain.

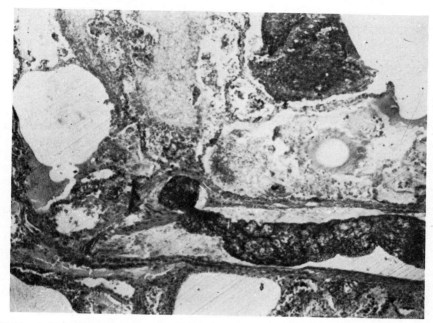

FIG. 15.2A. A fibrin "thrombus" in a branch of the pulmonary artery. × 100 approx. P.T.A.H. stain.

In a most unusual case seen by the author the pulmonary veins throughout the lungs were filled and occluded with a fibrin network which was absent in other organs (Fig. 15.2A). The appearances were quite unlike post-mortem blood clot and the cause was unknown.

Pulmonary Embolism

The term pulmonary embolism has by common use come to mean impaction of thrombus which has been transported in systemic veins to the pulmonary arteries. Pulmonary embolism in the strict pathological sense may be produced by any substance which is capable of being transported in the blood to the pulmonary arteries. There is, however, one fundamental difference between embolism caused by thrombus and the other forms of embolism to be considered in this section: thrombotic emboli are formed initially within the vascular system, whereas the other kinds of

emboli all originate outside vessels and subsequently, owing to extravascular tissue pressure changes or growth of cells, are sucked, spread or are injected into vessels and subsequently transported to the lungs in the usual way.

THROMBOTIC EMBOLISM (PULMONARY EMBOLISM)

The dramatic clinical picture of a patient dying suddenly from a massive pulmonary embolus a few days after an operation has been recognized since the early days of major surgery and was well known to the early pathologists. The frequency with which symptomless emboli occur in both adult medical and surgical patients is not, however, so generally appreciated. Several series of post-mortem statistics on the true incidence of pulmonary emboli have been published and in Table 15.1 figures from some of the larger series have been included. The quoted figures vary widely depending upon the nature of the cases examined and the ages of

TABLE 15.1

	Total post-mortems	Percentage with pulmonary emboli
Belt (1934a)	567	10
McCartney (1934) (injury cases)	1604	3·8
McCartney (1936) (post-operative cases)	2058	5·1
Hunter et al. (1941)	350	14·5
Hampton and Castleman (1943)	3500	9
Spain and Moses (1946)	1000	10·9
Raeburn (1951)	130	15·3
Morrell and Dunnill (1968)	263	51·7
MacIntyre and Ruckley (1974)	2291	13.2

the patients. Very high incidence rates occur among persons over the age of 45 who survive major lower limb fractures for a week or longer (Sevitt and Gallagher, 1959).

The incidence of proved pulmonary emboli varies directly with the amount of care that is taken in the examination of the lungs at autopsy. All observers, however, are agreed that the incidence in children and young adults is very much lower than the figures quoted in Table 15.1 which largely reflects the post-mortem findings in middle-aged and elderly persons. Haber and Bennington (1962), however, found that in Chicago 1·25 per cent of fatal pulmonary emboli occurred in children less than 10 years old. The incidence of pulmonary embolism fell during the two World Wars only to rise again to higher levels than previously during the subsequent peace-time (Zietlhofer and Reiffenstuhl, 1952).

The rising incidence of pulmonary thromboembolism is reflected in the Registrar-General's statistics for England and Wales where despite inaccuracies due to under-reporting and changes in the method of classification introduced in 1967, the death rate per million of the population has shown a continuing rise since 1945 and a rapid rise since 1953. Freiman et al. (1965) found the overall incidence of pulmonary thromboembolism in 2319 adult autopsies at the Beth Israel Hospital in Boston amounted to 33 per cent and in a specially examined series of sixty-one consecutive post-mortems to 64 per cent (39 cases). A similar post-mortem study by Morrell and Dunnill in which they examined only the right lung showed emboli present in 51·7 per cent. The usual frequency, however, is considered to be between 11 and 25 per cent. The continuous rise in incidence since the end of the Second World War has not been affected by the introduction of anticoagulant therapy as was shown by Barritt and Jordan (1961) who described seventy-two cases of pulmonary embolism in patients receiving such therapy. The introduction of subcutaneous low dosage heparin and dextran-70 regimens may hopefully reduce post-operative and puerperal thrombotic risks in the future. Furthermore, the introduction of the ^{125}I-fibrinogen test and venography while confirming and re-emphasizing the great frequency of peripheral vein thrombosis in adult surgical, medical and obstetric patients enables those at risk of developing pulmonary emboli to be more readily detected and treated.

The rising incidence of thromboembolism seen in the developed nations has not been paralleled by a similar increase in the underdeveloped nations (Hassan et al., 1973; Chumnijarakij and Poshyachinda, 1975). Significant differences have been found in the levels of plasma fibrinogen and prothrombin concentrations in age and sex matched groups of North American and African negro patients.

Healthy adults seldom develop massive or clinically detectable thrombotic pulmonary emboli. Breckenridge and Ratnoff (1964) and Fleming and Bailey (1966), however, both collected series of cases of massive and fatal pulmonary embolism occurring in previously healthy people under the ages of 45 and 70, respectively. Loehry (1966) also drew attention to the rising incidence of pulmonary embolism in young adults under the age of 40 who showed no apparent cause and were not taking contraceptive pills. The risk of thromboembolism developing under the age of 50 is greater in women but in older age groups men preponderate. A review of the whole subject has been presented by Nicolaides (1975).

Following the introduction and widespread use of the contraceptive pill many cases of fatal and non-fatal pulmonary embolism have been reported in healthy young women (Leather, 1965). The Boston Collaborative Drug Surveillance Program (1973) showed that women taking oral contraceptive pills had an eleven times greater risk of developing venous thrombosis and its embolic complications. Vessey *et al.* (1970) also showed that the same group of women were exposed to a greater risk of developing similar complications if they underwent surgery. High-oestrogen-content contraceptive pills are more likely to induce venous thrombosis and similar complications occur among both men and women receiving oestrogen-like compounds for the treatment respectively of prostatic carcinoma and the suppression of lactation (Daniel, 1969). Both pregnancy and the puerperium are associated with a slightly greater likelihood of thromboembolism and persons with blood group A are more susceptible. The rapid rise in the incidence since 1953 has corresponded with the introduction of muscle-relaxant drugs as supplements to inhalational anaesthetics, and these may be an additional important factor in causing post-operative thrombosis in the lower limb veins due to complete loss of the muscle pump action.

The majority of emboli originate in the veins of the leg. Rossle (1937), Hunter *et al.*, Raeburn, and Sevitt and Gallagher (1961) examined the leg veins post-mortem in adults and found the incidence of thrombosis to be 27, 52·7, 26·9 and 65 per cent, respectively. Venous thrombosis in the leg veins may start in six main sites (Sevitt and Gallagher) and sometimes simultaneously in more than one place. Thrombosis originating in and involving the calf veins (sural veins) is the most common, but the massive and fatal emboli arise from the large iliac and femoral veins and in these thrombosis may start in a venous valve pocket. A detailed description of the pathogenesis of deep vein thrombosis in the leg has been given by Sevitt (1973). The calf veins, which are by far the most common starting-point for phlebothrom-

bosis, enlarge and increase in size after the age of 40 and both the incidence of phlebothrombosis and pulmonary emboli increase with advancing age.

Recent studies *in vivo* using the ^{125}I-fibrinogen test have shown a high incidence of deep leg vein phlebothrombosis especially in patients with fractures of the large lower limb bones, and in those who suffer from "shock". Other conditions which especially predispose to thrombosis include post-operative surgical states, following myocardial infarction and post-prostatectomy cases (Nicolaides and Gordon-Smith, 1975).

Deep vein thrombosis is a bilateral condition in about one-third of all detected cases, an important fact to bear in mind should prophylactic surgical ligation of a vein be considered to prevent the occurrence of fatal embolism. Other much less common sources of emboli include the pelvic veins, particularly during the puerperium; the internal jugular veins in conjunction with septic states in the middle ear, pharynx and neck; and the right atrial appendage of the heart in conditions of atrial fibrillation. Thrombosis of peripheral veins is particularly liable to occur in conjunction with any cause in which there is a low output type of heart failure, in conditions associated with peripheral circulatory failure due to the sluggish venous flow, in post-operative and post-partum states and in some forms of malignant disease, notably carcinoma of stomach and pancreas. In post-operative and post-partum states it is partly due to the increase of circulating blood platelets and the release of thromboplastic substances often combined with the existence of other circulatory abnormalities already mentioned. The high incidence of peripheral venous thrombosis in heart disease was described by Axhausen (1929) and Rosenthal (1930–1).

Pulmonary emboli are almost always multiple and their distribution within the lungs of man and animals has been studied by Macleod and Grant (1954) and Pryce and Heard (1956); these workers have shown that the majority lodge in the lower lobes, particularly in the posterior basal and apical segments of the lower lobes. They are more common on the right side than on the left (Hamilton and Angevine, 1946).

The distribution of emboli in the posterior basal segments was attributed by Pryce and Heard to the direction of blood-flow into the main axial arteries supplying the lower lobes which terminate in the posterior basal segments. Although undoubtedly axial flow plays a part in their distribution, the much greater blood-flow through the lower halves of the lungs is probably a more important factor as evidenced by the paucity of emboli in collapsed lung in which there is a much-diminished blood-flow.

Massive pulmonary embolism proves fatal within the first hour in 34 per cent of cases, within 24 hours in 39 per cent and the remaining 27 per cent of the patients die within 2 to 5 days (Fowler and Bollinger, 1954). Sudden death from pulmonary embolism occurs when the pulmonary blood-flow is reduced by about 70 per cent. The coiled-up mass of thrombotic embolus frequently breaks into two portions each of which straddles the bifurcation and impacts itself where the respective left and right pulmonary arteries divide into their lobar branches. In some cases massive pulmonary emboli may not prove immediately fatal and should the patient survive over 12 hours signs of acute right ventricular strain and ischaemia develop. The deficient cardiac output coupled with decreased arterial oxygen saturation may lead to microscopic areas of infarction in the greatly strained right ventricle and this may give rise to an acute pericarditis confined only to the surface of this ventricle. Unusual electrocardiographic changes may occur and were first described by McGinn and White (1935). At the site where emboli lodge, the walls of the pulmonary arteries may undergo necrosis or show an acute focal arteritis (Meyer, 1960), and these changes may be mistaken for acute polyarteritis nodosa. The changes are considered to be caused by the embolus interfering with the nutrition of the arterial wall, resulting in an acute form of ischaemic arteritis.

Massive pulmonary emboli are usually preceded by several clinically undetectable smaller emboli. The majority of pulmonary emboli probably undergo complete lysis and even very large emboli, if not immediately fatal, may disappear within 2 or 3 weeks (Sautter et al., 1964). Alternatively, the organization of large and medium-sized emboli may later give rise to single or multiple delicate fibrous bands or webs which span the lumen of the vessel (Korn et al., 1962) (Fig. 15.5). Further fresh, small emboli may become entrapped within these intravascular fibrous meshes. As Korn et al. pointed out, proximal to the intravascular bands the walls of the pulmonary arteries often show extensive atheroma and intimal fibrous plaques which cease abruptly at the level of the bands. They believed that three factors were involved in the formation of intravascular bands, namely the initial arrest of an embolus, propagation of thrombus distal to the impacted embolus and the frequent incorporation of further thrombi at the sites of organization, causing a change in size and shape of the original lesions.

Under certain conditions repeated showers of small emboli may occur and if maintained can eventually lead to severe pulmonary arterial obstruction, a condition described more fully in Chapter 16 which is devoted to chronic pulmonary hypertension. Some of the large and small emboli eventually organize resulting in small plaques of intimal fibro-elastoid tissue. The "elastic" fibres in these plaques because of their abnormal staining reaction with Mallory's PAH stain are referred to as elastoid tissue and are probably related to or derived from collagen fibres (Gillman et al., 1955). During this process of organization, endothelium grows over the surface and into spaces formed by the retraction of the thrombus, resulting in the formation of new thin-walled sinusoidal channels. Later reticulin and collagen fibres appear between the newly formed spaces and the volume of the thrombus becomes greatly reduced. No revascularization occurs from the wall of the pulmonary artery unless the wall is damaged and the media is breached (Fig. 15.3). Incorporation of small emboli into the wall of the vessel by the overgrowth of endothelium proceeds very rapidly and is well advanced after 72 hours (Fig. 15.4). At the end of the first week the

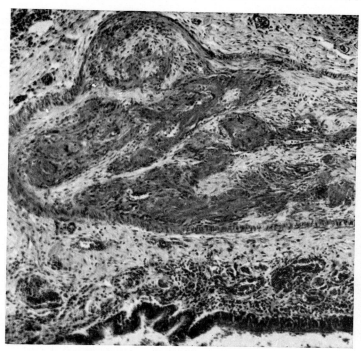

FIG. 15.3. A major branch of a pulmonary artery filled with organizing ante-mortem thrombus (embolus) showing recanalization which is proceeding from within the lumen. The medial muscle coat is intact and is not traversed by new capillary channels. × 100 H and E.

embolus may already be well anchored to the arterial wall and can only be detached by force. Small emboli, if not completely lysed, heal and leave very little to mark their previous existence except a plaque of intimal fibrous thickening.

Arrest in the lung of a large pulmonary embolus may cause a rise in the serum level of lactic dehydrogenase. When the embolus gives rise to a pulmonary infarct or to right heart failure, the serum bilirubin and glutamic pyruvic transaminase levels also rise (Wacker and Snodgrass, 1960; Klaus and Zeh, 1961).

Much discussion has centred around the mechanism of death in massive pulmonary embolism. Some believe that vasomotor tonus throughout the pulmonary arteries increases when an embolus lodges in the arterial tree and this together with the embolic occlusion of the pulmonary vessels leads to fatal right ventricular

strain coupled with anoxaemia. It has also been postulated that 5-hydroxytryptamine liberated from entrapped blood platelets within the embolus leads to pulmonary arterial vasconstriction but there is no evidence to support this view. Others believe there is no increase of the vasomotor tonus in the pulmonary arterial tree and that death is due entirely to the acute right ventricular failure. The evidence in support of these views is conflicting and has been mainly derived from animal experiments. Marshall *et al.* (1963) produced thrombotic pulmonary emboli in dogs but found no evidence of any cardiovascular or respiratory reflex effects and Knisely *et al.* (1957) using direct quartz rod illumination of animal lungs saw no evidence of vasospasm following the lodgement of an embolus within the pulmonary arteries. Viswanathan (1964), however, found that in man occlusion of a lobar

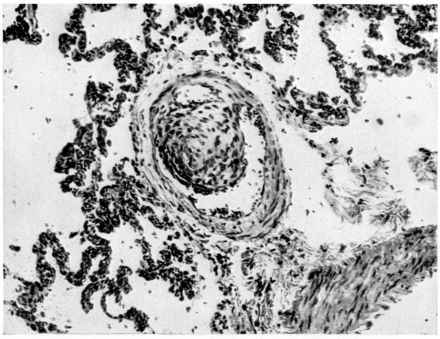

FIG. 15.4. A small pulmonary artery containing organizing ante-mortem thrombus (embolus). The thrombus is reduced to a small mass of intimal connective tissue. Note the atrophy of the medial coat beneath the organized thrombus. × 250 H and E.
(Reproduced by courtesy of the Editor of *J. Path. and Bact.*)

branch of the pulmonary arteries by an intraluminal balloon led to an immediate but transient rise in the blood-pressure proximal to the site of the occlusion. The mechanical obstruction was judged to have been insufficient by itself to have caused the rise in pressure and it was therefore concluded that simultaneous increased vasomotor tonus in the remainder of the unobstructed pulmonary arterial tree must have played a more important role. Transection and ligation of a pulmonary artery in man is usually unassociated with any rise in the pulmonary blood-pressure at rest, but intraluminal obstruction of comparable extent is usually associated in man with a transient reflex increase of vasomotor tonus. Krahl (1963) suggested that the sphincter-like muscle present in some of the small muscular pulmonary arteries which was first depicted by von Haÿek (1953), and which is known to be under vagal control in

rabbits, might be responsible for causing the transient rise in pulmonary arterial pressure. Tuller *et al.* (1961) in experiments designed to observe rabbits' lungs directly found that variation in perfusion rates of lung lobules followed the introduction of a cardiac catheter into the right side of the heart. They concluded the small branches of the pulmonary arteries were probably controlled by a neurohumoral mechanism. In many cases of massive human pulmonary embolism the total pulmonary blood-flow, and hence the cardiac output, may be reduced in excess of 70–75 per cent and this alone ensures a fatal outcome.

Emboli of Extravascular Origin

GENERAL FEATURES

The group of rarer emboli about to be described originate outside blood-vessels and

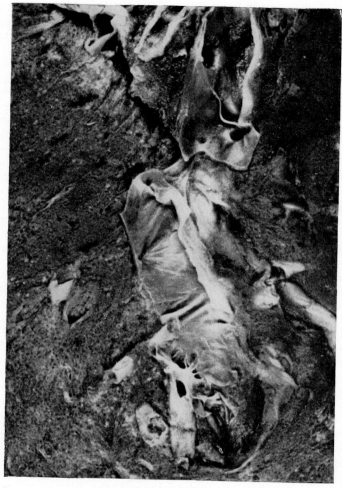

FIG. 15.5. A branch of the pulmonary artery traversed by fibrous bands caused by organized thrombotic emboli. × 2½ natural size.

become secondarily sucked into veins, often as a result of trauma. Almost any type of tissue may be aspirated into blood-vessels and give rise to emboli. Ribbert (1900) believed that when extravenous pressure exceeded intravenous pressure the veins were completely collapsed. This view was generally held until Young and Griffith (1950) described some simple hydraulic experiments *in vitro*, and showed that a nonrigid tube surrounded by fluid does not immedi-

ately collapse as the pressure outside it begins to rise. At one stage when the external pressure is only slightly in excess of the intratubal pressure it pulsates, and small bodies suspended in the liquid surrounding the tube may be sucked through a hole in its wall. These experiments may provide an explanation for the wide variety of tissues and substances that can find their way into the pulmonary circulation: these emboli of extravascular origin include (a) malignant cells (considered in Chapter 24), (b) fat, (c) air, (d) bone-marrow, (e) amniotic fluid, (f) trophoblast and decidual tissue, (g) brain and other body tissues, (h) bile thromboemboli, (i) cotton-fibre, (j) parasites, (k) vegetations, (l) emboli associated with drug addiction, (m) cardiac catheter emboli and (n) mercury emboli.

FAT EMBOLI

Fat emboli in the lung were first described by Zenker (1862) in a crushed railway worker and since then fat embolization has become well recognized. During both World Wars and especially during the Second, there was a renewed interest in the condition; Robb-Smith (1941), Wilson and Salisbury (1943–4) and Sevitt (1962) among many others have studied fat embolism, the earlier studies being mainly based on air-raid and battle casualties. Wilson and Salisbury defined fat embolism as a "state in which globules of fat are present in the circulation of sufficient size to cause blockage of the intimate vascularities of various organs".

Much misunderstanding has arisen because fat embolism has been confused with and regarded as being the same as traumatic lipaemia. These two conditions are different and should be regarded as two separate conditions with a different aetiology (Sevitt, 1966). Traumatic lipaemia is part of the complex of body reactions commonly encountered after haemorrhage, burns and cold injuries. It is a condition in which there is an alteration in the levels of finely particulate fat composing the plasma chylomicrons and serum lipoproteins, and a similar alteration occurs in diabetes mellitus. Chylomicron and serum lipoprotein fat forms a very stable emulsion and is not responsible for causing fat embolism in man.

True fat embolism follows the release of fat stored in body cells due to traumatic injury of the investing cell wall. Hallgren et al. (1966) showed that fat emboli produced experimentally in dogs by fracturing their legs had a triglyceride composition similar to normal bone marrow and depot fat, and was different from chylo-

micron fat. Most human fat emboli are derived from fatty bone marrow, less commonly from subcutaneous fat and very rarely follow injury to a fatty liver, acute pancreatitis, osteomyelitis, diabetes mellitus, burns or prolonged steroid therapy (Lancet, 1972). Lipid microembolism may follow lymphangiography.

Concussion and fracture injuries of bone are the usual cause of fat embolism but fat and bone marrow emboli may be released as a sequel to bone infarction, complicating such blood diseases as sickle-cell anaemia (Shelley and Curtis, 1958). It may also follow very rarely the use, for diagnostic and therapeutic reasons, of parenteral and intraurethral injections of oily substances, and it has been reported following lymphangiography. Fat "embolism" has also been reported following poisoning with carbon tetrachloride and phosphorus (MacMahon and Weiss, 1929), chronic alcoholism (Lynch et al., 1959), intrauterine douching employed for abortifacient purposes (Vance, 1945) and as a complication of the establishment of a temporary extracorporeal circulation (Adkins et al., 1962). In almost all of these cases, however, traumatic fat lipaemia has probably been mistaken for true fat embolism. Fat embolism may, however, complicate gas gangrene (Govan, 1946) and this results from the action of lipoproteolytic enzymes formed by Cl. welchii which split lipoproteins (α-toxins) and aggregates, thus liberating free fat (MacFarlane et al., 1941).

The incidence of fat embolism in battle and civilian injuries has been estimated at between 35 and 88 per cent, whilst the incidence in autopsy lungs following routine surgical operations was found by Whiteley (1954) to amount to 100 per cent. The incidence of fat embolism increases with the length of survival after injury and with the age of the patient, and although much of the fatty material is rapidly removed, traces of fat emboli may still be detected up to 15 days after injury (Palmovic and McCarroll, 1965). Fat emboli are not usually found in the lungs if the injury was immediately fatal.

Experimental fat embolism in animals results in a sudden rise of pulmonary arterial pressure followed by a rapid fall in the systemic pressure probably due to transient decreased cardiac output. In man, the occurrence of fat embolism in the lungs is not in itself of serious import and the acute stage passes unrecognized, being confused in most cases with the other changes due to shock caused by the injuries. When, however, there is massive pulmonary fat embolism, there is a danger that emboli may be released into the systemic circulation and transported to the brain, plugging many of the cerebral capillaries. Such a condition may occasionally prove fatal 24–48 hours after injury.

Fat emboli can readily be detected in the capillaries of the lungs and kidneys following fractures and other injuries but in most instances they cause no serious vascular obstruction, and are slowly eliminated from the circulation in various ways, including excretion in the urine, expectoration in the sputum, metabolization by the liver and destruction by the action of serum lipases. The distensible and extensive pulmonary capillary network usually acts as a trap for the majority of fat emboli and prevents their rapid discharge into the systemic circulation with consequent danger of fatal cerebral complications.

Green and Stoner (1950) showed that when fat is injected intravenously into animals together with adenosine triphosphate, a small dose of fat which alone would normally produce no symptoms will lead to the death of the animal. The fat is aggregated into larger globules which lodge in small pulmonary muscular arteries and arterioles but fail to pass into the alveolar capillaries.

Moore *et al.* (1958) showed that embolic blockage of the small muscular pulmonary arteries and arterioles caused far more serious embarrassment to the right ventricle than a similar amount of embolic fat lodged in the alveolar capillary network. Robb-Smith, however, suggested fatal fat embolism resulted from hydrolysis of the neutral fat in the emboli by lipase yielding the more toxic fatty acids, a view supported by Peltier (1957). Experimentally, neutral fat fails to cause a haemorrhagic pneumonia like that seen in fatal human fat embolism but there is no evidence that enzy-

matic digestion in the latter gives rise to fatty acids (Reidbord, 1974). A stress-induced fatty-acid lipaemia is thought to be a more likely cause of the heart failure. Unsaturated fatty acids cause severe vesicular swelling in the alveolar capillary endothelial cells in experimental animals with accompanying damage to the alveolar epithelium. Benatar *et al.* (1972) considered that similar changes were responsible for causing the lung damage in human fat embolism. Widespread endothelial vesicular swelling in other contexts is known to cause acute right ventricular failure, e.g. in experimental acute hypoxia, and may be responsible for the cardiac embarrassment in human fat embolism.

Macroscopically, the lungs in fatal cases of fat embolism are very haemorrhagic, oedematous and stiff, and the cut surfaces resemble liver and usually show multiple haemorrhages. Minute fat globules exude from the cut surface and the blood in the major pulmonary arteries often shows oily globules on its surface.

Microscopically, the emboli can be readily and quickly demonstrated in frozen sections of lung tissue stained to show fat but may be missed unless thick sections are employed (Fig. 15.6). Fat emboli are readily demonstrated in the post-mortem room by cutting very thin slices of lung with scissors and heating them gently on a slide with 5 per cent potassium hydroxide, later adding a drop of Scharlach R solution. The emboli can be seen as orange-red drops within alveolar capillaries and some of these small fat emboli normally drift into the systemic circulation. Apart from the presence of fat emboli the principal change is one of extensive pulmonary oedema.

Severe fat embolism results in severe hypoxia and a low arterial blood oxygen content (i.e. 40 mm Hg) together with acute right heart failure. The blood platelet and serum calcium levels are both greatly reduced, the calcium combining with fatty acids to form calcium soaps.

AIR EMBOLISM

Wepfer, in the seventeenth century, "by inflating with his mouth the jugular vein, did

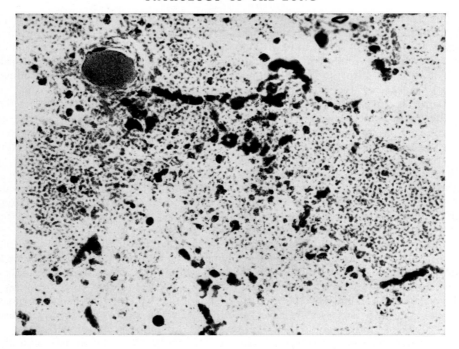

FIG. 15.6. A frozen section showing scattered fat emboli (black) lying in the alveolar capillaries which cannot be clearly seen. × 140 stained with Scharlach R and haematoxylin.

once on a time, lay prostrate and kill an ox, of a stupendous size" (quoted by Morgagni, 1769). Camerarius (quoted by Morgagni) confessed that "he was stirred up to make the like experiments with Wepfer, from the example of the sheep and the cow, that were then killed by him". From these early writings it is apparent that air embolism was recognized by the early anatomists and pathologists as a method for dispatching large animals. Later Virchow (1846) conducted similar experiments in cats and dogs, describing the appearance of air bubbles in the inferior vena cava and coronary sinus.

Air embolism has long been recognized as an accident that may occur particularly during operations on the head and neck following injury to the great veins, but it may also occur during other diagnostic and therapeutic procedures. Straus (1942) has listed some of the other commoner causes, which include:

(a) Intrauterine and intravaginal injection of soap (frothy) solutions and intravaginal manual manipulations forcing air into lacerated cervical or placental veins.

(b) Injection of solutions for therapeutic and diagnostic X-ray purposes into the nasal sinuses and urinary bladder.

(c) Inadvertent injection of air during the course of an intravenous infusion or following disengagement of the tubing connections.

(d) Therapeutic insufflation of the Fallopian tubes.

(e) Air introduced accidentally into the lung during pneumothorax.

(f) Decompression (dysbarism).

(g) Forced positive-pressure ventilation for hyaline membrane disease in the neonate.

In most of these conditions, air may be sucked into partly collapsed, large veins. The rate of infusion of the air into the veins is all-important; provided air is injected slowly experimental animals will survive, but if the same dose is injected quickly they succumb almost immediately after exhibiting generalized convulsions.

Although rapid decompression leads to the release of gas dissolved in the blood and tissues, this seldom causes air embolism. Rapid decompression, however, can lead to the formation of air-filled retention cysts under pressure in the lung if the bronchi are obstructed by mucus or other cause. Failure of the expanding air in the lung to escape naturally during decompression may lead to tearing of the lung tissues and air may be forced into the pulmonary vessels. This causes fatal air embolism mainly due to air emboli lodging in the systemic circulation, especially in the coronary vessels.

The mechanism which leads to death has been investigated by Cameron et al. (1951) who have found that the electrocardiographic changes in several species of animals show evidence of myocardial ischaemia.

Following massive embolism blood-pressure rises abruptly in the right atrium and very soon falls in the systemic circulation. There is no evidence from experimental studies that the presence of air within the pulmonary vessels initiates a vagal reflex resulting in coronary artery constriction as vagotomy does not affect the outcome. When air is introduced gradually, though the systemic pressure fluctuates, it does not fall precipitously. This has been regarded as evidence to support the view that collateral vessels open up within the lung. As Holden et al. (1949) have shown, the same response follows other forms of pulmonary embolic blockage.

The presence of air bubbles within the narrow alveolar capillaries breaking up the normally continuous column of blood causes surface tension forces to be released at blood–air interfaces; these are greater than the normal intracapillary pressure and lead to arrest of the flow. Small amounts of air are gradually absorbed and the flow is soon re-established, but sudden large volumes of air lead to gross mechanical obstruction of the pulmonary capillary circulation. A state of shock can follow decompression and was stated by Brunner et al. (1964) to be partly due to loss of plasma from the blood into the extravascular compartment resulting from capillary damage. This may cause severe pulmonary oedema and large pleural effusions.

The principal changes in air embolism are found in the right side of the heart where the right atrium is dilated and filled with bloody froth which not only fills the larger pulmonary arteries, coronary sinus, and inferior vena cava, but may even extend through the lung to the pulmonary veins. The left ventricle is found to be contracted and empty. Although the clinical history, together with the characteristic post-mortem findings, usually renders the diagnosis obvious, this condition must be distinguished from generalized gas gangrene septicaemia with gas formation. The presence of foul-smelling gas in all chambers of the heart, great vessels and viscera, together with intimal haemolytic staining of vessels in clostridial infections, usually leads to little difficulty in differentiating the two conditions. Pulmonary oedema and bilateral pleural effusions are found in all fatal cases of decompression sickness.

BONE-MARROW EMBOLISM

Bone-marrow emboli in the pulmonary arteries were first demonstrated by Lubarsch (1898); since then further cases have been described by Schenken and Coleman (1943), Warren (1946), Rappaport et al. (1951) and Tierney (1952).

Pulmonary emboli of bone marrow usually follow some form of direct or indirect injury to the skeleton. Direct fracture of a bone in addition to liberating particles of fat may also release marrow tissue. Indirect injuries to the skeleton resulting in fractures may occur during an epileptic attack, following electric shocks and electro-convulsion therapy, and nowadays frequently complicates resuscitative external cardiac massage owing to concurrent fracture of the ribs.

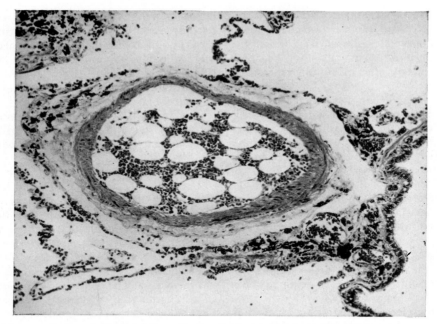

FIG. 15.7. A bone marrow embolus consisting of both active haemopoietic tissue and fatty marrow. × 140 H and E.

In osteoporotic states such as occur in rheumatoid arthritis, skeletal injuries are more likely to result in fractures and dissemination of marrow tissue. Bone-marrow emboli have occasionally followed diagnostic bone-marrow puncture. Also infarction of the bone and marrow tissue caused by sickle-celled anaemia may lead to the release of marrow emboli (Shelley and Curtis, 1958).

Microscopically the emboli can be recognized within the pulmonary arteries and consist both of fatty and haemopoietic bone-marrow tissue (Fig. 15.7). They are usually discovered during examination of routine sections and cause no recognizable macroscopic changes in the lungs.

The fate of bone emboli is uncertain as in most instances they are a terminal condition. Occasionally opportunities arise to study lungs after an interval of time has elapsed and in such cases the pulmonary arterial branches are found to contain amorphous haematoxyphil and partly calcified debris similar to that caused by embolization of vegetations from a bacterial endo-

carditis. The calcium salts probably combine with fatty acids released from the fatty marrow tissue.

AMNIOTIC FLUID EMBOLUS

Amniotic fluid embolism is a rare and fatal complication of the second and third stages of labour and has been estimated to occur once in every 26,000 deliveries. The presence of amniotic contents in the pulmonary arteries was first described by Meyer (1926). Steiner and Lushbaugh (1941) described a further eight cases, and the pathology of amniotic embolism has been further discussed and reviewed by Gross and Benz (1947), Wyatt and Goldenberg (1948), Attwood (1956 and 1958) and Scofield and Beaird (1957).

Amniotic embolism accounts for approximately 5·75 per cent of maternal deaths (H.M.S.O., 1969).

The mechanism of amniotic embolism has

been likened by Reid *et al.* (1953) to an intravenous infusion of amniotic fluid. It occurs following rupture of the membranes and partial detachment of the placenta with continuing powerful contractions. Usually the foetal head is wedged tightly in the pelvis and during strong uterine contractions amniotic fluid is forced either into open veins at the placental site or into veins exposed in a ruptured or surgically incised uterine wall. Amniotic fluid may also enter the maternal veins through small and often incomplete lateral tears in the lower uterine segment and cervix, especially if these extend into the broad ligament. Following massive injection of amniotic fluid into the veins the patient suddenly becomes dyspnoeic, cyanosed and collapsed and usually dies within a few seconds or minutes. Minor degrees of amniotic embolism may be an incidental finding unrelated to the main cause of death (Attwood, 1972). Survival may be followed by later complications described briefly below.

Macroscopically, the lungs show no characteristic changes, though patchy collapse, oedema and emphysema may be found.

Microscopically, the discovery of amniotic contents in the blood-vessels in the lungs removed from maternal fatalities is to a great extent dependent on the number of sections examined from the different lobes of the lungs. Attwood and Fortune (quoted by Attwood, 1972) found amniotic fluid emboli in 23 per cent of a small series of maternal deaths. The pulmonary arteries, arterioles and capillaries contain all or any of the following amniotic contents; epithelial squames which are occasionally nucleated, lanugo hairs, mucin and bile from the foetal meconium, and fat. These various constituents of an amniotic embolus may only be recognized with difficulty in routine haematoxylin and eosin-stained sections, but some are seen to best advantage in sections stained with phloxine and alcian green using the method described by Attwood (1958). By this method the epithelial squames stain red and the stringy mucus greenish-blue as seen in Fig. 15.8. Fat stains are required to demonstrate the fat droplets which are mainly found in the capillaries in the lung. Lanugo hairs are best demonstrated by their anisotropic properties in polarized light but may be recognized in stained sections viewed with ordinary light because of the melanin they contain. Amniotic embolism resulting from spontaneous abortion during the second trimester may show an absence of all constituents other than squames.

If the patient survives for a brief period, the embolic constituents cause the circulating polymorph leucocytes to become aggregated and the capillaries appear to be filled with such cells, a process described by Wyatt and Goldenberg as the "combing out" (of white cells) phenomenon. In experimental embolism produced in animals, the "combing out" of circulating polymorph leucocytes occurs within the first 24 hours.

A late complication, the first description of which was given by Ratnoff and Vosburgh (1952) and which was also further reviewed by Tuller (1957), is the development of afibrinogenaemia following extensive intravascular fibrin thrombosis. At about 8 days post-partum bleeding occurs at various sites and the sputum becomes blood-stained. This condition results from the liberation of thromboplastic material present in the amniotic contents (see fibrin thrombosis). Weiner *et al.* (1949) showed that amniotic fluid when added in a dilution of 1 part in 20 to blood reduced the clotting time by a half.

Experimental amniotic embolism in dogs results in very little pulmonary arterial reaction and the same probably applies to the human condition. For this reason it is very improbable that amniotic embolism is a contributory cause to the later onset of primary "idiopathic" pulmonary hypertension in women (Attwood, 1964). Experimental evidence points to the probability that amniotic fluid contains pulmonary vasoconstrictory substances which lead to transient but often fatal haemodynamic changes.

TROPHOBLAST EMBOLI

Emboli of trophoblast were first discovered in the pulmonary arteries in the lung by Schmorl (1893). Later, Schmorl (1905) found that the incidence of trophoblastic emboli in

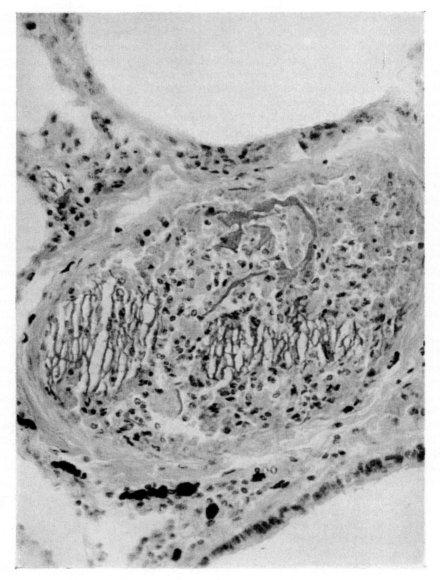

FIG. 15.8. An amniotic embolus consisting of vernix squames (red) and mucin (green). Section stained by Attwood's method. × 400 approx. (Reproduced by courtesy of Prof. H. D. Attwood.)

lungs of women dying not only during eclampsia but also during ordinary deliveries reached about 80 per cent. Novak (1952) regarded the finding of trophoblastic tissue in branches of the pulmonary artery as a normal occurrence in pregnancy and Park (1957), after carefully examining the lungs of women dying during pregnancy or the puerperium, found a minimum incidence of 44 per cent.

Although the violent uterine contractions associated with normal parturition are an important factor in dislodging trophoblastic tissue, this also occurs in the earlier stages of pregnancy, particularly in patients with hydatidiform moles and dying from eclampsia (Attwood and Park, 1961).

In most cases the fragments of trophoblast are lysed and within a week of delivery they are difficult to detect in the pulmonary vessels. Phagocytosis appears to play little part in their removal judged by the evidence gained by experiments on the placental destruction in mice (Park, 1958). According to Attwood and Park trophoblastic emboli behave as immunologically inert material despite the fact that they are of foetal origin.

Occasionally, emboli of trophoblastic tissue have been responsible for causing fatal embolism and pulmonary infarction, as in the cases reported by Hughes (1930) and Arnold and Bainborough (1957). Hydatidiform moles have been mainly responsible for these severe and eventually fatal cases.

In the case described by Arnold and Bainborough, the pulmonary arteries contained thrombi which resembled packed pulmonary emboli, and small amounts of trophoblast tissue were only identified by microscopical examination. The possibility that the trophoblast initiated clotting within the lung vasculature was not considered.

The emboli of placental tissue consist mainly of multinucleated plasmodial masses; many of the nuclei show degenerative changes and only rarely active mitoses (Fig. 15.9). All gradations exist between the highly malignant emboli found in chorion-carcinoma and the harmless fragments of trophoblast that occur in normal pregnancy which usually disappear without trace. Among the cases with an intermediate grade of malignancy was that described by Savage (1951). In this case the trophoblastic emboli produced lesions visible radiologically which finally regressed and disappeared many months after the uterus had been emptied.

DECIDUAL EMBOLI

The presence of ectopic endometrial tissue in many sites, including the lung and pleura, although of rare occurrence is now well recognized. The growth of decidual cells alone within the lung is very rare but examples have been described by Park (1954) and Wagenvoort et al. (1964). In the case described by Park, a small clump of decidual cells, a few of which showed active mitosis, lay within the alveolar spaces but outside blood-vessels (Fig. 15.10). In the example depicted by Wagenvoort et al. the decidual cells lay within a small branch of a pulmonary artery and were considered to be embolic; they were also associated with amniotic embolism.

There is no evidence to support the view that in these cases the decidual cells arose by a metaplastic change in existing lung tissue and the second case quoted above shows that blood-borne spread of decidual cells from the uterus can occur. Park suggested that in his case a single or a clump of cells had been carried in the blood-stream to the lung earlier in pregnancy, and these cells had continued to multiply, eventually bursting out of the vessel in which they had lodged initially.

BRAIN EMBOLI

Emboli composed of brain tissue were found by Krakower (1936) obstructing the pulmonary vessels in a case of a pulmonary infarct following a head injury.

Since this case was reported other similar examples have been described by Gruenwald (1941), Potter and Young (1942), Oppenheimer (1954), McMillan (1956) and Levine (1973).

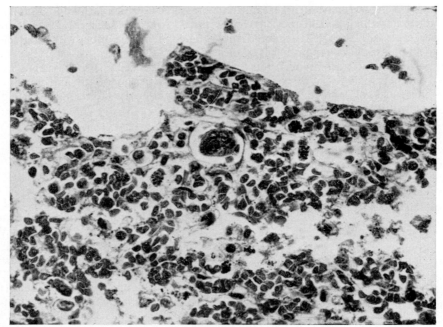

FIG. 15.9. A placental embolus consisting of a multinucleated mass of trophoblast. × 500 H and E.

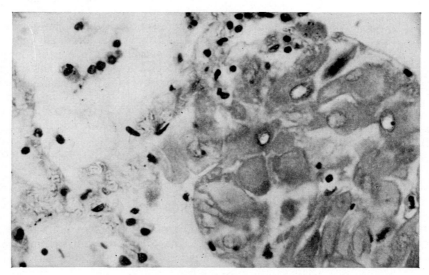

FIG. 15.10. A clump of decidual cells lying in the lung. The clump of cells was considered to have resulted from an embolus of decidual cells which had continued to grow in the extravascular tissue plane within the alveoli. × 400.

(Reproduced by courtesy of Dr. W. W. Parks.)

McMillan described four cases of traumatic cerebral embolism caused by accidental and surgical injuries to the brain, and all three examples seen by Levine followed gunshot wounds of the head. McMillan found cerebral emboli present in approximately 2 per cent of all cases of severe head injury. The emboli consist of viable glial tissue, blood-vessels or nerve cells together or separately (Fig. 15.11), and should be clearly distinguished from fibrin thrombi which they closely resemble and from which they can be distinguished by special staining methods for myelin (luxol fast blue and Weigert stains).

Cerebral emboli may also occur in the lung in neonates in association with major cerebral malformations, and in patients of any age following accidental or surgical trauma to the brain. The cases described by Gruenwald, Potter and Young and Valdes-Dapena and Arey (1967) all occurred in neonates. The infant described by Gruenwald was born with a frontal meningocoele and a severe brain defect, the two infants reported by Potter and Young

were both anencephalic and the two cases described by Valdes-Dapena and Arey occurred in neonates following brain trauma sustained during delivery.

Two theories have been advanced to explain the occurrence of cerebral emboli in conjunction with major cerebral abnormalities. Firstly, that brain tissue is carried to the lungs either through the lymphatics or blood-vessels and secondly, that brain tissue is aspirated into the lungs. Anencephaly arises very early in foetal life before lung alveoli have developed and it is most unlikely that aspiration could account for brain tissue reaching the lung during the period when anencephaly was developing. The vascular route was favoured by Potter and Young, and Gruenwald considered that a blow sustained by the mother during the second month of pregnancy had been responsible for causing the brain emboli in his case. The brain tissue may continue its organized growth within the lung but there is no evidence to show that it behaves as a neoplasm. McMillan described four cases

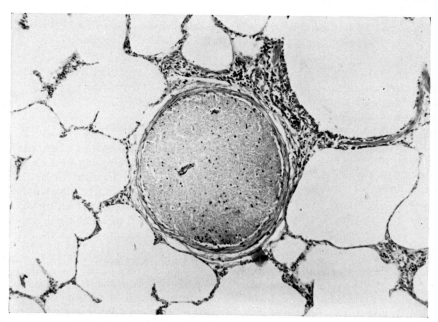

FIG. 15.11. A cerebral pulmonary embolus. The tissue occupying the lumen of the pulmonary artery is brain tissue containing a small capillary. × 100 H and E.

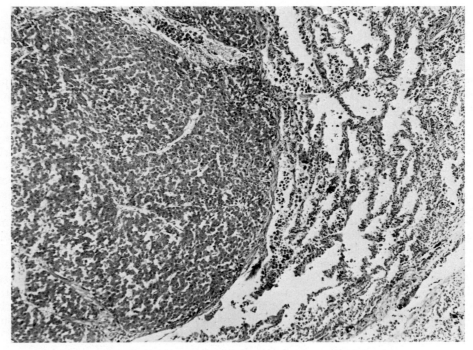

FIG. 15.12. A liver cell embolus in a branch of the pulmonary artery. × 75 H and E.
(Reproduced by courtesy of Dr. C. B. Chatgidakis, Johannesburg.)

of traumatic cerebral embolism caused by accidental and surgical injuries to the brain. He recorded the incidence of cerebral emboli in severe head injuries as approximately 2 per cent. In Oppenheimer's case the emboli were massive and filled both large and small pulmonary arteries and resembled pale blood clot. Persons surviving for a short interval following extensive lacerations or contusions of the cerebrum are more likely to show emboli of brain tissue.

LIVER EMBOLI

An example of a liver tissue embolus was first reported by Schmorl (1888) and more recent examples by Straus (1942) and Dunnill (1968). The first two cases and the unrecorded case (Fig. 15.12) all followed traumatic injury to the liver, while the most recently reported example

occurred in a patient dying of massive hepatic necrosis.

FATTY-TISSUE EMBOLI

Emboli of lipomatous tissue may be released into the venous circulation during the course of operations involving fatty tissue. The author has seen an embolus of fatty tissue in the lung derived from the periadrenal fat following an operation for the removal of an adrenal gland.

BILE THROMBOEMBOLI

Although jaundice following cholestasis is a not uncommon condition due to a variety of causes, bile emboli are rare. Mehta and Rubenstone (1967) described two cases in a series of 194 patients suffering from jaundice all of whom were shown to have cholestasis after death. In these cases bile thromboemboli were present in the lung in the small branches of the pulmonary artery and in alveolar capillaries which were found to be partially or completely filled with golden-yellow pigmented material which stained positively with Fouchet's reagent. The bile pigment is frequently associated with thrombus. The sequence of events leading to bile emboli are firstly, rupture of bile canaliculi allowing the escaping bile pigment to enter perivascular lymphatic channels and secondly, passage of the bile into the venous circulation. Bile pigment may be found in the thoracic duct before entering the blood-stream.

COTTON-FIBRE EMBOLI

The great increase in the use of parenteral therapy during the past 40 years, including the use of blood transfusion, has created many new pathological lesions. Needles wiped with sterile gauze pads are liable to retain fine filaments of cotton fibre and these may be introduced subsequently into the subcutaneous tissues or into veins. In the latter circumstance, the cotton fibres are carried to the pulmonary arteries and give rise to a characteristic granulomatous reaction within the walls of these vessels.

The first case of cotton-fibre embolus was reported by von Glahn and Hall (1949), and further cases have been described by Konwaler (1950) and Jaques and Mariscal (1951). Examples of subcutaneous cotton-fibre granulomas have been described by Amromin et al. (1955). von Glahn et al. (1954) injected cotton-fibre emboli into guinea-pigs and showed that the subsequent arterial changes bore many of the features of a hypersensitivity angeitis.

Although the cases previously described as examples of cotton-fibre emboli are probably genuine, it is unfortunate that micro-incineration studies were not carried out to prove the organic nature of the fibres. Following their introduction into the pulmonary circulation they tend to lodge in arteries between 500 μm and 150 μm in diameter, and eventually become incorporated

in the intima. According to von Glahn and Hall, the blood monocytes first collect around the fibres within the lumen of the artery and the whole collection adheres to the endothelium. The monocytes gradually fuse to form multinucleated giant cells and the endothelial cells proliferate; the result is an intimal granuloma. The giant cells enclose the refractile, curved fibres which vary in diameter and often show more highly refractile slits separating adjacent component parallel fibres (Fig. 15.13). In course of time, possibly aided by the arterial pressure, the internal elastic lamina disrupts beneath the granuloma and gradually the whole lesion is forced through the media and into the perivascular tissues as far as the surrounding alveoli. Once the granuloma has been extruded from the vessel it is soon surrounded and destroyed by further macrophage cells. The gap in the media is repaired by fibrous tissue but in time the muscle coat is reconstituted. These changes occur in the absence of vascular thrombosis.

The process whereby inert emboli are ejected through the vessel walls in the lung is also seen in human pulmonary schistosomiasis, in which the eggs excite a granulomatous reaction and eventually become extruded through the vessel walls. Similar granulomas occur in drug addict's lungs. The same change has also been observed by the author following the use of lycopodium spores to produce experimental pulmonary hypertension in animals and it has been observed following the experimental introduction of carbon particle emboli into the lungs of rabbits (MacCallum et al., 1966). It is essential, however, that a continuous blood-flow should circulate through the affected vessel as these changes do not appear to take place if thrombosis has supervened.

PARASITIC EMBOLI

Occasionally parasites may form emboli, but because of their small size they cause no symptoms and are only discovered in routine sections of lung tissue. Although many nematode

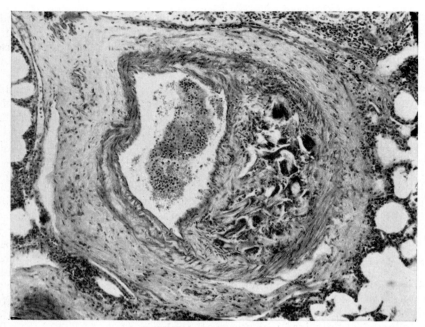

FIG. 15.13. An intra-arterial granuloma containing giant cells and refractile fibres (cotton fibres). Some of the fibres show longitudinal slits. × 100 H and E.

(Reproduced by courtesy of Dr. J. E. Todd.)

and trematode larvae are conveyed to the lung during the course of their life-cycle they are normally extruded into the alveoli and subsequently make their way through the air passages to reach the larynx and oesophagus.

The adult worms of various species of human parasites may find their way by accident into the pulmonary arteries. This subject is considered further in Chapter 10 which is devoted to parasitic diseases. Rupture of a hydatid cyst in the liver into a major branch of the hepatic vein may release scolices and fragments of cyst wall into the circulation which are then carried to the lungs, appearing in the branches of the pulmonary arteries where they cause thrombosis and where a considerable foreign-body giant-cell reaction ensues in the intima (Fig. 15.14).

Adult worms of the various species of human schistosomes may also find their way by accident into the pulmonary arteries. This

usually occurs when there is severe schistosomal portal fibrosis (pipe-stem liver) causing the establishment of alternative venous vascular pathways in the posterior mediastinum.

VEGETATION EMBOLUS

Very occasionally subacute bacterial endocarditis of the aortic valve due to the *α-hae-molytic streptococcus* may involve the region of the pulmonary conus and erode into the pulmonary artery, and portions of vegetation are released into the pulmonary circulation (Fig. 15.15). An example of this condition was described by Rhodes (1945).

EMBOLI IN DRUG ADDICT'S LUNGS

The progressive and downward course pursued by the drug addict is often marked by the victim switching from the oral to the intravenous route

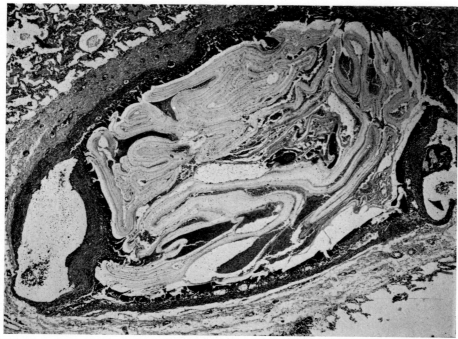

FIG. 15.14. A hydatid pulmonary embolus showing ectocyst twisted up and lying within a branch of the pulmonary artery. One surface of the parasitic embolus is covered with fused platelets. At the intimal surface of the artery there is considerable giant-cell formation. × 30 H and E.

(Reproduced by courtesy of Dr. V. McGovern, Sydney, Australia.)

for the administration of his drugs of addiction (main lining). The high market and scarcity value of illicit drugs leads to their adulteration with inert and often potentially harmful ingredients such as starch, lactose, talc and other insoluble "fillers". The intravenous injection of solutions of the product consequently results in a variety of emboli becoming impacted in the lung microvasculature (Fig. 15.16 A, B).

Starch (often corn starch in North America) leads to the formation of granulomas consisting of phagocytic cells in the interstitium of the alveolar wall. The starch particles are birefringent and appear as Maltese-cross-like bodies lying both in the lumen of the alveolar capillaries and outside within macrophages (Fig. 15.16 B). The starch granules are PAS-positive and are potentially or completely digested by diastase (Johnston and Waisman, 1971). In common with all forms of inert emboli

many of the particles are gradually extruded into the alveolar lumens whence they are removed by phagocytes.

Talc emboli follow intravenous injections of drug solutions prepared from dissolving tablets adulterated with talc powder, i.e. blue velvet (paregoric and tripelennamine hydrochloride), red devil (seconal) and methadone (Zientara and Moore, 1970; Groth et al., 1972). The lungs contain scattered but extensive greyish-white nodules up to 1 cm in diameter consisting of macrophages, foreign-body giant cells and fibrous tissue which is often laid down in parallel bands in which is embedded numerous particles of talc. The talc particles show as brilliant, birefringent specks and plate-like particles. Electron microscopy reveals small talc particles contained both within macrophages and lying in the alveolar capillary endothelium, while many escape through the vessel walls and lie free in

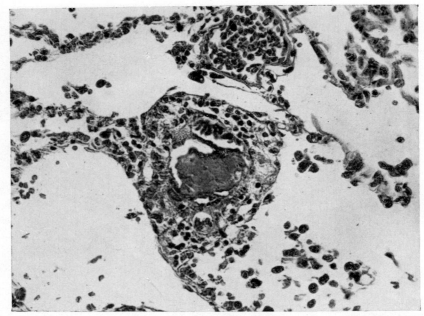

Fig. 15.15. An embolus of vegetation from a case of subacute bacterial endocarditis which had eroded into the pulmonary conus' liberating emboli.

the alveolar spaces where they provoke a fibrous reaction. Groth *et al.* considered the accompanying emphysematous changes resulted from ischaemic atrophy of alveolar walls.

Cardiac Catheter Embolus

Very rarely a small portion of a cardiac catheter may break off and embolize and lodge in a major branch of a pulmonary artery (Fig. 15.17). Thrombus tends to build up around the foreign body within the vessel lumen.

Mercury Embolism

An extremely rare form of pulmonary embolism may result from the accidental or intentional introduction of metallic mercury into the subcutaneous tissues. In a case described by Haubrich and Schuler (1949–50) a woman employed making thermometers developed an abscess in the forearm which was incised and a short while later a radiograph of her chest showed mercury emboli in her lungs. In one of two more recent cases seen by the author, a young woman with hysterical tendencies employed as a laboratory assistant injected mercury into the subcutaneous tissues of her forearm and developed a deep infection and injury to the median nerve. An operation was carried out using an elastic bandage to prevent haemorrhage and some of the mercury was evacuated. Unfortunately, the pressure exerted by the elastic bandage accidentally massaged some of the mercury into lymphatic channels whence it escaped into the general circulation. Numerous mercury emboli then appeared in the lungs (Fig. 15.18) and the patient later died of mercury poisoning.

Buxton *et al.* (1965) described nine cases of mercury embolism following the introduction of mercury into a vein whilst obtaining blood samples for gas analysis. To avoid the danger of this happening they advised that three-way stop-cocks should be used with the apparatus.

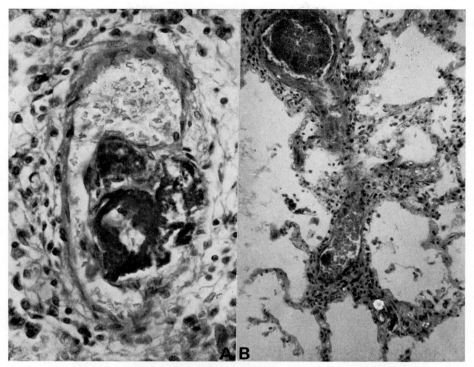

Fig. 15.16. (A) A drug addict's lung showing a small branch of a pulmonary artery partly filled and obstructed by debris from injected impure drug. (B). A drug addict's lung showing a branch of the pulmonary artery containing starch particles seen by polarized light. One of the particles in a capillary shows the typical Maltese cross appearance. A × 300, B × 120 approx.
(Reproduced by courtesy of Professor J. M. Cameron.)

One of the cases they reported died from mercury poisoning and serial chest radiographs in the others showed the gradual disappearance of the emboli within the lungs. When mercury poisoning follows the parenteral injection of mercury it is thought to be caused by the gradual conversion of metallic mercury into mercuric salts with the formation of the toxic mercuric ions.

Pulmonary Infarction

EXPERIMENTAL PATHOLOGY

Organs with a double blood-supply infarct less readily than those with a single supply, and this was shown to apply to the lung by Virchow (1847, 1851). He first showed that ligation of the pulmonary artery in animals failed to cause necrosis of the lung, provided the bronchial arterial circulation remained intact. He further showed that the bronchial arteries later dilated and concluded, therefore, that the increased bronchial arterial-flow had prevented infarction.

Since then much experimental work has been carried out in animals to determine the conditions under which infarction will occur. Karsner and Ash (1912), Ghoreyeb and Karsner (1913) and Karsner and Ghoreyeb (1913) repeated the earlier experiments and confirmed that pulmonary artery ligation alone failed to produce infarction. In addition, they concluded that the circulation distal to the obstruction of a lobar branch resulted from collateral anastomosis with adjacent intact branches of the

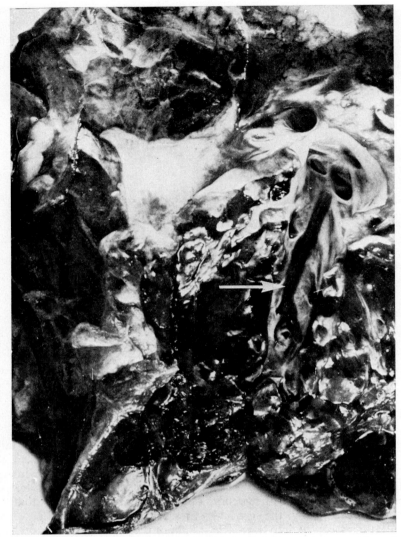

FIG. 15.17. A pulmonary embolus formed from a detached portion of a cardiac catheter (arrowed).
(Reproduced by courtesy of Dr. H. R. M. Johnson.)

pulmonary artery rather than from the opening up of bronchopulmonary anastomotic channels. They also showed that obstruction to the venous return was an important factor in causing pulmonary infarcts. Confirmation of the importance of interpulmonary arterial anastomoses in maintaining a circulation distal to an obstruction in a major pulmonary artery was provided by Steinberg and Mundy (1936); they injected a radiopaque oil into the unobstructed pulmonary circulation and found that it reached the capillaries which were normally supplied by the blocked vessel.

Bloomer *et al.* (1949) studied the bronchial arterial circulation following ligation of a pulmonary artery in a dog. They found that

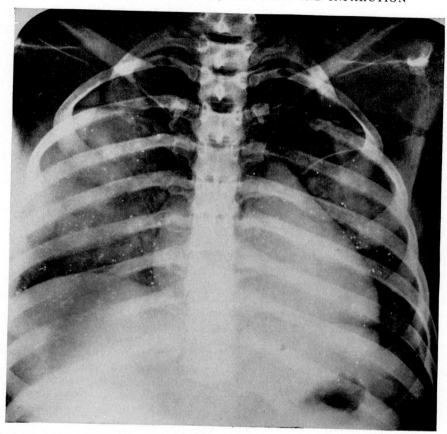

Fig. 15.18. A radiograph of the lungs showing multiple minute opacities due to metallic mercury emboli. (Reproduced by courtesy of Dr. H. R. M. Johnson.)

within the first few weeks an enormously increased blood-flow occurred in them, and that the blood reached the alveolar capillary bed through the opening-up of the pre-capillary anastomoses which they considered normally existed in the walls of the respiratory bronchioles. That the increased flow was very considerable was evidenced by the hypertrophy of the left ventricle which occurred following the raised output from the left side of the heart. Despite this increased bronchial blood-flow the pressure did not rise within the pulmonary capillaries, presumably because the enormous volume of the capillary bed exerted a cushioning and dissipating effect on the high-pressure bronchial

flow. Bloomer et al. also remarked upon the absence of any oedema in the alveoli which further confirmed the absence of any great changes in the alveolar capillary pressure.

Ellis et al. (1952) confirmed the results of Bloomer et al., but in addition drew attention to the fact that when small branches of the pulmonary artery were embolized following obstruction of the bronchial arteries, the animals survived. Only small areas of lung were deprived of a direct pulmonary circulation and these very soon filled with blood which seeped into them from adjacent non-obstructed branches of the pulmonary artery. Following this the capillaries in the ischaemic zone became turgid with blood that circulated sluggishly. Because of capillary congestion, the cushioning effect of the largely empty pulmonary capillary and arterial bed which was present when the main artery was ligated was lost. They believed that under natural conditions with unobstructed arteries the establishment of an increased bronchial supply to the infarcted area resulted in capillary rupture, intra-alveolar oedema and haemorrhage; these usually occurred about 4 days after the embolic obstruction.

PULMONARY INFARCTION IN MAN

As with all information gained from experimental pathology, allowance must be made for species differences when applying the knowledge to man. Although it rarely happens, pulmonary embolism may cause infarcts in healthy persons and ten cases have been reported by Hampton *et al.* (1945) in patients who were at work. The possibility exists that they may have suffered from some cardiac or respiratory disorder of which they were unaware at the time of the embolism, but no subsequent examination revealed its presence. Usually, when infarcts occur in healthy persons a very large branch is obstructed. Hampton and Castleman (1943) and MacLeod and Grant (1954) have closely investigated human pulmonary infarction. The former authors examined 370 cases and found that 40 per cent occurred post-operatively, 30 per cent in patients with some form of cardiac disease and 30 per cent were found in non-cardiac medical patients; very similar findings have been reported by MacLeod and Grant.

In addition to conditions resulting in a low output type of heart failure, Torack (1958) has listed other factors which predispose to lung infarction. These include: (1) collapse of lung, (2) any condition which mechanically interferes with the oxygenation of the alveolar capillary blood, such as bronchial obstruction or pneumonic consolidation, and (3) conditions of shock.

Emboli, however, are found less frequently in collapsed lung presumably due to the decreased blood-flow through such lung, but if they do occur, infarction is more liable to ensue due partly to the diminished pulmonary blood-flow, the failure to establish an adequate collateral circulation, and partly to the absence of aeration of the infarcted tissue. Both (2) and (3) increase the likelihood of infarction resulting from pulmonary embolism.

Pulmonary infarction occurs equally in both sexes and most cases are found in persons over the age of 40; it rarely occurs under the age of 20. The localization of infarcts corresponds with the distribution of emboli, and approximately 75 per cent occur in the lower lobes, often involving the costophrenic margin of the lung. Multiple infarcts are twice as common as single ones (Spain and Moses, 1946).

The changes following pulmonary embolization have been studied both radiologically and by the usual histological methods; they may terminate in three ways, namely, (a) without alteration in the lung, (b) by the development of incipient infarction, and (c) by the development of a true pale infarct.

Following obstruction of a branch of the pulmonary artery the related capillary bed becomes congested with blood flowing in from patent branches in the adjacent lung. This flow may be sufficient to prevent necrosis of the alveolar capillaries and the lung remains intact. The bronchial arterial flow scarcely increases around a small infarct and, following recanalization of the embolus, a pulmonary blood-flow is re-established. A few hours after embolization such lungs may appear more congested but they rapidly return to normal. Radiologically this condition gives rise to transient shadows in the lungs.

Incipient infarction is characterized by haemorrhage and oedema in the ischaemic zone (Figs. 15.19, 15.20). The pulmonary collateral circulation is insufficient initially to prevent stagnant congestion within the capillaries, which themselves suffer damage in the infarcted zone, although complete necrosis does not ensue. The secondary establishment within 12–24 hours of an increased bronchial artery-flow results in rupture of the damaged alveolar capillaries with leakage of oedema fluid and

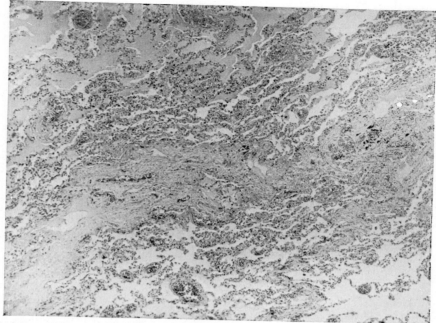

FIG. 15.19. Incipient infarction showing a local area of pulmonary oedema in the region of the infarcted tissue. × 40 H and E.

blood into the alveoli; later, with the establishment of a better circulation, reversal of these changes occurs and the lung is gradually restored to normal. A pale infarct is likely to occur in low cardiac output heart failure and is in part caused by the diminished blood-flow and also by the frequently associated pulmonary venous congestion. A pale infarct very infrequently follows a massive embolus obstructing a segmental artery in a patient with a normal cardiac output and no heart failure.

A pale infarct results when alveolar capillary walls necrose and the subsequent increase in the bronchial blood-flow results in gross haemorrhage and complete disruption of the affected area. The red cells later disrupt and haemosiderin-laden phagocytes appear in the infarcted zone. Subsequently, complete necrosis of all the parenchymal tissues occurs in the centre of the infarct, though ghost outlines of septa and vessels remain identifiable until a late stage. Later, organization proceeds from the marginal vascular zone and eventually results in scar formation. The final scar may be quite small and gives no indication of the volume of lung tissue that was originally infarcted. The scars are often linear in shape and confined to the subpleural tissues; within these scars bronchiolectatic bronchioles may survive because of their independent bronchial artery supply. The edges of a healing infarct often show considerable bronchiolar and alveolar epithelial cell proliferation (Fig. 15.21) which very rarely may later give rise to neoplastic changes resulting in a bronchiolar adenocarcinoma.

The differences between a scar resulting from an infarct and other forms of lung scar may be difficult to determine, but the presence of a recanalized branch of a pulmonary artery a short distance beyond the edge of the scar often provides a clue to the aetiology. Elastic fibres persist in a healed infarct although crowded together in a haphazard fashion if the infarct

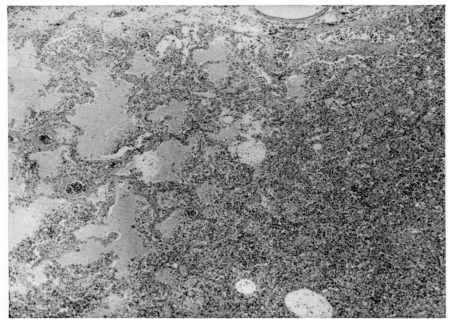

FIG. 15.20. Incipient infarction showing both pulmonary oedema and intra-alveolar haemorrhage but no actual necrosis of tissue. This stage is reversible as the circulation becomes re-established. × 40 H and E.

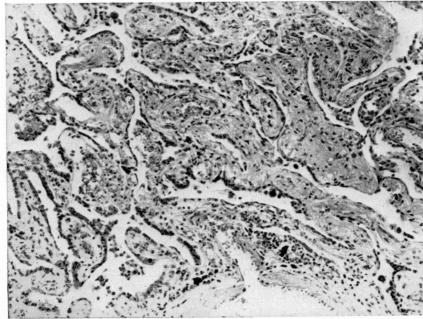

FIG. 15.21. The edge of a healing infarct showing epithelium-lined, thickened and fibrosed alveoli. × 140 H and E.

was previously infected (Castleman, 1940). The process of repair is usually retarded in patients suffering from heart failure.

Lung infarcts cause characteristic radiological changes in life, and angiograms may localize the arterial obstruction after death. Within 12–24 hours after embolization has occurred the infarcted lung shows as a blurred shadow. Later the shape of the infarct becomes more clearly discernible. As Hampton and Castleman (1943) have shown in post-mortem studies of lungs before removal from the body, the inner margin of many infarcts presents a convex surface. The popular concept of a lung infarct as an inverted cone with its base situated on the pleural surface is only partially correct—many infarcts involve the costophrenic margin of the lung and involve two pleural surfaces. Such infarcts are not

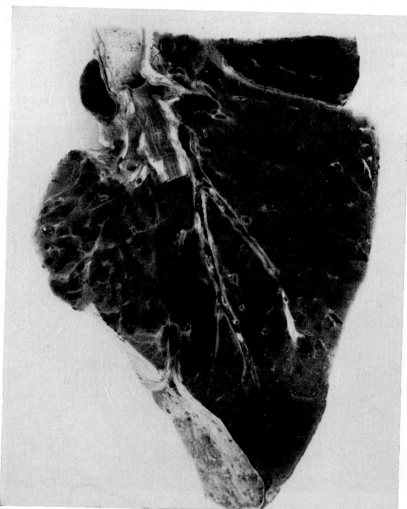

FIG. 15.22. A pulmonary infarct situated in the base of the lung and involving both the posterior and diaphragmatic surfaces of a lower lobe. The causative embolus is visible wedged astride a proximally situated bifurcation in the supplying pulmonary artery. Natural size approx.

pyramidal but wedge-shaped, extending beneath both pleural surfaces, and the base is situated centrally, presenting as a convexity towards the hilum (Fig. 15.22). Infarcts involving only one pleural surface conform to the cone shape with their bases situated externally on the pleura. Some aseptic infarcts may later undergo softening and simulate an infected infarct.

Injection of a contrast medium into the bronchial arteries after death often shows a greatly increased vascularity in the region of an infarct but the infarcted area fails to fill through the pulmonary arteries (Fig. 15.23).

Spain and Moses (1946) found that in a series of 100 pulmonary infarcts, thirteen were accompanied by a serosanguineous pleural effusion, thirty-four by a serous effusion and in the remainder there was no effusion. The thirty-four cases with a serous effusion were all associated with heart failure. The majority of pulmonary infarcts are associated with some degree of fibrinous pleurisy which may occasionally be extensive.

Very rarely a non-infected pale infarct may undergo softening and cavitation (Grieco and Ryan, 1968) and a further very rare complication—a bronchopleural fistula—may result, a complication carrying a very high mortality rate (McFadden and Luparello, 1969).

PULMONARY VENOUS INFARCTS

Pulmonary infarction is usually regarded as a sequel to embolic obstruction of the pulmonary

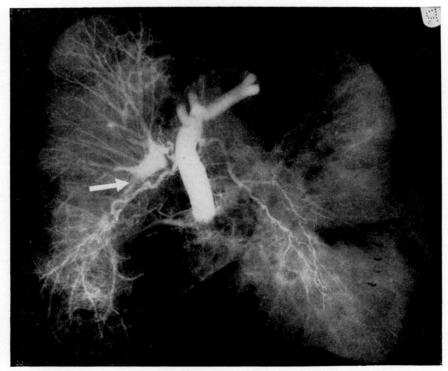

FIG. 15.23. An angiogram of the pulmonary artery and both bronchial arteries showing a major branch of the pulmonary artery blocked by an embolus (arrow) and compensatory increase in the bronchial artery blood-flow to the infarcted portion of the lung.

(Reproduced by courtesy of Dr. L. Cudkowicz and the Editor of *Brit. J. Tuberc.*)

arteries by a thrombus, but rarely it may follow pulmonary venous obstruction. Many of such venous thromboses have occurred in association with very extensive pulmonary tuberculosis but they may also complicate primary lung cancer. It has also followed segmental resection of the lung which causes thrombosis of the intersegmental veins. This in turn may result in small areas of infarction which may later give rise to bronchopleural fistulae and empyemas (Salyer and Harrison, 1958).

INFECTED PULMONARY INFARCTS

The incidence of lung abscess following pulmonary embolism and infarction has been estimated by Coke and Dundee (1955) to be about 2·7 per cent. The majority of such abscesses remain undiagnosed during the lifetime of the patient and are only discovered post-mortem. Infection may reach an infarct either from the embolus itself, from the blood circulating in the surrounding lung, or from the bronchial passages. Infarcts involving pneumonic lung are often infected from their inception and *Ps. pyocyaneus* is often the infecting agent. Rarely a septic infarct can follow acute or subacute bacterial endocarditis of the tricuspid or aortic valve. In the latter, infection may spread through the pars membranacea into the infundibular portion of the pulmonary conus. Infected emboli may be released from any vein, but are particularly liable to occur following septic thrombophlebitis of the lateral sinus or internal jugular vein which in turn is caused by middle-ear and peritonsillar infections. Septic infarcts are often multiple and small and undergo central liquefaction. The contents may consist of recognizable creamy-yellow pus or greyish-pink necrotic material. Liquefaction in an infarct does not always indicate infection and this change can result from autolytic softening. In both infective and simple autolytic change the patient is usually febrile and shows a leucocytosis in the peripheral blood. In recent years the incidence of saprophytic mycotic infection in pulmonary infarcts has increased; this has probably followed the widespread use of antibiotics, cytotoxic drugs and cortisone. *Aspergillus* sp. is the most frequently encountered secondarily infecting fungus, but in a few cases *Mucor* sp. have been found.

CHAPTER 16

Chronic Pulmonary Hypertension

GENERAL FEATURES AND PHYSIOLOGY OF THE PULMONARY CIRCULATION

Before considering the structural changes that occur in the pulmonary vessels in pulmonary hypertension it is first necessary to recall briefly some of the salient facts that are at present known about the physiology of the pulmonary circulation.

During foetal life when both cardiac ventricles are working in parallel, the systemic and pulmonary arterial blood-pressures are equal. Following birth when the two ventricles start to work in series, the pulmonary arterial pressure falls following the functional closure of the ductus arteriosus and the decrease in flow resistance offered to the pulmonary circulation by the smaller muscular pulmonary arteries, arterioles and the alveolar capillaries. The pulmonary blood-pressure does not reach its adult levels (25/8 mm Hg) until about 4 years of age.

Unlike all other organs in the body except the heart, the lungs, save in foetal and the immediate post-natal period, normally transmit the entire cardiac output which in turn is related to the surface area of the body and the age of the individual. Normally in young adults of average build at rest, the lungs transmit approximately 5–6 litres of blood per minute, but this amount may be increased threefold or more during exercise. With advancing age the cardiac output falls (Brandfonbrener et al., 1955) and in the elderly the pulmonary flow may only amount to about half that in the young adult.

The circulation time through the adult lung has been estimated to be about 0·75 second at rest, falling during exercise to 0·34 second (Roughton, 1945). These rapid rates of flow are normally achieved by the existence of an extensive system of readily distensible elastic arteries and by the low resistance offered by the pulmonary muscular arteries together with the short course and wide calibre of the very numerous alveolar capillaries. Increased pulmonary blood-flow also occurs in pregnancy and in such pathological conditions as pyrexia, severe anaemia, hyperthyroidism and Paget's disease of bone.

The pulmonary vascular flow in the erect position varies considerably from one part of the lung to another. In addition to the pumping pressure provided by the right ventricle, decreased pulmonary venous pressure occasioned by gravitational forces and the intrapulmonary gas tensions play an important part in determining regional blood perfusion within the lungs. In the upper third of the lungs in the erect position the pulmonary arterial pressure acts against gravity while the reverse applies to the pulmonary venous flow. During inspiration the intra-alveolar pressure rises above the pulmonary venous pressure and probably above the alveolar capillary pressure, leading to compression of the alveolar capillaries and an intermittent pulsatile blood-flow in this region of the lungs. If the body assumes a horizontal position a normal continuous flow occurs. The importance of gravitational forces in the distribution and flow of blood through different regions of the lungs was first suggested by Orth (1887) and has received confirmation in recent years from the gaseous and blood-perfusion studies on the lungs by the Hammersmith Hospital group of workers in England, using the radioactive gaseous isotopes $^{15}O_2$ and ^{133}Xe (West, 1963), and crystal scintillation counters applied to the outside of the chest wall.

In the basal portions of the lungs in the erect position both the pulmonary arterial and venous pressures are increased by gravitational force and the alveolar intracapillary pressures remain greater than the intra-alveolar gas pressures throughout the respiratory cycle. This leads to continuous perfusion of the alveolar capillaries and normally a greater minute blood-flow through the lower parts of the lungs. This in turn slightly increases the compliance of this part of the lung. In the middle field of both lungs the vascular flow is intermediate between the upper and lower regions.

The variations in ventilation : perfusion ratios in different regions of the lung are reflected in the clearance rates for alveolar carbon dioxide which are minimal at the apices, increasing in linear fashion towards the bases of the lungs. Nevertheless, under normal conditions of erect posture, the volume of gas exhaled from the upper lobes is slightly greater than that inhaled (1%), due to a greater volume of exhaled CO_2 compared with the volume of O_2 absorbed from the inhaled air. The reverse occurs in the lower lobes due to a greater volume of O_2 being absorbed from inspired air than the volume of CO_2 exhaled. The pulmonary arterial blood-flow, however, is normally greater in the upper lobe of the left lung compared with the upper lobe of the right lung owing to the normally axial flow of the pulmonary arterial blood towards the left side. Exercise or any pathological state that increases the pulmonary arterial flow and raises the pulmonary venous pressure abolishes the pulsatile flow found at rest in the erect position in the apical regions of the lungs, and leads to a considerable increase in the blood-flow through these regions.

Roughton (1945) calculated that the pulmonary capillary blood volume at rest was approximately 60 ml rising to 95 ml during exercise. Normally the distribution of the blood volume in the pulmonary blood-vessels varies with the phase of the respiratory cycle. During inspiration the elastic traction exerted by the distending alveolar walls dilates the more fixed pulmonary vessels such as the arteries and veins but the distension and consequent stretching of the alveolar walls constricts the alveolar capillaries. The pulmonary arteries and veins and left atrium consequently distend and fill but during expiration the process is reversed and leads to capillary filling and venous compression. These changes are most in evidence in the upper regions of the lungs in the erect position. The importance of intra-alveolar pressures in controlling pulmonary blood-flow was shown experimentally by Banister and Torrance (1960) who found that if the lungs of cats were perfused at a steady rate, the left atrial pressure being maintained at a constant level, the pulmonary arterial pressure varied in a linear fashion with the intratracheal pressure.

Normally the evolution of carbon dioxide and the absorption of oxygen in the lung capillaries takes a very small but finite time to accomplish but both are almost instantaneous, though oxygen diffusion is slightly slower. The term alveolar-capillary block syndrome has been widely used to describe the functional disability found in many forms of lung disease. In almost every instance this syndrome results from disorders of pulmonary blood perfusion and lung ventilation in various combinations and not from any failure of the blood gases to diffuse through the alveolar-capillary barrier.

For those who seek further information about the modern concepts of pulmonary physiology and their importance in lung pathology, reference should be made to the collected papers on these subjects presented at the Ciba Symposium (Ciba Symposium, 1962).

COMPARISON OF STRUCTURE OF THE PULMONARY ARTERIES DURING INTRA- AND EXTRAUTERINE LIFE

Before considering the pathology of the pulmonary arterial tree it is first necessary to understand the differences that exist between the adult and foetal pulmonary arterial circulations and especially the manner in which these differences affect the development and structure of the pulmonary vasculature.

During intrauterine life the left and right ventricles function mainly as two pumps working in parallel situated between the venous and arterial circulations. Only a very small fraction of the total right ventricular output passes through the lung, the remainder being shunted into the aorta through the ductus arteriosus. Although the foetal lungs are unexpanded and their contracted vessels transmit only a very reduced volume of blood, the pulmonary vasculature according to Smith *et al.* (1963) may still be capable of undergoing further vasoconstriction after appropriate stimulation.

During the latter stages of intrauterine life, the foetal arterial oxygen saturation begins to fall and reaches about 60 per cent (Dawes, 1958) and falls still lower during periods of active foetal movements. The foetus during this stage of pregnancy becomes adapted to an environment with a low oxygen tension and probably this largely accounts for the sometimes prolonged immediate post-natal survival of infants who are born with respiratory insufficiency or who develop hyaline membrane disease shortly after birth.

Dawes found that in lambs at the time of birth when the umbilical vein is interrupted the peripheral systemic vascular resistance rises abruptly following the fall in blood-pressure in the vena cavae. The state of anoxaemia that almost immediately results stimulates inspiratory respiratory movements and the subsequent attempted expiration of this air against a partially closed glottis expands the lung alveoli. The expansion of the lungs with gas (air) leads to a sudden diminution of the pulmonary vascular resistance (Dawes *et al.*, 1953) and a fall in the pulmonary arterial pressure together with a rise in the left atrial pressure. Whereas the pulmonary arterial pressure before birth exceeded the aortic blood-pressure, following the inauguration of successful respiratory movements the reverse obtains though the pulmonary arterial pressures only fall slowly in the human neonate (Harris and Heath, 1962). Before birth, Morris *et al.*

(1963) found that in lambs the blood flowed from the pulmonary artery to the aorta through the ductus arteriosus but after birth the flow, if any, occurred in the reverse direction. Normally the ductus arteriosus though not structurally obliterated becomes functionally closed within a few hours of birth. Rudolph *et al.* (1961) found that in human neonates the ductus arteriosus became functionally shut during the first 12 hours, a finding first demonstrated angiographically by Lind and Wegelius (1954). The stimulus to closure of the ductus arteriosus is thought to be the rise in arterial oxygen saturation following the commencement of respiration.

As a result of the changes in the pulmonary vascular physiology described above, the pulmonary arteries and capillaries undergo considerable alterations between the last stages of intrauterine life and early childhood at which time they assume their adult characteristics (Figs. 16.1, 16.2, 16.3, 16.4, 16.5). These changes have been studied in detail by Civin and Edwards (1951), O'Neal *et al.* (1957), Rosen *et al.* (1957) and Harris and Heath (1962).

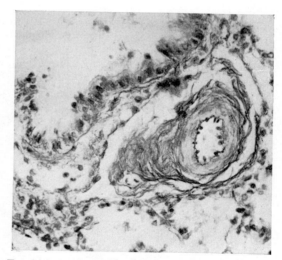

Fig. 16.1. Transverse section through a muscular artery 120 μm approx. in diameter shortly before *birth*, showing the thick muscular media at this age. The internal elastic lamina is scarcely visible but the swollen endothelial cells are visible as black dots. × 140 Verhoeff–Van Gieson. Mural thickness: total diameter 1 : 4·4.

During the last trimester of pregnancy the lungs are lobular organs with easily recognizable lobules of lung tissues separated by well-formed interlobular septa. The elastic arteries begin with the main pulmonary arteries and accompany the bronchi as far as the point where the latter become intralobular structures.

The *large pulmonary elastic arteries* during the last trimester of intrauterine life and at the time of birth closely resemble both in appearance and

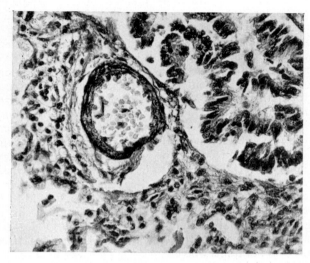

FIG. 16.2. Transverse section through a small muscular artery approx. 100 μm in diameter showing the rapid reduction in thickness of the media that has occurred in 4 *days* since birth and the rapid dilatation that has taken place. × 280 H and E. Mural thickness: total diameter 1 : 8.

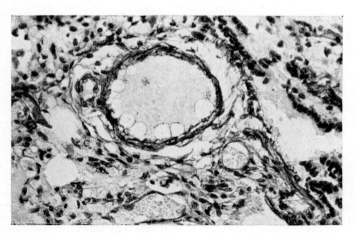

FIG. 16.3. Transverse section of a pulmonary artery 110 μm in diameter at the age of 3 *months* showing still further thinning of the wall and dilatation of the lumen. × 280 H and E. Mural thickness: total diameter 1 : 15.

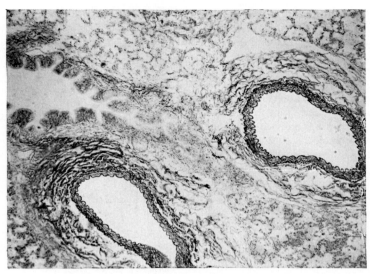

FIG. 16.4. Large elastic pulmonary arteries 800 μm in diameter showing internal elastic lamina with well-marked medial elastic tissue but ill-defined external elastic lamina; taken from a *full-term infant*. × 40 Verhoeff–Van Gieson.

structure the large elastic systemic arteries. The medial thickness of the main pulmonary artery equals that of the aorta at birth. Although the medial elastic fibres are arranged in a similar regular fashion and are found throughout the media, they nevertheless are rather fewer in number and slightly less regularly disposed than in the aorta (Fig. 16.6). Distal to the level of the segmental arteries the lumens vary from a half to three-quarters of the diameter of the adjacent bronchi from which the arteries are separated by loose connective tissue. The flattened endothelium rests almost directly on the subjacent innermost medial elastic fibres which are continuous with the medial muscle fibres. At about the time of birth an ill-defined external elastic lamina appears blending both with the media and periadventitial connective tissue. After birth, the lumen and external diameter of the small elastic (interlobular) arteries and to a lesser extent the large elastic pulmonary arteries increase considerably in size. At the age of 6 months the medial elastic tissue although reduced in amount is still mainly composed of parallel bundles of fibres which are beginning to

break up into numerous short irregular fibres. By the age of 2 years the elastic fibres have assumed their normal adult pattern though occasional areas of a foetal elastic pattern persist. By the age of 4 years the normal adult mural structure is achieved and the only further change is an increase in size of the arteries that occurs with the continued growth of the lungs up to the age of about 20 years. The changes in structure of the pulmonary artery trunk that occur at different ages have been described by Heath *et al.* (1959).

The *muscular pulmonary arteries* are mainly intralobular vessels but some of the largest branches form the terminations of the interlobular septal elastic vessels. The intralobular arteries accompany the bronchi as they enter the lobules and are separated from them by loose connective tissue. During the last trimester of pregnancy the lumens of all the muscular arteries are very narrow, being occluded by the swollen cuboidal endothelial cells which rest directly on the internal elastic lamina. Both the internal and external elastic laminae are well developed at the time of birth in the large muscular arteries and the medial muscle fibres are much more swollen than in adult vessels. The internal elastic lamina is less well developed in the smaller muscular arteries and may not be demonstrated until the fourth to sixth month of post-natal life.

Wagenvoort *et al.* (1961c) found that the amount of arterial medial muscle increased during the last half of pregnancy due mainly to an extension of the muscle coat into the walls of the small developing arteries. Almost immediately after birth both the size of the lumen and the external diameter of the muscular arteries increase rapidly with thinning of the media. Also after birth from the age of 6 months to a year the muscular arteries grow extensively in company with the rest of the lung and by the age of 20 years attain full development. During growth there is a progressive increase in the diameter of these arteries relative to the thickness of their walls.

At the time of birth and during late foetal life focal intimal proliferation occurs in some of the

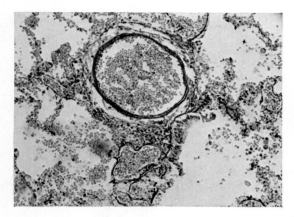

FIG. 16.5. A small muscular pulmonary artery in the lung of a child 7 *years* old showing two separate elastic laminae separated by a very thin layer of media. The vessel is 155 μm in diameter and is normal for this age. × 140 modified Sheridan's elastic stain and haematoxylin. Mural thickness: total diameter 1:23.

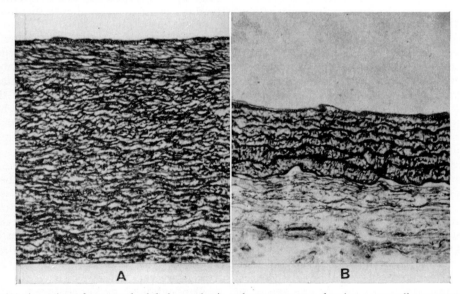

FIG. 16.6. (A) A section of a normal adult large elastic pulmonary artery showing a generally concentric arrangement of the elastic fibres which are finer than those found in the aorta and branch repeatedly. × 100. (B) A section of a large elastic pulmonary artery in the first day of neonatal life showing an "aortic" arrangement of thick concentric elastic fibres with less branching than is normally seen in the adult vessel. × 100 both stained with Sheridan's elastic stain.

smallest muscular arteries leading to constriction of the lumens at points lying immediately distal from their origins from the parent vessels. Similar constrictions have been found to occur in the developing pulmonary muscular arteries in a variety of mammalian species (Fig. 2.25).

The *pulmonary arterioles* are not clearly recognizable until about the third month after birth. At first they consist of little else than endothelium-lined tubes with a minimal amount of connective tissue forming their outer walls. Elastic tissue is slow to develop and is not completely formed before the age of 20 years. When fully developed they are almost indistinguishable in appearance from small venules.

The *pulmonary capillaries* during intrauterine life are largely collapsed and the lining endothelial cells present a swollen appearance. At the time of birth they become distended and first transmit a recognizable stream of blood whereas in foetal life they appear to be almost bloodless.

From the foregoing description of the development of the pulmonary arterial tree it will be appreciated that the contracted, thick-walled, small, muscular pulmonary arteries offer considerable resistance to the pulmonary blood-flow during intrauterine life, resulting in the diversion of the main pulmonary arterial flow into the ductus arteriosus and thence to the aorta. This shunt was demonstrated angiographically in the living foetus by Lind and Wegelius (1949). In addition to the restricted pulmonary arterial blood-flow some blood is shunted through bronchopulmonary arterial anastomoses that exist in foetal life. At the time of birth the lungs assume their aerating function and the extensive pulmonary capillary bed, hitherto collapsed, fills rapidly with blood and within a few hours almost the whole of the pulmonary arterial blood-flow passes through the lung. According to Wagenvoort *et al.* (1961c), the increase in the pulmonary blood-flow is mainly attributable to dilatation of the smallest muscular arteries.

Following upon the diversion of the main blood-flow through the ductus arteriosus to the pulmonary arteries at the time of birth and the establishment of alveolar capillary flow where previously such capillaries had been mostly collapsed and almost bloodless, intra-alveolar fluid is absorbed and its production greatly reduced following the commencement of respiration (Strang, 1973). The pulmonary lymphatic flow, which is considerable throughout the last few weeks of intrauterine development, causes distension of the very conspicuous intrapulmonary lymphatic channels at the time of birth and during the first few hours of post-natal life. This leads to an increase of lung compliance which may be responsible for the tachypnoea often observed during the first day of neonatal life (Avery, 1966).

The changes that occur in the pulmonary arteries after birth are adaptations to facilitate these changes and the muscular pulmonary arteries assume their normal thin-walled adult character by the age of 6–10 months (Valenzuela *et al.*, 1954; Civin and Edwards, 1951).

In the presence of certain congenital cardiovascular defects the pulmonary blood-flow may become very excessive after birth (hyperkinetic pulmonary hypertension). This results in persistence of the neonatal elastic structure in the large pulmonary arteries which fail to regress and adopt the normal adult pattern of elastic tissue. Furthermore, the muscular coat in the smaller muscular arteries fails to thin and undergoes still further hypertrophy and actual hyperplasia. These changes are described in greater detail in the section on chronic pulmonary hypertension.

AGEING OF THE PULMONARY ARTERIES

From early childhood the large *elastic pulmonary arteries* show increasing concentric fibrous thickening of the intima. At first minimal in amount, in old age it may equal half the thickness of the media. In late middle life a few fatty atheromatous flecks normally appear in the large elastic pulmonary arteries even in the absence of pulmonary hypertension. In the elderly, pools of acid mucopolysaccharide accumulate in the major pulmonary arteries between the fibres of the media and are accompanied by a reduction in the amount of elastic tissue. Rarely partial calcification of medial elastic fibres may occur in aged persons.

Age changes in the *pulmonary muscular arteries* were studied by Wagenvoort and Wagenvoort (1965). They found that the thickness of the media remained unchanged but the intima underwent Intimal fibrous thickening. The intimal changes were usually most pronounced in the upper lobes and consisted either of concentric thickening or segmental fibrous plaques. It has been suggested that the diminished blood-flow through the upper lobes may predispose the arteries in these lobes to the risk of spontaneous thrombosis and that the intimal plaques are a sequel to this change. Later both the intima and media may undergo hyaline change and the whole arterial wall may become hyalinized and structureless (Fig. 16.7). These changes are accelerated if pulmonary hypertension is present.

Chronic Pulmonary Hypertension

Arterial blood-pressure is determined by the interaction of three variable factors, namely: (1) the volume of blood transmitted through the vessels (the cardiac output), (2) the cross-sectional area of the vascular pathway through which the blood circulates (the vascular resistance), and (3) the viscosity of the blood.

Under normal conditions the pulmonary systolic blood-pressure does not exceed 25 mm Hg and the diastolic 8 mm Hg (Cournand, 1950–1), and this despite the fact, as already stated, that the lungs normally transmit the entire output of a ventricle. The lungs can normally transmit about three to four times the basal flow of 3·1 litres per square metre of body surface without any rise in pressure occurring. This large flow is accommodated in the distensible system of elastic arteries and in the readily dilated pulmonary capillary bed. The reserve capacity of the pulmonary capillary bed is very great and Wearn et al. (1934) showed by direct observation that, under resting conditions, only part of the capillary bed is perfused at any one time.

Although the pulmonary vasculature is admirably adapted to cope with normal physiological conditions, pathological demands, if of a gradual onset, produce a series of changes in the pulmonary vessels which reflect the general principle stated by Thoma (1893), that the

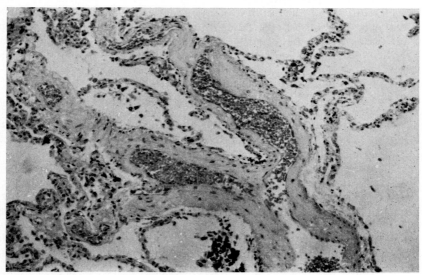

Fig. 16.7. Hyalinization of pulmonary arteries. A common change in old persons. × 130 H and E.

structure of a vessel reflects the volume of flow and pressure within it. The changes that follow a rise in pulmonary arterial blood-pressure depend to a great extent on the rapidity with which the changes occur. As a general rule, most living organisms can adapt to a gradual change but sudden variations often lead to very severe and fatal disturbances of function, and the lung is no exception to this principle. In the ensuing account only the changes that follow a chronic rise in pulmonary blood-pressure will be considered. In measuring and describing changes in the thickness of the vascular mural components allowance must always be made for the apparent thickening that follows contraction of an empty vessel wall.

Chronic pulmonary hypertension may be caused by the circulation of a persistently excessive volume of blood through the lung vessels, a condition referred to as hyperkinetic pulmonary hypertension. Alternatively it may result from mechanical obstruction causing an increased resistance to the circulation of the normal or diminished volume of blood or it may follow a chronic and considerable rise in pressure in the pulmonary venous circulation (chronic passive pulmonary hypertension or the post-capillary resistance group). Under normal conditions the pulmonary intracapillary pressure is much less than the plasma osmotic pressure (25 mm Hg) and there is no danger of pulmonary oedema. If, however, either the pulmonary arterial or the venous pressures are allowed to rise beyond a critical level pulmonary oedema immediately occurs and would rapidly lead to death unless adaptive changes took place to prevent its occurrence. It is with these adaptive vascular changes that prevent the onset of pulmonary oedema that the ensuing account is mainly concerned.

GENERAL CAUSES OF CHRONIC
PULMONARY HYPERTENSION

As the changes produced in the pulmonary arterial tree in most forms of chronic pulmonary hypertension are similar, a general description of these changes will first be given and later will be followed by a fuller description of each individual cause. The classification of the causes of chronic pulmonary hypertension is as follows:

(1) A group of conditions in each of which initially there may be pulmonary arterial or arteriolar vasoconstriction followed later by structural arterial and other changes in the lungs.
(a) Primary (idiopathic) pulmonary hypertension.
(b) Hyperkinetic pulmonary hypertension due to an increased pulmonary blood-flow.
(c) Chronic hypoxic pulmonary hypertension (high-altitude pulmonary hypertension, emphysema).
(d) Chronic passive pulmonary hypertension (post-capillary resistance group).

(2) Conditions which primarily cause mechanical obstruction to the pulmonary blood-flow.
(a) Emboli including malignant cell emboli.
(b) Thrombosis of small branches of the pulmonary arteries and capillaries.
(c) Pneumoconioses and other causes.
(d) Vascular obstruction due to progressive systemic sclerosis (PSC) and Raynaud's syndrome.

(3) Chronic pulmonary hypertension due mainly to hypoxia or in which the mechanism of production is not fully understood.
(a) Emphysema (see Chapter 14).
(b) Kyphoscoliosis.
(c) Pickwickian syndrome.
(d) Idiopathic pulmonary haemosiderosis (see Chapter 19).

Before considering the arterial changes that occur in chronic pulmonary hypertension the reader should be acquainted with the structure of the normal neonatal and adult pulmonary arteries, as a knowledge of the structure of these vessels is essential if the abnormalities which may develop in some cases of chronic pulmonary hypertension are to be appreciated.

An account of the pulmonary arterial changes common to many of the above forms of pulmonary hypertension will first be given, though it should be emphasized that considerable individual variation in the extent of the response to the hypertension occurs.

THE PULMONARY ARTERIAL CHANGES THAT OCCUR IN RESPONSE TO CHRONIC PULMONARY HYPERTENSION

The pulmonary arterial structural changes following a raised pressure may be subdivided into:
(i) those involving the elastic pulmonary arteries,
(ii) those involving the muscular pulmonary arteries and arterioles.

Although no two patients dying from the effects of chronic pulmonary hypertension exhibit all or the same combination of arterial changes because of individual variation in tissue response, an increasing pulmonary blood-pressure often results in a series of changes of increasing severity.

An increase in blood-flow accompanied by a rise in intraluminal pressure may at first lead to pulmonary arterial and capillary dilatation, but overdistension of the arterial wall provokes medial muscular contraction which, if sustained, leads to a work hypertrophy of the medial muscle. This change is usually the first to become recognized in all types of arteries.

CHANGES IN THE ELASTIC PULMONARY ARTERIES

The media of the elastic pulmonary arteries first undergoes hypertrophy, and probably hyperplasia, and the intimal connective tissue becomes thicker. When pulmonary hypertension commences at birth or very soon after, the regressive changes that normally affect the elastic tissue in the large elastic neonatal pulmonary arteries and which lead to the differences in structure between the adult aorta and major pulmonary arteries fail to take place, and the aortic pattern and quality of the elastic fibres persist (Heath et al., 1959).

Medial hypertrophy may later be followed by a form of mucoid degeneration characterized by the interstitial accumulation of excessive quantities of an acid mucopolysaccharide. This form of medial degeneration was produced experimentally by Liebow et al. (1950) in young puppies by artificially stenosing a pulmonary artery. This resulted in thickening and dilatation of the walls proximal to the obstruction and thinning with loss of elastic fibres in the post-stenotic segments. The elastic tissue in the pre-stenotic segments showed "mucoid" degenerative changes and partial fibrous replacement. In many cases of severe chronic pulmonary hypertension, the main and lobar branches of the pulmonary artery become dilated and this can lead occasionally to compression of the left recurrent laryngeal nerve as in the case of idiopathic pulmonary hypertension described by Brinton (1950). The mucoid degenerative change and atrophy of the medial elastic tissue may lead to aneurysmal dilatation and even rupture of a pulmonary artery. Rupture of a pulmonary artery has been described in pulmonary hypertension associated with a patent ductus arteriosus by Favorite (1934) and Whitaker et al. (1955), and in chronic mitral stenosis by Thomas et al. (1955). An example of aneurysmal dilatation occurring in chronic mitral stenosis is shown in Fig. 16.8. Rupture of the artery may be preceded by dissection of its coats. Dissecting aneurysms complicating chronic pulmonary hypertension were described by Durno and Brown (1908), D'Aunoy and von Haam (1934) and Ravines (1960) in patients with a patent ductus arteriosus, and by Brenner (1935, last paper) and Gold (1946) in cases of idiopathic pulmonary hypertension. Medial mucoid change is more liable to occur in older persons with pulmonary hypertension (Tredall et al., 1974) as well as in children with cystic fibrosis (mucoviscidosis) (Oppenheimer and Esterly, 1973). Rupture of a pulmonary artery may also follow the onset of chronic pulmonary hypertension induced by a Blalock–Taussig operation for the relief of Fallot's tetralogy (Ferencz, 1960b). The dissection of the pulmonary artery is usually a single focal lesion but occasionally may involve several vessels and leaves mural scars which become re-endothelialized (Ravines, 1960).

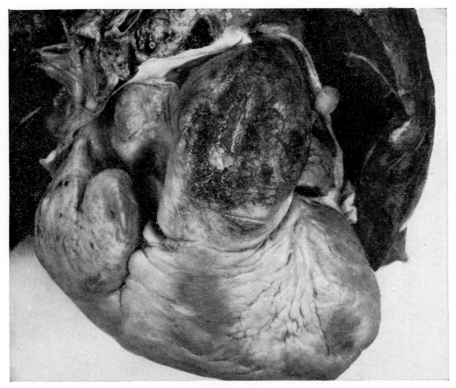

FIG. 16.8. Specimen of the heart and pulmonary artery showing aneurysmal dilatation of the latter which extended into the main branches of the artery. The patient died from chronic mitral stenosis and the aneurysm was partially filled with ante-mortem thrombus. Two-thirds natural size.

A similar form of medial degeneration occurs in the pulmonary arteries in the hereditary disease known as Marfan's syndrome (Fig. 16.9), a condition unassociated with pulmonary hypertension and which results in the formation of dissecting aneurysms usually in the aorta, but occasionally in the pulmonary artery as in the case described by Tung and Liebow (1952).

Chronic pulmonary aneurysms may later fill with laminated blood clot as occurred in the case of chronic pulmonary hypertension caused by schistosomiasis which was reported by Bedford *et al.* (1946).

In addition to the medial changes that occur in the elastic arteries, these vessels usually show excessive and sometimes very severe intimal atheromatous change (Fig. 16.10), especially at points of division. Normally, a few small yellow flecks of atheroma are found in the larger elastic pulmonary arteries in persons over 40 years of age and are often associated with a slight increase in the calcium content of the arterial wall. In the presence of chronic pulmonary hypertension these changes become very much more severe and widespread and even appear in young children. Calcification of the largest elastic pulmonary arteries when very extensive may become visible in chest radiographs (Philp, 1972). "Ulceration" of the endothelial surface covering the plaques, although rare, can lead to overlying thrombus formation which may in turn give rise to emboli which become impacted in the smaller, distal branches of the pulmonary arteries.

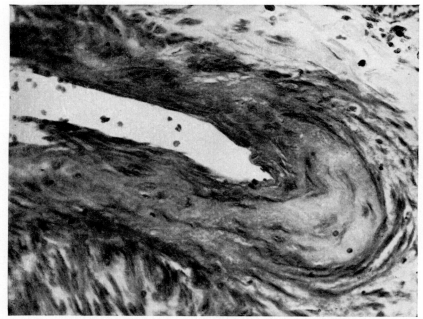

FIG. 16.9. A medium-sized pulmonary artery showing medial "mucoid" change in the wall. × 280 H and E.

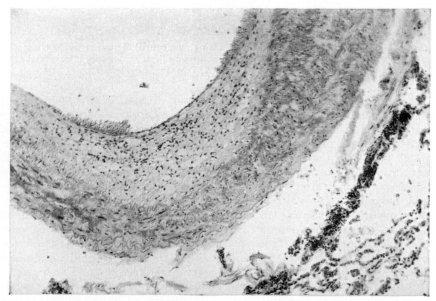

FIG. 16.10. A large elastic branch of the pulmonary artery from a case of primary pulmonary hypertension showing intimal atheroma and medial mucoid change. × 90 H and E.

(Reproduced by permission of the Editor of *J. Path. Bact.*)

Although microscopical examination confirms the presence of generalized medial hypertrophy throughout the elastic vessels, it may also show local patches of medial aplasia in the walls of both the large and small elastic arteries. The intima overlying the atrophic medial patches usually shows considerable fibrous thickening and this change was regarded wrongly by Gilmour and Evans (1946) and Spencer (1950) as the primary cause of the arterial obstruction. Both considered that the medial atrophy resulted from a congenital deficiency whereas it is now recognized that any lesion leading to thickening and fibrosis of the intima of an artery, i.e. atheroma, will *ipso facto* interfere with the nutrition of the underlying media causing it to atrophy. Furthermore, medial weakening tends to result in aneurysm formation, the reverse of arterial obstruction.

Two suggestions have been proferred in explanation of focal medial aplasia, firstly, that it results from organized mural thrombi caused by a turbulent blood-flow induced by the hypertension (Edwards, 1957), and secondly, that the medial deficiencies are residual scars following mural necrosis due to hypertensive pulmonary polyarteritis (Liebow, 1959b). Although the latter explanation probably accounts for medial scarring and damage in the muscular pulmonary arteries it is less likely to provide the explanation of the changes in the elastic arteries which are seldom affected by this condition.

Aschoff (1909) first stated that a turbulent blood-flow within blood-vessels causes mural thrombosis, and Harrison (1948) has shown that such thrombi in the pulmonary arteries become converted into segmental plaques of intimal fibro-elastoid tissue in the smaller branches. The presence of such intimal plaques splits the underlying part of the media and interferes with its nutrition leading to secondary atrophy of the muscle fibres (Figs. 16.10, 16.11). A similar change is frequently observed beneath the thick atheromatous plaques in systemic arteries and

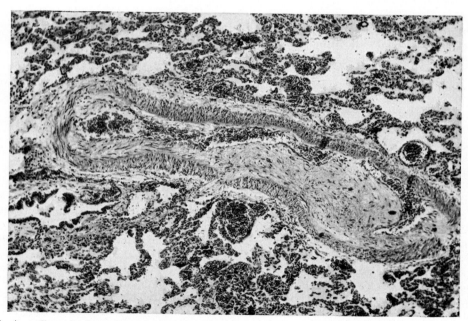

FIG. 16.11. A medium-sized muscular pulmonary artery showing an area of medial "hypoplasia" underlying a patch of intimal fibrosis. × 100 H and E.
(Reproduced by permission of the Editor of *J. Path. Bact.*)

beneath intimal plaques at the sites of branching of arteries forming the circle of Willis (Stehbens, 1959).

The localization of the pulmonary arterial medial deficiencies to the neighbourhood of vascular bifurcations is more readily explained on the basis of either local turbulence leading to local mural thrombosis or to the organization of small thrombotic emboli proceeding to intimal fibrosis.

CHANGES IN THE MUSCULAR PULMONARY ARTERIES

The muscular pulmonary arteries are the principal regulators of the pulmonary arterial blood-flow. Vasoconstriction of these vessels may be caused by anoxaemia, noradrenaline, 5-hydroxytryptamine (serotonin), as well as in response to an increase of the blood-flow above a critical level and in response to a gradual rise in pressure in the left atrium and main pulmonary veins. Rarely it occurs in response to an unknown cause in primary idiopathic pulmonary hypertension. In the first instance the vasoconstriction is probably initiated by either or both humoral and neural reflex mechanisms. Kay and Grover (1975) considered that mast cells played an important role in causing pulmonary arterial vasoconstriction in response to anoxaemia.

Vasoconstriction of the muscular pulmonary arteries leads to a rise of blood-pressure in the major pulmonary arteries and if maintained leads to structural changes in the muscular arteries themselves. The arterial lesions at first are of a reversible nature, but later irreversible structural changes result. Edwards (1957) divided the pulmonary vascular changes into three groups based on the reversibility or otherwise of the arterial changes. The groups were as follows: (a) a high-resistance/high-reserve type, (b) a high-resistance/low-reserve type, and (c) an intermediate type. By the term "low reserve" is meant a permanent state of restricted blood-flow due to irreversible structural changes in the smaller muscular arteries. By a "high reserve"

it is understood that although the blood-flow is restricted due to excessive vasoconstriction, the resulting structural changes, if any, are still potentially reversible and are capable of allowing a return to a normal flow. The importance of these subdivisions is reflected in the prognosis, as no operative measures can hope to fully restore the normal flow through vessels which are permanently sealed by irreversible structural changes.

The changes which occur in the muscular pulmonary arteries and arterioles in response to chronic pulmonary hypertension, when this is due entirely or in part to underlying pulmonary vasoconstriction, have been divided by Heath and Edwards (1958) into six grades. They classified each grade according to the changes found in the intima and media. In general, the grade of the arterial change corresponds to the severity and chronicity of the pulmonary hypertension. They are seen in their purest forms in primary (idiopathic) pulmonary hypertension, hyperkinetic pulmonary hypertension, and in a modified form in chronic passive pulmonary hypertension. The changes in chronic hypoxic pulmonary hypertension are mainly confined to the pulmonary arterioles and are restricted to grade 1 changes. Grades 1–3 more often correspond to the high-resistance/high-reserve group and grades 3–6 to the high-resistance/low-reserve category.

The six grades of lesions are as follows (from Heath and Edwards):

Grade 1. In this category there is medial hypertrophy in those arteries varying in diameter between 1000–100 μm and the development of the medial muscle extends into arterioles below 100 μm in size especially when chronic hypoxia is the underlying cause.

Grade 2. The lesions include those in the previous category together with intimal concentric fibrous proliferation in arteries less than 300 μm diameter (Fig. 16.12).

Grade 3. In addition to the changes found previously there is also included an extension of the intimal firobsis to arteries up to

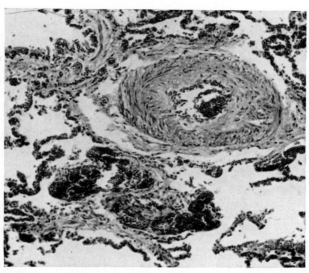

Fig. 16.12. A small muscular pulmonary artery showing medial hypertrophy and concentric intimal connective tissue proliferation. × 150 H and E.
(Reproduced by courtesy of the Editor of *J. Path. Bact.*)

500 μm diameter. The intimal fibro-elastosis may begin to undergo hyaline change due to ageing, and dilatation of the affected arteries may begin to occur when there is severe hypertension.

Grade 4. In this grade the arterial lesions are characterized by increasing dilatation of the affected arteries with in addition focal exaggerated dilatation. In addition in some of the larger muscular arteries increasing intimal thickening may lead to almost complete occlusion (Figs. 16.13, 16.14). Other characteristic lesions first seen at this stage include plexiform lesions (Figs. 16.15, 16.16), angiomatoids, and the appearance of vein-like arterial branches arising from greatly thickened and largely obstructed muscular pulmonary arteries.

Grade 5. In addition to all the previously described lesions this grade includes the changes due to pulmonary haemosiderosis.

Grade 6. In addition to all the previously described lesions this grade of lesion also includes pulmonary arteritis (pulmonary polyarteritis) (Figs. 16.17, 16.18).

The individual lesions included in the above grades of muscular arterial changes will now be described in greater detail.

In *grade 1* lesions medial hypertrophy is accompanied by slight thickening of the peri-adventitial connective tissue and prominence of the two elastic laminae. The muscular medial coat which normally ceases close to the junction of an arteriole with the parent muscular artery extends distally for some distance along the arteriole. This latter change is characteristic of chronic hypoxic pulmonary hypertension and most other changes may be absent. Although these changes occur during the earlier stages of all forms of pulmonary hypertension, they are often best seen in association with post-tricuspid congenital cardiac abnormalities such as a patent interventricular septal defect (IVSD). Post-natal development of an excessive pulmonary blood-flow leads to hypertrophy of the media throughout the pulmonary arterial tree. Wagenvoort *et al.* (1961a), however, have shown that for a brief period after birth the pulmonary arteries undergo the normal post-natal reduction in thickness but

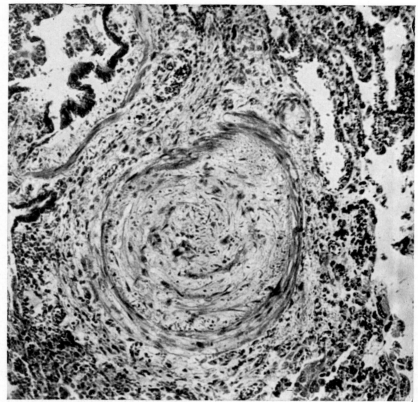

FIG. 16.13. The lumen of a small muscular pulmonary artery completely blocked by intimal connective tissue proliferation. × 150 H and E.
(Reproduced by permission of the Editor of *J. Path. Bact.*)

these changes are soon reversed in the case of an IVSD and are followed by hypertrophy of the muscle coat. In other forms of cardiac defect, notably atrial septal defects, the increased pulmonary blood-flow leads to hypertrophic changes even during foetal life which persist and increase in intensity after birth (Wagenvoort *et al.*, 1961b).

In *grades 2 and 3* lesions the previous changes continue to advance and in addition increasing intimal concentric fibrosis takes place. At first the intimal thickening consists of fibrous tissue alone but this is superseded by fibro-elastoid thickening and in time may lead to complete obliteration of the lumen of the smaller muscular arteries (Figs. 16.12, 16.13, 16.14). The most

striking obliterative changes affect segments of muscular arteries ranging in size from 300 to 40 μm in diameter. Such vessels may become completely occluded by concentrically arranged layers of loose fibrous and muscle tissue in the intima and they then bear a close resemblance to systemic small arteries in malignant hypertension. Later hyaline change may occur. The cause of the intimal change is uncertain but may be an exaggerated mural muscle hyperplasia which may follow excessive vasoconstriction of an artery in malignant hypertension. Hutchins and Ostrow (1973) regarded intimal fibro-elastosis as a sequel to mural tension causing arterial intimal layer separation with subsequent intimal

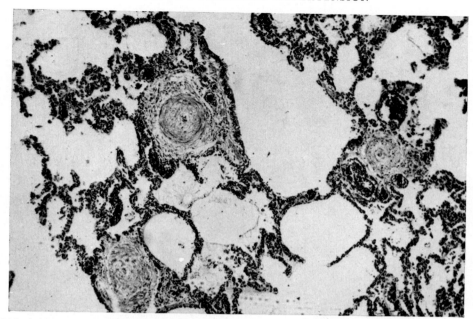

FIG 16.14. A low-power view showing three completely blocked muscular pulmonary arteries. × 90 H and E.

cellular proliferation. These obliterative changes when present are responsible for causing a high-resistance/low-reserve type of chronic pulmonary hypertension and are mainly found in idiopathic pulmonary hypertension. When fatal chronic pulmonary hypertension occurs in infancy due to any of the causes listed in the classification of chronic pulmonary hypertension, the pulmonary arterial changes are usually confined to medial hypertrophy and hyperplasia and permanent obliterative changes are less commonly seen than in adults.

Additional changes that may be associated with any of these grades of arterial lesions include the development of longitudinal muscle bands both inside the internal and outside the external elastic laminae of the muscular arteries (Figs. 16.19, 16.20). In the pulmonary arterioles in the chronic hypoxic group (less than 100 μm in diameter) the longitudinal muscle bands are restricted to the subendothelial plane. The presence of such longitudinal spiral muscle bands may be unconnected with hypertension but may be indicative of an increased longitu-

dinally applied stress to the vessel, because as Weibel (1960) showed, arteries subjected to intermittent excessive longitudinal stretch (i.e. in emphysema) develop longitudinal mural muscle bands. Turnbull (1914–15) considered the presence of such intimal muscle bands was due to hypertrophy of normally inconspicuous muscle fibres present in the subintima. That longitudinal stretching of the pulmonary arteries occurs is shown by the tortuosity of these vessels in many cases of chronic pulmonary hypertension which can be demonstrated by angiography after death (Fig. 16.21). In angiograms the affected arteries present a pig-tail appearance.

Grade 4 arterial lesions are characterized by the onset, in some cases only, of dilatation of the pulmonary arteries. Arterial dilatation may be regarded as the stage of vascular failure analogous to cardiac dilatation that follows initial cardiac hypertrophy. The dilatation of the arterial tree, though of general distribution, may be focally exaggerated giving rise to angiomatoid lesions. In other cases the arterial changes noted previously become exaggerated leading in some

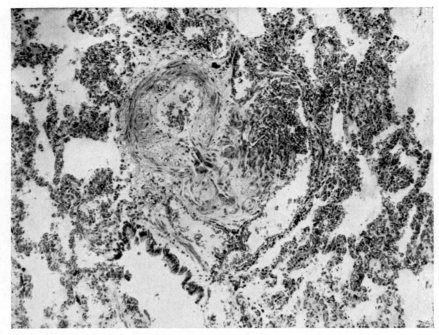

Fig. 16.15. A plexiform lesion showing greatly hypertrophied muscular pulmonary artery proximal to obstruction, with sinuous capillary channels traversing the obstruction which open into a dilated thin-walled continuation of the same vessel. × 140 H and E.

instances to obliteration of the lumen of small muscular arteries.

Plexiform lesions occur in small muscular branches immediately beyond their origin from greatly hypertrophied, parent, pulmonary muscular arteries. The parent arteries show in addition very extensive cellular intimal fibroelastosis. The media of the affected branch artery is lost, and the plexiform lesion consists of loose fibrous tissue continuous with the thickened intima. Within this tissue there is a mass of proliferating endothelial cells forming fine clefts connecting the arterial lumen proximally with the thin-walled, vein-like continuation of the vessel on the distal side. Some of the endothelium-lined channels lie within remnants of platelet thrombi. The whole lesion measures up to 700 μm in diameter. Many of the small sinuous endothelium-lined channels transmit blood and some of them connect through the damaged

arterial wall with capillaries in the neighbouring bronchial wall (Figs. 16.15, 16.16).

Plexiform lesions occur in severe pulmonary hypertension due to post-tricuspid causes and are rarely, if ever, found in association with chronic passive pulmonary hypertension and do not occur in chronic hypoxic hypertension. In the only two cases so far recorded of plexiform lesions in association with chronic passive pulmonary hypertension there have been other complicating lesions present such as a patent ductus arteriosus. Plexiform lesions are found most commonly in young children and are only rarely seen in infants (Wagenvoort, 1962). They were described by Shaw and Ghareeb (1938) in the lungs of patients dying of pulmonary schistosomiasis and have been found in patients dying of a variety of both pre- and post-tricuspid congenital cardiac and vascular defects including Eisenmenger's syndrome by Taussig (1947) and

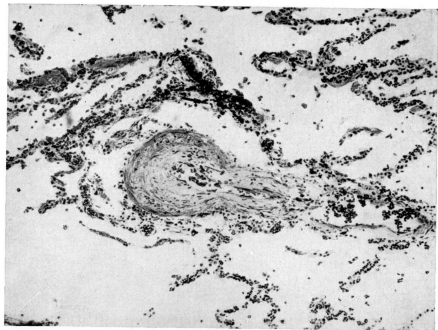

Fig. 16.16. A plexiform lesion showing the proximal hypertrophied segment of a muscular pulmonary artery and the obstructing mass containing thin-walled vascular channels and thin-walled continuation of the vessel beyond the obstruction. × 140 H and E.

Brewer and Heath (1959), in patent ductus arteriosus by Dammann *et al.* (1953), and in primary idiopathic pulmonary hypertension by Chiara (1950), Feuardent (1953) and Kuida *et al.* (1957). Although, as already stated, they have been described in association with mitral stenosis by Gordon *et al.* (1954) and Kanjuh *et al.* (1964), in both cases other congenital defects, including a patent ductus arteriosus, were present as well.

According to Liebow (1959), plexiform lesions only occur after pulmonary hypertension has become well established and they consist partly of the remnants of platelet thrombi filling the lumen and attached to the inner surface of arteries overlying local zones of medial necrosis. He concluded that the lesions were organizing thrombi which occluded previously damaged arteries and that the bronchial arteries played some part in the process of organization. Harrison (1958) found that plexiform lesions lay

entirely within the pulmonary arterial circulation and concluded that they were due to organized intravascular thrombi. A few of the capillaries forming these lesions drain through the damaged wall of the pulmonary artery to empty into pulmonary venules lying outside, but these communications never attain any functional importance. Such communications may be formed from venules draining repair tissue formed in the damaged segments of arterial wall overlying sites of platelet thrombus deposition. The platelet thrombi were judged by Kanjuh *et al.* (1964) to be caused by the fine jet of blood ejected under pressure through a narrowed branch orifice impinging on the endothelium of the branch vessels leading to platelet deposition.

The term *angiomatoid lesion* has been applied to a further type of vascular lesion found in the lung in advanced pulmonary hypertension. In this type of lesion thin-walled branch arteries arise from parent pulmonary muscular arteries

3

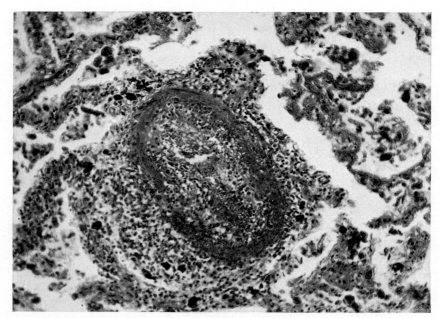

Fig. 16.17. A muscular branch of the pulmonary artery showing changes of pulmonary polyarteritis from a case of chronic mitral stenosis. × 140 H and E.
(Reproduced by permission of the Editor of *Brit. J. Tuberc.*)

showing severe hypertensive changes and expand to form cavernous sinusoidal vessels (Fig. 16.22). The cavernous sinus-like vessel may be subdivided internally by septa into several compartments through each of which blood flows eventually to drain into pulmonary capillaries. Although very similar to the sinusoidal thin-walled vessels that are found distal to plexiform lesions, no plexiform lesion is found in conjunction with an angiomatoid. This type of lesion may result from the venturi effect being an example of a post-stenotic dilatation lesion. The sinusoidal vessel contains elastic fibres in its walls and empties into the pulmonary capillaries.

In *grade 5* lesions the pulmonary muscular arteries show all the previously described changes but in addition both the dilated arteries and the thin sinusoidal vessels formed previously may rupture and give rise to focal haemorrhages and siderotic nodules (Figs. 16.23, 16.24). The latter, however, are usually encountered only in chronic passive pulmonary hypertension.

In *grade 6* the changes are distinguished by the presence of pulmonary hypertensive polyarteritis. In the past this condition was confused with the generalized disease of acute polyarteritis nodosa, but for the reasons stated below it is now regarded as a separate entity (Figs. 16.17, 16.18).

In acute systemic polyarteritis nodosa there is often a complete absence of hypertension and the disease is often accompanied by a blood eosinophilia and hyperglobulinaemia. Pulmonary hypertensive polyarteritis, however, is a localized disease confined to the pulmonary circulation, is always associated with a very high pulmonary arterial blood-pressure and is unassociated with any blood eosinophilia or hyperglobulinaemia. The blood-pressure in the systemic circulation is usually normal or below normal.

Pulmonary hypertensive polyarteritis was first described by Old and Russell (1950) in a case of

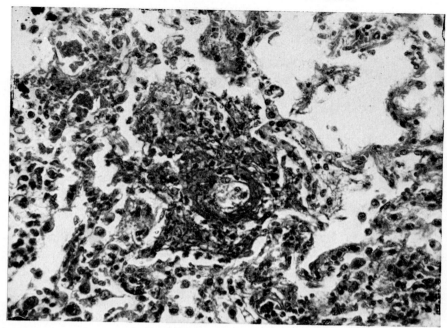

FIG. 16.18. A small muscular pulmonary artery showing fibrinoid changes due to pulmonary polyarteritis. It is not common for such small arteries to be involved. × 240 H and E.
(Reproduced by permission of the Editor of *Brit. J. Tuberc.*)

Eisenmenger's syndrome but since then further examples have been described by Spencer (1950 and 1957), Kipkie and Johnson (1951), Symmers (1952), McKeown (1952), Hicks (1953), Lopes de Faria (1954b) and Braunstein (1955). It has been found in a wide variety of conditions, but the common factor to them all has been the existence of a very high pulmonary arterial blood-pressure. The causes of the pulmonary hypertension have included mitral stenosis, massive cardiac fibrosis causing chronic left ventricular failure, multiple pulmonary emboli causing chronic pulmonary hypertension, primary pulmonary hypertension (idiopathic), conditions causing hyperkinetic pulmonary hypertension and pulmonary schistosomiasis.

Experimental proof that rapidly raising intra-arterial pressure can cause acute arteritis has been provided by Byrom (1954) and Wolfgarten and Magarey (1959). The former produced identical arterial lesions in rats during the experimental production of malignant hypertension, whilst the latter by forceful intra-arterial injections of saline into the carotid artery of rats caused similar changes. The initial reaction to a rapid rise of intra-arterial pressure is vasoconstriction which, if sustained, causes necrotic changes to appear later in the affected arteries.

Microscopically, the changes seen in pulmonary hypertensive polyarteritis are confined to the muscular arteries of all sizes and seldom involve the elastic arteries or pulmonary arterioles (Figs. 16.17, 16.18, 16.25). Often the fibrinoid change and most of the accompanying acute inflammatory cell reaction are confined to the intima and only involve the media in the smallest branches. The internal elastic lamina usually remains intact in the larger arteries and in these vessels there is little perivascular cellular reaction. Apart from these minor differences the changes closely resemble those of acute generalized polyarteritis nodosa.

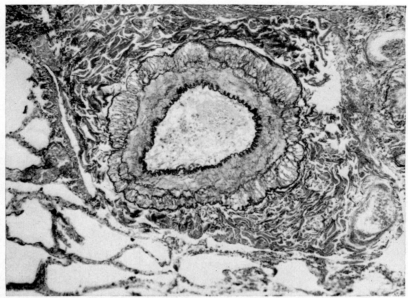

FIG. 16.19. A small muscular pulmonary artery showing considerable longitudinal muscle development outside the external elastic lamina in primary pulmonary hypertension. × 280 Verhoeff–Van Gieson.
(Reproduced by courtesy of Dr. J. Ball.)

Healing of the lesions may occur in subacute cases and often results, as in systemic polyarteritis nodosa, in segmental scarring of the whole thickness of the arterial wall and partial or complete destruction of the elastic laminae.

Among other pulmonary arterial changes that may often be encountered in chronic pulmonary hypertension is tortuosity of the arteries. Tortuosity of the pulmonary arteries, like tortuosity of the systemic arteries, occurs following fibrosis of the media in both elastic and muscular arteries. The change mainly involves the smaller elastic and muscular arteries which assume a corkscrew shape. The tortuosity is not usually seen in chronic passive pulmonary hypertension but may become well developed in the hyperkinetic group of pulmonary hypertension and in idiopathic primary pulmonary hypertension. It is best demonstrated in arteriograms taken either before or after death (Fig. 16.21).

In some cases of chronic pulmonary hypertension, particularly when this follows primary obstruction of the pulmonary arteries, the bronchial arteries may become enlarged and this may rarely lead to left ventricular enlargement if the volume of blood they transmit becomes considerable.

An additional method of examination that may prove of considerable value and interest in the elucidation of the pulmonary vascular changes in the different groups of pulmonary hypertension is post-mortem angiography. In this procedure a barium sulphate–gelatin mixture, warmed to body temperature, is injected into the pulmonary arteries at a pressure not exceeding the normal pulmonary systolic blood-pressure. The pulmonary arteries should first be washed out with lukewarm water to remove blood clots, and the lung to be injected should be immersed in warm water to maintain the fluidity of the injection medium. The resulting arterial patterns displayed by radiography differ according to the variety of chronic pulmonary hypertension. The different patterns of angiogram are described under the respective causes.

Hitherto only the vascular changes have been considered in connection with the pathology of chronic pulmonary hypertension. In some very

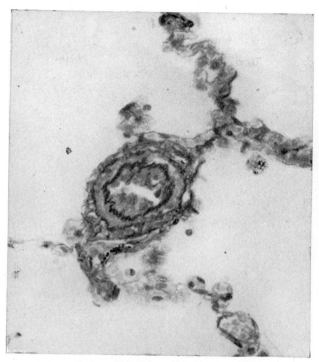

FIG. 16.20. A small muscular pulmonary artery showing intimal longitudinal muscle bundles and medial muscular hypertrophy. × 500 Verhoeff–Van Gieson.

(Reproduced by courtesy of Dr. J. Ball.)

advanced cases of obstruction to the pulmonary circulation the alveolar capillaries may become collapsed and ischaemic and this may lead to alveolar wall disruption.

In the later stages of all forms of chronic pulmonary hypertension the narrowing of the smaller pulmonary arteries leads to a reduction in the volume of the pulmonary blood-flow and a consequent increased tendency for autochthonous thrombosis to occur. Furthermore, the reduced cardiac output that inevitably ensues in many instances predisposes to systemic venous thrombosis with a consequent increased tendency for pulmonary embolism to occur. Pulmonary infarction is common and may follow the impaction of even quite small emboli in the pulmonary circulation. Reduction of the pulmonary blood-flow may become so excessive that it may result in transient attacks of cerebral

ischaemia (faints) and to Raynaud's syndrome.

Numerous attempts have been made to reproduce in animals the vascular lesions seen in human pulmonary hypertension. The results of these experiments show clearly that while the vascular changes parallel those seen in naturally occurring human disease, considerable individual variation in response occurs to an increased pulmonary blood-flow.

Heath *et al.* (1959 b) produced aorta-pulmonary anastomoses in dogs and showed that large anastomoses led to rapid death from cardiac failure, while smaller anastomoses were followed by survival and the development of chronic pulmonary hypertension. At the site of the anastomosis the media of the pulmonary artery became thinned and even totally destroyed. In some of the smaller muscular arteries plexiform lesions developed near to points of

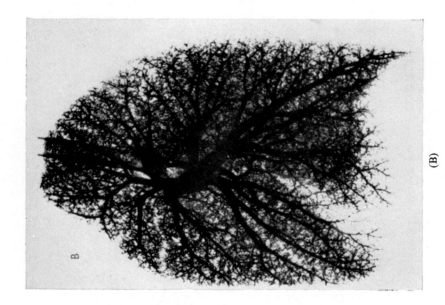

(B)

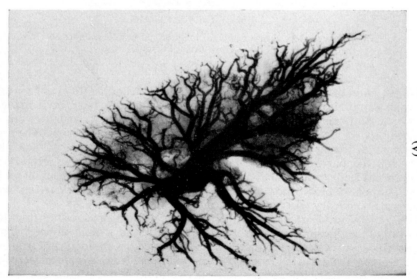

(A)

Fig. 16.21. (A) A pulmonary arteriogram from a case of primary pulmonary hypertension showing failure of the smaller branches to fill and increased tortuosity of the branches that have filled. (B) A normal pulmonary arteriogram. (Reproduced by courtesy of Drs. D. S. Short, W. Evans, and D. E. Bedford, and the Editor of *British Heart Journal*.)

branching, while others were occluded by intimal fibrosis. Dammann *et al.* (1957) anastomosed the left upper lobe branch of the pulmonary artery to the left subclavian artery in dogs. After initial dilatation, engorgement and haemorrhagic leakage from the pulmonary vessels, the pulmonary muscular arteries underwent hypertrophy and occlusion and some of the smaller branches developed plexiform lesions. In addition many of the alveoli became ischaemic and thin-walled.

Downing *et al.* (1963) also succeeded in reproducing plexiform and angiomatoid lesions in a dog by implanting the left lower lobe pulmonary artery into the thoracic aorta, after first removing the upper lobe on the same side. By this procedure the blood-flow through the surviving lobe of the left lung was increased about sixfold. Plexiform lesions occurred in the small, branch, muscular pulmonary arteries near their origins. The sequence of events leading to their formation was judged to be thrombosis caused by the local turbulent blood-flow which in turn led to damage of the underlying media and intima which became necrosed or atrophied. Subsequent revascularization of the damaged arterial wall and the superimposed thrombus followed the ingrowth of systemic capillaries and the formation of primitive angioblastic tissue, and resulted in the formation of a plexiform lesion. The same authors were also able to show that the bronchial arteries circumvented the obstruction in the pulmonary arteries.

The reversibility of pulmonary hypertensive arterial lesions has been proved in man. The pulmonary arterial changes induced by living at high altitudes may be reversed and the pulmonary blood-pressure restored to normal if the individual lives at sea level. Ramirez *et al.* (1968) similarly proved that pulmonary hypertension and the associated pulmonary arterial changes regressed following a successful valvuloplasty in mitral stenosis. Similarly, experimentally produced pulmonary hypertensive changes in both dogs and cattle may be reversed if the cause is removed as was shown by Ferguson and Varco (1955) and Wagenvoort *et al.* (1969), respectively. In some animals (dogs), however, there was less regression of the pulmonary vascular changes

and pulmonary hypertension. From these and other results it becomes evident that individual vascular response to a fall in intravascular pressure varies but it is improbable that the severe grades of arterial lesions (grades 3–6) are reversible to any great extent.

Lillehei (1958) stated that in children closure of congenital interventricular septal defects could result in the restoration of the pulmonary arterial blood-pressure to normal within 1–2 years even though the pulmonary arterial lesions were initially shown by lung biopsy to be of grade 3 severity. Intimal fibrous proliferation, even though of a severe grade, is a reversible change in young persons. According to Lillehei new capillary channels form in the intimal tissue and lead to recanalization of the obstructed lumen.

GROUP 1

Primary (Idiopathic) Pulmonary Hypertension (PPH)

Chronic pulmonary hypertension and cor pulmonale have long been recognized as sequels to many forms of lung damage, but in 1907 Mönckeberg showed that chronic pulmonary hypertension could also result from a primary arterial disorder confined to the lung. Since then an increasing number of such cases have been described and the term primary pulmonary hypertension has been applied to the condition. Unfortunately primary pulmonary hypertension was often loosely referred to in the past as Ayerza's disease, following a lecture delivered by Ayerza in Buenos Aires in 1901 in which he described a clinical syndrome characterized by cyanosis (cardiacos negros), dyspnoea and precordial pain and from which most of the patients died with right-sided heart failure. He gave no account of the underlying pathology and many of these cases were later wrongly attributed by Arrillaga (1913 and 1924) to pulmonary syphilitic endarteritis. It is probable that the majority of the cases described by Ayerza were examples of congenital heart diseases with associated pulmonary hypertension and cyanosis, while others may have been attributable to forms of obstructive lung disease (emphysema) and a few possibly to primary pulmonary hypertension.

An outbreak of unexplained pulmonary hypertension occurred in Switzerland, West Germany and Austria between 1965 and 1968 and was described by Gurtner *et al.* (1968) and Kaindl (1969). Its occurrence coincided with the release and use of a new slimming drug, aminorex(2-amino-5-phenyl oxazoline). Many of the

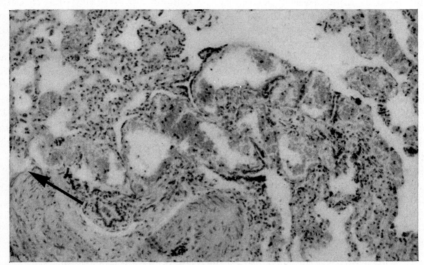

FIG. 16.22. An angiomatoid type of lesion showing a dilated sinusoidal branch artery arising from a stem vessel showing severe pulmonary hypertensive changes. There was no plexiform lesion present and the origin from the parent artery is arrowed. × 100 H and E.

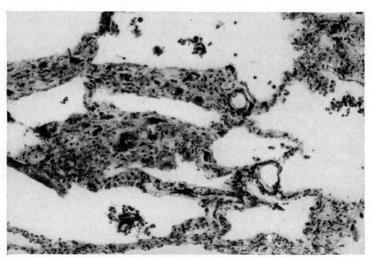

FIG. 16.23. A siderotic nodule in the lung. This condition is sometimes referred to erroneously as "endogenous pneumoconiosis". × 100 H and E.

patients were older than the usual age of victims of idiopathic primary pulmonary hypertension (PPH) and were taking the drug for obesity. The unexplained rise in incidence of PPH ceased when the drug was withdrawn in 1968 but all attempts failed to produce pulmonary hypertension with it in experimental animals. It is therefore still uncertain whether the drug was responsible for the vascular changes observed in man.

That drugs may produce pulmonary hypertension similar to PPH in man is supported by the experimental work of Valdivia *et al.* (1967) and Kay *et al.* (1967) both of whom showed that monocrotaline, a pyrolizidine alkaloid derived from the seeds of *Crotalaria spectabilis*, caused severe pulmonary hypertension in rats. Also extracts of fulvine derived from the plant *Crotalaria fulva* produces similar changes in rats (Wagenvoort *et al.*, 1974). The means whereby such drugs produce their effects is uncertain but they may act through their effects on the mast cells in the lung.

Primary pulmonary hypertension like its systemic counterpart essential hypertension may display a familial incidence (Dresdale *et al.*, 1954; Coleman *et al.*, 1959; Hood *et al.*, 1968). Not only may the condition occur among siblings, often twins, but it may affect several generations as shown by Melmon and Braunwald (1963). It is conveyed by a Mendelian dominant gene with variable penetrance (Porter *et al.*, 1967). It occurs more often in women and the majority of cases are discovered before the age of 35. It may occur in infants and very young children and the rate of onset and duration of the disease vary considerably. Death may ensue within 4 months of diagnosis but survival for 2 years or more is not uncommon. Sudden death may follow cardiac catheterization and during the induction of anaesthesia.

Clinically, the disease is characterized by increasing dyspnoea, precordial discomfort, syncopal attacks, anginal pain, occasional haemoptysis, vasospasm of the digital arteries and evidence of right heart failure.

In the early stages before the onset of recognizable structural changes in the pulmonary arteries, the pulmonary hypertension may be temporarily relieved by pulmonary vasodilator drugs such as acetyl choline and priscoline (Dresdale *et al.*, 1951 and 1954; Editorial, 1959) which result in a temporary increase in the cardiac output. Similar vasospasm affecting the pulmonary arteries may occasionally occur in rheumatoid disease and other "collagen" diseases (Walcott *et al.*, 1970).

As already stated primary pulmonary hypertension can occur very early in life (Berthrong and Cochran, 1955; Husson and Wyatt, 1959) and some of these cases were regarded by Goodale and Thomas (1954) as due to a persistence of the thickened and narrow pulmonary arteries found in the foetal lung. Confirmation of this view is provided by the fact that in some cases seen in infancy, the foetal structure of elastic fibres persists in the large pulmonary arteries. In older children suffering from this disease the large elastic pulmonary arteries show an adult pattern of elastic tissue, thus indicating that the disease was not congenital in nature, and started after the age of 4 years.

All grades of pulmonary arterial lesions occur in primary pulmonary hypertension and grades 3–6 lesions are not uncommonly found. A few fatal cases of this disease have been reported in which no structural changes beyond a slight medial hypertrophy were found (grades 1 and 2 lesions) (De Navasquez *et al.*, 1940; East, 1940).

Meyrick *et al.* (1974) investigated by electron microscopy what they claimed were the primary lesions responsible for PPH. They found that the smallest pulmonary arterial branches, those less than 140 μm in diameter, and the alveolar capillaries showed pinocytotic endothelial swelling. The same cells also showed an increased number of intracytoplasmic organelles (ribosomes and Golgi apparatus) and these changes were accompanied by swelling and thickening of the capillary basement membrane. As a result of these changes the capillary lumens were obstructed. The alveolar capillary pericytes proliferated and fine collagen and elastic fibres were laid down around and within the vessel wall which assumed a whorled onion-like appearance causing closure of the lumen. Such changes are

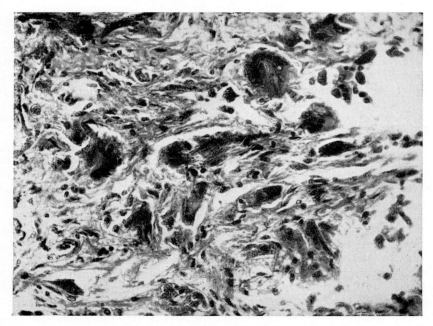

Fig. 16.24. A high-power view of siderotic nodule in lung showing foreign-body type of giant cell engulfing iron-encrusted elastic fibres and surrounding fibrosis. × 280 H and E.

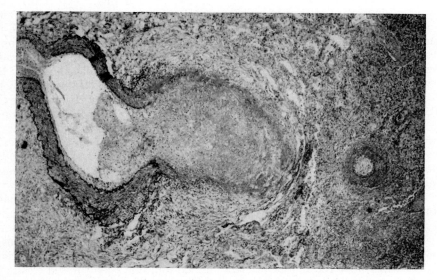

Fig. 16.25. Pulmonary arteritis and thrombosis affecting a muscular branch of a pulmonary artery due to hyperkinetic pulmonary hypertension caused by a patent interatrial septal defect. × 40 Verhoff–Van Gieson stain.

(Reproduced by courtesy of Dr. R. Salm.)

similar to those found in malignant hypertension in the systemic circulation and are responsible for the obliterative type of lesion shown in Fig. 16.13. Anderson *et al.* (1973) considered that the obliteration of the pulmonary arteriolar bed might be responsible for causing the pulmonary hypertension and the subsequent changes that occur in the more proximally situated arteries. That endothelial cells possess myofibrils and independent powers of contraction is well recognized but it is uncertain whether the changes described were responsible for or resulted from PPH. Both the pharmacological evidence and the pathology of the disease indicate that initially reduction of the pulmonary arterial blood-flow results from a functional vasoconstriction, a view which was supported by Kuida *et al.* (1957). Daly (1958) showed experimentally in animals that both vasoconstrictor and vasodilator nerve fibres supplied the pulmonary arteries, and Spencer and Leof (1964) have demonstrated the anatomical innervation of the pulmonary arterial tree in man.

The generalized character of the pulmonary arterial changes in primary pulmonary hypertension may be well displayed by post-mortem angiography. Evans *et al.* (1957) (Fig. 16.21) showed that in this disease the smaller muscular pulmonary arteries, which normally form a fine tracery of branches, completely failed to fill with the radiopaque medium. The difference in appearance between the abnormal and the normal arteriogram was likened to the difference between the leafless and pruned appearance of a shrub in winter and the same plant in summer. In addition to the occlusion of the finer branches of the pulmonary arteries, the same authors were also able to show that anastomoses formed between the bronchial and the pulmonary arterial systems.

The pulmonary arterial changes are those common to most forms of chronic pulmonary hypertension which have already been described. All grades of muscular arterial lesions are found in the lung and acute pulmonary polyarteritis may be found. The reader is referred to the preceding description of the pulmonary arterial lesions that occur in chronic pulmonary hypertension for a detailed account of these changes.

Primary pulmonary hypertension may be very difficult to differentiate clinically from pulmonary recurrent thromboembolic disease and pulmonary veno-occlusive disease. Dilatation lesions in the muscular pulmonary arteries (grade 4) as well as grades 5 and 6 lesions are encountered in PPH but are not usually found in thromboembolic disease and pulmonary veno-occlusive disease (Wagenvoort, 1964 and 1974). Thromboembolic vascular obstruction is usually more severe and widespread in the lower lobes of the lungs whereas in primary pulmonary hypertension the arterial lesions are more evenly distributed throughout all lobes of the lungs. Despite these differences, however, the not infrequent occurrence of autochthonous thrombosis in primary pulmonary hypertension may render the differentiation of these two conditions very difficult.

Hyperkinetic Pulmonary Hypertension

The pulmonary arteries and capillary bed are capable of accommodating treble the normal blood-flow without causing any rise in the pulmonary arterial pressure (Hickam and Cargill, 1948; West and Dollery, 1965). The increased blood-flow is accommodated partly by an increase in the radius of the pulmonary muscular arteries and partly by opening up of capillary pathways (reserve vessels) which under resting conditions are incompletely perfused. A further increase in the blood-flow when the limits of pulmonary vascular distensibility are reached leads to a rise in pulmonary arterial pressure. This type of secondary hypertension is referred to as hyperkinetic pulmonary hypertension and complicates a variety of forms of congenital heart disease.

All the conditions responsible for this form of pulmonary hypertension result initially in blood being shunted from the left side of the heart or aorta to the pulmonary circulation, and the pulmonary arteries consequently convey a greater volume of blood than the aorta. When the rate of flow exceeds the limit that can be conveyed by the pulmonary vessels without causing a rise in pressure, it is probable, though unproven, that vasoconstriction of the smaller muscular arteries occurs. This suggestion receives support from the observation made by Harris (1955) and Shepherd *et al.* (1959) that acetylcholine will temporarily reduce the ensuing pulmonary hypertension in the early stages, an effect enhanced by inhaling oxygen. The mechanism responsible for the vasoconstriction has not been determined but afferent impulses arising from the pulmonary arteries and conducted by the vagus were demonstrated by Walsh and Whitteridge (1945) in animals. As the resistance rises progressively in the pulmonary circulation, the pressure in the lesser circulation will equal that in the systemic circulation and a reversal of flow may result. This will cause deoxygenated blood to enter the systemic circulation and reduce the oxygen saturation of the arterial blood which in turn stimulates further pulmonary vasoconstriction. Later, secondary polycythaemia caused by prolonged arterial oxygen desaturation may still further raise pulmonary pressure by increasing the viscosity and volume of the blood.

In hyperkinetic pulmonary hypertension as opposed to passive pulmonary hypertension, the pathological changes in the pulmonary arteries affect all the branches throughout the lung and are not only restricted to those in the lower lobes (Heath and Best, 1958) as in chronic passive pulmonary hypertension.

The congenital cardiovascular abnormalities that may result in the development of hyperkinetic pulmonary hypertension are customarily divided into pre-tricuspid and post-tricuspid varieties. The only advantage to be gained from this subdivision is that pre-tricuspid abnormalities are less likely to develop pulmonary hypertension, and cyanosis is of late onset in this group. Pulmonary arterial changes of all grades of severity may be encountered in both varieties of lesions.

Among the congenital cardiovascular anomalies that may lead to the development of chronic pulmonary hypertension are the following:

Pre-tricuspid lesions
  (a) Interatrial septal defects (IASD).
  (b) Partial or total anomalous pulmonary venous drainage into the right atrium.

Post-tricuspid lesions
  (c) Interventricular septal defects (IVSD).
  (d) Eisenmenger's syndrome.
  (e) Patent ductus arteriosus (i) alone, (ii) with other defects.
  (f) Aorto-pulmonary shunts.
  (g) Transposition of the great vessels.
  (h) Persistent truncus.
  (i) Cor triloculare (bilateral atria with a single ventricle).
  (j) Surgical shunt operations for the relief of Fallot's tetralogy.

(a) Only about a quarter or less of all cases of interatrial septal defects (IASD) develop pulmonary hypertension and the condition has been reviewed by Massee (1947), Wood (1956) and Besterman (1961). Pulmonary hypertension may result from a greatly increased pulmonary blood-flow (hyperkinetic pulmonary hypertension) or it may be partly the result of increased mechanical obstruction of the pulmonary arteries brought about by secondary thrombosis.

Dexter (1955) showed that large interatrial septal defects resulted in blood passing from the left to the right atrium because of the greater distensibility of the latter chamber. This can lead, as was proved by McDonald *et al.* (1959), to an increase of over three times the normal pulmonary blood-flow causing the onset of hyperkinetic pulmonary hypertension. When this occurs a vicious cycle may become established leading to right ventricular failure, and to reversal of the interatrial shunt causing cyanosis and liability to the development of peripheral venous thrombosis with the consequent danger of pulmonary embolism. These changes predispose the pulmonary arteries to the development of thrombotic changes and three cases in which this complication occurred were described by Canada *et al.* (1953).

(b) Hickie *et al.* (1956) described thirteen cases with anomalous pulmonary venous drainage: in ten pulmonary hypertension developed. Uncomplicated anomalous pulmonary venous drainage does not usually result in pulmonary hypertension unless 50 per cent or more of the venous drainage is deflected to the right side of the heart.

(c) The occurrence of pulmonary hypertension in patients with an interventricular septal defect (IVSD) depends mainly on the size of the defect and partly on the existence of other cardiac deformities. The proportion of patients with an uncomplicated IVSD who develop pulmonary hypertension is still uncertain. Stanton and Fyler (1961) who followed a series of such patients found that only about 4 per cent ultimately developed pulmonary hypertension with advancing age but others have quoted a higher figure.

Brotmacher and Campbell (1958) analysed 175 cases of IVSD and found that when the defect was small (probably less than 0·5 cm) the normal ventricular pressure differences were maintained but became reduced with the increasing size of the defect. Large defects result in pulmonary hypertension which dates from birth. This is caused by the greatly increased pulmonary blood-flow which causes a persistence of the foetal pulmonary arterial constriction (Edwards, 1950). When smaller defects are present the pulmonary arterial blood-pressure first falls after birth, following the normal post-natal development and dilatation of the pulmonary arteries and their branches, but this may be followed by later development of pulmonary hypertension (Ferencz and Dammann, 1957). In such cases the moderate rise in the systemic blood-pressure which usually occurs about the time of puberty induces a left to right shunt of blood through the defect which may precipitate the onset of hyperkinetic pulmonary hypertension. The constriction of the pulmonary arteries and rise in pulmonary arterial pressure may cause a reversal of the shunt flow resulting in the onset of cyanosis which in turn may favour increased pulmonary vasoconstriction.

(d) Eisenmenger's syndrome (1897), as it was originally described, consisted of the following cardiac deformities: an interventricular septal defect, dextraposition of the aorta, dilatation of the pulmonary artery and hypertrophy of the right ventricle. As Hudson (1965) has stated, the changes in the pulmonary arterial circulation associated with Eisenmenger's syndrome may equally well occur as a result of a patent ductus arteriosus, a patent interventricular septum or an interatrial septal defect. These changes are due entirely to the increased volume of the pulmonary blood-flow which leads to hyperkinetic pulmonary hypertension. The increased resistance to the pulmonary arterial blood-flow in turn causes right ventricular hypertrophy and an equalization of the ventricular pressures which may cause shunting of blood from the right to the left side of the heart with consequent cyanosis. The pulmonary arterial changes in Eisenmenger's syndrome are identical with those found in other causes of hyperkinetic pulmonary hypertension and were described by Brewer and Heath (1959).

(e) Hyperkinetic pulmonary hypertension is an uncommon accompaniment of uncomplicated patent ductus arteriosus (DA) (Welch and Kinney, 1948), and depends initially on the size of the shunt and the additional volume of blood which flows through the lung. A patent DA complicated by pulmonary hypertension has been described by Cosh (1953), Hultgren et al. (1953) and Yu et al. (1954). This condition may cause death in young infants, and Heath et al. (1958) have attributed this to an unduly rapid transition from the thick-walled foetal pulmonary arteries to the thin-walled adult type of vessels resulting in fatal pulmonary oedema.

Edwards et al. (1949) described four cases of patent DA in association with coarctation of the aorta; in two of these the ductus joined the aorta proximal to the coarctation (type A), and in the other two below the coarctation (type B). In both instances pressure was raised in the pulmonary arteries.

In type A the pulmonary artery pressure equalled that in the proximal aorta, whereas in type B the right ventricle had been responsible from earliest life for maintaining a blood-flow to the distal aorta and lower limbs in addition to the lungs. After birth the vasomotor tonus of the peripheral limb vessels resulted in an increased blood-flow through the lungs; this in turn was responsible for a persistence and increase in the thickness of the medial coat of the pulmonary arteries that occurred after birth (Campbell and Hudson, 1951).

Hyperkinetic pulmonary hypertension may complicate aortic atresia unassociated with an IVSD. This anomaly may be combined with a persistent patent ductus arteriosus and Wagenvoort and Edwards (1965) collected twelve cases of this combined anomaly. Although many infants born with these abnormalities die at or soon after birth, a few may survive for 3 months or more, as in a case seen by the author. In such cases both the large and small branches of the pulmonary arteries show medial thickening and, less frequently, intimal fibrous thickening which in some cases antedated the birth of the child. The pulmonary hypertension results partly from exposure to excessive pulmonary blood-flow which is accentuated by chronic passive pulmonary hypertension resulting from the obstruction to the outflow of the left side of the heart.

(f and g) These congenital abnormalities are usually found in conjunction with other cardiac deformities. Aorto-pulmonary shunts are rare but occasionally the aorta and pulmonary artery may arise from the same ventricle (usually the right) and the resulting changes are similar to those occurring with septal defects in a common but incompletely divided truncus arteriosus. Kirklin et al. (1964) collected seventeen cases with a single ventricle in which nine had associated pulmonary stenosis and three an IVSD. All the patients developed pulmonary hypertension.

The author has recently seen severe hyperkinetic pulmonary hypertension complicating a case of complete transposition of the great vessels at the age of 3 months (Fig. 16.26). A similar case was also described by Moschcowitz and Strauss (1963). In both cases the elastic pulmonary arteries were greatly thickened due to intimal fibrosis and the muscular pulmonary arteries by medial hypertrophy.

(h) Christeller (1916–18) has shown that such patients may reach adult life with very little

disability and chronic pulmonary hypertension rarely occurs. A case with pulmonary hypertension was described by Heath and Best (1958) (Case 7).

(i) Two examples of pulmonary hypertension due to a trilocular heart (biatrial univentricular) with subaortic stenosis were reported by Edwards and Chamberlain (1951), both occurring in boys who survived to the ages of 6 and 8 years. The haemodynamic changes in this rare condition are almost identical with those found in Eisenmenger's syndrome. A further case was described by Chambers et al. (1961) in which the patient survived to the age of 41 as the pulmonary circulation was protected from an increased blood-flow by coincident pulmonary stenosis.

(j) Following the introduction of the earlier surgical measures to ameliorate the condition of children with Fallot's tetralogy, some of the patients later developed chronic pulmonary hypertension which ultimately proved fatal. Before the advent of the operation designed to repair directly the cardiac defects, it was customary to perform such pulmonary-systemic arterial by-pass operations as the Blalock–Taussig and Pott's procedures. In some instances these operations resulted in a striking but only temporary improvement in the patient's condition and after an interval, often of several years, the increased pulmonary blood-flow caused the onset of hyperkinetic pulmonary hypertension. The condition has been fully reviewed by Ferencz (1960b) and McGaff et al. (1962). Some patients died immediately after the operation due to the onset of acute pulmonary hypertension which caused rupture of the previously thinned pulmonary vessels. In these cases the adventitia of the medium and small pulmonary arteries became surrounded by blood which also suffused into the alveoli and filled the air passages. An example of this acute form of pulmonary hypertension accompanied by intrapulmonary haemorrhage is seen in Fig. 3.20. In patients who survived for a longer period the increased pulmonary blood-flow prevented the further occurrence of any pulmonary thrombosis which characterized advanced cases of untreated Fallot's tetralogy. It furthermore appeared to

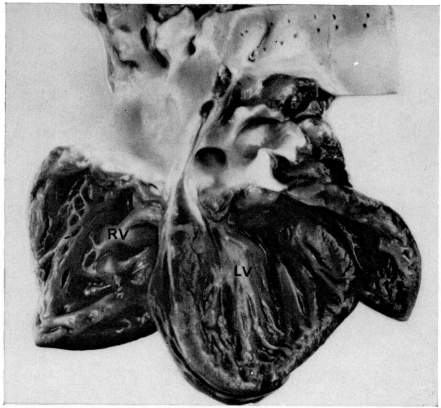

FIG. 16.26. Complete transposition of the great vessels. RV (the right ventricle) gives rise to the aortic arch, LV (the left ventricle) to the pulmonary arteries. Note the thickened pulmonary artery associated with hyperkinetic pulmonary hypertension.

lead to the dissolution of previously existing thrombi and some of the lesions resulting from them.

If the anastomosis proved eventually to be too large, pulmonary arterial changes occurred similar to those found in patients dying from other causes of hyperkinetic pulmonary hypertension. These changes included aneurysmal dilatation of the large elastic arteries, and arterial changes ranging from grades 1–3 in the muscular branches of the pulmonary arteries.

Microscopical changes found in the pulmonary arteries in chronic hyperkinetic pulmonary hypertension are similar to those already described and in most of the causative conditions all grades of arterial lesions may be

encountered. The changes in the large elastic pulmonary arteries were investigated by Heath *et al.* (1959) who found that the structure of the media was very closely related to the volume of the blood-flow passing through these vessels. In children with congenital cardiac disease causing a persistence of pulmonary hypertension after birth, the aortic pattern of elastic fibres normally found in the media of foetal pulmonary elastic arteries persisted. At the same time the medial muscle coat in both the elastic and muscular arteries underwent further hypertrophy and equalled in thickness the medial coat in the systemic arteries (Figs. 16.27, 16.28). The other changes are similar to those found in other forms of chronic pulmonary hypertension.

Chronic Pulmonary Hypertension due to Living at High Altitude

Synonyms: Monge's disease,
Chronic mountain
sickness

Chronic pulmonary hypertension may occur in people living at very high altitudes, usually above 12,000 feet (4000 metres) and less frequently at altitudes between 10,000 and 12,000 feet (3200–4000 metres). It results when the normal adaptive physiological changes found in persons living at very high altitudes fail and it was first described by Monge (1928).

There is considerable individual variation in the circulatory response to living at high altitude, and similar variability in adaptation is seen among different animal species transported to high altitudes. The indigenous animals to these regions such as the llama and vicuna are physiologically perfectly adapted and appear never to develop chronic mountain sickness.

Pulmonary hypertension sometimes develops in persons temporarily resident at altitudes as low as 11,000 feet (3600 metres) and it usually occurs between 5 months and 3 years after arrival. Pulmonary hypertension of this type was seen in some members of the Indian Armed Forces stationed for varying periods in the Himalayas and was described by Inder Singh *et al.* (1965). Studies of the effects of living at very high altitudes have mostly been made in the High Andes by Hurtado (1942), Penaloza and Sime (1971) and Arias-Stella *et al.* (1973) in addition to the original description by Monge after whom the disease was named.

Various types of chronic mountain sickness are now recognized (Arias-Stella *et al.*). Type 1 is seen in persons who move from a low altitude to live at a high altitude but never adapt fully to the change. A similar condition occurs in cattle moved for summer pasturage to high altitudes in the Rocky Mountains and is called brisket disease. Type 2 chronic mountain sickness occurs in persons who move from sea level to a high altitude and adapt satisfactorily but owing to the onset of some disorder which subsequently impairs their respiratory function, i.e. obesity,

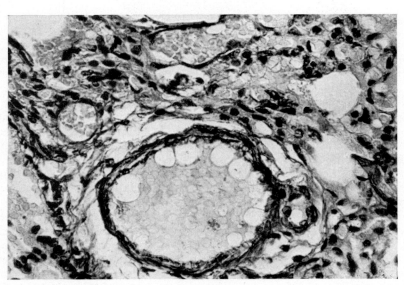

FIG. 16.27. A normal branch of the pulmonary artery of about 110 μm in diameter, from a child aged 3 months, showing an ill-formed internal elastic lamina and minimal medial muscle tissue. The external elastic lamina is not identifiable. × 450 stained with modified Sheridan's elastic stain and haematoxylin.

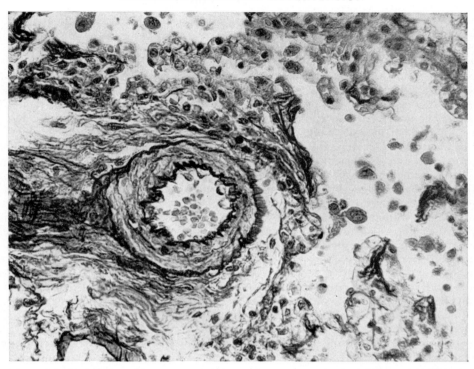

FIG. 16.28. A branch of the pulmonary artery about 100 μm in diameter from a case of patent ductus arteriosus in a child aged 3 months dying with severe pulmonary hypertension. The media is very hypertrophied and the internal elastic lamina is well developed. There are a few elastic fibres in the media itself. × 450 stained with modified Sheridan's elastic stain and haematoxylin.

kyphoscoliosis, emphysema, etc., they become hypoxic and develop chronic pulmonary hypertension. The third type (type 3) occurs in members of the indigenous population who occasionally develop chronic mountain sickness in adult life for reasons which are as yet uncertain but which Heath (1971) believes is precipitated because of the onset of centrilobular emphysema. Sorensen and Severinghaus (1968), however, attributed it to lessened sensitivity of the respiratory centre or chemoreceptor system acquired at an early age.

Rotta (1947) found that among the dwellers in the High Andes there was radiographic evidence of right heart enlargement and electrocardiographic evidence of right ventricular preponderance. Some of these persons showed a slightly decreased pulmonary circulation time and lowered systemic blood-pressure. Peñaloza et al. (1962a), in a series of cardiac catheterization studies in persons living above 14,900 feet in the Andes (4900 metres, barometric pressure 445 mm Hg), found that in a few people the pulmonary systolic and diastolic blood-pressures reached 41 and 15 mm Hg, respectively, with mean pressures at rest of 29 mm Hg rising on exercise to 60 mm Hg (normal 22/6 mm Hg, mean 12 mm rising to 18 mm Hg on exercise). The pulmonary wedge pressures were usually normal. In most people living in the same regions, however, the pulmonary blood-pressures lay within the normal ranges, thus demonstrating that considerable individual variation occurs in response to living at high altitudes. Below the age of 5 years the pulmonary blood-pressure in the high-altitude group was consistently higher than in similar children living at sea level, and the pulmonary blood pressures were often higher than in persons over the age of 5 living at the same altitude. Willerson *et al.* (1971) found the same results in a boy who was cardiac catheterized at the age of 2 in Denver (altitude 5200 feet, 1600 metres) because of a suspected interatrial defect, the pulmonary arterial pressure being found to be 75/35 mm Hg. Four years later when re-catheterized in Boston situated at sea level, the pressures were 38/7 mm Hg and the cardiac output had nearly doubled. They concluded that pulmonary hypertension developed prematurely in cases of atrial septal defect living at high altitudes. Similar pulmonary hypertension and accompanying vascular changes developed in a group of cattle kept at Denver in which one pulmonary artery was ligated. When the animals were subsequently removed to Houston at sea level both the pulmonary hypertension and accompanying pulmonary arterial changes diminished, indicating that pulmonary vasoconstriction in the patent vessels was primarily responsible for causing the changes (Wagenvoort *et al.*, 1969).

The high pulmonary blood-pressures recorded during the early years of life are probably due to persistence of foetal pulmonary vasoconstriction and of the foetal structure of the pulmonary arteries during early childhood. Despite the raised pulmonary arterial pressures found in normal children below the age of 5, their resting cardiac outputs and heart rates remain normal or may be raised (Theilen *et al.*, 1955).

Among the other normal adaptive changes found in high-altitude dwellers are polycythaemia and an increase in the plasma volume, together with an increase in the maximum breathing capacity due partly to the reduced muscular effort required to inhale a more rarefied atmosphere. The chest consequently becomes more barrel-shaped and the excessive removal of alveolar CO_2 tends to cause a respiratory alkalosis with subsequent excretion of an alkaline urine.

Post-mortem studies of infants and adults resident and dying at very high altitudes have been described by Fernan-Zegarra and Lazo-

Toboada (1961), Arias-Stella and Recavarren (1962), Arias-Stella and Saldana (1963), Reategui-Lopez (1969) and Arias-Stella (1971). Newborn and stillborn infants are found to have consistently heavier right ventricles than similar infants born at sea level. Furthermore, the right ventricle remains persistently heavier than normal in older children and adults and in some instances becomes heavier than the left ventricle of the heart. The basal region of the right ventricle is principally involved in the hypertrophic changes. The right ventricular hypertrophy results from changes in the small pulmonary muscular arteries and arterioles, mainly vessels less than 140 μm in diameter. The changes are identical with those described by Haselton *et al.* (1968) which are found in any chronic hypoxic state. They include the appearance of medial muscle in the pulmonary arterioles which is normally absent from these vessels. In addition, spiral longitudinal muscle bands are found in the intima of the smallest muscular pulmonary arteries and arterioles caused probably by the axial stretch applied to these vessels by the hyperinflation of the lung. Recent and organizing thrombi may be present at all levels in the pulmonary arterial tree (Fig. 14.19).

Persistence of pulmonary vasoconstriction and consequently of the foetal type of muscular pulmonary arteries into adult life accounts for the high pulmonary blood-pressures recorded in infancy. Such high pulmonary pressures lead to a greater likelihood of a persistent ductus arteriosus and the incidence of this type of congenital abnormality is greater in communities living at high altitudes than among those living at sea level. Other changes found in those who live at a high altitude include an increase in the size of the carotid bodies, a change common to other chronic hypoxic states, i.e. emphysema. The increased blood volume that accompanies the polycythaemia is mainly accommodated in the lung capillary bed as was shown by Campos and Iglesias (1957). Administration of oxygen to persons with chronic high-altitude pulmonary hypertension fails to reduce the hypertension. The hypertension is maintained by the arterial structural changes in the lung which are only slowly reversible and which lead to constriction of the smaller muscular pulmonary arteries and arterioles. Permanent residence at a lower altitude, however, usually leads to a gradual fall in the pulmonary blood-pressure due to reversal of the arterial changes in the lung. Inder Singh *et al.* found in a fatal case that thrombosis had occurred in many of the smaller branches of the pulmonary arteries and he believed that this was largely responsible for the pulmonary hypertension. It had resulted from sludging of intra-capillary red blood-cells and the formation of fibrin thrombi in alveolar capillaries. They suggested that the natural fibrinolytic system in the body had become disordered, but this as yet remains unproven.

Although some aetiological factors concerned in the causation of high-altitude pulmonary hypertension are still matters of debate, hypoxia probably plays the dominant role, a finding first demonstrated in animals by von Euler and Liljestrand (1946) and confirmed by Duke (1957). Both investigators showed that hypoxia is a powerful stimulus to pulmonary arteriolar vasoconstriction, which may be mediated by mast cells.

The changes described above are those found in persons long adapted to living at a high altitude and should be distinguished from the more dramatic changes that may sometimes befall persons newly arrived at a high altitude. In the latter group severe and often fatal pulmonary oedema may occur and is described in Chapter 17.

Chronic Passive Pulmonary Hypertension (Post-capillary Resistance Group)

Chronic passive pulmonary hypertension follows a *gradual* rise in pressure in the pulmonary veins. This in turn is caused by an incomplete obstruction of the venous outflow. Sudden complete blockage of the pulmonary veins or any cause of a sudden rise in the pulmonary venous pressure leads to pulmonary oedema when the intravenous pressure exceeds plasma osmotic pressure. This occurs particularly in sudden left ventricular failure from any cause which if severe results in extensive pulmonary oedema.

Gradual obstruction leading to complete blockage of a pulmonary vein causes the establishment of an abundant collateral venous circulation, and the intravenous pressure usually remains within normal limits.

Included among the causes of chronic passive pulmonary hypertension are:

(a) A chronically failing left ventricle.
(b) Chronic mitral stenosis.
(c) Congenital incomplete atresia of the pulmonary vein and cor triatrium.
(d) Incomplete obstruction of the pulmonary vein by growth or granuloma.
(e) Intermittent obstruction of the mitral valve orifice by a myxomatous tumour arising in the left atrium.
(f) Intermittent obstruction of the mitral valve by a ball-valve thrombus.
(g) Idiopathic thrombosis of the pulmonary veins.

As all these conditions may cause similar changes in the lungs, a detailed account of the vascular and other lesions found in mitral stenosis is applicable to them all to a greater or lesser degree. In passive pulmonary hypertension, changes are not only confined to the pulmonary arteries but also involve the capillaries, veins and lymphatics in the lungs, and it is with these latter changes that the ensuing account is mainly concerned.

The pulmonary arterial lesions are similar in most respects to those found in grades 1–3 previously described. There is medial hypertrophy in the smaller pulmonary arteries together with concentric and sometimes eccentric intimal fibrosis, the latter probably resulting from an organized thromboembolus.

EXPERIMENTAL PATHOLOGY

In acute experiments, Lasser and Loewe (1954) have shown that passive pulmonary hypertension maintains a direct relationship to left atrial pressure. Sarnoff and Berglund (1952a) and Guyton and Lindsey (1959), however, found that the pulmonary veins and the left atrium in dogs could distend to accommodate a large volume of blood with only slight rise in pressure proximal to a stenosed mitral valve. When the elastic limit of these structures was reached the pulmonary venous pressure rose abruptly following any further increase in volume, and pulmonary oedema occurred when the plasma osmotic pressure was exceeded. Borst *et al.* (1956) have also shown that within certain physiological limits, increased left atrial pressure may actually result in pulmonary capillary dilatation, although this stage is seldom seen in patients presenting with clinical evidence of mitral stenosis.

The diminishing output of blood from the venous side of the lungs in chronic pulmonary venous congestion would, in the absence of any compensating mechanism, lead to a rise in the capillary pressure and in turn to fatal pulmonary oedema. However, Sanger *et al.* (1959) have been able to show that in dogs raising the pulmonary venous pressure leads to an immediate constriction of the pulmonary arteries if the vagus nerves are intact. In the majority of the animals the pulmonary pressure rose but the cardiac output was not altered appreciably. The site of the baroceptor nerve endings in the pulmonary veins and the left atrium responsible for this reflex has not yet been established. Furthermore, in man the pulmonary veins are relatively indistensible structures and are incapable of expanding beyond a very minor degree when the intraluminal pressure rises.

That an intimate and very localized reflex exists in man is shown by the fact that if one pulmonary vein escapes obstruction by a mediastinal tumour, structural changes compatible with chronic vasoconstriction only occur in those branches of the pulmonary artery which supply the portion of lung drained by the obstructed veins, but are not found in that part of the lung drained by the unobstructed vein (Edwards and Burchell, 1951).

Chronic Left Ventricular Failure

Many causes of gradually increasing left ventricular failure result in pulmonary hyper-tension. As the vascular changes which ensue in the lung are similar to those found in chronic mitral stenosis no separate description will be given. Among the causes of chronic left ventricular failure that may cause pulmonary hypertension are: aortic stenosis, essential hypertension, severe myocardial fibrosis due to coronary artery occlusion, Fiedler's myocarditis (Heath *et al.*, 1957), endomyocardial fibroelastosis (Bedford and Konstam, 1946), sclerodermatous myocarditis, and cardiac amyloidosis. The decreasing cardiac output from the left ventricle results in a gradual rise of left atrial pressure with resulting secondary reflex pulmonary arterial changes similar to those seen in chronic mitral stenosis and described by Smith *et al.* (1954).

Mitral Stenosis

The term "brown indurated lung" has been given to the collective changes that occur in the lungs in chronic passive pulmonary hypertension. Such lungs retain their shape on removal and in patients who received adequate treatment oedema is usually absent or confined mainly to the upper lobes. Unlike normal lungs, the indurated lung has a reddish-brown tinge and brick-red flecks are often visible beneath the pleural surfaces. Infarcts, mainly confined to the lower lobes, may be found in all stages of development. When the lungs are compressed or cut they have a firmer texture than normal; the cut surfaces are often dry and brick-red in colour and thick pouting branches of the pulmonary arteries project from the surface. Scattered segments in the lower lobes of the lungs may be blanched and anaemic. In these areas the alveolar tissue is of normal texture but perivascular fibrosis may be apparent naked-eye. The main branches of the pulmonary artery are often visibly distended and thickened and show excessive atheromatous fatty flecking.

The main bronchi, in addition to congestion of the mucosa, may show conspicuous bronchial varices (Fig. 16.36) which rapidly collapse and empty when the lung is cut or detached from the other thoracic viscera.

VASCULAR CHANGES

Angiographic Studies

The pulmonary arteries have been studied *in vivo* by Goodwin *et al.* (1952), who carried out angiographic studies using 70 per cent diodone. The severity of the mitral disease was graded according to the New York Heart Association's Classification of 1939. In grade 1 lesions no alteration in the calibre of the pulmonary arterial tree was observed. In grades 2 and 3 there was dilatation of the main and lobar arteries followed by premature narrowing of the branches situated more distally. The segmental branches were irregular and tortuous and the shadows suggested obstruction caused either by vasoconstriction or organic obstruction. In the most severe cases the narrowing spreads to involve the major arteries themselves, and the circulation time through the lungs was prolonged in proportion to the severity of the occlusive changes. In early cases the peripheral arterial narrowing can be temporarily abolished by acetylcholine and the pulmonary hypertension reduced (Wood *et al.*, 1957), thus proving that it is at least partly due to functional vasoconstriction.

Charms *et al.* (1959) have also shown that temporarily blocking one pulmonary artery in patients with mitral stenosis known to have pulmonary hypertension results in no further rise in pressure. They attributed the failure of the pressure to rise still further in the non-occluded lung to the opening up of the pulmonary capillaries following the release of pulmonary arterial tonus in response to raised blood-flow. Their experiments showed that pulmonary arterial vasomotor tonus was an important factor in causing pulmonary hypertension even in well-established cases of mitral stenosis.

In mitral stenosis the critical pressure at which the alveolar capillaries open is raised and only when one pulmonary artery is blocked, causing a release of pulmonary arterial vasomotor tonus and increased blood-flow, is the critical pressure exceeded and a normal alveolar-capillary flow resumed. The cause of the raised critical opening pressure of the alveolar capillaries is unknown but may be due to increased bronchomotor tonus which causes a raised intra-alveolar air pressure and collapse of the capillaries.

Post-mortem angiographic studies on the intact lungs show narrowing of the smaller arteries, mainly localized to the dependent parts of the lungs (Fig. 16.29). Such angiographic studies have been reported by Short (1956) and Harrison (1958). In severe and chronic cases of mitral stenosis the main lobar and the commencement of the segmental arteries are usually dilated, whereas the distal parts of the segmental arteries and the branches beyond are narrowed, also the fine terminal branches often fail to fill completely. Similar pulmonary angiographic changes were found in living patients by Aber *et al.* (1963) who showed that the size of the pulmonary artery shadow could be correlated with the degree of intimal thickening in the smaller branches and with the level of the pulmonary blood-pressure. The larger the main pulmonary arterial shadows the greater was the pulmonary arterial pressure. In both the living and the dead lung these abnormalities are mainly confined to the lower halves of the lungs. The predominant localization of the hypertensive arterial and pulmonary arterial occlusive changes to the lower lobes of the lungs is attributed to the additional increment of venous pressure in the pulmonary veins draining these lobes caused by the weight of the column of blood extending to the heart. This is absent in the erect position in the veins draining the upper lobes. This small additional increment of pressure in the pulmonary veins draining the lower lobes is considered to raise the intravenous pressure above the critical level required to cause reflex arterial vasoconstriction (Harrison, 1958–9). The ensuing reduction in the blood-flow through the lower lobes of the lungs is followed by an increase of up to five times in the flow through the upper lobes of both lungs, and the blood-flow through the upper lobes may equal that through the normal lower lobes. These changes were demonstrated by Dollery and Hugh-Jones (1963) using radioactive isotopes to measure regional vascular perfusion and regional ventilation, the gas samples used to measure the latter being obtained through a bronchoscope. The increase in the regional

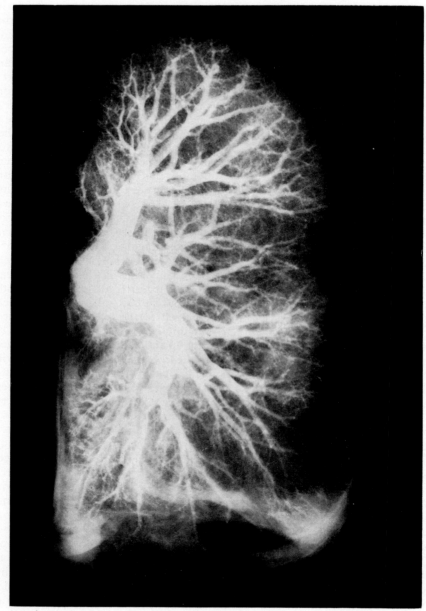

FIG. 16.29. An angiogram of the pulmonary arteries in a case of very severe and chronic mitral stenosis made shortly after death. For explanation see text.

blood-flow in the upper parts of the lungs in the less severe cases of mitral stenosis follows a lesser increase in the pulmonary venous pressure. West *et al.* (1964 and 1965) suggested, from the results of their experimental work in dogs, that the reduction of the pulmonary blood-flow seen in the lower lobes of the lungs was due to chronic perivascular oedema resulting from the raised pulmonary venous pressure. Their experimental work was repeated by Ritchie *et al.* (1969) who were unable to confirm their findings and conclusions which in any case would not explain the medial hypertrophy observed in the pulmonary muscular arteries in the lower lobes of the lungs in chronic mitral stenosis.

The *major elastic pulmonary* arteries are dilated in most long-standing cases and the size of these vessels is proportional to the intraluminal blood-pressure. Meyer and Richter (1956) found that in young persons both the circumference and weight of these vessels were appreciably increased and that the increase was related to the degree of hypertension. The structural changes in the major arteries are similar to those already described in primary idiopathic pulmonary hypertension.

The *smaller muscular pulmonary arteries.* The changes in these vessels are similar to those already described but they seldom progress beyond the changes found in stage 3. Concentric intimal fibroelastosis in addition to medial hypertrophy is characteristically found in established cases (Wagenvoort, 1973). The pulmonary arterial changes are mainly localized to and are far more severe in the lower lobes of the lungs (Fig. 16.30). MacKinnon *et al.* (1956) found that occasionally the pulmonary arterial pressure became so great that clinically the case might simulate one of primary pulmonary hypertension. Later mural hyaline change often supervenes. In cases with exceptionally high levels of pulmonary hypertension, pulmonary polyarteritis may occur, and a condition sometimes referred to as fibrinous vasculosis occurs in which

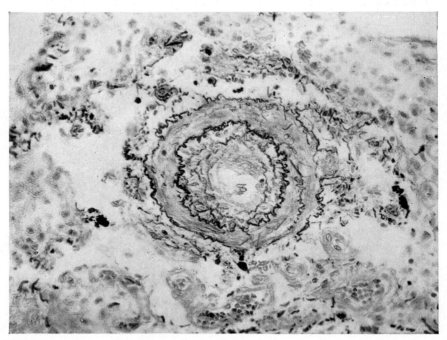

FIG. 16.30. Transverse section of a muscular artery from a lower lobe showing medial thickening and fibro-elastoid thickening of the intima. × 280 Verhoeff–Van Gieson.

plasma suffuses through the vessel wall. In many cases the arterial obstruction in the later stages of mitral stenosis is further increased by the presence of organizing thrombotic emboli and autochthonous thrombi. Histological studies of the vasculature have been made both in biopsy material obtained *in vivo* and in post-mortem tissues by Parker and Weiss (1936), Larrabee *et al.* (1949), Henry (1952) and Heath and Whitaker (1955). Heath and Best (1958) showed, however, that biopsy material obtained from the lingula segment is unsuitable for assessing pulmonary arterial changes as these vessels normally contain more medial muscle in their walls than vessels elsewhere in the lung, possibly due to their close proximity to the hilum.

In cases of mitral stenosis associated with aneurysmal dilatation of the left atrium there may be no pulmonary hypertension and very few structural changes occur in the pulmonary arteries (Best and Heath, 1964).

The *capillaries* may show a variety of abnormalities in addition to the arterial changes.

In normal alveolar capillaries only about two red cells flow abreast, but in some cases of mitral stenosis the lumen of capillaries in the upper lobes may be greatly widened (Fig. 16.31). This possibly reduces the efficiency of the oxygenation of those red cells in the centre of the capillary blood-stream. This change may represent the initial stage of capillary dilatation which is seldom seen in more chronic and advanced cases. The capillary dilatation together with the dilatation of the large elastic pulmonary arteries account for the increased blood-volume that was shown to be accommodated within the lungs of patients with mitral stenosis by McNeill *et al.* (1958) and Bates *et al.* (1960).

Aneurysmal dilatations of the capillaries were described by Weiss and Parker, but these were partly artefacts due to the plane of sectioning. Other changes induced by chronic alveolar oedema include thickening of the capillary basement membrane, the development of pericapillary reticulin and collagen fibres together with epithelialization of the related alveolar walls

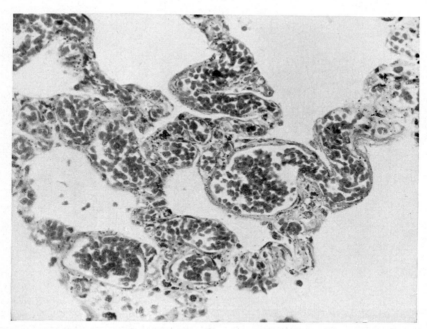

FIG. 16.31. Alveolar capillaries showing marked dilatation and aneurysmal bulging from the upper lobe of patient with mitral stenosis. × 140 H and E.

(Figs. 16.32, 16.33, 16.34). The association of chronic tissue oedema and the subsequent formation of fibrous tissue was described by Eppinger (1949), and this change in the case of the alveolar wall tends to restrict the development of further oedema (Hayward, 1955).

The ultramicroscopic changes in the alveolar-capillary wall in mitral stenosis were studied by Kay and Edwards (1973) who showed that the alveolar interstitial fibrosis occurred mainly in the normally wider, collagen-containing portions of the alveolar interstitium where it led to displacement of some of the alveolar capillaries. Collagen fibres were not usually formed in the thinnest portions of the alveolar walls where the alveolar endothelial and epithelial basement membranes are normally fused together. In these regions, however, the fused basement membranes were greatly thickened due to oedema and there was severe intracellular oedema of the alveolar capillary endothelial cells. Fragments of disintegrating, extravasated,

red blood-cells were found in the oedematous fused basement membranes, and the overlying alveolar epithelium underwent a metaplasia to the type 2 form of pneumocytes.

Though much of the alveolar fibrosis undoubtedly results directly from chronic oedema, Moolten (1962) showed that fibrosis may also result from inflammatory damage in the lung (Fig. 16.35). The chronic lymphoedema in the lung predisposes it to bacterial and viral infections and furthermore helps to disseminate such infections. The resulting fibrosis develops especially around the smaller bronchi and extends into the alveolar interstitial plane. The narrowing, obstruction and thickening of the capillaries resulting from these changes can cause an alveolar capillary block to oxygen diffusion (Carroll et al., 1953) causing a fall in the arterial oxygen saturation in advanced mitral stenosis. The capillary turgidity and particularly the increased fibre content of the alveolar wall together cause the induration of the lung and are partly responsible

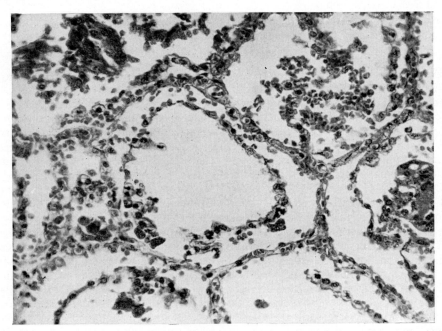

FIG. 16.32. Alveoli in a case of mitral stenosis showing thickening and epithelialization of the walls. Although a space now exists between the epithelium and the alveolar capillary this was formerly occupied by oedema fluid. × 140 H and E.

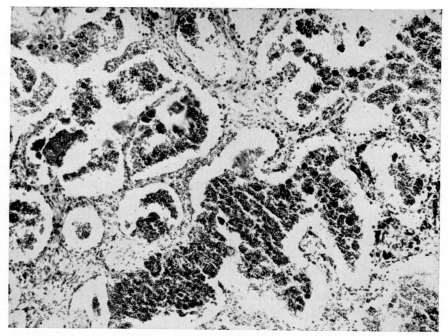

FIG. 16.33. Fibrous thickening of the alveolar walls and collections of haemosiderin-laden phagocytes in alveolar lumens from a case of advanced mitral stenosis. × 140 H and E.

for the reduction in lung compliance. Holling and Venner (1956) considered that the excessive lymph formation and pulmonary hypertension *per se* also played a considerable part in the alteration of the lung compliance.

The *pulmonary veins* though less strikingly affected than the pulmonary arteries may nevertheless undergo medial hypertrophy and intimal fibrosis at all levels (Fig. 16.37). The medial hypertrophy is less conspicuous than in comparably sized arteries because the muscle tissue is normally interspersed with collagen and elastic tissue, and is less conspicuous. Intimal fibrosis and oedema may in some cases become very severe and this occurs especially in the condition of a mediastinal granuloma resulting in very severe pulmonary venous obstruction. De Bettencourt *et al.* (1953) have shown by tomography that pulmonary veins are capable of active contraction when the pressure within them is raised. Also Burch and Romney (1954)

referred to a "pulmonary venous throttle mechanism" caused by the contraction of the sleeve of left atrial muscle which ensheaths the central ends of the pulmonary veins, extending for a distance of up to 15 mm from the atrial wall.

It has long been known that chronic mitral stenosis complicated by an interatrial septal defect (Lutembacher's syndrome) carries a better prognosis than uncomplicated mitral stenosis. Bronchial varices, which have been fully described by Ferguson *et al.* (1944) and Bland and Sweet (1949), may function in a similar manner to an atrial septal defect. Communications normally exist between the bronchial and pulmonary venous systems in the walls of the main and lobar bronchi (Miller, 1947) and if these dilate, a pathway is provided for blood to return from the lung parenchyma through the azygos and hemi-azygos veins to the right atrium. The greatly increased volume of blood carried by the bronchial veins leads to the formation of bronchial varices which

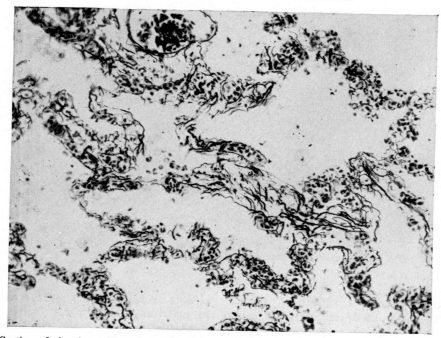

FIG. 16.34. Section of alveolar wall to show the increase in reticulin fibres that occurs in advanced mitral stenosis. × 280 stained with modified Foot's reticulin stain.

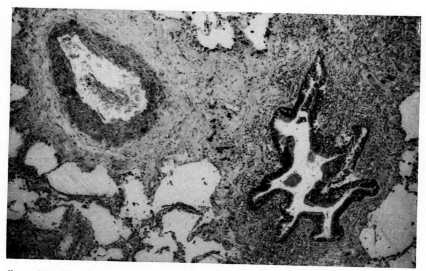

FIG. 16.35. Peribronchial fibrosis and chronic inflammation in the lung from a patient with chronic mitral stenosis. The fibrosis probably resulted very largely from previous inflammation of the air passage.

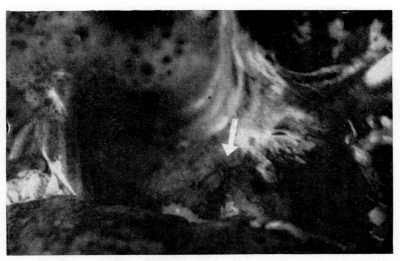

Fig. 16.36. A lobar bronchus opened to show a leash of bronchial varices indicated by the arrow. × 4.

may rupture, causing extensive haemoptysis (Fig. 16.36). The return of part of the pulmonary venous blood to the right atrium adds a further burden to the right side of the heart and its output will then exceed that of the left (Nakamura, 1958).

Inadequately treated patients suffering from mitral stenosis may die with extensive pulmonary oedema due in part to the bottle-neck formed by the *pulmonary lymphatics*. These can drain up to several times their normal flow, but beyond their limits of expansion fluid accumulates first in the wider parts of the alveolar interstitium and later may begin to accumulate in the alveolar lumens. Lymphatic dilatation is stated to occur when the pulmonary arterial blood-pressure exceeds 32 mm Hg or when the diastolic pressure exceeds 25 mm Hg. Hayward (1955) regarded the fibrosis and epithelialization of the alveolar walls as a mechanism rendering the passage of oedema fluid into the alveolar space more difficult, but as already stated these changes are a sequel to chronic alveolar wall oedema. For this reason pulmonary oedema may be encountered early in the course of mitral stenosis but disappears when the secondary adaptive changes have occurred. The raised pulmonary venous pressure accompanying mitral stenosis, in controlled cases,

seldom reaches a dangerous level likely to result in massive pulmonary oedema. Normally the interstitial fluid comprises about 40 per cent of the total lung fluid, a much higher proportion than in other body tissues. In pathological states such as mitral stenosis where there is excessive production of tissue fluid and hence lymph, the pulmonary lymphatics dilate and accommodate many times their normal flow. Such adaptation, however, requires time for its achievement. Heath and Hicken (1960) showed that the pulmonary lymphatic distension seen in lung biopsies correlated with the level of left atrial pressure and the presence of distended lymphatics also indicated increased flow rates. The dilatation of the interlobular lymphatics together with the accompanying oedema of the septal tissues are responsible for the linear shadows seen in radiographs of the lower lobes known as Kerley's B lines (Fig. 16.39).

Although serious pulmonary oedema is usually controlled by adequate treatment, intermittent pulmonary oedema occurs and can result in intra-alveolar fibrinous exudate which may later organize to form foci of intra-alveolar bone. The latter may measure up to 2 mm in diameter and are visible in lung radiographs. They were described by Elkeles and Glynn (1946) and

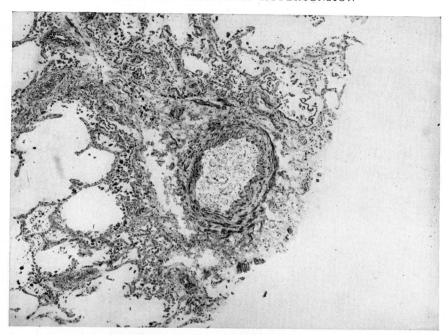

FIG. 16.37. Section showing a greatly hypertrophied medial coat of a branch of the pulmonary vein in a case of chronic mitral stenosis. × 280 H and E.

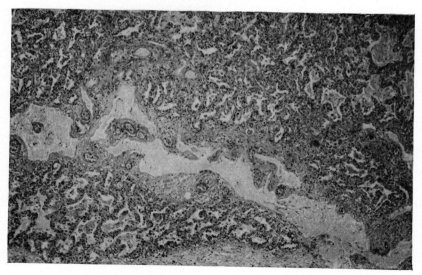

FIG. 16.38. Dilatation of a pulmonary lymphatic channel in chronic mitral stenosis. × 40 H and E. (Reproduced by courtesy of Dr. A. A. Liebow.)

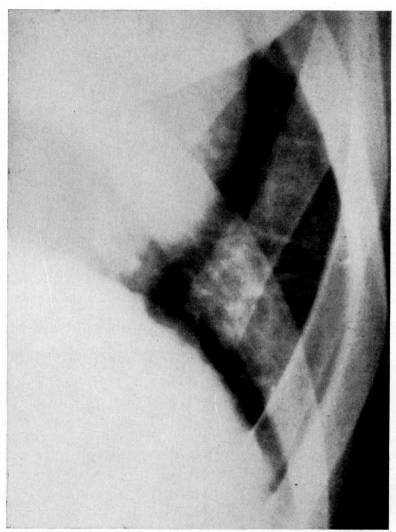

FIG. 16.39. X-ray of phreno-costal margin of lung showing transverse linear shadows caused by oedematous interlobular septa in a case of mitral stenosis, Kerley's B lines.

Yesner (1956) who have shown that the alveolar elastic fibres persist within the larger bony masses (Fig. 16.40).

Rarely pulmonary alveolar microlithiasis may complicate mitral stenosis.

PULMONARY HAEMORRHAGES

Among the other radiological changes found in the lungs in some cases of mitral stenosis is a miliary mottling caused by collections of haemosiderin-laden macrophages (Fig. 16.41). The same nodules are seen as brick red dots 2–3 mm in diameter on the cut surfaces of the lungs postmortem. The application of the Prussian-blue test, either to the gross specimen or to histological sections, displays the siderotic nodules very clearly and they are found scattered through the lung parenchyma and beneath the pleura. These nodules consist of intra-alveolar collections of haemosiderin-laden macrophage cells, often referred to as "heart-failure cells" (Fig. 16.42), together with local fibrous thickening of

the neighbouring alveolar walls. In addition, there is fragmentation and encrustation with iron salts of damaged elastic fibres in the wall of the adjacent alveolar ducts and alveoli; some of these fibres closely resemble the "bamboo fibres" found in siderotic nodules in the spleen in haemolytic anaemias. The iron-encrusted fibres cause a foreign-body giant-cell reaction (Fig. 16.24) and eventually the whole lesion organizes to become a small mass of fibrous tissue. The origin of siderotic nodules is not known with certainty, but Wood et al. (1937) showed that in mitral stenosis with right ventricular hypertrophy, capillary anastomoses were established between the pulmonary and bronchial circulations in the walls of the alveolar ducts and in the pleura. Lendrum et al. (1950) believed that in mitral stenosis and chronic passively congested lungs from any cause, the greatly raised pulmonary blood-pressure might nearly equal the systemic pressure, causing rupture or stasis of the flow in dilated capillaries at the points of anastomoses between the two circulations. Haemorrhages then occur and are responsible for siderotic nodules.

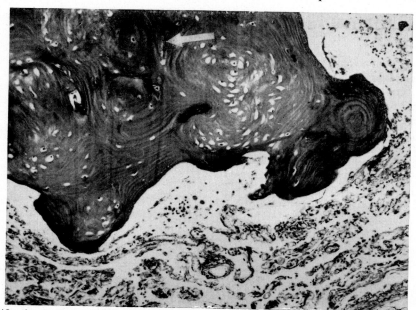

FIG. 16.40. Ossification in the alveoli of the lung in a case of chronic mitral stenosis. The arrow points to the remains of the alveolar elastic framework deep in the ossified mass. × 140 Verhoeff–Van Gieson.

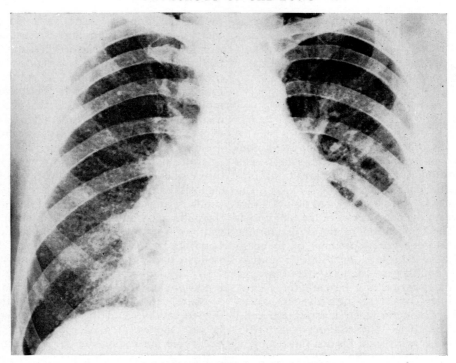

FIG. 16.41. A radiograph of the lungs showing siderotic nodules due to chronic mitral stenosis.
(Reproduced by courtesy of Dr. J. W. Pierce.)

Deposited iron pigments may also occur in the elastic laminae of branches of both pulmonary arteries and veins. The iron-encrusted elastic laminae excite a foreign-body giant-cell reaction. Siderotic pigment in addition to forming siderotic nodules may be deposited on and within the lobular septa and beneath the surface of the pleura. Septal deposition of iron pigment may be visible in lung radiographs and gives rise to Fleischner's lines (Fleischner and Reiner, 1954).

Heath and Whitaker (1956), however, found that siderotic nodules were more commonly found in patients with a high pulmonary venous pressure, a condition known to predispose to the formation of bronchial varices. Rupture of such varices may result in brisk haemoptyses and blood may be inspired into the alveoli. Magarey (1951) has shown that in rats, inhaled blood is gradually taken up by phagocytes which collect in aggregations resembling siderotic nodules,

and the same mechanism may be responsible for siderotic changes in human lungs.

Gough (1960) showed that in some patients with well-developed pulmonary siderosis the lower lobes and lingula might be much less involved than the upper portions of the lungs.

PULMONARY EMBOLI IN MITRAL STENOSIS

A further common vascular complication in mitral stenosis is the presence of ante-mortem thrombus in branches of the pulmonary artery. It is usually impossible to determine whether this is due to primary autochthonous thrombosis or embolism. Such thrombi, however, are found three times as often in the lower lobes, and are frequently located at the points of division of the arteries which makes it more likely that they are embolic in nature. Pryce and Heard (1956)

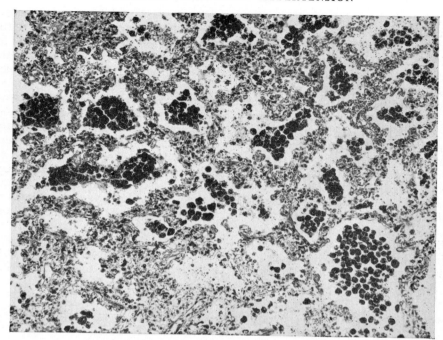

FIG. 16.42. Collections of intra-alveolar haemosiderin-laden phagocytes and thickening of the alveolar walls in a case of chronic mitral stenosis. × 140 H and E.

showed in rabbits emboli are more frequently diverted to the lower lobes of the lungs than elsewhere, and the same distribution seems to occur in man. Many of the emboli cause infarction and, as Harrison (1958–9) has stated, may decrease the already reduced pulmonary blood-flow by up to a further third. This presents an insuperable mechanical burden to the right ventricle and may cause it to fail rapidly. Should the thrombi subsequently canalize, the new lumen may remain permanently narrowed and in severe cases it resembles an atheromatous coronary artery.

OTHER CHANGES

Attention thus far has been mainly devoted to a description of the vascular changes but parenchymal changes are also found, though they are rarely encountered in the more rapidly progressive cases of chronic passive congestion. Rindfleisch (1872) and Davidsohn (1907) both described the generalized increase of muscle tissue that is often found in the respiratory bronchioles and alveolar ducts, and the latter referred to the "musculare Lungencirrhose" of mitral stenosis (Fig. 16.43). More recently Harkavy (1924) and Rodbard (1950 and 1953) have again drawn attention to the muscular overgrowth; the latter suggested that contraction of the muscle by obstructing expiration maintains intra-alveolar pressure and helps to impede the outflow of capillary oedema fluid. Further confirmation that the hypertrophied muscle helps to reduce fluctuations of the intra-alveolar gas tension in chronic passive congestion received some support from the observation of Christie and Meakins (1934) who showed that the pleural pressure remains raised throughout the respiratory cycle in congestive heart failure. This

4

may, in part, result from decreased lung compliance as already mentioned.

Since the introduction of surgical measures to alleviate mitral stenosis it has now become possible to assess the results of mitral valvotomy. A minority of cases, mainly those classified in clinical groups III and IV, failed to show the anticipated improvement. The reasons for this failure are in part due to pulmonary vascular changes and in part to the fibrous thickening of the alveolar walls which cause an alveolar-capillary block to oxygen diffusion.

Hickam and Cargill (1948) have shown that obstructive changes in the pulmonary artery may be reversible, particularly where these are due mainly to vasoconstriction. Braunwald *et al.* (1965) studied the functional effects on the pulmonary circulation of substituting artificial valves (Starr–Edwards prosthesis) for chroni-

cally stenosed and incompetent mitral valves. They found that in almost all the thirty patients they examined there was a rapid fall in the pulmonary arterial systolic pressure, and the average pressure gradient between the pulmonary arteries and the pulmonary veins fell by approximately 40 per cent. These changes were accompanied by an increase in the cardiac output, indicating dilatation of the smaller pulmonary arteries and the reversible nature of some of the arterial changes. It is unlikely, however, that permanent obstructive changes in the pulmonary arteries resulting from organization of emboli can be entirely reversed by any such operative measures, and obliteration of the capillary bed once established probably remains unaltered.

MacIntosh *et al.* (1958) have shown that the oxygen-diffusing capacity through the alveolar walls is unaffected by the operation although an

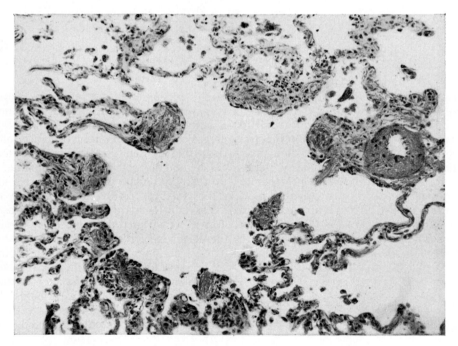

FIG. 16.43. Respiratory bronchioles showing hypertrophy of the muscle which also extends into alveolar ducts. The terminal branch of the accompanying pulmonary artery is greatly thickened. From a case of chronic mitral stenosis.
× 140 H and E.

improvement in the oxygen ventilatory equivalent occurred *pari passu* with the clinical improvement.

Rare Causes of Chronic Passive Pulmonary Hypertension

Certain rare conditions tend to produce pulmonary venous obstruction which may cause chronic passive pulmonary hypertension that is often of a more severe grade than that found in chronic mitral stenosis. Although the resulting pulmonary vascular and parenchymal changes resemble those found in chronic mitral stenosis, owing to the extent of the pulmonary venous obstruction which may become extreme, and the rate at which it develops, chronic pulmonary oedema and its sequelae may be more severe and further thrombosis can lead to complete occlusion of the very narrowed pulmonary veins.

The rarer causes of chronic passive pulmonary hypertension include:
(1) Congenital atresia of the pulmonary veins and cor triatrium.
(2) Obstruction of the pulmonary veins by a mediastinal neoplasm or granuloma.
(3) Myxoma of the left atrium.
(4) A ball-valve thrombus in the left atrium.
(5) Idiopathic thrombosis of the pulmonary veins.

1. CONGENITAL ATRESIA OF THE PULMONARY VEINS AND COR TRIATRIUM

In this condition the pulmonary veins may be completely obstructed or narrowed at the point of entry into the left atrium. A considerable collateral circulation may be established through the bronchopulmonary hilar venous anastomosis. An example of this condition was described by Bernstein *et al.* (1959) but usually only cases causing an incomplete obstruction result in pulmonary hypertension.

Cor triatrium is a condition caused by persistence of the common pulmonary vein which distends to form an atrial-like cavity. This "third" atrium communicates with the true left atrium through a narrow orifice which functions in the same manner as a stenosed mitral valve and leads to obstruction of the blood-flow in the pulmonary veins. The condition was described by Belcher and Somerville (1959) who reviewed the literature. A further case was described by Bernstein *et al.* (1959). In a recent case seen by the author (Fig. 16.44), the changes in the lungs resembled those seen in very severe mitral stenosis but were not restricted to the lower lobes of the lungs, the upper lobes sharing in the changes to a greater extent than in mitral stenosis.

In some of the smaller pulmonary veins medial hypertrophy and severe widespread intimal thickening with concentric layers of oedematous connective tissue reduce the lumen to a slit. Recent and recanalized thrombi and newly formed adventitial and periadventitial capillaries may also be present in the pulmonary veins. The severe alveolar oedema which ensues from these changes leads to extensive interstitial alveolar fibrosis of a more severe degree than that usually seen in uncomplicated chronic mitral stenosis. There may also be considerable hyperplasia of smooth muscle in the interlobular septa and pleura.

The severe obstruction offered to the pulmonary venous circulation can lead to the establishment of large bronchopulmonary venous anastomoses.

2. OBSTRUCTION OF PULMONARY VEINS BY A MEDIASTINAL NEOPLASM OR GRANULOMA

Very rarely a mediastinal mass of growth or a collagenous granuloma may compress one or more branches of the pulmonary veins. The gradual and incomplete obstruction results in adaptive changes in the branches of the pulmonary arteries supplying the same portion of the lung. Cases of pulmonary hypertension due to this cause were described by Edwards and Burchell (1951) and Inkley and Abbott (1961). Two further cases have recently been seen by the author (Fig. 16.45). A mediastinal granuloma

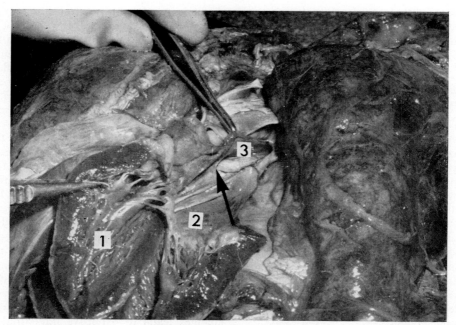

FIG. 16.44. Cor triatrium showing (1) left ventricle, (2) left atrium and (3) the persistent common pulmonary vein. The arrow points to the narrow opening between (2) and (3) which has been opened.

may readily be mistaken on macroscopic examination for a neoplasm and only microscopical examination reveals its true nature.

The granulomatous process, which sometimes follows the use of methysergide, is usually of unknown aetiology and is similar to and may occasionally be associated with retroperitoneal idiopathic fibrosis (Ormond's disease, Raper's idiopathic retroperitoneal fibrosis). It consists of hyaline collagenous fibrous tissue with a resemblance to a keloidal scar and contains eosinophils, lymphocytes and occasional foreign-body type of giant cell. The lesion usually starts around the adventitia of the venae cavae, innominate veins or pulmonary veins within the chest (Fig. 16.46). It starts most commonly in the middle mediastinal compartment and leads to very severe venous obstruction and lymphoedema in the regions drained by the obstructed veins. The affected segments of the lungs drained by obstructed pulmonary veins develop severe lymphoedema, extensive interstitial alveolar fibrosis and septal and pleural fibrosis

(Fig. 16.47). The alveolar fibrosis may lead to alveolar obliteration. The intrapulmonary branches of the pulmonary veins show medial hypertrophy, intimal fibrosis and luminal thrombosis (Fig. 16.48), and pulmonary arterial changes are found similar to those seen in chronic mitral stenosis (Fig. 16.49). All the vascular changes are strictly limited to those areas of the lungs drained by the obstructed vein or veins and only if several pulmonary veins are involved will the corresponding pulmonary arterial changes become sufficiently widespread and severe enough to cause pulmonary hypertension (Evans, 1959, case 3). The fibrosis of the lung is accompanied by hyperplasia of septal muscle tissue (Fig. 16.50).

Very rarely a major branch of a pulmonary artery may be obstructed and involved in the spread of a mediastinal granuloma. Intimal thickening occurs inside the compressed segment of the vessel which may be very narrowed though rarely completely obstructed. A case of this nature was described by Nelson et al. (1965).

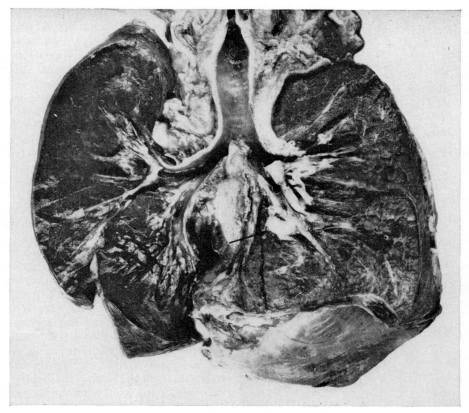

FIG. 16.45. A mediastinal granuloma (arrowed) surrounding a main lobar bronchus. One-third natural size.

3. MYXOMA OF THE LEFT ATRIUM

This rare cause of chronic, intermittent mitral valve obstruction may lead to changes similar to those encountered in the lungs in chronic mitral stenosis. An example of this condition is shown in Fig. 16.51. Examples of left atrial myxomas which led to pulmonary parenchymal and vascular changes were described by Jones and Julian (1955), Harvey (1957) and Goodwin *et al.* (1962). The last-named authors described four cases all of which developed pulmonary hypertension. In three of the cases pulmonary infarction followed the occlusion of the pulmonary vein draining a lower lobe. Studies of the pulmonary blood-flow using $C^{15}O_2$ showed that, as in chronic mitral stenosis, there was a diminished perfusion of blood through the lower lobes, and angiography revealed constriction of the pulmonary arteries supplying these lobes. Left atrial cardiac myxoma may be associated with pyrexia and clinically may closely resemble bacterial endocarditis.

4. A BALL-VALVE THROMBUS IN THE LEFT ATRIUM

Like a myxoma arising in the left atrium, a ball-valve thrombus may very rarely cause intermittent obstruction of the mitral valve and result in pulmonary hypertension. Such a case was described by Nanson and Walker (1952).

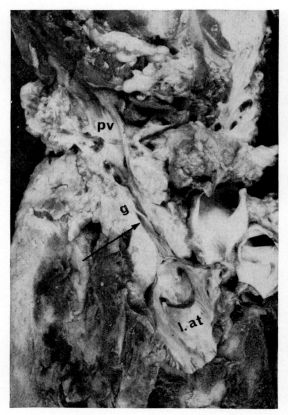

FIG. 16.46. A mediastinal granuloma compressing and largely obliterating a main pulmonary vein ; pv = pulmonary vein within the lung, g = mediastinal granuloma lying outside and infiltrating the wall of the pulmonary vein which it has practically obliterated, and l.at. = left atrium.

5. PULMONARY VENO-OCCLUSIVE DISEASE

Pulmonary veno-occlusive disease was first described by Höra (1934) and until recently was a little-known disease. In 1957 Walford and Kaplan described twelve cases of an unusual pulmonary disorder which they named "endogenous pneumoconiosis". Several of their cases were almost certainly unrecognized examples of thrombotic pulmonary veno-occlusive disease.

Since the first description further cases have been reported by Brewer and Humphreys (1960),

Crane and Grimes (1960), Burki (1963), Stovin and Mitchinson (1965), Brown and Harrison (1966), *Brit. Med J.* (1972), Liebow *et al.* (1973) and Wagenvoort and Wagenvoort (1974). Thadani *et al.* (1975) reviewed the published literature and added further cases.

Pulmonary veno-occlusive disease occurs at any age but the majority of cases have been found in persons under the age of 20. Wagenvoort *et al.* (1971) reported it in two infants who were siblings at the ages of 2 and 3 months, respectively; both of the children subsequently died. Like primary pulmonary hypertension it shows a familiar tendency.

Clinically, pulmonary veno-occlusive disease closely resembles and may be confused with primary pulmonary hypertension and pulmonary thromboembolic disease. It presents with increasing dyspnoea, syncopal attacks, evidence of increasing right heart failure and weight loss. As the disease progresses evidence of right ventricular strain can be detected electrocardiographically and Kerley's B lines are frequently seen in chest radiographs. Angiocardiography may provide valuable information inasmuch as the pulmonary arterial tree usually fills normally thus excluding pulmonary thromboembolic disease and primary pulmonary hypertension. Cardiac catheterization usually shows a moderate degree of pulmonary hypertension but the capillary wedge pressures are very variable depending upon whether the venous obstruction affects the large central pulmonary veins or the small peripherally situated veins and venules. If the former are the site of obstruction the capillary wedge pressure is usually raised but if the peripheral veins are mainly involved, collateral drainage through unaffected branches prevents a rise in the capillary pressure. Usually the pulmonary veins are not obstructed uniformly, some being severely affected, others unaffected. Polycythaemia is commonly present.

Macroscopically, the lungs are firm, indurated and the surface lymphatic channels are distended and very prominent. The cut surface shows multiple small haemorrhages and the larger venous channels if obstructed are surrounded by a greyish-white mantle of fibrous tissue. The small

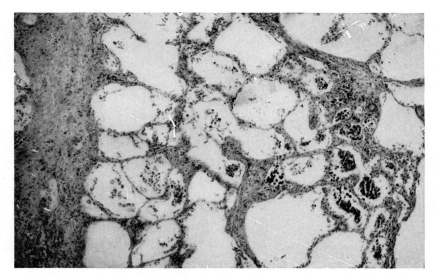

FIG. 16.47. Interstitial fibrosis of the alveoli and septal fibrosis due to obstruction of the pulmonary veins by a mediastinal granuloma. These changes are a sequel to chronic oedema. × 40 H and E.

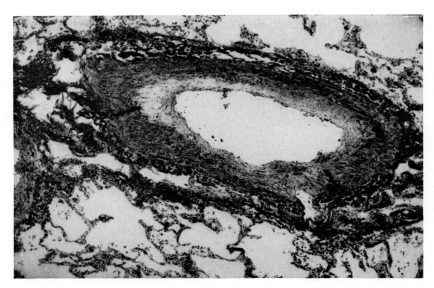

FIG. 16.48. A greatly thickened wall of a pulmonary vein lying in the lung. The main branches had been obstructed in the hilar region by a mediastinal granuloma. Note the medial hypertrophy and intimal fibrosis. × 40 Verhoff–Van Gieson stain.

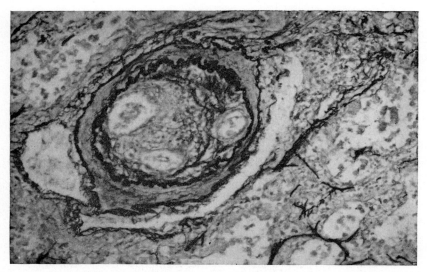

Fig. 16.49. A muscular branch of a pulmonary artery in a lung the pulmonary venous drainage of which was obstructed by a mediastinal granuloma. Note the medial hypertrophy, intimal fibroelastosis and lumen occupied by recanalized thrombus. Also note the greatly dilated periarterial lymphatic channels. × 100 Verhoff–Van Gieson stain.

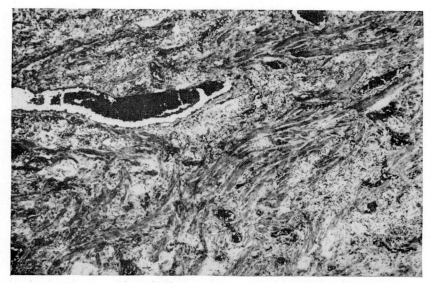

Fig. 16.50. Interlobular septal muscular hyperplasia following pulmonary venous obstruction by a mediastinal granuloma. × 40 H and E.

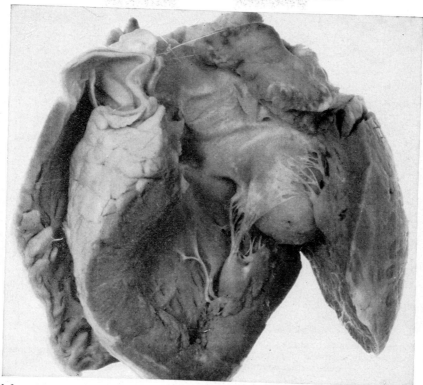

FIG. 16.51. A left atrial myxoma projecting through a tightly stretched mitral valve and obstructing its lumen. The left ventricle has been opened to expose the tumour. × 2 natural size.
(Reproduced by courtesy of Dr. H. R. M. Johnson.)

pulmonary arteries are thickened and atheromatous change is commonly found in the large elastic arteries due to pulmonary hypertension.

Microscopically, the pulmonary veins in the interlobular septa are partially or totally occluded by loose-textured intimal fibrous tissue. Recanalization of the obstructed lumen is observed and dilated venous channels are often present in the perivenous septal tissues outside the obstructed vein (Fig. 16.52). Recently deposited thrombus may be found in the pulmonary veins and less frequently in the small pulmonary arteries. The bronchial veins are very prominent and may resemble angiomatous lesions (Carrington and Liebow, 1970).

In addition to the pulmonary venous changes which are the most severe, the pulmonary arterioles and small muscular pulmonary arteries also show medial thickening and patchy intimal fibrosis, while in the larger elastic arteries fibroelastic tissue surrounds new vascular channels formed from recanalized thrombi. The staining reaction of the newly formed fibroelastic tissue often indicates more than one previous thrombotic episode.

As a consequence of the venous obstruction the intrapulmonary lymphatics are dilated, the septal tissues oedematous and thickened, and the alveolar walls adjacent to the interlobular septa show interstitial mural fibrosis consequent on the persistent interstitial alveolar oedema. Focal areas of alveolar capillary overdistension, varicosity and congestion resulting in intra-alveolar haemorrhages are present near the septa. In these areas haemosiderin-laden intra-alveolar phagocytes are present and there is siderotic

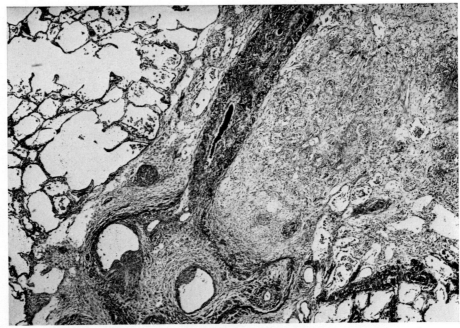

FIG. 16.52. Pulmonary venous thrombosis showing a recanalized branch of a pulmonary vein surrounded by newly formed anastomotic capillary channels in the perivenous tissues. The alveoli adjacent to the interlobular septum show well-marked interstitial alveolar fibrosis. × 75 H and E.

encrustation of the alveolar elastic fibres which provoke a foreign-body type of giant-cell reaction (Fig. 16.53). It was this latter change which led Walford and Kaplan to regard the disease erroneously as "endogenous pneumoconiosis". Electron-diffraction spectroscopy confirms the presence of iron and calcium salt deposits on the elastic fibres. In some cases obliteration of the alveolar capillaries by sludged red blood-cells may lead to alveolar wall atrophy and disruption.

The aetiological cause of pulmonary veno-occlusive disease remains unknown, but in some reported cases there has been both clinical and subsequent pathological evidence of an antecedent infective illness. Wagenvoort et al. (1971) reported an infective illness during the 35th week of pregnancy in the mother who subsequently gave birth to infants who developed the disease.

Corrin et al. (1974) found ultramicroscopic and immuno-fluorescent evidence of deposits of IgG together with complement (B_1) in the capillary basement membrane in the alveolar wall.

They believed such antigen–antibody deposits could have resulted from an antecedent infection of viral nature.

In the author's experience pulmonary veno-occlusive disease usually affects either mainly the larger branches of the pulmonary veins or predominantly the small veins and venules. The latter form of the disease is particularly associated with capillary engorgement and alveolar siderotic changes and the small venules are found to be filled with sludged debris of red blood-cells. Alveolar siderotic changes tend to be less severe when the larger pulmonary veins are maximally involved.

In addition to the pulmonary arterial and other vascular changes that complicate long-standing chronic passive congestion, Andrews (1957) has described advanced alveolar structural changes that may rarely complicate very chronic and extreme degrees of pulmonary venous obstruc-

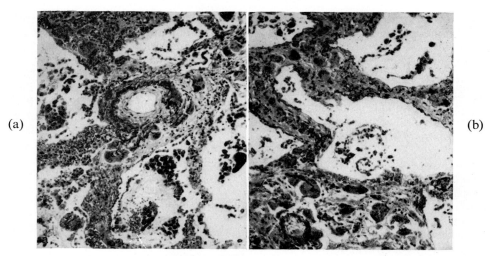

(a)

(b)

FIG. 16.53. Chronic pulmonary venous thrombosis showing in (a) interstitial giant cells ingesting remnants of iron-impregnated elastic fibres and iron-impregnation of the internal elastic lamina of a pulmonary artery. In (b) there is widespread interstitial alveolar fibrosis and numerous giant cells and fragments of disrupted impregnated elastic tissue. In (a) some of the very distended alveolar capillaries can be seen filled to bursting point with red blood-cells; in (b) they have become obliterated. Both × 100 stained H and E.

tion. He described five such cases occurring in association with the following conditions: a left atrial myxoma, thrombotic occlusion of the pulmonary veins, obstruction of the pulmonary veins by mediastinal collagenization and an extreme degree of congenital stenosis of the left pulmonary veins.

In the area of lung drained by the obstructed veins there was widespread interstitial fibrosis.

Microscopically, considerable alveolar capillary congestion and oedema was followed by fibroblastic proliferation and alveolar epithelialization on a scale greater than that already described as accompanying chronic mitral stenosis and shown in Fig. 16.33. The changes, although widespread, were not evenly distributed throughout the affected venous territory. In addition to the above changes there were also siderotic nodules and similar arterial changes to those already described. The interstitial fibrosis that resulted from this extreme form of pulmonary venous occlusion can usually be differentiated from idiopathic interstitial fibrosis by the absence of inflammatory cells and the presence of haemosiderin-laden macrophages (heart-failure cells). Although the alveoli become lined by cuboidal epithelium this does not assume such bizarre metaplastic changes such as mucus-secreting epithelium sometimes seen in idiopathic interstitial alveolar fibrosis.

GROUP 2

Chronic Pulmonary Hypertension due to Primary Mechanical Obstruction of the Pulmonary Arteries

Many diseases are included in this group all of which are caused by a primary mechanical obstruction of the pulmonary arterial or capillary blood-flow.

In the varieties of chronic pulmonary hypertension previously described, the hypertension was caused initially by a functional vasoconstriction of the muscular pulmonary arteries. In the group about to be described the arteries or capillaries are either obstructed by emboli or compressed from without, but if arterial vasoconstriction plays any part at all, it is only transient and occurs at the moment of embolic impaction.

The conditions responsible for causing this form of chronic pulmonary hypertension include:

(1) Embolic group consisting of:
    (a) recurrent thrombotic emboli,
    (b) malignant cell emboli,
    (c) parasitic emboli (schistosomiasis),
    (d) trophoblastic emboli,
    (e) pulmonary hypertension in drug addicts.
(2) Thrombosis of small branches of the pulmonary artery and capillaries.
(3) Vascular compression and obliteration due to:
    (a) pneumoconioses,
    (b) widespread alveolar destruction and capillary compression,
    (c) the pulmonary histiocytoses.
(4) Vascular obstruction in progressive systemic sclerosis (PSC), Raynaud's syndrome and rheumatoid arthritis.

1. EMBOLIC GROUP

Recurrent Pulmonary Emboli

An increasing number of cases of chronic pulmonary hypertension have been attributed to obstruction of the pulmonary arterial tree and capillaries by recurrent showers of pulmonary emboli. Despite the increasing number of published cases of chronic obstructive pulmonary hypertension attributed to this cause, relatively few withstand critical examination. Many of the reported cases remain unproven and may well be cases of primary (idiopathic) pulmonary hypertension associated with autochthonous thrombosis. The majority of cases of obliterative pulmonary disease due both to thromboembolic obstruction or to primary (idiopathic) primary pulmonary hypertension occur in women under the age of 40 and often start during pregnancy or soon after delivery.

Proof of the embolic nature of the thrombi is made more difficult because thrombi in the pulmonary arteries may form secondarily in any condition where the pulmonary blood-flow is seriously reduced. In the past it was mistakenly believed that pulmonary thrombi were always embolic in nature and that they were responsible for the vascular obstruction. Primary pulmonary thrombosis may complicate many pulmonary vascular abnormalities including "primary" pulmonary hypertension (Wade and Ball, 1957), pulmonary stenosis (Rich, 1948; Heath et al., 1958) and mitral stenosis (Thomas et al., 1956).

Before considering the acceptable cases the experimental pathological evidence bearing on the subject will first be considered.

Experimental Pathology

Harrison (1948) injected fibrin emboli into the pulmonary circulation of rabbits and found that they rapidly became incorporated within the intima, finally resulting in the production of small eccentric intimal plaques of fibrous tissue; these caused either little or no appreciable mechanical obstruction to the blood-flow. He later (1951) repeated the experiment using repeated showers of small emboli and found that right ventricular hypertrophy resulted; it increased while the emboli continued to be injected, but disappeared when injections ceased.

Haynes et al. (1947) had shown earlier that obstruction of a main pulmonary artery in dogs by a balloon failed to alter the pulmonary arterial blood-pressure, whereas injection of minute emboli caused an immediate rise in pressure. This was attributed by Daly et al. (1947), following experiments in dogs using microemboli, to a vasoconstrictory mechanism and unless repeated emboli were administered the pressure rapidly returned to normal levels. Moore et al. (1958) repeated the experiments in dogs but used micro-emboli of barium sulphate. They found that small emboli became lodged in arterioles unless an anti-pressor drug such as hexamethonium was first given, in which case the emboli passed on into the capillaries. Owing to the much greater cross-sectional area of the capillary compared with the arteriolar bed, no hypertension resulted when the emboli lodged in the capillaries. Barnard (1954) injected intravenously minute autogenous, clot emboli into rabbits and produced pulmonary hypertension. Later microscopical examination of the pulmonary arteries showed two distinct changes: (a) eccentric intimal fibrous plaques and (b) concentric intimal fibrosis. The former resulted from organized blood clot; the latter was thought to be an intimal reaction to a reduced pulmonary blood-flow resulting from prolonged vasoconstriction.

Certain facts emerge from these experiments. Firstly, that because of the enormous reserve capacity of the pulmonary arteriolar system, if emboli are to cause persisting pulmonary hypertension they must be released continuously into the circulation and, secondly, that normal vaso-constrictory tonus plays a part in retaining the emboli in the arterioles and may be increased following injection of intravascular emboli.

Human Pathology

The first full account of this condition was given by Ljungdahl (1928, quoted by Wilhelmsen et al., 1963). The clinical signs and symptoms shown by patients with chronic thromboembolic pulmonary hypertension differ very little from those found in patients with primary pulmonary hypertension except that a history of major embolic incidents and signs and symptoms due to resulting pulmonary infarcts are more likely to be elicited in recurrent thromboembolic disease.

Pulmonary angiograms carried out before or after death may sometimes be of help in distinguishing embolic obstruction of a pulmonary artery by showing an abrupt filling defect commencing in a segmental or occasionally larger artery and involving all its branches. In primary pulmonary hypertension there is more likely to be a generalized failure of the small muscular pulmonary arteries to fill throughout the lung.

Castleman and Bland (1946) described a case in a woman where post-mortem showed occlusion of most tertiary branches of the pulmonary artery by organized and recanalized emboli. Other accounts of cases, probably acceptable as examples of recurrent thromboembolic pulmonary disease, have been given by

McMichael (1948), Owen et al. (1953), Keye and Thomas (1955), Hollister and Cull (1956) and Goodwin et al. (1963).

Muirhead et al. (1952) also described two cases, one of which they regarded as certainly embolic, but in both there was an associated cryoglobulinaemia and an increased tendency to intravascular thrombosis. The final diagnosis depends upon excluding other causes of chronic pulmonary hypertension, a consideration of the history in which embolic episodes may be recognized, the detection of emboli in all stages of organization, and the finding of a source from which emboli could have arisen.

Anderson et al. (1973) in a careful quantitative study following injection of the pulmonary arterial tree at hypertensive pressures with radio-paque medium claimed to be able to differentiate between primary pulmonary hypertension and thromboembolic pulmonary hypertension. In primary pulmonary hypertension the pulmonary arterioles around 40 μm in diameter (pulmonary arterioles) were reduced in number due to obliteration and the walls of the small pulmonary muscular arteries were thickened. In thromboembolic disease the number of small pulmonary arteries appeared to be increased at first in the non-obstructed portions of the pulmonary arterial tree, probably due to the opening up of physiologically temporarily closed branches.

Occasionally, total occlusion of the main pulmonary arteries or their major branches occurs due to repeated embolic episodes which gradually increase the extent of the obstruction and result in the rapid onset of right heart failure (Hollister and Cull, 1956).

Chronic pulmonary hypertension may rarely complicate cirrhosis of the liver and chronic thrombosis of the portal vein. Six cases were described by Naeye (1960), and Levine et al. (1973) described three further cases of progressive pulmonary hypertension in children associated with portal hypertension. Haworth et al. (1974) have reviewed the subject. The pulmonary hypertensive changes, which are very rare, are quite separate from the pulmonary vascular changes sometimes associated with primary liver disease which causes secondary portal hypertension and which were described by Berthelot *et al.* (1966) but which were not associated with pulmonary hypertension. The pulmonary hypertensive changes associated with liver disease were considered by Levine *et al.* to have resulted from initial vasoconstriction but the agent responsible for causing this was not identified. It was suggested that the vasoconstrictory agent might have either been produced by or failed to be metabolized by the diseased liver. Alternatively, it has been suggested that the small pulmonary arteries are filled with thrombotic emboli which reach them from the portal venous system by way of collateral perioesophageal and mediastinal veins, or alternatively that thromboplastic substances are diverted from the liver to the systemic veins subsequently causing thrombosis in the small pulmonary arteries.

Macroscopically, the lungs may show little abnormality apart from dilatation of the major elastic arteries together with intimal atheromatous flecking, changes that are common to all forms of chronic pulmonary hypertension. Recognizable ante-mortem thrombus may extend into the branches of the segmental arteries, but often the occluded vessels are much smaller. The lungs are dry in contra-distinction to the oedema and ascites present in other parts of the body due to the accompanying right heart failure. Arteriograms of the pulmonary arteries after death may enable the blocked segments of arteries to be visualized (Short, 1956) and the extent of the obstruction noted, and this is usually more marked in the lower lobes.

Microscopically, in the majority of cases the small elastic and the larger muscular pulmonary arteries are filled with thrombi in all stages of organization, but many of the smaller muscular branches are not involved and may in fact be thinned. In primary pulmonary hypertension the maximal changes occur in the smaller muscular pulmonary arteries below 500 μm in diameter, whereas in recurrent thromboembolic pulmonary hypertension the changes are usually maximal in arteries between 1500 and 500 μm. The dilatation (grade 4) lesions found in the pulmonary arterial system in other forms of chronic

pulmonary hypertension are not seen in recurrent pulmonary embolic disease apart from the thinned vessels distal to an embolic blockage.

When thrombotic emboli are present two forms of intimal changes are recognizable: in the first the thrombus is reduced to an eccentric intimal plaque of fibro-elastoid tissue containing sinuses and endothelium-lined spaces, in the second the whole intima is concentrically thickened by similar tissue which encroaches upon and may completely obliterate the lumen. The medial coat may atrophy beneath the thickened intima. Other changes include dilatation of branches of the bronchial artery in the neighbourhood of occluded branches of the pulmonary artery, and the establishment of subpleural pulmonary arterial anastomoses between the ischaemic and adjacent lobules and segments (Short, 1956). Many of the subpleural anastomoses connect with the bronchial arterial system. In the later stages of embolic obstruction, when the pressure starts to rise, secondary changes identical with those already described under primary pulmonary hypertension occur, and the differentiation of embolic pulmonary hypertension from other causes becomes extremely difficult.

Recurrent pulmonary embolism leads to a reduction in the vital capacity probably due to restriction of lung movement and a considerable reduction in the blood-flow through the lungs. The reduced perfusion:ventilation ratio results in impairment of diffusion of oxygen but only rarely of carbon dioxide. The arterial oxygen saturation may become reduced during exercise but the arterial pCO_2 usually remains normal and a respiratory alkalosis may even ensue due to the tachypnoea that is induced. The functional changes resulting from this condition were described by Heilman *et al.* (1962) and Goodwin *et al.*

Subacute Pulmonary Hypertension due to Pulmonary Carcinomatosis

Unlike some of the forms of chronic pulmonary hypertension so far discussed, the course of this disorder is usually much more rapid, the patients rarely surviving longer than 3 months.

A case showing the features of this condition was described by Bristowe (1868), and Girode (1889) was the first to draw attention to the accompanying fibrous endarteritis. Later Schmidt (1903) published a further series of cases, and more recently the condition has been reviewed by Morgan (1949) and Storstein (1951).

The primary growth is usually located in the stomach, though it may arise in the breast, lung, prostate and pancreas and is often so small as to escape clinical detection.

Macroscopically, the lungs, if not collapsed by pleural effusions, are firm, greyish-white and the pleural surfaces may be seeded with secondary growth or show diffuse subpleural lymphatic carcinomatous permeation (Fig. 20.59). When withdrawn from the chest the lungs retain their shape well, partly due to the accompanying pulmonary oedema. The cut surface shows the smaller transected branches of the pulmonary

arteries surrounded by a cuff of greyish-white tumour tissue. Some of these vessels are completely occluded forming greyish-white cords. The small bronchi may be similarly cuffed with tumour tissue. The large branches of the pulmonary arteries seldom show longstanding chronic pulmonary hypertensive changes because of the rapid course of the disease.

Microscopically, many of the muscular arteries are completely or partially filled with loose fibroelastoid tissue, in which may be found either occasional clumps or solitary carcinoma cells. Outside, the perivascular lymphatic channels are either totally obliterated by fibrous tissue or may still contain viable carcinoma cells (Fig. 16.54). The medial muscle and elastic laminae are often entirely destroyed in the affected arteries, being replaced by fibrous tissue, and capillary vessels may invade the wall from without.

Much controversy has arisen about the nature of the intravascular changes, some considering them a reaction to the perivascular lymphatic

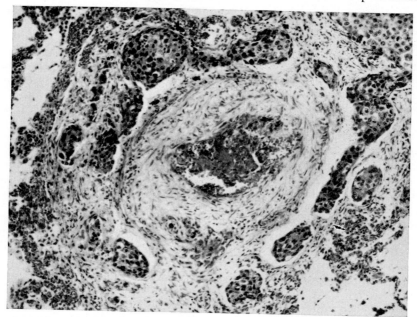

FIG. 16.54. Lymphangitis carcinomatosa in the lung showing secondary carcinoma cells filling the perivascular lymphatics together with recent thrombus and early endarteritis obliterans in a small branch of the pulmonary artery. The lumen of the artery shown actually contains viable cancer cells, though these are difficult to identify. × 140
H and E.

growth (Greenspan, 1934), whilst others regard them as being caused by intravascular tumour emboli. Intravascular metastatic cancer cells when about to spread out of the vessel first adhere to the endothelium, acquiring a fibrin covering, and later migrate between the endothelial cells often in the wake of polymorph leucocytes. Following this, should the carcinoma cells fail to survive, the fibrin later becomes replaced with fibro-elastoid tissue. The second view was supported by Schmidt and Morgan, and is that to which the author subscribes as a result of his own observations. The widespread nature of these changes may lead to very rapid production of pulmonary arterial obstruction and a rapidly rising pulmonary arterial blood-pressure. Also the lymphatic oedema coupled with widespread lymphatic and vascular fibrosis causes an increase in the lung compliance. Reference should be made to Chapter 24 in which the subject of secondary tumours is discussed in greater detail.

Chronic Pulmonary Hypertension due to Schistosomiasis

Although the incidence of lung lesions in schistosomiasis is high, only a very small proportion of such patients develop cor pulmonale. According to Richert and Krakaur (1959) 65 per cent of patients harbouring *S. mansoni* have lung lesions but less than 1 per cent of these lesions are of serious import and lead to serious pulmonary vascular obstruction. In the case of *S. haematobium* infections 33 per cent of patients have pulmonary involvement and about 2 per cent of these cases are associated with cor pulmonale. The vascular changes occurring in schistosomiasis are considered in detail in Chapter 10.

The lung may become involved in schistosomiasis during the migratory adult-fluke stage, or eggs may be carried to the pulmonary circulation following their entry into the vesical and pelvic venous plexus. Eggs may also reach the pulmonary arteries from the portal venous system following the establishment of anastomoses between the portal and systemic mediastinal veins as a consequence of schistosoma hepatic portal fibrosis (pipe-stem liver).

Trophoblastic Emboli

Among the rare forms of pulmonary emboli, trophoblastic emboli may very rarely cause obstruction and thrombosis in the pulmonary arteries and result in a subacute form of cor pulmonale (Arnold and Bainborough, 1957). This condition is described in greater detail in Chapter 15.

Pulmonary Hypertension in Drug Addicts

Pulmonary hypertension may occasionally follow intravenous injection of a variety of adulterated drugs by addicts. Tripelennamine hydrochloride tablets include in their composition a variable quantity of such binding agents and adulterants as talc, magnesium stearate and corn starch to mention but a few. The resulting emboli become trapped in the smaller branches of the pulmonary arteries and capillaries where they cause an arteritis and thrombosis. Granuloma formation subsequently occurs and if the changes are very extensive pulmonary hypertension can occur (Wendt et al., 1964).

2. THROMBOSIS OF SMALL BRANCHES OF THE PULMONARY ARTERY AND PULMONARY CAPILLARIES

Thrombosis may complicate the state of chronic pulmonary hypertension with decreased cardiac output, the thrombi forming autochthonously as stated previously. Thrombosis may occasionally complicate primary blood disorders and has been reported in cases of sickle-celled anaemia by Yates and Hansmann (1936). The thrombosis involves the smaller muscular arteries and capillaries throughout the lung and may result in pulmonary hypertension and cor pulmonale.

3. VASCULAR COMPRESSION AND OBLITERATION

Obstruction or ligation of one pulmonary artery normally fails to raise the pulmonary blood-pressure provided the pulmonary arteries in the opposite lung are normal. Until the pulmonary blood-flow has been restricted by about 60 per cent the pressure at rest remains at the normal levels. Although many forms of chronic pulmonary disease lead to very extensive damage to both the pulmonary arteries and the alveolar capillaries, it is common in such conditions for bronchopulmonary anastomoses to be established. The resulting rise in pulmonary arterial blood-pressure is partly due to the obliteration of the pulmonary vessels by the disease process, and to the remaining patent portion of the pulmonary arteries filling with blood conveyed by the bronchial arteries at systemic pressure. Among the diseases which may lead to severe pulmonary arterial obstruction is massive fibrosis of the lungs due to silicosis. It may complicate coal-miner's lung and haematite-miner's lung; the former condition has been described in detail by Wells (1954). The other causes of vascular obliteration are described under their respective diseases.

A rare cause of chronic pulmonary hypertension follows obliteration of the smaller branches of the pulmonary arteries by lipid-laden granulomatous tissue found in the Hand–Christian–Schuler and eosinophil granulomatous forms of pulmonary histiocytosis X disease; the condition is described in Chapter 22.

4. PULMONARY HYPERTENSION IN PROGRESSIVE SYSTEMIC SCLEROSIS (PSC)
Raynaud's syndrome and Rheumatoid arthritis

The relationship between pulmonary hypertension and this group of disorders is still uncertain, but Raynaud's syndrome has been

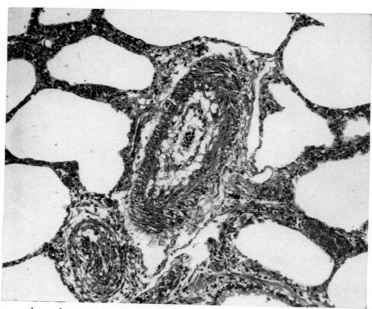

FIG. 16.55. A small muscular pulmonary artery from a case of Progressive Systemic Sclerosis showing subendothelial oedema with lifting off of the endothelial cells from the subjacent tissue of the intima. × 140 H and E.
(Reproduced by courtesy of Dr. J. Ball.)

reported in association with pulmonary hypertension by Matsui (1924), Linenthal and Talkov (1941), Taft and Mallory (1946) and Wade and Ball (1957). Raynaud's syndrome frequently heralds the later development of Progressive Systemic Sclerosis, and some of the cases quoted by these authors (those of Matsui and Linenthal, and one of the two cases of Wade and Ball) belonged in this category.

Pulmonary hypertension may also be associated with rheumatoid arthritis and was described by Gardner *et al.* (1957). Changes in the digital arteries and arterioles have also been reported.

Increasing evidence points to the existence of a form of pulmonary arterial lesion associated with this group of disorders in addition to the other pulmonary changes which occur affecting the parenchyma and pleura (described in Chapter 19).

The arterial changes accompanying Progressive Systemic Sclerosis (PSC) were first reported by Masugi and Yä-Shu (1938) and more recently by Wade and Ball (1957).

The arteries principally affected range in size from 300 to 100 μm, but some doubts were raised by Ellman and Cudkowicz (1954) as to whether the vessels involved were branches of the pulmonary or the bronchial arteries. The occurrence of pulmonary hypertension in some patients proves, however, that the pulmonary arteries are involved. The endothelium of the affected artery is first separated from the underlying intima by oedema-filled spaces (Fig. 16.55); this is later followed by a fibroelastoid intimal proliferation which reduces or completely blocks the lumen of the artery (Fig. 16.56). Perivascular fibrosis accompanies these changes as part of the general thickening of the pulmonary connective tissues but this is not responsible for arterial obstruction.

In the earlier stages, some of the smaller arteries may show an extensive perivascular

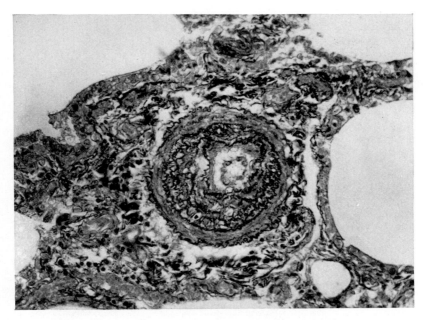

Fig. 16.56. A later stage than is shown in Fig. 16.55 showing fibroelastoid thickening of the intima and persisting oedema beneath the endothelium. × 280 Verhoeff–Van Gieson.

(Reproduced by courtesy of Dr. J. Ball.)

cuffing with lymphocytes before the onset of the intramural changes.

GROUP 3

Chronic Pulmonary Hypertension of Uncertain or Ill-understood Causation

PULMONARY HYPERTENSION DUE TO KYPHOSCOLIOSIS

Kyphoscoliosis is a well-recognized but still poorly understood cause of pulmonary hypertension. Kyphoscoliosis involving the upper thoracic spine and due to structural abnormalities of the vertebrae rather than to postural causes is more likely to result in cor pulmonale. The condition has been investigated by Bergofsky et al. (1959), Abrahamson (1959) and Dollery et al. (1965).

In severe kyphoscoliosis involving the upper thoracic spine, the chest wall is rotated by the scoliosis and deformed due to the approximation and crowding together of the ribs. Both deformities lead to a reduction in the volume of the pleural spaces. Furthermore, the pleurae are further reduced by the upward thrust of the diaphragm caused by compression of the abdominal contents resulting from the posture of the patient.

The structural changes in the chest wall result in a considerable diminution of the lung volumes and particularly in the vertical dimensions of the lungs. Dunnill (1965) showed that usually there is a normal complement of lung alveoli unless the thoracic deformity was present from birth and interfered with post-natal alveolar development, when the number of alveoli was reduced in the adult lung. He found that the main abnormality in the lungs was reduction in size of the alveoli. It is usually difficult to demonstrate any structural changes in the lungs apart from mild chronic pulmonary hypertensive arterial changes (grades 1–2). The small muscular pulmonary arteries and arterioles often show intimal longi-

tudinal muscle and may contain recanalized thrombi, both changes being identical with those found in other hypoxic states such as emphysema and chronic mountain sickness. Tortuosity and kinking of the pulmonary arterial tree including the stem and main branches was described by Abrahamson and also by Dunnill (1970).

Respiratory failure and changes due to cor pulmonale usually first appear in middle age due in many instances to the development of emphysema and when they are present they often increase rapidly. Although the respiration rate is usually increased, the shallow breathing results in hypoventilation of alveoli, each breath sufficing in the main to change merely the dead space air. Furthermore, the reduced vertical dimensions of the lungs result, as Dollery et al. have shown, in a more even blood perfusion throughout all regions of the lung compared with that which occurs normally. This change results partly from the reduced vertical dimensions of the lungs and partly from a raised pulmonary venous blood-pressure that is frequently encountered in kyphoscoliosis. More even blood perfusion, although of no significance at rest, results in a more rapid rise of the pulmonary arterial pressure during exercise in the vertical position due to loss of the buffering action normally provided by the intermittently collapsed but readily distensible alveolar capillaries in the upper third of the lungs. In time this will lead to sustained pulmonary hypertension and the development of cor pulmonale. Also the low maximum breathing capacity may result in the terminal stages in a lowered arterial oxygen saturation and to alveolar CO_2 retention. Normally, however, as the lung function studies of Dollery et al. (1965) with ^{133}Xe have shown, although ventilation is reduced it is fairly even throughout the lungs and gaseous exchange is usually adequate. Terminal alveolar retention of CO_2 may lead to loss of respiratory drive due to increasing insensitivity of the respiratory centres which become adapted to a high level of retention. Severely kyphoscoliotic patients are particularly liable to develop respiratory infections including chronic bronchitis, and this may lead to air-trapping and cause superadded anoxaemia. Anoxaemia further stimulates pulmonary vasoconstriction, increasing the condition of cor pulmonale.

PULMONARY HYPERTENSION DUE TO THE "PICKWICKIAN" SYNDROME

In recent years a new cause of pulmonary hypertension and cor pulmonale has been recognized which is associated with an extreme degree of obesity. The term "Pickwickian" syndrome has been applied to this condition because of the physical similarity of these subjects to the literary character portrayed by Dickens. The first description of this syndrome was given by Auchincloss et al. (1955) and further examples have been described by Burwell et al. (1956), Carroll (1956) and Fadell et al. (1962).

In every case the patients have been very grossly overweight, often of a low intelligence and become increasingly dyspnoeic. In some of the cases there has been no evidence of pulmonary hypertension and cor pulmonale has not developed, but the enormous mechanical burden provided by the excessive subcutaneous and lipomatous tissue in other sites leads to increasing respiratory embarrassment. Respiratory movements are hindered by the subcutaneous adipose tissue which acts as a binder to the chest wall, mechanically preventing its expansion.

At post-mortem examination the extreme obesity of the patients contrasts strangely with the often diminutive pleural spaces, and the lungs tend to be small. The lipomatous tissue infiltrates the musculature of the body including the respiratory muscles and according to Fadell et al. impairs their mechanical efficiency. Also the excessive omental and retroperitoneal abdominal fat leads to elevation of the diaphragm which still further restricts the available space for lung movements.

Dunnill (1970) found a normal complement of alveoli, but the greatly increased oxygen uptake and CO_2 output occasioned by the obesity requires increased alveolar ventilation and hypoxia then results on exertion. The respiratory centre in time becomes less sensitive to the high arterial levels of CO_2 as in emphysema and oxygen therapy given for hypoxia may then prove very dangerous.

IDIOPATHIC PULMONARY HAEMOSIDEROSIS

This condition which is discussed in detail in Chapter 19 is a rare cause of chronic pulmonary hypertension. As the condition is considered in detail elsewhere it will not be further described.

Aneurysms of the Pulmonary Artery

Aneurysms of the pulmonary arteries are extremely rare and are customarily divided into those involving the trunk and main pulmonary arteries and those involving the small peripheral arterial branches.

Aneurysms involving the extrapulmonary portions of the pulmonary arteries are very uncommon, and descriptions of these lesions have been given by Boyd and McGavack (1939) and Deterling and Clagett (1947). The latter authors analysed the findings in a large series of postmortems and found the incidence of pulmonary artery aneurysms involving the major pulmonary arteries amounted to eight cases among a total of 109,571 post-mortem examinations. They were able to collect 142 examples of pulmonary arterial aneurysms from the literature.

A true aneurysm should be distinguished from general dilatation of the pulmonary arteries that not uncommonly complicates chronic pulmonary hypertension. Excluding such dilatations, the causes of true aneurysms include syphilis, mycotic aneurysms, trauma and chronic pulmonary hypertension, particularly those forms of chronic pulmonary hypertension associated with congenital anomalies of the heart and great vessels. Unlike aortic aneurysms which occur five times more commonly in males, pulmonary artery aneurysms occur equally in both sexes. Aneurysms are seldom found in the trunk of the pulmonary artery but occur more commonly in the main and segmental arteries in each lung. Pulmonary artery dissecting aneurysms may develop in Marfan's syndrome and subsequently rupture (Natelson et al., 1970; Thomas, 1971).

In approximately 66 per cent of all cases there are associated congenital cardiovascular abnormalities and in about 20 per cent a patent ductus arteriosus is present. Other congenital anomalies associated with these aneurysms include unequal division of a truncus arteriosus and a patent interventricular septum. Ruptured aneurysms have also been described in association with chronic pulmonary hypertension due to primary (idiopathic) pulmonary hypertension and pulmonary schistosomiasis, although some of these latter varieties of aneurysms have followed initial dissection of the walls of the artery.

Syphilitic mesarteritis of the pulmonary arteries is very rare but Deterling and Clagett considered that it was responsible for about 33 per cent of pulmonary arterial aneurysms. It may be associated with thrombosis of the affected arteries.

Mycotic aneurysms are usually secondary to subacute bacterial endocarditis affecting a heart valve, a cardiac anomaly or a congenital vascular lesion such as a patent ductus arteriosus. Subacute bacterial endocarditis may spread from an aortic valve lesion into the pulmonary conus of the right ventricle with subsequent liberation of infected emboli into the lungs. Similar aneurysms occasionally follow the release of infected thrombi from a peripheral vein.

The association of systemic venous thrombophlebosis with single or multiple aneurysms in the major pulmonary arteries was first described by Beattie and Hall (1911). Hughes and Stovin (1959) found that in such cases the bronchial arteries may also show degenerative changes, and the occurrence of venous thrombophlebosis, aneurysms of the pulmonary arteries, pulmonary emboli and degenerative changes in the bronchial arteries is now often referred to as the Hughes–Stovin syndrome. The bronchial arterial changes are thought to affect the nutrition of the pulmonary arteries and this in turn interferes with the normal process of organization of the thrombotic embolus resulting in weakening of the arterial wall and the production of an aneurysm. Most of the cases of this syndrome so far reported have occurred in young men and present with haemoptysis.

Dilatation of the pulmonary artery may occur distal to a congenital stenosis but this type of dilatation is not usually regarded as an aneurysm.

Traumatic aneurysms have also been described but their existence remains unproven.

Aneurysms involving the distal branches of the pulmonary or bronchial arteries are mostly either mycotic aneurysms or result from tuberculous infection (aneurysms of Rasmussen). The former result from infected emboli and the latter are found in the walls of tuberculous cavities. Very rarely intrapulmonary deposits of secondary malignant hydatidiform mole or chorion carcinoma may lead to erosion of branches of the pulmonary arteries and lead to aneurysm formation.

CHAPTER 17

Pulmonary Oedema and Its Complications
and
the Effects of Some Toxic Gases and Substances on the Lung

Pulmonary Oedema

The importance of pulmonary oedema as a potentially fatal process was first recognized by Laennec (1834); he described the changes in the oedematous lung and drew attention to the respiratory embarrassment that the condition causes. Since then much experimental work has been done in an attempt to find out precisely how the oedema is produced, and although many of the experiments have been carried out under highly artificial conditions, the results obtained have undoubtedly helped towards a better understanding of the condition.

Before discussing the causes of pulmonary oedema it is important to recall some of the relevant physiological and anatomical factors concerning the drainage and formation of tissue fluid and lymph in the lung. A summary of the present knowledge of these factors and a review of the experimental work bearing on the subject have been given by Staub (1974).

The alveolar capillaries and venules are generally regarded as the source of tissue fluid in the lung and because of the increased hydrostatic pressure within the capillaries at the base of the lung, amounting to 30 cm H_2O in the erect position, both normal and excessive production of tissue fluid (pulmonary oedema) occurs mainly and at first in the lower halves of the lungs. The capillary endothelial cell junctions lack the "tight" junctions present between the alveolar type 1 pneumocytes which normally prevent the ready access of fluid into the alveolar lumen.

The normal drainage of the tissue fluid is through the alveolar interstitium to gain entry to the pulmonary lymphatics which are larger than lymphatic channels elsewhere in the body. The basement membrane of the pulmonary lymphatics forms a discontinuous structure as was shown ultramicroscopically by Lauweryns and Boussauw (1969) and the endothelium lining the lymphatics also forms a discontinuous layer with gaps between neighbouring cells. The lymphatic endothelial cells are tethered to the extralymphatic tissues by fine fibrils which normally permit the gaps to open to allow the free entry into the lymphatic lumen of tissue fluid and its contained protein molecules together with some red blood-cells (Drinker, 1942). The route and means whereby tissue fluid flows through the alveolar interstitium to gain the lymphatic channels, which are mainly situated in the loose perivascular interstitium, is still conjectural. Staub has postulated, and there is supporting evidence, that fluid and protein molecules follow preferential routes of flow and move as free fluid films along the surfaces of the interstitial collagen and reticulin fibrils rather than through the gel compartment of the interstitial space. Furthermore, as Haas and Johnson (1967) have shown in a foam model, solutions travel rapidly through such models at fluid–gas interfaces. Staub found that injected dye-containing fluids travelled through the alveolar interstitium in the lung, which behaves as a semisolid foam, to gain the loose and more capacious portion of the interstitial compartment situated perivascularly where the principal lymphatic pathways are located. The movement is facilitated by the rhythmical respiratory movements of the alveoli. Propulsion of lymph within the lymphatics is effected partly by the respiratory movements causing intra-alveolar pressure changes, and probably in part by active contraction of the lymphatics themselves when there is excessive lymph production.

The interstitial fluid comprises about 40 per cent of the total lung fluid, a much higher proportion than in other body tissues. In pathological states in which there is excessive production of lymph the pulmonary lymphatics dilate and can accommodate many times their normal flow, but such adaptation requires time for its achievement. Heath and Hicken (1960) showed that the pulmonary lymphatic distension seen in lung biopsies correlated with the levels of left atrial pressure, and the distended lymphatics probably also indicated increased flow rates.

Before discussing the causes of pulmonary oedema, it is important to consider the differences which exist between the capillaries of the lung and those in other parts of the body. Unlike any other organ, the lungs normally transmit the entire output of the heart—the resting flow has been calculated by Drinker (1945) to amount to about 4 litres per minute, reaching a maximum flow of 30 litres per minute after violent exercise. The additional flow is partly accommodated in the very extensive and distensible alveolar capillary bed which is 120–140 m² in extent (Cameron, 1948), and which can accommodate 60 ml of blood at rest and 95 ml during hard work (Roughton, 1945). It may also result in the transudation of fluid through the walls of the capillaries which is removed through the lymphatic network. In

extremely severe exercise frothy fluid may be expectorated (Drinker and Warren, 1943), the surplus intra-alveolar fluid being rapidly reabsorbed when the exercise has ceased.

Pulmonary capillary transudation is subject to the same laws as apply to capillaries elsewhere. The intracapillary pressure has been estimated to be about 9–10 mm Hg (Cameron, 1948) and plasma osmotic pressure is the same as in the rest of the body; thus under physiological conditions the tendency to oedema formation is negligible except during violent exercise, when the greatly increased flow results in raised intracapillary pressures. Under the latter conditions the cardiac reflexes (Bainbridge, 1914) increase the cardiac output, but pulmonary oedema and pleural effusions may occur rarely as transient conditions (Yamada, 1933). Pleural effusion in the past was regarded as a sequel to pulmonary oedema following transudation of intrapulmonary oedema fluid through the visceral pleural surfaces. It is now thought to be mainly caused, however, by a raised systemic venous pressure and transudation through the parietal pleura and consequently complicates chronic heart failure.

Under resting conditions intra-alveolar liquid is absorbed rapidly as was demonstrated by Courtice and Phipps (1946) in animals and by Potter (1952) in babies. The internal surfaces of the alveoli are normally moist and the fluid is thought to be actively secreted by the alveolar cells together with surfactant.

Unlike capillaries elsewhere in the body, those in the lung lack external supporting tissue pressure, but they are cushioned to some extent by the intra-alveolar air pressure which varies during the different phases of the respiratory cycle. In inspiration the pressure normally falls 10 mm below atmospheric pressure and in heavy breathing 70 mm, and in this phase fluid tends to be sucked out of the capillaries, whereas in expiration the position is reversed.

Although the lung possesses a great reserve capacity which prevents excessive tissue-fluid production, should this fail and fluid accumulate in the lung, the consequences are very much more serious than in other tissues as they interfere with normal oxygenation which may further increase capillary permeability, thus establishing a vicious circle.

Among the causes of pulmonary oedema are included:

(1) Capillary pressure alterations.
(2) Capillary damage leading to increased permeability.
(3) Neurogenic oedema.
(4) Oxygen and ozone toxicity oedema.
(5) Transfusional oedema and lowered plasma protein content.
(6) Bronchomotor tonus alterations.
(7) Lymphatic obstruction.
(8) High-altitude oedema.

CAPILLARY PRESSURE ALTERATIONS

For over 50 years it was unquestioningly accepted that interference with the output from the left ventricle in the presence of a normal output from the right ventricle was responsible for raising the pulmonary capillary pressure and causing pulmonary oedema. This simple explanation followed from a series of experiments on rabbits and dogs by Welch (1878) in which the aorta and innominate arteries were tied and the left ventricle damaged by crushing.

This widely accepted view was assailed by Cameron (1948) who pointed out that the distensibility of the pulmonary capillaries coupled with baroceptor reflexes prevent the onset of oedema. He believed that when oedema occurred in the lung following left ventricular failure additional factors were responsible.

One of the factors proposed was the existence of a coronary artery–pulmonary capillary reflex whereby failure of an adequate coronary artery blood-flow caused reflex pulmonary oedema. This suggestion was not supported by the work of Paine et al. (1950), who showed in animals that obstructing the aorta distal to the openings of the coronary arteries did not prevent them from developing pulmonary oedema.

Pulmonary oedema may occur despite the distensibility of the pulmonary capillary bed when the volume of blood in the lungs is very excessive and the intracapillary pressure rises. A shift in the blood volume may occur from the systemic capillary circulation to the lung capillary bed following excessive liberation of catechol amines of nor-adrenaline type, an occurrence that may be stimulated by the decreased cardiac output and by the anxiety caused by an ischaemic cardiac episode causing a diminished cardiac output from the left ventricle.

Recent experiments, in which the left atrial pressure was raised in dogs by narrowing the mitral valve, have shown that the occurrence of pulmonary oedema is very closely related to the left atrial pressure (Guyton and Lindsey, 1959). Also Gorlin et al. (1951) found that in mitral stenosis in man the onset of pulmonary oedema is dependent on the level of the lung capillary pressure, appearing when the latter exceeds 35 mm Hg at rest.

In infarction of the left ventricle in man, in which condition the output from this chamber is known suddenly to cease or fall to very low levels, there invariably follows severe congestion and oedema of the lungs. In cases of fatal quinidine poisoning, in which sudden total cardiac arrest occurs, the lungs remain dry and are not congested, whereas in infarction the right ventricle probably continues to contract in a dissociated fashion for a few beats, thus producing the oedema.

CAPILLARY DAMAGE

Capillary damage is the most important factor in the development of pulmonary oedema (Drinker, 1945) and of all the causes of such damage anoxia is the most important. Anoxia alone is probably incapable, except when present in extreme degree, of causing pulmonary oedema (Gibbs et al., 1943), but it is often associated with cardiac failure which causes a rise of intracapillary pressure, and these two factors combined produce the oedema (Drinker).

Many chemical agents may damage the alveolar capillaries, rendering them extremely permeable; such substances include phosgene, chlorine, irritating smoke gases, ammonia, α-naphthyl thiourea (ANTU), nitrogen dioxide, ozone, cadmium fumes and many others.

Although the ultralight microscopic changes induced in the lungs of experimental animals by many of these agents have not been studied, those caused by ANTU have been investigated by Meyrick et al. (1972). They found that oedema fluid first appeared in the perivascular connective tissue sheaths, later spreading to form blebs beneath and outside the capillary endothelium which it displaced. The change is later followed by intra-alveolar oedema and greatly increased pinocytosis in both the alveolar epithelial cells and capillary endothelium. Individual idiosyncrasy to some drugs, notably nitrofurantoin (Israel and Diamond, 1962), may cause pulmonary oedema which can be produced by giving small doses of the offending agent. A possible cardiac cause for the oedema can be excluded as the circulation time remains unaltered. Likewise pulmonary oedema may also result from the direct toxic effects of circulating intravascular chemicals including organic iodine derivatives used for angiography (Sawyers et al., 1964). Some protection from the latter substances may be provided by the simultaneous injection of 20 per cent glucose or low molecular weight dextrans.

Capillary damage can also result from bacterial and viral agents. Foremost among the bacteria causing oedema is the β-Haemolytic streptococcus which gives rise to an abundant sero-sanguineous alveolar exudate. Influenza virus pneumonia, as distinct from the secondary bacterial pneumonias, also causes fulminating and very severe alveolar capillary damage which results both in intra-alveolar haemorrhage and extensive oedema.

Physical agents such as excessive X-irradiation are capable of producing severe capillary damage resulting initially in an area of pulmonary oedema commensurate with the irradiated area, and these changes may be observed in patients who die shortly after being irradiated for lung and breast tumours.

In acute rheumatic fever the extensive pulmonary oedema which sometimes occurs is partly the result of cardiac failure, but it is likely that a hypersensitivity capillaritis is also partly responsible.

In almost all the conditions described above, the increased capillary permeability, in addition to allowing the escape of oedema fluid, also allows the escape of more or less plasma protein, which by raising the osmotic pressure of the intra-alveolar contents still further increases intra-alveolar transudation.

NEUROGENIC OEDEMA

Pulmonary oedema frequently occurs in conditions of raised intracranial pressure in man, and was formerly known as hypostatic oedema. This so-called "neurogenic oedema" has been investigated by Cameron and De (1949), Cameron and Sheikh Hamid (1951), Harrison and Liebow (1952), Sarnoff (1951) and Sarnoff and Berglund (1952b).

Neurogenic pulmonary oedema follows almost instantaneously upon brain injury and is now thought to be associated with hypothalamic disturbance and is independent of cardiac disease or disturbance. It can be inhibited in experimental animals by the previous administration of α-adrenergic blocking drugs.

Two forms of experimental "neurogenic" pulmonary oedema are recognized. Cameron and De and Sarnoff (1951) found that pulmonary oedema could be caused in dogs by the intracysternal injection of fibrin. This procedure brought about a reflex constriction of the peripheral systemic arteries which caused a shift of the blood volume to the pulmonary circulation, resulting in a secondary rise of the pulmonary capillary pressure. This form of oedema was prevented by vagotomy and is probably related to the human neurogenic form of pulmonary oedema that follows a rapid rise in intracranial pressure especially when the latter is associated with head injuries. This form of pulmonary oedema in man does not occur if there is simultaneous accidental cervical spinal cord transection.

The second form of "neurogenic" oedema

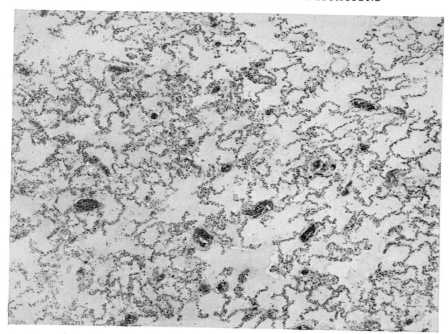

FIG. 17.1. Section of lung showing acute pulmonary oedema. × 40 H and E.

was demonstrated by Sarnoff and Kaufman (1951) who administered ammonium chloride to cats and found that two out of the twelve treated animals developed pulmonary oedema. The oedema followed a rise in the left atrial and vena caval pressures which were caused by the failure of the left ventricle; this change was mediated through the sympathetic nervous system and could be abolished by sympatholytic drugs such as dibenamine. This form of oedema may also be partly caused by the ammonium ion which causes some local toxic effect on alveolar structures similar to that caused by ANTU (Hayes and Shiga, 1970).

OXYGEN AND OZONE TOXICITY OEDEMA

Pulmonary oedema following exposure to high concentrations of oxygen was first produced in animals by Boycott and Oakley (1932, 1933) and later by Pratt (1958). It was originally considered to be caused by vagal stimulation but it is now regarded as a consequence of the direct toxic action of oxygen on the alveolar wall components.

Oxygen Toxicity

During the past 30 years it has gradually been appreciated that exposure to concentrations of oxygen in inspired gases in excess of 40 per cent (Sevitt, 1974), far from being beneficial, may if prolonged for more than a few minutes result in permanent damage to the lung or death. The lungs of different animal species vary in their susceptibility to oxygen toxicity. According to Kapanci *et al.* (1969) the susceptibility of the monkey approximates to that of man and this species has been used to study the early stages in the evolution of oxygen toxicity in the human lung. The effects of oxygen on the human lung have been studied by Pratt (1958, 1965) and

Bonikos *et al.* (1975). The lesions comprise the early onset of a severe form of fibrinous oedema, the formation of hyaline membranes, thickening of the alveolar walls as well as loss of bronchial cilia and foci of bronchiolar mucosal necrosis. Later interstitial alveolar fibrosis supervenes, the alveoli become lined by light microscopic visible alveolar epithelium and focal dilatation of alveolar capillaries occurs. Death may take place within the first 10 days from acute respiratory failure and if the patient survives some permanent impairment of respiratory function may result.

The ultramicroscopic changes in the lung caused by breathing pure oxygen have been described by Kapanci *et al.* and Pariente (1975). The latter also studied the effects of other oxidizing agents on the lungs. The earliest change seen after 48 hours was swelling of the alveolar capillary endothelial cells with some interstitial oedema. At 96 hours both the types 1 and 2 pneumocytes showed extensive damage, especially the former, many being shed to expose the underlying basement membrane which was covered with a fibrin containing exudate to form hyaline membranes. At the same time the interstitial oedema became greatly accentuated and focal vesicles formed beneath the alveolar capillary endothelium. The endothelial cells themselves swelled and some were later destroyed. By 7 days the alveolar epithelial repair by type 2 pneumocytes was well advanced though these cells were swollen and contained few lamellated bodies. The capillary endothelial cells remained very swollen and also the interstitium into which red blood-cells had extravasated. During the subsequent days focal cellular proliferation and interstitial alveolar fibrosis appeared and the less-affected remaining interstitium was thickened. Some alveolar capillaries were destroyed by the initial damage, others by the subsequent interstitial changes, while the volume of the remainder was increased and was the probable explanation of the pulmonary capillary "proliferation" previously noted by Pratt in human lungs. Accompanying the alveolar changes, the bronchiolar epithelium is also damaged though to a lesser extent.

Ozone and Nitrogen Dioxide

The action of ozone and nitrogen dioxide has been investigated experimentally in animals and the changes followed by electron microscopy by Plopper *et al.* (1973) and Bils (1970).

Ozone causes maximal damage in the more central parts of the acini where it damages both the alveolar epithelial and capillary endothelial cells, resulting in interstitial and intra-alveolar oedema. In the less-damaged peripheral portions of the acini the damage is confined to the alveolar capillary endothelium and interstitial oedema precedes the onset of intra-alveolar oedema. Ozone in addition affects the intracytoplasmic organelles in the type 2 pneumocytes resulting in a reduction of lamellar bodies and consequent atelectasis.

The effects of nitrogen dioxide are considered under "silo-filler's lung".

TRANSFUSIONAL OEDEMA AND LOWERED PLASMA PROTEIN PRESSURE

The possibility that intravenous administration of fluids may, when carried to excess, cause pulmonary oedema has become a question of more than academic importance now that the parenteral injection of fluids is so widely used. Clinical experience has taught that intravenous infusions both of fluids and whole blood in cases of chronic severe anaemia may be fraught with danger. In such persons acute left ventricular failure may be precipitated partly due to the fatty degenerative changes present in the heart muscle interfering with the normal function of the muscle.

Altschule *et al.* (1942) investigated the effect of large intravenous fluid injections in normal persons and found that up to about 2 litres of saline, administered at 185 ml per minute, was incapable of causing pulmonary oedema. These authors felt, however, that in patients with chronic heart failure or renal disease the results would have been different.

Cutting *et al.* (1939), working on the same lines, found that when animals were transfused until the venous pressure rose, cardiac decompensation and death occurred. Post-mortem, the bowel and peritoneum of the animals were found to be filled with fluid and there was extensive pulmonary oedema. The raised venous pressure in the great veins may also have obstructed lymphatic drainage from the thoracic duct.

Opdyke *et al.* (1948), in similar experiments, found that massive intravenous infusions of saline raised the left atrial pressure more than the right; they considered this to be the principal mechanism responsible for causing pulmonary oedema. They also found that a close relationship existed between the level of plasma protein and the pulmonary venous pressure in the production of oedema of the lungs, animals with low plasma protein developing pulmonary oedema with a correspondingly lower rise of pulmonary pressure.

BRONCHOMOTOR TONUS AND OEDEMA

The part played by bronchomotor tonus in regulating alveolar capillary flow and the control of pulmonary oedema has hitherto been neglected, and is still little understood. Rodbard (1950 and 1953) emphasized its importance and drew attention to the fact that compared with other tissues the intra-alveolar air pressure replaces tissue pressure as a means of alveolar capillary support.

Nissen (1927) reported a case of acute massive oedema of the lung which followed immediately upon the resection of a chronic tracheal stenosis. It is clear that in this case the tracheal narrowing must have produced a chronic rise in the intra-alveolar air pressure and it is possible that with the sudden release of the obstruction, alveolar pressure fell, thus diminishing the support of the capillaries and producing the oedema.

Also, Burford and Burbank (1945), discussing the nature of acute massive collapse of the lung—a rare complication of either injuries to the chest or upper abdominal operations—showed that this condition is preceded by severe pulmonary oedema which in turn is followed by bronchospasm and finally massive absorption collapse. Clinically, such patients have been described as suffering from bronchial asthma. Teleologically, the bronchospasm could be regarded as an acute reaction to reduce the pulmonary oedema. Bronchospasm has also been reported by Rodbard (1953) in the early oedematous phase of pulmonary infarction due to embolism.

In chronic passive pulmonary hypertension where there is a persistent tendency to pulmonary oedema, hypertrophy of the muscles of the bronchioles and alveolar ducts often occurs. Clinically, this condition shows itself as cardiac asthma.

LYMPHATIC OEDEMA IN THE LUNG

Although the lung is richly supplied with lymphatic vessels, the lymph-flow is normally small. Courtice (1963) found that in dogs the lymph-flow from the lungs amounted to about only one-tenth of the flow in the thoracic duct. This may in part be attributable to the normally low pulmonary capillary pressure. An acute rise in pulmonary venous pressure, however, leads to the very rapid onset of pulmonary oedema and owing to the relatively indistensible nature of the pulmonary lymphatic channels, pulmonary oedema rapidly ensues. In acute experimental pulmonary oedema provoked by compression of the pulmonary veins, red blood-cells extravasate and reach the lymphatic channels in addition to lymph (Warren and Drinker, 1942), and the alveolar basement membrane becomes considerably thickened (Kisch, 1958) with accumulation of tissue fluid in the interstitial plane when viewed electron microscopically.

The pulmonary lymphatic system also plays an important part in the resorption and drainage of fluids from the alveolar lumen as is seen in the lungs of drowning persons and animals.

HIGH-ALTITUDE PULMONARY OEDEMA

Synonyms: Mountain sickness, Soroche

Since the close of the last century many cases of non-fatal and occasional cases of fatal pulmonary oedema have been reported among climbers and visitors to places above an altitude of 9000 feet (3000 metres). The dyspnoea associated with rapid ascent to a high altitude by non-acclimatized persons may be of immediate or delayed onset. Heath (1975) considers the former to be due to the immediate effects caused by hypoxia. The better-known mountain sickness, however, takes up to 6 hours to develop and is often most severe and noticeable at night and results from acute pulmonary oedema. It is with this condition that this account is concerned.

The first case of high-altitude pulmonary oedema was described by Mosso (1898) in a physician who died climbing in the Alps. Later Hurtado (1937) described the first case of a similar nature in the Andes and during the past 15 years an increasing number of cases of high-altitude pulmonary oedema have been reported from Peru and Bolivia. A few cases have also occurred in North America in the higher ranges of the Rocky Mountains and a large series of cases occurred among Indian troops engaged in mountain warfare in the Himalayas in northeast India. The same condition was also experienced by some of the Chinese troops operating in Tibet (Li and Liu, 1958). These examples illustrate the importance of acute mountain sickness in military medicine when troops are rapidly transported to a high altitude. Unless some days are allowed for physiological adaptation to living at a high altitude the military effectiveness of such troops will be gravely impaired.

The condition mainly affects young, healthy, active persons shortly after arrival at a high altitude or may affect the normal inhabitants at a high altitude if they live at sea level for a short time and later return to their original domicile. It has also been recognized as a hazard of high-altitude flying when pressurization is inadequate or faulty. Insufficient time for physiological adaptation to living at a high altitude, often coupled with a familial tendency to the condition, are the principal factors in the development of this form of oedema. Active exercise at a high altitude is often a precipitating factor.

Recent accounts of high-altitude pulmonary oedema have been given by Houston (1960), Hultgren et al. (1961 and 1962), Arias-Stella and Kruger (1963), Nayak et al. (1964) and Inder Singh et al. (1965).

Clinically, the condition may present acutely as "mountain sickness" characterized by increasing dyspnoea, a dry cough, vomiting and weakness or may follow premonitory headache and insomnia due to dyspnoea. This may develop into a clinical picture of severe pulmonary oedema characterized by cough, cyanosis, chest pain, vomiting, hallucinations and the production of copious watery and frothy sputum. Death may follow in unrelieved cases in from 12 hours to 1 week. Relief is provided by the administration of oxygen and removal to a lower altitude.

Post-mortem, the lungs are bulky and plum-coloured due to very severe pulmonary oedema which is usually more severe in the lower lobes. Areas of haemorrhage and foci of bronchopneumonia are common and pleural effusions are usually present. Very rarely death may occur suddenly due to a shock-like condition in persons recently arrived at high altitudes and is thought to be caused by cardiac arrhythmias.

Microscopically, the lungs show very severe and generalized oedema with deposition of fibrin and collections of red blood-cells within the alveoli and considerable lymphatic distension. Hyaline membrane formation is common and sludging of red blood-cells causes thrombi in the alveolar capillaries, small pulmonary arteries and arterioles. Numerous intrapulmonary magakaryocytes have been described within the alveolar capillaries. Evidence of bacterial pneumonia may be found in cases where death was delayed but is often absent. In some cases the pneumonia is related to aspirated stomach contents. The cause of this form of pulmonary oedema in man is still unknown, but as was shown by Hultgren et al. (1961) blood is redistributed from the systemic to the pulmonary circulation. This results from the peripheral systemic vasoconstriction that occurs during the early stages of acclimatization to a high altitude. The redistribution of the blood volume leads to distension of the large pulmonary arteries and alveolar capillaries though the left atrial pressure remains normal. For this reason pulmonary venous obstruction (vasoconstriction) was thought to play an important role in causing the pulmonary oedema.

Heath *et al.* (1973) exposed rats to a hypobaric atmosphere of 265 torr (approximately 29,000 feet or 9800 metres altitude) and then examined their lungs electron microscopically. They found that after exposure for 12 hours to the low-pressure focal vesicular collections of fluid were present in the basement membrane outside and beneath the alveolar capillary endothelial cells. The latter became greatly stretched over the surface of the vesicles and were invaginated into and in places almost blocked the capillary lumens. Similar vesicular collections of fluid in the alveolar capillary basement membrane occur in other forms of pulmonary oedema and it is likely that they cause mechanical obstruction to the capillary blood-flow as well as being associated with coincident alveolar oedema. Removal to a lower altitude and rectification of the hypoxia would be expected to reverse such changes.

Alkalosis induced by the hyperpnoea may retard haemoglobin oxygen desaturation promoting tissue anoxia and pulmonary venous spasm. Failure to inactivate serotonin due to anoxia has also been postulated as a cause of pulmonary arterial vasoconstriction and of the consequent pulmonary hypertension that occurs. The increase in atmospheric ozone found at such high altitudes and at certain seasons in the High Andes may also contribute to the pulmonary oedema.

No evidence has been found to show that any form of myocardial damage plays a part in the onset of high-altitude pulmonary oedema although the electrocardiographic and radiographic changes may sometimes simulate myocardial infarction at the height of the illness.

Other cases may show electrocardiographic and radiographic evidence of pulmonary hypertension probably caused by hypoxia.

A similar state to human high-altitude pulmonary oedema has been described in cattle grazed at high altitudes during the summer months in the Rocky Mountains in Colorado and Utah. Oedema fluid collects in the dependent subcutaneous tissues of the trunk especially in front of the forelegs, a condition known as brisket disease (Hecht *et al.*, 1962). Although of slower and more insidious onset than the human disease the condition is probably of similar aetiology and often proves fatal.

PLEURAL EFFUSIONS IN PULMONARY OEDEMA

Pulmonary oedema is often accompanied by pleural effusion. The mechanism of its production is uncertain and it probably results from several causes.

In chronic heart failure transudation into the pleural sac arises mainly from systemic venules lying beneath the parietal pleura. In neoplastic disease and radiation pneumonitis obstruction to the free flow of lymph both in the intrapulmonary lymphatics and through the mediastinal lymph glands is an important factor but lymphatic obstruction has to be widespread as the system of pulmonary lymphatics is both very extensive and intercommunicating.

Brock and Blair (1931–2) found that in dogs pleural effusion depended partly on the respiratory movements; during respiration the alveolar interstitium gained fluid from the alveolar capillaries which entered the lymphatics of the lung and pleura. From overdistended pleural lymphatics it then flowed into the pleural cavity on account of the cyclical pressure changes during respiration. Pleural effusions may occur rarely in physiologically produced pulmonary oedema caused by very severe exercise as was found by Yamada (1933) in violently exercised soldiers.

Obstruction of the pulmonary lymphatics may

occasionally result in oedema of the lungs. Because the pulmonary lymphatic system is so extensive and intercommunicating, the obstruction has of necessity to be widespread throughout the lymph-vessels. In the later stages of such cases the pulmonary oedema is often not apparent, as the lung remains collapsed by the gradually increasing pleural effusion which accompanies the pulmonary oedema.

Macroscopically, the lungs in acute pulmonary oedema are plum-coloured and retain their shape well in the oedematous areas. They are heavier than normal and the oedematous tissue sinks in water; the cut surface pours frothy, blood-stained fluid on gentle pressure and similar frothy fluid fills the air passages.

Acute Pulmonary Oedema due to Chemical Agents

Although numerous gases can cause pulmonary oedema, phosgene and chlorine, because of their use as war gases, merit particular attention. More recently the gaseous oxides of nitrogen have claimed increasing attention as causes of pulmonary oedema leading to permanent pulmonary damage, and are considered in some detail.

PHOSGENE

This gas has a direct action on pulmonary capillary endothelium, as shown by Short (1942) in animal experiments. It is readily hydrolysed and damages the capillary endothelial mitochondria before the capillaries begin to leak fluid. At first this fluid is mainly a transudate but later all the plasma proteins escape. Very few opportunities occur for studying the effects of this gas on man but a few specimens of lungs from soldiers gassed in the First World War still exist.

Occasionally accidental inhalation of phosgene may occur in chemical laboratories, and can also result from the use of methylene chloride paint stripper when it is used in confined spaces in the presence of a naked flame. Traces of carbonyl chloride (phosgene) are formed and a fatal case of poisoning due to this cause was described by Gerritsen and Buschmann (1960). The effects caused by phosgene may be immediate or delayed in onset for 2 to 3 hours. The alveoli fill with oedema fluid, including some fibrin, which is mainly responsible for forming hyaline membranes found in such cases (Fig. 17.2). The epithelium lining the air passages also suffers damage and in the smaller bronchi it may be completely destroyed (Fig. 17.3). Later, infection supervenes and should the patients survive, widespread bacterial pneumonic consolidation occurs. Accompanying the initial changes there is oedema of the interlobular septa, distension of the lymphatic channels throughout the lung, and the frequent co-existence of a pleural effusion.

Similar changes occur in the lungs of persons exposed to chlorine.

Gross *et al.* (1965) showed experimentally in rats that inhalation of very small amounts of phosgene irritated the alveolar epithelium but was insufficient to damage the alveolar capillaries. As a result there was proliferation of the alveolar and bronchiolar epithelium and a chronic pneumonitis developed with production of reticulin fibrils. The damage was maximal in the respiratory bronchioles where the walls were unprotected by a film of mucus and because there was less exudation of fluid from underlying capillaries to dilute the noxious agent. It resulted in an obliterative bronchiolitis. These changes are not seen in human phosgene poisoning as the condition usually proves fatal and if recovery occurs the lung pathology remains unknown. These experimental findings are of importance as they help to explain the changes seen in human lungs following exposure to other irritant gases such as nitrogen dioxide.

NITROUS FUMES

Synonym: Silo-filler's disease

Nitrogen forms a series of gaseous oxides of which nitric oxide (NO) and nitrogen dioxide (NO_2, N_2O_4) are the most likely to cause

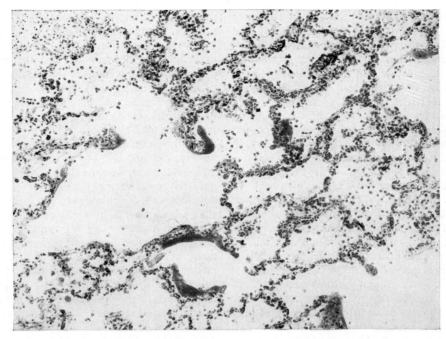

FIG. 17.2. The early oedematous phase showing hyaline membranes in the lung from a fatal human case of phosgene poisoning. × 140 H and E.

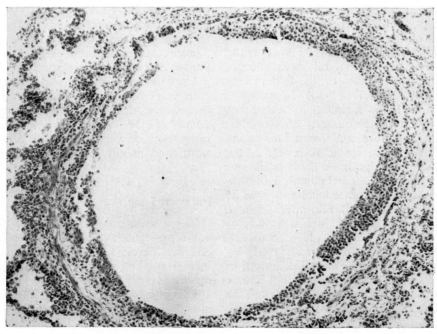

FIG. 17.3. Acute bronchitis showing destruction of the bronchial epithelium due to phosgene poisoning. × 140 H and E.

damage to human lungs. Nitric acid is used very extensively in industry for copper, brass and silver dipping, the preparation of nitrocellulose, collodion and methyl nitrate, and the production of sulphuric, chromic and picric acids, together with many other substances. It may also in the future be increasingly used as an oxidizer in rocket fuel. In addition to the nitrous fumes which may arise in any of the above processes, they are also liberated in electric arc welding, in mines using explosives with inadequate ventilation and in silos filled with cattle forage collected during times of partial drought. In conditions of drought, the forage (corn) is often harvested before it is fully grown and the young plants contain a higher nitrate content. Later during storage, anaerobic fermentation of potassium nitrate liberates nitrates which, as the temperature rises with fermentation, liberate various oxides of nitrogen. Lethal concentrations of nitrogen dioxide are not usually reached before a week to 10 days after filling the silo and may take as long as 6 weeks to be generated.

Hayhurst and Scott (1914) gave an account of four silage workers who sat on top of fresh silage and became unconscious after a few minutes and all of whom subsequently died. It is uncertain, however, whether these fatalities were caused by the inhalation of nitrous fumes and whether this account was the first record of silo-filler's disease. The first full description of the effect of nitrous fumes on the human lung was given by Nichols (1930) following a disastrous fire in a hospital in Cleveland, U.S.A., in which a large quantity of X-ray film made of nitrocellulose underwent slow combustion in a confined space. A review of the literature and a description of the bronchiolitis obliterans that may follow inhalation of sub-lethal amounts of the fumes has been given by Darke and Warrack (1958). Silo-filler's disease, due to the same cause, has been described by Delaney et al. (1956), Grayson (1956), Rafii and Godwin (1961) and Moskowitz et al. (1964).

The outcome following the inhalation of nitrous fumes (mainly nitrogen dioxide) in man depends on the concentration of the gas and the length and frequency of exposure. The maximum safe industrial concentration is generally considered to be about 10 parts per million of air. Following inhalation of nitrogen dioxide, which is readily soluble in water, the respiratory epithelium is extensively damaged and often becomes totally destroyed, being completely shed especially in the small bronchi and bronchioles. Coincidentally, the alveolar capillaries are damaged and oedema fluid rapidly seeps into the alveoli, welling up as a frothy, watery, mucoid and blood-tinged fluid into the air passages. A latent interval of several hours may occur between exposure to the gas and the development of pulmonary oedema similar to that which may follow exposure to phosgene, cadmium and mercury vapours.

Exposure to sub-lethal doses may at first cause pulmonary oedema, but later bronchopneumonia and bronchiolitis obliterans develop. *Macroscopically*, in this latter condition the lungs contain palpable small nodules and show haemorrhages together with increasing interstitial fibrosis. *Microscopically*, a well-developed bronchiolitis obliterans is found together with generalized interstitial lymphocytic infiltration and slight fibrosis, and the alveoli become filled with macrophages (Fig. 17.4). Serial lung biopsies have shown that most of these changes are reversible and may have disappeared within 6 months. In most severe cases some interstitial fibrosis and centrilobular emphysema persist.

Lung function studies show that the initial pulmonary oedema causes a reduction of lung compliance, decreased vital capacity, and the arterial oxygen saturation may be reduced. Long-term recovery of function depends on the severity of the initial structural damage.

Paraquat Lung

The introduction of paraquat (1,1'-di-methyl-4,4'-bipyridylium dichloride) in 1964 as a very effective herbicide has been followed by an increasing number of fatalities in man due to its ingestion accidentally, suicidally and occasionally from homicidal intent. Fatal poisoning has also occurred following its absorption through

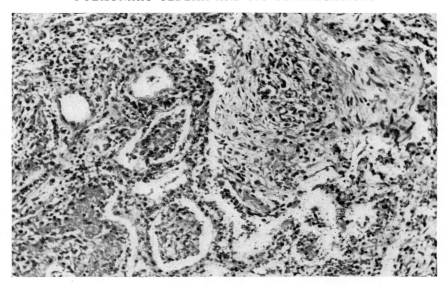

FIG. 17.4. Organizing pulmonary oedema and bronchiolitis obliterans resulting from silo-filler's disease (nitrogen dioxide inhalation). × 90.
(Reproduced by courtesy of Dr. A. A. Liebow.)

the skin. Fatal human cases have been described by Bullivant (1966), Matthew et al. (1968) and Smith and Heath (1974). Of the 474 deaths caused by paraquat up to the beginning of 1975, 102 had occurred in Japan and 60 in Great Britain. Absorption occurs very rapidly and can occur in the mouth where it causes ulceration and may prove fatal. Paraquat dichloride is the active ingredient of the herbicides "Weedol" and "Gramoxone". Following its ingestion it causes diarrhoea and in addition damages the renal tubules and centrilobular cells in the liver, but it mainly accumulates in the lungs and muscles from which it is only slowly eliminated over a period of 3 weeks or longer.

Damage to the lung becomes apparent about the fourth day and leads to increasing dyspnoea, cyanosis and death from respiratory failure from 4 days to 5 weeks following ingestion. The appearance of the lungs depends on the length of survival following ingestion. If death occurs within a week the lungs are heavy and oedematous and are similar to lungs following oxygen poisoning. If survival continues beyond this

time the lungs assume the character of honey-comb lungs common to most forms of interstitial lung fibrosis.

Microscopically, in the first few days the alveolar capillaries are congested and the alveoli are filled with a fibrinous exudate containing desquamated alveolar epithelial debris, macrophages, red blood-cells and contain hyaline membranes. The epithelial destruction extends into the distal respiratory bronchioles.

Depending on the length of survival varying degrees of organization of the intra-alveolar exudate by plump, strap-like fibroblasts is seen and the alveolar interstitium becomes thickened due to mononuclear cell infiltration, haemorrhage and interstitial fibroblastic proliferation. The longer the period of survival the more extensive becomes the alveolar and interstitial fibrosis and consequent bronchiolectasis. The interstitial elastic tissue may persist almost undamaged as the interstitial fibrosis results primarily from organization and incorporation of intra-alveolar exudate. The small muscular pulmonary arteries begin to show medial thickening and with the development of a honeycomb change intimal longitudinal muscle bands appear in these vessels. Regeneration of type 2 pneumocytes occurs to cover the alveolar surfaces but owing to the severity of the alveolar damage almost the whole of the alveolus may come to be filled with young connective tissue.

The ultramicroscopic changes seen in experi- mentally induced lung lesions in rats have been described by Vijeyaratnam and Corrin (1971) and Smith and Heath. Both reported that the earliest changes were observed in the type 1 alveolar epithelial cells which became swollen. This was followed very shortly by hydropic degeneration in both types 1 and 2 pneumocytes which both proceeded to undergo necrosis at the end of 48 hours, exposing the underlying

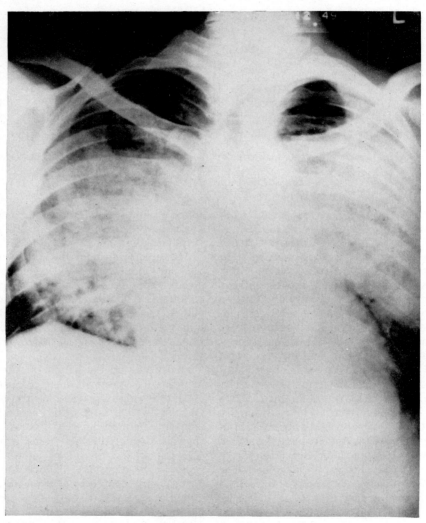

Fig. 17.5. An X-ray of the lungs showing "batswing" type of shadow seen in uraemic lung. (Reproduced by courtesy of Dr. R. C. Grainger and the Editor of *Brit. J. Radiol.*)

basement membrane. The subsequent intra-alveolar exudation observed by light microscopy then followed and the fibrinous exudate was rapidly invaded by immature fibroblasts. The earliest changes have been claimed by others to occur in the alveolar capillary endothelial cells, and the alveolar epithelial changes though early and much more striking are considered to be secondary to the initial endothelial changes. If this view is correct then paraquat lung resembles the changes seen in oxygen and ozone toxicity.

The slow elimination of paraquat from the lung and muscles should be borne in mind if early lung transplantation is considered, as the transplanted lung may suffer the same paraquat damage as the one it is replacing.

Both paraquat and ozone cause the production of superoxides which are the agents responsible for causing cellular damage. For this reason superoxide dismutase has been used with some success in the treatment of paraquat poisoning.

"Uraemic" Lung (Fibrinous Pulmonary Oedema)

Synonym: Azotaemic lung

This unfortunately named condition was first recognized by radiologists because it causes a characteristic butterfly or "batswing" type of shadow in antero-posterior X-rays of the lung (Fig. 17.5). The periphery of the lung remains normally translucent, whilst the more central areas are opaque and display a waist-like constriction of the shadow (Day et al., 1929).

The pathological change underlying the radiological shadows remained unknown until it was described by Doniach (1947), who recognized that it was a form of pulmonary oedema. As it frequently complicates advanced renal failure it was originally referred to as "uraemic" lung, and this widely applied name will still be used to describe this condition.

Ehrich and McIntosh (1932) described changes in the lungs of patients with renal failure but failed to recognize the nature of the process

and referred to it as bronchiolitis obliterans, noting that no micro-organisms were found, that there was no destruction of bronchial walls, and that the condition was frequently associated with serous effusions caused by coexistent renal failure.

In 1937 Masson et al. described an unusual form of chronic intra-alveolar fibrosis resulting from organization of intra-alveolar fibrinous exudate in acute rheumatic fever, and Hadfield (1938) and Neubuerger et al. (1944) also described similar changes which they called "rheumatic pneumonia". In retrospect, it is obvious that all these earlier descriptions referred to a similar condition to that now under discussion.

Now that "uraemic" lung is considered a form of pulmonary oedema, its occurrence in non-uraemic conditions has been recognized; these include left ventricular failure from essential hypertension (Hodson, 1950) and acute rheumatic fever. It had also been reported by Heard et al. (1958) in acute tubular nephrosis due to surgical shock following an operation. It is encountered not in conditions of acute left ventricular failure such as accompany coronary insufficiency, but in the more chronic forms of failure which accompany the conditions described above.

PATHOLOGY

Macroscopically, the lungs are plum-coloured, bulky, of a soft rubbery texture and retain their shape; the cut surface exudes a little frothy fluid but only on firm pressure. The distinction between the oedematous and non-oedematous parts of the lung which is so clear in radiographs cannot usually be appreciated on direct examination.

Microscopically, the lung is very congested, the interlobular septa are swollen and oedematous, and the alveoli and alveolar ducts are filled with eosinophilic proteinaceous fluid (Fig. 17.6). The quality of the oedema fluid varies from one case to another, but there is often an abundance of fibrin and hyaline

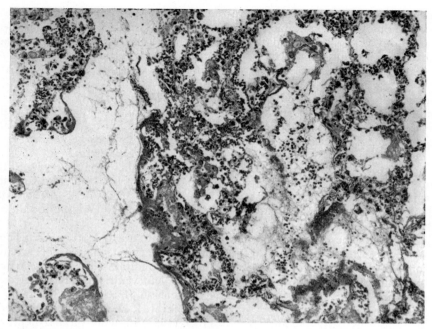

Fig. 17.6. The early oedematous phase of uraemic lung. The alveoli contain fibrinous oedema fluid and the inter-lobular septum is very oedematous and swollen. × 140 H and E.

membrane formation is common. In such cases the appearances suggest a lobar pneumonic exudate without any polymorph leucocytes and very few red blood-cells.

In many cases the oedema fluid is completely absorbed, but in more chronic persistent cases in which the oedema fluid is rich in fibrin, the exudate organizes and is gradually replaced by intra-alveolar and intraductular whorls and plugs of oedematous fibrous tissue—the *bourgeons conjonctifs* described by Masson *et al.* (1937) in "rheumatic pneumonia" (Fig. 17.7). This condition constitutes bronchiolitis obliterans. The newly formed fibrous tissue may contain macrophage cells laden with haemosiderin and intact red blood-cells and the surface of the whorls may later become covered with alveolar epithelial cells. Organization of the intra-alveolar exudate proceeds from the walls of the alveolar ducts. In addition to the intra-alveolar changes described above, the more chronic cases show epithelialization of the

alveolar walls and often the alveolar capillaries are separated from the newly developed epithelial cells by a clear zone occupied in life by oedema fluid.

Morrow *et al.* (1953), Doniach *et al.* (1954) and Freiss (quoted by Pokorny and Hellwig, 1955) described a form of chronic interstitial fibrosis of the lung which followed treatment of essential hypertension with I-hydrazinophthalazine (apresoline) and hexamethonium chloride. Morrow *et al.* believed that the lung changes that sometimes followed apresoline were directly attributable to the drug which is known occasionally to cause diffuse lupus erythematosus. As chronic interstitial lung fibrosis may follow both in diffuse lupus erythematosus and in the closely related rheumatoid disease their belief may be correct but it does not account for similar changes that may follow prolonged hexamethonium therapy for essential hypertension. Doniach *et al.* considered that hexamethonium therapy by controlling the

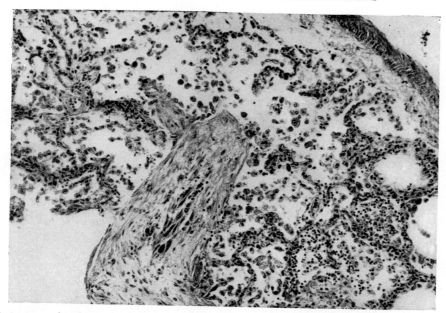

FIG. 17.7. Later stage showing organization of the intra-alveolar exudate and filling of an alveolar duct with a plug of young connective tissue containing haemosiderin-laden macrophages. This stage has been described as non-inflammatory bronchiolitis obliterans. × 140 H and E.

hypertension prolonged life and increased the chronicity of the terminal oedema. Similar lung changes have also been reported following the use of pentolinium (Hildeen *et al.*, 1958) and mecamylamine (Rokseth and Sorstein, 1960). All these drugs cause a fall in pulmonary arterial pressure due to pulmonary vasodilatation, and Viersma (1955) (quoted by Heard, 1962) believed that it was the excessive pooling of blood in the alveolar capillaries which was responsible for causing pulmonary oedema.

Heard (1962) suggested that deficient fibrinolysis predisposed to the later production of chronic interstitial fibrosis of the lung, a finding supported by MacLeod *et al.* (1962) who showed that there was deficient activator of fibrinolytic enzyme in the lungs of patients dying of renal failure. The chronic oedema of the alveolar walls and the organization of the fibrinous intra-alveolar exudate both result in interstitial pulmonary fibrosis with consequent extensive damage to the lungs as seen in Fig. 17.8. Heard (1962) described in detail the post-mortem changes found in the lung of one of the patients of Doniach *et al.* original series of cases. In both of the lungs of this patient, who survived 8 years, the parenchyma was fibrosed in the perihilar regions and in the fibrotic areas the small bronchi and bronchioles were compensatorily distended and the smaller branches of the pulmonary arteries were crowded together. Microscopical examination showed conchoid bodies growing from the surfaces of disrupting alveolar elastic fibres and the appearances were similar to those described by Walford and Kaplan (1957) as "endogenous pneumoconiosis". The conchoid bodies stained strongly for iron. The changes seen in chronic "hexamethonium" lung resemble those described by Andrews (1957) in patients dying from severe chronic pulmonary venous obstruction. The series of changes described above form a sequence which may be interrupted at any stage by the death of the patient and only rarely are the most advanced stages encountered.

It is now generally agreed that uraemic lung

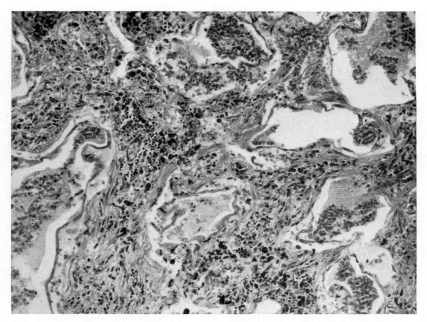

Fig. 17.8. "Hexamethonium lung", the final stage in chronic pulmonary oedema. The fibrosis occurs both inter-stitially in the alveolar walls, and also results partly from the organization of the alveolar exudate as in uraemic lung. The remains of the alveolar spaces are partly lined by alveolar epithelium. × 140 H and E.
(Reproduced by courtesy of Professor I. Doniach.)

is a form of pulmonary oedema caused primarily by left ventricular failure and pulmonary engorgement, but in which in addition a local capillary toxic factor and possibly deficient fibrinolysis also play a part. The additional toxic factor may be either a substance elaborated in renal failure or a hypersensitivity capillaritis in rheumatic fever; the two factors together so increase the permeability of the alveolar capil-laries that macromolecular proteins such as fibrinogen and even red blood-cells may escape into the alveoli.

The localization of the oedema to the central zones of the lung has attracted much attention. Herrnheiser and Hinson (1954), as a result of an anatomical investigation of the pulmonary arterial branching in the lung, have shown that each lobe consists of three zones which they have named (from within outwards): (1) the central or lobar root zone, (2) the medulla and (3) the cortex or peripheral parts of the lung (Fig. 2.2).

The central zone contains no lobules of lung tissue but only the large bronchi and vessels; the medulla contains poorly developed lobules supplied by branch arteries which are dis-proportionately small compared with the parent vessel but the arterial pathway to the medullary lobules is shorter than that supplying the better-developed cortical lobules. The cortex, composed mainly of lung lobules, is supplied by proportionately sized arterial branches, the sum total of the diameters of the branch arteries approximating to that of the parent vessel. These anatomical differences are considered to play an important part in the localization of the oedema to the inner two-thirds of the lung. Also, Prichard et al. (1954), in experiments on small animals, showed that if the lungs were rendered anoxic, injection of methylene blue into the pulmonary artery at the hilum dis-coloured the central areas but left the peripheral zones uncoloured. This they attributed to diversion of the blood-flow through the shorter blood vascular channels in the more central parts of the lung. As the blood-supply to the more central areas of the lung has a shorter course than that supplying the periphery it is unnecessary to postulate any central anastomotic mechanism between arteries and veins. The anoxic constriction of the longer arterial pathways would tend to divert the blood-flow through the more central parts of the lung.

In conditions where there is raised pulmonary venous pressure due to a failing left ventricle, reflex vasoconstriction occurs in the pulmonary arteries which in turn will tend to restrict the flow to those parts of the lung supplied by the shortest arterial pathway.

This diversion of flow, together with additional factors causing toxic capillary damage, may largely explain the characteristic distribution and nature of the oedematous exudate found in "uraemic" lung.

Alveolar Lipo-proteinosis (Alveolar Proteinosis)

This condition was first described in man by Rosen *et al.* (1958) and was named by them alveolar proteinosis. It was regarded at first as a disease *sui generis* but following its experimental production in animals exposed to silica, coal dust and drugs and its occurrence in patients exposed to various dusts and fumes or receiving treatment for leukaemia with busulphan (Doyle *et al.*, 1963), it is now more properly regarded as a form of response of the lung to a variety of noxious agents. Although it is often impossible to determine the cause in many human cases, Davidson and MacLeod (1969) found that a considerable number of their patients had a past history of exposure to a dust, a finding confirmed by Heppleston and Young (1972) and by Buechner and Ansari (1969). The author has seen its occurrence in a patient exposed to a very high concentration of fine silica dust.

Human idiopathic alveolar lipo-proteinosis though first described in the U.S.A. has now been reported from many countries. It is about three times more common in males and affects persons of all ages.

Clinically, the disease in man is often asymptomatic and may only be discovered following a routine chest radiograph. Some patients, however, complain of chest pain, fatigue, loss of weight, persistent pyrexia and expectorate sputum consisting of chunks of gelatinous material which may be blood-streaked. The chest radiograph frequently shows an appearance similar to the "butterfly" type of shadow seen in uraemic lung. Sputum culture is unhelpful but the diagnosis may usually be established either by light microscopic or electron microscopic examination of sputum or bronchial washings. Alternatively a lung biopsy may provide evidence of alveolar lipo-proteinosis. The peripheral blood picture often shows a moderate polymorph leucocytosis.

Macroscopically, the affected lungs contain greyish-white consolidated areas which often fuse to involve much of a lobe. The septa are prominent and the intervening lung between the consolidated areas is dark red, collapsed and of softer consistency.

Microscopically, the respiratory bronchioles and alveolar walls are usually of normal thickness but may be slightly thickened due to a mainly lymphocytic infiltration. The alveoli are lined by flattened epithelial cells which, in the more chronic cases, may partly or completely disappear (Figs. 17.9, 17.10). The lining cells secrete the proteinaceous material which fills the alveoli and some of the respiratory bronchioles, and which is the most striking feature of the disorder (Fig. 17.11). The proteinaceous material stains positively by the PAS method, gives a metachromatic reaction with toluidine blue, but fails to stain with alcian blue. It is probably a mucopolysaccharide with a few acidic radicals attached to the carbohydrate moiety and it is also coupled with much lipid. It usually stains homogeneously and resembles other colloid secretions, but in some cases it has a distinctly granular appearance. The granular alveolar epithelial cells (type 2 pneumocytes) first fill with PAS-positive granules and later the nuclei of these cells disappear by a process of either

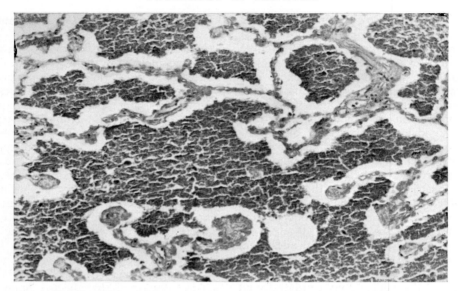

FIG. 17.9. Alveolar proteinosis showing the PAS-positive material filling the alveolar lumens and the swollen alveolar epithelial cells. Approx. × 150.
(Reproduced by courtesy of Dr. A. A. Liebow.)

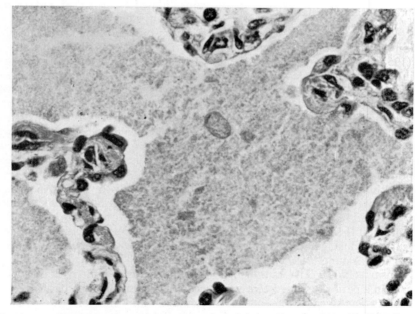

FIG. 17.10. Alveolar proteinosis showing a higher-power view of the swollen alveolar epithelial cells and proteinaceous material filling the alveolar lumen. × 500 H and E.
(Reproduced by courtesy of Dr. A. A. Liebow.)

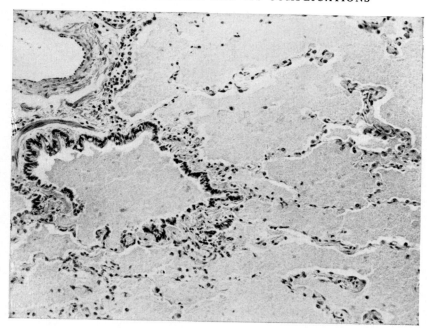

FIG. 17.11. Alveolar proteinosis showing the alveolar contents extending into the smaller bronchi and bronchioles.
× 140 H and E.
(Reproduced by courtesy of Dr. A. A. Liebow.)

karyolysis or pyknosis. Some of the non-nucleated remains of cells may be found amidst the material filling the alveoli (Fig. 17.12) and others are transformed into laminated bodies. Kuhn et al. (1966) found that despite the probable derivation of much of the lipid found in the proteinaceous material from type 2 pneumocytes, extracts of the lipid failed to show surfactant properties. In addition to the PAS-staining material, intracellular lipid, identified as cholesterol or cholesterol ester by a positive Schultz test, is also found in the alveolar walls. Acicular spaces filled with similar lipid are found in the alveolar lumens amidst the proteinaceous material. Proteinaceous material similar to that filling the alveoli may, in some cases, be found within the sinus-lining cells of the regional lymph glands.

The course of the disease is variable; amongst the cases so far observed the majority have resolved, though eight of the original twenty-seven cases have died (Fraimow et al., 1960). Minor degrees of interstitial fibrosis may occur in portions of the lung formerly the seat of proteinaceous changes.

Its course is uninfluenced by steroid or antibiotic therapy. In the majority of cases there is little functional disability. The principal changes found have included a slight reduction of the vital capacity and the pulmonary compliance, and a moderate reduction of oxygen uptake leading in some cases to some reduction of the arterial oxygen saturation principally at rest. The changes in lung function were studied by Slutzker et al. (1961).

The diagnosis of the disease may be made by light microscopic examination of sputum but this may prove difficult and lung biopsy may be required. Endobronchial instillation of trypsin solution or saline at pH 7·4 results in expectoration of the characteristic intra-alveolar contents which may be identified histologically

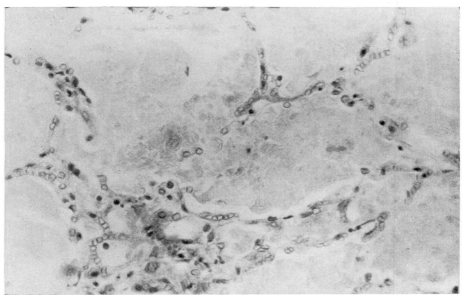

Fig. 17.12. Alveolar proteinosis showing the ghost outline of desquamated alveolar epithelial cells (type 2 pneumocytes) lying amidst the proteinaceous material filling the alveolar lumen. Approx. × 200.
(Reproduced by courtesy of Dr. A. A. Liebow.)

using the PAS stain. Biochemical tests on the blood are unhelpful though the serum lactic dehydrogenase content may be found increased. Other common serum enzyme levels are unchanged.

Electron microscopy also provides an additional method of diagnosis using material obtained by bronchial washing techniques (Basset *et al.*, 1973; Costello *et al.*, 1975). Bronchial lavage as a therapeutic measure was introduced by Ramirez-Rivera (1966). If a lung biopsy is first attempted time should be allowed for healing to occur before bronchial lavage is resorted to for either diagnostic or therapeutic purposes.

Electron microscopic examination of the alveolar contents shows that it consists of alveolar macrophages containing numerous cytoplasmic membrane bound bodies (lipidic phagolysosomes). These membranous structures contain lamellar osmiophilic lipoidal bodies which either have a concentric arrangement or are amorphous. The constituent lamellae have a similar appearance to the osmiophilic lipid

material shown in Fig. 2.13 and consist of two electron dense lines separated by an electron-lucent band. Lamellar material (phospholipid) is present both within macrophage cells and lying free amidst amorphous debris. Although the phospholipid material found in alveolar lipo-proteinosis morphologically resembles surfactant found in normal type 2 pneumocytes it lacks its surface activity.

Alveolar lipo-proteinosis was first produced in rats exposed to silica dust by Corrin and King (1966) and their observation was confirmed by Heppleston (1967) and Gross and de Treville (1968). Human alveolar lipo-proteinosis has been reported in sandblasters by Buechner and Ansari (1969). Heppleston and Young have also produced alveolar lipo-proteinosis in rats by exposing them to coal dust.

Vijeyaratnam and Corrin (1973) induced a condition in rats similar to alveolar lipo-proteinosis by giving them the drug iprindole by mouth and Heath *et al.* (1973) have produced the same result by giving rats chlorphentermine hydrochloride intraperitoneally.

These varied methods of producing alveolar lipo-proteinosis in animals confirm the conclusion reached by Heppleston and Young that both the human disease and the induced disease in rats result from an initial type 2 granular pneumocyte hyperplasia which causes excessive production of a phospholipid material combined with protein, similar to but lacking surfactant properties. This overwhelms the phagocytic potential of the alveolar phagocytes. The result is an accumulation within the alveolar lumen as well as within intra-alveolar macrophages of the phospholipid material which both fill it and destroy it as an effective aerating unit. There is no evidence to support the view that the alveolar macrophages lack their normal motility or that alveolar lipo-proteinosis results from disordered bronchiolar and bronchial clearance mechanisms. There is also no evidence that it is due to *Pneumocystis carinii* infection.

Adult Hyaline Membranes

Hyaline membranes may be found in the distal air passages and alveoli of the adult lung in many conditions of varied aetiology. Among these conditions are included "uraemic" lung, phosgene poisoning, acute influenzal virus pneumonia, acute haemolytic streptococcal pneumonia, pneumonic plague, acute tuberculous pneumonia and many other forms of acute bacterial and chemical pneumonitis. The common factor to all is partial destruction of alveolar epithelium and the presence in the lungs of abundant oedema fluid containing fibrinogen, both of which are essential for hyaline membrane formation. Hyaline membrane disease may occur in persons of all ages. Neonatal hyaline membrane disease because it is related to the problem of neonatal lung collapse is described separately in Chapter 13 and will not be further considered here.

Hyaline membrane disease in adults often complicates severe pulmonary oedema from which it is inseparable. The greater lung surface in the adult, the greater size of the alveolar sacs and the more forceful respiratory movements render the obstructing action of hyaline membranes of much less importance than in the neonate.

CHAPTER 18

Degenerative and Metabolic
Disorders of the Lungs

Amyloidosis of the Lung

Although it has been customary in the past to subdivide amyloid disease into primary, secondary (classical) and para-amyloid forms, more recent studies based on the chemistry of amyloid and distribution of amyloid deposits in the tissues have shown that the subdivisions are artificial. In the ensuing description most cases of pulmonary amyloidosis fall within the category previously referred to as primary amyloidosis.

Early descriptions of amyloidosis of the respiratory tract were given by Lesser (1877) and Wild (1886) and the latter described in detail "primary" amyloidosis, a disease unassociated with any previous chronic suppurative infection. "Secondary" or classical amyloid disease which usually followed chronic suppurative infection or tuberculosis had been described earlier by Rokitansky (1842).

Although lung involvement in "secondary" amyloidosis and para-amyloidosis is usually regarded as of rare occurrence, recent studies have cast doubt on this view as small deposits of amyloid are sometimes found in post-mortem tissues related to capillaries and small arteries in the lung. Also deposits of para-amyloid and amyloid are found in other forms of lung disease such as diffuse lymphocytic interstitial pneumonia (LIP) and in plasma cell granulomas. In the ensuing description of pulmonary amyloidosis the account mainly refers to those forms of "primary" amyloidosis which occur in the lung and respiratory tract.

The nature of the amyloid material found in the lung deposits was investigated by Page *et al.* (1972). They examined amyloid material taken from a nodular intrapulmonary deposit and found that it bore a resemblance chemically to the lambda Bence–Jones proteins and was probably derived from the lambda light chain protein of an immunoglobulin. Amyloid has a fibrillar structure and is probably produced in the lung from plasma cells. This view received support from the fact that plasma cell granulomas in the lung contain abundant intercellular amyloid material (Carrington and Liebow, 1966); Glenner *et al.* (1971) had also suggested that pulmonary amyloid deposits were nodules of a modified immunoglobulin protein produced by intra-pulmonary collections of plasma cells.

The literature concerning pulmonary amyloidosis is now extensive but many cases go unrecorded. The reader is referred to the accounts given by Sappington *et al.* (1942), Weismann *et al.* (1947), Whitwell (1953) and Lee and Johnson (1975).

Primary amyloidosis in the lung may be divided into the following forms: (a) a localized deposit in the bronchus, (b) multiple or diffuse bronchial deposits, (c) localized or multiple parenchymal deposits and (d) diffuse (parenchymal) amyloid infiltration of the alveolar walls and blood-vessels in the lung.

Pulmonary amyloidosis occurs mainly in persons over the age of 60 and affects both sexes equally. It may occur in persons with a neoplasm elsewhere in the body.

(a) **Localized bronchial deposits** are usually found in the large lobar or segmental branches and project as rounded, smooth, greyish-white

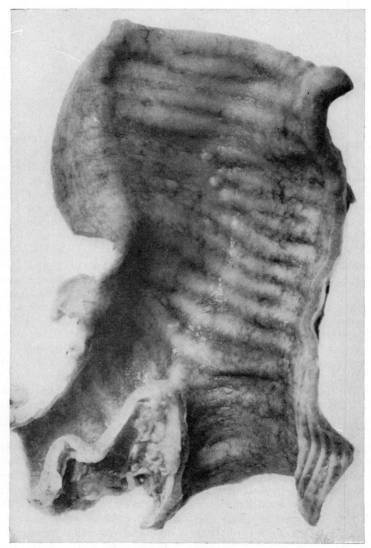

FIG. 18.1. Submucosal nodules of amyloid material projecting into the lumen of the trachea and main bronchi. × 1¼ natural size.

sessile tumours into the lumen, which they may occasionally block, causing secondary obstructive pneumonitis in the lung beyond. An amyloid tumour of this type was described by Weismann et al., and Prowse (1958) reviewed two further reported cases. All these cases occurred in males.

(b) **Diffuse bronchial deposits** are commoner and were first described by Balser (1883) and more recently by Antunes and Vieira da Luz (1969), Mainwaring et al. (1969) and Attwood et al., (1972). The majority of cases have occurred in men in whom numerous sessile smooth nodules varying in size up to 1 cm in diameter are found either on bronchoscopic or post-mortem examination projecting into the lumen of the trachea and larger bronchi. The entire wall may become diffusely infiltrated with amyloid causing appreciable stenosis of the smaller bronchi (Fig. 18.1).

Microscopically, the mucosal epithelium remains intact, though it may undergo squamous metaplastic change. The submucous layer is diffusely filled with amyloid which encases the mucous glands and cartilages causing the former to atrophy (Fig. 18.2). Extension of the amyloid occurs into the peribronchial connective tissue, but the lung parenchyma is seldom involved in diffuse bronchial disease though single bronchial deposits may extend more widely. Foci of calcification and giant-cell formation are found at the edges of deposits and the amyloid in most cases stains metachromatically with methyl violet and selectively with Congo red. Most cases of diffuse bronchial amyloidosis are associated with deposits in the larynx and trachea which cause hoarseness.

Single amyloid tumours in the bronchi may be successfully resected without recurrence, but diffuse bronchial involvement often leads to obstruction and consequent obstructive pneumonitis.

(c) **Parenchymal deposits (nodular pulmonary amyloidosis)** in the lung may be single or multiple and are often discovered at a post-mortem examination. They exist independently of bronchial deposits. The first case was described by Hallermann (1928) and further examples and reviews of the literature have been added by Glauser (1955), Duke (1959), Becker

(1961), Condon et al. (1964), Dyke et al. (1974) and Lee and Johnson (1975). Parenchymal deposits may occasionally occur in conditions of dysproteinaemia such as Waldenström's macroglobulinaemia. This form of pulmonary amyloidosis has not been reported, however, in multiple myelomatosis though the chemical structure of the amyloid is a modified lambda light chain protein derived from an immunoglobulin and similar to Bence–Jones protein.

The amyloid deposits may be single or multiple appearing usually as translucent, grey tumour masses varying in size up to about 8·0 cm in diameter. In the cases described by Bergman and Linder (1958) and Becker, the deposits contained either chalky spots or presented as hard calcareous masses. The tumour masses usually shell out readily from the surrounding compressed lung. Diffuse parenchymal deposits of amyloid may produce a radiological picture which can be confused with that caused by multiple secondary tumour deposits. Although the clinical progress of these cases may be slow and for a long time they lead to little disability, eventually in most cases the disease proves fatal.

The deposits are found mainly beneath the pleural surfaces originating from the muscle walls of small blood-vessels and they spread into the interstitial tissues (Fig. 18.3). At their edges they are surrounded by plasma cells, lymphocytes and occasional foreign-body giant cells which may contain ingested amyloid (Fig. 18.5). Calcification and bone formation within the deposits also occur. The staining reaction of the amyloid material is inconstant and a number of staining methods should be used including methyl violet, Congo red, silver stains and thioflavine T. The last named gives a bright blue fluorescence with amyloid.

Surgical excision of a pulmonary amyloid nodule may be followed by its recurrence after a period of several years (Dyke et al.).

(d) **Diffuse parenchymal amyloidosis.** This is the least common form of pulmonary amyloidosis and the subject was reviewed by Poh et al. (1975). It is usually associated with widespread amyloidosis elsewhere in the body and usually involves most of both lungs though local

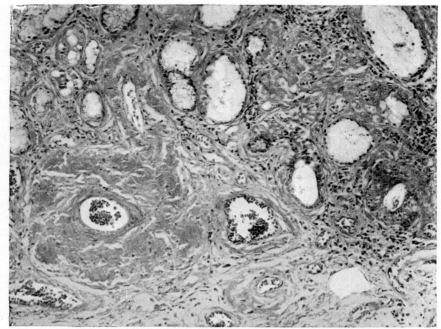

FIG. 18.2. Amyloid deposits in a main bronchial wall showing some atrophy of the mucous glands. × 140 H and E.

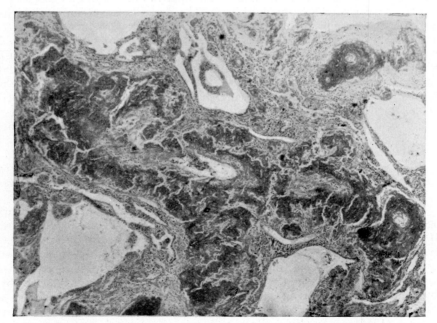

FIG. 18.3. A local deposit of amyloid in the lung parenchyma. × 140 Congo red.

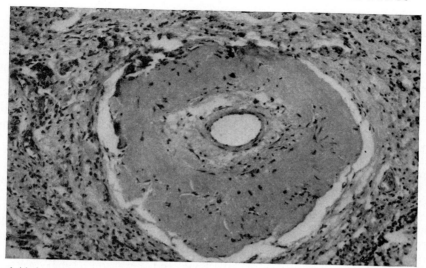

FIG. 18.4. Amyloid change in the wall of a small branch of a pulmonary artery showing mainly periarterial deposition with foreign-body giant-cell reaction at the edge of the amyloid deposit. × 100 Congo red and haematoxylin.

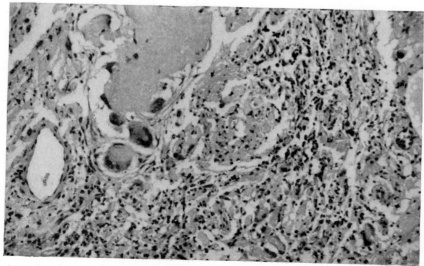

FIG. 18.5. An intrapulmonary amyloid deposit showing surrounding giant-cell reaction. × 100 stained with Congo red and haematoxylin.

variations in the intensity of the changes are common. Post-mortem the lungs are bulky, pale and rubbery in texture and microscopically diffuse amyloid deposition occurs in the alveolar walls and around the smaller branches of the pulmonary arteries and veins (Fig. 18.6) (Dahlin, 1949). The changes originate in the alveolar capillary basement membranes. As a result of the amyloid deposition, capillary obstruction and possibly an alveolar–capillary block to gaseous diffusion occur and lead in extreme cases to a reduction of the arterial oxygen saturation on exertion. Calcification and ossification may occur in the amyloid deposits (Zundel and Prior, 1971).

Alveolar septal amyloidosis was investigated electron microscopically by Rajan and Kikkawa (1970) who found that the first deposits occurred in relation to the alveolar capillary and epithelial basement membranes and later spread to fill most of the interstitial space in the alveolar wall.

Pulmonary Corpora Amylacea

The term corpora amylacea is applied to rounded, eosin staining, acellular structures found in the alveoli of lungs removed post-mortem, and in sputum before death. They were first seen in sputum and described by Friedrich (1856), and more recently by Michaels and Levine (1957) who examined them with the electron microscope.

In ordinary sections of lung they appear as rounded, homogeneous masses varying in size between 30 and 200 μm though usually they measure 60–100 μm.

The centre may consist of a small amount of black pigment or a doubly refractile rod. The outer part is concentrically ringed, and between the rings fibres are arranged radially (Fig. 18.7). Baar and Ferguson (1963) investigated their composition and concluded that glycoproteins were the principal constituents. Their stratified

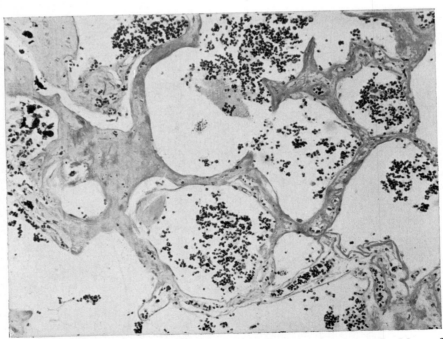

FIG. 18.6. Diffuse pulmonary amyloidosis involving most of the alveolar capillary walls. Many of the alveolar capillaries are almost totally obliterated. × 140 H and E.

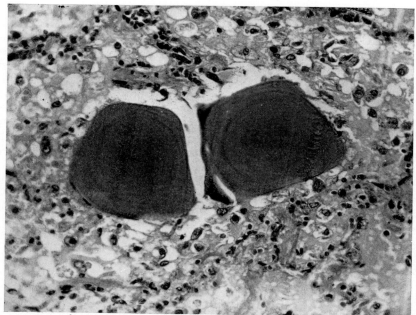

FIG. 18.7. Two corpora amylaceae showing their concentric ringed appearance and slight peripheral radial striation. They have excited minimal cellular reaction but are partly surrounded by alveolar phagocytic cells on one side. × 280 H and E.

structure was attributed to precipitation of protein substances due to the Liesegang phenomenon. Histochemical tests often give variable results but the bodies stain pink with eosin, lilac with haematoxylin and blue with aniline blue. Some have found them to react positively to the PAS and Hale's stains but in the author's experience neither method yields a convincing positive result. Lubarsch and Plenge (1931) found a variable response to iodine, the bodies staining anything from a bluish-green or green to yellow in colour. Iodine may produce a non-homogeneous staining response, the periphery staining greenish, the centre remaining almost colourless. In the author's experience the majority fail to stain at all. Crushing the bodies causes fissuring in a radial pattern.

Electron-microscopic examination by Michaels and Levine showed that they consisted of fibrils 100–150 Å in thickness, with striations at 200 Å periodicity. The electron-microscopic appearances closely resemble those of the bodies found in microlithiasis alveolaris pulmonum, except

that in the latter condition the structures are also calcified.

The cause of corpora amylacea in the lung is unknown but they are found most frequently in diseases in which there is persistent but moderate pulmonary oedema. Such conditions include pneumonia, infarction, collapse, and conditions in which persistent pulmonary oedema follows chronic left-sided heart disease. Reabsorption of water from the alveolar exudate and diffusion of molecules into the protein residuum were considered by Baar and Ferguson to account for their stratified structure. The presence of corpora amylacea in the alveoli excites very little cellular reaction and they usually lie free in the alveoli, though occasionally macrophage cells may be applied to their surfaces and they may be adherent at some point to the alveolar wall.

Pulmonary Alveolar Microlithiasis

This rare and unusual condition was first described by Harbitz (1918), but was named by

Puhr (1933) who called it "mikrolithiasis alveolaris pulmonum". Since then further individual and small series of cases have been reported by Sharp and Danino (1953), Kent *et al.* (1955), Sosman *et al.* (1957), Thomson (1959), Ravines (1969) and Coetzee (1970). According to Sears *et al.* (1971) over ninety cases have now been reported and the disease is of a world-wide distribution.

It may show a familial tendency and nine cases occurring in three related families were described by Ravines. In such cases the diagnosis may be able to be made at an early stage before the onset of symptoms.

Although the majority of the reported cases have occurred between the ages of 20–40, it has occurred as early as 6, and in an adult as late as 66. In most cases the onset has been insidious, and in the case reported by Manz (1954), it was observed to spread gradually over a period of 25 years. A case with a rapid course was reported by Biressi and Casassa (1956), development occurring in as short a period as 1 year.

Pulmonary alveolar microlithiasis may be associated with chronic mitral stenosis and pulmonary fibrosis.

Clinically, the diagnosis has usually been first established following a routine chest radiograph taken as part of a general examination in persons complaining of respiratory symptoms such as shortness of breath. The radiographs show dense miliary mottling or confluent shadows mainly confined to the lower two-thirds of the lung fields, although in the case described by Schild-knecht (1932), the apical portion of the lungs was also involved.

When the lungs are examined at autopsy they are remarkable for their weight and hardness. Their shape is exceptionally well maintained, and the lower lobes are pale and stony hard; even the margins of the lung may not be able to be cut in the normal fashion. The lungs are often sawn open only with much difficulty, and thick mounted sections of such lungs prepared by the Wentworth and Gough technique resemble very coarse sandpaper.

Microscopical examination of the lung after decalcification by 20 per cent nitric acid, lesser concentrations being useless, show that the alveoli and occasionally a bronchiole are filled with psammoma-like bodies termed calcospherites (Fig. 18.8). In the less severely affected parts of the lung the alveoli contain concentrically laminated calcospherites which are completely devoid of all nuclear structure. The appearance of the calcospherites is best appreciated in non-decalcified, thick, frozen sections (Fig. 18.11), and in radiographs of such sections.

Where these changes are most severe the calcospherites are separated from each other by remnants of alveolar walls, consisting of avascular strands of fibrous tissue containing monocytes, lymphocytes and occasional giant cells (Fig. 18.9). Calcospherites also occur within thickened alveolar walls and in the walls of bronchi but do not damage the overlying mucosa of the latter. They may also occur in extrapulmonary sites such as the lumbar sympathetic chain and the testes. The calcospherites, in addition to the concentric laminations, show fine radial striation but do not contain a central black core like the non-calcified corpora amylaceae. In places lamellar bone may form around the calcospherites, and nodules up to several millimetres in diameter occur (Fig. 18.10) which are visible through the pleura as white spots. Cleared, thick sections of less-damaged parts of the lung show that the calcospherites are often deposited in relation to the finer blood-vessels, and vary in diameter up to about 1 mm, being roughly spherical.

With haematoxylin-eosin stain the centre of the spherite usually stains darker than the periphery, and it stains intensely with the PAS and colloidal iron stains.

Chemical analyses show that the calcospherites contain calcium, phosphorus mainly as phosphates, a small amount of iron and traces of magnesium (Leicher, 1949). Frozen sections show that they also contain sudanophilic and doubly refractile fatty material.

Calcospherites bear a superficial resemblance to the commonly occurring corpora amylaceae found particularly in conditions of heart failure,

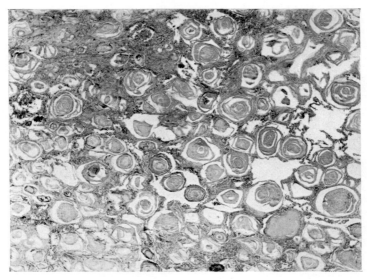

Fig. 18.8. Low-power view of the lung in microlithiasis alveolaris pulmonum showing the alveoli filled with typical concentrically laminated intra-alveolar calcospherites. × 32 H and V.G.
(Reproduced by courtesy of Drs. Mary Sharp and E. A. Danino and the Editor of *J. Path. Bact.*

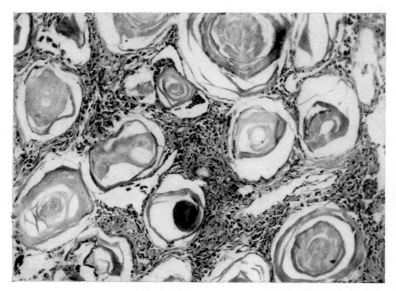

Fig. 18.9. Multiple intra-alveolar calcospherites showing chronic alveolar wall damage. × 130 H and E.
(Reproduced by courtesy of Drs. Mary Sharp and E. A. Danino.)

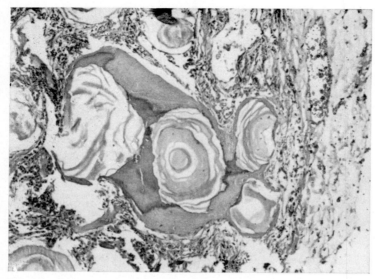

FIG. 18.10. Section showing commencing ossification around the calcospherites and severe damage to the adjacent alveolar walls. × 84 H and V.G.
(Reproduced by courtesy of Drs. Mary Sharp and E. A. Danino and the Editor of *J. Path. Bact.*)

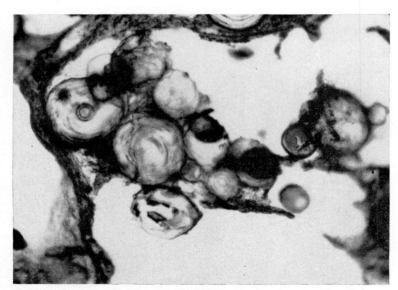

FIG. 18.11. Non-decalcified calcospherites in a thick 100 μm section. The black linear structure on the left of the photomicrograph is an alveolar capillary. × 85 H and E.
(Reproduced by courtesy of Drs. Mary Sharp and E. A. Danino and the Editor of *J. Path. Bact.*)

pulmonary infarction and chronic bronchitis. Corpora amylaceae, however, are not calcified, are more regular in shape, often have a central core of black pigment, and are lamellar structures. Like calcospherites they also show radial striation.

The relationship of alveolar microlithiasis to the osseous nodules found in persons dying from rheumatic mitral stenosis is undetermined. Such intra-alveolar ossification in chronic rheumatic heart disease was described by Elkeles and Glynn (1946), and had been observed by Wells and Dunlap (1943).

Sharp and Danino remarked upon the uniformity of size and appearance of the calcospherites, which they considered developed in both lungs following a single inflammatory or vascular episode causing a generalized alveolar exudate.

The functional changes caused by this disease have been studied by Thomson and Lebacq et al. (1964) who both found a slight reduction of the vital capacity which gradually increased the longer the condition persisted. Arterial oxygen desaturation occurs late in the disease due to the uneven distribution of air and blood in the affected lung. The hyperpnoea seen in the last stages is also partly related to the very gross disturbances of lung compliance.

Alveolar Calcification in the Lung

Synonym: Calcinosis of lung

Calcification occurs in the lung in two forms, the dystrophic and the metastatic varieties. In dystrophic calcification deposits of calcium salts are formed in damaged or necrotic lung tissue but the change is unassociated with any alteration of the blood calcium and phosphorus levels. Dystrophic calcification will not be considered further as it is described under the various causative conditions.

Metastatic calcification which is the form responsible for alveolar calcification or calcinosis of the lungs generally follows alterations of the serum calcium and phosphorus levels which lead to the precipitation of calcium salts in many tissues including the lungs. The principal groups

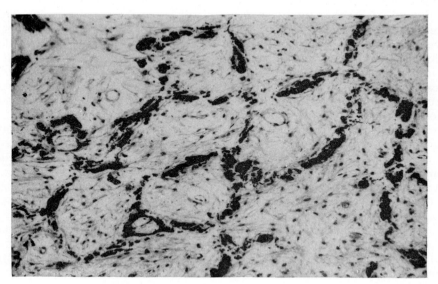

FIG. 18.12. Alveolar wall calcification in a chronically fibrosed lung. This is probably a metastatic form of pulmonary calcification. The intra-alveolar fibrosis represents organized fibrin-rich transudate. × 100 H and E.
(Reproduced by courtesy of Dr. P. D. Byers.)

of disorders responsible for this form of calcification include primary and secondary diseases of the skeleton, chronic renal diseases, hypervitaminosis D and a miscellaneous group of diseases in which the reasons for metastatic calcification are ill understood.

Alveolar calcification was first described by Virchow (1855), but many further examples have since been described and the whole subject of metastatic calcification was reviewed by Mulligan (1947), who also discussed the current views on the pathogenesis of the condition. In most cases the calcific deposits have not been confined to the lungs but are also found in the myocardium, kidneys, cardiac end of the stomach and less frequently in other sites.

The principal primary skeletal disorder responsible for calcinosis of the lungs is osteitis fibrosa caused either by a parathyroid adenoma or parathyroid hyperplasia. In this condition increased urinary excretion of phosphate leads to subsequent osteoclastic resorption of the skeleton, particularly of the trabecular bone. As a consequence the serum calcium level rises and the increased excretion of calcium and phosphorus by the kidneys leads to the precipitation of stones in the renal tract and in the interstitium of the kidneys. Later nephrocalcinosis leads to renal failure and the serum phosphorus levels rise, resulting in deposition of calcium salts in many sites including the lung.

Widespread metastatic growth in the skeleton due to secondary carcinoma, malignant lymphomas including chronic myeloid leukaemia, multiple myelomatosis and more rarely extensive inflammatory lesions in bone may all cause metastatic calcification. The blood levels of calcium and phosphorus may be slightly increased but are more often within normal limits.

Among the renal diseases responsible for metastatic calcification are chronic glomerulonephritis, chronic pyelo-nephritis and polycystic disease of the kidneys. Among the many biochemical changes that may occur in the later stages of renal failure is a raised serum inorganic phosphorus level, the serum calcium being variable and often reduced. Secondary parathyroid hyperplasia may follow and lead to absorption of bone, causing a secondary rise in the serum calcium levels. This may be followed by metastatic calcification in many sites including the lung.

Hypervitaminosis D following excessive intake of dietary vitamin D (calciferol) may rarely lead to metastatic calcification. Excessive vitamin D in the diet of rats can lead to a rise in the level of both serum calcium and phosphorus following osteoclastic resorption of the skeleton, a change enhanced by giving an alkaline diet. Excessive quantities of intravenously administered calcium gluconate together with excessive oral intake of vitamin D led to alveolar calcification in the case described by Cooke and Hyland (1960).

Among the miscellaneous disorders responsible for causing metastatic calcification is pyloric stenosis which may cause an alkalosis (Volland, 1941).

The ages of patients developing alveolar calcification has varied between 8 weeks and 70 years and both sexes are equally affected. The most common cause is the skeletal group of disorders followed by the nephropathic conditions, though in the author's experience the nephropathies are equally responsible. The distribution of metastatic calcific deposits in the lungs, stomach and kidneys was attributed by Hueper (1927) to the fact that local tissue alkalosis probably existed at such sites and facilitated the deposition of calcium salts. Alveolar calcification is not only found in man but was reported in animals, including birds, by Barnard (1946a).

Patients with widespread metastatic (hypercalcaemic) alveolar calcification sometimes develop severe terminal pulmonary oedema in the absence of heart failure. It has been suggested that the calcification destroys the normal alveolar capillary wall, allowing excessive transudation to occur.

Macroscopically, the lungs when cut feel gritty and may resemble in texture an extremely fine mesh calcified sponge. The calcification can be demonstrated radiologically in lungs removed at post-mortem (Fig. 18.13), and it has been seen in radiographs of the chest taken during life (Hild, 1942).

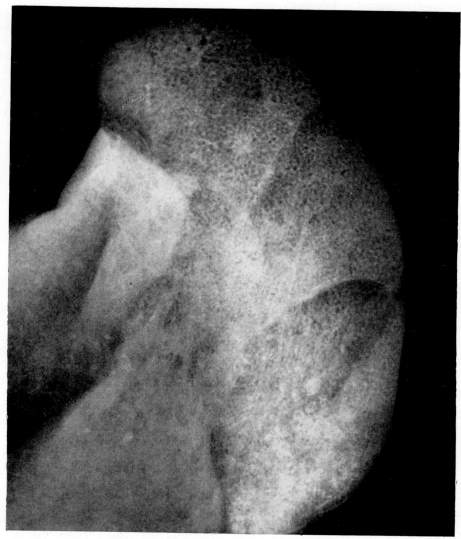

FIG. 18.13. A radiograph of a lung with generalized alveolar calcification taken post-mortem after removal from the body, showing the fine filigree of calcified alveolar spaces with some denser aggregations of calcium salts.

Microscopically, the calcium salts are first deposited in a fine, light haematoxyphil line in the alveolar and bronchiolar basement membranes. Later as the proportion of calcium salts increases, the basement membranes in the alveoli, bronchioles, and later the vessels' walls become clearly outlined and are best stained by von Kossa's method (Fig. 18.14).

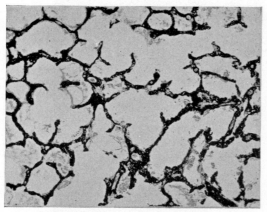

FIG. 18.14. Section of lung taken from a case with generalized alveolar calcification. × 60 stained by von Kossa's method.

Pulmonary Ossification

Osseous tissue is not infrequently formed in areas of dystrophic calcification and may also replace normal or neoplastic cartilage tissue in the lung. Ossification of bronchial cartilage rings, which is common in the elderly, is preceded by calcification and invasion of the dead cartilage by capillaries. Later, fatty and haemopoietic tissues usually fill the interstices between the newly formed bone trabeculae.

Two further rare forms of pulmonary ossification occur, the first involving the walls of the large bronchi known as bronchopathia osteoplastica and the second in which rods of trabecular bone radiate peripherally through the lung lying outside but in relation to the air passages and branches of the pulmonary arteries (diffuse parenchymal ossification of the lung).

TRACHEO- AND BRONCHOPATHIA OSTEOPLASTICA

Although this rare condition affects principally the trachea it frequently spreads to involve the larger bronchi, and the condition is characterized by the formation of osseous tissue beneath the mucous membrane of these air passages.

The condition was first clearly described by Wilks (1857) and was given its present name by Aschoff (1910). In 1947 Dalgaard was able to collect ninety cases from the literature and discussed its pathogenesis. Recently, further cases have been described by Carr and Olsen (1954), Bowen (1959) and Ashley (1970). Bowen reviewed the cases in the French and German literature and Ashley considered that long-standing bronchial inflammation was a causative factor.

The disease is more common in males and with a few exceptions has occurred mainly in persons over the age of 50. When the changes are confined to the trachea and bronchi the patients suffer little inconvenience, and the condition is often first recognized in a routine chest radiograph or at bronchoscopic examination when it causes a characteristic grating as the instrument is introduced and passed down the lumen. Many cases, however, remain undiagnosed until post-mortem examination is performed. If the changes spread to involve the smaller cartilaginous bronchi the latter become obstructed and changes due to obstructive pneumonitis result.

Macroscopically, the large bronchi and the trachea form a rigid system of tubes, the mucosal surfaces of which are raised by rounded bosses and plaques of ossified and cartilaginous tissue. The nodules are mainly confined to the portions of the walls normally bound by cartilage, being almost entirely absent from the membranous segments of the bronchial and tracheal walls (Fig. 18.15). The mucosal surfaces remain intact.

Microscopically, numerous irregular deposits of cartilage and bone appear immediately below the basement membrane of the surface epithelium and mostly deep to the muscularis submucosa (Fig. 18.16). Many of the newly formed

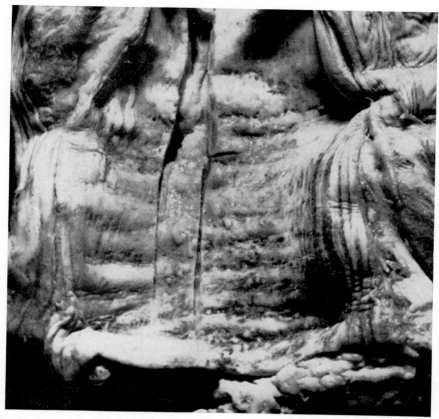

FIG. 18.15. Trachea opened longitudinally to show submucosal nodules and bosses of bone and cartilage in bronchopathia osteoplastica. The membranous portion of the trachea on the right side is free of submucous nodules. × 2½.
(Reproduced by courtesy of Dr. E. H. Bailey.)

osseous nodules are continuous with the perichondrium of the bronchial cartilages but others appear to be completely independent, occurring between the cartilage bars. In the centre of many nodules, fatty and actively haemopoietic bone marrow tissue is found.

Although many views have been expressed concerning the pathogenesis of this condition, observations have shown that the bone and cartilage form by a process of metaplasia in hyalinized fibrous connective tissue, the latter including the perichondrium of the normal cartilage bars. The presence of ectopic bone and cartilage leads to secondary atrophy of the submucosal glands and their ducts, whilst the surface epithelium remains intact and is usually unaltered.

Bronchopathia osteoplastica is associated with no alteration in the blood calcium and phosphorus levels and the condition appears to be unrelated to calcinosis of the lungs, chronic inflammation or any other form of lung disorder.

DIFFUSE PARENCHYMAL OSSIFICATION OF THE LUNG

Nodular heterotopic ossification may occur within the lung alveoli in mitral stenosis, in organized inflammatory tissue, occasionally in the walls of a chronic cavity and very rarely by a

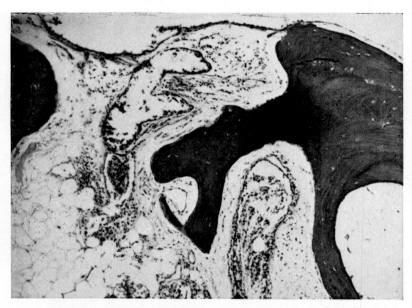

FIG. 18.16. Bronchopathia osteoplastica showing the submucous masses of bone which extend up to the overlying epithelium. The submucous glands and the duct epithelium undergo atrophy. The mass of bone is independent of the true bronchial cartilaginous rings. × 60 H and E.

(Reproduced by courtesy of Dr. E. H. Bailey.)

process of metaplasia within normal lung. Metaplastic ossification and bone marrow formation is also recognized as taking place within bronchial and tracheal cartilages with increasing age, and in the centres of chondromatous hamartomas. Idiopathic heterotopic ossification unrelated to any obvious pathological change within the lungs is very rare but a radiograph of the condition is shown in Fig. 18.17. A similar case was described by Green *et al.* (1970). It may affect all lobes or may be restricted, as in the case illustrated, to one lobe only. Discrete deposits of osseous tissue are laid down mainly in the perivascular and peribronchial connective tissue and these later develop metaplastic deposits of fatty bone marrow and occasional intraosseous foci of lymphocytes. Later the osseous deposits fuse to form long radiating rods which grow outwards from the hilar region and were described by Forster (1856, quoted by Ewing, 1940). Occasionally a whole lobe may become transformed into a mass of osseous

tissue. The nature of the change is uncertain but it does not behave like a neoplasm and may only be discovered at a post-mortem examination.

The Pulmonary Lipoidoses

Among the conditions known collectively as the lipoidoses, those which most commonly affect the lungs are Gaucher's disease and Niemann–Pick's disease. The whole group of lipoidoses are characterized by the intracellular deposition of a variety of lipid substances in both fixed and mobile macrophage cells, and in certain specialized cells, particularly in the central nervous system. In Gaucher's disease, which is often familial, the stored lipid is a cerebroside known as kerasin and is usually accompanied by an increase of acid phosphatase in the affected storage cells. In Niemann–Pick's disease the lipid material has been identified as the phospholipid sphingomyelin and five

FIG. 18.17. An X-ray of a lobe of a lung showing diffuse shadows due to intrapulmonary ossification. (Reproduced by courtesy of Dr. C. F. Ross.)

main clinical subvarieties of this disorder are recognized.

Although the brain is principally affected in the juvenile form of the disease, the lungs may be involved both in children and in adults.

In common with the other members of the lipoidoses Niemann–Pick's disease is often a familial condition. Clinically it usually becomes manifest in early infancy and leads to the death of the child before puberty. Survival beyond 20 years of age is unusual but a few adult cases have occurred. In the adult case described by Terry et al. (1954) the patient died from respiratory insufficiency due to the lung changes caused by the disease.

The tissues most frequently involved include the central nervous system, liver, spleen, lymph glands, adrenals and lungs, though most organs are found to contain a few typical lipid-laden phagocytes. The pathology of the disease has been described in detail by Crocker and Farber

(1958) and McCusker and Parsons (1962). Only the respiratory changes will be discussed in this text.

Macroscopically, the lungs are characteristically pale, airless and rubbery in texture. *Microscopically*, characteristic lipid-laden macrophages are found both in the interstitial tissues and lying free in the alveoli (Figs. 18.18, 18.20). The interstitial collections are found principally in the alveolar walls around the bronchi and blood-vessels, and in the subpleural and interlobular connective tissue septa. The cells vary in size from 20 to 90 μm and the nuclei tend to be small with thick nuclear membranes containing up to four nucleoli. The nuclei are found anywhere in the cell, sometimes being displaced to the outer wall, and in others occupying a central position. The cytoplasm is finely vacuolated, and giant-cell forms of the cells occur containing up to twenty nuclei often arranged around a central, more deeply eosinophilic staining and homogeneous mass (Touton cells) (Fig. 18.19).

The staining properties of the intracellular contents have been studied by both Crocker and Farber and McCusker and Parsons. In frozen sections the lipid stains orange with Scharlach R, black with Sudan Black B, bluish-black with Baker's phospholipid haematein stain, pale blue or lavender with nile blue sulphate and dark blue to black with phosphomolybdic acid stains. The lipid gives a positive reaction with the Schultz cholesterol test but a variable response with the PAS staining methods. Sphingosin, a product of hydrolysis of sphingomyelin, reacts with periodic acid to form an aldehyde and if this is formed a positive PAS reaction ensues.

The electron microscopic structure of the sphingomyelin deposits in alveolar macrophages was described by Lynn and Terry (1964) and Skikne et al. (1972). They occur as membrane-bound cytoplasmic bodies with a vacuolated structure. At very high magnification they show a lamellar structure with alternating osmiophilic and osmiophobic layers and a periodicity of 5 nm. The periodicity measurement enables sphingomyelin to be distinguished from other forms of lipid storage disease.

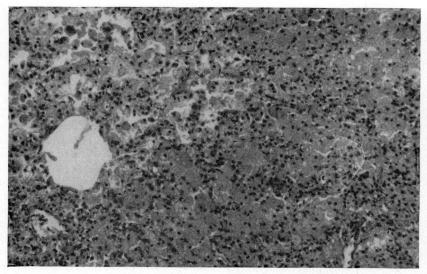

FIG. 18.18. Niemann–Pick's disease showing pulmonary changes resulting in diffuse intra- and interalveolar infiltration with the characteristic lipid-laden cells. × 40 H and E.
(Reproduced by courtesy of Dr. A. A. Liebow.)

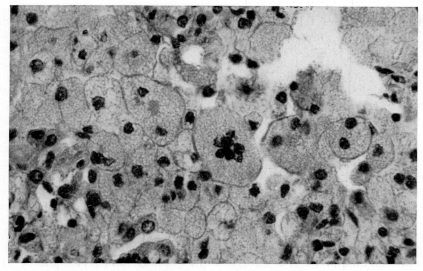

FIG. 18.19. Lipid-laden cells in the intra-alveolar spaces in a case of Niemann–Pick's disease. Note the Touton type of giant cell. × 400 H and E.
(Reproduced by courtesy of Dr. A. A. Liebow.)

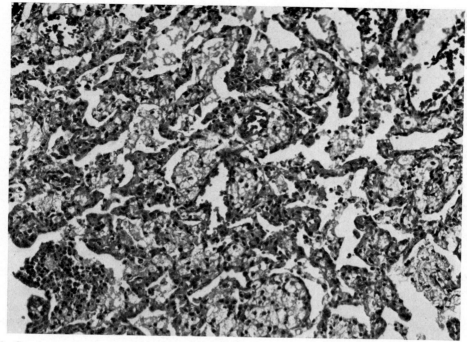

FIG. 18.20. Gaucher's disease showing lipid-laden cells in the interstitium of the alveolar walls. × 180 H and E.

The cause of the accumulation of the abnormal lipid is not known but the evidence does not support the idea of an overproduction of a normal lipid substance or a deficiency of a sphingomyelinase or co-factor. It is currently thought that there is a defect in intracellular protein synthesis resulting in the formation of a protein with an abnormal affinity for lipids.

von Gierke's Disease

Among the rare metabolic disorders that cause changes within the lungs is the type 2 variety of von Gierke's disease.

An atypical case was reported by Caplan (1958) in which the stored polysaccharide was closely related to glycogen except that it was not digested by diastase.

The glycogen or related polysaccharide is found within macrophage cells, many of which are shed into the alveolar lumens (Fig. 18.21).

Small amounts are present in bronchial cartilage cells, bronchial mucous gland epithelium, and in the bronchial epithelial cells.

Cystine Storage Disease (Lignac–Fanconi Disease)

Among the inborn errors of metabolism that may give rise to widespread pathological changes in many organs is cystine storage disease. The first cases were described by Lignac in 1924 and later Fanconi (1931) drew attention to further biochemical changes in this condition including the non-diabetic glycosuria. The disease is extremely rare, occurring among siblings of one generation only, and according to Bickel and Harris (1952) it is transmitted by a simple Mendelian recessive character. Baar and Bickel (1952) have discussed the various theories of its pathogenesis and concluded that cystine was elaborated within the reticulo-endothelial

6

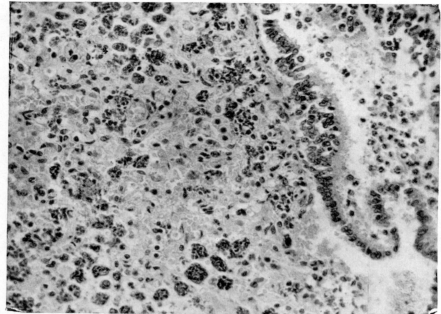

FIG. 18.21. Section of a lung from a case of von Gierke's disease showing intra-alveolar and interstitial macrophage cells laden with glycogen granules (black granules). × 280 Best's glycogen stain.

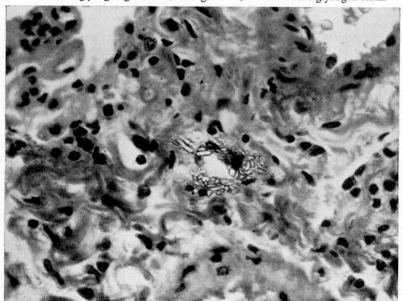

FIG. 18.22. Cystinosis of the lung showing crystals of cystine contained within reticulo-endothelial cells surrounding pulmonary vessels. A branch of the pulmonary artery has been cut across in the upper right-hand corner and along the lower margin. × 500 H and E. Specimen fixed in 95 per cent alcohol.

cells following a complex congenital disorder of protein anabolism. Biochemical investigations have shown that the abnormality in metabolism involves many amino-acids in addition to cystine.

Cystine storage disease principally affects the bones, kidneys, lymph glands, spleen and liver and leads to dwarfism and renal failure, though cystine may also be found within the reticulo-endothelial cells in other organs, including the lung (Fig. 18.22). The deposits provoke no cellular reaction or alteration of lung function and are readily mistaken for calcific deposits, an error which is encouraged by the positive reaction given by cystine-crystals with von Kossa's stain. Normally, pure cystine will not reduce silver salts, but the presence of traces of cysteine is responsible for the positive reactions. In tissues in which the cystine is present in only small amounts it may readily be missed owing to its solution in aqueous fixatives, and alcohol fixation is therefore to be recommended. In tissue sections the crystals are birefringent and when treated with concentrated sulphuric acid and phosphotungstic acid they form clumps of radiating needle-like crystals.

When crystals are found in the lung they are mainly distributed within the peribronchial and periarterial reticulo-endothelial cells, though they have also been reported within reticulo-endothelial cells contained in the alveolar walls.

For a more detailed description of the histochemical identification, crystallography, and methods for extraction of cystine from the tissues the reader is referred to the papers of Baar and Bickel (1952), Sullivan (1926) and Gatzimos et al. (1955).

CHAPTER 19

Pulmonary Diseases
of Uncertain Aetiology

Rheumatic Pneumonitis

The existence of a special form of "pneumonia" in acute rheumatic fever has been debated since the condition was originally described by Cheadle (1888). He attributed it to chills contracted through overventilation of the wards during the hot weather, but he also added a note to the effect that cardiac depressant drugs such as sodium salicylate and aconite should not be used in these cases. Following this description, which was based entirely on clinical observations, several accounts have been recorded of the pathological changes in the lungs of patients dying of acute rheumatic fever and occasionally of adults with recurrent rheumatism, including those by Masson *et al.* (1937), Hadfield (1938), Rich and Gregory (1943), Scott *et al.* (1959) and Grunow and Esterly (1972). Earlier, von Glahn and Pappenheimer (1926) described pulmonary vascular changes in rheumatic fever which included periadventitial collections of chronic inflammatory cells, fibrinoid necrosis of the walls of small arteries and concentric fibrous thickening and increased vascularity of the intima.

In recent years, the term rheumatic pneumonitis has fallen into disfavour, the majority of the changes found in such lungs having been attributed entirely to pulmonary oedema caused either by left heart failure or possibly by a hypothetical toxic factor. Careful examination of sections from the lungs of children dying of rheumatic fever leaves little doubt, however, that in some cases specific changes in addition to pulmonary oedema may be found; therefore it is justifiable to refer still to rheumatic pneumonitis.

Macroscopical Appearances

The pleural surfaces of the lungs are usually covered with a thin film of fibrin. The parts most commonly affected are the lower lobes which become bulky, rubbery in texture and plum-coloured. On section the lung is dark red, glairy and airless, but sticky oedema fluid may be expressed on firm pressure. Small, darker scattered areas of haemorrhage can be seen against the general background but none of these gross features is distinctive (Fig. 19.1).

Microscopical Appearances

The alveoli and alveolar ducts are partly filled with lightly-staining, eosinophilic, fibrin-containing pulmonary oedema fluid, and in focal areas much darker staining, brick-red masses of exudate intermingle with large mononuclear cells many of which are phagocytes. Hyaline membranes may be present and the alveolar surfaces are lined with a single layer of cells formed by proliferating metaplastic type 2 pneumocytes. The alveolar septa necrose in some places and focal intra-alveolar haemorrhages are common. Later the exudate organizes resulting in interstitial alveolar fibrosis and intra-alveolar plugs of loosely arranged fibroblasts which partly fill the alveolar ducts, the "bourgeons conjonctifs" described by Masson

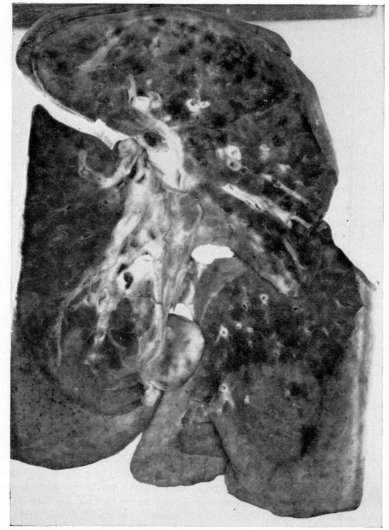

FIG. 19.1. Specimen of a lung from a case of acute rheumatic fever showing focal haemorrhagic areas. Five-sixths natural size.

et al., and thought wrongly by them to be characteristic of rheumatic pneumonitis. The dense intra-alveolar exudate is often separated from the walls of the alveoli by a space or is invested with a single layer of mononuclear cells similar to those found in the interstitium of the alveolar wall (Fig. 19.2). Focal intra-alveolar haemorrhages are common. Any alveolar exudate containing fibrin, including pneumonic exudates, and fibrinous oedema fluid such as may be found in chronic pulmonary oedema due to the left ventricular failure, produces the same appearance when it becomes organized (see Chapter 7).

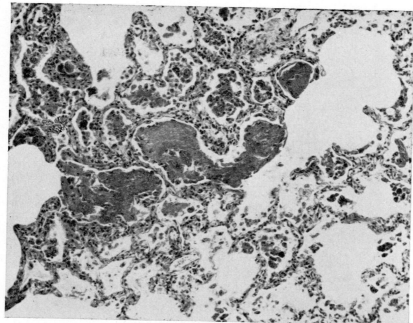

Fig. 19.2. Acute rheumatic fever showing the highly proteinaceous exudate in the alveoli which has excited a surrounding cellular reaction. The exudate appears as brick-red material in the H and E stained sections. × 140 H and E.

In addition to the intra-alveolar exudative changes, an interstitial mononuclear cell proliferation occurs in the alveolar walls; where this takes place the alveolar capillaries become almost bloodless and may be difficult to identify.

Many vascular changes have been described in acute rheumatic fever, but only two seem to occur with any frequency. The smaller branches of the pulmonary arteries may contain hyaline thrombi and show fibrinoid mural changes; whilst rarely alveolar capillaries undergo focal fibrinoid changes resulting in their rupture. Where this occurs the capillary walls appear to have ruptured outwards allowing the escape of red cells (Fig. 19.3).

The association of acute rheumatic fever with changes of generalized acute polyarteritis nodosa as reported by Friedberg and Gross (1934) is more than coincidental and has been seen by the author.

Although the changes in rheumatic pneumonitis are similar to those found in association with left ventricular failure (uraemic lung), differences exist. Many of the changes occurring in the lung in acute rheumatic fever are entirely attributable to pulmonary oedema resulting from left ventricular failure, and hence Cheadle's advice to avoid the use of cardiac depressant drugs has now been shown to be based on a sound pathological reasoning. The presence, however, of arterial and capillary damage, interstitial cellular reaction and the nature of the alveolar exudate distinguishes it from uncomplicated left ventricular failure.

Rich and Gregory, partly as a result of their experimental studies on hypersensitivity changes in animals, regarded acute rheumatic fever as a form of acute hypersensitivity reaction in man. This suggestion receives support from the occasional but more than coincidental association of acute rheumatic fever and acute polyarteritis nodosa, the latter of the type that Zeek (1952) and Knowles et al. (1953) described as hypersensitivity angeitis. It has long been considered

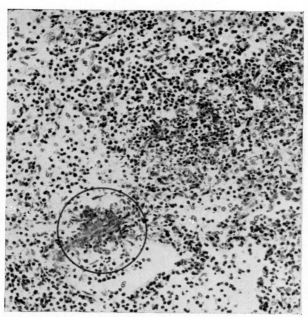

FIG. 19.3. An acute alveolar capillaritis in a case of acute rheumatic fever. The affected area is ringed and shows fibrinoid change affecting the alveolar wall. × 140 H and E.

that acute rheumatic fever is related to an abnormal immune response of the body to infection by the β-*Haemolytic streptococcus*; the changes occurring in rheumatic pneumonitis may therefore be regarded as partly due to pulmonary oedema resulting from left ventricular failure, and partly to a hypersensitivity vasculitis sometimes affecting arteries and including the alveolar capillaries.

Bronchial Asthma

Bronchial asthma alone rarely causes death. When it does prove fatal it is due either to the concurrent bronchitis and bronchiolitis or to heart failure (cor pulmonale). For this reason few opportunities occur to study the lung pathology due to asthma alone. Occasionally, however, death may take place during a state of unremitting asthmatic attacks (status asthmaticus) before infection has had time to occur and characteristic changes are found in the lung.

Although the first attack of status asthmaticus may prove fatal, in the majority of cases the final attack has been preceded by others. Asthma beginning after the age of 40 usually runs a more rapid and often fatal course than the same condition starting in childhood. The majority of fatal cases of status asthmaticus occur in women.

The changes in status asthmaticus have been described by Kountz and Alexander (1928), Houston *et al.* (1953) and Cardell and Pearson (1959).

Macroscopically, the lungs are overaerated and pale, but despite the bronchial obstruction, emphysematous change in status asthmaticus is rare as was shown by Gough (1955 and 1961). Whole lung sections fail to show centrilobular or other forms of destructive emphysema provided permanent lung damage has not resulted from concurrent bronchitis. The lungs usually remain very distended after removal from the thorax.

On section both the large and small bronchi are thickened and those between 0·2 and 1 cm in diameter are filled with very thick, yellowish plugs of mucus (Fig. 19.4). These plugs may be expressed from the cut surface and are of semisolid consistency. Leopold (1959) also drew attention to the presence of small subpleural scars resulting from intrapulmonary lesions which mainly occurred in the upper lobes.

Microscopically, the changes affect all the bronchi down to a size of 1 mm in diameter, beyond which point mural cartilage ceases and the remaining mucous glands become small and soon disappear.

The mucus filling the bronchi contains numerous eosinophils arranged in layers, Charcot–Leyden crystals, together with desquamated bronchial epithelial cells, and is directly continuous with mucus filling the ducts of the mucous glands. The bronchial epithelium is mostly shed save for the basement layer, but where still intact it shows extensive mucous metaplasia, the ciliated epithelium being replaced almost entirely by goblet cells. Clusters of shed bronchial epithelial cells may be found in the sputum from asthmatic patients and they were originally described by Vierordt (1883) who referred to them as *Epithelialzellballen*. Dunnill (1960) considered that shedding of bronchial epithelium was caused by subepithelial oedema.

Mucous glands in the bronchial wall are increased in size and number and their lumens are filled with thick mucus. Beneath the surface epithelium the basement membrane is greatly thickened and may be ruptured in places allowing herniation of the underlying tissue into the bronchial lumen (Fig. 19.5). Bohrod (1958) considered this thickening was due to the deposition of some unidentified material on its deep surface, now known to contain IgA and IgG. The entire wall of the bronchus including the peribronchial tissues is infiltrated with eosinophil cells.

The muscle coat in the bronchi is greatly hypertrophied and the bronchial cartilages may undergo partial atrophy (Fig. 19.6). The presence of bronchospasm during life is evidenced by the exaggerated plication of the soft tissues of the bronchial wall and the formation of diverticula of the mucous membrane seen after death. In addition to the characteristic changes in the bronchi a proteinaceous type of oedema fluid may sometimes be found filling the alveoli (Bohrod).

Leopold (1959) described a form of interstitial pneumonitis found beneath the pleural surface mainly in the upper lobes which was accompanied by an infiltration of eosinophils and polymorph neutrophils and in which the alveoli are filled with macrophage cells and small multinucleated cells. The changes appear to originate in the respiratory bronchioles, and the lesions bear a close resemblance to those found in eosinophilic pneumonia, and terminate in patches of interstitial fibrosis and localized patches of saccular bronchiectasis (Dunnill, 1960).

The changes described above go far to explain the clinical symptoms which result mainly from bronchospasm and blockage of the bronchi by mucus casts and these together prevent effective respiration.

Moore (1925) believed that loss of ciliated epithelium was a major factor in causing the mucus stagnation, and Hilding (1943) stressed the importance of mucous metaplasia of bronchial epithelium as a cause for the reduction in the number of ciliated cells. In addition, the quality of the mucus differs markedly from that found in normal bronchi, and this is an important factor in causing bronchial obstruction.

The aetiology of bronchial asthma is still little understood, but the association of asthma with other allergic states such as infantile eczema and hay fever, together with the characteristic infiltration of the walls of the respiratory tract by eosinophils, are reasons for regarding it as a hypersensitivity type of reaction. The localization of the changes to the respiratory tract remains unexplained, as the changes are not only confined to respiratory tissue in the lungs but have been found by Thomson (1945) in unorganized respiratory epithelium contained within an ovarian teratoma in a woman dying from status asthmaticus. Experimental studies in animals using isotope-labelled trace elements have failed to show the localization of antigens or antibodies in the

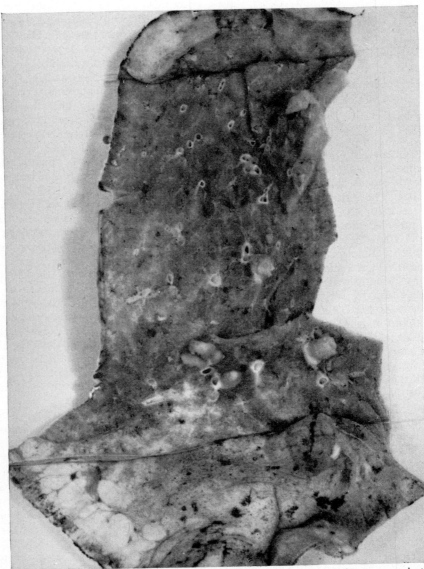

FIG. 19.4. Lung from a fatal case of status asthmaticus showing the plugs of very viscid mucus projecting from the cut
edges of the bronchi. The lung was not emphysematous. Approx. natural size.

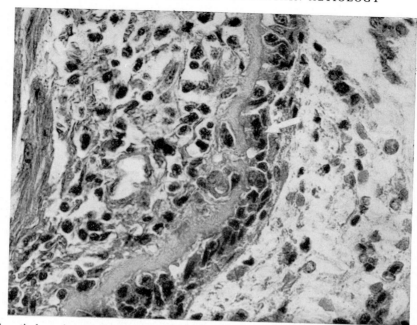

FIG. 19.5. Asthmatic bronchus showing hyaline thickening of the basement membrane and shedding of all except the basement layer of bronchial epithelium. The bronchial epithelium is indicated by an arrow. × 280 H and E.

bronchial musculature (Warren and Dixon, 1948), but they are found in the surface basement membrane.

The close relationship of asthma and eosinophilic pneumonia was illustrated in the case described by Danziger (quoted by Harkavy, 1941), in which a patient dying of status asthmaticus showed in addition an eosinophilic arteritis of the pulmonary vessels, together with an eosinophilic, lymphocytic and polymorph leucocytic infiltration in the walls of the alveoli. The changes described resembled those reported in eosinophilic pneumonia and those found by Leopold in asthma and described above.

An unusual syndrome occasionally encountered in chronic asthmatic patients is the development of focal granulomatous lesions and diffuse interstitial eosinophilic infiltration in various organs, especially the prostate, in addition to the lungs. The prostatic lesions were described in detail by Stewart et al. (1954), though Löffler (1932) had earlier drawn atten-

tion to the association of eosinophilic epididymitis in a chronic asthmatic patient who had also shown fine nodular opacities scattered throughout his lung radiographs.

In addition to the usual bronchial asthmatic changes described above, there is a very extensive diffuse and focal peribronchial, periarterial, alveolar interstitial and pleural infiltration with eosinophils together with some lymphocytes and a few plasma cells. Scattered irregularly in the bronchial walls, in the periarterial connective tissues and in the pleura, and usually corresponding to the most intense regions of eosinophilic infiltration, are small miliary granulomatous lesions the centres of which contain deeply eosinophilic but granular material often erroneously described as fibrinoid change (Figs. 19.7, 19.7A). Closer examination of the "fibrinoid" material shows that it consists very largely of granular material formed by degenerate eosinophil cells. These lesions known as "eosinophil abscesses" are also found in eosinophilic pneumonia and are described in detail under that

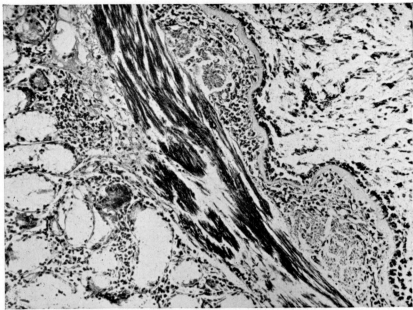

FIG. 19.6. Section of a bronchial wall in asthma showing the hypertrophy of the muscle, hypertrophy of the mucous glands, and denudation of the surface layers of bronchial epithelium into the lumen of the bronchus which is filled with cell-laden mucus. × 140 H and E.

section. The prostatic lesions described by Stewart *et al.* were of similar nature.

Any patient suffering from chronic asthma and signs or symptoms of prostatic involvement should be examined for the presence of excessive eosinophils in their prostatic secretions as this syndrome should be capable of being correctly diagnosed in life.

In many cases of chronic bronchitis there is an element of bronchospasm but after death there is little evidence of the asthmatic changes described above. Occasionally, infiltration of the walls of the smaller bronchi by eosinophils and thickening of the epithelial basement membrane suggests a background of allergic asthma. In the majority of such cases, however, the findings are those of chronic bronchitis, emphysema and varying degrees of inflammatory damage to the lung.

Chronic Asthma

In 1951 Shaw drew attention to an unusual condition occurring in bronchial asthma in which mucus of a peculiar and semisolid consistency became impacted distal to the division of the segmental bronchi and usually in bronchi in the upper lobes of the lungs. Some of the patients suffered from idiopathic bronchial asthma but others as was shown by Hinson *et al.* (1952) followed allergic bronchopulmonary aspergillosis. Accounts of this disorder have been given by Harvey *et al.* (1957), Sanerkin *et al.* (1966), Morgan and Bogomoletz (1968) and Katzenstein *et al.* (1975).

Much of the bronchial epithelium undergoes almost total mucous metaplasia and many of the cells, except the basal layer, are shed into the bronchial lumen which is filled with yellowish-green, very tenacious, glairy, ropy, mucus plugs similar to those found in fatal status asthmaticus.

Microscopical examination of the mucus shows that they are laid down in lamellae and contain collections of disintegrating eosinophils and darker patches which stain positively with phosphotungstic acid–haematoxylin stain and consist of nuclear debris. In places the bronchial

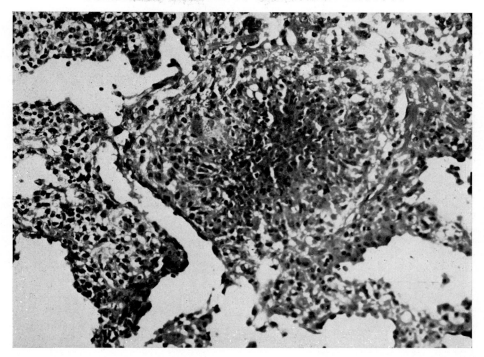

FIG. 19.7. An eosinophil "abscess" in interstitium of the lung in a case of eosinophilic pneumonia. The contents of the abscess consists of eosinophil granules released from the disintegration of the parent cells. × 180 H and E.

epithelium undergoes squamous metaplasia and the affected mucus-filled bronchi in chronic cases become stretched and dilated causing the bronchial cartilages to atrophy. The walls of the affected bronchi show changes identical with those present in bronchial asthma and these as Dunnill (1960) showed include hyperaemia, oedema and an infiltration with eosinophils and lymphocytes. In severe cases the bronchial obstruction was followed by suppurative changes in the bronchi distal to the obstruction. The abscesses, however, were non-putrid and were usually confined within the bronchial walls resulting in a cystic form of bronchiectasis. Bronchi in other parts of the lung also showed plugging with mucus containing eosinophils, together with the other characteristic changes found in bronchial asthma.

An earlier stage of this condition was probably observed radiologically by Cole (1951), who described reversible changes of bronchiectasis in

persons suffering from chronic asthma. These chronic asthmatic changes are similar to if not identical with those seen in eosinophilic pneumonia and bronchocentric granulomatosis and are described below.

Eosinophilic Pneumonia and the Pathergic Angeitides

The conditions about to be discussed share features common to those described previously in this chapter. Although differing in their clinical presentations and pathology they nevertheless merge one into the other and each may show features common to the others. The occasional presence of sarcoid-type granulomas in eosinophilic pneumonias, a type of lesion which is found in other presumed immunological (hypersensitivity) disorders, and also the occurrence of acute vasculitis lend support to the view that probably most of these disorders

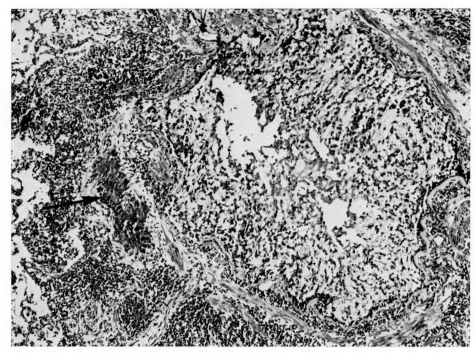

FIG. 19.7A. Bronchial asthma showing the bronchial lumen filled with desquamated lining cells, mucus and eosinophils and an eosinophil "abscess" (arrowed) within the bronchial wall. These appearances are identical with those seen in allergic bronchitis caused by *Aspergillus* sp. infection, but no fungus can be identified. × 30 H and E.

have an immunological basis and that Good-pasture's syndrome falls within the type 2 category (Gell and Coombs). The pulmonary changes are associated with a form of vasculitis characteristically found in hypersensitive states and described by Zeek *et al.* (1948). Fienberg (1955) referred to the whole group of disorders as "Pathergic Granulomatoses" following Rössle's (1933) definition of pathergy as "any lesion caused by an altered immune reactivity of body tissues".

For descriptive purposes the group will be divided into the following:

(1) Eosinophilic pneumonia and broncho-centric granulomatosis.
(2) Allergic granulomatosis.
(3) Classical Wegener's granulomatosis.
(4) Localized Wegener's granulomatosis.

(5) Sarcoidal angeitis.
(6) Lymphomatoid granulomatosis.
(7) Goodpasture's syndrome.

1. Eosinophilic Pneumonia and Broncho-centric Granulomatosis

Synonyms: Pulmonary eosinophilia, Pul-monary infiltration with eosino-philia (PIE syndrome)

Eosinophilic pneumonia is a descriptive name applied to a form of pneumonia characterized by an eosinophil infiltration of lung tissues. It comprises a heterogeneous group of diseases some of which, if not all, are probably basically caused by immunological disorders.

Löffler (1932, 1936) described a series of cases all showing a characteristic clinical, radiological and haematological picture; all the patients complained of respiratory symptoms, particularly asthma, which were accompanied by pyrexia, and the disease usually followed a mild course, recovery invariably occurring. Radiographs of the lungs showed numerous irregular opacities scattered through all the lung fields varying in size up to 0·5 cm in diameter. In addition the blood count showed an eosinophilia which in some cases reached 50 per cent of the total white cells.

Löffler described this clinical syndrome but it lacked any pathological basis. The term Löffler's disease should be abandoned and the more general term eosinophilic pneumonia substituted, denoting as it does an eosinophil infiltration of the lung which is the dominant feature of this form of pneumonia. Although the pulmonary eosinophil infiltration is usually accompanied by a blood eosinophilia this is not invariably so. A variety of causes of eosinophilic pneumonia are recognized and in most instances the clinical features conform to those described by Löffler. Eosinophilic pneumonia has been reviewed by Crofton et al. (1952), Liebow and Carrington (1966 and 1969), Scadding (1971) and Katzenstein et al. (1975) and its principal causes include:

(a) Drug reactions.
(b) Asthma including allergic pulmonary aspergillosis and its most severe form—bronchocentric granulomatosis.
(c) Parasitic infections.

(a) DRUG REACTIONS

The revolutionary changes that have occurred in therapeutics during the past 40 years following the introduction of antibiotic and synthesized chemical drugs in place of the older pharmacological preparations, as well as the greatly increased number of chemical environmental pollutants, has resulted in the appearance of a variety of new drug and chemically induced iatrogenic diseases. Among these diseases are included the occasional appearance of a blood eosinophilia accompanied by eosinophilic pneu-

monic infiltrates. The drugs and chemicals responsible for these changes are thought to behave in occasional individuals as haptenes though the drugs themselves are not antigenic. Among the drugs known to cause such reactions are nitrofurantoin, furazolidine, penicillin, para-amino-salicylic acid, hydralazine, chlorpropamide, mephenesin, the sulphonamide group of drugs and probably many other substances in an occasional sensitized individual. Very rarely some of these drugs, i.e. penicillin, may cause a dramatic type 1 immunological response (anaphylactic shock). In less severe reactions pulmonary oedema associated with hyaline membrane disease and a variable interstitial and intra-alveolar eosinophil and mononuclear cell reaction occurs which later may be followed by interstitial lung fibrosis. In most instances the nodular opacities seen in lung radiographs at the height of the drug reaction disappear rapidly following drug withdrawal only to be followed by their reappearance if it is given again. Drug reaction to nitrofurantoin was described by Muir and Stanton (1963), to furazolidine by Cortez and Pankey (1972), to penicillin by Reichlin et al. (1953), to p-amino-salicylic acid by Warring and Howlett (1952) and to sulphasalazine (Lancet, 1974). Pulmonary angeitis may occasionally occur in and complicate drug-induced eosinophilic pneumonia as in the case which was biopsied and described by Isenberg et al. (1968). In this case fibrinoid vascular necrosis was present in some of the small pulmonary arteries and arterioles accompanied by foci of lung necrosis and extensive infiltration of neutrophil and eosinophil leucocytes in the surrounding lung. The cause for these changes was not established with certainty but following a total colectomy for chronic ulcerative colitis and the withdrawal of the previously administered sulpha-drugs the pulmonary radiological changes resolved within 10 days. Sarcoid-type lesions may also sometimes be found in such a drug-induced eosinophil pneumonia similar to those found in extrinsic allergic alveolitis. Pepys et al. (1959) also showed that some patients harbouring A. fumigatus in their sputum exhibited transient pulmonary eosinophil infiltrations as evidenced by the presence of these cells in their sputa. The eosinophil reaction was considered to be a response to a precipitin-type antibody–antigen complex.

(b) EOSINOPHILIC PNEUMONIA ASSOCIATED WITH ASTHMA, ALLERGIC PULMONARY ASPERGILLOSIS AND BRONCHOCENTRIC GRANULOMATOSIS

Clinically, it is convenient to divide asthma into those cases occurring in response to a recognizable allergen which were referred to by Scadding (1971) as extrinsic asthma and those cases occurring in the absence of any recognizable allergen which have been called intrinsic asthma. Among the well-known allergens that reach the lung through the air passages are the large group of organic dusts responsible for causing extrinsic allergic alveolitis which purely for the sake of descriptive convenience have been described in Chapter 11. Other inhaled allergens include pollen grains responsible for hay fever, other plant pollen allergies and bronchial mycotic infections especially those due to various *Aspergillus* sp. Although extrinsic allergens act primarily on the bronchial and alveolar surfaces, proteins and possibly haptenes are readily absorbed into the circulation and produce immunological disturbances of the type 1 and 3 varieties (Gell and Coombs). In intrinsic asthma no allergen can be detected and the symptoms tend to be of a more continuous nature rather than the episodic changes following exposure to extrinsic allergens.

In the ensuing account the changes to be described are characteristically seen in extrinsic asthma but many of the pathological changes encountered in the intrinsic form are indistinguishable and no subdivision will be made when considering their pathology.

At least two of the fatal accident cases, the post-mortem findings in which were described by von Meyenburg (1942a, b) and in which the patients had suffered from Löffler's syndrome in life, showed an eosinophilic pneumonia.

Bayley *et al*. (1945) also described a case of eosinophilic pneumonia associated with pulmonary vasculitis and Smith (1948) described a very similar case with pulmonary vasculitis, eosinophilic pneumonia and early necrotic lesions identical with those seen normally in Wegener's granulomatosis. More recent studies and examples of eosinophilic pneumonia in association with asthma have been presented by Leopold (1959), Liebow and Carrington (1969), Scadding (1971) and Katzenstein *et al*. It has become increasingly realized that many cases of eosinophilic pneumonia and asthma are associated with *Aspergillus* infections and these changes will be described in greater detail.

Hinson *et al*. (1952) first drew attention to the association of severe asthma in England with pulmonary infiltrations mainly of eosinophils, a peripheral blood eosinophilia and sticky mucoid sputum containing *Aspergillus fumigatus*. Pepys *et al*. (1959) also showed that some patients harbouring *A. fumigatus* in their sputum exhibited transient pulmonary eosinophil infiltrations as evidenced by the presence of these cells in their sputa. The eosinophil response was considered to be caused by a precipitin antigen–antibody reaction. The characteristic viscid mucoid plugs of sputum found in this condition had formerly been described as typical of "plastic bronchitis".

Opportunities occasionally arise to examine the histology of such an eosinophilic pneumonia and the alveolar interstitium, bronchial walls and peribronchial tissues are found to be heavily infiltrated with eosinophils, lymphocytes, a few plasma cells and plasmacytoid cells, while the alveoli are filled with eosinophils, macrophages and desquamated alveolar epithelial cells. A desquamative interstitial pneumonia (DIP) may also be found (Fig. 19.8). In the case described by Scadding (case 2) there was in addition a small necrotic focus of eosinophils surrounded by histiocytes in the lung parenchyma. This lesion appeared from the illustration to be an example of an "eosinophil abscess", other examples of which have been seen by the author (Fig. 19.7A). Steroid therapy successfully reverses these changes in the great majority of

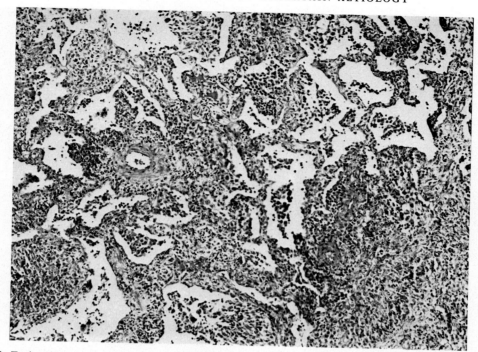

FIG. 19.8. Eosinophilic pneumonia showing in addition to severe interstitial eosinophilic pneumonia a desquamative pneumonic type of picture in some areas. Most of the smaller cells are eosinophils. × 75 H and E.

radiologically diagnosed cases. Although pneumonic infiltrates are present, the principal changes are found in the smaller bronchi which are often dilated and filled with thick viscid mucus containing many eosinophils, nuclear debris and hyphae of non-invasive *Aspergillus* sp. The walls of such bronchi are very heavily infiltrated with eosinophils, lymphocytes and some plasma cells (Fig. 19.9A). Damage to the bronchial epithelium may result in granulation tissue formation and bronchiolitis obliterans (Fig. 19.9B,D). The changes described are often referred to as *allergic pulmonary aspergillosis*. In many severe cases "eosinophil abscesses" also occur in the walls of the affected bronchi, peribronchial tissues, pleura and scattered through the alveolar interstitium. Similar but fewer "abscesses" are also found in non-allergic asthma as previously described. The abscesses, which also occur in other organs, consist of a central mass of disintegrating eosinophils surrounded by one or more layers of radially arranged histiocytes some of which ingest the eosinophilic granular material from the disintegrating eosinophils, while others form small giant cells which contain ingested crystalline bodies resembling minute Charcot–Leyden crystals. Acute vasculitis may also be present in addition as in a case described by Chan-Yeung *et al.* (1971) though vasculitis, apart from perivenous lymphocytic cuffing, is not usually found in uncomplicated eosinophilic pneumonia. Although *Aspergillus* sp. have been mainly incriminated as a cause of eosinophilic pneumonia associated with asthma, other fungi including *Candida albicans* are probably also capable occasionally of behaving in a similar manner.

Associated with the changes described above are positive skin antigen tests of the immediate type together with the presence of precipitin antibodies in the patients' sera. Such patients

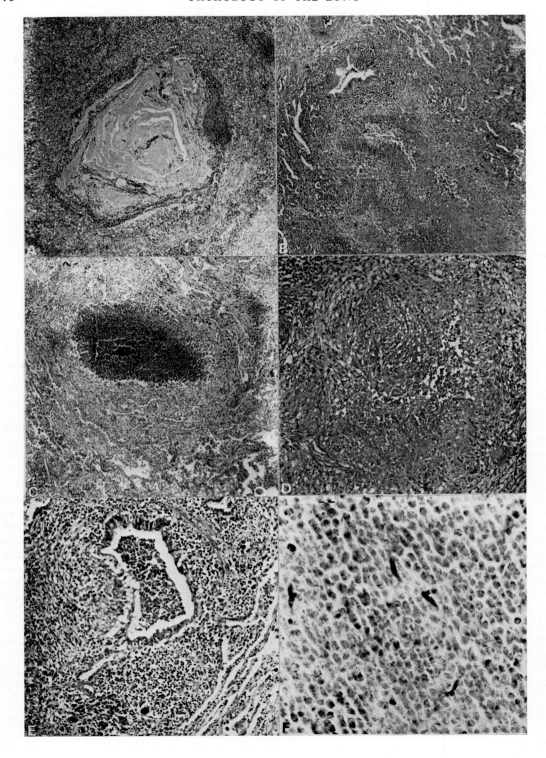

are very hypersensitive to *Aspergillus* antigen which should be used with caution.

The most severe cases of eosinophilic pneumonia caused by allergic pulmonary aspergillosis result in the condition reviewed by Katzenstein *et al.* and called by them *bronchocentric granulomatosis*, the condition having been so named originally by Liebow (1973). In bronchocentric granulomatosis the changes already described as allergic pulmonary aspergillosis are far more severe and result in a variable and often extensive necrosis of the lining of the bronchial wall with spread of the necrotic change and inflammatory granulomatous reaction into the adjacent peribronchial lung tissues. Twenty-three cases were reviewed by Katzenstein *et al.* Ten patients had suffered from chronic asthma but the remainder gave no history of asthma. Some of the cases showed evidence of coexistent eosinophilic pneumonia and two non-asthmatic cases had rheumatoid disease. So far most of the reported cases have occurred in North America and Australia despite the greater frequency of *Aspergillus* infection in Britain, though a similar case was seen by Hinson (1975) which was known to have had an *Aspergillus* infection many years earlier. Clinically, the patients complain of nasal congestion, fatigue, cough, haemoptysis and general malaise with fever. Patients may show a positive precipitin test for *Aspergillus* infection and for *Candida albicans*. Blood eosinophilia is, however, a very inconstant finding. *Macroscopi-*

cally, many of the smaller bronchi are dilated, saccular and filled with caseous-like material while the more proximal larger segmental bronchi may show mucus impaction (Fig. 19.10). The affected bronchi are surrounded by a mantle of greyish-white tissue which extends a variable distance into the surrounding lung. The lesions may be mistaken for a neoplasm or caseous bronchial tuberculosis. *Microscopically*, the bronchial epithelium and submucosal tissues are destroyed and the lumen filled with necrotic material and nuclear debris, dead and viable polymorph leucocytes and occasional eosinophils. The necrotic material resembles that found in the necrotic areas of Wegener's granulomatosis but contains small abscess-like collections of polymorph leucocytes. Surrounding the necrotic tissue are radially arranged spindle-shaped histiocytic cells some of which are multinucleated. Outside the histiocytic layer are large numbers of lymphocytes, plasma cells, occasional eosinophils and giant cells. The granulomatous reaction may involve the bronchial cartilage plates which are eroded and destroyed and it can spread extensively into the surrounding peribronchial tissues (Fig. 19.9 C, D, E). The pulmonary arteries are not primarily involved and angeitis is not a feature of the disease though the arteries can become involved in the outward spread of the destructive process. Later fibrosis supervenes and bronchial stenosis and periarterial fibrosis are notable features. Widespread interstitial alveolar lymphocytic

---

FIG. 19.9. A–E show successive stages in the evolution of eosinophilic bronchopneumonia due to *Aspergillus* sp. infection.

A. Severe eosinophilic bronchitis with outpouring of thick mucus, destruction of most of the lining cells and a very heavy eosinophil infiltration of the entire bronchial and peribronchial tissues.
B. A later stage showing necrosis of the bronchial lining.
C. Similar to B but the lumen is now occupied by a mass of darkly staining eosinophilic granules resulting from necrosis of the previously intact eosinophils.
D. Total destruction of the bronchial wall with giant-cell granulomatous reaction in the bronchial mural tissues. This stage corresponds to bronchocentric granulomatosis.
E. Bronchocentric granulomatosis showing a still recognizable bronchus with granulomatous reaction in its walls. Many of the cells in the wall are eosinophils, others mainly lymphocytes.
F. Bronchial contents from the case of bronchocentric granulomatosis shown in Fig. 19.10. The bronchial contents consisting mostly of necrotic debris also contains hyphae of *Aspergillus* sp.
A, B, and C × 25, D and E × 65, F × 250. A–E stained H and E, F stained Grocott. (The author is indebted to Dr. K. F. W. Hinson for A and C and to Dr. F. Whitwell for D, E, and F.)

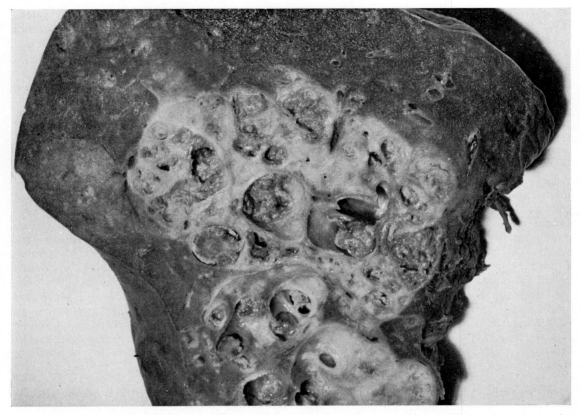

FIG. 19.10. Bronchocentric granulomatosis caused by *Aspergillus* sp. infection. The debris-filled cystic spaces are the dilated and damaged remains of bronchial walls.
(Reproduced by courtesy of Dr. F. Whitwell.)

and plasma cell infiltration are present in the lung outside the bronchi. In a high proportion of cases scattered hyphae of the causative fungus are found in the necrotic debris filling the destroyed bronchi (Fig. 19.9 F).

(c) PARASITIC INFECTIONS

The parasitic causes of eosinophilic pneumonia are considered in Chapter 10. Several common human helminth infections notably those caused by *Ascaris lumbricoides*, *Ankylostoma duodenale*, *Strongyloides stercoralis* and probably *Toxocara canis* undergo larval stages in their life-histories that involve the lungs. Similarly the larval stages of the blood nematode infections caused by *Brugia malayi* and the feline *B. pahangi* are

thought to be responsible for causing tropical eosinophilic pneumonia.

Two of the fatal accident cases examined by von Meyenburg (1942) showed intestinal ascariasis and the pulmonary eosinophilic infiltration may have been caused by migrating larvae. Eosinophilic infiltration in the lungs occurs in response to migrating larvae and Symmers (1954) showed that pulmonary arteritis could also be caused by the presence of migrating ascaris larvae.

2. Pulmonary Allergic Granulomas

Some uncertainty surrounds this condition which occupies a position intermediate between eosinophilic pneumonia and classical Wegener's granulomatosis. The clinical history and often prolonged course relate it more closely to eosinophilic pneumonia. The author regards it as an intermediate stage in a disease process with a wide spectrum of changes. It can be found in some varieties of eosinophilic pneumonia as well as in generalized Wegener's granulomatosis.

The clinical course of the illness is more severe than in the cases originally described by Löffler but less dramatic than Wegener's syndrome, and the majority of the patients die after a variable period. Two of the patients described by Churg and Strauss (1951) survived up to 5½ years but in the case described by Ehrlich and Romanoff (1951) death followed within 10 days of the onset of the disease. The patients complain of asthma, which becomes very severe in some cases, associated with a low-grade fever and a very high blood eosinophilia which may reach to over 80 per cent of the total white cell count. Patients may also suffer from recurrent seasonal attacks of hay fever. A raised blood-pressure occurred in about half the cases after the illness had been in progress some time, and a mild haematuria and albuminuria were noted in a few instances.

Macroscopically, the lungs are either bound by adhesions to the chest wall, or the pleural sacs may be partly filled with sero-sanguineous fluid. Scattered through the lungs are greyish-yellow nodules varying in size from a miliary tubercle to lesions 1·5 cm in diameter. The nodules may vary from two to three in number being situated in one or both lungs to innumerable miliary nodules scattered throughout both lungs. The centre of the lesions is filled with yellowish-white caseous-like contents which is bounded in the larger lesions by a distinct margin. The walls of the bronchi are thickened and the lumens often filled with thick viscid mucus. Some branches of the pulmonary arteries may be thickened and narrowed. Granulomatous lesions are often found in other organs especially the spleen.

Microscopically, the nodules consist of a central necrotic zone of fibrinoid or frankly infarcted alveolar tissue surrounded by a layer of radially arranged endothelioid cells, together with small foreign-body type of giant cells (Figs. 19.11, 19.11A). Numerous eosinophils and lymphocytes surround the pallisaded layer of endothelioid cells and the whole lesion bears a superficial resemblance to a rheumatoid necrobiotic nodule. Other nodules consist of focal areas of interstitial alveolar fibrosis accompanied by masses of eosinophils and chronic inflammatory cells filling damaged alveoli. The walls of the bronchioles also become heavily infiltrated with eosinophils and lymphocytes, and their basement membranes are thickened and hyalinized. The interlobular septa are also thickened, oedematous and occasionally show focal patches of fibrinoid change.

In the majority of cases the above lesions are accompanied by peri-arterial and peri-capillary inflammation and in the most severe cases by typical lesions of acute granulomatous type of polyarteritis nodosa. The walls of some of the small arteries, and more rarely veins, are replaced in whole or in part by a mass of histiocytes, lymphocytes and occasional polymorph leucocytes but there is usually no associated mural fibrinoid change. Other arteries show destruction and replacement of the intima by loose connective tissue, and a layer of small foreign-body giant cells attached to the inner surface of the disintegrating internal elastic lamina.

In addition to the pulmonary lesions, several, but not all, of the reported cases have been accompanied terminally by a thrombotic and haemorrhagic form of acute glomerulitis. This change, which is typical of fully developed cases of Wegener's syndrome, is an inconstant finding in the condition under discussion.

3. Generalized (Classical) Wegener's Granulomatosis

Generalized Wegener's granulomatosis was defined by Fahey *et al.* (1954) as a condition characterized by the presence of granulomas in

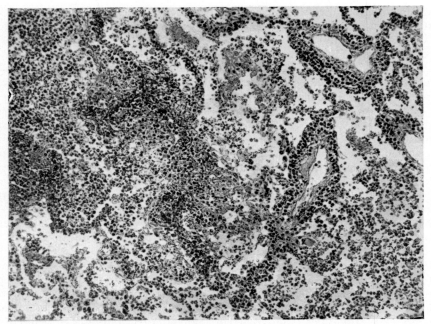

FIG. 19.11. A low-power view of a pulmonary allergic granuloma showing the changes related to the alveolar capillaries. × 140 H and E.

the nasopharynx or antra, lesions in the respiratory tract, a generalized arteritis and a focal form of glomerulitis; it is a more severe and extensive form of pulmonary allergic granuloma.

The clinical condition was originally described by Klinger (1931) but Wegener (1936, 1939) reported further cases and first included an account of its pathology.

The pathology of this disorder has been described by Godman and Churg (1954), Fahey *et al.* (1954), Walton and Leggatt (1956), Rose and Spencer (1957), Spencer (1957) and Walton (1958). A localized form of the disease has been described by Carrington and Liebow (1966).

Walton (1958) carried out a retrospective survey of fifty-six cases including ten of his own. He found that two-thirds had suffered from persistent purulent rhinorrhoea together with nasal obstruction, antral pains, epistaxis and proptosis. In the remaining third, pulmonary

signs and symptoms had been the presenting feature and these included a chronic cough, haemoptysis and pleurisy. During the latter part of the illness, most patients pass a blood-stained urine loaded with albumen and casts. The progress of the disease once established is rapid and death invariably occurs, often within as short a time as a month. The most common cause of death was found by Walton to be renal failure, which occurred in 83 per cent of his series.

Radiologically, irregular shadows come and go in the lungs but tend to be very much larger than those found in allergic granulomas and eosinophilic pneumonia. Approximately 45 per cent of cases show a high eosinophil count, amounting in some cases to 80 per cent of the total white cells and hyperglobulinaemia is commonly found.

Macroscopically, the lesions in the nose and sinuses may progress and lead to palatal and orbital perforation with destruction of the facial

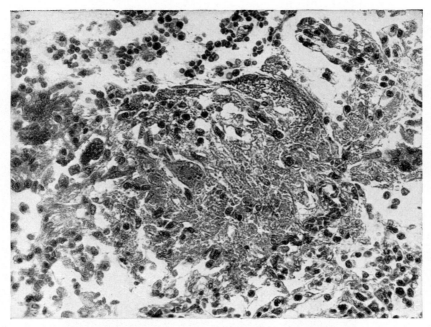

FIG. 19.11A. Pulmonary allergic granuloma showing a small fibrinoid necrotic focus with surrounding endothelioid cells and an occasional small giant cell. There were also eosinophils in the alveolar lumens. × 280 H and E.

tissues near to the root of the nose (midline granuloma). The lesions in the respiratory tract involve the air passages throughout their extent as well as the lung parenchyma and its vasculature.

Ulceration of the trachea and bronchi is very common and extends into the intrapulmonary branches (Fig. 19.12).

In the lung, creamy-yellow necrotic lesions and haemorrhagic infarcts are found. The former vary in size from small nodules resembling caseous tuberculous lesions about a centimetre in diameter to massive involvement of most of a lobe. In massive lesions, the affected lobe is greyish-white and airless at the periphery but it cavitates or is filled with necrotic material in the centre (Fig. 19.13). The peripheral portions of the lesion closely resemble unresolved pneumonia. Smaller nodules in other lobes may coexist with the massive lesions. The pleural surface overlying the massive lesions

becomes fibrotic and usually densely adherent to the chest wall. Pleural effusions, often haemorrhagic, occur in the unobliterated portions of the sac.

Microscopically, the smaller lesions are similar to those described as allergic granulomas. The massive lesions present a histological picture which partly resembles both pulmonary tuberculosis and infarction, conditions with which it may readily be confused at first sight. At the edge of the lesions, characteristic vascular changes are found together with severe chronic interstitial fibrosis in the surrounding lung.

In the central necrotic zones it is often possible to recognize the ghost outlines of the destroyed tissues and the dead tissue contains much nuclear debris; scattered haphazardly through it are areas where the tissues are still viable though filled with a variety of acute and chronic inflammatory cells. The lesions differ both in shape and appearance

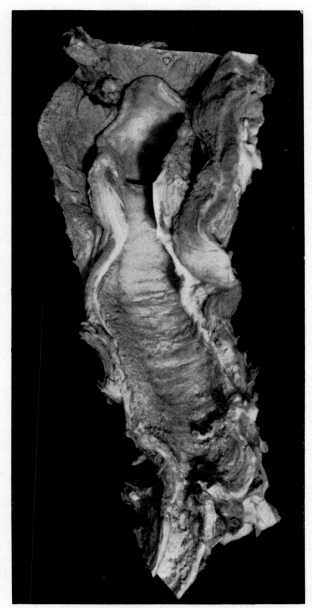

FIG. 19.12. Ulceration of the trachea and larger bronchi in a case of generalized Wegener's granulomatosis. (Reproduced by courtesy of the Editor of *Quart. J. Med.*)

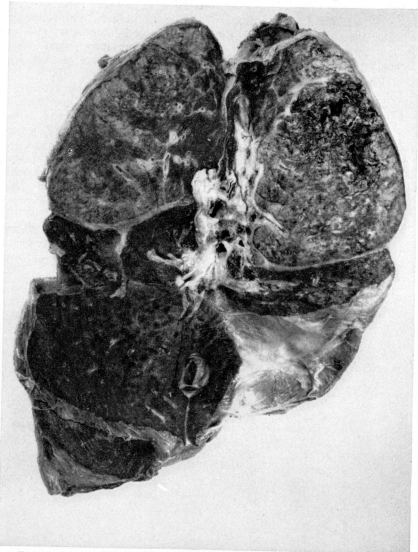

FIG. 19.13. Generalized Wegener's granulomatosis showing extensive destruction of the upper lobe of a lung with cavitation. Almost the entire lobe is consolidated and replaced by necrotic tissue. One-half natural size. (Reproduced by courtesy of the Editor of *Quart. J. Med.*)

from classical pulmonary infarcts, being found anywhere within the lung and possessing a completely irregular shape.

At the edges of the necrotic tissue are numerous giant cells of Langhans' type and the alveolar walls and spaces are filled with a mixture of acute and chronic inflammatory cells in varying proportions, but eosinophils are scanty in the cellular parts of the lesion outside the edges of the necrotic tissue (Figs. 19.14, 19.15).

In addition to the destructive and granulomatous pulmonary lesions, characteristic changes are found in the blood-vessels. In the unaffected parts of the lung, and in many other organs, widespread changes of acute polyarteritis nodosa occur. These changes are identical with but more widespread than those found in allergic granuloma. At the edges of the massive lesions the arteries and the veins become largely replaced by granulomatous tissue which narrows or occludes the lumen and replaces most of the wall (Figs. 19.16, 19.17). The granuloma consists of a loose form of connective tissue, rich in fibroblasts, and containing both acute and chronic inflammatory cells and giant cells. In the intact lung, small focal patches of fibrinoid change are present in the alveolar capillaries.

The walls of the large bronchi show extensive superficial mucosal ulceration together with polymorph leucocytic and lymphocytic infiltration, and changes of acute polyarteritis nodosa frequently involve the bronchial arteries.

In addition to the pulmonary changes, some patients may show chronic spreading granulomas composed mostly of plasma cells, lymphocytes and polymorph leucocytes eroding the walls of the nasopharynx, antra or middle ear. Granulomatous lesions with many plasma and plasmacytoid cells and features more akin to a malignant lymphoma may lead to extensive nasal and

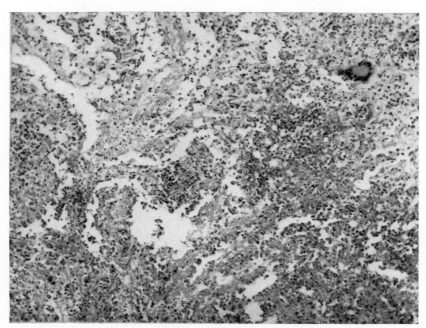

FIG. 19.14. Largely necrotic (infarcted) lung with giant cell reaction in a case of generalized Wegener's granulomatosis syndrome. × 140 H and E.

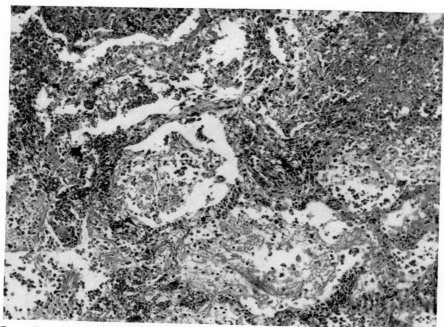

FIG. 19.15. Generalized Wegener's granulomatosis showing diffuse alveolar capillaritis with a larger necrotic focus in the upper right-hand corner. × 140 H and E.

orbital destruction. Such lesions are similar to those seen in pulmonary lymphoid granulomatosis and only occur in association with diffuse or localized Wegener's granulomatosis. This type of change only occurs in very longstanding nasal lesions which often antedate the appearance of the other changes found in generalized Wegener's granulomatosis by many years.

Other changes occurring in generalized Wegener's granulomatosis include multiple splenic infarcts of unusual shape and distribution, together with a thrombotic and haemorrhagic form of acute glomerulitis which is also seen, though less frequently, in allergic granuloma.

Aldo *et al.* (1970) studied the glomerulonephritic lesions electron microscopically and claimed to have found immune complexes.

Eosinophilic pneumonia, allergic granulomatosis and classical generalized Wegener's granulomatosis form a triad of conditions with overlapping features and of increasing severity.

Eosinophilic pneumonia due to varied causes is usually followed by recovery while the fully developed classical form of Wegener's granulomatosis usually proves fatal though azathioprine, methotrexate and alkylating agents may favourably affect the prognosis.

Wegener (1939) regarded the lung lesions as infarcts, but Walton and Leggatt found that the initial lesion was a peribronchial granuloma which led to secondary involvement of the vessels and other lung structures. Spencer (1957) regarded the massive pulmonary lesions as the sequel of a pan-vasculitis which included the capillaries. The presence of generalized arterial changes due to acute polyarteritis nodosa has already been noted in pulmonary allergic granuloma and in generalized Wegener's granulomatosis. Similar small focal lesions may sometimes occur in eosinophilic pneumonia.

It is very necessary, however, to draw a clear distinction between true generalized polyarteritis nodosa, in which the pulmonary

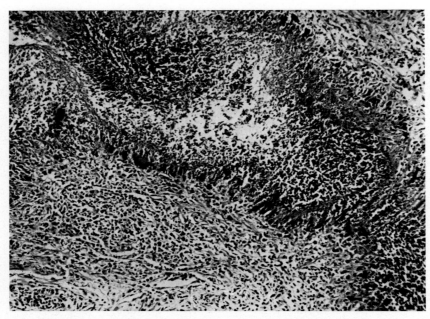

FIG. 19.16. Very severe pulmonary arteritis from a case of generalized Wegener's granulomatosis. The vessel wall has been totally destroyed and there are numerous giant cells engulfing destroyed medial elastic tissue. × 100 H and E.

arteries and the systemic vessels are involved, and hypertensive pulmonary arteritis which is strictly confined to the pulmonary arteries and which is discussed in Chapter 16. The latter is directly related to the presence of hypertension in the lesser circulation whereas generalized polyarteritis nodosa is frequently unassociated with hypertension.

Generalized polyarteritis nodosa involving the lung is usually preceded by some form of respiratory illness which has often been present for several years before the appearance of generalized arterial disease (Rose and Spencer). Also it is often associated with a high blood eosinophilia, hyperglobulinaemia and with a granulomatous type of lesion in the smaller arteries.

The aetiological causes of allergic granuloma and classical Wegener's granulomatosis are unknown but are likely to be similar. Rich (1946–7) considered that polyarteritis nodosa was a hypersensitivity reaction to histamine while others believed that drugs or the products of bacterial or viral infections behaving as haptenes caused antigen–antibody reactions which in turn led to local release of histamine. Support for the latter view was provided by Harkavy (quoted by Reichlin et al., 1953) who found lung changes with proved pulmonary angeitis together with renal changes following the use of penicillin. Also Citron et al. (1970) described four cases of non-pulmonary necrotizing angeitis which they attributed to the addictive use of methamphetamine. Other drugs that have been reported as causing these reactions are the sulphonamides (Walton and Leggatt, 1956), thiouracil and its derivatives, and a variety of other drugs including sera (Lichtenstein and Fox, 1946). The β-Haemolytic streptococcus, long recognized as being in some unknown way connected with the aetiology of acute rheumatic fever, has also frequently been found in the respiratory tract preceding the onset of acute polyarteritis nodosa involving the lung. As Rose and Spencer have implied, both antibiotic drugs and sulphonamides are

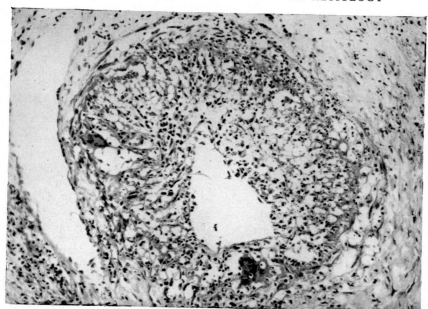

FIG. 19.17. Granulomatous vasculitis (probably an artery) in which the intima is greatly thickened and replaced by oedematous granulation tissue containing giant cells. This type of vascular lesion is widespread in the affected parts of the lung. × 140 H and E.

(Reproduced by courtesy of the Editor of *Quart. J. Med.*)

frequently employed to treat such infections and many have been erroneously regarded as the cause of the ensuing arterial disease which may have been related to the bacterial infection.

4. Localized Wegener's Granulomatosis

Synonym: Pathergic granulomatosis

Carrington and Liebow (1966) described sixteen cases, ten women and six men, in whom a lung biopsy had shown a localized pulmonary lesion with features similar to those seen in generalized Wegener's granulomatosis but which lacked the other components of this syndrome. Various names have been applied to similar cases described in the past. In a later review (Liebow, 1973) eighty-five cases were presented and the localized form of the disease is now more commonly seen than the classical generalized variety. It predominantly affects women (8 : 5). A few patients are asympto-

matic but the majority present with a history which includes any of the following symptoms or signs: cough, chest pain, dyspnoea, sore throat, painful subcutaneous nodules progressing to deep penetrating skin ulcers, pyrexia and loss of weight. Chest radiographs are normal in all cases except for discrete and usually multiple rounded or nodular opacities varying in size up to 5·0 cm diameter which are more commonly found in the lower halves of the lungs. Cavities are commonly seen in the larger opacities. Renal function tests show no evidence of renal disease and there is no evidence of the characteristic thrombotic type of acute glomerulo-nephritis seen in Wegener's syndrome.

Very few of the patients gave an unequivocal history of drug sensitivity. Of the sixteen original cases, five died within a year from generalized vascular involvement and from the granulomatous lesion in the lungs, nine survived for longer than a year and of these two had received no steroid therapy, and two died from unrelated diseases.

Macroscopically, the lesions are yellowish-white, rounded or pyramidal in shape, varying in size up to 6·0 cm diameter and are found to be distributed throughout the lung parenchyma. When cavitated the cavity is lined by shaggy necrotic tissue and the centres of some of the lesions are filled with liquid contents. Occasional superficial ulcers are found in the bronchi and trachea.

Microscopically, the lesions present similar features to those seen in the pulmonary lesions in generalized Wegener's granulomatosis only on a smaller scale. The wall of the lesions is formed of fibrous tissue and a dense underlying zone of lymphocytes, plasma cells, histiocytes and occasional foreign-body type of giant cells which surround necrotic, infarct-like tissue. Sarcoid-type granulomas are sometimes found in the larger lesions in the surrounding parenchyma and fibrous septa. A panangeitis is a conspicuous feature and is best seen in the vessels at the periphery of the lesions and is similar to the changes found in the pulmonary vessels in generalized Wegener's granulomatosis.

The ulcerating lesions found in the upper respiratory tree and those in the subcutaneous lesions are both attributable to the vasculitis, and evidence of acute arteritis can sometimes be found in muscle biopsy specimens.

The pulmonary lesions in localized Wegener's granulomatosis closely resemble those found in the classical and generalized form of the disease but the most notable difference is the usual absence of both thrombotic acute glomerulo-nephritis and midline nasopharyngeal granulomas, and the not infrequent self-limited nature of the disease. Pseudotumours of the orbit occasionally occur in this form of pulmonary granulomatosis (Cassan *et al.*, 1970).

The differential diagnosis includes the specific forms of chronic granulomas due to mycotic and mycobacterial infections, sarcoidosis and rheumatoid necrobiotic lesions.

Localized Wegener's granulomatosis carries a better prognosis than the generalized form of the disease, and even in the absence of treatment with immuno-depressive and cyto-toxic drugs, spontaneous recovery sometimes occurs.

5. Sarcoidal Angeitis

Synonym: Necrotizing sarcoidal angeitis

Necrotizing sarcoidal angeitis was first named and described by Liebow (1973) who reviewed eleven cases in patients of all ages. The author has also seen three cases which qualify to be included in this group. It may occur at any age and in either sex. One patient in Liebow's series had been exposed to beryllium and a second may have had idiopathic sarcoidosis. One of the author's patients subsequently died at a later date from lung cancer. The lesions may be unilateral or bilateral and the patients present with pyrexia, cough and pleuritic-type pain. Radiologically the lesions are nodular and usually subpleural causing an overlying pleural fibrosis.

Microscopically, giant-cell granulomatous lesions superficially resembling sarcoidal granulomas are present around the outside and in the walls of the muscular pulmonary arteries and veins. The granulomas differ from Boeck's sarcoidal granulomas inasmuch as the lesions are more ill-defined, the giant cells are smaller and the cell populations more heterogeneous consisting of histiocytic cells, lymphocytes, plasma cells, eosinophils and occasional polymorph leucocytes (Fig. 19.18). The affected vessels are often occluded by intimal proliferation or by extension of the intramural granuloma, and the media shows variable segmental necrosis as evidenced by elastic stains. Other vessels may lack giant cells but are diffusely infiltrated with lymphocytes which extend outwards into the extravascular tissues. The centres of the larger granulomas sometimes undergo necrosis and necrosis may extend into the extravascular tissues (Fig. 19.19). Liebow has drawn attention to the presence of similar granulomas in the mucosa of the

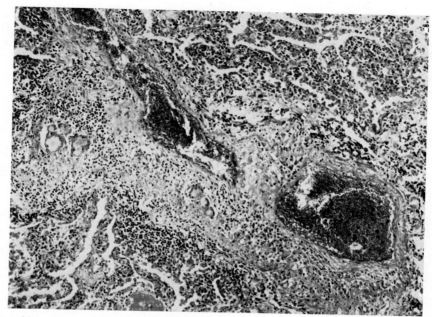

FIG. 19.18. Sarcoidal angeitis showing a giant-cell granulomatous lesion outside the adventitia of a small pulmonary artery. × 100 H and E.

bronchi which may lead to extensive bronchiolitis obliterans with resulting obstructive pneumonitis. Massive giant-cell response to destruction of vessel walls with accompanying histiocytic reaction is characteristically seen in generalized Wegener's granulomatosis with giant-cell formation (Fig. 19.16), but massive and necrotic angeitis tends to form one end of a spectrum of angeitic change which at its other extreme blends with sarcoidal angeitis.

6. Lymphomatoid Granulomatosis

In 1972 Liebow *et al.* described at length a new form of pulmonary angeitis based on a study of forty cases. They described it as "an angiocentric and angiodestructive lymphoreticular proliferation and granulomatous disease involving predominantly the lungs". In 1973 Liebow again reviewed the condition and added a further thirty-four cases. Almost all cases of the disease so far reported have occurred in North America, a few in Australasia but so far the author has not seen any cases originating in Great Britain. The disease is twice as common in men and the majority of cases occur between 25 and 50 years of age. The mode of clinical presentation is very similar to that found in localized Wegener's granulomatosis, namely pyrexia, cough and dyspnoea. Radiologically nodular opacities very similar in appearance to metastatic growth are found mainly in the lower lobes. These nodular shadows come and go and are unassociated with enlargement of the hilar lymph glands. Like localized Wegener's granulomatosis to which the condition bears many resemblances, it may present initially with cutaneous ulceration which can precede the lung changes by several years. Other signs and symptoms result from involvement of the nervous system (central and peripheral) or to joint disorder (arthralgia).

Macroscopically, the pulmonary lesions usually present as rounded, greyish-pink, solid masses unless abutting upon the pleura. Some are

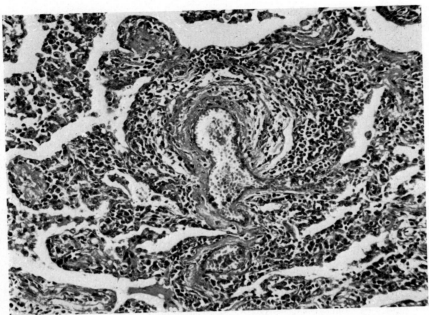

FIG. 19.19. Sarcoidal angeitis showing arteritis involving mainly the outer coats of a small branch of a pulmonary artery together with extensive chronic interstitial pneumonia in surrounding lung. × 160 H and E.

related to bronchi, others to lobular septa depending on whether the angeitis affects mainly the pulmonary arteries or veins. Large nodular lesions may undergo central and often extensive necrotic change and cavitate, while bronchial obstruction leads to areas of obstructive pneumonitis.

Microscopically, the lesions are centred on either the pulmonary arteries or veins, the walls of which are infiltrated with lymphocytes, plasma cells, plasmacytoid cells and large mononuclear cells of undetermined nature referred to as reticulum cells (Fig. 19.20). The microscopical picture consists of primitive lymphoid series of cells together with their more mature derivatives, lymphocytes and plasma cells. The more immature cells show numerous mitoses and form bizarre multinucleated cells (Fig. 19.21). Necrosis is a feature of the diffusely infiltrating areas of lymphomatoid granulomatosis.

The picture presented by this condition has features consistent with both an angeitis and a lymphomatous or prelymphomatous process. Ten of the original series of patients developed a malignant reticulosis in the central nervous system similar in structure to the tumours that may follow renal transplantation. Five of the original series of patients developed malignant lymphomas. When reviewing his collected cases, Liebow (1973) found that twenty-eight of the forty patients were dead, the majority having died from intrapulmonary extension of the lung lesions. Both the intracerebral reticuloendothelial neoplasms and the malignant lymphomas that follow the lung disease proceed to grow autonomously despite the healing of the lung lesions.

Unlike classical Wegener's granulomatosis, dysproteinaemia was not discovered in any of the patients reviewed by Liebow *et al.* though it was not always sought.

Other organs and tissues that may become the sites of an angeitis and associated lymphomatoid infiltration include the kidneys, liver and skin. Lymph glands, spleen and bone

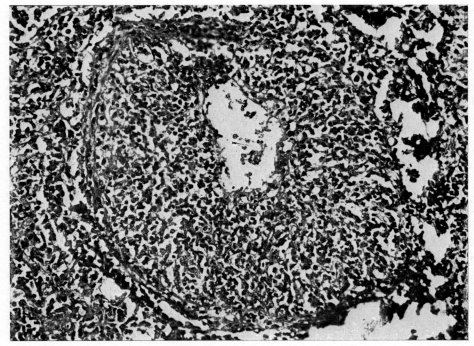

FIG. 19.20. Lymphomatoid granulomatosis showing severe chronic arteritis with infiltration of the wall of a pulmonary artery with lymphoid and plasma cell series of cells. × 180 H and E.
(Reproduced by courtesy of Dr. A. A. Liebow.)

marrow also show atypical proliferative changes with many plasma and plasmacytoid cells and in one case a malignant lymphoma developed.

Liebow *et al*. have discussed at length the nature of lymphomatoid granulomatosis, a lesion presenting several features of great biological interest. The reader is strongly recommended to read their review in its entirety as it is impossible to fully cover the subject of its aetiology and nature in this text. Many of the features of the disease bear a close similarity to localized Wegener's granulomatosis. Both conditions may be preceded by cutaneous lesions and both present with similar respiratory signs and symptoms.

Some cases of eosinophilic pneumonia caused by drugs, as already stated, may be associated with an acute angeitis. A vasculitis affecting arteries, veins and capillaries of increasing severity is a feature of both localized Wegener's granulomatosis and classical Wegener's granulomatosis. In all three conditions the angeitis has the character of an acute and rapid process in response to an acute injury of a presumed immunological nature.

A widespread granulomatous vasculitis which occasionally shows florid necrosis is the principal feature of sarcoidal angeitis. In this disorder lesions of a granulomatous type, which are also found in more chronic (type 4) forms of immunological disorders, i.e. extrinsic allergic alveolitis or Boeck's sarcoid, are present in and around vessel walls in the lungs. Lymphomatoid granulomatosis, a prominent feature of which is angeitis, may also represent the outcome of a previously existing state of low-grade immunological damage of long duration. It is well recognized that chronic inflammation of varied aetiology together with the immunological changes it provokes can occasionally predispose

7

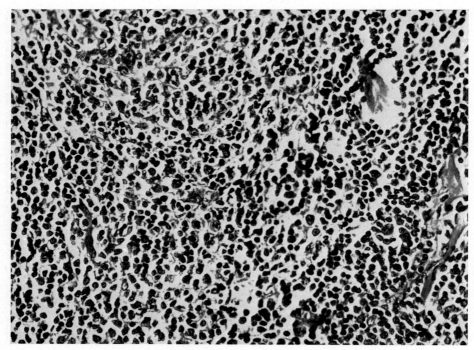

FIG. 19.21. Lymphomatoid granulomatosis showing the character of the pulmonary infiltrate which consists mainly of lymphocytes, plasmacytoid cells, a few mature plasma cells and histiocytic cells. × 300 H and E.
(Reproduced by courtesy of Dr. A. A. Liebow.)

to the development of malignant lymphomas. This is evidenced by the greater incidence of such tumours in underdeveloped countries where so many of the population are exposed to a variety of recurrent infections from earliest infancy. In addition many of the people suffer from alterations of both humoral and cellular immunity following malnutrition in early childhood and subsequently. In some respects the high incidence of malignant lymphomas and prelymphomatous states in such populations bears similarities to the high incidence rates of reticulo-endothelial proliferations encountered in several human genetically linked immune deficiency diseases, especially in the Wiskott–Aldrich syndrome (Brand and Marinkovich, 1969).

Liebow et al. in their review of the possible aetiological causes of lymphomatoid granulomatosis also drew attention to its similarities with such prelymphomatous lesions as those which follow allogeneic tissue grafting in young mice, in which the animals are injected with parental cells which provoke a chronic host–graft reaction (Armstrong et al., 1970).

An alternative view proposed by Liebow et al. was that an oncogenic virus infection might follow a disordered and reduced immune state which subsequently directly induced lymphoproliferative changes by altering tissue cell membranes and thus causing an autoimmune response.

At present the aetiology of lymphomatoid granulomatosis and the other forms of angeitis described above remain uncertain, but lymphomatoid granulomatosis must be regarded as a pre-lymphomatous condition occurring in the lung analagous to other similar conditions in other organs such as Sjøgren's syndrome, Waldenström's macroglobulinaemia, the changes associated with chronic coeliac disease of the small bowel, so-called "pseudolymphoma" of the lung, lymphadenoid goitre and angio-immunoblastic lymphadenopathy (Frizzera et al., 1974).

7. Goodpasture's Syndrome

In 1919 Goodpasture described several cases of influenza that he had seen during the 1918–19 pandemic, including an unusual case in which a naval rating developed massive pulmonary haemorrhages and died 3 days after readmission to hospital following an attack of influenza 6 weeks earlier. A post-mortem examination showed that he suffered from acute glomerulo-nephritis in addition to numerous intrapulmonary haemorrhages.

Little attention was paid at the time to Goodpasture's observation but during the last 20 years several further examples of the syndrome have been reported. Further cases have been described by Rusby and Wilson (1960), Lundberg (1963), Canfield et al. (1963) and Benoit et al. (1964).

The majority of cases have presented with the symptom of haemoptysis which may be slight or massive and this is usually followed by anaemia and the development of retrosternal pain. At the same time urine examination has revealed a proteinuria and haematuria consistent with an acute glomerulonephritis, but in the majority of cases the pulmonary changes have preceded and overshadowed in importance the renal changes. In several cases a Group A, Type 12 β-Haemolytic streptococcus (Strep. pyogenes) infection preceded the onset of the pulmonary and renal changes. Most of the patients have died after a short illness lasting less than 6 weeks and very few have survived for longer than a few months.

Goodpasture's syndrome may bear a close clinical resemblance to idiopathic pulmonary haemosiderosis and Benoit et al. have summarized the principal difference between the two diseases. In Goodpasture's syndrome the patients tend to be males and over 20 years of age, the

disease is more rapid in onset and in its clinical course it may not cause manifest haemoptysis although extensive intrapulmonary haemorrhage is present, and there is usually no hepato-splenomegaly.

Macroscopically, multiple reddish-brown patches are found throughout both lungs due to recent and organizing haemorrhages.

Microscopically, the intra-alveolar spaces, pulmonary septal tissues and hilar lymph glands are filled with red blood-cells and haemosiderin-laden macrophage cells. Nodular collagenous lesions appear later in the alveolar septa in those patients who do not succumb in the first 2 or 3 weeks of the illness. The renal lesions are characterized by focal intracapillary thrombotic change and endothelial cell proliferation, crescent formation and later increasing glomerular and periglomerular fibrosis. These changes are similar to those found in the acute and early subacute stages of proliferative (Ellis type 1) glomerulonephritis. Both the pulmonary and renal lesions result from a type 2 immunological reaction in which an IgG globulin autoantibody to alveolar and glomerular basement membranes coupled with complement B 1C component is deposited on the basement membranes of the alveoli and glomeruli as was shown by Beirne et al. (1968) and Duncan et al. (1965), respectively, using immunofluorescent techniques (Fig. 19.22). In man both the alveolar and glomerular basement membranes react similarly to the circulating antibody and possess a common antigenic component. Oliveira and Laus-Filho (1962) produced experimental lesions in the lungs of rats, resembling those of human Goodpasture's syndrome, by injecting intravenously a nephrotoxic serum. By absorption testing they showed that the antigen concentration was similar in the lungs, kidneys and placenta, indicating a common antigenic component which was probably derived from basement membrane.

Benoit et al. postulated that a viral infection might first destroy the alveolar epithelium and expose the underlying basement membrane. Both light and electron microscopy, however,

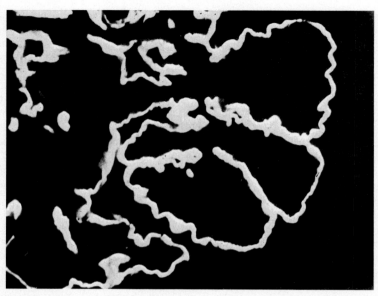

FIG. 19.22. Goodpasture's syndrome showing a glomerulus treated with antihuman IgG fluorescent-coupled serum. The entire basement membrane of the gomerulus fluoresces and the pulmonary basement membrane would respond likewise.

(Reproduced by courtesy of Dr. David Davies.)

fail to reveal any evidence of damage to any component of the alveolar wall other than to the basement membrane which is much thickened.

Doubt has been expressed as to whether idiopathic pulmonary haemosiderosis is the same condition as Goodpasture's syndrome but there is no present evidence to support this suggestion as very few iron-pigment impregnated and disintegrating elastic fibres are found in the alveolar and pulmonary vascular walls in Goodpasture's syndrome.

Although Goodpasture's syndrome is uncommon, McCaughey and Thomas (1962) found that in about 15 per cent of 252 fatal cases of acute and subacute proliferative glomerulonephritis there were old or recent pulmonary haemorrhages. The cause of the condition is unknown but it has been suggested that it may be related to an autoimmune reaction. Animal experiments have shown that antiserum prepared against human whole lung may also induce an acute glomerulonephritis, thus showing that there is a common antigen in both types of tissue. Cross antigenic reactivity of the lung and kidney basement membranes, however, is seen only in man (Poskitt, 1970).

Idiopathic Interstitial Fibrosis of the Lung (IIFL)

Synonyms: Chronic interstitial fibrosis of the lung, usual interstitial pneumonia (UIP), fibrosing alveolitis, sclerosing alveolitis, cryptogenic fibrosing alveolitis, Hamman–Rich lung

During the past 30 years an increasing number of cases of an idiopathic form of progressive pulmonary interstitial fibrosis have been reported. As already described in Chapter 7, interstitial fibrosis of the lung may be brought about in several ways and may result from a variety of causes which include bacterial and viral pneumonias, physical agents such as radiation, chemical fumes and dusts, familial diseases and probably disordered states of body immunity including autoimmunity. The ensuing account is concerned only with the idiopathic group of cases.

Although the term "usual" interstitial pneumonia has been widely adopted in North America it is a meaningless term and no indication has been given as to what should be regarded as the "unusual" form. The term fibrosing or sclerosing alveolitis though favoured by others is a purely descriptive term describing the principal pathological end product and would be equally applicable to other causes of interstitial lung fibrosis where the aetiology is known. The term Hamman–Rich lung, much employed in the past, should be discontinued and if used reserved only for the occasional case of idiopathic interstitial lung fibrosis of rapid onset and course and for which the aetiological cause is unknown. Future research if it results in a better understanding of the probable multiple causes of the present "idiopathic" group will eventually result in a classification based on aetiology. Because this type of lung lesion represents a common end product resulting from many causes of lung damage it is often impossible to determine how or why it started.

In 1935 Hamman and Rich briefly presented to the American Clinical and Climatological Association four cases of an unusual pulmonary disorder. Although each case varied slightly in the character and duration of its symptoms and signs, all the patients suffered from severe dyspnoea and cyanosis, developing right ventricular hypertrophy and failure from which they died within 3 months of the onset. In one case, the disease pursued a very rapid course, the patient dying within 3 days of admission to hospital, and the authors expressed some doubts as to whether this could be classed with the other three cases, but considered that the lung changes found post-mortem possibly represented the early acute stage of the disease.

In 1944 Hamman and Rich published details of four fatal cases which included three of those briefly described in 1935, and they discussed in detail the pathological changes observed in the lungs. Since this report the condition has often been referred to as either the Hamman–Rich lung or chronic idiopathic diffuse interstitial fibrosis of the lung. Of the four cases originally described, three were women and three of the patients were negroes.

Many further cases have since been described and many remain unreported and the disease appears to be becoming more common. In a comprehensive review in 1957, Rubin and Lubliner collected sixty-four cases, including sixteen of their own. Livingstone et al. (1964) described a personal series of forty-five cases and reviewed the world literature while more recent accounts have been given by Liebow and Carrington (1969), Gaensler et al. (1972) and Liebow (1975).

The outstanding pathological change is the chronic interstitial fibrosis which occurs in a patchy fashion throughout the lungs. It is usual to find early, intermediate and terminal stages of lung damage in the same lung thus supporting the view that it is a sporadically active but continuing cause for the lung damage. Occasionally the changes may be restricted to a lobe of a lung and in such cases the lesions may become centrally calcified. The increasing pulmonary fibrosis interferes with lung function by interfering with respiratory gas exchange due to ventilation–perfusion imbalance. In addition the fibrosed and inelastic lungs impair the normal compliance thus restricting ventilatory movement. The latter is compensated for by hyperventilation which in turn can lead to respiratory alkalosis. Despite the thickening of the alveolar capillary membrane gaseous diffusion is seldom impaired until the final stages of the disease. In the late stage of the disease cor pulmonale develops.

Similar changes occur in the lungs in other conditions, notably the "collagen diseases" such as rheumatoid disease and progressive systemic sclerosis (scleroderma), and it has followed the use of an increasing number of drugs which include hydrazinophthalazine (apresoline), bleomycin, nitrofurantoin, cyclophosphamide and para-aminosalicylic acid.

A considerable number of cases of IIFL show positive serological tests for rheumatoid disease and eventually display some of the clinical features of this and other "collagen" diseases. Because of these similarities it has been suggested that IIFL should be included in that remarkable assortment of conditions referred to as the "collagen diseases", and the evidence for this will be discussed under theories of causation.

IIFL occurs slightly more often in males and is encountered at all ages from childhood to the eighth decade, although the majority of cases have occurred in persons between the ages of 30 and 50. It has been reported in twins (Peabody et al., 1950) and frequently occurs in siblings. Solliday et al. (1973) described a family in which three generations developed the disease including a father and son suggesting that in some cases it has an autosomal dominant pattern of transmission. The disease often does not manifest itself until middle age.

Although usually a widespread and bilateral disease, it may occasionally be confined to one lobe as shown by Geever et al. (1951). Clinically, the majority of cases have followed the pattern described by Hamman and Rich. There is a rapid onset of progressive dyspnoea, orthopnoea, minor degrees of haemoptysis, constricting type of chest pain, increasing cyanosis and a persistent unproductive cough. Later, loss of weight and clubbing of the fingers occurs, and cor pulmonale develops. Despite the rapid onset of symptoms the majority of the patients are afebrile at first though pyrexia occurs when pneumonia supervenes. The duration of life in acute cases may extend from a few days to 2 months, but in the more chronic cases survival has been reported for several years.

The radiological changes are very variable and according to Rubin and Lubliner only about a third of all cases show diffuse shadows due to interstitial fibrosis and oedema (Fig. 19.23). Livingstone et al., however, found mottling in lung radiographs of all their cases at some stage in the disease. When the mottling was very fine it gave rise to a "ground glass" appearance. Translucencies were also found to

be common and varied in size up to 5 mm and were sometimes associated with coarse mottling. The translucent areas were usually basal in position and when confluent denoted a honeycomb type of change in the lung. In all cases the maximal changes were found in the lower lung fields.

No specific diagnostic laboratory test for this disease has so far been discovered. The haemoglobin levels frequently exceed 17·0 g per 100 ml when the patients are cyanotic, but the white blood-cell counts in the absence of pulmonary infection (pneumonia) are usually normal. The sedimentation rate is often considerably raised and may mirror alterations in the plasma proteins. Although the plasma protein levels often remain within normal limits, electrophoresis may show an abnormal composition of the globulins.

Macroscopical appearances. Very few descriptions exist of the changes that are found in the early stages. If the fourth case originally described by Hamman and Rich in 1935 is accepted, the appearances are similar to those found in acute influenzal virus pneumonia, the lungs being plum-coloured due to severe oedema.

Usually by the time the condition proved fatal the pleural surfaces of the lungs are thickened and sometimes bound by adhesions to the chest wall. Non-adherent areas of the thickened pleural surface present a "hobnailed" appearance, the whole lung being reduced in size and feeling rubbery. On section it is reddish-grey and presents a fine honeycomb of smooth-walled air-filled cysts, a few of which may contain inspissated pus or mucus. The cysts vary in size up to about 5 mm in diameter and are most numerous in the lower and outer zones of the lungs (Fig. 19.24). Very little normal lung tissue may remain in the later stages, and that which is present is intersected with numerous bands of fibrous tissue. Small but dilated bronchi can usually be traced to within about 2 mm of the pleural surface when "honeycomb" changes are well established. The majority of the cystic spaces are thin-walled emphysematous bullae but some are dilated

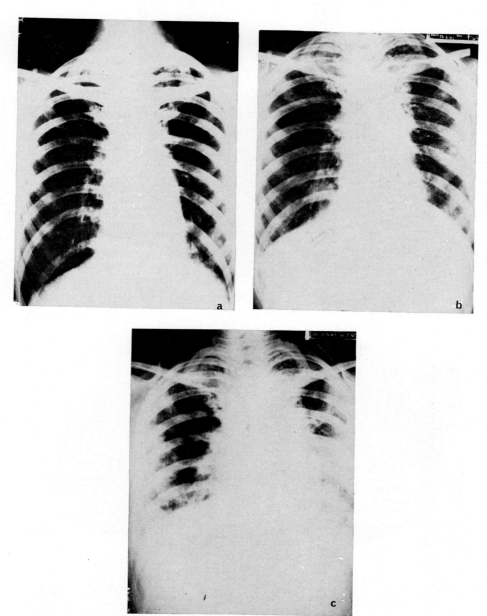

FIG. 19.23. A series of lung radiographs showing the progressive nature of the lesions over a period of 2 years in idiopathic interstitial lung fibrosis. The changes have increased in rapidity towards the end of the period of survey as seen in (c).

(Reproduced by courtesy of Dr. E. Davies.)

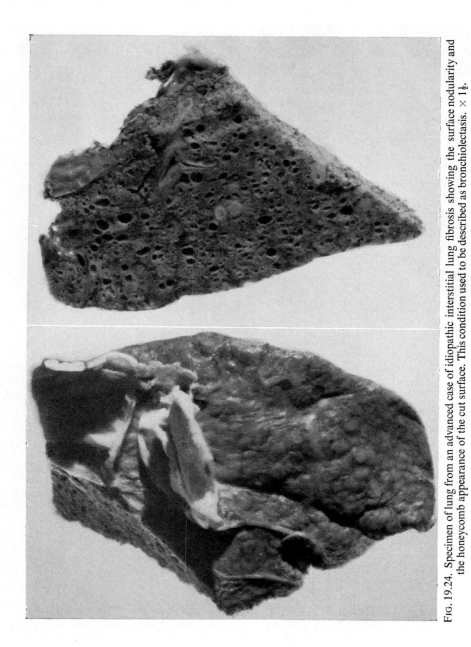

Fig. 19.24. Specimen of lung from an advanced case of idiopathic interstitial lung fibrosis showing the surface nodularity and the honeycomb appearance of the cut surface. This condition used to be described as bronchiolectasis. × 1½.

bronchioles and small bronchi. A terminal pneumonia is of frequent occurrence.

Microscopical appearances. Attempts have been made to grade the severity of the lesions in this disease but as the extent of the changes vary within the same lung little is achieved by such classifications.

Initially, there is an intra-alveolar exudate of fibrinous oedema fluid, red blood-cells and desquamated macrophage cells often accompanied by the presence of hyaline membranes (Fig. 19.25). The maximal damage occurs in the alveolar walls and involves the alveolar ducts, and the interlobular septa become very swollen due to oedema. During these early stages the alveolar capillaries become very swollen and congested and the changes appear to be mostly attributable to increased capillary permeability. This stage merges with the next in which large numbers of histiocytic cells accumulate in the interstitial plane of the alveolar wall though many desquamate into the alveolar lumens.

The source and origin of these cells is uncertain but may be either from pre-existent interstitially located histiocytes or from the blood monocytes. Reticulin fibrils soon begin to appear around fibroblasts situated in the walls of the alveoli and lead to interstitial mural thickening. The further maturation of the reticulin into collagen leads to a state of chronic interstitial fibrosis of the lung (Figs. 19.26, 19.27, 19.28). Accompanying these intramural changes, some of the damaged alveolar walls may undergo partial dissolution and this later change contributes to the macroscopic honeycomb appearances of the lungs.

During the course of these changes the affected lungs may pass through a stage where they show all the features characteristic of desquamative interstitial pneumonia (DIP). As already stated DIP is a response of the lung to many forms of injury and it is often a stage in the evolution of IIFL (Scadding and Hinson, 1967).

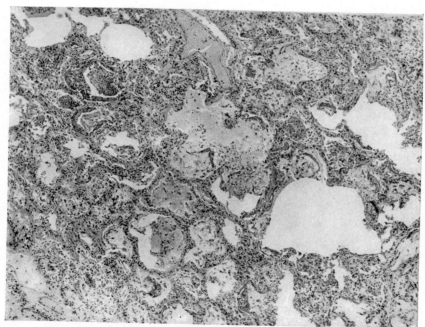

FIG. 19.25. Idiopathic interstitial fibrosis of lung showing the initial stage of pulmonary oedema, interstitial oedema and epithelialization of the damaged alveolar walls. × 40 H and E.

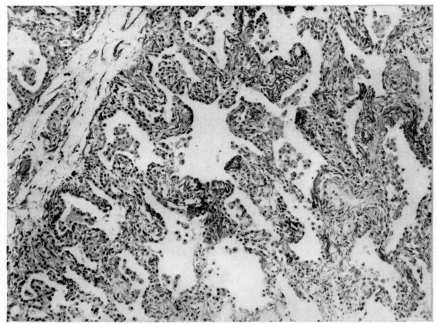

FIG. 19.26. A slightly later stage in the same disease showing commencing interstitial fibrosis and alveolar epithelialization. × 140 H and E.

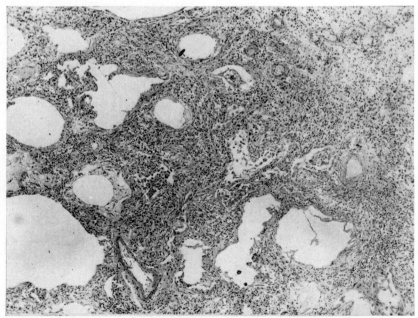

FIG. 19.27. A later stage in the same process showing more advanced interstitial fibrosis and destruction of the lung. × 40 H and E.

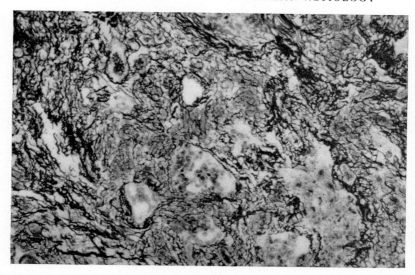

FIG. 19.28. Idiopathic interstitial lung fibrosis showing reticulin fibrils being laid down in relation to the intramural histiocytic cells. The fibrils are formed around each cell and are mainly responsible for the interstitial mural thickening. × 100 stained by modified Foot's reticulin stain.

The damaged but unobliterated alveoli become lined by cuboidal epithelium which may often undergo metaplasia into either a columnar and mucus-secreting or a stratified, squamoid type of epithelium. Included among the regenerating alveolar epithelial cells are occasional large, atypical alveolar epithelial cells showing both normal and abnormal mitotic activity including nuclear triploidy. The regenerating epithelial cells may be derived from either hyperplastic surviving alveolar epithelial cells or from bronchiolar epithelium which has spread through Lambert's canals as shown in Fig. 12.6. The extent of the epithelial metaplasia may cause parts of the lung to resemble the appearances seen in pulmonary adenomatosis. The extension of the fibrotic process may lead ultimately to considerable fibrosis causing obliteration of alveoli, alveolar ducts and bronchioles (Figs. 19.29, 19.30). Although much of the chronic interstitial fibrosis occurs in the manner outlined above it also results partly from organization of intra-alveolar fibrinous and pneumonic exudates and incorporation of the connective tissue into the alveolar wall (Gross, 1962).

No characteristic type of cellular reaction accompanies the above changes, though interstitial collections of eosinophils are found in some cases, and lymphocytes and other types of chronic inflammatory cells are present in all cases.

Although eventually most of the lungs become extensively fibrosed, it is usually possible to find some areas which still show the earlier exudative type of reaction. Also it is possible, as judged from animal experimental lesions, that some of the newly formed reticulin fibrils may be resorbed accounting for the occasional spontaneous remission of symptoms noted in some patients.

The bronchioles share in the general alveolar fibrotic changes, and while some become occluded, in others the muscle coat undergoes hypertrophy and the lumen becomes lined by a multilayered flattened epithelium. Cyst production also occurs in the absence of bronchiolar obstruction due to emphysema. Damage

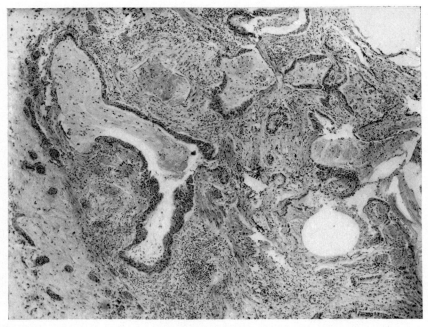

FIG. 19.29. Advanced stage of idiopathic interstitial lung fibrosis showing almost total destruction of the alveoli, those that remain being partly epithelialized by proliferating bronchiolar epithelial cells and hyperplastic alveolar lining cells. There is a moderate interstitial chronic inflammatory cell infiltration, and bronchiolectasisis present. × 40 H and E.

and fibrosis of the alveolar tissue results in compensatory bronchiolar dilatation. Occasionally either the emphysematous areas or the bronchiolectatic bronchioles may rupture causing spontaneous pneumothorax.

Following the extensive obliteration of much of the alveolar capillary bed the alveolar capillaries in the less-damaged portions of the lungs become very distended and in places aneurysmally dilated. The extensive alveolar capillary and pulmonary arteriolar obliteration results in the establishment of bronchopulmonary arterial precapillary anastomoses which may readily be demonstrated by injecting the bronchial arteries with radiopaque substances which quickly appear in the pulmonary arteries.

Other microscopical changes that may occur include foci of calcification and ossification, the formation of intra-alveolar giant cells and microlithiasis pulmonum (Leicher, 1949).

Hyperplasia of plain muscle tissue throughout the destroyed lung is very commonly present. Some of the muscle tissue is related to small bronchioles, elsewhere it occurs where alveoli have become obliterated and it also develops in relation to the walls of lymphatics.

The ultramicroscopic changes found by Corrin and Spencer (1975) in lung biopsy material include an initial increased pinocytotic activity in the alveolar capillary endothelium, swelling of the alveolar capillary and epithelial basement membranes due to oedema which then together form a very much thickened homogeneous membrane. Also early replacement of the type 1 pneumocytes by either type 2 cells or more primitive cuboidal progenitor cells. No immune complexes were seen. In longstanding cases increasing numbers of histiocytes, lymphocytes and fibroblasts together with increasing amounts of reticulin and collagen

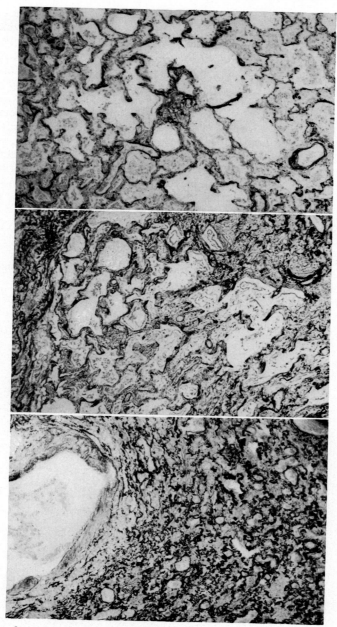

Fig. 19.30. Sections taken from the same area of lung showing the progressive obliteration of the lung alveoli in idiopathic interstitial fibrosis of the lung. Despite the increasing interstitial and intra-alveolar fibrosis the elastic fibre framework of the lung remains visible though the elastic fibres become coarser and thickened. The elastic framework of the lung is still visible in (3) although almost all the alveoli have been obliterated. All × 40 Sheridan's elastic stain and neutral red.

fibrils appear in the interstitial plane of the alveolar wall. Finally collagen fibres replace and fill the interstitium of the alveolar wall and lead to compression and obliteration of many of the alveolar capillaries.

GIANT-CELL INTERSTITIAL PNEUMONIA (GIP)

Reddy *et al.* (1970) and Liebow (1975) both briefly described the microscopical features of this condition. Liebow described five cases of idiopathic interstitial alveolar fibrosis in which the alveoli were partly filled with large, bizarre giant cells. He provisionally called the condition giant-celled interstitial pneumonia (GIP). The clinical course differed in no respect from other cases of idiopathic alveolar fibrosis. In no case was it possible to grow or identify a virus or other causative micro-organism. Microscopically, numerous, large, multinucleated giant cells partly filled many of the alveolar

spaces and there was diffuse interstitial fibrosis of the alveolar walls which also contained mononuclear cells (Fig. 19.31). None of the giant cells contained inclusion bodies as in measles and their cell of origin was uncertain. The giant cells were actively phagocytic for other cells and a few contained asteroid bodies (Fig. 19.32). Unlike measles pneumonia in which the majority of the giant cells form part of the alveolar epithelial lining, the giant cells in GIP lie free in the alveolar lumen. GIP is probably not a specific form of pneumonia but forms part of the spectrum of changes that may be found in idiopathic interstitial alveolar fibrosis (IIFL).

Lung Function Changes

As a result of the changes described above, certain functional changes occur and these were described by Luchsinger *et al.* (1957), Livingstone

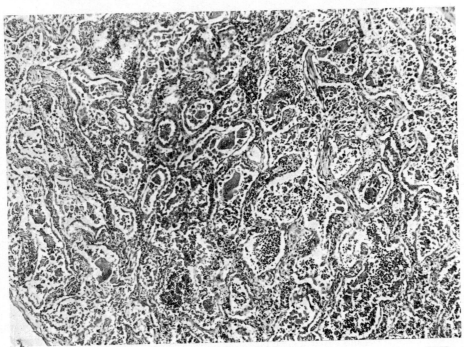

FIG. 19.31. Low-power view of lung from a case of giant-cell interstitial pneumonia. × 75 H and E. (Reproduced by courtesy of Dr. A. A. Liebow.)

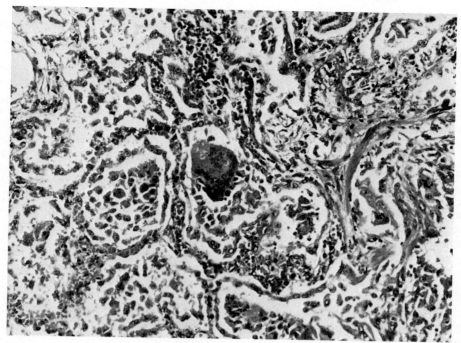

FIG. 19.32. High-power view of giant-cell interstitial pneumonia showing an intra-alveolar giant cell containing a small early asteroid body. × 180 H and E.
(Reproduced by courtesy of Dr. A. A. Liebow.)

et al. and Baglio *et al.* (1960). Destruction and fibrosis of the lung tissue with consequent reduction of lung volume and compliance impose limitations on the chest and diaphragmatic movements and also lead to reduction of the vital capacity and to hyperpnoea. Even in the later stages of the disease there is little or no obstruction to the air passages but arterial oxygen saturation falls and pulmonary hypertension becomes increasingly evident. These latter changes result from alveolar fibrosis which causes obliteration of many alveolar capillaries and small pulmonary arteries resulting in interference with oxygenation. Furthermore, diversion of the blood-flow through the remaining patent alveolar capillary networks leads to gross overdistension and probably less efficient oxygenation of their contained blood. Eventually increasing anoxaemia and vascular obliteration leads to right heart failure, a process

that may be accentuated by superimposed attacks of pneumonia.

THEORIES OF CAUSATION

(a) *Infective*

Although it has usually been impossible to prove a bacterial, viral or mycoplasmal cause of IIFL it is known that viral and mycoplasmal infections may both cause an interstitial form of pneumonia in which fibroblasts begin to appear in the interstitium as early as one week (Liebow and Carrington). Lourdes *et al.* (1974) also followed by serial lung biopsy the changes that might follow primary influenzal virus pneumonia and showed that it could progress over a period of time to cause widespread interstitial lung fibrosis and chronic interstitial inflammation. A similar type of change also occurs in giant-cell pneumonia due to the measles virus and in some adenovirus pneumonias. Thus the view expressed earlier by Golden and Tullis (1949) and Callahan *et al.* (1952) postulating a persistent primary "atypical" pneumonia eventually involving the whole of both lungs as a cause of some cases of IIFL is both possible and likely. The pleuro-pneumonia group of organisms was suggested as one of the possible aetiological agents largely on the basis of experimental work in animals but there is no evidence that they are responsible for human cases. Pokorny and Hellwig (1955) also suggested that a disturbance of fibrinolysis might be responsible for causing the fibrosis. Following the more recent work it now appears that some cases of IIFL undoubtedly represent the burnt-out residue of an earlier viral or mycoplasmal infection.

(b) *Drug Sensitivity*

Some cases of IIFL have been attributed to allergy partly because a proportion of patients give a personal or family history of an allergic disorder and partly because in some cases there may be numerous eosinophils in the lung. It has become evident that certain drugs may cause severe reactions in occasional individuals resulting in the eventual production of interstitial fibrosis in the lungs. Among drugs which may cause this complication are nitrofurantoin (Rosenow *et al.*, 1968; Israel *et al.*, 1973), bleomycin (Luna *et al.*, 1972), cyclophospha-mide (Topilow *et al.*, 1973), *p*-amino salicylic acid, and busulphan (Oliner *et al.*, 1961). It is likely that many other drugs, environmental pollutants and occupational chemical hazards of unrecognized nature may produce the same result. It is thought some drugs may behave as haptenes and that the lungs for reasons which are not clear are the target organs for the resulting immunological damage.

(c) *Collagen Diseases*

Peabody *et al.* (1953) considered that inter-stitial lung fibrosis was related to the group of disorders which includes rheumatoid disease, progressive systemic sclerosis (scleroderma), diffuse lupus erythematosus and dermatomyosi-tis. Interstitial lung fibrosis has been reported as a complication of dermatomyositis by Frazer and Miller (1974) and in diffuse lupus erythe-matosus by Rakov and Taylor (1942) and both disorders are considered separately. Most of these conditions are usually generalized disorders involving many organs whereas IIFL is usually a localized disorder confined to the lungs. It is now recognized, however, that in some of these unfortunately named "collagen" diseases the lesions, before they become generalized, may be restricted to one organ or system. In common with the other conditions comprising this group of disorders, cases of IIFL are becoming more frequently encountered and cases have been reported in siblings. The disease behaves as a Mendelian autosomal dominant character and may also be associated with dysproteinaemia (Bonanni *et al.*, 1965).

Rubin and Lubliner have drawn attention to the frequency with which arthralgia occurs in the past history of patients dying of IIFL. It is now also recognized that very similar changes may occur in the lungs in scleroderma and rheumatoid arthritis. The same authors described a female patient who, 5 years after the onset of rheumatoid arthritis, was submitted to a lung biopsy which showed chronic interstitial fibrosis. Later, changes compatible with diffuse lupus erythematosus occurred but when, 7 years after the onset of the illness, the patient died, post-mortem findings were confined to lung changes of the IIFL type. Muschenheim (1961) reported a case in which there was a necrotizing glomerulonephritis and arteritis, with changes in the blood-vessels characteristic of polyarteritis nodosa. The merging of the clinical and patho-logical changes characteristic of one "collagen" disease with another is by no means uncommon, and the cases quoted above may represent such a change.

Since the introduction of hydralazine hydro-chloride (apresoline) as a drug used for the treatment of hypertension, several patients treated with it have developed clinical dissemin-ated lupus erythematosus, and after death have shown a diffuse interstitial fibrosis of the lung. The clinical sequence of events in these cases closely resembles the case described by Rubin and Lubliner quoted above.

The use of hexamethonium, another popular hypotensive drug, may be followed by a chronic diffuse interstitial fibrosis of the lung, but systemic lupus erythematosus has not been reported. The lung fibrosis in this instance is usually regarded as the result of repeated non-fatal episodes of pulmonary oedema.

Abnormal immune reactions involving the absorption of complement may lead to increased capillary permeability and chronic inflammatory reactions at the sites where they occur. Turner-Warwick (1966) using a fluorescent–antibody coupling technique showed that anticomplementary antibody may be found attached to the alveolar capillary walls, and Turner-Warwick et al. (1971) stated that immunoglobulin and complement became attached to the alveolar walls when antinuclear factor was present, a finding independently confirmed by Nagaya et al. (1973). Turner-Warwick (1966) also found that IgG and IgM globulins were localized within lymphoid foci in the damaged lung and IgA proteins within some of the alveolar macrophage cells in IIFL. Turner-Warwick and Doniach (1965) also found that in 27 per cent of cases rheumatoid factor was present in the patients' sera and Gottlieb et al. (1965) discovered latex-fixing antibodies (macroglobulins) in the same proportion of cases. Turner-Warwick et al. also found anti-nuclear factor present in a number of their patients and suggested that this factor might become attached to the nuclei of dead and dying cells both on the walls and in the lumen of the alveoli. Despite these immunofluorescent findings Spencer and Corrin (1975) have been unable to find any evidence of immune complexes in lung biopsy material removed and examined electron microscopically from cases of IIFL.

The use of cortisone and adrenocorticotrophic hormones in cases of IIFL has occasionally resulted in arrest of the disease and the extension of fibrosis. Such drugs, in addition to inhibiting fibrosis, also interfere with antigen–antibody

reactions and, as Peabody et al. (1953) have shown, the withdrawal of the drugs may precipitate a violent exacerbation of the disease. Acute exacerbations following the withdrawal of steroid drugs is a well-recognized risk in many forms of systemic "collagen diseases".

(d) Congenital Disease Theory

Donohue et al. (1959) have claimed that about a quarter of all cases of idiopathic pulmonary interstitial fibrosis result from an inherited defect. Careful questioning of the patients has revealed that it is a familial disease in which there is a failure of the foetal alveolar epithelium to undergo normal post-natal attenuation. They claim that interstitial fibrosis and cyst formation follow the persistence of the foetal type of alveolar epithelium. For this type of idiopathic interstitial lung fibrosis they have proposed the name of familial fibrocystic pulmonary dysplasia.

Although there is no general support for their views, it is now recognized that such familial diseases as tuberose sclerosis and generalized neurofibromatosis (von Recklinghausen's disease) can both be associated with a form of chronic interstitial lung fibrosis which is genetically determined.

Lung Changes in Progessive Systemic Sclerosis (PSC)

Synonym: Scleroderma lung

Scleroderma was first described by Curzio in a dissertation delivered in Naples in 1752, but it was not until the close of the nineteenth century that it was realized that it might involve tissues other than the skin.

Since Finlay (1889) first described lung fibrosis in a patient with scleroderma, it has gradually become realized that many organs may become involved in this disease process, and for this reason the term progressive systemic sclerosis (PSC) will be used in preference to scleroderma. The latter term is reserved for the cutaneous lesions.

Radiological changes in the lung may antedate the cutaneous lesions (Hayman and Hunt, 1952) but the lung damage seldom leads to disturbance of respiratory exchange though reducing the vital capacity.

In common with scleroderma the lung changes in PSC are seen more commonly in women over 40 and lead to increasing exertional dyspnoea; they are very often associated with Raynaud's syndrome.

The pulmonary changes in PSC have been described by Matsui (1924), Goetz (1945), Getzowa (1945), Spain (1948), Aronson and Wallerstein (1950) and D'Angelo *et al.* (1969). The last named found that interstitial lung fibrosis was present in 74 per cent of fifty-eight cases of systemic sclerosis that they examined post-mortem.

Macroscopical appearances. In PSC—in which the lungs are affected—the pleura becomes greatly thickened, though this change is confined to the visceral layer and adhesions are often completely absent. In the lungs, large air-filled cysts occur which project from the surface and occasionally rupture to cause a pneumothorax (Israel and Harley, 1956).

Parts of the lung show a honeycomb of small, air-filled cysts, a change largely confined to the outer parts of the lower lobes (Fig. 19.33). The cyst walls are smooth and little may remain of the surrounding lung except thickened interlobular septa. The demarcation between the cystic and non-cystic lung appears abrupt, though careful examination will usually show that the adjacent lung on the hilar side of the cystic zone is firmer and less well aerated.

Microscopical appearances. In the transitional zone between normal and cystic lung, the alveolar walls at first become lined by a visible layer of alveolar epithelium and the interstitium becomes oedematous. Many fine reticulin fibres are laid down interstitially and these mature to form collagen (Fig. 19.34). Some of the damaged alveolar walls become disrupted at an early stage while others gradually become lost in the expanding fibrosis. In the early stages many

dilated alveolar capillaries are present (Fig. 19.35) but later most of them are obliterated by dense collagenous tissue which also causes the disruption of the alveolar elastic fibres, only coarse fragments of which remain. Many of the alveoli are eventually completely obliterated while the wall of others after first undergoing hyaline fibrosis rupture to form cysts or large air-filled sacs (Fig. 19.36). Calcification is an uncommon sequel of the fibrotic changes.

The alveolar changes are not accompanied by any characteristic cellular reaction though lymphocytes are scattered in the interstitial tissue, and desquamated alveolar epithelium and lipid-laden phagocytes fill many of the remaining alveolar spaces. The interference caused in this disease to normal oesophageal function by mural fibrosis can lead to chronic regurgitation of oesophageal contents into the air passages and lungs resulting in superadded pneumonic changes.

There is usually considerable pleural, interlobular and peribronchial fibrosis in the later stages which is preceded by earlier oedema of these structures. The peribronchial fibrosis may lead to obstruction of the lumen and atrophy of bronchiolar muscle. Although some of the bronchioles are obstructed, as was shown bronchographically by Church and Ellis (1950), others undergo dilatation but fail to show mural muscular hyperplasia.

Among the more constant changes that occur in PSC is the occurrence of pulmonary hypertension. This may precede other structural changes in the lung (Sackner *et al.*, 1962), and is sometimes associated with Raynaud's syndrome in the hands. At first the pulmonary hypertension may be temporarily relieved by giving acetyl choline but later the occurrence of structural changes in the small muscular branches of the pulmonary arteries leads to an irreversible condition. The pulmonary arterial changes are further described in Chapter 16.

Electron-microscopic studies of the lung changes in PSC were carried out by Wilson *et al.* (1964). Although their observations were made on post-mortem material they found that the earliest changes consisted of a thickening of the

FIG. 19.33. "Scleroderma" lung from a case of progressive systemic sclerosis (PSC) showing honeycomb appearance of the outer and lower part of the lung together with overlying pleural fibrous adhesions. Carcinomatous changes had occurred and the hilar gland is replaced with growth. Two-thirds natural size.

(Reproduced by courtesy of Dr. K. F. W. Hinson.)

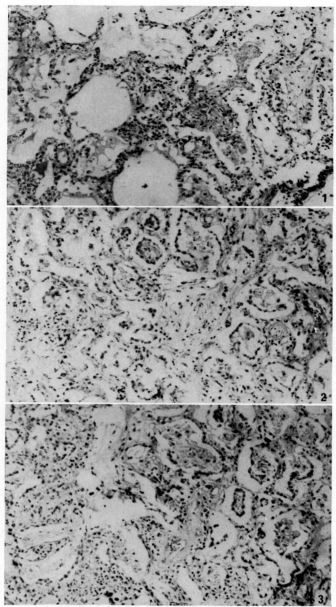

FIG. 19.34. Sections taken from the same area of the lung showing the different stages in the early development of lung fibrosis in progressive systemic sclerosis (scleroderma). The earliest stage (1) is found in the more centrally situated regions of the lung whilst (2) and (3) are found in the subpleural regions. In (1) the changes are mainly those of wide-spread intra-alveolar and intramural oedema. In (2) and (3) progressive intra-alveolar and interstitial fibrosis have occurred due to organization in the oedematous regions. All × 100 H and E.

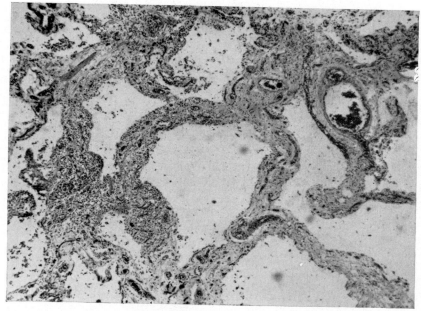

Fig. 19.35. The earlier vascular stages in the development of the pulmonary changes in progressive systemic sclerosis (PSC) showing the prominent alveolar capillaries in the thickened and cellular alveolar walls. × 140 H and E.

alveolar basement membrane and of the membrane underlying the endothelium lining the small pulmonary arterioles and venules.

As in idiopathic fibrosis of the lung, bronchiolar and alveolar cell hyperplasia is prone to occur, often the alveoli becoming lined by tall, columnar, mucus-secreting epithelium. Because of the chronicity of PSC, the alveolar cell hyperplasia may develop insidiously into malignant change, and adenocarcinomas of the lung can result. This change has been reported by Zatuchni et al. (1953), Richards and Milne (1958) and Collins et al. (1958) and is described in Chapter 20.

As a result of the structural changes that occur in the lungs considerable functional derangement occurs in the later stages of the disease. Lung function was investigated by Leinwand et al. (1954), Sackner et al. (1964) and Wilson et al.

Impaired respiratory function is mainly caused by ventilation–perfusion mismatch due to vascular obliteration, reduced vital capacity and reduced compliance. Arterial oxygen saturation is usually near normal at rest but may reach low levels during exercise, though carbon dioxide diffusion is usually unimpaired. Hyperventilation may lead to respiratory alkalosis. The arterial oxygen desaturation has been shown to be in part caused by pulmonary veno-arterial shunts.

Respiratory function may also be impaired by sclerodermatous changes affecting the skin of the chest wall causing accompanying atrophy of the intercostal muscles; both of these changes lead to restriction of respiration and lung ventilation.

The aetiology of PSC is unknown but it is usually classed as a "collagen disease". The pulmonary changes are similar to those found in IIFL and the chronic rheumatoid lung, but the progress of the disease is slower and the patients often survive for many years after its discovery. The early interstitial oedematous phase seen in the other two diseases is not

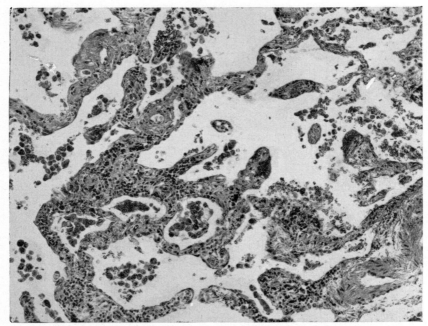

Fig. 19.36. A later stage in the evolution of the pulmonary changes showing increasing alveolar fibrosis and disruption of some of the walls. There are still many lymphocytes present in the damaged tissue. × 140 H and E.

so conspicuous, and there is an absence of any interstitial eosinophils. Malignant change occurs more commonly in PSC, but occurs only occasionally in IIFL probably because of its more rapid course but may be expected to occur in the future in cortisone-controlled cases which survive longer.

Pulmonary Changes in Rheumatoid Disease

Lung involvement in rheumatoid arthritis is of comparatively recent recognition and was first reported by Ellman (1947). Its recognition was due to the greater attention paid to the non-articular changes occurring in rheumatoid arthritis, since it was recognized that the latter disease was a generalized disorder of which the joint changes form but one manifestation. Also the discovery of a "rheumatoid factor" in the blood of patients suffering from this disease has provided a laboratory test for the recognition of

this and closely allied disorders. The realization that joint changes are only one of the lesions caused by the diseases led Christie (1954) to introduce the term rheumatoid disease.

The pulmonary changes may conveniently be divided into: (a) pleural lesions, (b) pulmonary interstitial changes (c) modified pneumokoniotic lesions (Caplan's lesions), and (d) lung lesions in ankylosing spondylitis.

(a) PLEURAL LESIONS

Baggenstoss and Rosenberg (1943) examined the lungs in thirty cases of rheumatoid arthritis and found that in twenty-two of them pleural adhesions were present. Cruickshank (1957), in a further series of 100 cases, found that 42 per cent had adhesions, an incidence greater than the 22 per cent found in control series by Sinclair and Cruickshank (1956). Pleural effusions are not infrequently found to be present at some

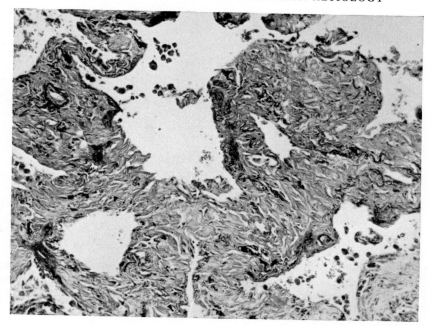

FIG. 19.37. Lung alveoli in advanced progressive systemic sclerosis (PSC) showing collagenous thickening of alveolar walls. There is almost complete absence of any interstitial cellular reaction. × 140 H and E.

stage during the course of rheumatoid disease.

Rheumatoid nodules. Among the least common of the pulmonary changes found in rheumatoid disease are necrobiotic lesions similar in structure to those commonly found in the subcutaneous tissues over the olecranon and forearm bones. Pulmonary necrobiotic lesions are associated with pleural effusions and were described by Gruenwald (1948), Ellman *et al.* (1954) and Robertson and Brinkman (1961). Their discovery usually follows a chest radiograph in which they present as single or multiple nodularities in the lungs. In most instances exploratory thoracotomy is required to elucidate their nature. On opening the chest they present as raised single or multiple greyish-white masses up to 2·0 cm in diameter in the visceral pleura and the subjacent interlobular septa (Fig. 19.38). The centres of the nodules are filled with cheesy contents and the nature of the lesions may be

unsuspected unless the patient shows other and more characteristic changes of rheumatoid arthritis.

Occasionally following steroid therapy absorption of the necrotic contents and healing occur and the lesions later present as fibrous-walled cavities (Dumas *et al.*, 1963). Such cavities may rupture and cause a pneumothorax.

Microscopically, they show typical necrobiotic changes in the centre, surrounded by a palisade layer of spindle-shaped endothelioid cells, some of which are phagocytic, and outside these are numerous lymphocytes, plasma cells and fibroblasts (Fig. 19.39). The necrotic central areas often contain much pyknotic nuclear debris, whilst at the edge of the lesions the adjacent lung collapses, becoming incorporated in the outer fibrotic tissue layers of the lesion.

Silver stains show a severance of reticulin fibres at the edge of the necrotic zones. This is

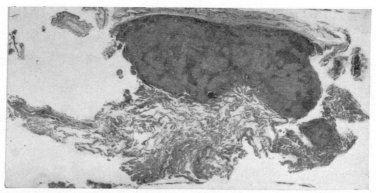

FIG. 19.38. A pleural necrobiotic nodule in rheumatoid arthritis showing necrotic areas scattered throughout its substance. The lesion is entirely restricted to the pleural tissue. × 7.
(Reproduced by courtesy of Dr. L. Cudkowicz and the Editor of *British Journal of Tuberculosis*.)

caused by spread of the fibrinoid change which as Collins (1937) had shown starts in the proliferating mesenchymal cells which initially form the nodules. Kellgren (1952) considered fibrinoid change was a chemical loosening of the collagen–mucopolysaccharide complex, causing the collagen molecule to disintegrate into simpler products, leaving in the affected zone mucopolysaccharides and glycoproteins which were responsible for the fibrinoid changes. Similar necrobiotic changes may occur occasionally in the interlobular septa within the lung.

(b) INTERSTITIAL PNEUMONITIS AND FIBROSIS

In 1948, Ellman and Ball described a form of interstitial pneumonitis which they considered

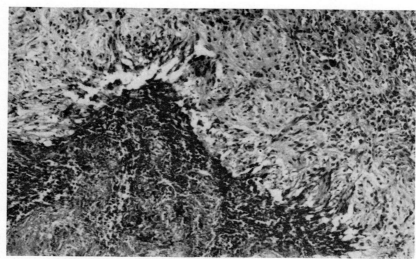

FIG. 19.39. A rheumatoid necrobiotic nodule in the pleura showing in the lower left-hand corner the fibrinoid necrotic centre surrounded by a palisading layer of histocytic cells. × 100 H and E.

was caused by rheumatoid arthritis, and since then similar cases, some of which have shown additional features including vascular and bronchial lesions, have been described by Rubin (1955), Price and Skelton (1956), Edge and Rickards (1957) and Cruickshank (1959).

In the most advanced stages the periphery of the lower halves of the lungs may be replaced by a honeycomb of fibrous-walled spaces, and this condition constitutes one variety of "honeycomb lung". The sequence of changes occurring in rheumatoid interstitial lung fibrosis is similar to those described previously for IIFL and some cases of the latter are of undoubted rheumatoid aetiology.

Initially, the alveolar capillaries are congested and the alveolar walls are oedematous and both of these, together with the peribronchiolar tissues, are filled with lymphocytes. At this stage the alveoli contain fibrinous exudate, red blood cells and macrophages, but intra-alveolar exu-

date as seen in the bacterial pneumonias is entirely absent in uncomplicated cases (Fig. 19.40). Later, the interstitial histiocytic and lymphocytic reaction in the affected parts of the lung is replaced by an increasing amount of fibrous tissue which ultimately obliterates many of the alveoli except for some epithelialized alveolar remnants and bronchiolectatic bronchioles. The latter are mainly responsible for the honeycomb appearances seen in gross specimens.

Foci of lymphocytes, some of which contain germinal centres, persist in the fibrosed lung and are often a distinctive feature (Fig. 19.41). Elastic tissue stains, whilst revealing disruption of the elastic pattern in the fibrotic areas, do not show the bizarre fragmentation and clubbing of these fibres seen in radiation fibrosis. The fibrosis results mainly from organization of the interstitial pneumonitis, and there is no evidence that it is caused by organization of intra-alveolar exudate.

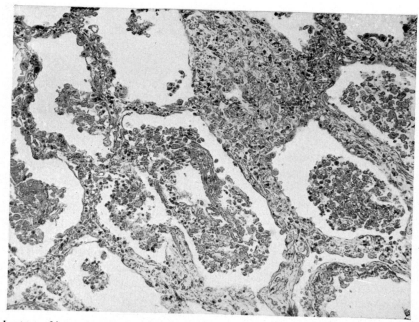

FIG. 19.40. Early stage of interstitial alveolar oedema and thickening which is the earliest abnormality found in cases of rheumatoid lung. The alveoli contain desquamated alveolar epithelial cells, red blood cells and fibrinous exudate. × 140 H and E.

(Reproduced by courtesy of Professor B. Cruickshank.

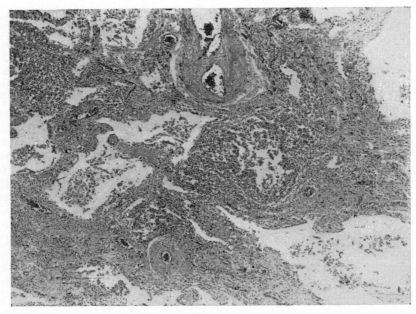

Fig. 19.41. Later stage in the development of interstitial pneumonitis due to rheumatoid disease showing severe alveolar fibrosis and interstitial collections of lymphocytes. Many of the alveolar walls have been destroyed by the advancing interstitial alveolar fibrosis. × 40 H and E.

In addition to the changes described above granulomatous lesions involving the bronchiolar walls and resembling in miniature the necrobiotic nodules seen in the pleura were first reported by Price and Skelton, and have been further described by Cruickshank (1959).

Lesions of the smaller muscular pulmonary arteries leading to their obstruction and causing chronic pulmonary hypertension were reported by Gardner et al. (1957) and have been found in some cases.

(c) CAPLAN'S LESIONS

In 1953, Caplan described unusual radiological changes in coal-miners suffering from rheumatoid arthritis, and subsequently radiologists in Germany, France, Belgium and Holland reported similar findings. The lung X-rays showed large, rounded and well-defined shadows mainly in the peripheral parts of the lung fields, but unlike the shadows caused by massive fibrosis they were not restricted to the upper lobes and tended to enlarge more rapidly. The appearance of the pulmonary lesions sometimes preceded the onset of arthritic changes and Caplan et al. (1962) found that 65 per cent of such patients showed positive serological tests for rheumatoid disease.

The pathological changes responsible for the radiological appearances were investigated and described by Gough et al. (1955). The lesions vary in size from discrete rounded nodules 1 cm in diameter to larger confluent masses and closely resemble tuberculous silicotic nodules. The centre may be partly cavitated and shows concentric black and yellowish rings, the pale zones frequently being liquefied (Fig. 19.42). The centres of some nodules may subsequently calcify, a change which occurs more frequently than in simple silicotic nodules.

Microscopically, the centre of the mass is composed of collagen, anthracotic pigment and necrotic debris. At the edge of the necrotic zone

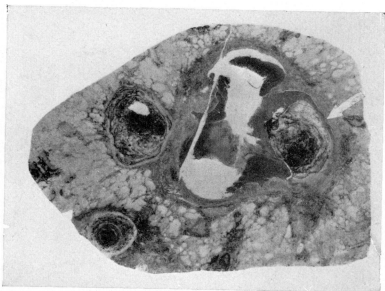

FIG. 19.42. A "Caplan" node. A low-power view of pneumoconiotic lesions showing the split which has appeared between the central contents and the outer fibrous wall. × 7.
(Reproduced by courtesy of Professor B. Lennox.)

a cleft separates the dead tissue from an adjacent surrounding layer of palisaded spindle-shaped fibroblasts. Between the cellular and necrotic tissue is a narrow belt of polymorph leucocytes and nuclear debris derived from macrophage and other cells (Fig. 19.43). Much of the pigment formerly carried by macrophages is deposited as free granules and forms the black rings seen macroscopically. Sections stained to show reticulin fibres reveal an abrupt destruction of the fibres as they pass from the necrotic to the peripheral cellular areas. Outside the palisade layer are numerous lymphocytes and plasma cells, beyond which lies a further concentric layer of collagen fibres. The arteries lying beyond the lesion are usually blocked by endarteritis accompanied by much lymphocytic and plasma cell infiltration of the vessel walls.

Despite its similarity to a tuberculous silicotic nodule, both macroscopically and microscopically, it can be distinguished by the cellular zone consisting of polymorph leucocytes and a palisaded layer of fibroblasts that surrounds the necrotic centre and separates the latter from the outer collagenous layers. Also central calcification occurs more readily than in a simple silicotic nodule, and tests fail to reveal the presence of tubercle bacilli.

A similar type of change has been reported in silicotic lesions by Clerens (1953), in foundry workers (silico-siderosis) by Caplan *et al.* (1958), in boiler scalers by Campbell (1958) and in asbestosis by Rickards and Barrett (1958).

In Caplan's lesions complicating asbestosis the particles of asbestos fibres are scattered throughout the lesions and are not deposited in concentric rings.

(d) LUNG CHANGES IN ANKYLOSING SPONDYLITIS

A disease identical with or very similar to rheumatoid disease and which is also associated with interstitial lung fibrosis is ankylosing spondylitis. Unlike classical rheumatoid disease the most seriously affected parts of the lungs are

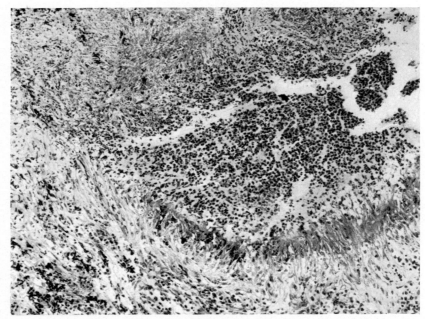

FIG. 19.43. A high-power view of the boundary zone between the central part of the lesion and the surrounding wall showing the layer of "palisaded" endothelioid cells and the intervening collection of polymorph leucocytes.
× 40 and E.
(Reproduced by courtesy of Professor B. Lennox.)

the upper lobes and the pulmonary changes may not start until 7 to 10 years after the appearance of the joint changes in the spine (Davies, 1972). The interstitial fibrosis is similar to that seen in rheumatoid interstitial lung fibrosis. Bronchiolitis obliterans also commonly occurs and causes obstructive airways disease and accumulations of foamy macrophage cells in the alveoli.

To summarize the changes occurring in rheumatoid disease, both the necrobiotic pleural and septal lesions and those described by Caplan in pneumoconiotic nodules are two distinctive rheumatoid lesions which are not encountered in other diseases. Interstitial lung fibrosis due to rheumatoid disease is not, however, a specific change and cannot be distinguished histologically from the idiopathic form of interstitial lung fibrosis. The coincident presence of conspicuous lymphoid follicles, often containing germinal centres, and the occasional presence of necrobiotic lesions involving the walls of small bronchi in an interstitial fibrotic lung are, however, helpful differential histological features which suggest a rheumatoid aetiology and may enable it to be distinguished from other causes of pulmonary interstitial fibrosis.

THE RELATIONSHIP OF THE THREE DISORDERS

The inter-relationship of these three diseases, IIFL, PSC and rheumatoid disease, is still uncertain, although an increasing volume of evidence points to their similarity if not actual identity. It is singularly unfortunate that the term "collagen" disease has come to be applied to this group of disorders as it is both misleading and fails to indicate the probable underlying abnormality, which is likely to be a disordered immunity reaction. If these diseases are caused by disorders of the immune mechanism they are disorders of the delayed type of response (type 4) rather than of the acute Arthus (type 2) reaction. The acute necrotizing Arthus type of reaction in the lung results in eosinophilic pneumonia and allergic granulomatous lesions. Although important clinical differences exist between the pulmonary manifestations in these three conditions, notably the duration of the diseases, IIFL tends to pursue a more rapid course than the other two. Laboratory tests reveal certain similarities. Tests for the detection of "rheumatoid factors" in sera have shown that the factors are present in highest titres in patients with rheumatoid disease but that they are present in 61 per cent of patients with idiopathic intersitital fibrosis of the lungs and are found in low dilution in patients suffering from PSC (Tomasi et al., 1962).

All three conditions show oedema of the interstitial planes of the lung at an early stage though this is least well developed in PSC. The suggestion has recently been proposed that in rheumatoid disease depolymerization of the normal macromolecular mucopolysaccharides filling the tissue spaces causes attraction of fluid into this tissue plane. In PSC, the pulmonary and other changes tend to be very chronic and, by the time lung changes are diagnosed, they have often been present for a long time. Their chronicity is reflected in the pulmonary histology which usually shows a greater degree of fibrosis, with less evidence of the earlier cellular and oedematous phases seen in the other two disorders. This difference probably accounts for the radiological differences between PSC and the other two conditions. The distribution of the pulmonary lesions tends to be localized to the periphery and lower lobes in both rheumatoid arthritis and particularly PSC, whereas in idiopathic interstitial lung fibrosis the changes are more generalized.

Occasionally, cases are reported with clinical and pathological features common to two disorders. Gardner et al. reported a patient with features of both active rheumatoid arthritis and acrocyanosis of the fingers. Later, the same patient developed pulmonary hypertension which proved fatal, and both the pulmonary and digital arteries showed changes similar to those seen usually in PSC. Also, Rubin and Lubliner, and Katz and Auerbach (1951) have both drawn attention to the occurrence of articular and muscular changes in IIFL patients resembling the joint changes normally found in rheumatoid arthritis.

The pulmonary response to steroid therapy has been very disappointing in PSC and not very effective in rheumatoid disease, but, as the lung damage is usually well established and permanent before clinical symptoms due to pulmonary fibrosis appear, this would be anticipated.

Ellman and Cudkowicz (1954) drew attention to the similarity in the distribution of the fibrotic changes in the lung in so-called "collagen" diseases and the territory normally supplied by the bronchial arteries. They considered that these disorders were ischaemic in nature and followed a reduction in the bronchial artery flow. They also thought that the arterial changes had been wrongly localized to the pulmonary circulation, when they in fact involved the terminal divisions of the bronchial arteries, a confusion which is likely to occur in the absence of injection of one or other arterial system with a contrast medium. The occurrence of pulmonary hypertension in all three conditions indicates, however, that the pulmonary arterial system is involved.

Despite the clinical differences which exist between these three conditions, the similarity of the pulmonary lesions, the occurrence of cases with mixed clinical features and the not infrequent presence of rheumatoid and antinuclear factors in the sera in all three disorders support the view that each may be a different manifestation of a related disorder which has a wide spectrum of clinical presentation. To these three conditions may also be added diffuse lupus erythematosus and dermatomyositis each of which may also cause interstitial lung fibrosis. Interstitial lung fibrosis may also occur together with chronic liver disease (Turner-Warwick, 1968) of unknown aetiology but which is also thought to be a manifestation of a "collagen" disease. In these disorders the pulmonary damage ultimately causes a decrease of the vital capacity, decreased lung compliance, interference with oxygen diffusion, ventilation–perfusion disorder and obliteration of much of the pulmonary capillary bed. The clinical features and pathophysiology associated with this group of interstitially fibrosed lungs were discussed by Austrian et al. (1951) and Gaensler et al. (1972).

Pulmonary Changes in Diffuse Lupus Erythematosus and Dermatomyositis

Many cases of diffuse lupus erythematosus (DLE) are unassociated with any identifiable changes in the lung.

Teilum (1946), however, described changes which he regarded as a form of "focal allergic pneumonia". These were found chiefly in the subpleural and interlobular connective tissue, though focal granulomatous lesions also occurred in the alveolar walls and in the walls of veins. In the septal tissues, focal fibrinoid changes occurred which progressed to form small areas of complete necrosis accompanied by accumulation of histiocytes and the formation of focal granulomatous lesions, but no eosinophils or giant cells were found. Vascular lesions resembling those seen in polyarteritis nodosa occurred in both arteries and subpleural veins. The venous lesions resulted in the development of intraluminal fibrous polyps.

Rakov and Taylor (1942) described the occurrence of a chronic interstitial pneumonia leading to alveolar thickening with chronic inflammatory cells and particularly mononuclear cells with dark-staining nuclei; this change led to alveolar obliteration. Baggenstoss (1952) noted an interstitial mucinous oedema in the alveolar walls, peribronchial and perivascular tissues, a change that led to gradual obliteration of the alveolar spaces.

Lyons et al. (1964) carried out lung function tests on twenty-two cases of diffuse lupus erythematosus and found that there was a reduced vital capacity, arterial oxygen desaturation on exercise and hyperventilatory respiratory alkalosis. The functional derangements were usually unassociated with any radiological evidence of lung damage.

Interstitial lung fibrosis mainly confined to the subpleural portions of the lower lobes has been seen by the author and described by Frazier and Miller (1974) in dermatomyositis and polymyositis. According to Frazier and Miller about 5 per cent of the cases show radiological evidence of lung involvement.

In dermatomyositis the damaged alveolar interstitium is occupied by oedematous connective tissue and the alveolar capillaries are reduced in number.

Pulmonary Changes in Sjøgren's Syndrome

The first clear clinical description of this condition was given by Sjøgren (1933, 1951) who was the first to recognize and describe the clinical syndrome which is now named after him. In two-thirds of all cases there is a chronic polyarthritis of the rheumatoid type and serum flocculation tests for rheumatoid factors are positive in between 83 and 100 per cent of all cases (Cardell and Gurling, 1954).

The lungs are rarely affected but may show atrophic changes in the bronchial mucous glands. The acinar cells atrophy and later become replaced by fibroadipose tissue. These changes are accompanied by periglandular lymphocytic and plasma cell infiltration which extends into the lamina propria (Fig. 19.44). Interstitial alveolar fibrosis and pleural fibrosis may accompany these changes.

Sjøgren's syndrome may occur in association with lymphomatoid granulomatosis and lymphocytic interstitial pneumonia (LIP) and is itself now regarded as a pre-lymphomatous condition.

Idiopathic Pulmonary Haemosiderosis

Synonym: Ceelen's disease

Siderosis of the lungs may result from intrapulmonary haemorrhages which may be caused

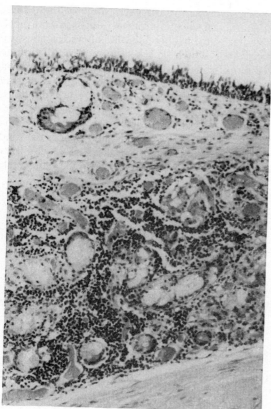

FIG. 19.44. A large bronchus showing atrophy of the mucous glands in Sjøgren's disease. The atrophic changes are accompanied by lymphocytic infiltration. × 100 H and E.
(Reproducd by courtesy of Dr. B. S. Cardell.)

by a variety of different diseases, notably chronic mitral stenosis, chronic pulmonary venous occlusion, Goodpasture's syndrome, allergic vasculitis (pulmonary allergic granuloma), and idiopathic pulmonary haemosiderosis which was first described by Ceelen. It is with the last-named condition that the present account is concerned. Much confusion exists about the pathology of the disease under discussion as some of the other conditions named above have been confused with Ceelen's disease.

In 1931 Ceelen demonstrated before the Berlin Pathological Society the condition which now bears his name. He subsequently described it in greater detail (Ceelen, 1931).

Although it is an uncommon disease, an increasing number of cases have been reported in the past 20 years from many parts of the world. Full descriptions of the clinical aspects and pathology have been given by Wylie et al. (1948), Soergel (1957) and Soergel and Sommers (1962a, b). Approximately four-fifths of all cases have occurred in children below the age of 10 and it has been reported occasionally in children during the first year of life. Wynn-Williams and Young (1956) described the findings in fifteen adult cases including one of their own.

Idiopathic haemosiderosis of the lung when it occurs in childhood affects both sexes equally, but when it occurs in adults is more prone to affect males. Clinically, it is characterized by sudden attacks of dyspnoea, coughing and the vomiting of blood, the blood appearing both in the stomach contents and in the sputum. These attacks lead to increasing fatigue and pallor, and the attacks, which cause the patient great alarm, may last up to a week. As the attack progresses the patients frequently become jaundiced and both the spleen and liver enlarge. Repeated attacks lead to increasing dyspnoea on exertion and progressive pulmonary hypertension which usually causes death within 3 years of the onset. Clubbing of the fingers is marked in the later stages. Fulminating cases have also been reported and these may prove fatal within 10 days of the onset (Grill et al., 1962).

The changes in the lung radiographs are similar to those found in lung siderosis accompanying mitral stenosis (Fig. 16.38), and consist of diffuse miliary mottling in the lung fields. In the more chronic cases respiratory function tests show a reduction of the vital capacity when the lungs become fibrosed, and the arterial oxygen saturation may be reduced below normal.

Macroscopically, the lung surfaces are smooth, brownish-purple in colour and recent petechiae may be visible beneath the pleural surfaces. On section the lungs are dry and present the characteristic brick-red colour and indurated texture seen in the later stages of mitral stenosis.

The changes involve all lobes of the lungs, and the hilar lymph glands are a similar colour. Both the lungs and the hilar glands give a strongly positive Prussian blue reaction indicating the presence of much haemosiderin. Large thick sections of the lung show that the haemosiderin is deposited in the subpleural region, in the septa, around arteries and to a lesser extent around the bronchi (Fig. 19.45) (Grainger, 1958). Focal siderotic nodules are present throughout the lung (Fig 19.46) and constitute the principal abnormality.

Microscopically, the early stage of the disease is characterized by numerous diffuse focal intra-alveolar and intrabronchiolar haemorrhages, no source for which can be found. In addition to red blood-cells there are numerous haemosiderin-laden macrophages filling the alveoli together with hyperplasia of the alveolar epithelial cells. The hyperplastic alveolar epithelium is a conspicuous feature and the hyperplastic cells may show finger-like cytoplasmic processes and vacuoles. Thickening of the alveolar capillary basement membrane may also be an early

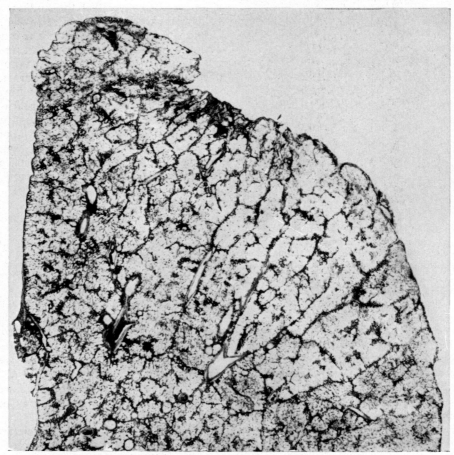

FIG. 19.45. Section of lung showing the distribution of haemosiderin pigment in Ceelen's disease. The pigment occurs in relation to the tissues surrounding the vessels, septa, and bronchi and is also found in the pleural and subpleural tissues. The distribution corresponds to that of dusts in many forms of pneumoconioses.

(Reproduced by courtesy of Dr. R. C. Grainger and the Editor of *Brit. J. Radiol.*)

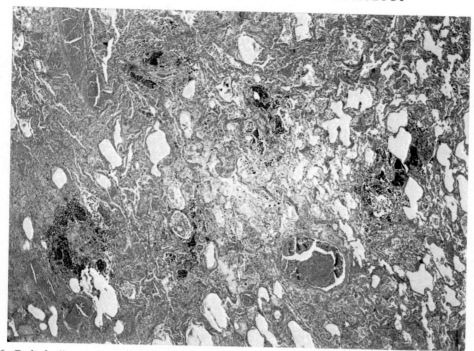

FIG. 19.46. Ceelen's disease (idiopathic haemosiderosis) showing black (siderotic) lesions scattered through the lung. × 30 Perl's stain.

feature. The repeated focal and sometimes extensive intrapulmonary haemorrhages eventually lead to interstitial alveolar fibrosis causing obliteration of many of the alveolar capillaries (Fig. 19.47). The repeated pulmonary haemorrhages also cause disruption of alveolar elastic fibres and encrustation of the fragments with a mixture of iron and calcium salts. These impregnated fibres give rise to siderotic nodules consisting of foreign-body giant cells, fibroblasts and haemosiderin-laden phagocytes in the walls of the alveolar ducts and the terminal respiratory bronchioles. Similar changes affect the elastic laminae of the smaller branches of the pulmonary arteries and veins. In addition many of the small muscular pulmonary arteries undergo progressive intimal fibrosis and hyalinization which may lead to pulmonary hypertension and cor pulmonale. Many of these changes are identical with those seen in pulmonary venous thrombosis.

Other changes include the presence of perivascular collections of lymphocytes and foci of haemosiderin-laden phagocytes around the perivascular lymphatic channels. The hilar lymph glands draining the lungs are also filled with haemosiderin-laden phagocytes.

The extent of the interstitial alveolar fibrosis varies considerably and seems to bear little relationship to the duration of the disease; it is seldom found in children under 1 year. Death results in the majority of cases from haemorrhage and not from cor pulmonale, and during the earlier stages of the disease patients frequently suffer from hypochromic and microcytic anaemia due to the lung haemorrhages which may amount to a considerable volume of blood. The electron microscopic features were studied by Hyatt et al. (1972). They found breaks in the continuity of the alveolar capillary basement membrane but could not find any evidence of immune complex deposition, cf. Goodpasture's syndrome. Their

8

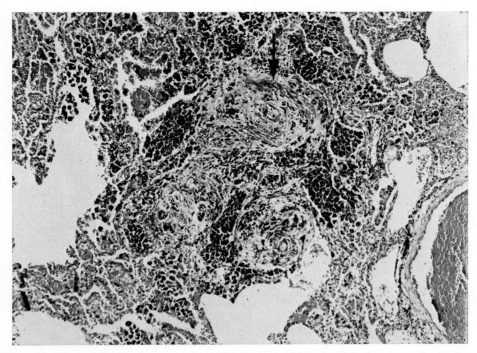

FIG. 19.47. A siderotic nodule showing numerous siderophages and elastic fibres impregnated with iron pigment (arrowed). × 75 H and E.

findings have been confirmed by Donald *et al.* (1975) who considered that the disorder primarily affected the alveolar epithelial cells causing secondary disruption of the underlying basement membrane.

Diagnosis of the condition rests largely on the clinical history, radiological changes in the lungs causing miliary mottling and, in some cases on the results of a lung biopsy. In some cases a blood eosinophilia occurs, and Wylie *et al.* found that occasionally cold agglutinins and antibodies giving a positive Coomb's test may be present in the blood. The plasma proteins remain normal but the iron content of the lungs may become increased up to 2000 times (Soergel and Sommers, 1962a). Haemosiderin-laden macrophages swallowed in sputum may be recovered from gastric washings. All attempts to demonstrate antibodies to human lung tissue using precipitin tests, fluorescein antibody coupling techniques

and the Ouchterlony agar double diffusion plate technique have proved repeatedly negative.

Pathogenesis

The cause of this disease is unknown but some consider it is the same as Goodpasture's syndrome. Steiner (1954) suggested that idiopathic pulmonary haemosiderosis was caused by an auto-immune reaction in which the lungs were the principal target organ. Chikamitsu (1940) (quoted by Lundberg, 1963) showed that anti-rabbit lung serum damaged not only the basement membranes in the lung but also the glomerular basement membrane causing an acute glomerulonephritis. The reverse effect, namely that antiglomerular basement membrane serum could damage alveolar basement membrane, was shown by Markowitz (1960). This and other experimental work shows that both the lung and renal glomeruli share a common basement membrane antigen (see Goodpasture's syndrome). Zollinger and Hegglin (1958) considered the pulmonary changes were due to an allergic capillaritis and that the disease was allied to Schönlein–Henoch purpura in the skin and the renal glomerulitis seen in polyarteritis nodosa. This view was supported by Heptinstall and Salmon (1959) and MacGregor et al. (1960), who both described cases of idiopathic pulmonary haemosiderosis accompanied by acute glomerulonephritis in which fibrinoid change was present in the glomeruli.

These cases would probably now be regarded as examples of Goodpasture's syndrome or qualify for inclusion as pulmonary allergic granuloma or even generalized Wegener's granulomatosis.

For the present idiopathic pulmonary haemosiderosis is still considered in this text as a separate entity but the future may well show that it is the same as Goodpasture's syndrome or is being confused in some cases with primary pulmonary venous thrombosis.

Relapsing Polychondritis

Synonyms: Systemic chondromalacia, Chronic atrophic polychondritis, Diffuse perichondritis, Dyschondroplasia

This rare but interesting disease first became recognized following the descriptions given by Altherr (1936) and von Meyenburg (1936), although it had been originally described by Jaksch-Wartenhorst (1923) who called it "Polychrondroplasia". The cartilage tissue in the trachea and large bronchi is often involved together with the larger cartilage masses forming the nasal cartilages and septum, parts of the internal and external ears, and joint surfaces. The changes in the bronchi may lead to bronchial collapse and obstruction which can be observed in lung tomographs. Secondary pulmonary complications are common especially in the later stages of the disease. Recent reviews of the condition, including accounts of the histo-

chemical studies of the degenerating cartilage, have been given by Pearson et al. (1960), Verity et al. (1963) and Kaye and Sones (1964).

As its name implies it is a relapsing disease associated with episodes of painful inflammatory swelling of the cartilage in the soft tissues of the nose, eyelids and ears and with painful swelling, particularly of the cartilage in the peripheral joints. The condition often occurs associated with diffuse lupus erythematosus, rheumatoid disease and Sjøgren's syndrome. Hypergammaglobulinaemia is commonly present.

Microscopically, the cartilage in the sites mentioned, including the trachea and large bronchi, undergoes progressive destruction. The normal acid mucopolysaccharide ground substance becomes less metachromatic, staining poorly with toluidine blue, by Hale's method and with alcian blue. Simultaneously the cartilage ground substance loses its normal haematoxyphil properties and stains more strongly by the PAS-method. These changes cause a narrowing of the interlacunar spaces and result in secondary degeneration of many of the chondrocytes whose nuclei become pyknotic. These destructive processes usually start beneath the perichondrium and spread into the interior of the cartilage. Accompanying these changes, the cartilage is invaded and replaced by vascular granulation tissue together with lymphocytes and plasma cells (Figs. 19.48, 19.49, 19.50). An acute polymorphonuclear leucocytic reaction with local abscess formation may also be encountered. The loss of the normally inhibitory acid mucopolysaccharide ground substance allows calcium salts to be deposited in the degenerating cartilage. Dissolution of the bronchial cartilage rings and their replacement by granulation and fibrous tissue leads to bronchial expiratory collapse and narrowing, and later to changes of obstructive pneumonitis in the lung.

The cause of this rare disease is unknown but its association with other so-called "collagen diseases" leads to its present inclusion among this group of disorders. Depolymerization of ground substance in connective tissue has been regarded as one of the most important changes in this group of disorders, and it is the most

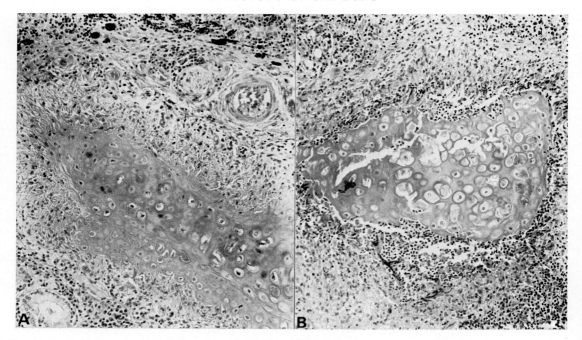

FIG. 19.48. Relapsing polychondritis. In (A) the cartilage plate in the bronchus is undergoing perichondral destruction with accompanying histiocytic reaction and in (B) there is also considerable polymorphonuclear leucocytic reaction around the disintegrating cartilage. Both × 160 H and E.
(Reproduced by courtesy of Dr. W. M. Thurlbeck, Winnipeg, Canada.)

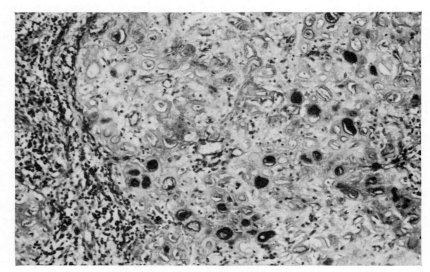

FIG. 19.49. Relapsing polychondritis. × 100 PAS and haematoxylin.
(Reproduced by courtesy of Dr. C. C. S. Pike.)

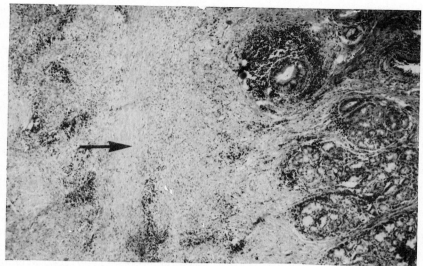

FIG. 19.50. Dyschondroplasia of the bronchus showing complete disappearance and destruction of the bronchial cartilage. The arrow marks the former site of the destroyed cartilage. × 40 H and E.
(Reproduced by courtesy of Dr. A. A. Liebow.)

characteristic change found in relapsing polychondritis. The changes found in the affected cartilages resemble those seen following experimental destruction of cartilage in animals by local injections of papain or overdosage with vitamin A.

Pulmonary Sarcoidosis

Synonyms: Boeck's sarcoid,
Besnier–Boeck–Schaumann's disease
and uveoparotid tuberculosis

The skin lesions were first described by Hutchinson (1875); later Boeck described the histology of the nodules and the condition was subsequently known as Boeck's sarcoid. Schaumann (1917) first showed that the cutaneous lesions were only part of a generalized disorder involving many tissues, basing this discovery on the similarity of the histological changes in the different organs. Sarcoidosis is a difficult condition to define. It may be defined according to either clinical, pathological or combined criteria.

Mitchell and Scadding (1974) defined it as "a generalized, non-caseating, epithelioid and giant-cell granulomatosis".

The incidence of idiopathic (Boeck's) sarcoidosis, with which this account is concerned, varies considerably in different parts of the world and between different races and even among peoples of the same race.

The highest incidence of the disease occurs in Sweden where it is estimated that 64 per 100,000 of the population show radiographic evidence of the disease. This received post-mortem confirmation from the findings of Hagerstand and Linell (1964) in Malmö where between 0·64 and 1·0 per cent of all autopsies show evidence of the disease. The comparative incidence of the disease in Norway, England and Wales, Holland, Switzerland, France, Canada and Denmark is respectively 26·7, 20, 21·6, 16·3, 10, 10 and 8 per 100,000 of the population (Siltzbach, 1965). The incidence of sarcoidosis in the U.S.A. was estimated by Freiman (1948) to amount to 13 per 100,000 but recent studies in New York have shown an incidence of 39 per 100,000 of the population. Mallory (1948) found twenty-three cases among 6000 post-mortems at the Massachusetts General Hospital confirming the almost invariably greater number of cases revealed at autopsy. Many cases discovered post-mortem have previously been in good health and died as a result of accidents or unrelated causes. Recent surveys in the U.S.A. have shown that the disease is more common in the south-eastern states and that the incidence is greater among Negroes than Caucasians. In England and Wales the estimated incidence, as gauged by mass miniature radiography, is about 20 per 100,000 of the population but amounts to 40 per 100,000 in women of child-bearing age. The incidence rate is lower in the north of England than in the south and the peak incidence in both sexes was reached between 25 and 35 years of age. Among persons of Irish descent living in London the incidence was as high as 220 per 100,000 among women, and 110 per 100,000 in men (James, 1962). Among the white race the Scandinavians (Swedes and Norwegians) are most commonly affected but the disease is also common among the Irish.

In most large series of cases there is a predominance of female cases and the female to male incidence ratio is about 1·7:1·0. The disease has also been described several times among siblings.

The aetiology of sarcoidosis remains uncertain but it is currently believed to be a type of tissue response caused by an altered immune state and similar histological lesions may be caused by a variety of infective and non-infective agents. Some of the causes result in generalized lesions resembling idiopathic sarcoidosis, including involvement of the lungs, but in other causes the condition may be restricted to the skin, regional lymph glands or other tissue. Among the causes of a sarcoid-type of lesion other than idiopathic (Boeck's) sarcoid are included the following: (a) tuberculosis, (b) brucellosis, (c) leprosy, (d) histoplasmosis, (e) coccidioidomycosis, (f) berylliosis, (g) silica particles buried in tissues, (h) pine pollen, (i) a variety of organic vegetable protein containing dusts including those responsible for farmer's lung, bagassosis and bird

fancier's lungs and (j) in lymphatic glands draining a cancerous region. In many of these conditions phospholipid substances are liberated and are thought to play an important part in provoking the typical histological sarcoid lesions.

Cummings and Hudgins (1958) claimed that in the U.S.A. most cases of sarcoidosis occurred in areas where there were coniferous forests and attributed the disease to the inhalation and subsequent sensitization of the body to pine pollen, particularly to a phosphatide and an amino acid fraction containing α-ε-diaminopimelic acid. Later, however, Cummings concluded that this view of the aetiology of sarcoidosis was incorrect.

Many clinicians in Europe believed that sarcoidosis was related to tuberculosis, and Nethercott and Strawbridge (1956) claimed to have demonstrated mycolic acid (a constituent of tubercle bacilli) in sarcoid lesions, but their findings were disputed by Cummings and Hudgins. The occasional transition of sarcoidosis into tuberculosis and the appearance of tubercle bacilli in the sputum led Scadding (1950) to state: ". . . perhaps if I had the courage of my convictions . . . (I would) label my cases (of sarcoidosis) non-caseating tuberculosis." Pinner (1946) had earlier reported that acid-fast bacilli were occasionally found in sarcoidosis but that normally they were rapidly destroyed. American negroes treated by steroids for pulmonary sarcoidosis may develop pulmonary tuberculosis. The evidence in these cases strongly favours transition of the sarcoidal to a tuberculous lesion and is against the latter being an independent coincident infection. Renewed interest in the infective nature of sarcoidosis followed the discovery by Vaněk (1968) and Vaněk and Schwarz (1970) of acid-fast bacilli either as isolated bacilli or in small groups in histologically proven and acceptable cases of sarcoidosis. Their discovery received support from the electron-microscopic studies of Greenberg et al. (1970) who found that the epithelioid cells in the centres of the sarcoidal granulomas contained cytoplasmic inclusions with a structure compatible with an acid-fast organism. Mitchell and Rees (1969) also succeeded in producing sarcoid lesions in the footpads of mice by injecting a homogenate prepared from human sarcoid lesions and were able to use the mouse lesions to reproduce further lesions in other mice. These latter results were confirmed by Taub and Siltzbach (1974). Earlier Chapman and Speight (1967) had detected significant quantities of mycobacterial antibodies against unclassified mycobacteria in 80 per cent of patients with Boeck's sarcoid compared with their presence in only 31 per cent of control cases. Sarcoid lesions have also been produced in mice by injecting unclassified mycobacteria. Doubt still surrounds the nature of the infective agent which conceivably might be an ultralight microscopic (protoplasmic) form of classical *M. tuberculosis* or a different acid-fast organism. There is little evidence, however, to support the view proposed by Mankiewicz and Van Walbeek (1962) that the absence of antibodies to mycobacteriophages led to bacteriophages modifying *M. tuberculosis* resulting in the formation of a sarcoid type of tissue response.

Formerly, the presence of a positive tuberculin skin reaction was regarded as a contraindication to the presence of sarcoidosis, but Hoyle *et al.* (1954) found that 28 per cent of their patients diagnosed as sarcoidosis on histological evidence had a positive skin reaction. That profound changes occur in cellular immunity is evidenced by lack of skin sensitization in sarcoidosis to dinitrochlorobenzene (DNCB), the lack of response to *Candida albicans* antigen, the lack of lymphocyte transformation in culture of such cells treated with phytohaemagglutinin and the reversion of positive tuberculin tests following BCG vaccination.

The lung is the most common site for sarcoid lesions and Mather *et al.* (1955) found that in a series of ninety-three cases, 42 per cent involved both lungs and the hilar lymph nodes, and in 28 per cent the lungs alone were involved. James (1961) reported that some form of intrathoracic lesion occurred in 84 per cent of a series of 261 cases that he examined and diagnosed radiologically. Snapper and Pompen (1938) had shown previously that radiological involvement of the hilar lymph glands often preceded the changes in the lungs. Radiological changes in the lungs consist of either a fine miliary mottling or more irregular coarse nodules scattered throughout both organs.

The natural history of the pulmonary form of the disease shows a strong tendency to natural regression and probably the majority of cases never seek treatment. Of those that are clinically diagnosed, James and Thomson (1959) have shown that approximately 50 per cent regress when followed over a long period. King (1941) also found the same tendency and found that the average period for lesions to regress was 25 months. Approximately a quarter of all clinically diagnosed cases become progressive and according to Siltzbach (1962) about 7 per cent eventually die. Meier and Wurm (1960) found that the prognosis was better in women and the ultimate prognosis was better in those cases which initially pursued a rapid course.

Macroscopically, the changes depend upon the stage to which the disease has progressed. In early cases, such as may be found in persons killed accidentally, the lesions usually escape detection as they are unsuspected and often submiliary in size. If they are larger they resemble miliary tubercles, consisting of small greyish-white nodules up to 1 mm in diameter scattered throughout many organs, including the lung.

In chronic pulmonary disease widespread fibrosis ensues which is mainly confined to the upper lobes; very rarely the lesions may calcify. The lungs are "honeycombed" with air-filled cysts and this condition cannot be distinguished, in the absence of lesions elsewhere, from the other forms of honeycomb lung (Fig. 19.51). Larger cavities due to sarcoid were reported by Ustvedt (1948) in which no evidence of tuberculous infection could be proved. Such cavities are liable to become infected with *Aspergillus* sp. resulting in an aspergilloma. Severe emphysematous change may occasionally follow bronchial stenosis caused by healed sarcoid lesions in the walls of the larger bronchi though the majority of such lesions heal without causing stenosis.

Although pulmonary sarcoidosis is almost invariably a bilateral disease, a unilateral case

FIG. 19.51. Boeck's sarcoidosis of the lung showing the sarcoid nodules scattered throughout the lung together with more generalized fibrosis near the periphery of the upper lobe of the lung, resulting in a fine honeycomb of air spaces. Approx. one-half natural size.

(Reproduced by courtesy of the late Professor H. A. Magnus and Dr. B. S. Cardell.)

involving only the lower lobe of the left lung was described by Balboni and Castleman (1963).

Microscopically, in early nodules the centre is occupied by a collection of pale staining endothelioid cells, some of which fuse to form giant cells of Langhan's type. These cells are in turn surrounded by a ring of lymphocytes and the whole lesion is sharply defined from the surrounding tissue (Fig. 19.52). True caseous change is absent though the centre of the nodules may contain a small amount of eosinophilic necrotic material but acid-fast organisms are almost invariably absent. Nickerson (1937) using silver impregnation stains found fine reticulin fibres in structureless areas of sarcoid lesions but these were absent in areas of tuberculous caseation. Later an amorphous hyaline substance is laid down between the endothelioid cells which resembles amyloid but differs from it in its staining properties. Because of these differences it was called paramyloid by Teilum who regarded sarcoidosis as a systemic granulomatous condition caused by abnormal activity of the body immunity mechanism (Teilum, 1964). In the later stages of the disease the lesions are converted into concentric whorls of collagenous tissue.

The giant cells in the later stages may contain a variety of intracytoplasmic inclusions which include Schaumann's bodies, asteroid bodies and centrospheres (Fig. 19.53). In 1941 Schaumann redescribed the doubly contoured, laminated, haematoxyphil bodies which now bear his name. He considered, as did Metchnikoff previously (quoted by Schaumann), that the bodies were related to the presence of tubercle bacilli. Zak (1964) in a later study of the formation of these bodies concluded that the first step in their formation was the accumulation of a mucopolysaccharide substance which had an affinity both for calcium and ferric ions and this gradually developed into calcified and iron salt-impregnated bodies. After an interval, transformation of the amorphous structure into a laminated body occurred due to physico-chemical changes analogous to the formation of Liesegang rings

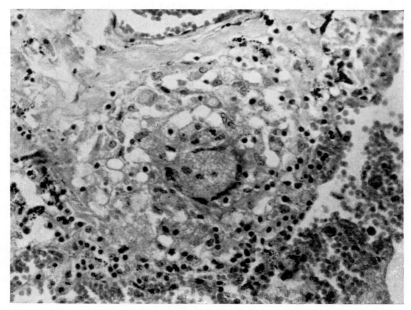

FIG. 19.52. Boeck's sarcoid in the early cellular phase showing centre of the lesion occupied by endothelioid cells and a giant cell, and a peripherally situated ring of lymphocytes. Note the comparatively well-defined margin to the lesion. × 280 H and E.

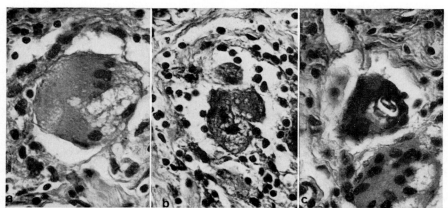

FIG. 19.53. Illustrating the different types of inclusions found in the giant cells in Boeck's sarcoid: (a) centrospheres, (b) asteroid body and (c) Schaumann's body. All × 400 H and E.

In many instances, however, Schaumann bodies undoubtedly appear to start as calcium and iron impregnation of elastic fibres. Schaumann bodies are also found in tuberculosis, berylliosis, atypical mycobacterial pneumonias and in "endogenous pneumoconiosis" (pulmonary siderosis) as seen in mitral stenosis, cortriatrium and chronic pulmonary venous occlusion.

Schaumann bodies persist long after the sarcoid lesions have healed.

Asteroid bodies are spider-shaped structures found almost invariably within the Langhans giant cells. From the central core of the body radiating spicules project, the whole measuring about 5·0–20 μm in diameter. Azar and Lunardelli (1969) examined asteroid bodies electron microscopically and found that they consisted of criss-crossing collagen fibre bundles. The experimental work of McDougall and Azar (1972) in rats has shown that macrophages (from which the asteroid-containing giant cells are derived) possess the ability to metabolize proline and hence may themselves produce collagen.

Centrospheres are ill-defined clusters of vacuoles found within the giant cells. Uehlinger (1964) considered that all three types of cellular inclusions were related and developed from each other, the Schaumann body representing the final mature product.

Sarcoid lesions are found throughout the interstitial tissues of the lung including the peribronchial, perivascular, subpleural and alveolar interstitial tissues. In chronic cases the nodules become increasingly fibrosed and much of the lung is transformed into dense collagenous fibrous tissue and, as Heppleston (1956) observed, it may be difficult to distinguish respiratory from non-respiratory bronchioles (Fig. 19.54). The cystic changes that occur are the consequence of interstitial fibrosis and the cystic spaces are formed partly by bronchiolectatic bronchioles and partly by compensatory distension of uninvolved alveolar tissue. The elastic fibres are totally destroyed within the sarcoid lesions and the alveolar capillary bed becomes progressively destroyed in advanced cases.

Sarcoid lesions are frequently found in the submucosa of the larger bronchi, especially in the proximal end of the segmental and subsegmental bronchi (Fig. 19.55). Usually they heal without causing subsequent bronchial stenosis but stenosis may occur and often involves a length of the wall (Honey and Jepson, 1957). Collapse of the middle lobe following bronchial stenosis was reported by Goldenberg and Greenspan (1960).

The diagnosis of pulmonary sarcoidosis rests on the clinical and radiographic findings coupled

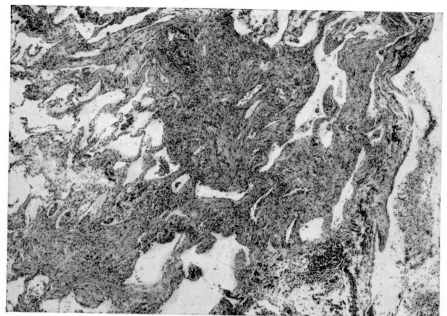

FIG. 19.54. Healed Boeck's sarcoids in the lung leading to extensive fibrosis and the formation of a honeycomb of air spaces. × 40 H and E.

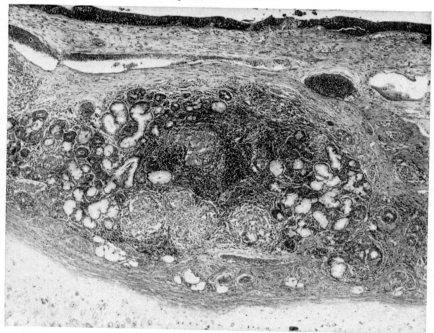

FIG. 19.55. Boeck's sarcoids in the wall of a large bronchus. × 40 H and E.

with the discovery of histologically acceptable lesions in biopsy material. Biopsies may be taken from any of the following sites: liver, lymph glands, nasal and palatal mucous membrane, scalene and mediastinal lymph glands, lung or bronchus. In difficult cases resort may have to be made to either bronchial or open lung biopsy, and Reid and Lorriman (1960) showed this procedure yielded a positive diagnosis in almost all suspected cases. The right middle lobar bronchus is very commonly affected and a bronchial biopsy yields positive confirmation in about 50 per cent of cases despite the absence of suggestive clinical signs or symptoms of bronchial disease.

The Kveim test, which has been used extensively as a diagnostic test for sarcoidosis, consists of the intradermal injection of a saline extract of sarcoid tissue obtained from either an involved lymph gland or spleen. A positive reaction results in an indurated skin lesion with the histological features of a Boeck's sarcoid. James and Thomson (1955) found that about 75 per cent of cases of sarcoidosis in which no suitable lesion was available for biopsy gave a positive result. Siltzbach (1964) found that the Kveim test was positive in 85 per cent of histologically confirmed cases of sarcoidosis and that there were less than 1 per cent of false positive results among patients free of the disease. False Kveim tests may occur in leprosy.

Sarcoidosis associated with a severe fibrosis of the lungs leads to functional alterations including hyperventilation at rest, a reduced maximum breathing capacity and diminished total lung volume. In addition, Riley et al. (1952) have shown that the alveolar-capillary-block syndrome occurs and is the most serious defect in the late stages of the disease. Progressive impairment of lung function occurs mainly during the first 5 years of the disease but little change occurs after this period (Hamer, 1963). Among the later complications that may occur are pneumothorax and the development of cor pulmonale. For other studies on the functional disabilities that result from sarcoidosis of the lungs reference should be made to the work of Tammeling et al. (1964) and Holmgren and Svanborg (1964). Steroid therapy has a variable effect on the progress of the disease, some cases appearing to be markedly improved while others are unaffected. James and Thomson (1959) showed that corticosteroid therapy could modify early (subacute) sarcoid lesions restoring the tissues to normal or resulting in a lesser degree of fibrosis, but they found antituberculous drugs had no such effect.

Desquamative Interstitial Pneumonia (DIP)

Liebow (1962) first drew attention to and briefly described an unusual form of chronic pneumonia and named it desquamative interstitial pneumonia (DIP) after its most characteristic feature. Herbert et al. (1962) also described a series of cases of chronic interstitial fibrosis of the lungs of differing aetiology, and referred to two of the cases which showed unusual features including large numbers of desquamated alveolar cells. The two latter cases from their descriptions and illustrations were probably cases of DIP. Liebow et al. (1965) collected and described at length eighteen cases of DIP all of which had occurred in the U.S.A. They regarded DIP as a disease sui generis but since their description it has now become generally recognized that DIP is a response of the lung to injury and has been encountered in many conditions including asbestosis (Corrin and Price, 1972), following exposure to silica, graphite and talc dusts (Gaensler et al., 1972), to tungsten carbide dust (Coates and Watson, 1971; Patchefsky et al., 1973) and in idiopathic interstitial fibrosis of the lung (IIFL) by Scadding and Hinson (1967). In many cases, however, a causative agent cannot be identified despite a careful history. The lung may remain for long periods in the stage of DIP but lung biopsies and post-mortem evidence show that ultimately in many cases interstitial fibrosis supervenes with loss of the DIP features (Scadding and Hinson, Patchefsky et al.).

DIP affects both sexes equally and occurs at any age including infancy, and is usually seen at an earlier age than the majority of cases of idiopathic interstitial fibrosis of the lung as would be anticipated. Although it usually involves the lower fields in both lungs as judged radiographically, it may occasionally present as localized subpleural nodules as in the case described by Kinjo *et al.* (1974).

The patients complain in almost every instance of increasing dyspnoea and the illness pursues a chronic, afebrile course complicated by a non-productive cough, hoarseness, fatigue and weight loss. Cyanosis, finger-clubbing and variable chest signs develop later and are sometimes accompanied by pleural effusions. Exploratory thoracotomy is usually required to confirm the diagnosis. The pulmonary arterial pressure usually remains within normal limits at rest but rises on exercise. Only in one case of the series described by Liebow *et al.* was cor pulmonale present. The lung compliance is reduced

and the principal functional disability results from inefficient ventilation of alveoli in which the capillary circulation is still unimpeded. The vital capacity usually remains unchanged or is only very slightly reduced.

Radiographically the lungs show a characteristic ground-glass opacity mainly confined to the basal parts of the lungs (Fig. 19.56). This opacity may clear following the introduction of steroid therapy leaving a reticular pattern in the lung fields.

Macroscopically, the affected lung is firm, airless and greyish-yellow in colour. *Microscopically*, the most striking feature, and the one from which the name of the condition is derived, is the large number of desquamated granular alveolar epithelial cells lying within the alveolar lumens (Fig. 19.57). The cells vary in size from 7 to 18 μm in diameter, some are spindle-shaped

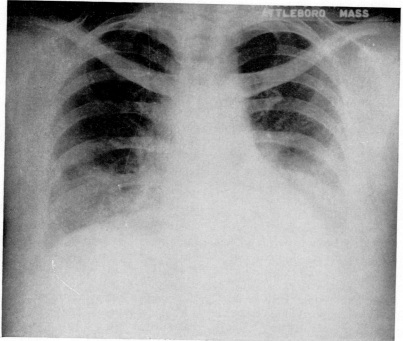

FIG. 19.56. Radiograph of the lung in a case of desquamative interstitial pneumonitis showing the characteristic ground-glass wedge-shaped opacities in the lower lung field. The bases of the opacities are situated peripherally and the apices are directed towards the cardiophrenic angles.

(Reproduced by courtesy of Drs. H. S. W. Udis and A. A. Liebow and the Editor of *American Journal of Medicine*.)

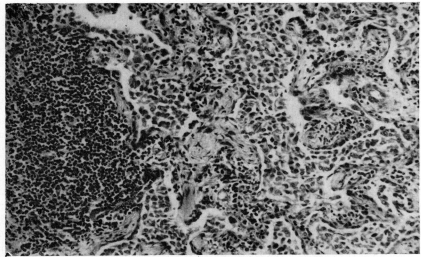

FIG. 19.57. Desquamative interstitial pneumonitis showing desquamated granular cells in the alveolar lumens, slight interstitial alveolar fibrosis and a lymphocytic focus on one side of the field. There is slight thickening of the walls and narrowing of the lumens of some of the smaller branches of the pulmonary arteries. × 100 H and E.
(Reproduced by courtesy of Dr. A. A. Liebow.)

and multinucleated forms occur. The desquamated cells may contain a few fine vacuoles though they seldom contain any anthracotic or other ingested particles. Their cytoplasm contains numerous PAS-positive, diastase resistant granules and often a few iron-containing pigment granules, but lipid stains are usually negative. Electron microscopy shows that the desquamated cells though predominantly type 2 pneumocytes also contain large numbers of alveolar macrophages. Liebow et al. found that both the lining alveolar epithelium and the desquamated cells underwent mitotic division and the former showed both hyperplasia and hypertrophy. Despite the intraluminal collections of desquamated cells which extend into the bronchiolar lumens, very few of these cells are found in the interstitial plane. Unlike many forms of chronic interstitial pneumonia there are no hyaline membranes present in DIP.

A variable amount of interstitial fibrosis and reticulin fibre formation accompany the other changes and the newly formed tissue sometimes displays a myxomatous character. Interstitial muscle fibres are found as in other pulmonary conditions associated with rigidity of lung tissue, and interlobular, septal and pleural oedema and fibrosis are common. Despite all these changes, however, alveolar obliteration is uncommon. Pulmonary obliterative endarteritis occurs in the most severely affected regions of the lungs and a further striking feature is the presence of focal collections of lymphocytes which may contain germinal centres.

DIP as already stated has multiple aetiological causes among which a viral cause has also been suggested as the desquamated alveolar cells are similar to those seen in "grey lung virus disease" of rats. Liebow et al. and Patchefsky et al. have drawn attention to the presence of intranuclear eosinophilic inclusions in some cases, the inclusion bodies being surrounded by a clear halo. Although suggestive of a viral inclusion body this has not been confirmed by electron microscopy.

The changes in DIP may remain unchanged for many years and many cases respond favourably to steroid therapy provided interstitial fibrosis has not supervened.

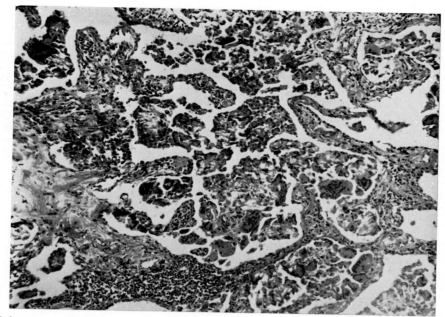

FIG. 19.58. Pulmonary malakoplakia showing numerous intra-alveolar giant cells and histiocytes as well as widespread interstitial pneumonia. × 90 H and E.

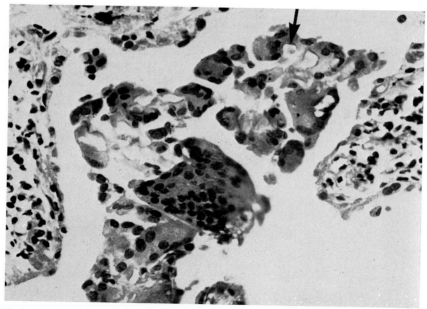

FIG. 19.59. Pulmonary malakoplakia showing an intracellular calcospherite (arrowed). × 300 H and E.

Malakoplakia of Lung

Malakoplakia is customarily only associated with chronic infection in the urinary tract, but as Gupta *et al.* (1972) showed a similar condition may occasionally be found in the lung probably in response to a haematogenous spread of *E. coli*. Bodies similar to Michaelis–Gutmann bodies, which appear very pale blue in ordinary stained sections, occur in both macrophages and giant cells in the alveolar lumens (Figs. 19.58, 19.59). The surrounding lung also shows severe chronic interstitial pneumonic change. The bodies are of glycolipid nature and autofluoresce in unstained sections exposed to ultraviolet light. The overall pulmonary histological appearances are at first sight suggestive of desquamative interstitial pneumonia (DIP) and illustrate the fact the DIP is a response of the lung to varied insults including chronic infection. It is highly probable that the few reported cases of localized DIP were cases of a similar nature.

CHAPTER 20

Carcinoma of the Lung

Incidence Rates

At the beginning of this century, primary lung cancer was generally regarded as one of the less common forms of malignant growths, and in his monograph in 1912 Adler reviewed 374 cases collected from the world literature. Since the end of the First World War the great increase in the number of cases of primary lung cancer and the interest displayed in this condition has been evidenced by the ever-growing number of published accounts of personally collected and analysed series of cases, among which may be mentioned those by Seyfarth (1924), Fried (1927, 1932), Bonser (1929), Frissell and Knox (1937), Ochsner and De Bakey (1941–2), Bryson and Spencer (1951) and Galluzi and Payne (1956) and others.

Many of these published statistics are open to serious errors, and it is difficult to obtain accurate and comparable national statistics, except in a few countries where medical care and death certification are of a high standard. Hospital post-mortem statistics, though often showing an apparent increase in a disease, may only reflect the type of disorders in the class of patient admitted to the institution. Also, greater efforts may be made to obtain autopsies on certain classes of disease in which the staff are personally interested, thus invalidating any comparative value of the statistics obtained. Despite these difficulties and inaccuracies, however, many large series of post-mortem statistics are now available from large hospitals throughout the world. The majority, obtained from Western Europe, North America and Australia, all show an increasing incidence of lung cancer.

For a full resumé of the comparative statistics the reader is referred to the paper by Steiner (1953) and a comprehensive account of the incidence in five continents (U.I.C.C., 1970, 1976).

In Tables 20.1 and 20.2 are listed the crude death rates from respiratory tract cancers in several countries, and without exception all show an unmistakable rise in incidence, which becomes more apparent when a comparison is made between the figures obtained before and after the Second World War. The greatest rise has occurred in the highly industrialized and populated countries, and particularly among the population of the larger cities, where the increase over the past 40 years can only be described as startling.

The validity of statistical conclusions was questioned in earlier years by Willis (1948) who believed that improved methods of diagnosis, errors in death certification and incomplete or mistaken interpretation of pathological findings accounted for much of the apparent increase.

If this argument of incorrect death certification were in fact true, the Registrar-General (1957) has shown that in the decade 1911–20 the proportion of wrongly certified deaths in this country would have amounted to 38 per cent of the certified deaths, a figure which is most unlikely to be correct. After allowance has been made for many fatal cases of lung cancer having been wrongly certified in earlier years, there has been a steady rise causing at least a twenty-five-fold increase in the number of cases since the beginning of the century.

Gilliam (1955), using similar arguments, found a comparable increase in the U.S.A.

TABLE 20.1

TOTAL DEATHS FROM ALL FORMS OF MALIGNANT DISEASE AND NUMBERS OF DEATHS FROM ALL FORMS OF RESPIRATORY TRACT CANCER IN BOTH SEXES

Country	Year	Deaths from all forms of respiratory tract cancer		Total deaths from cancer and all malignant tumours
		Male	Female	
Great Britain (England and Wales)	1928	1991	760	56,896
	1949	10,163	2232	80,663
	1954	14,888	2681	90,095
	1962	20,278	3501	—
	1972	25,762	5897	—
Scotland	1929	238	114	7108
	1949	1038	287	9587
	1954	1626	339	10,505
	1962	2375	397	—
	1972	2862	711	—
Denmark	1931	66	47	5024
	1949	356	92	6634
	1954	569	146	8370
	1962	964	213	—
	1972	1743	411	—
Spain	1931	908	158	16,757
	1949	1493	389	20 552
	1961	2655	671	—
	1972	4672	960	—
France	1943	2541	775	57,752
	1949	4344	1210	69,315
	1954	6578	1515	78,063
	1962	6948	1463	—
	1972	11,174	1673	—
Eire	1929	62	30	3116
	1949	229	67	4161
	1954	385	138	4594
	1962	497	142	—
	1972	769	253	—
Italy	1931	599	287	30,342
	1949	3029	878	47,376
	1954	4689	1197	59,339
	1962	6971	1391	—
	1972	13,517	2123	—
Norway	1929	24	12	3340
	1949	144	59	4765
	1954	225	79	5417
	1962	298	75	—
	1972	555	129	—
Holland	1929	258	85	9400
	1949	1214	221	13,997
	1954	1887	267	16,588
	1963	3390	275	—
	1972	5298	393	—
Switzerland	1929	222	29	6006
	1949	589	111	8366
	1954	872	139	9271
	1962	1085	111	—
	1972	1699	212	—

TABLE 20.1 (contd.)

Country	Year	Deaths from all forms of respiratory tract cancer		Total deaths from cancer and all malignant tumours
		Male	Female	
South Africa (Whites only)	1932	58	15	1656
	1949	221	61	2959
	1954	374	78	3683
	1962	446	95	
Canada	1929	177	91	8792
	1949	1130	306	16,732
	1954	1692	328	19,694
U.S.A.	1930	2660	1147	114,186
	1949	15,822	3696	206,325
	1954	23,257	4370	234,669
	1971	54,931	13,686	
Australia	1929	149	72	6256
	1949	676	142	9930
	1954	1023	206	11,611
	1962	2026	254	
New Zealand	1929	42	9	1467
	1949	191	34	2472
	1954	285	38	2879
	1962	439	63	
	1972	636	172	

(Reproduced from *Epidemiological and Vital Statistics Report*, W.H.O. (1952) 5, 1–144, and from *Annual Epidemiological and Vital Statistics*, W.H.O., for 1954, 1962 and 1971–2)

The figures quoted for the years up to 1949 were based on the International Classification modified by the revisions made in 1900, 1909, 1920, 1929 and 1938. The figures quoted for 1954 comprise tumours classified under the locations 160–165 as given in the International Detailed List of 1948.

The figures quoted for 1962 were based on the Intermediate List (List A) of the International Classification of Diseases and comprise only groups A 49 and A 50. As a result, the increase in the incidence for the year 1962 and subsequently is greater than appears from the figures quoted in the above table.

Doll (1953) drew attention to the fact that he greatest rise in incidence had occurred between 1930 and 1950, a period during which modern diagnostic aids had become available, and during which clinicians and pathologists were made continuously aware of its increasing importance. Although conclusions drawn from some of the earlier post-mortem figures, notably in Leeds, failed to show any rise in incidence (Bonser, 1929), figures from the same city revised to include later years confirmed that it had increased (Watkinson, quoted by Doll,

TABLE 20.2

DEATH RATES FROM PRIMARY MALIGNANT TUMOURS OF THE TRACHEA, BRONCHI AND LUNGS PER 100,000 OF THE POPULATION

Country	1950	1954	1958	1960	1962
Great Britain (England and Wales)	27·8	36·7	43·9	48·1	51·0
Scotland	26·8	35·4	43·1	46·5	50·1
Denmark	11·5 (1951)	13·9	18·2	23·1	25·3
Spain	4·5 (1951)	6·6	8·7	10·9	
France	8·7	12·3	14·7	16·3	17·9
Eire	7·5	14·3	17·0	20·0	22·6
Italy	6·4 (1951)	8·9	12·1	14·4	16·4
Holland	13·3	18·3	21·8	25·2	28·7
Switzerland	16·2 (1951)	17·9	19·5	21·3	21·4
Canada	9·0	11·6	13·3	14·3	16·3
U.S.A.	12·2	15·4	18·7	20·3	22·3
Australia	9·2	11·9	14·5	16·2	18·9
New Zealand	11·0	14·5	17·5	17·2	20·1
Japan	1·3	2·6	4·6	5·5	6·5
German Federal Republic	16·0 (1952)	18·5	23·2	25·8	28·3

(Reproduced from *Epidemiological and Vital Statistics Report* of the W.H.O. (1965) **18**, 319.)

1953). It is also difficult to believe that the eminent pathologists who practised 50 or 70 years ago were unable to recognize lung cancer, and that the standard of histological diagnosis since has so greatly improved.

The rise in incidence has been more rapid and apparent among males but nevertheless it has also risen considerably among women. Doll (1953) analysed the statistics for lung cancer in Great Britain and found that in 1915 the male:female ratio was 1·5:1, in 1925 it was 1·9:1 and in 1951 5·7:1. The same rise in the incidence of male to female cases of lung cancer was found to a lesser extent by Rigden and Kirchoff (1961) when they analysed the sex ratio data of 17,609 collected cases from the literature. In 1900 the ratio of males to females was 4:1 and by 1955 this had changed to between 5 and 6:1. These findings were broadly confirmed by Wynder *et al.* (1973) who found that in the U.S.A. since 1960, when

the ratio of male to female deaths was 6·6:1, the ratio has shown a progressive change reaching 5:1 in 1969. A similar trend has been noticed in Great Britain where the ratio in 1970 was 4·6:1 and this is attributed to the greatly increased habit of cigarette smoking among women dating from the time of the Second World War. Since the First World War there was a twelvefold increase of the disease in women in Great Britain and in 1955 lung cancer accounted for 1 in every 100 deaths among women and the proportion is still increasing.

Although the increase in the disease has been greatest in the United Kingdom, the same trend has been noticed in all highly industrialized countries. The rising incidence has been less apparent in Asiatic, African and South American countries where, however, the vital statistics are often less complete and therefore less accurate. The rise is apparent especially in the more rapidly developing and industrializing countries in all continents, but in the less developed non-industrialized countries in the same continents this form of cancer is still rare.

The differing incidence, while partly attributable to increasing industrialization in the Western nations and Japan, is largely attributable to different customs and habits, and also to the lower life expectancy in many of the underdeveloped areas of the world. Davies (1948) in Uganda and Gharpure (1948) in Bombay, India, both found no increase in their autopsy figures of this disease, but more recently Viswanathan *et al.* (1962) found that the number of patients attending several teaching hospitals in India with lung cancer has been increasing but at a much slower rate than in Western countries. Kageyama (1960) found that the incidence of lung cancer in Japan had doubled in both sexes between 1947 and 1956 and that lung cancer now ranked as the third most common cancer in Japanese males. Strachan (1934) found in the first survey of lung cancer in South Africa that it was confined to the white South African population but more recent studies of the disease in Johannesburg miners have shown that it now occurs and is becoming more common among the Bantu miners (Hurwitz, 1964). In Ceylon only 0·2 per cent of all post-mortems in Colombo hospitals were due to lung cancer (Cooray and Leslie, 1958), whilst in Morocco 2 per cent of all cancers reported upon at the Central Tumour Laboratory (1958) from both white and coloured patients originated in the lung. In Malaysia the incidence of lung cancer, though low, is nearly five times as great in both Chinese males and females as compared with both sexes of Malays (Marsden, 1958). This latter finding is probably a reflection of the fact that the Chinese are largely urban and city dwellers and the statistics reflect the fact that the disease is more common in the large cities, especially Singapore. A further factor that may partly account for the different racial distribution in Malaysia is the habit of opium smoking, which is indulged in mainly by the Chinese population and which is connected with the high incidence of nasopharyngeal cancers in this population group. During the period 1957–9 lung cancer accounted for approximately 12·0 per cent of hospital admissions for all forms of cancer in Singapore and about 7·0 per cent in women (Muir, 1962). Kozlowa and Rodionow (1963) found that in the Ukraine Republic (U.S.S.R.), the incidence of lung cancer per 100,000 of the population in 1948 and 1954 was 7·3 and 16·1 in men and 1·5 and 3·0 in women, whilst in 1960 the incidence had risen to 21·6 and 4·8 respectively.

The rising incidence of lung cancer began to be noticed at the end of the First World War in the industrial cities of eastern Germany. Since then it has been repeatedly noticed that the frequency of its occurrence correlates very closely with the degree of atmospheric pollution and the population density and is always lower in rural areas. Stocks (1959) found that the standardized mortality ratio (see Chapter 4) for males in Great Britain increased from 64 in rural districts to 126 in the large industrial centres, and for females from 76 to 121. In the rural areas it also increased with the population density (Curwen et al., 1954). Bonser and Thomas (1955) also found a similar trend in their comparative survey of the incidence of lung cancer in a rural area of Aberdeen and the City of Leeds. Hoffman and Gilliam (1954) reported that in the U.S.A. in 1948–9 the rate was consistently higher in urban residents; also Clemmesen and Nielsen (1957), in a comparative study in Denmark, found that the incidence of the disease in both sexes was nearly twenty times as great in the population of Copenhagen as in a Danish rural area, the provincial towns occupying an intermediate position. In Belgium the greatest incidence of lung cancer is found in the most highly industrialized areas of Liege, Namur and Charleroi and is lowest in the rural areas of the Ardennes. Since 1942 the age specific death rate from lung cancer has shown a ninefold increase in the Province of Ontario in Canada and the rise has been greatest in the largest cities, falling proportionately with diminishing population density (Burr et al., 1963). Although lung cancer death rates are always highest in very large industrial cities, the incidence of the disease within such cities may vary from district to district depending on the extent of the atmospheric pollution. These district variations are further discussed under the section on atmospheric pollution as a cause of lung cancer.

Recent statistical evidence collected in both Great Britain and the U.S.A. shows that the disease is more common in Social Class V (the poorest) and decreases in the higher social classes (Registrar-General, 1951; Cohart, 1955). This may be partly explained by the fact that the poorer sections of an industrial society tend to live in the cities nearer where they work, rather than in the less-populated suburbs.

Aetiological Factors

The undoubted and striking increase in lung cancer that has occurred during the past 40 years, and which has not yet reached its peak, has stimulated a search for environmental causes which might be responsible for the change. These are now known to include occupational factors, general environmental causes and especially personal habits such as cigarette smoking. To these might be added an hereditary factor for which, however, there is no proof in man, although it is of proved importance in certain mouse pulmonary tumours, as was shown by Strong (1936).

OCCUPATIONAL CAUSES

Radiation: The Schneeberg and Joachimstal (Jachymov) Miner's Lung

The oldest-recorded occupational lung cancers were those described in the Joachimstal and Schneeberg miners. For over four centuries a variety of minerals have been mined from the Erzgebirge mountains which separate Saxony from Czechoslovakia. Among the metals extracted from the ores were silver, copper, cobalt, bismuth, nickel, zinc, lead and, in recent years, uranium.

Agricola in the sixteenth century, in his book *De Re Metallica*, described the respiratory ailments from which these miners suffered, although he did not know the causes. In 1879 Härting and Hesse first recognized that the disease from which many of the Schneeberg (Saxony) miners were dying was lung cancer, a finding later confirmed histologically by Rostoski *et al.* (1926). Since this observation, over 600 cases have been reported (Hueper, U.S. Public Health Service Mem. 452).

Although Agricola had originally described the malady among the Bohemian miners, it was not until 1929 that Löwy reinvestigated the Joachimstal (Czechoslovakian) miner's lung, and described the first two confirmed cases of lung cancer, regarding it as an industrial disease in that area. In 1932, Pirchan and Sĭkl published the results of their extensive investigation into the nature of the Joachimstal miner's lung, confirming that it was due to lung cancer. Peller (1939) showed that the miners in the "radium mines" were thirty times more liable to develop lung cancer than adult men of comparable age groups in Vienna; and Sĭkl (1950) reported that the average induction time for these tumours was 17 years after the victim

had first started working in the mines, no case having occurred in less than 13 years. Evans (1950) estimated that during 17 years' employment in the Erzgebirge mines, a miner would have inhaled radon gas emitting the equivalent of 3000 Roentgens of radiation and probably received a much higher dose.

As already stated, minerals, such as pitch-blende, yielding many radioactive elements including uranium are mined from this mountain range, and furnished the Curies with the raw material from which they extracted and first isolated the element radium. The miners working underground, besides breathing radon gas, inhaled other dusts and gases notably silica, particles of radioactive minerals and traces of arsenious oxide, which may have formed nuclei upon which radioactive material could aggregate. The importance of the inhaled radioactive dust particles lies in the fact that if they are retained in the lungs they continue to emit radiation long after the miners have left the mine workings and even after they cease to be miners. Furthermore, Burrows and Clarkson (1943) showed that in some animals cancer can be induced more readily by irradiation when the tissues are damaged by non-specific causes, and thus silica by causing silicosis might have predisposed the miners to a greater likelihood of developing cancer. Also arsenious oxide is itself a potentially carcinogenetic substance.

Lung cancer among the Schneeberg miners is now becoming increasingly rare because of the improved ventilation and safety measures employed (Sepke, 1957). Since Schneeberg and Joachimstal miners' lung was shown to be lung cancer caused by the chronic inhalation of radioactive gases, further examples of this form of occupational lung cancer have been described in other areas of the world. Archer *et al.* (1962, 1973) and Wagoner *et al.* (1963, 1964 and 1965) found that uranium miners, but not surface workers, working in the Colorado uranium mines showed at least a three times greater incidence of lung cancer, compared with a matched group of non-mining, male population living on the surface of the

Colorado plateau. They carefully excluded other possible causes for the higher incidence in the group of miners. It has been found that the increased incidence of lung cancer occurred among the Colorado miners employed from 1950 to 1968 who had been exposed for more than 120 months to working levels of radon daughter products (Archer *et al.*, 1973a, b). The incidence of lung cancer was even greater in those miners who smoked or who had been exposed to diesel fumes or suffered damage to their lungs from pneumoconiotic disease. The higher incidence of lung cancer continues to appear among the miners even though the initial dangerous levels of radioactivity were reduced in the mines and many of the miners had stopped mining. Saccomanno *et al.* (1964) found that 79 per cent of the lung cancers found in Colorado uranium miners were oat-celled cancers. Webster (1969) also found a similar slight increase of oat-celled cancers in South African gold-miners who were exposed to traces of uranium and radioactive elements in some of the deep Rand gold-mines.

De Villiers and Windish (1964) reported a very high incidence of lung cancer, amounting to twenty-nine times the expected rate, among a small group of fluorspar miners living in an isolated community in Newfoundland. The miners during the course of their work were exposed to between 2·5 and 10 times the suggested safe working level of radon in the mines. The average exposure to this gas amounted to about 19 years which was considerably less than the average exposure of 25–35 years needed to induce lung cancer quoted by Hueper (1962). No reports have so far appeared of an increased incidence of lung cancer among miners employed in the uranium mines in Canada, Australia and Zaire, but an insufficient interval of time may have as yet elapsed and some of these mines are open-cast workings with consequent diminished risk to those employed. Furthermore, the average expectation of life of some African miners is such that there may be insufficient time for lung tumours to develop.

The importance of the radioactive gas and dust hazard in certain mines may be judged from the fact that, according to Hueper (1962), 70 per cent of the Schneeberg and 45 per cent of the Joachimstal miners used to die from lung cancer.

In all the examples of radioactive-induced lung cancers quoted above, the source of the radioactivity was mainly inhaled radon gas and its daughter products. Court-Brown and Doll (1965) have also reported an increased incidence of lung cancer among patients receiving externally applied radiation for anky-losing spondylitis. This occurs in addition to the better-recognized increased incidence of myeloid leukaemia. Wanebo *et al.* (1968) and Cihak *et al.* (1974) also found a slight increase of oat-celled lung cancers among survivors of the Hiroshima atomic explosion who received a dose in excess of 130 rads at the time of the explosion. Both the latter examples of lung cancer followed externally applied radiation. The lung damage results in fibrosis which also exerts a synergistic effect in addition to the direct effect of the radiation.

Tumours may be induced in the lungs of several species of animals including mice, rats and beagles by inhalation of particulate radio-active beads, plutonium aerosols and externally applied X-rays. The importance and significance of radioactive elements as potential causes of lung cancer arises because of the ever-increasing risk of accidents to atomic reactors with release of plutonium and other highly radioactive elements and isotopes.

Many experimental radioactively induced tumours have occurred in the periphery of animal lungs and have been either benign adenomas or malignant lymphomas, though cancers were described in 100 per cent of rats exposed to radon for 11 months (Perraud *et al.*, 1972).

Arsenic

Arsenic has long been recognized as a substance capable of producing hyperkeratoses in the skin and ultimately cancerous lesions.

Only recently has it been recognized that the inhalation, and possibly the ingestion, of inorganic arsenical compounds may also lead to lung cancer and other internal cancers.

The manufacture of arsenical sheep dips in a small English factory was found by Hill and Fanning (1948) to be associated with a seven times greater incidence of lung cancer among the workers when compared with the incidence in the general male population of the same age. Also, the higher lung cancer mortality among residents living down-wind of copper smelting plants in Wyoming was also attributed to the fumes of arsenious oxide liberated into the atmosphere (Lull and Walker, quoted by Hueper, 1957).

Blot and Fraumeni (1975) confirmed these earlier findings and have shown that both sexes have a higher incidence of lung cancer in those areas of the United States where lead, copper and zinc smelting and refining occur. This they attribute to the release of arsenic trioxide into the atmosphere.

Lung cancer due to the inhalation of arsenical insecticidal sprays has been reported by Hess (1956) and Roth (1960) among both male and female vineyard workers, and Braun (1958) found sixteen vintners with arsenical keratoses nine of whom later developed bronchial carcinomas. The therapeutic use of arsenic compounds may be followed after an interval of many years by the development of both skin and bronchial carcinoma. Sommers and McManus (1953) described the first case, and the author has seen further cases of bronchial carcinoma in chronic epileptics who had been treated for many years with sedative bromide mixtures containing small amounts of Fowler's solution (Liq. Arsenicalis B.P.C.). Other cases of a similar nature were described in Norway by Danbolt and Foss (1958) and more recently Robson and Jelliffe (1963) described a further six examples. The sex incidence and age distribution of the patients were often atypical for bronchial carcinoma and in most instances a long latent period of 20–30 years elapsed between the first ingestion of the arsenic and the appearance of the bronchial tumour.

The importance of arsenic as a potential lung carcinogen was increased following the introduction of arsenical parasiticidal mixtures to spray tobacco crops. Both Deschreider (1957) and Weber (1957) estimated the arsenical content of tobacco and found that it varied between 7 and 55 parts per million, being highest in Virginian leaf, but since the use of arsenical sprays has been discontinued the level has dropped considerably; it is not suggested, however, that arsenic is the principal carcinogen liberated by burning tobacco.

Holland et al. (1960) found that the arsenic content of urban air was greater than rural air due to the combustion of large amounts of fuel in towns which released traces of arsenious compounds. They also found that the arsenic content of the bronchial mucosa and submucosa was greater in cases of lung cancer than in a control group of lungs.

Nickel

Stephens (1933) was the first to recognize that there was an increased incidence of cancer of the paranasal sinuses and lungs among nickel workers at the Clydach refinery in South Wales.

Later, Morgan (1958) collected and described 131 cases of lung cancer among the same group of nickel workers that had occurred between the years 1927 and 1957. The pathology of some of the lung tumours was described in detail by Williams (1958). Doll (1958) found that the incidence of lung cancer was five times greater among those employed refining nickel than among a comparable group of the general population. These findings, however, only applied to men over the age of 50 who had been employed in the industry for many years. The increased incidence among the nickel refiners amounted to fourteen times that in the general population between the years 1938 and 1947 but had dropped to three times the incidence during the period 1957–64. This change was partly due to a fall in the incidence in nickel workers and partly to the general rise which had occurred in the general population. The recent figures indicate that the cause of the risk in the nickel industry had been removed at an earlier date (Doll et al., 1970).

Almost all the nickel refining in Great Britain is carried out at the Clydach refinery at Swansea by the Mond process which was reorganized in 1924 to reduce exposure of the workers to arsenic-containing acids and fumes (sulphuric acid) and to the fumes of nickel carbonyl. At the same time most of the initial stages of ore refining were transferred to the region of the nickel mines in Ontario, Canada, and only the further refining was carried out in this country. Following these changes the incidence of lung cancer at the Clydach works has fallen progressively but there is still no evidence to suggest that the original factor was nickel or nickel carbonyl. Lung cancer still remains a recognized hazard at the Canadian and Norwegian nickel ore refineries (Hueper, 1962).

Passey (1962) re-analysed the statistics for lung cancer among nickel workers in South Wales and concluded that neither metallic nickel nor nickel carbonyl was carcinogenic but caused lung cancer indirectly by chemically damaging the lungs. This damage acted as a promoting factor and excited secondary malignant changes at a later date. In support of this view he pointed to the fact that the duration of employment in the industry did not alter the age at which lung cancer appeared. Those who entered the industry at an early age developed their lung cancers at about the same age as those who had been employed for only a few years. Passey's suggestion gains support from the observation of Williams (1960) that nickel refiners may show chronic interstitial fibrosis in their lungs, a condition known to predispose in other circumstances to the development of lung cancer.

Saknyn and Shabynina (1973) have more recently reported an increased incidence of lung cancer in four nickel refineries in the Ural region of Russia. The workers had been exposed to aerosols containing nickel sulphide and nickel oxide as well as metallic nickel. They also noticed an increased lung cancer rate among workers in an electrolytic works where they had been exposed to high concentrations

of nickel sulphate and nickel chloride. Most of the affected persons had worked for more than 7 years in the refineries.

Hueper (1958) succeeded in producing a single lung sarcoma among a group of thirty-four rats by directly implanting nickel powder into their lungs. The nickel powder prepared from nickel carbonyl caused local fibrosis and led to considerable bronchiolar epithelial hyperplasia, the change extending to involve the alveolar epithelium. Many of the multicentric lesions that resulted bore some resemblance to human pulmonary adenomatosis.

Sundermann et al. (1959) also succeeded in producing lung tumours in rats using nickel carbonyl.

Metallic Iron and Iron Oxides

Turner and Grace (1938) found that foundry workers, including metal grinders, showed a higher mortality rate from lung cancer than any other occupational group in Sheffield, a finding that was corroborated by McLaughlin and Harding (1956). Bonser et al. (1955) have also shown that haematite miners in Cumberland show a seventeen times greater incidence of lung cancer than comparable groups of non-miners in the same county, and it is concluded that iron, probably in the Fe^{3+} form, in high concentration in the tissues is potentially carcinogenic. More recent investigations, however, have shown that these miners are also exposed to increased radioactivity, which is probably of greater importance.

Chromates

Machle and Gregorius (1948) and Brinton et al. (1952) in the U.S.A., Bidstrup and Case (1956) in Great Britain and Spannagel (1953) in Germany have all found a higher incidence of lung cancer among those employed in the manufacture or handling of chromates. In the U.S.A. chromate workers have shown a thirty times greater incidence, but in Great Britain it was reported to be only four times the rate in the general population. This disparity is to some extent explained by the higher overall mortality rate from lung cancer in Great Britain compared with the U.S.A.

Bourne and Rushin (1950) showed that the increased incidence of lung cancer in workers in factories using chromium salts not only involved those who actually came into personal contact with the substances but also involved the other employees not actually engaged in handling these materials. Furthermore, the incidence of lung cancer may be increased in populations living downwind of factories refining and extracting chromium and its salts. Negro workers appear to be more susceptible to the development of lung cancer than the white employees.

Passey (1962) claims that, like nickel and its compounds, chromium and its salts are not themselves carcinogenic but cause damage to the lung which later induces malignant changes. He claimed to show that length of exposure to chromium compounds did not influence the age at which lung cancer occurred. The hexavalent compounds of chromium appear to be particularly carcinogenic, however, and experimental proof of this property was provided by Hueper and Payne (1959). Among the industries in which the risk from chromate cancer is greatest is chrome pigment manufacture, chrome steel and alloy production and chrome electroplating.

Asbestos

The occurrence of lung cancer among those manufacturing or handling asbestos was noted by Gloyne (1935), and many additional cases have since been reported including those by Homburger (1943), Bonser *et al.* (1955) and Howard *et al.* (1975). The incidence of lung cancer in patients with asbestosis has been estimated at 20 per cent, though as Doll (1955) has shown, only those employed in the industry for longer than 20 years are liable to develop this complication. The incidence of lung cancer, which depends on the type of asbestos used and con-

sequently the physical and chemical properties of the individual fibres, is consequent upon the development of asbestosis in the lung. Crocidolite is the most dangerous form as it penetrates most readily into the terminal air spaces whereas chrysotile and amosite which penetrate less readily are associated with less extensive damage to the lung and a much lower risk of cancer. McDonald *et al.* (1971) have shown that the risk of Quebec asbestos-miners developing lung cancer (chrysotile asbestos) was only about twice that of the general population and that it occurred almost entirely in those miners who were additionally heavy cigarette smokers. A similar low increased incidence has been reported in the asbestos (chrysotile) mines around Sverdlovsk in Western Siberia (Kogan quoted by Wagner, 1972). A low incidence of lung cancer is also found among anthophyllite asbestos miners in eastern Finland. On the other hand, very high incidence rates which are increasing have been reported from the U.S.A. by Selikoff *et al.* (1970). This is probably due to the continuing, unrestricted and increasing use of the dangerous crocidolite asbestos in that country. A similar situation occurs in other highly industrialized nations such as Western Germany and Japan where there is unrestricted use of crocidolite.

The carcinoma usually arises in the lower lobes where the changes due to asbestosis are most severe (Jacob and Anspach, 1965). The tumour is preceded by considerable alveolar epithelial hyperplasia induced by the presence of asbestos fibres and by the chronic interstitial fibrosis. The resulting cancers tend to be peripherally situated adenocarcinomas or solid undifferentiated alveolar cell cancers.

The carcinogenic properties of asbestos have been attributed to its iron content but Harington (1962) proved the existence of traces of benzo-(a)pyrene and related aromatic, polycyclic, carcinogenic hydrocarbons in freshly mined and virgin South African crocidolite and amosite forms of mineral asbestos. These substances were presumably formed in geological times and were only found in sedimentary asbestos-containing rocks and were absent in mined asbestos igneous rocks. The carcinogen content of asbestos varies widely according to the geographical area in which it was mined. Experiments using asbestos from which the naturally occurring carcinogen had been removed showed that it was equally carcinogenetic.

In addition to causing lung cancer asbestos (crocidolite and amosite) may cause malignant pleural and peritoneal mesotheliomas which are considered separately. Hueper (1962) analysed 909 collected cases of asbestosis in the U.S.A. and found that the incidence of lung cancer was 18·4 per cent (171 cases). The death rate from lung cancer among males suffering from asbestosis and who smoke heavily in England and Wales is, however, in excess of 50 per cent.

The risk of asbestosis and the subsequent development of lung cancer not only involves those who actually mine, transport, process and use the mineral, but extends to persons living in the vicinity of asbestos-mines and waste heaps and anybody who may even remotely and by accident come in contact with the dust of this mineral. The increased use of asbestos has been incriminated as a major aetiological factor in the rising incidence of lung cancer in Great Britain but this is unlikely and remains unproven. Cauma et al. (1965) in Pittsburgh, U.S.A., found asbestos-like bodies in lungs of 41 per cent of a series of unselected autopsies on adults, but the presence of the bodies was not associated with asbestosis and lung cancer.

The majority of asbestos lung cancers are situated in the lower lobes of the lungs where the fibrotic changes caused by asbestos are most severe. The tumour is usually preceded by considerable alveolar epithelial hyperplasia induced by the chronic interstitial fibrosis. The resulting lung cancers tend to be of the peripheral adenocarcinomatous variety or solid alveolar cell growths.

Beryllium

Although beryllium lung disease (berylliosis) is not a proven cause of lung cancer, the two diseases may occasionally be associated. Further-more, beryllium salts are suspected of being carcinogenetic agents. Beryllium is a well-recognized cause of chronic interstitial lung fibrosis and this latter condition may lead in other situations to the formation of alveolar epithelial cell hyperplasia. Also as Schepers (1961, quoted by Hueper, 1962) showed, beryllium salts can induce multicentric malignant changes in the lungs of rats.

Gas-workers

Kennaway and Kennaway (1947) found that an increased risk of developing lung cancer existed among London coal gas employees who had worked in the retort houses and coke ovens, and Doll (1952) and Doll et al. (1965) found that the risk was twice that found in a comparable group of the male population in London. The concentration of benzo(a)pyrene found in the atmosphere in the vicinity of an emptying horizontal type of gas retort is about a hundred times that in the atmosphere of a large industrial city, but it is considerably less in the vicinity of a vertical type of retort. The concentration of benzo(a)pyrene in the air around a large gas works has been found to be five to ten times that found in the atmosphere of a busy industrial city.

Kuroda (1937), Kawahata (1938) and Kawai et al. (1960) have all reported a number of cases of lung cancer among a group of gas workers employed in gas generators in the Yawata steel works in Japan. All were exposed intermittently to pitch oil fumes emitted at high temperatures from the generators.

Other Industries Associated with a High Incidence of Lung Cancer

Other industries stated to be associated with a high incidence of lung cancer include the manufacture of isopropanol, though little evidence can be found to support this statement. Weil et al. (1952) were only able to find one case of lung cancer, though cancer of the nasal sinuses was more common. Recent evidence indicates that some other constituent than iso-propyl alcohol in the crude liquor was the carcinogenic agent (Hueper, 1962).

Mustard gas has been incriminated as a powerful carcinogen as well as being a well-known antimitotic (alkylating) agent. Some suggestive but not statistically confirmatory evidence was obtained from an analysis of the medical records of both British and American troops who were exposed to this gas during the First World War. Yamada (1963) reported a very high incidence of both lung and laryngeal cancer among a small group of Japanese chemical workers who had engaged in the manufacture of mustard gas during the Second World War. In common with lung cancer caused by other radioactive and radiomimetic substances there was a high proportion of oat-celled cancers among these mustard-gas workers.

Figueroa et al. (1973) have reported a high incidence of oat-celled lung cancers among chemical-workers exposed to chloromethyl-methyl-ether. The length of exposure varied from 3 to 14 years. Commercial chloromethyl-methyl-ether, which is used as an intermediate substance in the synthesis of organic chemicals including ion exchange resins, contains up to 7 per cent of bischloromethyl-ether, a substance which is highly carcinogenetic to mice and rats (Van Duuren et al., 1968). These halogenated ethers are alkylating agents which are a group of chemical substances known to be carcinogenetic and to have a radiomimetic and destructive effect on mitosing nuclei.

Monson and Peters (1974) claimed to have also shown a slightly raised incidence of lung cancer in persons who handle vinyl chloride.

ATMOSPHERIC POLLUTION

Although cigarette smoking is regarded by many as the only cause of importance responsible for the increased incidence of lung cancer, the author does not entirely subscribe to this view.

The rising incidence of lung cancer has coincided with a period of great industrial expansion, in Great Britain and other countries. This expansion, together with the introduction of the automobile, has added very greatly to the amount of atmospheric pollution, particularly in industrial areas. The increase of lung cancer was first noticed at the beginning of the present century in the industrial cities of eastern Germany, and since the end of the Second World War numerous statistical reports have appeared on the importance of atmospheric pollution as a cause of this disease.

Among the carcinogenic substances so far identified as atmospheric pollutants are 3:4-benzpyrene, 1:12-benzperylene, arsenious oxide, coal-tar fumes, petroleum oil mists, and traces of radioactive substances, the last having risen to surprisingly high levels during periods of "smog" in large cities such as London. Goulden et al. (1952) found that the benzo(a)pyrene content of the atmosphere in London varied between 7·1 and 5·8 μg per 100 cubic metres of air, the arsenic content reaching a mean value of 7·0 μg. The benzo(a)pyrene levels in the air around Liverpool and North Wales were shown by Stocks and Campbell (1955) to vary from 0·1 μg per 100 cubic metres of air in the countryside of North Wales to 4·75 μg per 100 cubic metres of air in an industrial town in east Lancashire. Hueper et al. (1962) found considerable variation in the amounts of polycyclic aromatic hydrocarbons present in the atmospheres of several large American cities and showed that the quantities present, however, bore little relationship to the local incidence of lung cancer. Although benzo(a)pyrene content of air is often used as an index of the carcinogenicity of a polluted atmosphere, recent work suggests that the levels of this substance may not always prove a reliable guide. Atmospheric pollutant content varies from one city to another and from one country to another. Other equally and often more potent carcinogenic pollutants may include ozone, asbestos dust, traces of nickel, chromium and arsenic compounds, uncombusted aliphatic hydrocarbons capable of conversion into unsaturated and polymerized carcinogenic compounds, and probably many as yet unidentified substances, the number of which is increasingly rapidly with the introduction of new industrial processes. Furthermore, the physical state in which pollutant substances are present in the atmosphere is of considerable importance. Some exist as aerosols, some in the gas phase and others as particles with adsorbed substances, though the latter may be less dangerous than they would appear at first sight as adsorption often renders the adsorbed substance innocuous. In some of these states pollutants may reach the terminal air passages and alveoli where they are capable of causing damage at these sites. Recent evidence supports the view that chronic bronchitis and the alveolar parenchymal damage resulting from inhaled irritants and infection may be of greater importance as carcinoma-promoting factors than any direct carcinogenic effect of any one individual substance. The bronchial epithelial changes are at first reversible if the individual is removed from the area, but if persistent exposure occurs the changes become irreversible and can lead to regenerative hyperplasia, metaplasia and ultimately malignant change in a few cases. Furthermore, chronic damage to the protective bronchial ciliary and mucus-producing mechanisms may enable viruses and bacteria to gain a foothold leading to further lung damage which may hasten the onset of malignant change.

The incidence of lung cancer increases with the number of inhabited dwellings per acre and Stocks (1959) compared the standardized mortality rate (SMR) for lung cancer with the smoke index, deposit index of the atmosphere, and the population per acre. The results have shown in a most convincing way that the SMR for lung cancer varies directly with the smoke index. Although population per acre tends to be greatest in areas with the most smoke, the relationship to lung cancer is not so close, non-industrial cities having a considerably lower incidence of lung cancer than industrial cities of comparable size.

Similar findings have been reported in the U.S.S.R. by Mats and Mizyak (1958) who found that in 1956 lung cancer accounted for 8·5 per cent of all forms of cancer throughout the country, but it was nearly twice as high in the industrial centres of Kharkov, Dnepropetrovsk and Kiev. Clemmesen and Nielsen (1957) also found that the mortality from lung cancer in both sexes in Denmark was highest in Copenhagen and lowest in the rural areas, and a similar finding was reported by Hoffman and Gilliam (1954) in the U.S.A. In London the highest incidence of lung cancer is found in the northeastern boroughs, which are downwind districts with the prevailing south-west wind. Skramovsky (1963) found that in Czechoslovakia the same direct relationship existed between the size of a town and the death rate from lung cancer. In cities with over a million inhabitants (Prague) there was the greatest death rate from lung cancer and the rate became progressively less with decreasing city size. The same relationship of lung cancer to increasing atmospheric pollution has been demonstrated in Japan. In general the death rate also mirrors the benzo-(a)pyrene content in the atmosphere.

Although the statistical evidence in many countries has shown that the increase in the habit of tobacco (cigarette) smoking correlates very closely with the rising incidence of lung cancer, nevertheless, the rising incidence of the disease in many of the developing countries as well as in the developed nations corresponds closely with the increase in the number of registered petrol- (gasoline) and diesel-driven road vehicles as well as industrially generated levels of pollution. In an endeavour rightly to focus attention on the dangers of cigarette smoking, too little attention has been focused on other rapidly growing sources of pollution, especially atmospheric pollution associated with massive industrial combustion of fossil fuels and the volatile products they release into the atmosphere. A close relationship exists between the density and number of registered road vehicles and the incidence of lung cancer. A great deal more attention needs to be paid to this factor than hitherto. Inhaled carcinogens whatever their source summate, and while every endeavour should be made to reduce the habit of tobacco smoking other major sources of carcinogens in the environment have attracted too little attention though equally deserving of similar study.

The largest sources of atmospheric pollution are industrial, especially installations such as electricity power stations using conventional methods of raising steam. In the larger cities in England domestic heating and to an increasing extent vehicle exhaust fumes, especially diesel fumes, have contributed to the extensive atmospheric pollution. Petrol engines, however, contribute more nitrogen oxides, formaldehyde, carbon monoxide and aromatic and aliphatic hydrocarbons to the overall pollution than well-tuned diesel engines both when idling and particularly when moving and decelerating (L.C.C. Report by M.O.H., 1964).

SMOKING

Since Müller (1939) first drew attention to a relationship between lung cancer and heavy smoking, innumerable investigations and statistical surveys have been carried out and have abundantly confirmed this finding.

Only a few of the better-known and pioneer investigations can be mentioned in the space available, including those by Doll and Hill (1952, 1954, 1964) in England, by Wynder and Graham (1950), Hammond and Horn (1954) and Hammond (1958) in the U.S.A., and by Kreyberg (1955, 1956) in Norway.

All the large-scale retrospective and prospective surveys have led to the same conclusion, namely that heavy smoking was more commonly indulged in by patients with lung cancer than a control group of patients suffering from other diseases. In the series of 1357 men with lung cancer reported by Doll and Hill (1952), only 0·5 per cent were non-smokers, whereas 25 per cent of this group gave a history of having smoked more than 25 cigarettes a day for many years in the past. A more recent survey by Doll and Hill (1954, 1956, 1964) of lung cancer deaths among doctors on the British Medical Register has revealed a direct relationship in this occupation group between the amount of tobacco smoked and the incidence of lung cancer. In both this and other previous surveys it has been noticed that cigarette smoking was of greater importance as a cause of lung cancer than the smoking of cigars or a pipe. Furthermore, the survey has shown that giving up the habit of smoking reduces the risk of the subsequent development of lung cancer and that the risk is reduced proportionately to the length of time smoking was abandoned.

Stocks and Campbell (1955) in a survey of lung cancer in the North Wales and Liverpool areas found that the rural death rate from lung cancer increased proportionately to the number of cigarettes smoked per week, and they considered that in cities such as Liverpool half the deaths from lung cancer in males were probably directly attributable to heavy smoking.

The lower incidence of lung cancer in women has been attributed in part to the past lower incidence of the habit of smoking in this sex. An outstanding exception to this general rule occurs in New Zealand where the Maori female from school age is often a heavy cigarette smoker, and two-thirds of Maori women smoke cigarettes; consequently the proportion of Maori male to female cases of lung cancer (1968–70) was of the order of about 2·4:1 (Department of Health, N.Z., 1973).

A similar association was found by Hammond and Horn (1958) in a 44-months follow-up study of 187,783 men in the U.S.A. The lung cancer death rate was highest among the heavy smokers living in large cities, and was consistently lower for all levels of smoking among those who lived in rural rather than urban areas.

Cigarette smoking in the U.S.A. appears to carry less risk of the subsequent development of lung cancer, and this has been attributed by Doll et al. (1959) partly to the fact that when cigarettes are discarded they are left with a longer butt than in this country. This may afford some protection to the American smoker, who is less likely to inhale high-temperature volatile products of combusted tobacco including fumes of arsenious oxide.

Apart from the statistical evidence the association between smoking and cancer of the lung rests upon the following evidence:

(a) The finding of benzo(a)pyrene and other carcinogens in combusted tobacco.

(b) The finding of histological changes in the bronchial epithelium of heavy smokers including pre-carcinomatous changes.

(c) The increased incidence of other diseases associated with smoking in patients with lung cancer.

(d) The experimental production of tumours in animals following the application to the skin and internally of tobacco condensates.

(a) Among the substances formed in the combustion products of cigarettes are traces of 3:4-benzpyrene, 1:12-benzperylene and other chemically similar carcinogenetic substances, together with traces of arsenic which volatilizes as arsenious oxide and radioactive polonium compounds. It appears that doses of individual carcinogens may summate, in the same manner as radiation, until the individual tolerance dose is exceeded, when pre-malignant and ultimately malignant changes may be induced. The country dweller, owing to the absence of heavy atmospheric pollution, would be able to absorb a far greater quantity of carcinogens from heavy smoking with less risk of inducing malignant change.

(b) The histological changes found in smokers' lungs are described in greater detail later, but Wynder and Graham (1950) in the U.S.A. have shown that the increased numbers of bronchial cancers seen in smokers were mostly squamous-celled or oat-cell cancers, whereas the adeno-carcinomas found in both sexes appear to bear less relationship to the smoking habits of the individual. Similar findings have been reported by Doll et al. (1957) in Britain.

(c) The incidence of chronic duodenal ulcer is higher in patients with cancer of the lung than in control series of patients.

(d) In 1937 Roffo painted rabbits' ears with the combustion products of tobacco and succeeded in producing skin carcinomas. Nearly 20 years later Wynder et al. (1953) reproduced the same changes in rabbits and mice using cigarette tar. The active principal was found to be contained in the neutral fraction, but the actual carcinogen or carcinogens responsible have not yet been isolated.

Blacklock (1957) injected cigarette filter tar into the hilar regions of rats' lungs after opening the chest. He dissolved the tar in olive oil and added dead tubercle bacilli, and was successful in producing cancers in the hilar region of the lungs.

Innumerable experiments have been carried out in attempts to show that tobacco smoke condensates contain carcinogenic substances. A very recent review of the many animal experimental models employed and the results obtained have been presented in a U.I.C.C. Technical Report (1976).

One such attempt, however, was made by Borisjuk (1967) who instilled whole tobacco tar concentrate in a protein blood substitute medium directly into the trachea of rats. Two squamous-cell cancers were produced in a batch of forty-three rats. Numerous "smoking machines" have been devised and animals made to inhale tobacco smoke but nearly all the experimental cancers produced have been adeno-carcinomas unlike most human cancers.

Combustion of tobacco results in traces of 3:4-benzpyrene being formed and this substance was identified by Cooper and Lindsey (1955) in cigarette smoke. That this is not the only potentially carcinogenic substance that is found in tobacco smoke was shown by the experiments of Lasnitzki (1958) when she applied different fractions of cigarette smoke condensates to organ cultures of human foetal lung and found that other substances present were also capable of inducing atypical hyperplastic changes.

Although there is undeniable statistical evidence to link tobacco smoking with lung cancer, the association may well not be so direct as has hitherto been assumed. Kotin and Galk (1960) and Passey (1962) have produced convincing arguments and statistical confirmation to show that tobacco may behave indirectly rather than directly as a cause of lung cancer. The principal arguments against the hitherto-accepted views may be briefly stated as follows:

(a) Human lung cancers occur at the same age whether the patients have smoked heavily since adolescence or have only been light, moderate or even non-smokers. This runs contrary to the evidence gained from experimental cancer research which shows that increasing the dose of a carcinogen hastens the appearance of the resulting tumour. The risk of lung cancer developing, however, varies with the duration of smoking and the amount smoked.

(b) It is now confirmed that if a heavy cigarette smoker ceases the habit the risk of developing the disease diminishes with every year since smoking stopped. After 10 years the risk is little more than in a non-smoker. This runs contrary to the information gained from all experimental cancer research which has shown that once a threshold dose of a carcinogen has been applied, withdrawal of further applications does not prevent the onset of the tumour.

(c) The consumption of cigarettes by the white South African male population is amongst the highest in the world but the incidence of lung cancer is still low compared with Western Europe and North America. This finding suggests that some factor other than cigarette and tobacco smoke plays an important part in the genesis of lung cancer.

(d) If tobacco smoke was directly carcinogenic it would be reasonable to expect that the incidence of buccal, oral, nasopharyngeal and laryngeal cancers should also have increased commensurately with lung cancer as these structures would be even more directly exposed to the action of any potentially carcinogenic substance. Furthermore, squamous epithelium, which lines many of these structures, is known to be a susceptible form of epithelium to directly applied carcinogens in animal experiments and in man (mule spinner's, lubricating oil scrotal cancer). Although laryngeal cancer has increased slightly in frequency this rise in incidence bears no comparison with the increase in lung cancer.

(e) If tobacco smoke were as carcinogenic as has been suggested it is surprising that the large number of animal experiments that have been conducted have yielded relatively few lung tumours when different species of animals are exposed to tobacco smoke in various ways.

From these arguments the author does not wish to convey any impression that tobacco (particularly cigarette smoking) and lung cancer are not statistically related, a fact which is of the greatest importance and which now appears undeniable from numerous investigations carried out in many countries throughout the world. What is less certain is the manner in which tobacco smoke contributes to the causation of lung cancer. Recently increasing evidence has accumulated to show that cigarette smoke, and to a less extent cigar and pipe smoking, act mainly by provoking a state of chronic irritation in the lungs and this may predispose not only to chronic bronchitis but also to parenchymal (alveolar and respiratory bronchiolar) damage. Jeffery (1973) and Graham and Levin (1971) have also shown that tobacco smoke inhaled by rats increases the cell turnover rate in the basal germinal layer of cells in the bronchi. Furthermore, the damage to the ciliary and mucus-producing cells in the larger bronchi may allow the more ready ingress of bacteria and viruses into the lungs. If cigarette smoking is allied with breathing heavily polluted and cold and damp air due to an inclement climate, the extent of the resulting bronchitis and often concurrent parenchymal damage may become considerable. In general the death rate from lung cancer parallels the incidence of chronic bronchitis, and all the factors mentioned above which are capable of causing bronchitis may summate. The resulting damage and reparative changes in the lung structures are thought to play an important part in initiating lung cancer.

Some confirmatory evidence for these views was provided by Eastcott (1956) and Dean (1959), both of whom found that British emigrants to New Zealand and South Africa respectively showed a higher incidence of lung cancer than the locally born white male population although matched for smoking habits and age. This difference may have resulted from lung changes already present in the lungs of the men when they left the United Kingdom induced by the causes mentioned above.

Passey has also pointed out that women normally suffer less from chronic bronchitis than men and therefore would be expected, if the foregoing arguments were correct, to show a lower incidence of lung cancer.

HEREDITARY FACTORS

The importance of hereditary and racial factors as a contributory cause of any form of cancer is extremely difficult to evaluate in human pathology.

In the U.S.A., the incidence of lung cancer in the American Indian, Negro and white populations shows considerable variation, the proportion being approximately 1:3:4. An accurate comparison is very difficult, however, as the Indians are almost entirely a rural people, whilst the Negroes and whites live under both urban and rural conditions. In the U.S.A., the Negro population show a very much greater incidence of lung cancer than their kinsmen in Africa, although both are of common descent. This variation is, therefore, likely to be due entirely to environmental differences.

Rakower (1957) reviewed the figures for lung cancer in Israel and showed that the incidence was much lower among Asiatic- and African-born Jews than among those born in Europe, and this despite the fact that the Jewish people have remained racially pure throughout the centuries. Although heavy cigarette smoking in the Western European Jew may to some extent explain this disparity, it is also partly attributable to the different environments in which the two types of Jews previously lived. Rakower also showed that the overall incidence of cancer was lower in comparable Asian/African-born age groups than among those born in Europe, lending support to the view that cancer occurs more commonly among highly industrialized communities. Buechley et al. (1957) showed that there was a twofold excess of lung cancer in Mexican women born in Mexico but resident in California over a similar group of Spanish-surname women born in California. This was attributed to the different environment under which the former group of women had been reared.

All the evidence gained from these and similar investigations tends to show that environmental factors are of greater importance than hereditary factors or racial susceptibility.

CHRONIC INFLAMMATION

In many sites, cancer is known to be preceded by chronic inflammatory changes and the same sequence of events is now known to precede some lung cancers.

As the subject of lung scar cancers is considered in more detail later, it will not be discussed further at this juncture.

Bryson and Spencer (1951) and later Finke (1956), Passey (1962) and Campbell and Lee (1963) have drawn attention to the importance of chronic bronchitis as an antecedent condition in some cases of lung cancer. A long-standing history of chronic bronchitis appears to be at least twice as common among patients with lung cancer as in a comparable group of control patients matched for smoking habits and age. Furthermore, as Stocks (1959) has shown, statistically the incidence of lung cancer parallels that of chronic bronchitis and examples of minute lung cancers may occasionally be discovered post-mortem in persons suffering from long-standing bronchitis. Campbell and Lee considered that chronic bronchitis might be a major factor in the aetiology of lung cancer in Australia.

Atypical Hyperplasias and Pre-carcinomatous Conditions in the Lung

During the past few years several extensive and detailed researches have added considerably to the knowledge of atypical proliferative lesions and precancerous states in the lung.

"TUMOURLETS"

In 1929 Gray and Cordonnier described what they considered was an example of an early carcinoma arising from alveolar ducts. Later, Womack and Graham (1941) noted similar microscopical tumours in each of three specimens of lungs removed for "congenital cystic disease". They regarded the tumours as atypical but non-malignant epithelial proliferations which were potentially capable of becoming malignant.

Since then further reports of similar epithelial proliferations, variously regarded as carcinomas or atypical bronchiolar or alveolar celled hyperplasias, have been described by Stewart and Allison (1943), Petersen *et al.* (1949), Spain and Parsonnet (1951), Prior and Jones (1952), King (1954), Spencer and Raeburn (1954) and Whitwell (1955). The last named called them "tumourlets" on account of their distinctive appearance and behaviour. In every instance the growths have consisted of uniformly staining, non-hyperchromatic, small, round, oat- or spindle-shaped cells, filling alveoli and growing mostly from respiratory bronchioles, alveolar ducts and alveolar epithelial cells (Figs. 20.1, 20.2, 20.3 and 20.4). The "tumourlets" are never found in the absence of damage to the lung, which is often severe and frequently associated with bronchiectasis. The changes are usually multifocal and several "tumourlets" arise separately throughout the damaged lung. Although the cells seldom show mitoses they invade lymphatic spaces but very rarely metastasize to the regional hilar glands, though an exception was demonstrated by Cureton and Hill (1955) (Fig. 20.6), and a few similar cases have been reported by Hansman and Weimann (1967) and Merle and Legg (1972). Similar alveolar- and bronchiolar-celled proliferations may occur in the recovery stages of influenza, giant-celled pneumonia and following paraquat poisoning, and there is evidence to suggest that bronchiolar obliteration due to scarring or collapse may be an aetiological factor in their formation. King regarded fixation of the lung and the subsequent growth of the epithelium of the terminal air passages but especially of the alveoli as the cause of the "tumourlets". Prior, however, regarded them as peripheral adenomas.

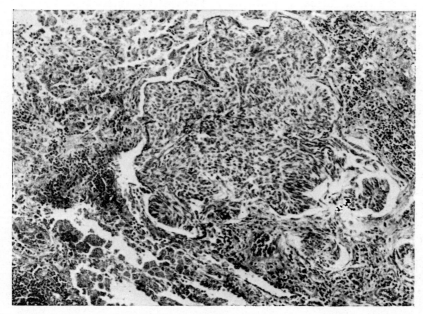

FIG. 20.1. A "tumourlet" arising in a damaged alveolus in a bronchiectatic lung and filling the alveolus. × 120 H and E. (Reproduced by courtesy of the Editor of *Journal of Pathology and Bacteriology*.)

9

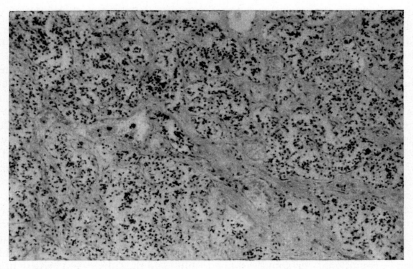

FIG. 20.2. A tumourlet showing typical clumps of small darkly staining cells and complete destruction of the lung alveolar structure. × 100 H and E.

Hoffmann (1962) studied thirty-one cases, the majority of which occurred in lungs containing pulmonary emboli and blocked pulmonary arterial branches. He divided them into three varieties and regarded them all as atypical proliferations of "pulmonary epithelium" resulting from injury to the lung caused by a variety of lesions. From a study of a number of these tumours the author believes that tumourlets are composed of two separate groups of tumours. The first, a large group, are peripheral carcinoid tumours and may on occasion cause endocrine

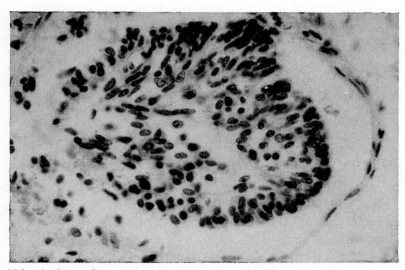

FIG. 20.3. A clump of tumourlet cells showing some spindling of the cells. × 400 H and E.

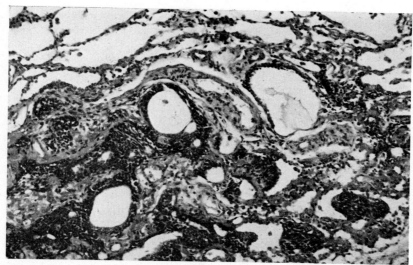

Fig. 20.4. A small tumourlet beneath the pleura showing canalization of some of the cell masses. In the region of the tumourlet the lung parenchyma has been destroyed by fibrosis. × 100 H and E.

disturbances such as Cushing's syndrome (Fig. 20.5). The component cells are larger and usually show a more compact arrangement similar to carcinoid tumors in the large bronchi. These form a less common group. The second group comprising the great majority of tumourlets are proliferations of either bronchiolar or alveolar epithelium in response to damage and irritation caused by a variety of conditions and are found equally in both sexes. Both bronchiolar and alveolar epithelia possess enormous regenerative capacity and in the presence of damage, following

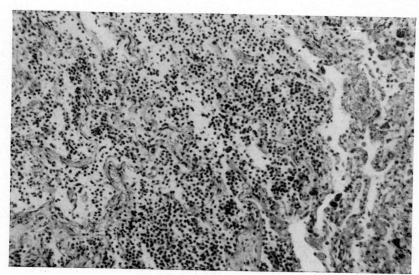

Fig. 20.5. A small peripheral bronchial carcinoid tumour resembling a tumourlet. × 100 H and E.

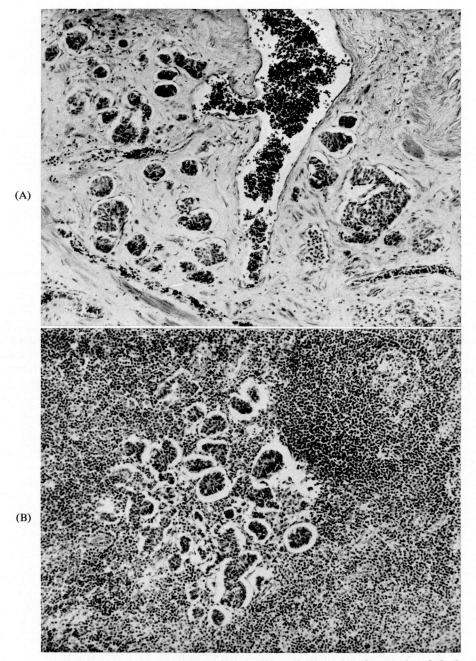

FIG. 20.6. A "tumourlet" arising in a bronchiectatic lung (A), and (B) metastatic deposits of similar tumour cells in a hilar lymph gland. × 140 H and E.

(Reproduced by courtesy of Dr. R. J. R. Cureton.)

upon injury, may proliferate to form solid buds of cells. The buds may show metaplastic squamoid features or the component cells may be loosely arranged and occupy fibrous-walled spaces. In almost every instance growth is self-limited and the tumours regress. Some of the solid buds of cells may cannulate to form epithelium-lined spaces (Fig. 20.4).

Tumourlets may be produced experimentally by the intratracheal instillation into the lungs of rabbits of 1 per cent nitric acid, the animals being protected from the onset of bacterial pneumonia by antibiotic cover (Totten and Moran, 1961; Moran, 1965). Their formation is stimulated by giving the animals cortisone.

The benign nature of the majority of these growths is attested by their subsequent non-recurrence and the survival of patients in whom they were discovered in the excised lung tissue. The patient originally described by Stewart and Allison was alive and well 10 years following the removal of the "tumourlet", and a similar history was obtained in the series of cases reported by Whitwell. MacMahon et al. (1967) followed a "tumourlet" by both macroscopic and microscopic examinations over a period of 12 years. At the time of the patient's death the lesion was too small to be seen macroscopically but was still present at the original site in the scarred lung. Their findings support the view that it represents an atypical form of predominantly bronchiolar cell hyperplasia with a very low neoplastic potential.

Unlike true lung cancer peripherally situated "tumourlets" have occurred rather more frequently in women. Tumourlets need to be distinguished from small pulmonary chemodectomas which they may closely resemble in appearance, the latter being unassociated with evidence of lung damage. Similarly multiple peripheral carcinoids are often unassociated with severely damaged and fibrotic lungs.

BRONCHIAL METAPLASIA AND THE EFFECTS OF SMOKING ON THE LARGE BRONCHI

Lindberg (1935), when examining the bronchi from thirty-nine cases of bronchial carcinoma, found that fifteen showed metaplastic change in the non-malignant parts of the bronchial epithelium. This occurred most commonly in association with centrally situated squamous-celled carcinomas but was never seen in adenocarcinomas of the lung.

Later, Black and Ackerman (1952) reported sixty cases of squamous and undifferentiated carcinomas of the bronchus and showed that 22 per cent had carcinomas in situ, and a further 13 per cent showed anaplastic hyperactivity of the bronchial epithelium adjacent to the growth. In fifty-three cases squamous metaplasia of the bronchial epithelium was present. Valentine (1957) found that the incidence of squamous bronchial metaplasia amounted to 36·8 per cent in normal lungs and to 61·9 per cent in those dying of carcinoma of the bronchus.

These findings indicate that metaplastic bronchial changes often precede and occur in association with squamous-celled bronchial cancers.

The recent incrimination of heavy cigarette smoking as probably the most important factor in the genesis of lung cancer has led to a study of the effects of the habit on the respiratory tract. Auerbach et al. (1956, 1957a, b, 1962a, b) and Cunningham and Winstanley (1959) have studied the bronchial epithelium from the lower part of the trachea and the larger bronchi removed from lungs as soon after death as possible. The former removed the bronchial tree after blunt dissection and fixation and divided it transversely into some 200 blocks of tissue from each of which five or six sections were prepared. The results showed a remarkable association between the smoking habits of the individual and the subsequent histological appearances of the bronchial epithelium. The principal changes include: (1) a basal-celled hyperplasia, (2) stratification of the epithelium, (3) squamous-celled metaplasia and (4) carcinoma in situ. In about 6 per cent of the slides from those known to be heavy smokers, unsuspected carcinomas in situ were present. An almost identical proportion of slides prepared from the bronchial tree removed from patients with lung cancer showed carcinomas in situ in areas remote from the invasive growth. The carcinomas *in situ* showed the same features as similar growths in other anatomical sites.

The bronchial epithelium at first may show only a goblet cell metaplasia indicative of chronic catarrhal inflammation. Later foci of squamous-cell metaplasia are of frequent occurrence. Both of these changes are reversible if smoking is discontinued. Continuation of heavy smoking may lead on to the stage of bronchial epithelial dysplasia in which the number of epithelial cells is increased and in which the cells assume a spindle shape and are arranged perpendicularly to the surface. The surface layer of cells may still, however, remain ciliated. Dysplasia is also a reversible state if the cause is removed. If heavy smoking continues the hyperplastic bronchial epithelium loses its resemblance to normal ciliated or squamous epithelium and consists entirely of hyperchromatic cells some of which enlarge and contain more than one nucleus (Fig. 20.7 A, B, C, D). At the same time irregular club-shaped downgrowths of epithelial cells project into the lamina propria but the underlying epithelial basement membrane remains intact and no true infiltration occurs. Similar carcinoma *in situ* changes affect the ducts of the bronchial mucous glands and the gland acini (Fig. 20.8). Carcinoma *in situ* is a multifocal change being found in scattered foci throughout the larger bronchi of both lungs. These findings were confirmed by Chang (1957) with the aid of a new technique which viewed whole fields of bronchial epithelium.

In a further investigation Auerbach *et al.* (1962a) studied 38,621 sections taken from 456 men and 302 women. They found, firstly, that atypical bronchial epithelial changes were more common in male cigarette smokers than in a comparable group of women smokers. Secondly, very few atypical cells were found in the bronchial epithelium of non-smokers of either sex. Thirdly, there were slightly more atypical bronchial epithelial changes in non-smoking women living in urban areas than in a comparable group living in rural areas. Age appeared to make little difference to the number and extent of the epithelial changes in the groups of women non-smokers. Among the group of male smokers, however,

increasing age resulted in a greatly increased number showing atypical changes. Pipe and cigar smokers showed fewer changes than cigarette smokers. Auerbach *et al.* (1962b) studied the bronchial epithelium in a group of men who had discontinued cigarette smoking for periods of between 5 and 45 years before death and compared the findings with those in a group of men who actively smoked until the time of death. Cells with atypical nuclear changes similar to those found in established bronchial cancer were found in 93·2 per cent of the active smokers but in only 6·0 per cent and 1·2 per cent of ex-smokers and non-smokers respectively. The ex-smokers sometimes showed bronchial epithelial cells with disintegrating nuclei surrounded by cells with vesicular nuclei. These findings corroborate the epidemiological studies which have shown a diminishing risk of lung cancer in ex-smokers.

Knudtson (1960) in a study of the effects of smoking on the respiratory epithelium extending from the trachea to the terminal bronchioles found that metaplasia was always more common at points of major bronchial bifurcation in smokers and was present to a much lesser extent in the lungs of urban non-smokers. He furthermore emphasized the importance of bronchial basal-cell hyperplasia and squamous metaplasia as antecedent changes in the development of atypical bronchial proliferative changes.

Although carcinoma *in situ* in other sites, notably the cervix uteri, is recognized as a reversible change, there is reason to believe that if the irritation that causes it is not removed the condition may progress, after a variable and sometimes long period of time, to an infiltrating carcinoma. Woolner *et al.* (1960) followed-up and described the subsequent course of fifteen postoperative cases of carcinoma *in situ* discovered by cytological and other methods of examination following resection of a lung for an undoubted lung cancer. Although some of the tumours they described would be regarded as infiltrating bronchial carcinomas, nine of the patients were still alive and well 4 to 8 years after the original resection. In eleven of the fifteen cases the presence of the tumours was revealed by cytodiagnosis and in one of these cases the sputum had

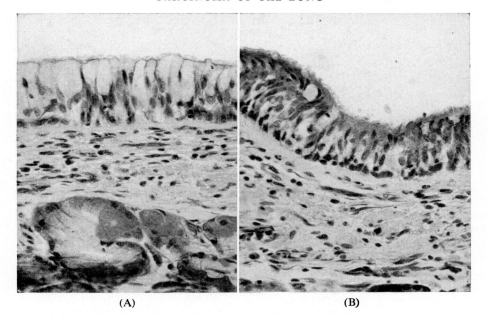

(A)   (B)

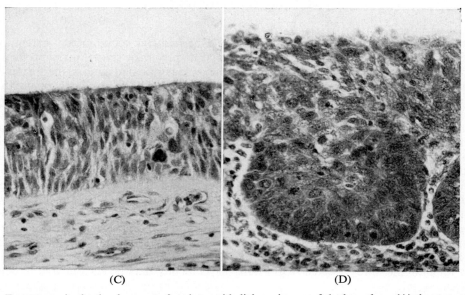

(C)   (D)

FIG. 20.7. Four stages in the development of an intraepithelial carcinoma of the bronchus: (A) the stage of mucosa metaplasia, (B) hyperplasia of the epithelium with loss of mucus-secreting cells (dysplasia), (C) still further epitheliul hyperplasia with large hyperchromatic cells and loss of a well-defined basal layer of cells, and (D) a fully developed intraepithelial carcinoma with downgrowths into the underlying lamina propria and numerous mitoses in the neoplastic epithelium. All these appearances were present in the same bronchus. All ×280 H and E.

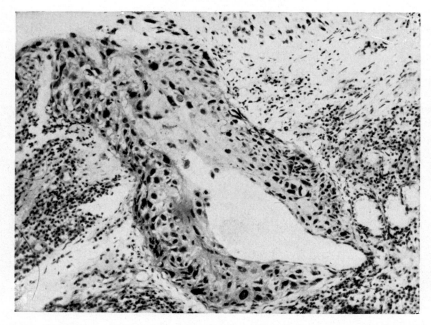

FIG. 20.8. An intraepithelial bronchial carcinoma showing the epithelial changes spreading down the walls of a duct of a bronchial mucous gland. × 280.

been continuously positive for 3 years. Auerbach *et al.* (1957) found carcinomas *in situ* present in 89 per cent of cases in the bronchi of lungs that had been resected for an invasive large bronchus cancer, and a comparable incidence occurred in the remaining lung, the carcinomas *in situ* being distributed evenly throughout the bronchial tree.

Mason and Jordan (1969) drew attention to the papillary nature of some carcinomas *in situ* and to the long-term risk of recurrent tumours. Even when a tumour which started as a carcinoma *in situ* shows very early invasion of the bronchial mucosa, making it a squamous-celled carcinoma, a favourable prognosis may still follow its complete excision. Ryan and McDonald (1956), also found a carcinoma *in situ* in the opposite lung to the one resected for established bronchogenic carcinoma.

Carcinomas *in situ* in the bronchi had been reported previously by Papanicolaou and Koprowska (1951), Umiker and Storey (1952) and Raeburn and Spencer (1953).

The frequent occurrence of carcinoma *in situ* and its precursor stages in the bronchial tree of heavy cigarette smokers and the greater frequency of the condition in the lungs of urban dwellers point to chronic irritation and contact with possible potential carcinogenic substances as the factors mainly responsible for its causation.

A wide spectrum of papillary carcinomas *in situ* exist. At one extreme of the spectrum is the benign bronchial papilloma considered in Chapter 21. At the other extreme is the papillary squamous-celled carcinoma with minimal bronchial invasion which develops from a carcinoma *in situ*, examples of which were described by Mason and Jordan (Figs. 20.9, 20.10, 20.11, 20.12).

Difficulty may be encountered in determining whether some of the intermediate grades of papillary tumours should be classified as benign or malignant, and a series of such tumours was described by Smith and Dexter (1963). They show all grades of histological de-differentiation

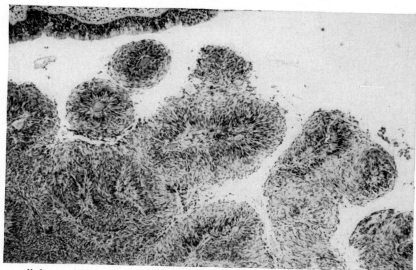

Fig. 20.9. A very cellular papilloma of the bronchus. Multiple tumours were present and these tumours, despite their histological appearances, may behave as invasive tumours. It may be difficult or impossible to be certain of their innocent or malignant nature from microscopical examination. × 40 H and E.

including tumours covered with a transitional type of epithelium examples of which have been seen by the author and by Assor (1971). Others are similar to the Ringertz tumours found in the upper respiratory tract. Smith and Dexter and the author regard many of these polypoid tumours as papillary bronchial carcinomas *in situ* and, like the more usual variety of bronchial carcinoma *in situ*, when they are discovered they show no invasion of the bronchial wall. They are usually multicentric and some of the undetected tumours may have already developed

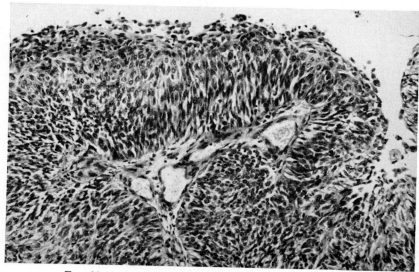

Fig. 20.10. Same tumour as Fig. 20.9. × 100 H and E.

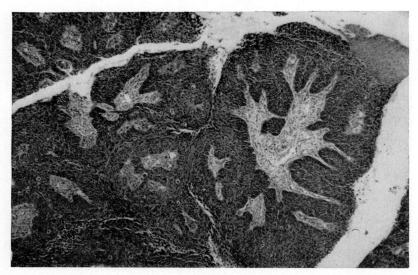

FIG. 20.11. A polypoid carcinoma *in situ*. Although the epithelium is fairly well differentiated it also shows the features in some places of a carcinoma *in situ*. Compare this tumour with those seen in the previous three figures. × 40 H and E.

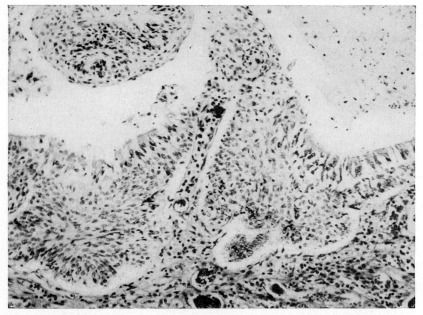

FIG. 20.12. An atypical intraepithelial bronchial proliferation still lined by ciliated columnar epithelium and beginning to form a polypoid type of epithelial tumour. This type of lesion probably represents the first stage in what may develop into a polypoid type of slow-growing but eventually invasive bronchial carcinoma. × 120 H and E.

(Reproduced by courtesy of Drs. C. Raeburn and H. Spencer and the Editor of *Thorax*.)

invasive features in other sites in the bronchial tree.

General Features and Histological Varieties of Lung Cancer

Sixty years ago this tumour showed an almost equal sex incidence but the proportion of male to female cases subsequently rose steadily in most developed countries and reached a 6:1 ratio in Great Britain. The last decade, however, has seen a reversal of this trend especially in England and Wales and the United States where the ratios have changed to 4·36:1 and 4:1, respectively. This has been attributed to the increase of cigarette smoking among the adult female populations.

The age distribution of a large post-mortem series of cases is shown diagrammatically in Fig. 20.13. A feature noted universally is the later age of the peak incidence of this tumour in women though this is becoming less obvious. Although lung cancer is not usually seen under the age of 20, a number of fatal cases were described in children under the age of 17 by Cayley et al. (1951) and Sawyer et al. (1967). The youngest seen by the author was a male aged 18.

Despite the appalling high incidence of lung cancer in Great Britain, recent evidence indicates that the peak-incidence of squamous and oat-cell lung cancer may diminish with the passing of the present over-60 generation and that thereafter it may begin to drop among the oncoming younger generations. The reasons for this are probably numerous but chiefly due to the reducing incidence and popularity of cigarette smoking (Springett, 1966).

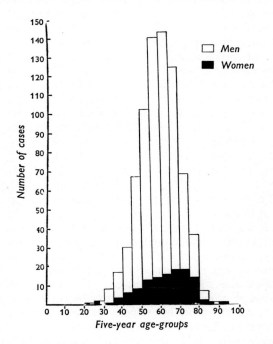

FIG. 20.13. Diagram showing the incidence of lung cancer in 5-year age-groups in both sexes. The peak incidence in women occurs approximately 10 years later than in men.

(Reproduced by courtesy of the Editor of *Quart. J. Med.*)

DISTRIBUTION

Most large series of lung cancers have shown a slight preponderance of growths in the right lung (Duguid, 1927; Simons, 1937; Mason, 1949; Bryson and Spencer, 1951), and in most a greater number of growths have started in the upper lobes, particularly the right upper lobe.

MULTIPLE LUNG CANCERS

In common with many forms of cancer, carcinoma of the lung is sometimes a multicentric disease in which occasionally more than one growth may arise. Billroth and Winiwarter (1887, quoted by Rohwedder and Weatherbee, 1974) established criteria for judging whether double primary growths in a single organ had occurred. The criteria were (a) that the tumours should be found in separate places, (b) that the tumours should be of differing histological patterns and (c) that each primary tumour should produce its own metastases. According to Rohweller and Weatherbee 155 examples have been reported but many of them do not fulfil all the above criteria. Multiple primary lung tumours are particularly associated with radioactive-induced lung tumours (uranium miners), asbestosis and exposure to chromates. The patients are almost always heavy cigarette smokers and the commonest association is a primary squamous-celled cancer with an oat-celled cancer. Examples of multiple lung cancers have been described by Nordmann (1938) in a case of asbestosis, Howard and Williams (1957), Payne *et al.* (1962) and Glennie *et al.* (1964). As already stated multiple carcinomas *in situ* may occur in association with established squamous carcinoma arising in major bronchi. These pre-infiltrative tumours may, if given sufficient time to grow, undergo changes resulting in a true infiltrative growth. The majority of multiple tumours, excluding multifocal alveolar-cell tumours, are of the squamous and oat-celled types, and each may metastasize separately to regional lymphatic glands (Newman and Adkins, 1958). A further unreported case of a double primary lung cancer is shown in Fig. 20.14.

ANATOMICAL LOCALIZATION OF LUNG
CANCER (WITHIN THE LOBE)

Lung cancer may arise in the central, intermediate or peripheral parts of the lung. These three regions of the lung were defined by Walter and Pryce (1955b) as follows: (a) the central zone including the main bronchus to the point of division of the segmental bronchi; (b) the intermediate zone from the outer edge of zone (a) as far as the smallest visible branches arising from a nominate bronchus; and (c) the peripheral zone comprising the remainder of the minute distal bronchi and bronchioles which cannot be seen with the naked eye. The classification employed here will include only two groups, namely (i) the central group comprising (a) and (b) of the previous classification, and (ii) the peripheral group consisting of the previous group (c).

The central growths account for 44·5 per cent of all lung cancers according to Hinson (1958b) and for 50·4 per cent according to Walter and Pryce (1955b). Occasionally the primary tumour may remain minute and may escape detection during a very careful post-mortem examination.

Such growths may be of microscopical dimensions and are only discovered by chance, a finding that particularly applies to some of the peripheral carcinomas. Distribution figures based on resected surgical specimens give a lower figure for peripheral growths. Distribution patterns based on surgical specimens have been criticized on the grounds that such cases are selected for their operability, and tend therefore to include a greater proportion of the earlier and more operable central growths which cause bronchial obstructive symptoms at an early stage in the disease. Radiographical and pathological investigations into the site of origin have shown that many more tumours than are generally recognized start in the periphery of the lung though the definition of "peripheral" differs from that adopted by Walter and Pryce.

The first radiographic evidence pointing to a more peripheral origin of many cancers outside the main, segmental and subsegmental bronchi was provided by Rigler *et al.* (1953). They showed that an undiagnosed peripheral shadow might precede the appearance of a central shadow by a period of 1 to 2 years. The first shadow was subsequently proved to have been a primary peripheral lung cancer and the later appearing central shadows were caused by secondary enlargement of the hilar lymph nodes by secondary growth. In subsequent radiographic studies this early observation has been confirmed in several countries by Rigler (1957, 1964 and 1975), Veeze (1968) and Tala (1967). The proportion of all lung cancers due to peripherally arising growths has been variously stated to be between 50 and 60 per cent. In a survey in which annual radiographic examinations were carried out on several thousand men over the age of 45 years in Philadelphia, and in which the radiographic onset of the disease could be observed, approximately 60 per cent of lung cancers arose in the periphery of the lungs. Meyer and Liebow (1965) also showed by careful examination of specimens of resected lung cancers that almost 20 per cent originated in the periphery of the lung in relation to scarred areas, and 60 per cent arose distal to a segmental ronchus.

FIG. 20.14. A lung showing double primary lung cancers both of squamous-celled type. The two growths were situated in different lobes and appeared to be completely independent one of the other. × two-thirds natural size.

MACROSCOPIC VARIETIES OF GROWTHS

The majority of central cancers arise as warty growths which rapidly invade the underlying wall of the bronchus in which they arise. They soon encircle the affected bronchus leading to obstructive pneumonitis in the area of lung they supply (Fig. 20.15). Some growths completely fill the lumen and grow along the parent bronchus and its immediate branches (Fig. 20.16), whilst others remain as flattened, ulcerated plaques in the wall of the bronchus from which they arise.

Peripheral lung cancers often arise in relation to scars, the centre of the growths being occupied by a dark zone of anthracotic scar tissue whilst the edges of the tumours grow into the surrounding lung presenting an irregular edge (Fig. 20.17), unlike the sharply defined margin of a secondary

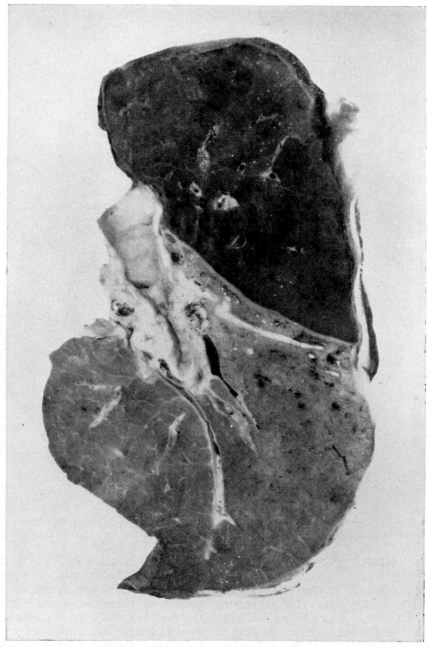

FIG. 20.15. An annular constricting type of bronchial carcinoma (squamous-celled type) showing early changes of obstructive pneumonitis in the lung beyond. Natural size.

devised. It includes the following groups:

		W.H.O. classification
(1a)	Squamous-celled carcinomas	i
(1b)	De-differentiated squamous-celled carcinomas or polygonal cell carcinomas	i
(2)	Small round- or oat-celled carcinomas	ii
(3)	Adenocarcinomas including scar cancers	iii
(4)	Giant-cell carcinomas	iv
(5)	Bronchiolar-alveolar cell carcinomas	vi
(6)	Clear-cell carcinomas	—
(7)	Carcino-sarcomas	v
(8)	Basal-celled carcinoma of the bronchus	—

A statistical relationship has been established between histological groups (1), (2) and (3) and heavy cigarette smoking by Doll *et al.* (1957) and Meyer and Liebow (1965). Adenocarcinomas are mainly found in the periphery of the lung, tending to become more frequent with increasing age but are not so closely related to the previous smoking history as groups (1) and (2). Adenocarcinoma may be associated in women with evidence of hyperoestrinism (Sommers, 1958) and with disturbances of the sex hormones in non-smoking males as reported by Kraus (1957). Group (2) are linked with more endocrine and secretory products than any other variety of lung cancer and these are considered later.

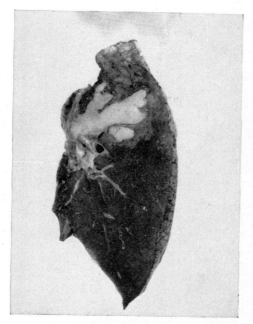

FIG. 20.16. A carcinoma arising in an upper lobe bronchus showing tumour tissue growing into and filling up the bronchial passages. × one-third natural size.

carcinomatous deposit. The appearances of bronchiolar-alveolar cell carcinoma are described separately.

HISTOLOGICAL TYPES OF LUNG CANCER

The histological classification recommended by the W.H.O. Committee included the following varieties of growths:

(i) epidermoid carcinoma; (ii) small-cell anaplastic carcinoma; (iii) adenocarcinoma; (iv) giant-cell carcinoma; (v) combined epidermoid and adenocarcinoma; and (vi) bronchiolar-alveolar carcinoma.

The classification that will be adopted, while similar to that proposed originally by the W.H.O., differs in detail and includes further subvarieties recognized since the W.H.O. classification was

(1a) SQUAMOUS-CELLED CARCINOMA

The proportion of squamous-celled lung cancers has been variously reported, and ranges from 4·9 per cent (Bryson and Spencer) to 61·2 per cent (Fried, 1938). This discrepancy may be largely explained by the more exacting criteria adopted by the former authors who regarded a tumour as a squamous-celled carcinoma only if it showed either intercellular prickles or cell-nests with keratin (Fig. 20.18). Other less exacting classifications have included any tumour remotely resembling a squamous-celled cancer. Using the more liberal definition the figures given by Bryson and Spencer would have been nearer 45 per cent.

Carlisle *et al.* (1951b) laid down four criteria for this group of tumours: (i) the presence of

FIG. 20.17. A specimen of a peripheral adenocarcinoma arising from a scar which may be recognized as the black anthracotic area occupying the centre of the growth. Natural size.
(Reproduced by courtesy of Professor H. Spencer and C. Raeburn and the Editor of *Journal of Pathology and Bacteriology*.)

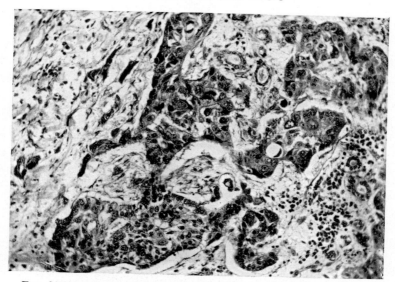

FIG. 20.18. A squamous-celled carcinoma of bronchus. × 140 H and E.

intercellular bridges; (ii) definite cell-nest formation; (iii) "squamatization" and (iv) polarization or whorling arrangement of the cells. Employing these criteria 37·8 per cent of their 849 lung cancers were of the squamous-celled type, whilst Hinson (1958b) found that 56 per cent of 211 cases in his series fell into this category.

Squamous-celled carcinomas occur mainly in the large bronchi and form a large proportion of central growths, although they may also occur in the periphery of the lungs arising in relation to scars; an example of the latter was described by James and Pagel (1944).

These tumours spread mainly in the extra-cartilaginous parts of the bronchial wall invading the adjacent lymph glands directly. Squamous-celled growths are particularly liable to undergo central necrosis and form one variety of abscess cavity; according to Carlisle et al. 13 per cent of the growths undergo this change (Fig. 20.19).

Squamous-celled carcinoma may arise occasionally in the walls of bronchiectatic and abscess cavities; examples of these growths were described by Mallory et al. (1948b), Bryson and Spencer and Konwaler and Reingold (1952) (Fig. 20.20).

The majority of patients with squamous-celled carcinoma present with symptoms and signs attributable to bronchial obstruction. This results in pneumonia and the development of bronchiectasis and lung abscesses in the obstructed portions of the lung (Fig. 20.21).

Untreated, this type of cancer has the longest life expectancy, but apparently successful surgical removal may be followed several years later by new growths which arise from further small undetected carcinomas *in situ* elsewhere in the large bronchi, or from a local or distant recurrence.

(1b) THE POLYGONAL AND UNDIFFERENTIATED LARGE-CELLED CARCINOMAS

The majority of these tumours are anaplastic squamous-celled carcinomas in which the characteristic features of squamous cells are lost (Fig. 20.22); they spread rapidly by both the lymphatics and the blood stream.

They comprise a variable proportion of the total lung cancers depending on the criteria used to define a squamous-celled cancer. Patton

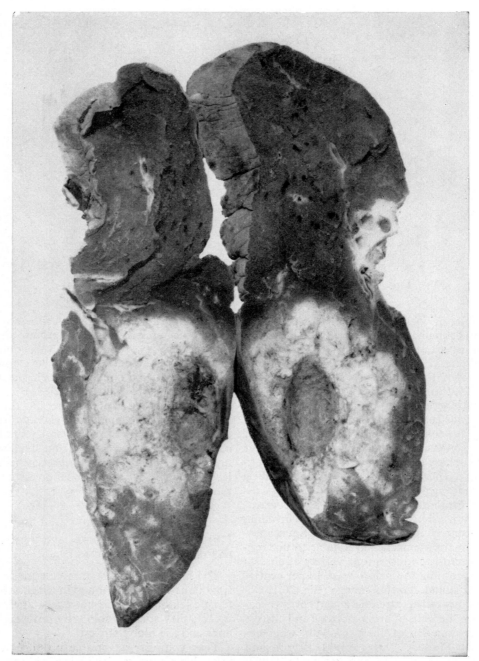

FIG. 20.19. A squamous-celled carcinoma arising in the lung showing cavitation of the primary tumour. This is not uncommonly seen in squamous-celled lung cancers. Nine-tenths natural size.

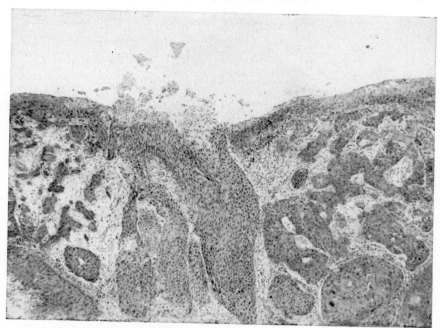

Fig. 20.20. A squamous-celled carcinoma arising in the wall of a bronchiectatic cavity. The growth is well differentiated. × 40 H and E.

(Reproduced by courtesy of the Editor of *Brit. J. Tuberc.*)

et al. (1951) found that 40·2 per cent of their tumours fell within this category when they used the same criteria for a squamous-celled cancer as was employed by Carlisle *et al.*

(2) SMALL ROUND- AND OAT-CELLED CARCINOMA

These tumours were formerly regarded as mediastinal lymphosarcomas until Barnard (1926) showed that they formed a group of bronchial carcinomas.

This group of tumours behave in a very uniform manner both macroscopically and in their microscopical appearances. Bryson and Spencer found that this group comprised 36 per cent of their series of 866 cases of lung cancer, a figure in close agreement with the 37·1 per cent found by Walter and Pryce (1955a). The proportion of these growths found in surgically excised specimens is much lower as they tend to be very malignant tumours and are usually beyond surgical aid when first seen. The primary tumours may sometimes be extremely small and escape detection unless the greatest of care is taken in the examination of the walls of the larger bronchi. These tumours undoubtedly account for a large proportion of undiagnosed primary growths. Death often occurs within 4 months of their first discovery as they grow rapidly to invade the mediastinum and distant lymphatic glands at an early stage (Fig. 20.23).

The majority of oat-celled carcinomas arise in the larger central bronchi and become progressively fewer in the smaller air passages. These tumours are now considered to be the most malignant form of neoplasm derived from the Kultschitzky-type cells present in the bronchial tree and are very closely related to the bronchial carcinoid tumours (Bensch *et al.*, 1968).

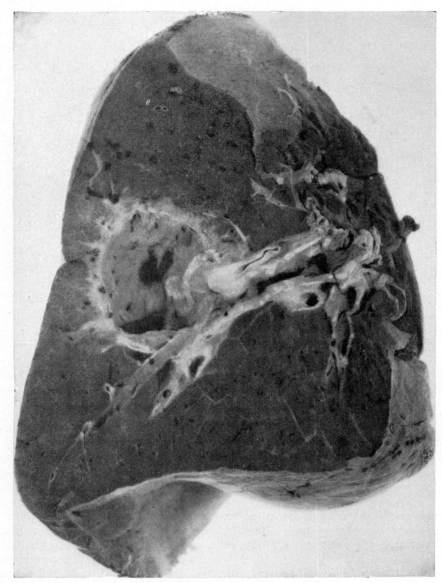

FIG. 20.21. An abscess cavity distal to a fairly small carcinoma of the bronchus of squamous-celled type. Growth is spreading partly around the walls of the cavity which is also partly lined by non-neoplastic squamous epithelium. Natural size.

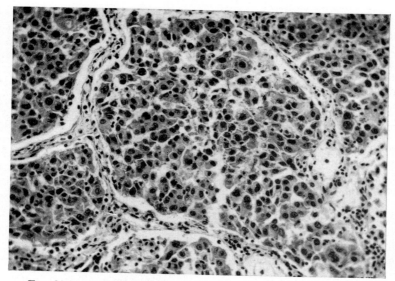

Fig. 20.22. A polygonal-celled carcinoma of the lung. × 190 H and E.

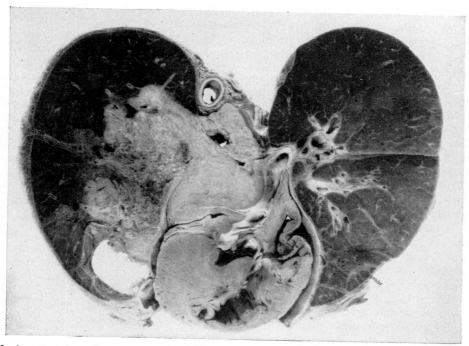

Fig. 20.23. A transverse section of the thorax showing an oat-celled carcinoma invading the hilar region of the lung and involving the mediastinal glands and back of the pericardial sac. The oesophagus is embedded in growth. × one-quarter approx.

Microscopically, this group of lung cancers are most distinctive and the tumour cells are either small round- or oat-shaped often superficially resembling lymphocytes (Fig. 20.24). Occasionally the cells are spindle-shaped and then resemble fibrosarcomas though no collagen or reticulin fibrils are ever produced by the tumour cells (Fig. 20.25). In all the tumour cells there is a very high nuclear : cytoplasmic ratio and the nuclei contain several nucleoli. The cells may be arranged in solid undifferentiated masses, ribbons, rosettes or ductules and the cells lining the latter may secrete a PAS-positive mucinous material. The arrangement of the cells in rosettes and ductules led to the confusion of these tumours by Walter and Pryce (1955a) with adenocarcinomas. There is, however, very little resemblance of the two types of lung cancers and the shape, size and arrangement of the tumour cells in the two types of tumours should prevent any confusion. Islands of larger cells with less dense nuclei may be found interspersed through an oat-celled tumour (Azzopardi, 1959). A striking feature sometimes seen in this group of lung cancers is the presence of darkly staining haematoxyphil vessel walls which may be mistakenly regarded as calcified vessels (Fig. 20.26). As Azzopardi showed, the dark staining was due to DNA and was thought to have arisen from the release of nuclear components by the surrounding tumour cells. Electron microscopic confirmation that the haematoxyphil material was caused by nuclear fragments was provided by Ahmed (1974) who postulated that the absence of DNAases in oat-celled carcinomas might account for the persistence of unchanged DNA.

A few oat-celled carcinomas may pursue a more indolent course but eventually metastasize. Such tumours form an intermediate group both in structure and behaviour between bronchial carcinoids and the very malignant oat-celled carcinomas (Fig. 20.27). Also occasionally the stroma of an oat-cell carcinoma may contain amyloid, confirmed by electron microscopy (Gordon et al., 1973), thus further resembling a bronchial carcinoid tumour.

The small round-celled variety of these tumours may occasionally spread to fill capillaries in several organs with the neoplastic cells and may then closely resemble a leukaemic infiltration. This type of spread was described by Watson (1955) and a further example is shown in Fig. 20.28.

Although oat-cell cancers were previously regarded as a variety of bronchial cancer arising from the germinal layer of cells in the bronchial epithelium, these tumours are now known to be of a different nature and arise from the Kultschitzky-like cells incorporated in the bronchial epithelium and bronchial mucous glands. The genesis and relation of these tumours to bronchial carcinoids is very close and is further considered in Chapter 21. To summarize the similarities:

(a) Unlike the majority of lung cancers which display a variety of histological structure from one part of the tumour to another, the oat-celled carcinoma tends to reproduce the same structure both throughout the primary growth and in its metastases.

(b) All stages in histological structure may be recognized between the benign carcinoid bronchial tumour and an oat-celled carcinoma (Fig. 20.27). Although histological similarities are notoriously unreliable as a means of proving the identity or relationship of different tumours to each other, the resemblances are so striking in this instance that when combined with other evidence they lend further support to the idea of a common origin. The ribbon-like cellular arrangement is common to both varieties of tumour.

(c) The distribution of both carcinoid tumours and oat-celled cancer in the bronchial tree is very similar, both becoming progressively less frequent in the peripheral bronchi.

(d) Both bronchial carcinoid tumours and oat-celled carcinomas alone produce similar endocrine disturbances comprising Cushing's syndrome, the carcinoid syndrome and hyponatraemia with excessive secretion of antidiuretic hormone.

(e) The carcinoid syndrome elsewhere in the body is caused only by tumours of the parocrine (APUD) system of cells which comprises

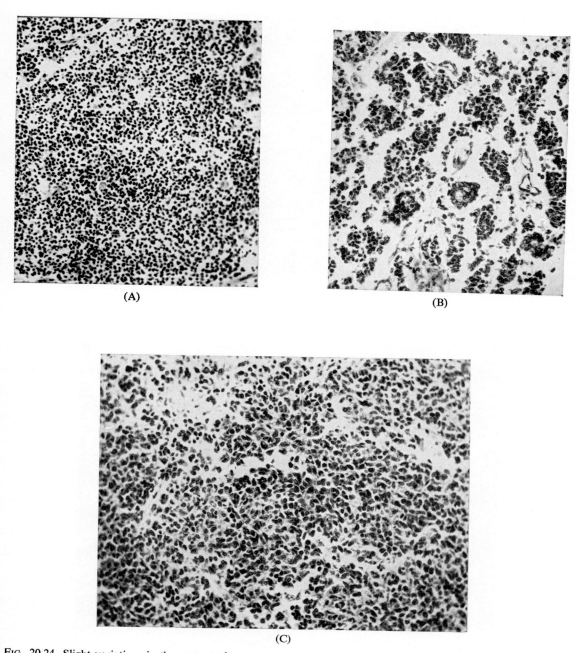

FIG. 20.24. Slight variations in the pattern of oat-celled carcinomas showing in (A) small round-celled type, (B) a tendency to tubule formation, (C) a larger-celled type of growth. × 190 H and E.

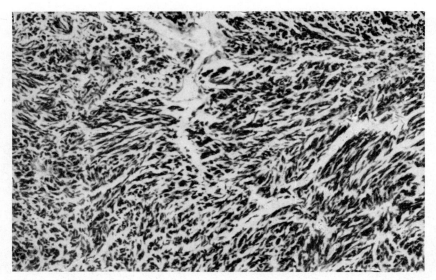

FIG. 20.25. Spindle-celled change in an oat-celled bronchial carcinoma. × 100 H and E.

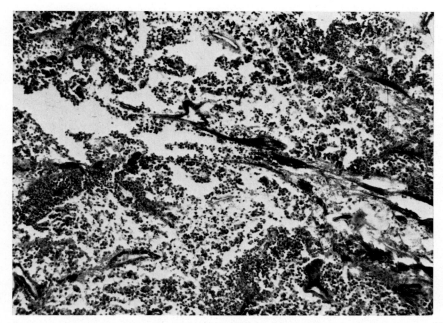

FIG. 20.26. An oat-celled lung cancer showing the walls of blood-vessels (black) impregnated with DNA released from necrotic tumour cells. × 100 H and E.

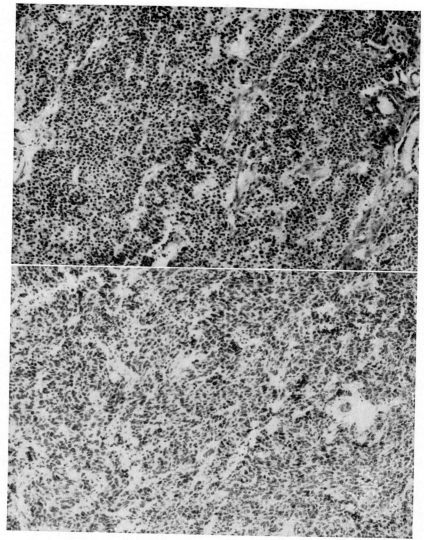

FIG. 20.27. A bronchial carcinoid tumour showing dedifferentiation. In the lower field the microscopical appearances resemble an oat-celled carcinoma.

the argentaffin cells arising in the alimentary canal. The tumour cells in both bronchial carcinoids and oat-celled cancers contain neurosecretory-type cytoplasmic granules when examined by E.M. (Fig. 20.29) (Stoebner et al., 1967; Pariente et al., 1967; Bensch et al., 1968; Hattori et al., 1972).

(f) Search for the earliest stages in development of an oat-celled carcinoma has occasionally revealed minute unsuspected bronchial carcinoid tumours in large bronchi with histological similarity in places to an oat-celled cancer. The tumours appear to be situated almost entirely in the submucosal plane and may, like oat-cell

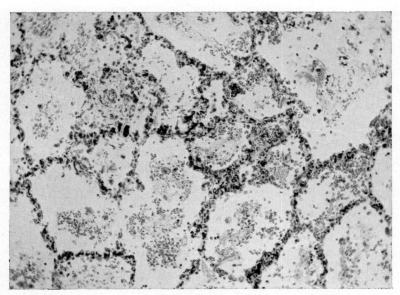

FIG. 20.28. Diffuse capillary infiltration by an oat-celled carcinoma of the bronchus. This type of spread may sometimes simulate a lymphatic leukaemic condition.
(Reproduced by courtesy of the Editor of *Quart. J. Med.*)

cancers, result in little or no disruption of the overlying mucous membrane.

(g) Oat-cell carcinomas have now been reported as arising from parocrine cells situated in other organs than the lung, i.e. pancreas.

(3) ADENOCARCINOMA INCLUDING SCAR CANCERS

A glandular carcinoma may occasionally arise from the mucous glands in the walls of the large bronchi (Fig. 20.30) but the majority of these growths occur in the periphery of the lung. According to Patton *et al.* (1951) they account for 13·2 per cent of lung cancers, and Walter and Pryce (1955a) and Hinson (1958b) found that they comprised 11 and 28·3 per cent of their series respectively. In recent years the incidence of adenocarcinoma appears to be increasing.

These tumours often remain clinically silent until signs and symptoms appear due to distant metastases and provoke a search for the primary growth. Some cases are first discovered during

the course of mass miniature radiographic surveys and others may ultimately be discovered when they invade the chest wall causing local pain and discomfort. In other cases the appearance of metastatic tumours may be the first indication of the existence of a primary growth within the lung.

Because many peripheral adenocarcinomas are related to and often arise in relation to lung damage, scar cancer will be described at length. Although bronchiolar-alveolar carcinomas are closely related to adenocarcinomas because of their distinctive features they are considered separately.

Although in the past the sex incidence of pulmonary adenocarcinoma was regarded as being nearly equal, Meyer and Liebow (1965) found that in their series of 153 resected lung cancer specimens, when an adenocarcinoma arose in relation to chronic interstitial fibrosis in the lung, it invariably occurred in men.

Approximately a third of the pulmonary squamous-celled carcinomas, three-quarters of

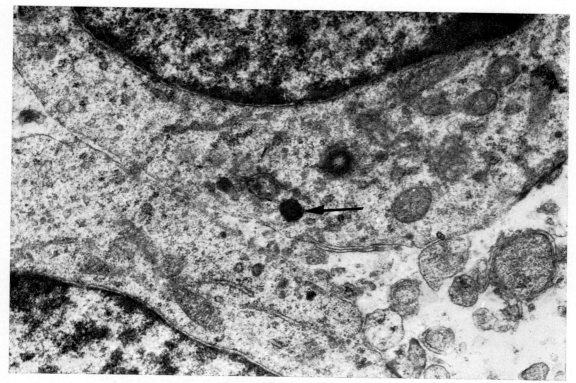

FIG. 20.29. An electron micrograph of an oat-cell lung cancer showing typical neurosecretory-type cytoplasmic body (arrowed). × 22,000.

the adenocarcinomas and one-fifth of the oat-celled tumours arise in the periphery of the lung. Some of the adenocarcinomas are mixed adeno- and squamous-celled cancers and many grow in relation to areas of scarring.

Fibrosis in the lung may follow focal productive lesions such as the pneumoconioses (e.g. silicosis) and tuberculosis, the resulting scar replacing the destroyed lung and extending as chronic interstitial fibrosis into the walls of the surrounding alveolar tissues. Alternatively, fibrosis of the lung may take the form of chronic interstitial fibrosis (honeycomb lung) resultant upon diverse conditions including asbestosis, infective pneumonias and diseases of unknown aetiology such as idiopathic interstitial fibrosis of the lung (IIFL), rheumatoid disease and

scleroderma. Damage to alveolar walls resulting in fixity of the walls results in alveolar epithelial-cell hypertrophy and hyperplasia. Under certain conditions, notably virus infections and exposure to paraquat, the subsequent hyperplasia and metaplasia of the alveolar epithelium that occurs may result in some of the alveoli becoming lined with a multilayered epithelium closely resembling squamous epithelium. Alveolar epithelium is very labile tissue endowed with multipotential properties which include the production of mucous and stratified epithelium and it readily undergoes hyperplasia in response to many forms of injury (Figs. 20.31, 20.32). Although such reactive hyperplasia may be excessive, after an initial burst of activity the growth rate subsides and eventually flattened

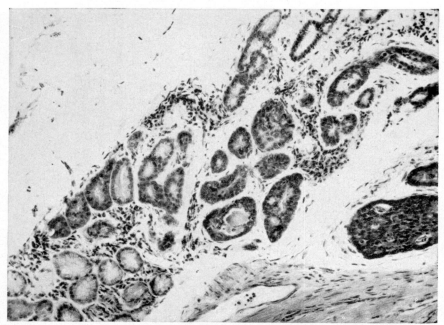

FIG. 20.30. A carcinoma of the bronchus (basisquamous) showing its origin not only from the actual bronchial lining epithelium but also from the subjacent mucous glands. Normal, hyperplastic and neoplastic mucous glands lie in the same field. × 140 H and E.

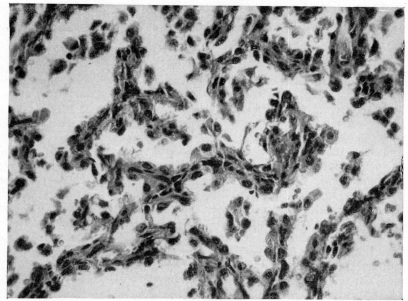

FIG. 20.31. Alveolar cell hyperplasia and hypertrophy found as an incidental lesion in an otherwise normal lung. × 280 H and E.

(Reproduced by courtesy of the late Dr. Sybil Robinson.)

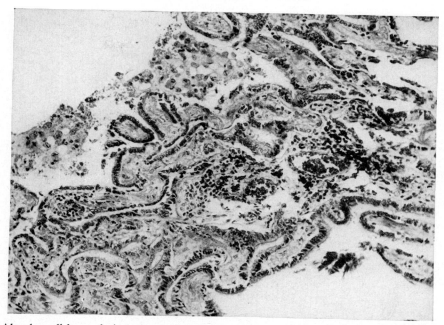

Fig. 20.32. Alveolar cell hyperplasia and metaplasia to a columnar celled type in a nonspecific scar in lung. × 140 H and E.

epithelium-lined spaces result. In a few instances, especially when a condition of chronic interstitial fibrosis occurs, regenerative hyperplasia may become excessive and can occasionally develop malignant changes. Atypical regenerative hyperplasia of alveolar epithelium was first described by Friedländer (1876).

Although both focal lung scars and chronic interstitial alveolar fibrosis (honeycomb lung) can give rise to lung cancer, the latter form of lung injury is much more prone to develop this change. In focal scars, if a cancer develops, it usually commences at the edge of the scarred area where the damage spreads into the surrounding alveoli causing an interstitial lung fibrosis (Fig. 20.33).

Meyer and Liebow in their study of 153 surgically resected lung cancers found that there was evidence of chronic interstitial fibrosis with atypical alveolar epithelial proliferation in 22 per cent. This type of lung fibrosis occurred more commonly in the upper lobes of the lungs

and was associated with the development of pulmonary adenocarcinomas. All the patients who showed both honeycombing of the lung and resulting cancers were males, all were cigarette smokers, and over half gave a history of previous attacks of pneumonia during the preceding 5 years.

A similar smoking history was found by Hinds and Hitchcock (1969) in their series of cases, but nevertheless a greater proportion of patients are non-smokers than in the other common histological varieties of lung cancer.

Lung scar cancers were first described by Friedrich (1939) and Rössle (1943). Raeburn and Spencer (1953) examined a large number of lung scars and found that in a few it was possible to find excessively active bronchiolar and alveolar epithelial hyperplasia and occasionally unsuspected malignant changes. In some scars the intact bronchioles at the edge become hyperplastic, whilst in others alveolar epithelial cells lining unobliterated spaces within the scarred

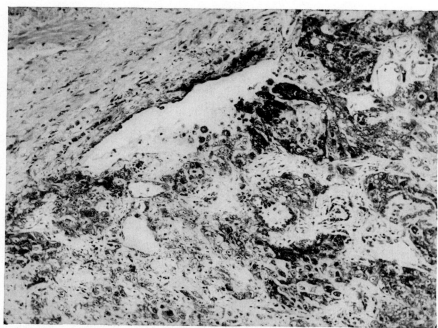

FIG. 20.33. Scar cancer growing around the edge of the tuberculous lesion shown in Fig. 20.36. × 120 H and E.

zone undergo similar changes. The cells in the interior of the scars tend to become undifferentiated, polygonal-shaped or giant cell in form. In a few instances they become indistinguishable from an infiltrating carcinoma. In other scars, particularly those resulting from interstitial fibrosis, the remains of the alveolar spaces become lined by cuboidal epithelium which undergoes further growth filling up the alveolar lumen and developing into a hilic type of carcinoma (Fig. 20.34). The hilic or solid type of carcinoma often consists of nests of swollen, vacuolated cells with ill-defined cell margins. Many such tumours have undoubtedly been confused with polygonal-celled undifferentiated squamous-celled growths. In other tumours the cells grow in lepidic fashion to line the intact alveoli bordering the scar and are identical with a bronchiolar-alveolar-celled carcinoma arising *de novo*.

Similar bronchiolar hyperplastic changes were seen by Montgomery (1942–3) following the repair of experimental wounds inflicted on the lungs of cats, though malignant change was absent.

The same sequence of events—regenerative hyperplasia developing into malignant change— is now well recognized to occur in many sites in the body.

Although described as lung scars, in most instances the scar is composed of elastoid tissue as defined by Gillman *et al.* (1955), who showed that collagen exposed to repeated stretching developed elastic properties, and that such tissue frequently formed the stroma of malignant tumours. Furthermore, the blockage of lymphatic channels by the scar tissue, together with trapping of dust-laden histiocytes already present within the scar, causes a concentration of anthracotic pigment with possible adsorbed pollutant carcinogens. Also cholesterol, already recognized as a carcinogen (Hieger, 1949), is often found within old lung scars, particularly those of tuberculous nature. Tuberculous scars especially were proved to be the site of origin of scar cancers by Themel and Lüders (1955). The author

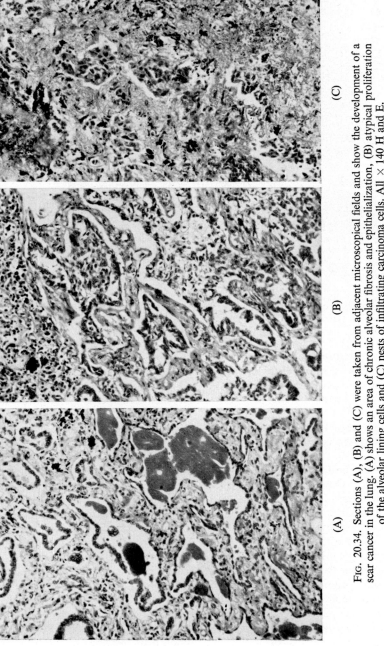

(A) (B) (C)

Fig. 20.34. Sections (A), (B) and (C) were taken from adjacent microscopical fields and show the development of a scar cancer in the lung. (A) shows an area of chronic alveolar fibrosis and epithelialization, (B) atypical proliferation of the alveolar lining cells and (C) nests of infiltrating carcinoma cells. All × 140 H and E.

(Reproduced by courtesy of the Editor of *J. Path. Bact.*)

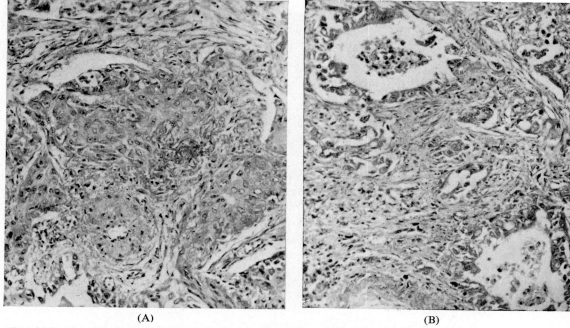

(A)   (B)

FIG. 20.35. A scar cancer in the periphery of the lung showing adjacent microscopical fields. In (A) the growth is of the squamous-celled type, and in (B) the appearances are those of an adenocarcinoma. × 140 H and E.
(Reproduced by courtesy of the Editor of *Thorax*.)

during earlier investigations also saw further cases (Fig. 20.36), and more recently confirmation of the occasional association of the two diseases has been provided by Campbell (1961) in Australia, Yesner (1961) in the U.S.A., Kheifets (1962) in the U.S.S.R., Crinquette (1962) in France and Mody and Poole (1963) in Great Britain.

Fischer (1922) was the first to investigate the epithelial proliferative changes that take place around the edges of lung infarcts involving both bronchiolar and alveolar epithelium. He thought the changes were caused by tissue anoxia. Balo *et al.* (1956), Raeburn and Spencer (1957), Galy *et al.* (1958), Schwarz (1958) and Touraine (1958) have all drawn attention to the hyperplastic and occasional malignant change that may affect bronchiolar and alveolar epithelium around infarcts. Raeburn and Spencer (1957) also showed that scars of varied aetiology may be responsible for the development of malignant

changes, including scars resulting from infarction, bullet wounds (Fig. 20.37), foreign bodies and certain forms of pneumoconiosis particularly asbestosis (Fig. 20.38) and haematite miner's lung. Whitwell *et al.* (1974) found that about 35 per cent of lung cancers associated with asbestosis were adenocarcinomas.

The majority of scar cancers are adenocarcinomas, but a few squamous-celled cancers may originate in this manner (Fig. 20.35), and an example was originally described by James and Pagel (1944). Psammoma bodies may also occasionally occur in the papillary variety of lung adenocarcinoma (Unterman and Reingold, 1972). Although scar cancers frequently remain small growths, they often invade both lymphatic channels and blood-vessels at a very early stage in their development, giving rise to distant metastases before the primary growth can even be detected radiologically. Peripheral growths situated near the apex of the lung or related to the

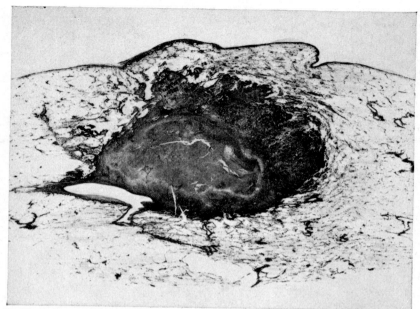

Fig. 20.36. A tuberculous caseous focus near pleura of lung with surrounding early carcinomatous change occurring round half of its periphery. × 4 natural size.
(Reproduced by courtesy of the Editor of *Brit. J. Tuberc.*)

paravertebral sulcus give rise to pleural adhesions and often spread directly through these into the overlying chest wall.

Among the conditions resulting in chronic interstitial fibrosis of the lungs that may cause later development of cancer is Progressive Sys-temic Sclerosis (PSC) (Fig. 20.39), idiopathic interstitial alveolar fibrosis and rheumatoid interstitial lung fibrosis (Fox and Risdon, 1968).

Primary scar cancers may readily be confused with secondary carcinomas which may show a predilection for growing in the vicinity of

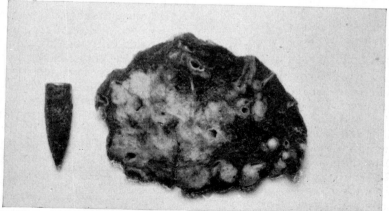

Fig. 20.37. A scar cancer occurring around the site of an embedded bullet. × nine-tenths.
(Reproduced by courtesy of the Editor of *Brit. J. Tuberc.*)

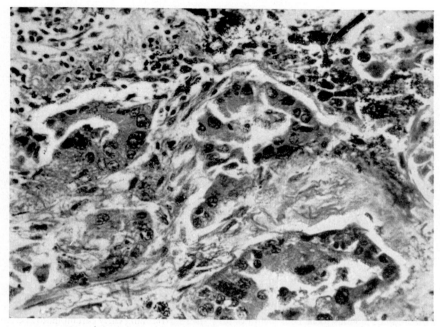

FIG. 20.38. An adenocarcinoma arising in an asbestos lung. The arrow points to an asbestos body. × 280 H and E.

damaged lung. The author has found that primary pancreatic cancer is particularly liable to behave in this fashion and great difficulty may sometimes be experienced in determining the site of the primary growth.

(4) GIANT-CELLED CARCINOMA

Although this variety of lung cancer is considered separately it is thought by some to be a variety of undifferentiated adeno-carcinoma. This type of tumour was first described by Nash and Stout (1958), and Flanagan and Roeckel (1964) and Guillan and Zelman (1966) described further examples. It was also seen by Bryson and Spencer (1951) in their survey of lung cancer (Figs. 20.40, 20.41). These tumours are rare, forming less than 1 per cent of lung cancers. The tumour cells are extremely pleomorphic, multinucleated, and resemble sarcomatous giant cells. The multinucleated cells may contain in-

gested polymorph leucocytes, red blood cells, tumour debris and anthracotic pigment but they show no evidence of striation as in a rhabdomyosarcoma. This tumour should be distinguished from an irradiated squamous-cell carcinoma of bronchus in which bizarre giant tumour cells occur (Fig. 20.42).

Ozello and Stout (1961) confirmed, by means of tissue culture, that the tumour cells were epithelial in origin but lost their cohesion.

The clinical course of the illness due to these growths is very short and the majority of the patients give a history of heavy cigarette smoking. A severe form of haemolytic anaemia may occur. Nash and Stout found that four out of their five cases arose in the periphery of the lung and carried an extremely poor prognosis as they soon invaded the pleura and chest wall to which they became attached.

Although the giant-celled tumours of the lung are still customarily regarded as a variety of lung cancer some of these tumours have proved to be caused by metastatic adrenal cortical

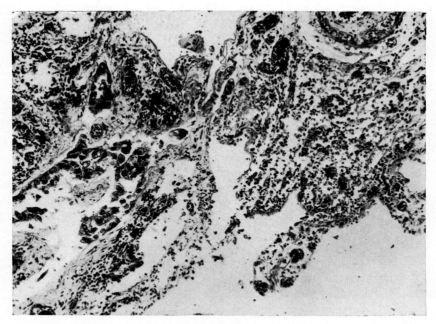

FIG. 20.39. A carcinoma developing in a scleroderma (PSC) lung. × 140 H and E.

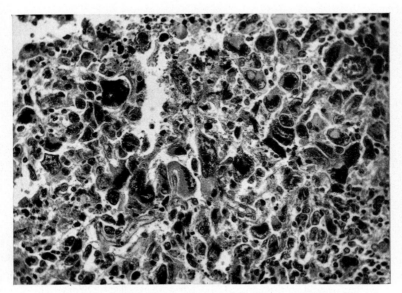

FIG. 20.40. A giant-celled carcinoma of the lung. × 190 H and E.

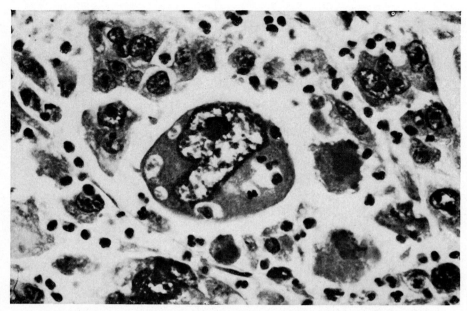

FIG. 20.41. A giant-celled carcinoma of the lung showing the phagocytic properties of the tumour giant cells. × 360 H and E.

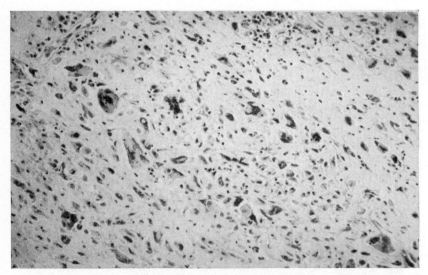

FIG. 20.42. An irradiated squamous-celled carcinoma of the bronchus showing atypical giant-cell forms of tumour cells. Such irradiated tumours should be distinguished from a true giant-celled carcinoma. The irradiated tumour cells are not phagocytic. × 100 H and E.

carcinoma. The phagocytic properties of the tumour cells is more in keeping with a mesenchymal tumour than a classical malignant epithelial tumour unless this was derived from mesenchymal derived epithelium.

Giant-cell carcinoma should be distinguished from giant cells formed following irradiation of anaplastic lung cancers. The latter are atypical tumour cells caused by radiation damage.

Recent electron microscopic studies by Wang *et al.* (1976) support the suggested origin of these tumours from epithelial cell or cells situated in the bronchioles or alveoli. Intranuclear and cytoplasmic inclusions were found in the tumour cells but their nature remained undetermined.

(5) BRONCHIOLAR-ALVEOLAR CELL CARCINOMA (BENIGN AND MALIGNANT PULMONARY ADENOMATOSIS)

Synonyms: Alveolar cell carcinoma, bronchiolar carcinoma, bronchiolar adenomatosis, bronchiolar apocrine tumour

Although the terms benign and malignant pulmonary adenomatosis have frequently been used for these tumours in the past, as a group they exhibit all gradations of malignancy from an apparently localized bronchiolar-alveolar cell carcinoma (benign pulmonary adenomatosis) to metastasizing tumours but all are now regarded as carcinomas.

The outstanding feature of this group of neoplasms is the lepidic growth characteristics of the tumour cells which grow outwards from their origin to line surrounding alveolar walls which they use both as a scaffolding and to serve as stroma.

In the past a great deal of discussion centred around the question of the cell of origin of these growths, some believing they arose from alveolar epithelium while others, who denied the existence of such epithelium, thought they arose from bronchiolar lining cells. This argument is now of historic interest only, since Low (1953) has shown by electron microscopy that alveoli are lined by an attenuated layer of ultramicroscopic epithelium which is continuous with the visible epithelium lining the bronchioles.

Some of the tumour cells have been shown by electron microscopy to show features compatible with an origin from type 2 pneumocytes, i.e. they have microvilli on their free surfaces and lamellated bodies within their cytoplasm (Coalson *et al.*, 1970; Basset *et al.*, 1974). Others hold a contrary view believing the tumour cells showed features linking them with bronchiolar epithelium (Geller and Toker, 1969). The tumour cells may then show E.M. features linking them with mucus-secreting or Clara cells (Greenberg *et al.*, 1974). The argument is academic as the tumours can arise from any of the epithelial cells in and distal to the terminal bronchioles, which as Low (1953) showed form one continuous layer of epithelium. As already stated, included among these cells are Clara cells, and Montes *et al.* (1966) reported a bronchiolar tumour with many of the features of Clara cells including club-shaped cytoplasmic processes on their free surfaces.

In the so-called benign form of the tumour (benign pulmonary adenomatosis) the growth remains confined to the lung and fails to metastasize. Malignant pulmonary adenomatosis (bronchiolar-alveolar cell carcinoma), however, behaves like any other form of lung cancer and spreads rapidly both by the lymphatics and by the blood-stream.

The first description of malignant pulmonary adenomatosis was given by Malassez (1876), who described his tumour as an example of "multiple, nodular, encephaloid carcinomata in the lung". Later Musser (1903) described the first example of a diffuse bronchiolar-alveolar cell carcinoma which involved most of the lung. Since these earliest descriptions many other cases have been described by Helly (1907), Oberndorfer (1930b), Dacie and Hoyle (1942), Neuberger and Geever (1942), Herbut (1944, 1946), Storey *et al.* (1953), Laipply *et al.* (1955) and Spencer and Raeburn (1956).

The incidence of bronchiolar-alveolar cell carcinoma has been estimated at between 3·9 and 5·0 per cent of all lung cancers by Frissell and Knox (1937) and Drymalski *et al.* (1948), though Beaver and Shapiro (1956) consider that this form of lung cancer is now more common, an opinion shared by the author who believes that if peripheral lung scar adenocarcinomas were included in this group of lung cancers as suggested by Geller and Toker, the incidence would be a good deal higher than 5 per cent.

Beaver and Shapiro suggested that the rising incidence of these growths might be related to the widespread use of antibiotic drugs in the treatment of pneumonia which has increased the number of lung scars. Antibiotic drugs by suppressing the leucocyte response in many forms of bacterial pneumonia, as well as destroying the causative organisms, lead to depression of fibrinolytic activity in the consolidated areas; this results in more residual fibrosis. Chronic lung damage, as already stated, probably plays an important part in the aetiology of this group of tumours.

Unlike central lung cancers, the tumours occur equally in both sexes and the average age of onset is 50 years. These growths in both their benign and malignant forms often occur multifocally in one or both lungs, the first case of bilateral tumours being described by Helly and further recent examples having been added by Laipply and Fisher (1949), Peterson and Houghton (1951) and Fisher and Holley (1953). Recently considerable doubt has been cast on the view that these are multifocal tumours. Liebow (1960) reviewed the evidence and claimed that they represented intrapulmonary extensions from a small and usually well-differentiated peripheral primary adenocarcinoma.

For convenience of description the earlier subdivision into local or benign and metastasizing or malignant tumours will be retained.

Local Bronchiolar-alveolar Cell Carcinoma

Synonym: Benign pulmonary adenomatosis

These tumours may be multiple or single, occurring in one or both lungs as greyish-white, often mucus-secreting growths, and when ex-

tensive almost the entire lobe may be replaced by tumour tissue.

Microscopically, the alveoli are lined by tall, columnar and often peg-shaped cells, with basally situated nuclei and eosinophilic cytoplasm (Fig. 20.43). The cells may secrete mucus (Fig. 20.44) or contain lipid and show no mitoses. Although the cells closely resemble normal bronchiolar epithelium, cilia are very rarely present though they were found in the cases described by Fisher and Holley (case 1) and Spencer and Raeburn (1956) (Fig. 20.45). The normal peg-shaped free borders of the cells are readily mistaken for cilia when sections thicker than 5 μm are examined.

Metastasizing Bronchiolar-alveolar Cell Carcinoma

Synonym: Malignant pulmonary adenomatosis

Malignant pulmonary adenomatosis may invade and replace a lobe or most of a lung with greyish-white, often very sticky mucoid growth, which bears a superficial resemblance to Type III pneumococcal pneumonia (Fig. 20.46). In other cases the tumours appear as multiple small, greyish-white, irregular nodules scattered through one or both lungs which are likely to be mistaken for areas of pneumonic consolidation or secondary growth. In addition, growth may be found in the hilar lymph glands and in distant sites.

Microscopically, the growths consist of tall columnar epithelium lining intact but often thickened alveolar walls. The cells are anaplastic, more hyperchromatic than those found in the "benign" lesions and often form papillary projections within the alveolar spaces (Figs. 20.47, 20.48). Outlying fragments of tumour cells are found attached to the alveolar walls in advance of the main edge of the growth (Fig. 20.49), and Hutchison (1952) considered these were airborne metastases. Despite the enormous size reached by many of these growths, the bronchioles are not compressed or destroyed, remaining patent and potentially capable of transmitting air-borne metastases. Storey et al. considered that airborne metastases disseminating through

(A)

(B)

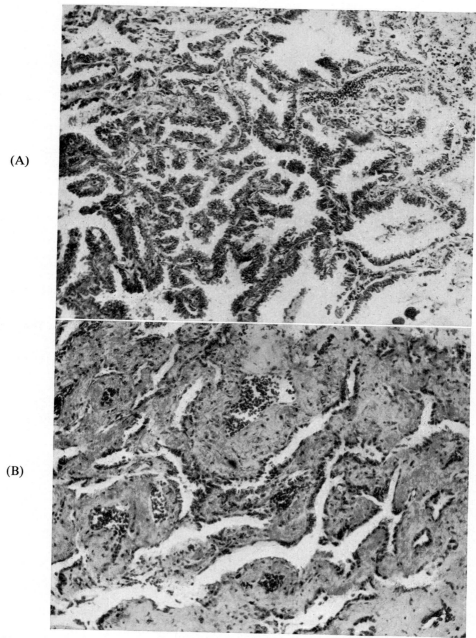

FIG. 20.43. (A) "Benign" pulmonary adenomatosis showing the alveoli lined by peg-shaped cells. The nuclei of the tumour cells are situated near their points of attachment to the alveolar walls. × 140 H and E.
(B) Another part of the same specimen showing a scarred zone of lung tissue with epithelium-lined clefts from which the tumour shown in (A) probably arose. × 140 H and E.
(Reproduced by courtesy of the Editor of *J. Path. Bact.*)

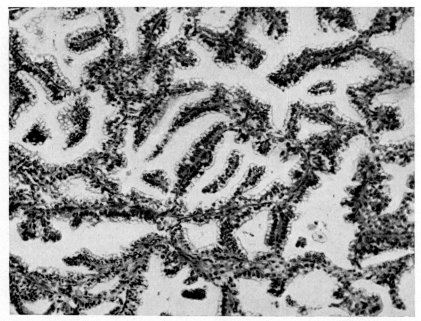

FIG. 20.44. "Benign" pulmonary adenomatosis in which the tumour cells are actively secreting mucus. There was no evidence of distant metastasis in this case. × 140 H and E.

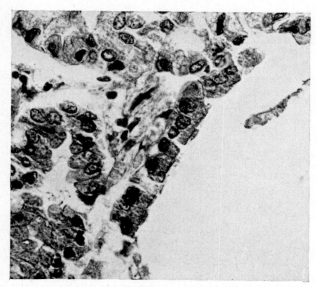

FIG. 20.45. Bronchiolar-alveolar cell carcinoma showing a small collection of ciliated tumour cells. × 280 H and E. (Reproduced by courtesy of the Editor of *J. Path. Bact.*)

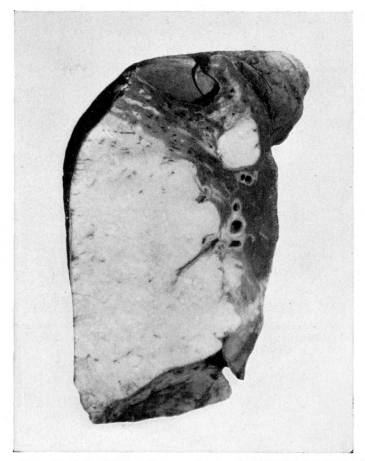

FIG. 20.46. Specimen of lung showing a diffuse malignant pulmonary bronchiolar-alveolar carcinoma replacing most of a lobe of a lung. One-third natural size.

the bronchial tree were responsible for the multi-centric nature of many of these growths.

The alveolar walls may be destroyed whilst in other cases there is a moderate lymphocytic infiltration of the thickened alveolar septa. In the more rapidly growing and anaplastic tumours the cells may grow in a hilic fashion.

The proportion of cases showing metastases was estimated by Neubuerger and Geever and Storey *et al.* to amount to 50 and 54 per cent respectively. Approximately 40 per cent metastasize to the hilar lymph glands, often at a very early stage, and the metastases in these

glands exhibit the same lepidic growth character-istics as the primary tumour.

Blood-borne metastases occur in those sites which are common to all forms of lung cancer.

Before any case of malignant pulmonary adenomatosis can be accepted as such, it is essential to exclude any other source of a primary tumour, as these growths may be exactly simu-lated by secondary pulmonary metastases. Even direct continuity of tumour tissue with normal bronchiolar epithelium does not constitute proof of origin from the bronchiole as was shown by Spencer and Raeburn (1956). Many cases of

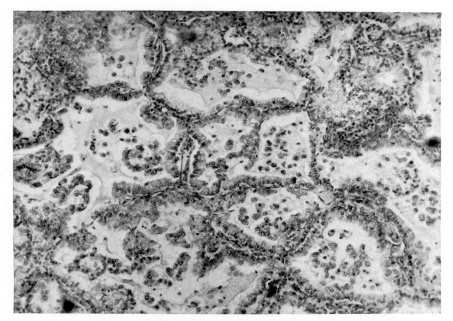

Fig. 20.47. Bronchiolar-alveolar cell carcinoma showing the lepidic nature of the tumour cells many of which are secreting mucus. The individual cells are rather less uniform than those seen in "benign" pulmonary bronchiolar-alveolar cell carcinoma. × 100 H and E.

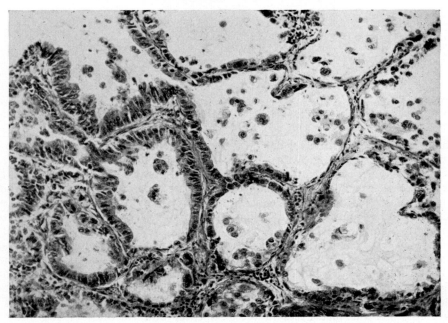

Fig. 20.48. Bronchiolar-alveolar cell carcinoma showing marked hyperchromatism and irregularity in the size of the tumour cells. Some of the tumour cells are detached and free in the alveolar lumen; these may be responsible for air-borne spread. × 140 H and E.

(Reproduced by courtesy of the Editor of *J. Path. Bact.*)

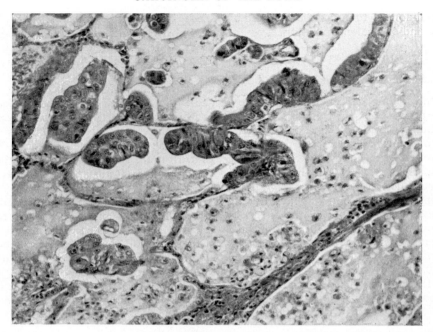

Fig 20.49. Bronchiolar-alveolar cell carcinoma showing clumps of tumour cells attached to the walls of alveoli in advance of the margin of the main growth. × 280 H and E.

bronchiolar-alveolar cell carcinoma diagnosed purely on clinical grounds are suspect, and only the most careful post-mortem examination followed by histological examination can exclude the possibility of the pulmonary growth being a secondary tumour. The diffuse type of bronchiolar-alveolar cell carcinoma which grows by direct continuity to involve most of a lobe or lung is so characteristic in appearance and behaves in a manner so unlike a secondary tumour, that a correct diagnosis may be hazarded on the pulmonary findings alone, provided a clinical examination has excluded a primary growth elsewhere.

More suspect is the multiple, discrete, nodular variety of bronchiolar-alveolar cell carcinoma, an appearance mimicked by secondary metastatic growth from an extrapulmonary source, the most common extrapulmonary carcinomas being situated in the pancreas, colon, breast, stomach and kidney. Occasionally eosinophilic intranuclear inclusions may be found in the tumour cells and have been thought to be possibly of viral nature. They were examined electron microscopically by Kuhn (1972) who found that the nuclear inclusions consisted of membrane-bound vacuoles and tubules and who regarded them as a form of degenerative change.

Basset et al. (1974) who studied the ultrastructure of these tumours found dendritic Langerhans-type cells similar to those normally present in the epidermis. The cells contained in their cytoplasm the characteristic rod-shaped cytoplasmic structures which often resemble tennis rackets in shape. The significance of these findings is uncertain but suggest that Langerhans cells remain to be discovered in the normal lung. Langerhans cells are currently regarded in other sites as cells of mesenchymal origin.

In bronchiolar-alveolar cell carcinoma, many of the malignant cells are exfoliated and can readily be detected by cytological examination of the sputum. A cytological diagnosis may sometimes be made many weeks or even months before the pulmonary tumours are detected by radiographic and clinical examination.

Aetiology and Comparative Pathology

Unlike central lung cancers which are very difficult to produce experimentally, tumours resembling human local and very rarely metastasizing bronchiolar-alveolar cell carcinoma occur spontaneously in many animals including the sheep, horse, guinea-pig, rat and chinchilla.

The disease in sheep is known as *jaagsiekte*, an Afrikaans term meaning "the driving sickness", and was first noticed among flocks driven from one pasturage to another; the affected animals became increasingly dyspnoeic, being forced to drop out of the flock, and eventually died (Fig. 20.50). M'Fadyean (1894, 1920) had applied the term verminous pneumonia to this sheep disease believing it was caused by the

sheep lung worm, whilst Theiler (1918) considered that the disease as seen in South Africa was due to the animals eating the plant *Crotalaria dura*. Dungal (1946) in Iceland considered that *jaagsiekte* was caused by a filterable virus acting together with the sheep lung worm and named it epizootic pulmonary adenomatosis. In Montana in the U.S.A. the disease is known as chronic progressive pneumonia in sheep. It has also been reported in sheep, lambs and goats from the Andean region of Peru by Cubo-Caparo *et al.* (1961) and by Enchev (1963) in sheep in Bulgaria.

Mackay *et al.* (1963) isolated pleuropneumonia-like organisms (PPLO) from three sheep suffering from the disease and found antibodies in the blood of four other sheep suffering from *jaagsiekte*. The importance of this observation awaits further confirmation.

From all observations on the disease in sheep it is apparent that it is related to lung

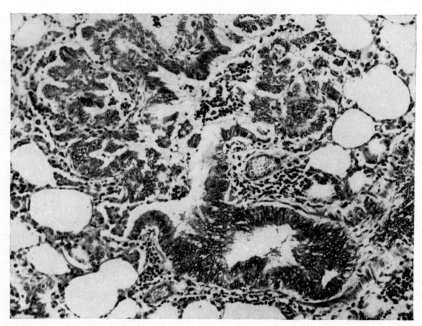

FIG. 20.50. Section showing the pulmonary changes in *jaagsiekte* in a sheep. The condition is very similar to "benign" pulmonary adenomatosis in man. × 140 H and E.

(Reproduced by courtesy of Professor E. Cotchin and the Editor of *J. Path. Bact.*)

damage. Bonne (1939) showed that interstitial pneumonia preceded the alveolar-cell proliferation. The same conclusions were reached concerning the pathogenesis of the human conditions of benign and malignant pulmonary adenomatosis by Drymalski *et al.*, Swan (1949), Beaver and Shapiro (1956), Spencer and Raeburn, Spain (1957) and Roelsen *et al.* (1959), who have all shown that many of these growths arise in relation to areas of scarring in the lung.

Unlike the human disease very few sheep affected with *jaagsiekte* develop a metastasizing form of the disease although rare exceptions in which metastases were found have been reported by Aynaud (1926) and Cubo-Caparo *et al.*

Eversole and Rienhoff (1959) found microscopical foci of alveolar cell hypertrophy and hyperplasia in lungs in which the cells closely resembled those found in benign pulmonary adenomatosis. They regarded them as benign bronchiolar cell tumours related to congenital malformations. Similar microscopical islets of hyperplastic alveolar cells have been found by the author in lungs removed routinely post-mortem and he concurs with the view that they are probably congenital lesions and not true neoplasms. Transient collapse of the lung, and possibly inhaled irritants, together with many other transient causes of lung damage, are all equally capable of evoking these changes, but all appear to cause fixity of the underlying alveolar wall.

(6) CLEAR-CELL CARCINOMA OF THE LUNG

This form of peripheral lung cancer is uncommon but several cases were described by Morgan and Mackenzie (1964) who found that it accounted for 2·8 per cent of a group of 380 lung cancers and was more commonly found in the upper lobes of the lungs.

These tumours macroscopically show few distinguishing features. Microscopically, they consist of large, clear, rounded cells containing small central nuclei (Figs. 20.51, 20.52). The cells are filled with PAS-positive material which fails to stain for glycogen. The tumour cells are arranged in clumps and sheets which

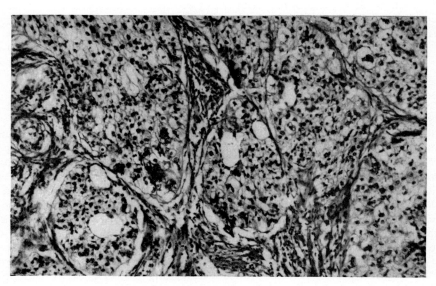

FIG. 20.51. Clear-celled carcinoma of the lung showing central vacuolation in some of the cell masses. × 100 H and E.

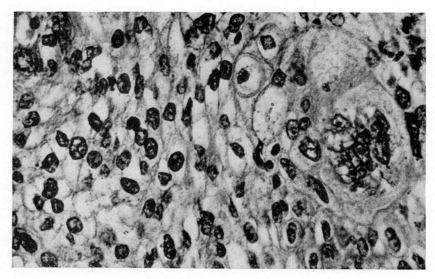

FIG. 20.52. Clear-celled carcinoma. × 400 H and E.

characteristically undergo central necrosis. The tumour is intersected by numerous well-developed septa of fibrous and elastic tissue.

The importance of this group of peripheral lung cancers lies in the fact that they may readily be mistaken for secondary deposits of renal cortical carcinoma.

(7) CARCINO-SARCOMA OF THE LUNG

The existence of such tumours anywhere in the body has been a subject of controversy since Virchow (1863–7) stated that they could result from the occurrence of simultaneous malignant change in stroma and epithelium.

Drury and Stirland (1959) reviewed the whole subject with particular reference to mixed tumours of the respiratory system. Similar tumours were described by Bergmann *et al.* (1951), Peabody (1959), Prive *et al.* (1961), Moore (1961), Chaudhuri (1971) and Diaconita (1975), and although definite proof of the dual nature of the malignant process has not always been obtained beyond doubt, nevertheless most of the tumours that have been described as

carcino-sarcomas of the respiratory tract have shown a close resemblance to each other, displaying certain characteristic features. Proof of a mixed tumour is provided by the production of both carcinomatous and sarcomatous secondary deposits but has seldom been found though described by Chaudhuri. Metastases of differing histology are only likely to be found when carcino-sarcoma results from a collision tumour as opposed to a true carcino-sarcoma. Some reports describe mixed metastases but most have been either carcinomatous or less frequently sarcomatous in nature.

Similar tumours occur in the nasopharynx, larynx and oesophagus and in all the situations where they are found they have identical characteristics. Carcino-sarcomas form a very distinctive group of tumours and are quite distinct from pulmonary blastomas with which they have been confused. They occur mainly in men and the growth rate of the tumours is often slow, Bergmann *et al.* reporting a 6-year survival in an untreated case. The majority grow initially for a long period only by endobronchial extension. The tendency to infiltrate the bronchial wall at the site of attachment is slight and if it does occur spread may only involve the sub-mucosa. If the tumours are left eventual spread to the regional lymph glands occurs.

Microscopically, they consist of well-differentiated squamous-celled carcinoma overlying a cellular, spindle-celled stroma which may contain small foci of bone, osteoid tissue and cartilage (Fig. 20.53). The pattern of reticulin fibres is characteristic of a sarcoma, and the material obtained by bronchoscopy usually consists of the stromal parts of the growth.

These tumours should be distinguished from some squamous-celled bronchial carcinomas in which the cells become spindle-shaped and separated from their neighbours by inter- and intracellular oedema (Fig. 20.54). It may be difficult or impossible to differentiate a true primary carcino-sarcoma from a pulmonary hamartoma which has undergone malignant change.

Bronchoscopic biopsy may result in severe and even fatal haemorrhage and necessitate urgent excision of the lung. The more peripherally arising carcino-sarcomas tend to spread more readily and soon involve the chest wall.

(8) BASAL-CELLED CARCINOMA OF THE BRONCHUS

This form of lung cancer is the least common form encountered. It is a very rare tumour and the microscopical appearances are similar to basal-celled carcinomas of the skin, consisting of nests of tumour cells bounded by an outer palisade layer of cells and surrounded by a characteristic layer of very loosely woven stromal tissue distinct from the normal denser bronchial connective tissue (Fig. 20.55). The bronchial epithelium overlying the tumour shows squamous metaplasia. Very few of these tumours have been recognized and little is known about the ultimate prognosis attaching to them. In the few cases known to the author removal of the tumour has been followed by long survival and distant lymphatic and blood-borne metastases are uncommon.

ENDOCRINE AND OTHER DISTURBANCES ASSOCIATED WITH LUNG CANCER

An increasing number of endocrine disturbances and paramalignant disorders are

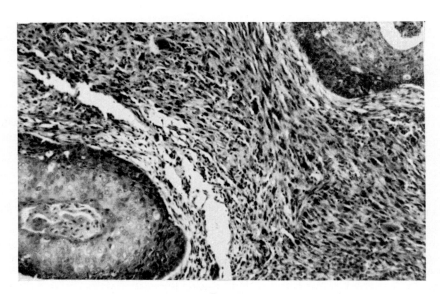

FIG. 20.53. A section of a carcino-sarcoma arising in a major bronchus. Parts of the tumour (not shown) contained bone. × 100 H and E.

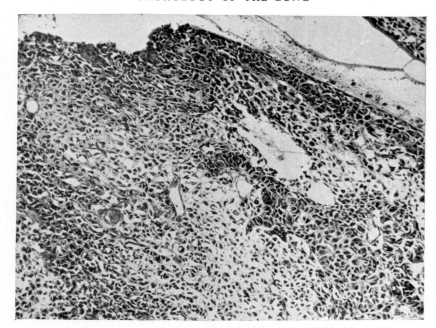

FIG. 20.54. A squamous-celled carcinoma of the bronchus showing intracellular hydropic change. This form of squamous-celled carcinoma is usually slow growing and may be confused with a carcino-sarcoma of bronchus. × 140
H and E.

known to complicate malignant disease in many sites and especially lung cancer. Of the different histological varieties of lung cancer, the oat-celled cancers are most commonly productive of endocrine disturbances.

Ever-increasing evidence now supports the view proposed by Bensch *et al.* (1968) and Hattori *et al.* (1972) that both bronchial carcinoid tumours (described in Chapter 21) and oat-celled cancers of the lung are very closely related and are the benign and malignant tumours respectively derived from bronchial Kultschitzky cells. These cells form part of the parocrine (APUD) system of cells now known to be present throughout the original entodermal canal and which secrete a large number of hormones. Because of their close histogenetic relationship the endocrine disturbances produced by both bronchial carcinoids and oat-celled carcinomas are similar. The following hormones have now been identified and de-

scribed in conjunction with oat-celled cancers:

(a) an adrenocorticotrophic-like (ACTH) substance,
(b) 5-hydroxytryptamine (5-HT),
(c) an anti-diuretic hormone-like substance (arginine vasopressin),
(d) growth hormone-like substance,
(e) β-melanocyte-stimulating hormone (MSH),
(f) insulin and glucagon.

(a) *Cushing's Syndrome due to Release of an ACTH-like Substance*

Brown (1928) described a case of Cushing's syndrome in association with an oat-celled lung cancer and since then and particularly during the past 10 years an increasing number of such cases have been reported. Lung cancer now accounts for more cases of Cushing's

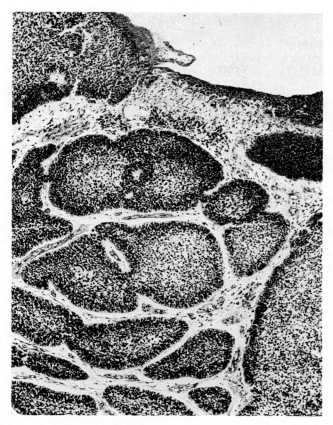

FIG. 20.55. A basal-celled type of bronchial carcinoma. The bronchus is lined by metaplastic squamous epithelium from which the tumour is arising. Note the palisade layer of cells around the tumour masses. × 100. (Reproduced by courtesy of Dr. G. Fitzpatrick, Memorial Hospital, New York.)

syndrome than either pituitary or adrenal adenomas. Parker and Sommers (1956) found that 12·3 per cent of patients dying from respiratory cancer showed hyperplasia of the adrenal glands. Neoplasms arising in the pituitary or adrenal glands were found by Riggs and Sprague (1961) to be responsible for only 5·6 per cent of 232 cases of Cushing's syndrome. The majority of the responsible malignant tumours were oat-celled lung cancers though occasionally primary tumours situated in the pancreas, thymus, thyroid and other sites were responsible.

The author agrees with Azzopardi and Williams (1968) that oat-celled carcinomas or bronchial carcinoids are the only primary lung tumours associated with Cushing's syndrome. Both of these tumours can secrete a corticotrophin-like substance which may be only a portion of the complete molecule but which nevertheless stimulates the adrenals and depresses the normal secretion of ACTH by the pituitary. The hormone-like substance is secreted on a cyclical basis but unlike natural ACTH formed by the anterior pituitary it is not inhibited by dexamethasone (Liddle, 1966). Examples of Cushing's syndrome due to oat-celled lung cancers have been described by Cheng et al. (1958), Vogel et al. (1961), Hymes *et al.* (1962), McDaniel *et al.* (1963), Ross (1966) and Azzopardi *et al.* (1970).

Hymes *et al.* investigated adrenal function in cases of lung cancer and in control groups of patients. They found that the level of plasma 17-hydroxycorticosteroids (17-OH-CS) was raised in two patients with Cushing's syndrome complicating lung cancer, and there was also increased excretion of conjugated 17-OH-CS in the urine. Increased ACTH activity may be demonstrable both in the plasma and in the lung tumour-tissue extracts but pituitary ACTH secretion is reduced. In several cases excessive secretion of aldosterone has also been found.

The incidence of Cushing's syndrome in oat-celled carcinoma has been found to be about 0·5 per cent (Azzopardi *et al.*). Electron microscopic examination has shown that the tumour cells contain more neurosecretory-type cytoplasmic granules than non-secretory tumour cells. Production of ACTH-like substances may be associated with the formation of other hormones as in the case described by Rees *et al.* (1974).

As a consequence of the secretion of an excess of an ACTH-like substance, the adrenal glands may undergo both cortical hyperplasia and hypertrophy and may reach a combined weight of 40 g (normal 12–16 g). Large eosinophilic, slightly vacuolated cells appear in the greatly hyperplastic zona fasciculata. Kennedy and Williams (1964) counted the number of Crooke cells (hyalinized basophil cells) in the anterior pituitary in fatal cases of lung cancer and in a control group of pituitaries taken from persons dying of other forms of cancer. The presence of Crooke cells was considered evidence of an increased amount of circulating cortisone-type hormone. Crooke cells were present in 80 per cent of cases dying from an oat-celled lung cancer but in only 6·6 per cent of cases dying from other forms of cancer.

The onset of Cushing's syndrome in lung cancer usually occurs late in the disease and is of sinister prognostic significance. It is usually associated with hypokalaemia, alkalotic hypochloraemia and the blood sugar level may be raised. Raised levels of plasma and urinary 17-OH-CS may also be found in patients suffering from oat-celled lung cancer who show no clinical evidence of Cushing's syndrome.

(b) *5-Hydroxytryptamine (Serotonin)*
5-HT Production

Harrison *et al.* (1957), Parish *et al.* (1964), Gowenlock *et al.* (1964) and Hattori *et al.* (1968) have all described cases of oat-celled lung cancer associated with the carcinoid syndrome. In two patients there was evidence that the primary tumour or its metastases secreted 5-hydroxytryptamine (5-HT). Hattori *et al.* examined nine cases of oat-celled carcinomas and one bronchial carcinoid tumour and claimed to have shown raised serum and, where estimated, tumour tissue levels of 5-HT. In the case described by Kinloch *et al.* (1965) there was a raised blood-level of 5-HT and an increased urinary excretion of both 5-HT and its metabolic product 5-hydroxyindole acetic acid (5-HIAA). 5-HT was also found in small quantities in the metastatic deposits in the liver. Blood-levels of 5-HT do not, however, correlate well with the presence of the clinical carcinoid syndrome (see Chapter 21). In the case described by Parish *et al.* approximately one-tenth of the tumour cells showed a golden-yellow granular fluorescence in unstained sections. A similar fluorescence is shown by dilute 5-HT solutions. The argentaffin, diazonium and other staining methods commonly employed to selectively stain carcinoid tumours fail to reveal granules similar to those seen in intestinal carcinoids. Although the tumours are rarely argentaffinic they are not infrequently argyrophilic (Gowenlock *et al.*).

(c) *Antidiuretic Hormone (Arginine Vasopressin AVT)*

In 1957 Schwartz *et al.* described a case of lung cancer associated with hyponatraemia, excessive urinary loss of sodium ions and in which the urine was of higher osmolarity than the blood plasma. Since their original description several other examples of this syndrome have been described (Amatruda *et al.*, 1963; Thorn and Transbøl, 1963; Williams, 1963; and Editorial, 1963). In almost every case the lung tumour was of the oat-celled type.

Schwartz *et al.* considered that the changes resulted from abnormal secretion of anti-diuretic hormone (ADH). In no instance has there been evidence of any adrenal or renal structural abnormality and the hyponatraemia, which results from fluid retention and continued excretion of sodium, is improved by restricting the fluid intake.

The cause of the syndrome is now attributed to the secretion of arginine vasopressin (AVP) or neurophysin, the latter being a collective name for polypeptides which bind both vasopressin and oxytocin. Examples of AVP production were reported by Barraclough *et al.* (1966), Hamilton *et al.* (1972), Epstein *et al.* (1973) and Rees *et al.* (1974). The last named have reviewed the previously recorded cases. Removal of the tumour or its treatment with cytotoxic drugs may restore the biochemical disturbances to normal. Ross (1963) considered that aldosterone was reduced following the increase of the plasma volume.

(d) *Growth Hormone and Anti-growth Hormone Inhibitor Production*

Evidence of abnormal growth hormone production in association with oat-celled cancers and bronchial carcinoids is of recent recognition. The acromegaly produced may disappear following successful treatment of the growth. Although it is claimed that immuno-reactive human growth hormone (RHGH) can be detected in cases of oat-celled cancers and Greenberg *et al.* (1972) were able to demonstrate synthesis of this hormone by cultured tumour cells, in some cases the hormone produced by the lung tumour acts by inhibiting the anti-growth hormone substance normally produced in the hypothalamus. This substance controls output of growth hormone by the anterior pituitary. This is evidenced by the enlargement of the pituitary gland which occurs and which disappears following successful treatment of the lung tumour.

In addition to the growth disturbances the patients also display a diabetic type of glucose-tolerance test which returns to normal when the tumour is removed or destroyed.

(e) *β-Melanocyte Stimulating Hormone (MSH)*

Occasionally patients with oat-celled carcinoma or a rapidly growing bronchial carcinoid may become hyperpigmented, especially on exposed surfaces, due to secretion by the tumour of MSH. Examples have been described by Shapiro *et al.* (1971) and Island *et al.* (1965).

(f) *Insulin and Glucagon*

Examples of insulin secreting oat-celled bronchial carcinomas have been reported by Unger *et al.* (1964) and Shames *et al.* (1968). The levels of tissue immunoreactive insulin may often be greater in the secondary metastases than in the primary tumour.

An example of an oat-celled carcinoma producing both insulin and glucagon was reported by Unger *et al.*

Plurihormonal secreting tumours are not common but examples occur in which various combinations of hormones can be identified using radioimmunoassay methods. An example of a plurihormone-producing tumour was described by Rees *et al.* (1974) who also reviewed the literature on similar cases.

Adenocarcinoma of the lung may cause *hyperoestrinism* especially in women (Sommers, 1958, and Weintraub and Rosen, case 1, 1971). These tumours and squamous-cell lung cancers may sometimes cause gynaecomastia in males due to secretion of human somato-mammotropin or gonadotropin which stimulate the interstitial cells of the testis to produce oestrin (Cottrell *et al.*, 1969).

Squamous-celled lung cancers can occasionally cause increased androgen secretion in males (Kennedy and Williams), and they may also cause *hypercalcaemia* (Azzopardi *et al.*). Hypercalcaemia occasionally occurs in conjunction with an oat-celled cancer and is then the result of skeletal destruction by secondary growth. Hypercalcaemia can result from the production of a parathormone-like substance by the tumour cells but more commonly results from massive destruction of the skeleton by metastatic growth. Higgins *et al.* (1974) have also described a patient who developed the nephrotic syndrome in association with an oat-celled lung cancer. It was suggested that this complication followed an immune-complex deposition due to host sensitization to tumour-produced antigenic products. These were possibly related to DNA released from necrotic tumour cells.

Other Para-malignant Conditions

Among the other para-malignant conditions found in association with lung cancer are dermatomyositis (Williams, 1963), acanthosis nigricans (Fox and Gunn, 1965), erythema gyratum repens (Greenberg *et al.*, 1964), leukemoid reactions, bone marrow aplasia (Entwistle *et al.*, 1964) and a variety of neuro-myopathic syndromes (Lennox and Prichard, 1950; Brain and Henson, 1958) which are not caused by metastatic tumour in the central nervous system.

The pathology of the neuromyopathic syndromes due to malignant disease was studied by Brain and Henson and they found that thirty-two of forty-three cases were caused by lung cancer. They found that the nervous symptoms sometimes preceded the discovery of the lung tumour by an interval of up to 3 years. They described five varieties of neuro-muscular lesions which included: (a) loss of Purkinje cells and sometimes of the granular cells in the cerebellum; (b) degeneration of the nerve cells in the cerebellar dentate nucleus and spinal anterior horn cells, coupled with degeneration of the lateral and posterior column

tracts; (c) degeneration of the posterior nerve root ganglion cells; (d) loss of the anterior horn cells; and (e) muscular lesions resembling dermatomyositis. Many of these lesions were associated with cerebral and posterior root ganglion lymphocytic infiltrations. Lennox and Prichard found that 1·7 per cent of 299 patients with lung cancer developed peripheral neuritis.

The size of the lung tumours bears no relationship to the severity of the neural changes and the cause of the neural and muscular lesions remains unknown. It has, however, been postulated that the presence of a tumour allows either superinfection with a virus or gives rise to a metabolite which poisons the nervous tissues.

THE BLOOD-SUPPLY OF LUNG CANCERS

The blood-supply of bronchial carcinomas was studied by Wood and Miller (1938), Cudkowicz and Armstrong (1953) and Wagenvoort and Wagenvoort (1965). It is now generally agreed that the blood-supply to a primary lung cancer is carried by the bronchial arteries which often become tortuous as well as becoming enlarged. Wagenvoort and Wagenvoort found that the smaller muscular branches of the pulmonary artery in the same lobe as the primary growth may show intimal fibrosis and undergo medial hypertrophy. The remainder of the pulmonary arterial tree is unaffected by these changes. They regarded the lesions in the pulmonary arterial branches as being either secondary to the increased blood-flow and consequent rise in blood-pressure resulting from the back flow of blood through broncho-pulmonary arterial anastomoses, or were due to organized thrombi. It is unlikely, however, that organization of thrombi would account for the medial hypertrophy.

The Spread of Lung Cancer

The spread of lung cancer follows the same general principles that govern the spread of cancer from any site, namely spread by direct infiltration, by lymphatic permeation and embolism, and by the blood-stream.

DIRECT SPREAD

The first structure to be involved in the direct spread is the bronchus of origin and the spread within this structure was investigated by Cotton (1959) in surgically excised specimens.

Growth may spread proximally from the main tumour in the bronchial wall both by direct infiltration and by filling the submucosal lymphatic plexus. In its further spread it often involves other neighbouring bronchi leading to their occlusion, and may also directly infiltrate the extrabronchial lymph glands. From these lymph glands growth frequently invades the adjacent pericardial wall both directly and by retrograde lymphatic spread (Fig. 20.56). According to Simpson (1929) 40 per cent of cases eventually show spread of growth to the pericardium.

Other bronchial carcinomas may spread directly along the lumen of the bronchus and its branches leading to bronchial obstruction.

Direct extension within the lung may frequently lead to invasion of branches of the pulmonary veins and to compression of the major divisions of the pulmonary arteries, though the latter structures are seldom infiltrated directly.

Intrapulmonary spread of some tumours may follow an interstitial plane and lead to compression and obliteration of the alveolar spaces (Fig. 20.57).

Other routes of spread are followed by peripherally situated growths. Growths arising in the apex of the lung soon invade the pleura leading to the formation of adhesions and the spread of cancer cells through the newly formed lymphatics to the overlying chest wall. Such growths also invade the perineural lymphatics in the brachial plexus, and the sympathetic chain and ganglia in the root of the neck. This variety of tumour was described by Pancoast (1924 and 1932) and in his second paper he referred to it as a "superior pulmonary sulcus tumour", an unfortunate term lacking any anatomical or embryological basis.

Peripheral lung cancers situated beneath the pleura in the region of the costovertebral sulcus may infiltrate the adjacent bodies of the vertebrae, even reaching the spinal canal through the intervertebral foramina and causing paraplegia. Extension of peripheral lung cancers situated in the lower lobes, such as may be found in asbestosis, lead to infiltration of the diaphragm and paralysis of the phrenic nerves.

LYMPHATIC SPREAD

Ochsner and De Bakey (1941–2) reviewed the extent of lymphatic spread of lung cancer in 1298 cases collected from the literature and found the major groups of glands involved as follows:

(1) Tracheobronchial glands 69·7%,
(2) Abdominal glands   20·7%,
(3) Cervical glands   17·4%,
(4) Retroperitoneal glands 8·1%,
(5) Axillary glands   6·6%·
(6) Peripancreatic glands   6·5%,
(7) Supraclavicular glands   4·2%.

The extent of the lymphatic spread of lung cancer depends partly on the histological variety of the growth and partly on the site of origin within the lung.

Most central lung cancers spread to involve the hilar lymph glands though the rate at which this occurs varies with the histological type of growth. The glandular involvement is often the result of direct extension of the growth rather than lymphatic spread. In surgically excised specimens, 100 per cent of oat-celled cancers and 34 per cent of squamous-celled growths had spread to the hilar lymph glands (Nohl, 1956); the remaining histological types of growth occupy an intermediate position. In post-mortem material, 97 per cent of all cases show metastases in the hilar lymph glands. Onuigbo (1963) investigated the extent of the spread of the growth in the hilar lymphatic glands in 100 cases of lung cancer. He found that in 84 per cent spread of the tumour occurred only or predominantly to the ipsilateral groups of mediastinal (hilar)

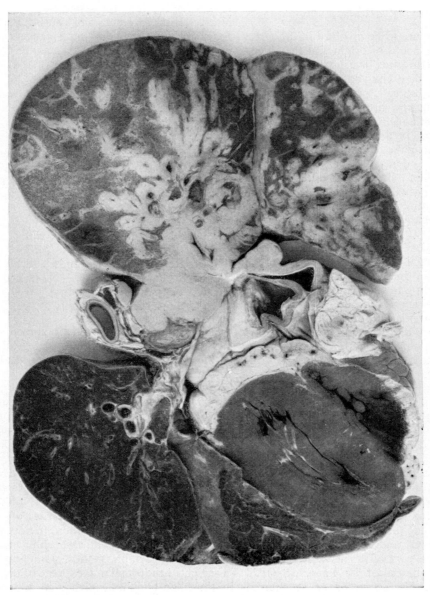

FIG. 20.56. A transverse section of the thoracic contents showing diffuse periarterial and peribronchial lymphatic infiltration by carcinomatous tissue, the primary growth being sited in the main bronchus. Note the enlarged malignant mediastinal lymph glands causing compression of the oesophagus and invasion of the pericardium. Approx. one-quarter size.

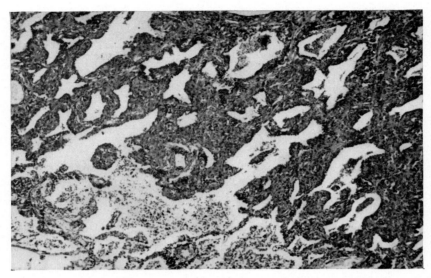

FIG. 20.57. Illustrates the interstitial mode of spread of a primary lung cancer at the edge of an advancing growth. ×100 H and E.

lymph glands, and only in 13 per cent were the contralateral lymph glands involved. In almost every instance one or more lymphatic glands contained metastases.

The later routes of spread in the hilar and mediastinal lymph glands have been studied by Borrie (1952) and Nohl, both of whom found that growths arising in the upper lobe of the right lung metastasized to the lymph glands lying in relation to the origin of the eparterial bronchus. In addition metastases may appear in the lymph glands situated between the middle lobe and the right eparterial bronchus and in the gland situated about the origin of the apical segmental bronchus to the lower lobe. From both of these glands further upward lymphatic spread occurs to the right paratracheal and anterior tracheal groups of glands, the inferior tracheobronchial glands often escaping involvement by growths arising in the right lung (Fig. 20.59).

Tumours arising in the right middle and right lower lobes spread to the glands situated at the root of the middle lobe bronchus and the lower lobe apical segmental bronchus, and the further upward spread follows the same course as that taken by growths arising in the upper lobe (Fig. 20.60).

The lymph glands situated about the root of the middle lobar bronchus and the apical segmental bronchus to the lower lobe may be involved in the spread of growths from all the lobes of the right lung and are therefore of great surgical importance. As they drain all lobes of the right lung they were referred to by Borrie as the "lymphatic sump" of the right lung.

Growths arising in the upper lobe of the left lung metastasize to the glands situated at the root of the left upper lobe bronchus and to the subaortic lymph glands. In addition they metastasize to lymph glands situated between the lingular segmental bronchus and the main left upper lobe bronchus, and a gland situated at the root of the apical segmental bronchus to the lower lobe. From these two glands upward spread to the inferior tracheobronchial glands occurs and eventually spread takes place from the inferior tracheobronchial (subcarinal) glands to the right paratracheal glands.

Tumours arising in the lower lobe of the left lung follow a similar path. The glands, situated at the root of the lingular and apical segmental

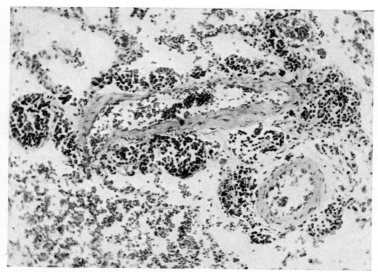

FIG. 20.58. Perivascular lymphatic infiltration in the lung by a small undifferentiated oat-celled carcinoma. × 140 H and E.

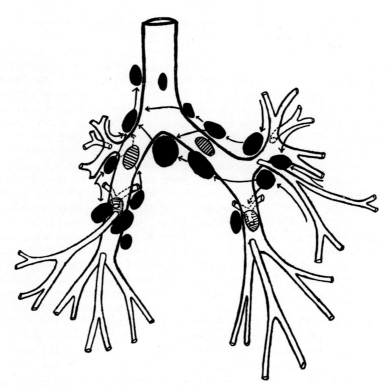

FIG. 20.59. Diagram showing the route of lymphatic glandular spread of lung cancers arising in the upper lobes of the lungs. For key to the names of principal lymph glands and segmental bronchi see Fig. 2.1.

bronchi, drain both lobes of the left lung and are therefore of considerable surgical importance, constituting the "lymphatic sump" of the left lung.

In addition to the pathways outlined above growths arising in the lower lobes not infrequently spread to the paraoesophageal (posterior mediastinal) glands, and growths situated near the anterior surface of the lungs may involve the paraphrenic (anterior mediastinal) glands.

Although the above description applies to the lymphatic spread of the majority of lung cancers, a few growths situated at the apices of the lungs and beneath the pleural surface may spread to the chest wall through newly developed lymphatic channels which arise in pleural adhesions. This may lead to early invasion of the supraclavicular and axillary groups of glands through lymphatic channels which traverse the chest wall to drain into the latter groups of glands, as described by Rouvière (1939).

According to both Bignall and Moon (1955) and Nohl, growths arising in the lower lobes tend to spread more widely by the lymphatic system than those situated elsewhere in the lung. Although in most cases the tumour spreads in order from one group of glands to another, in some cases several groups may be "skipped", only for growth to appear in a more distantly situated gland.

The later lymphatic spread involves the supraclavicular and upper deep cervical chain of glands and spread through the diaphragm involves the large group of para-aortic glands lying around the coeliac axis. The lymphatic pathways connecting the lower lobes of both lungs to the lymphatic glands in the region of the coeliac axis and on the posterior wall of the abdomen were investigated by Meyer (1958). He found that direct lymphatic channels connected the lower lobe of the right lung to the coeliac lymph glands, but that the pathways connecting the lower lobe on the left side were more devious and followed the oesophagus. The diaphragmatic lymphatics form a link between the thoracic and upper abdominal lymphatic glands. Infiltration of the thoracic duct was reported by Young (1956).

Lymphatic spread is a centrifugal process, the most distant nodes in the abdomen and neck being the least commonly affected; this confirms the opinion of Fried (1932) who stated that "the frequency of metastases in bronchiogenic cancer in man diminishes with the distance from the central axis".

Distant lymphatic spread may be responsible for some adrenal and renal secondary growths as often the larger metastasis or the only one present occurs on the same side as the primary tumour in the lung (Onuigbo, 1957 and 1958). Furthermore, Onuigbo has stated blood-borne metastases to these glands might be expected to lodge in the cortical capillaries whereas the majority of secondary deposits occur in the medulla.

Following obstruction by growth of the central hilar groups of lymph glands, retrograde lymphatic spread of tumour occurs within the lung filling the perivascular and peribronchial lymphatic channels (Fig. 20.58) and encasing these structures with growth. Eventually tumour cells reach the subpleural lymphatic plexus, and when this happens the growth-filled lymphatic plexus is readily seen on the surface of the lung (Fig. 20.61). Widespread lymphatic permeation may be responsible for some of the pleural effusions encountered in this disease.

During the course of the lymphatic spread, growth may involve a lymph gland situated behind and to the inner side of the superior vena cava immediately below the point of entry of the azygos vein. Growth may later spread from this gland into the wall of the superior vena cava; the latter being tethered by the azygos vein rapidly becomes occluded leading to obstruction of the venous return. Szur and Bromley (1956) found that 14·6 per cent of all cases of carcinoma of the lung developed superior vena caval obstruction, and that lung cancer was the commonest cause of this condition today. Other organs which may become involved by lymphatic spread include the lower poles of the thyroid gland following retrograde spread of tumour cells from the paratracheal group of glands.

Reference has already been made to submucosal lymphatic spread in the large bronchi from a primary tumour arising in these structures.

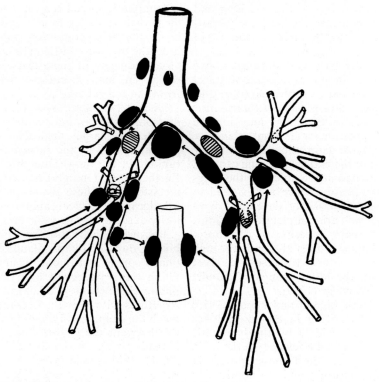

FIG. 20.60. Diagram showing the route of lymphatic glandular spread of lung cancers arising in both lower lobes and the right middle lobe. For key to the names of principal lymph glands and segmental bronchi see Fig. 2.1.

Peripherally situated lung cancers may spread rapidly by the lymphatic system to the glands lying outside the lobar bronchi, and subsequent retrograde spread involves the submucosal lymphatic plexus in the walls of these adjacent bronchi. When this occurs the secondary growth in the bronchus may very closely resemble a primary growth for which it is readily mistaken (Spencer, 1954). Rigler *et al.* (1953) and Rigler (1957, 1964 and 1975) observed the growth of peripheral lung cancers radiographically and found that in some cases the appearance of the secondary metastases simulated a primary central lung cancer. Kramer and Som (1936) also found that bronchoscopy tended to give a high figure for primary main and lobar bronchial growths compared with the actual number found subsequently at autopsy. This was probably explained by the fact that some of the growths considered to be primary tumours in the large bronchi were in fact secondary growths, as explained above.

A further possible and likely route of tumour spread to the hilar glands is by the blood-stream. Goldmann (1907) showed that injection of the bronchial veins also filled the adjacent extrabronchial lymph nodes, a finding confirmed by the author. This provides a possible alternative route by which a primary central bronchial cancer may spread to the hilar glands.

Extension of lymphatic tumour permeation may occasionally involve the liver and kidneys though both organs are usually involved by blood-stream dissemination.

Fig. 20.61. Diffuse carcinomatous infiltration of the subpleural lymphatics by a growth arising in a major bronchus.

BLOOD-BORNE METASTASES

No other form of cancer can disseminate so widely as lung cancer. Situated in close relation to the low-pressure pulmonary veins the opportunities for dissemination are only limited by the unsuitability of certain organs to provide a "soil" for tumour emboli to grow.

Wolf (1895) first described the invasion of pulmonary veins by lung cancer and in recent years the subject has been re-examined in an attempt to explain the unsatisfactory results of surgical treatment. Ballantyne et al. (1957) examined fifty-nine lungs or lobes of lung that had been excised for lung cancer, and found that fifty-two (88 per cent) showed invasion of the pulmonary veins. In ten, branches of the pulmonary artery had also been invaded (Fig. 20.62). Occasionally, a lung cancer may invade a major branch of a pulmonary vein, filling its lumen and extending along it as far as the left atrium (Fig. 20.63). Surgical removal of

FIG. 20.62. Invasion of a major branch of a left pulmonary artery by lung cancer. Natural size.

such a lung may result in the detachment of the intravenous extension which occasionally has resulted in serious embolic complications (Aylwin, 1951). To reduce the risk of detachment of the growth and subsequent embolic complications, it is recommended that during a pneumonectomy the section and ligation of the major pulmonary vessels should be performed within the pericardium.

Several attempts have been made to detect circulating malignant cells in the blood of patients during resection of a lung for lung cancer. Kangas and Viikari (1960) detected free circulating cancer cells in blood collected from the pulmonary veins in 30 per cent of patients at operation. Robinson *et al.* (1963) found single circulating malignant cells in samples of brachial artery blood in about 9 per cent of lung cancer patients during resection of the tumour. They considered this lower figure was due to the fact that the arterial blood has been derived from blood which had flowed through both the tumorous and non-tumorous portions

of the lungs. Some of the released tumour cells pass through the alveolar capillary bed into the systemic arterial circulation. Such showers of cancer cells which are released during the operative removal of a lung cancer are probably largely responsible for the crop of blood-borne secondary tumours that so often develop 6 to 12 months later. The advisability of giving pre-operatively alkylating or similar drugs to lessen the risk of survival of released cancer cells should be considered. Oat-celled cancers tend to spread most commonly in this manner and they are followed by adenocarcinomas and squamous-celled cancers.

Branches of the pulmonary veins are not infrequently invaded by tumour and may undergo thrombosis due to compression of the vein by extramural growth and Simpson (1929) found about a fifth of his series of 139 lung cancers ultimately showed this condition.

Invasion of branches of the pulmonary arteries, though less common, was found to have occurred in almost 16·5 per cent of specimens of resected

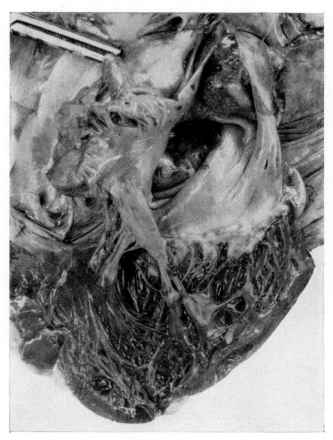

FIG. 20.63. The left ventricle and auricle opened to show a mass of carcinomatous tissue growing by continuity down the lumen of a pulmonary vein. The primary growth arose in a bronchus in the left lung. Three-quarters natural size.

lung cancers by Ballantyne *et al.*, when their tumours were examined microscopically. Pryce and Walter (1960) also claimed to have observed macroscopic invasion of large and medium-sized branches of the pulmonary artery in 43 per cent of 183 surgically resected specimens. Invasion and subsequent filling of a major branch of a pulmonary artery with growth occasionally results in extensive lung infarctions, an occurrence which is more likely to occur if there are coincident changes due to obstructive pneumonitis. The oat-celled and adenocarcinomas of the lung are mainly responsible for arterial invasion. Stevenson and Reid (1957)

drew attention to the compression and rarely invasion of the pulmonary artery that may occur following spread of a growth to the hilar glands. Bronchospirometry reveals a failure to take up oxygen although ventilation of the affected lung may still be unimpaired. Rarely massive and fatal haemoptysis may follow invasion of a lobar branch of a pulmonary artery (Fig. 20.64).

Invasion of the pulmonary blood-vessels, especially the veins, is associated with a very poor prognosis and Johnson (1957) found that only 6 per cent of such patients were alive after 5 years, whilst approximately 75 per cent without invasion of blood-vessels survived this period.

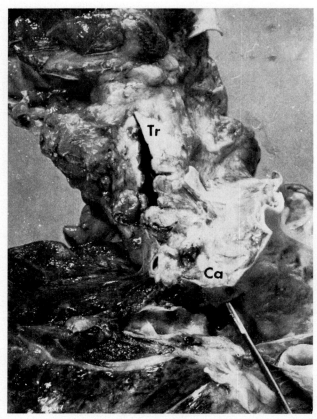

FIG. 20.64. The right pulmonary artery exposed to show a rupture of its wall due to an infiltrating carcinoma (Ca). The probe enters the hole in the arterial wall and the tip can be seen within the trachea (Tr). The patient died from a massive sudden haemoptysis.

These figures were in close agreement with those of Collier *et al.* (1958) who found that three-quarters of all patients with hilar lymph gland involvement also showed pulmonary vascular invasion if this was sought. Hinson and Nohl (1960), after filling the pulmonary veins of lungs resected for lung cancer, found that in 92 per cent malignant cell invasion of the smaller branches had occurred. They concluded that only invasion of the larger branches of the pulmonary veins was associated with a subsequent poor prognosis and that invasion of the small branches probably occurred in nearly 100 per cent of all lung cancers.

The most malignant growths, the oat-celled carcinomas, all show vascular involvement when first discovered, and the peripheral adenocarcinomas are notoriously liable to spread in the same way at a very early stage in their development. The well-differentiated squamous-celled, central carcinomas may only invade pulmonary blood-vessels at a late stage which probably partly accounts for their better prognosis.

Many authors have analysed series of cases of lung cancer to determine the incidence of blood-borne metastases in various organs, but only three of the larger series will be quoted (Table 20.3).

TABLE 20.3

Author and no. of cases analysed	Liver %	Adrenals %	Brain %	Skeleton %	Kidneys %	Spleen %
Ochsner and De Bakey (1941–2), 3047 cases (collected from literature)	33·3	20·3	16·5	21·3	17·5	—
Galluzi and Payne (1956), 741 cases	39·0	33·0	26·0†	15·0	15·0	5·0
Spencer (1954), 1000 cases	38·5	26·4	18·4	15·5	14·3	4·6

The figures quoted for the skeleton are approximate only, as complete examination of the skeleton at autopsy is not possible and only the accessible bones are likely to be examined.

† This figure was obtained from those cases in which the brain was examined, and does not represent the whole series of cases.

The liver was the organ most frequently involved in all three series, and Cameron (1954) discussed the possible reasons for the frequent involvement of this tissue in the metastatic spread of many varieties of cancer. The size of the secondary deposits of growth may reach enormous proportions and lead to the early development of jaundice due to intrahepatic bile-duct obstruction by the metastases, a subject further discussed by Berkowitz et al. (1942). Later tertiary spread from the liver metastases into the branches of the hepatic veins may lead to dissemination of growth to both lungs.

Although adrenal metastases are usually regarded as being blood-borne, the likelihood of lymphatic dissemination to these glands has recently been emphasized by Onuigbo (1957). The ipsilateral adrenal gland is more likely to be involved first when the growth reaches it by the lymphatic pathways. In many cases the first evidence of an adrenal metastasis is found in the medullary veins and not in the cortical capillaries where it might be expected to lodge and grow if it was blood-borne. Despite the enormous size attained by some adrenal metastases and the almost complete replacement of the glands that occurs, Addison's disease is very rarely encountered.

Metastatic spread to the brain and pituitary is common and the incidence rises in direct proportion to the care exercised in the examination. Many of the metastases are single and they are usually situated in the cerebral hemispheres.

Dissemination in most of the cases almost certainly occurs by the pulmonary veins and the great systemic arteries, but Wack et al. (1958) have shown that in dogs there is a free anastomosis between the pulmonary and vertebral veins. If such a communication is of importance in man, it might provide an alternative route by which metastases could reach the brain stem. The high figures for brain metastases quoted by some hospitals admitting only patients with nervous diseases are unrepresentative of the overall incidence, as such hospitals tend to attract patients with symptoms due to cerebral tumours.

The low figure found for metastases in the kidney is notable in view of the very great blood-flow to this organ and may be attributable to the disruptive influence of this factor. In animal experiments tumour emboli composed of clumps of cells are more likely to grow and give rise to metastases than individual malignant cells. Squamous-celled carcinomas seem more liable than the other varieties of lung cancer to give rise to secondary growths in this organ.

Although the spleen is generally considered an uncommon organ for secondary metastases, careful search often reveals small clumps of malignant cells as well as macroscopic deposits in other cases. Very occasionally the small round- and oat-celled cancers may diffusely infiltrate the pulp and resemble a leukaemic infiltration of this organ. The same change is also encountered in the liver.

Among the other less common sites for metastases are included the myocardium, pancreas, ovary, intestine and skin. The author has seen only one secondary deposit in the body of the testis and that from an oat-celled bronchial carcinoma.

According to Mead (1932–3) the commonest primary tumour responsible for secondary metastases in the heart is a carcinoma of the lung. Skin metastases are reported to occur in 1–4 per cent of all lung cancers, the majority being from oat-celled growths.

TRANSPLEURAL SPREAD

Both peripheral and central lung cancers may spread to involve the pleura, and result in the widespread dissemination of cancer cells over the surface of the lung, chest wall and diaphragm (Fig. 20.65). In the case of central growths this only occurs when widespread lymphatic dissemination has involved the subpleural lymphatic plexus. The resulting pleural effusion is usually blood-stained and leads to secondary pressure collapse of the lung.

Carcinoma of the lung, like most other forms of cancer, is a clinically silent disease until either the primary growth or its secondary metastases interfere with the function of an organ or the primary growth ulcerates causing haemoptysis. Because of the ubiquitous nature of its metastases, this form of cancer may present clinically in an almost unlimited number of ways. In the majority of cases, the symptoms and signs are related to the obstructive pneumonitis which occurs when the growth causes bronchial obstruction. Pleural effusions, pulmonary infarcts of unexplained nature, pneumothorax, jaundice and symptoms referable to damage of the central nervous system are also common ways in which this tumour may first present.

A less common but well-recognized manner in which these neoplasms first present is with obstructive emphysema and the formation of giant air cysts. Castex and Mazzei (1951) considered the air cysts were caused by bronchial obstruction following enlargement of the hilar glands by growth. Giant air cysts should be distinguished from lung abscesses due to obstructive pneumonitis, the latter being very commonly seen in the later stages of central growths causing bronchial obstruction.

Although lung cancer is usually regarded as a rapidly growing form of tumour which soon proves fatal, several cases have now been observed to grow slowly over a period of years, and Rigler et al. (1953) watched such a case which progressed slowly for a period of 9 years before proving fatal. According to Rigler (1957), more than 50 per cent of patients with lung cancer show radiological evidence of the disease 2 years before the onset of symptoms, and this particularly applies to peripheral growths. The same author also noted the variations that occur in the rate of growth at different stages in the life-history of the tumour. Laitinen et al. (1964) analysed 202 fatal, proven cases of lung cancer and found that approximately 10 per cent had pursued a slow course as judged by retrospective examination of the radiological evidence. In some cases tumours had been watched for up to 10 years before death occurred. These findings also supported the view that many growths started in the periphery of the lung and later involved the central lymphatic glands. In general, the degree of invasiveness and rate of growth increase with the age of the tumour. Garland (1966) studied the growth rate of different forms of lung cancer by means of serial radiographs. He found that most tumours have passed through three-quarters of their total lifespan when first discovered clinically. It was calculated that adenocarcinomas take 15 years to reach a size of 2 cm diameter, but squamous-celled carcinomas grow twice as fast. It has been computed that the volume doubling time of a well-differentiated squamous-celled carcinoma is between 80 and 150 days, a poorly differentiated oat-celled cancer about 48–55 days, and adenocarcinomas have a growth rate intermediate between the other two. It should be emphasized, however, that growth rates are variable and are not constant. Lung cancers developing in pre-existing bronchial carcinoids and adenomas also grow slowly (Goldman, 1942; Smith, 1954). Very rarely a primary

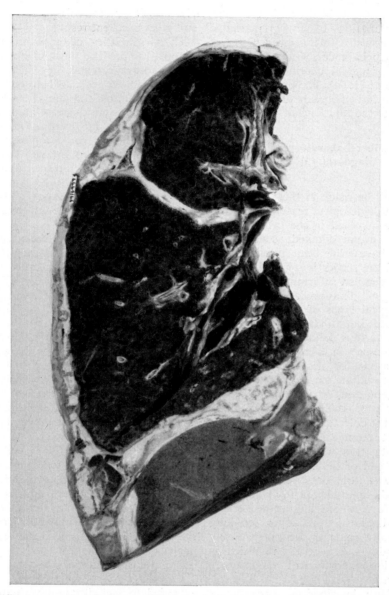

FIG. 20.65. A diffusely spreading carcinoma of the lung simulating a pleural mesothelioma. One-half natural size.

squamous-celled bronchial carcinoma may regress whilst its metastases continue to grow in the absence of any treatment (Blades and McCorkle, 1954). The explanation of this remarkable behaviour is unknown. A further proven case of spontaneous disappearance of an inoperable undifferentiated bronchial carcinoma was described by Bell *et al.* (1964). The patient was still surviving in good health 5 years after all radiological evidence of the tumour had disappeared.

Cytological and Other Laboratory Procedures in the Diagnosis of Lung Cancer

No attempt will be made in this text to describe in detail the cytological features and procedural minutiae involved in the cytodiagnosis of lung cancer. Several detailed accounts, including books, have now been published on this subject, and the reader is referred to these works for more detailed information. Only the principal cytological features of each of the common histological types of lung cancer will be briefly described, but an attempt will be made to evaluate the various laboratory diagnostic procedures at present in use and to discuss some of the reasons for the failure of these methods to diagnose lung cancer.

In 1935 Dudgeon and Wrigley published the results of the first large-scale cytological examination of sputa for the detection of malignant cells. In 1938 Barrett summarized the results of the earlier investigation which had been continued up to the end of 1937. For several years this procedure was little practised outside St. Thomas' Hospital, London, but following these early pioneer efforts which had demonstrated its value as a diagnostic procedure for lung cancer, it gradually came to be more widely accepted until today it is generally regarded as the most valuable diagnostic and confirmatory test for this disease.

In 1946 Bamforth described the appearances of the common forms of lung cancer cells in sputum, using the original method of fixation in Schaudinn's fluid and staining with Mayer's haemalum and eosin devised by Dudgeon. Although other staining methods have been introduced, including the stain devised by Papanicolaou (1942), the original method of Dudgeon still gives excellent results.

Of all varieties of growths, the oat-celled group provides the most characteristic cytological appearances and they may occur together with squamous cells. The oat cells usually occur in clumps and the individual cells vary much in size and shape. Compared with non-malignant cells in sputum, they have larger nuclei with little or no cytoplasm, each nucleus contains one or two nucleoli and the nuclear chromatin is stippled throughout the nucleus (Fig. 20.66). Similar cellular features are found in the secondary metastases and in malignant cells obtained from pleural effusions due to these growths.

Squamous-celled carcinoma cells are more difficult to find. The cells are mostly found singly though clumps occur but are less numerous than in the previous variety. The individual cells are larger, variable in size and shape, containing irregularly shaped nuclei together with eosinophilic cytoplasm, and multi-nucleated cells are found (Fig. 20.67). Unless clumps of cells are present, it may be difficult to identify this type of growth from the appearances of the individual cells. This group includes both the histologically well-differentiated squamous-cell cancers and the less-differentiated neoplasms included under the term polygonal-celled carcinoma.

In alveolar (bronchiolar) celled carcinoma some of the cells are vacuolated with large, irregularly shaped nuclei containing clumps of chromatin, whilst others appear to bud (Fig. 20.68). Intracellular vacuoles may stain positively for mucus. A subvariety occurs in which the cells are smaller, more uniform in size and nuclear staining pattern, and may show fewer or no mucus-containing vacuoles.

Cytodiagnosis of lung cancer has now been extended to include the examination of pleural fluids, bronchial washings, bronchial swabbings and bronchial brush biopsies. The highest figures for positive results is obtained from bronchial brush preparations and after that

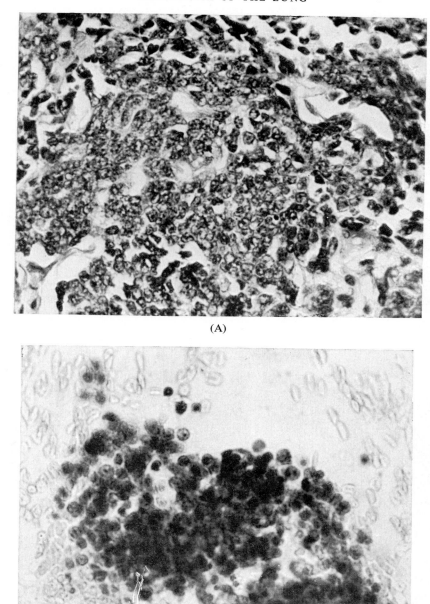

(A)

(B)

Fig. 20.66. An oat-celled carcinoma arising in the bronchus showing part of the growth prepared by the ordinary histological processes (A), and (B) the cytological features of the malignant cells from the same growth found in the sputum. × 550.

(Reproduced by courtesy of the late Dr. J. Bamforth.)

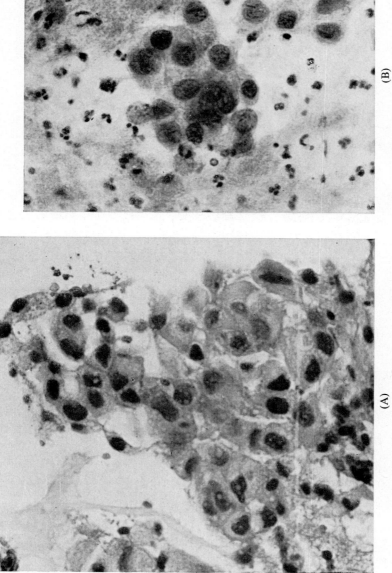

Fig. 20.67. A squamous-celled carcinoma of the bronchus showing part of the growth prepared by the ordinary histological processes (A), and (B) the cytological features of the malignant cells from the same growth found in the sputum. Both × 550. Note the presence of abundant polymorph leucocytes.

(Reproduced by courtesy of the late Dr. J. Bamforth.)

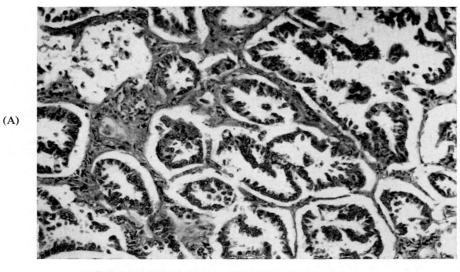

(A)

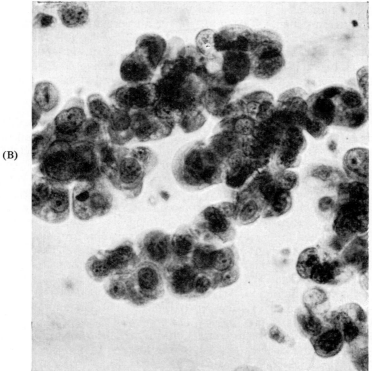

(B)

FIG. 20.68. A bronchiolar-alveolar celled carcinoma showing the general histological appearances of the growth (A), and (B) the cytological features of the malignant cells from the same growth found in the sputum. Note the foamy appearance in the cytoplasm due to mucus. (A) × 145, (B) × 550.
(Reproduced by courtesy of the late Dr. J. Bamforth.)

either from examination of mucus aspirated through a bronchoscope or from examination of sputum obtained by a "deep cough". If ordinary sputum samples are used, early morning specimens should be obtained.

The accuracy and value of sputum examination for exfoliated tumour cells has now been assessed. Spjut *et al.* (1955) found that among 501 cases of histologically confirmed lung cancer, 76·6 per cent had been diagnosed pre-operatively by cytodiagnosis. To obtain this percentage of positive results 1269 smears of sputum or bronchial washings were examined which involved at least two sputa or the products of three bronchial washings from each patient. Umiker (1961) found that among 152 consecutive histologically proven cases of lung cancer, 70 per cent had been diagnosed pre-operatively from examination of sputum or bronchial aspirate. Cytological examination of pleural fluid was positive in eight of fourteen patients. Teir *et al.* (1963) obtained a correct result by examination of sputa in 74 per cent of cases and by examination of bronchial aspirate in 72 per cent. The false negative results of 26 per cent and 28 per cent respectively, when analysed, showed that this had been due either to absence of any communication between the tumour and a bronchial lumen, or to stenosis of bronchi and to misinterpretation of the cytology.

The proportion of cytologically proven cases varies directly with the number of sputum examinations undertaken in each case, the attention to detail in the collection and preparation of the specimens and to the experience of the examiner. Hinson and Kuper (1963) have recommended that fresh sputum when received should be placed in a petri dish against a dark background and only white threads and bloodflecked streaks selected for examination. When five or more examinations are made in the case of each patient 96 per cent of cases may be accurately diagnosed (Umiker, 1960). Oswald *et al.* (1971) were able to confirm the diagnosis in 85 per cent of cases if four or more specimens were examined.

Cancers situated in the upper lobes of the lungs yield more positive cytological results from sputum examination than growths situated elsewhere, and more positive results are obtained from examination of bronchial washings in the case of tumours situated in the lower lobes. Sputum cytodiagnosis has proved the most valuable and by far the best presently available method of diagnosing peripheral lung cancers, though centrally situated growths yield a higher proportion of positive results than peripheral tumours. The greater the size of a tumour the more likely are malignant cells to be discovered in the sputum.

False positive results though fortunately rare occur in about 1·5 per cent of cases but in the series reported by Oswald *et al.* they amounted to only 0·7 per cent. They may result from examination of the sputum in any resolving inflammatory condition but especially in cases of active pulmonary tuberculosis, acute asthma and bronchiectasis. Particular care should be taken in the interpretation of sputum cytology during the first 2 weeks following a bronchoscopic examination (Hinson and Kuper). Cells very closely resembling malignant cells are also found in pleural effusions complicating pulmonary infarcts.

Cytological examination proves a more accurate and an earlier method of diagnosis for small peripherally situated lung cancers, including the alveolar (bronchiolar) celled carcinomas, than radiography, though it gives no information concerning the localization of the tumours. As might be expected the larger and more inoperable cases yield a higher proportion of cytologically positive sputa than the small but more operable growths.

For further information on the cytological methods of diagnosis of lung cancer the reader is referred to the reports by Richardson *et al.* (1949), Watson *et al.* (1949), Farber *et al.* (1950), Buffmire and McDonald (1951) and McCormack *et al.* (1955).

Cytodiagnosis yields fewer positive results when the lung tumours are secondary growths. Rosenberg *et al.* (1959) identified malignant cells in the sputum in only 38 per cent of such patients. In such cases needle aspiration using radiological image intensification for localization and cytological examination of the aspirate has yielded the best results.

Daniels (1949) introduced biopsy examination of scalene lymph nodes as a method for diagnosing lung cancer and other forms of cancer. A positive result showing secondary carcinoma cells in these glands is indicative of extensive spread and an inoperable state. This method of diagnosing lung cancer has little to recommend it and a positive result is a contraindication to surgical treatment of the primary tumour. The operation carries a 6 per cent risk of serious complications mainly due to haemorrhage and mediastinitis. Scalene lymph node biopsy as might be expected is more often positive with growths arising in an upper lobe of a lung and with a highly anaplastic oat-celled carcinoma than in the slower-growing squamous- and adeno-carcinomas. Positive results are seldom obtained in more than 9–10 per cent of proved cases and the merits of the procedure have been discussed by Lees and McSwan (1962) and Skinner (1963).

CHAPTER 21

Rare Pulmonary Tumours

THE rare pulmonary tumours consist of a heterogeneous collection of lesions, mostly neoplasms, but including some chronic granulomas. Some of the neoplasms are benign while others are malignant. Hamartomatous lesions, with the exception of angiomas, are considered separately in Chapter 23 together with pulmonary blastomas and pulmonary teratomas.

For the purposes of classification it is proposed to group the lesions considered in this chapter into four groups, although it should be emphasized that this classification will probably require future amendment in the light of further knowledge that will be gained about some of these conditions.

The rare pulmonary tumours are classified as follows:

(1) Tumours possibly or probably derived from cells of neural origin:
  (a) Bronchial carcinoid tumour.
  (b) Benign clear-cell tumour of lung.
  (c) Neurofibroma and neurogenic sarcoma.
  (d) Myoblastoma of the bronchus.
  (e) Primary melanoma of the bronchus.
  (f) Pulmonary chemodectoma.
(2) Tumours derived from the bronchial epithelium and bronchial mucous glands:
  (a) Papilloma of the bronchus.
  (b) Mucoepidermoid tumour of bronchus and cystadenoma of bronchus.
  (c) Bronchial cylindroma (adenoid cystic carcinoma).
(3) Tumours derived from the pulmonary and bronchial mesenchyme:
  (a) Chondroma of the bronchus.
  (b) Lipoma of the bronchus.

  (c) Fibromas and myxomas of the lung.
  (d) Leiomyofibromas.
  (e) Pulmonary sarcoma.
  (f) Rhabdomyosarcoma.
  (g) Bronchiolo-alveolar tumour (IVSBAT).
  (h) Angiomas.
  (i) Haemangiopericytomas.
  (j) Benign pleural tumours.
  (k) Malignant pleural tumours.
  (l) Pleural endometriosis.
(4) Tumours of chronic granulomatous or uncertain nature:
  (a) Plasma cell granuloma.
  (b) Sclerosing "angioma".

1 (a) Bronchial Carcinoid

Synonym: Carcinoid bronchial adenoma

In the past it was customary to divide bronchial adenomas into two varieties, the carcinoid adenomas and the cylindromatous adenomas. Although it was usual to regard both as bronchial adenomas their totally different microscopical structure and secretory properties indicated their origins from different types of cells.

The term carcinoid is now applied to the commoner type of tumour because of its close histological resemblance to the intestinal tumours of the same name. The first bronchial carcinoid was described by Müller (1882). Bronchial carcinoids, bronchial cystadenomas (cylindromas) and bronchial muco-epidermoid adenomas form respectively 90 per cent, 8 per cent and 2 per cent of the group of tumours previously collectively called bronchial adenomas. According to Burcharth and Axelsson (1972) they form 1·2 per cent of all lung tumours but account for 3 to 10 per cent of all *surgically* excised broncho-pulmonary neoplasms and in a mass miniature radiographic survey of 300,000 persons, two such tumours and seventy-one lung cancers were discovered (Zellos, 1962). Bronchial carcinoid tumours have been reviewed by Arrigoni *et al.* (1972) and Lawson *et al.* (1976).

With few exceptions bronchial carcinoids occur in the main, lobar or segmental bronchi in regions of bronchial bifurcations and 90 per cent are visible through a bronchoscope. The mucosal surface is seldom extensively ulcerated (Liebow, 1952), but nevertheless the very vascular stroma of these tumours is commonly responsible for causing small repeated haemoptyses. Relatively few bronchial carcinoids arise in the peripheral bronchi but examples were reported by Mayo (1942), Maier and Fischer (1947) and Pollard *et al.* (1962). Occasionally multiple bronchial carcinoids occur scattered through the bronchial tree, the central tumours being of macroscopic dimensions while the peripheral tumours are usually only microscopical in size. Careful search of the major bronchi as a routine measure at autopsies may occasionally reveal very small bronchial carcinoids varying in size up to 2·0 mm in the region of bronchial bifurcations.

Bronchial carcinoids are found at all ages and the average age at which they present clinically is between 44 and 48 years. Unlike the far more common lung cancer, bronchial carcinoids are slowly growing and symptoms referable to them often antedate surgical removal by 5 years or longer (Delarue, 1951). Some of these tumours may also cause endocrine disturbances, notably Cushing's syndrome, for which the patients may first seek advice.

Macroscopically, bronchial carcinoids are equally distributed between both lungs but show a slight predilection for the lower lobe lobar and segmental bronchi (Rabin and Neuhof, 1949). The majority of these tumours assume two distinct forms and grow either as a bronchial polyp or as an infiltrating growth or "iceberg" tumour. The polypoid variety grows by direct extension along the large bronchi as seen in Fig. 21.1, whilst the infiltrative type may barely raise the bronchial lining but extends deeply into the underlying bronchial and extrabronchial tissues (Fig. 21.2). Both varieties of tumour appear to grow in the submucous layer of the bronchi separate from the lining epithelium and sometimes separated from it by a well-developed layer of fibrous tissue. The surface bronchial epithelium frequently undergoes metaplastic change but is rarely extensively ulcerated. The cut surface of the tumours is greyish-white or pink and is often intersected by fibrous strands which may undergo calcification or even ossification and amyloid changes.

Microscopically, bronchial carcinoids consist of small, uniformly staining clear cells arranged either in solid clumps, trabeculae or poorly developed tubules (Figs. 21.3, 21.4). The stroma is often very vascular and may undergo hyaline change, calcification or even ossification (Figs. 21.5, 21.7).

Bronchial carcinoids like their intestinal counterparts can form acini and secrete mucus. Occasionally some of the tumour cells may be argentaffinic (Fig. 21.8), though more often they are argyrophilic. Very rarely cartilage-like tissue may be present (Fig. 21.6). Carcinoid tumours in the bronchi show a spectrum of microscopical appearances and clinical behaviour varying between a typical locally invasive tumour to the highly malignant metastasizing oat-celled carcinoma. It is sometimes difficult from the histological appearances alone to assess the degree of malignancy and the likely clinical course of the intermediate grades of tumours. Furthermore, the microscopical appearances in one part of a tumour may not be representative of the whole. In parts it may be consistent with a benign bronchial carcinoid while in others it may show a less-differentiated pattern akin to an oat-celled carcinoma. In undoubted carcinoid tumours mitoses are uncommon but tumour cells may infiltrate the immediately adjacent tissues including the bronchial wall and hilar lymph glands (Fig. 21.9). Distant blood-borne metastases are rare but occur in between 2 and 5 per cent of all bronchial carcinoids and are usually found in the liver, adrenals and bone (Fig. 21.10). An

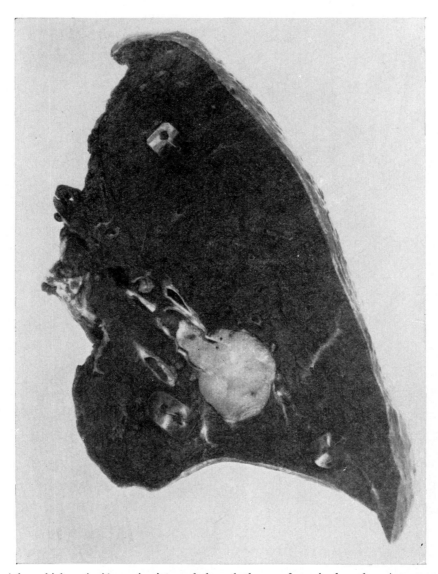

FIG. 21.1. A bronchial carcinoid growing into and along the lumen of a major bronchus. Approx. natural size

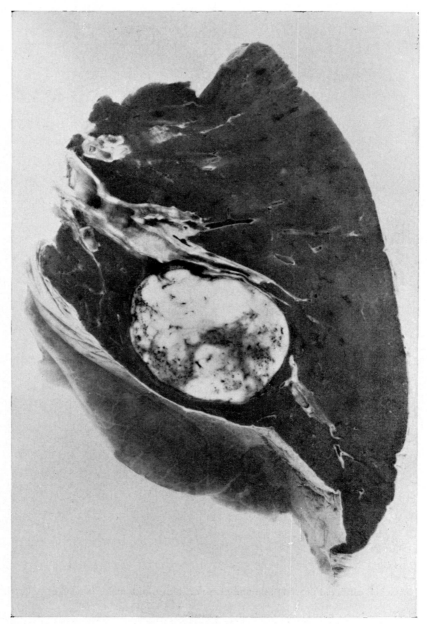

FIG. 21.2. A bronchial carcinoid of the "iceberg" type lying almost entirely below the bronchial surface. Nine-tenths natural size.

unusual subvariety of a bronchial carcinoid is the bronchial oncocytomatous carcinoid, examples of which were described by Black (1969) and Fechner and Bentinck (1973) (Fig. 21.11). The tumour cells are larger than those present in a typical bronchial carcinoid and have a pale, granular eosinophilic cytoplasm closely resembling the oxyphil cells of the parathyroid glands. The majority of bronchial oncocytomas are bronchial carcinoids (Fig. 21.11) but a few are probably true bronchial mucous gland adenomas (*see* **Bronchial Cystadenoma** on p. 887). The microscopic appearances of oncocytoma cells both in the bronchi and elsewhere is caused by an enlargement and increase in the number of mitochondria. Bronchial oncocytomas were first described by Hamperl (1937) and later by Stout (1943). The former considered that it arose from oncocytes that are present occasionally in the epithelium of the bronchial mucous glands and their ducts. The ultramicroscopic study of a bronchial oncocytoma by Black showed that the cells also contained the osmio-philic neurosecretory-type granules character-istically found in bronchial carcinoids (Fig. 21.13).

Histochemical studies have shown that the tumour cells are particularly rich in acid phosphatase and 5-nucleotidase (Willighagen *et al.*, 1963). As already stated the vascular stroma may undergo hyalinization, chondro-myxomatous change, calcification, ossification and amyloid change, the latter being common to many other tumours derived from the parocrine (APUD) system of cells, e.g. medullary carcinoma of the thyroid and thymic and intestinal carcinoids. Osteoblastic reaction to the tumour cells is also found in metastatic deposits in those carcinoids which metastasize. The stromal changes have been wrongly interpreted as evidence for the blastomatous nature of these tumours but are in reality changes induced in the stroma by the tumour cells.

Price-Thomas and Morgan (1958) considered that ossification was sometimes related to the proximity of bronchial cartilage, the cartilage

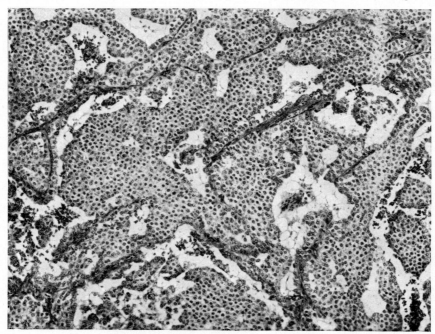

FIG. 21.3. A low-power view of a carcinoid adenoma of the bronchus showing nests of uniformly staining cells and no evidence of hyperchromatism. × 140 H and E.

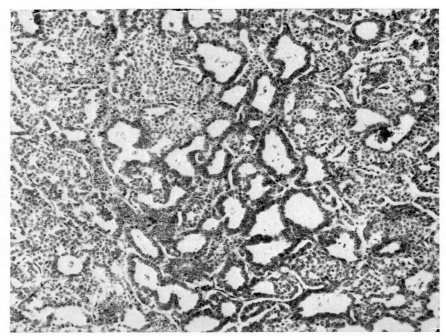

FIG. 21.4. A bronchial carcinoid showing attempt at tubule formation. × 140 H and E.

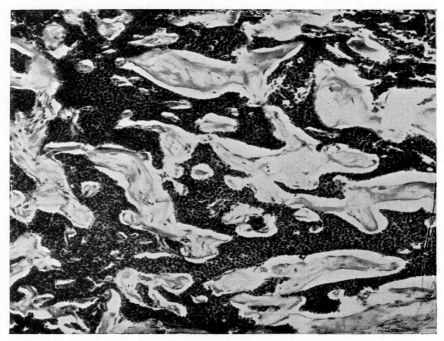

FIG. 21.5. Section showing hyaline changes in the stroma of a bronchial carcinoid. × 140 H and E.

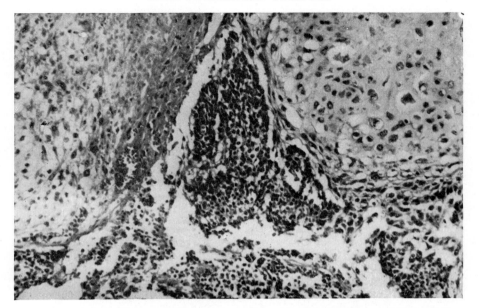

FIG. 21.6. A bronchial carcinoid tumour containing islands of cartilage. ×140 H and E.

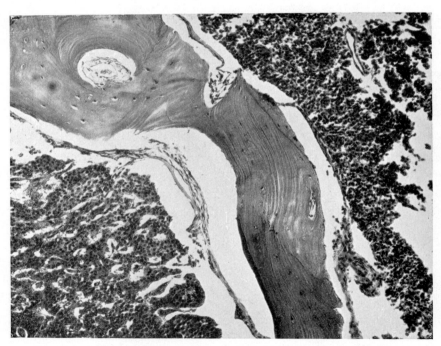

FIG. 21.7. Ossification of the stroma of a bronchial carcinoid. × 140 H and E.
(Figs. 21.7, 21.43, 21.44, 21.46, 21.47, 21.48 and 21.49 are reproduced by kind permission of the Thoracic Society of Great Britain.)

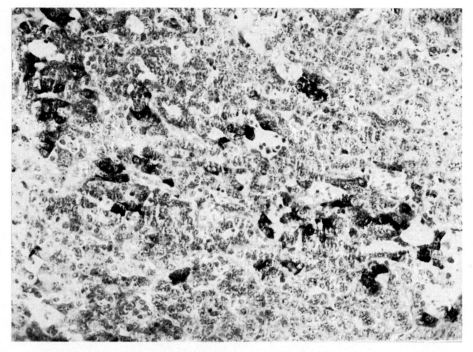

FIG. 21.8. An argentaffinic carcinoid tumour arising in a bronchus. Only a few of the tumour cells are argentaffinic (black-stained cells). × 300 Fontana stain.

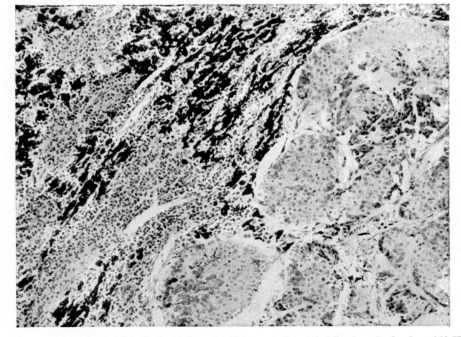

FIG. 21.9. A secondary deposit of a bronchial carcinoid in an anthracotic hilar lymph gland. × 140 H and E.

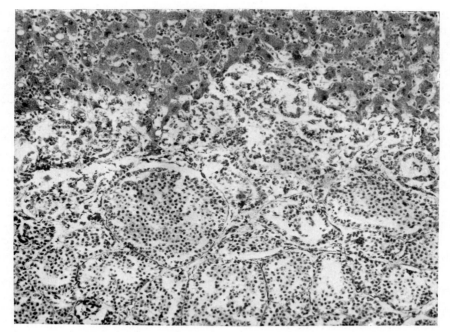

FIG. 21.10. A secondary metastasis of a carcinoid in the liver. × 140 H and E.

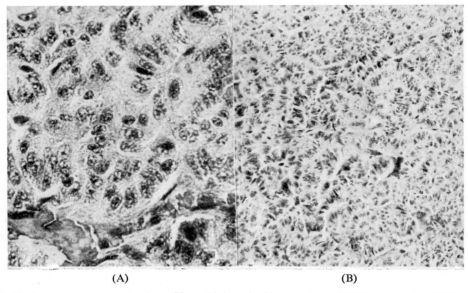

(A)   (B)

FIG. 21.11. An oncocytomatous variety of bronchial carcinoid tumour. A × 1000 approx., B × 300 approx.

first undergoing necrosis which later was followed by calcification that provided a medium in which later heterotopic ossification occurred. It is more likely, however, that ossification is caused by the influence of enzymes secreted by the carcinoid cells on the surrounding stroma as evidenced by the osteoblastic reaction provoked by metastatic deposits of malignant carcinoid cells in bone (Toomey and Felson, 1960). Peripherally sited bronchial carcinoids may be confused with tumourlets, but carcinoid tumours do not usually arise in relation to chronic lung damage and are often multiple and microscopic in size, cf. tumourlets. Occasionally the carcinoid nature of the tumour is evidenced by the development of secondary endocrine disturbances, the commonest of which is Cushing's syndrome.

THEORIES OF ORIGIN

For many years the nature and origin of bronchial carcinoids was uncertain and they were included with other bronchial tumours and referred to as bronchial adenomas. The similarity of bronchial carcinoids to growths of the same name arising in the bowel and other offshoots of the primitive endodermal canal led to attempts to prove that they had a similar ancestry. With few exceptions, notably the cases reported by Foot (1945), Holley (1946) and McConaghie (1962), bronchial carcinoids seldom stain positively for argentaffin granules though many such tumours are argyrophilic. The absence of argentaffinic properties occasions no surprise as Feyrter (1954, 1960) showed that argentaffin cells or the helle-Zellen-Organes can be found in the bronchial and bronchiolar epithelium in early childhood and infancy and sometimes in the presence of inflammation of the bronchial wall (Fig. 2.5), but they disappear normally soon after the first year of life. Furthermore, success in the demonstration of argentaffin granules requires rapid fixation of the tumour. As in the case of the intestinal arrcinoid tumours the granules may sometimes be demonstrated by the diazonium reaction and by autofluorescence. The argentaffin and clear

cells in the epithelium of the digestive and bronchial tracts represent part of a very widespread and an extremely important biological system of hormone-secreting cells named by Feyrter the "peripheren Endokrinen" or "parocryne" system of cells. The same system of cells is now often referred to as the APUD (amine precursor uptake and decarboxylase) system. Frohlich (1949) claimed to have shown neural connections with clear, non-argentaffinic cells in the basal layer of the bronchial tree. Bensch et al. (1965a) first showed the presence of Kultschitzky-type cells incorporated between the exocrine cells of the bronchial mucous glands and lying just within the limiting basement membrane. These cells possessed neurosecretory-type granules within their cytoplasm and had long dendritic processes insinuated between the adjacent exocrine cells. Similar cells were also found by Bensch et al. (1968) in the bronchial epithelium situated among the basal layer of cells and they possessed features similar to those of the Kultschitzky cells found throughout the intestinal tract and like those cells were capable of producing a variety of hormones.

Bensch et al. (1965b) investigated the electron microscopic structure of tumour cells removed from a bronchial carcinoid and found that all of them contained similar neurosecretory-type granules and that the tumour cells were of two types similar to the "light" and "dark" neurosecretory cells they had found within the bronchial mucous glands (Figs. 21.12, 21.13). The cytoplasm of the cells contained unusual arrangements of the endoplasmic reticulum causing adjacent cisternae to fuse to form structures with four membranes. The ultrastructure of bronchial carcinoids and their relationship to the histochemistry of the cells has been investigated further by Hage (1973). The similarity of bronchial carcinoid cells to the Kultschitzky-type cells present in the bronchial epithelium and the bronchial mucous glands, and the tumour distribution which corresponds to the numbers of such cells in the different parts of the bronchial tree confirms their origin.

Churg and Warnock (1976) claim to have shown that many tumourlets contain similar

(b) *Gastrin-like substances.* Bronchial carcinoids may occasionally occur as part of a pjuriglandular syndrome with multiple APUD system adenomas in several endocrine glands. These may occasionally be associated with recurrent peptic ulceration (Zollinger–Ellison syndrome) as in the case quoted by Williams and Celestin (1962).

(c) *Adrenocorticotrophin (ACTH-like substance)* was among the first hormones to be identified with bronchial carcinoids. Cohen *et al.* (1960) described cases of Cushing's syndrome and numerous examples have since been described. The hormone substance appears to act directly on the adrenals which undergo cortical hyperplasia often of an extreme degree. The pituitary does not enlarge and is probably not directly implicated.

(d) *Antidiuretic-like hormone (arginine vasopressin AVP)* is among the least common hormones produced by bronchial carcinoids but is more commonly found in oat-celled lung cancer. Bailey (1971) described AVP production in a "malignant bronchial carcinoid tumour (case 3)" which produced both AVP and an ACTH-like substance.

(e) *Insulin production* has been reported by Shames *et al.* (1968) in a "malignant bronchial carcinoid tumour".

(f) *An inhibitory substance to anti-growth hormone inhibiting factor* may be produced by bronchial carcinoids and is responsible for causing acromegaly. An example of such a tumour was described by Altmann and Schutz (1959) and three further examples have been seen by the author. Pituitary enlargement occurs because the hormone released by the bronchial carcinoid inhibits the naturally produced growth inhibiting hormone secreted by the hypothalamus thus allowing unrestrained growth hormone production by the anterior pituitary which consequently undergoes hyperplasia and enlarges. Acromegaly regresses and the pituitary reduces in size following removal of the bronchial carcinoid.

(g) *β-Melanin stimulating hormone (MSH)* is usually produced in association with other hormone products as in the cases described by Gramlick and Wiethoff (1960) and Melmon *et al.* (1965 case 3). Hyperpigmentation of the exposed areas and pressure points results.

The very close relationship that exists between bronchial carcinoids and oat-celled cancers is supported by the similar and merging light microscopic appearances of the two tumours, the presence of common electron microscopic features in both tumours and the fact that they both secrete an identical series of hormones. These several facts leave little doubt that they have a common ancestry and represent the two extremes of malignant potential, the non-metastasizing bronchial carcinoid being the benign form of tumour and the rapidly metastasizing oat-celled carcinoma representing the opposite extreme. Between these two extremes all degrees of clinical and histological evidence of malignancy may be encountered and light microscopy shows all gradations of changes, the same tumour often showing a mixture of histological appearances. This same spectrum of malignancy was found by Arrigoni *et al.* who reviewed 215 bronchial carcinoid tumours and found 23 had atypical histological features. Of the latter atypical cases almost a third had died following a mean survival time of just over 2 years.

Occasionally bronchial carcinoids may be associated with a tumour resembling a neurofibroma lending additional support to the idea that a bronchial carcinoid may be closely related to the peripheral part of the autonomic nervous system (Fig. 21.14). In this respect these tumours are analogous to certain pigmented naevi in the skin which may also be associated with neurofibromatosis.

1 (b) Benign "Clear-cell" Tumour of Lung

In 1963 Liebow and Castleman presented four examples of what they considered were new lung tumours. In each case the tumour was detected radiologically as a rounded opacity in the lung fields and required an open thoracotomy for its identification and removal. In 1971 Liebow and Castleman further reported a

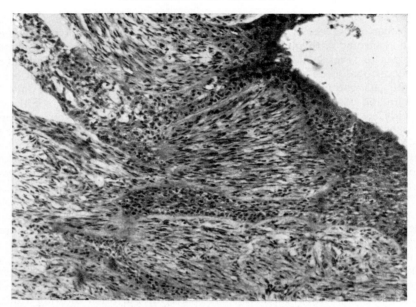

FIG. 21.14. Section of "neurofibromatous-like" tumour immediately adjacent to a bronchial carcinoid. × 140 H and E.

series of twelve cases. The tumours are sharply defined, non-encapsulated, reddish-brown and do not appear to arise from bronchi. They occur between the ages of 30 and 60 and are usually discovered as a result of a routine chest radiograph in which they appear as a coin shadow.

Microscopically, they consist of sheets of large, clear, vacuolated, rounded or spindle-shaped cells with some multinucleate Touton-type giant cells (Figs. 21.15, 21.16). The tumours are very vascular and tumour cells abut directly upon the walls of the capillaries which often undergo hyaline change. The component cells contain large intracytoplasmic masses of PAS-positive material which stains for glycogen and which is digested by diastase. Interstitial deposits of calcium salts may be found. Cells resembling nerve cells with acidophilic cytoplasm and containing a lipochrome pigment also occur. Mitoses are notably absent but occasional tumour cell nuclei contain acidophilic inclusions composed of glycogen.

The tumours are not bounded by a capsule but are nevertheless sharply differentiated from the surrounding lung.

The cell of origin and nature of clear cell tumour of the lung is uncertain. It is not related to clear cell lung cancer and unlike that tumour grows very slowly and does not metastasize.

The ultramicroscopic structure of a clear cell tumour was studied by Becker and Soifer (1971) and Hock *et al.* (1974). The former found that in some cells (2–5 per cent) there were membrane-bound cytoplasmic granules similar in size and appearance to the neurosecretory-type granules present in bronchial carcinoid cells. Because of this feature these tumours are tentatively regarded as a variant type of bronchial carcinoid. The intracytoplasmic accumulation of glycogen within the tumour cells has been attributed to an absence of α-glucosidase. Hoch *et al.*, however, concluded that these tumours were more likely to have arisen from plain muscle or from vascular pericytes.

1 (c) Neurofibroma and Neurogenic Sarcoma of Lung

These two tumours are among the rarest encountered in the lung and will be considered

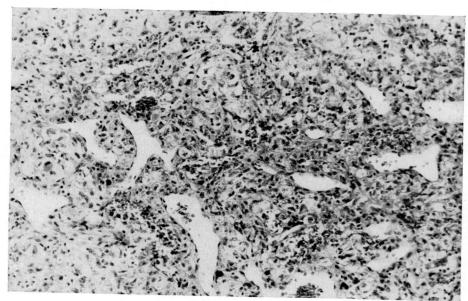

FIG. 21.15. A benign clear-celled tumour in the lung showing the character and arrangement of the cells. × 186 H and E.
(Reproduced by courtesy of Dr. A. A. Liebow.)

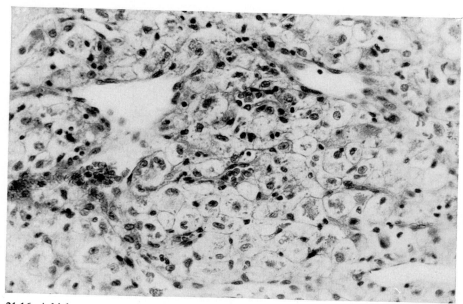

FIG. 21.16. A higher-powered view of the constituent cells of a benign clear-cell tumour in lung. × 372.
(Reproduced by courtesy of Dr. A. A. Liebow.)

together. Askanazy (1914) first mentioned the presence of bronchial neurofibromas in generalized neurofibromatosis, but the first case to be described in detail was published by Rubin and Aronson (1940). Neilson (1958) and Bartley and Arean (1965) reviewed the previously reported cases and both added further examples. Bartley and Arean traced twenty-four bronchial or intrapulmonary neurofibromas and eight neurilemmomas (schwannomas) in the literature which they regarded as acceptable examples. Their own case (1) was a further example of a schwannoma but their case (2) which they presented as a malignant schwannoma was of less certain nature. The author's own experience has been limited to two benign neurofibromas and one malignant neurofibroma. One of the benign tumours occurred in a patient with generalized neurofibromatosis and the other was a solitary extrabronchial lesion connected at one point with a nerve trunk.

The ages at which the reported cases have occurred range from 2½ to 57 years. The majority of schwannomas have occurred in women but the neurofibromas are equally distributed between the sexes. The majority of the tumours have not caused clinical symptoms though some have led to bronchial obstruction, and intrapulmonary tumours may sometimes cause pain radiating to the shoulders and interscapular regions as well as joint pains, stiffness, fever and dyspnoea.

Macroscopically, both types of neurofibromas are related to the bronchi and the schwannomatous variety tend to be smaller, seldom exceeding 3·0 cm in diameter. They form polypoid intrabronchial tumours or are attached to the outside of a bronchus by a broad pedicle (Fig. 21.17). The tumours are lobulated, usually well encapsulated, greyish-brown or pale yellow and often contain cystic areas, haemorrhage and rarely calcification (Pétriat et al., 1953).

Microscopical. The appearances depend on

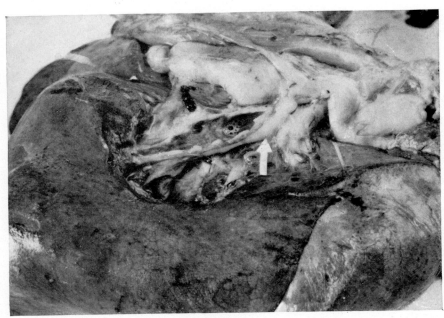

Fig. 21.17. Dissected lung from a case of von Recklinghausen's disease (generalized neurofibromatosis) showing a neurofibroma lying outside the opened bronchial wall; the arrow indicates the tumour. Approx. one-half natural size.

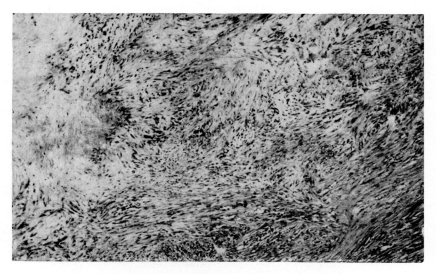

FIG. 21.18. A pulmonary neurofibrosarcoma. × 40 H and E.

whether the tumour is a fibromatous neuro-fibroma or a schwannoma. The former can only be distinguished from any other variety of fibroma when the tumour can be seen to arise in relation to a nerve. Schwannomas contain bundles of elongated fibres showing characteristic palisading of their nuclei (Antoni B tissue). These are separated by acellular but finely fibrillary substance, the Antoni A tissue. Cystic degeneration (myxoid change), hyaline vessel walls and haemorrhages into the tumour are common in the schwannomas. Occasionally Antoni B tissue may be replaced by lympho-cytoid-type cells. The arrangement of reticulin fibres around the cells rather than parallel to their long axes (Stout, 1935) has not been found of much help. Both types of neurofibroma may contain collections of foam cells which are responsible for the yellow zones seen in the gross specimen.

In malignant neurogenic sarcomas, nuclear hyperchromatism, abnormal mitoses, giant cells, and occasional invasion of the capsule serve to differentiate these tumours from the benign form of the growths (Fig. 21.18). The differentiation, however, from other forms of soft tissue sarcoma, such as fibrosarcoma, depends largely, as Willis (1953) has stated, on the identification of the characteristic features of a benign neurofibroma elsewhere in the same tumour.

The clinical course of both the benign and malignant growths will depend entirely on the degree of bronchial involvement and obstruction they cause, the occurrence of haemorrhage, the development of metastases, and in cases associated with the generalized disease the changes produced by tumours elsewhere in the body, i.e. cranial nerve neurofibromas.

A form of chronic interstitial lung fibrosis may occur in association with generalized neuro-fibromatosis and is described in Chapter 23.

1 (d) Myoblastoma of the Bronchus

Myoblastomas of the bronchus are amongst the rarest of the tumours encountered in the lung. Cases have been described by Kramer (1939), Kraus *et al.* (1948), Murphy *et al.* (1949), Hebert *et al.* (1957) and Peterson *et al.* (1957). The few tumours so far described have been

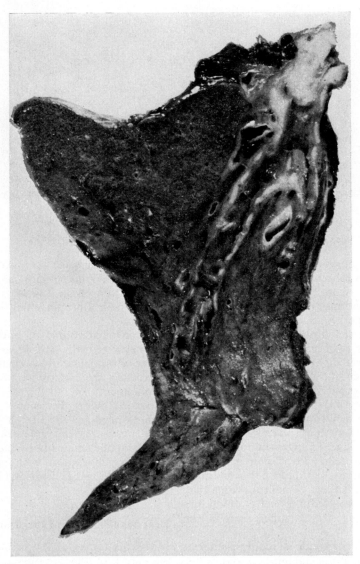

Fig. 21.19. A myoblastoma of the bronchus arising near the origin of a lobar bronchus. The lung distal to the growth shows changes of obstructive pneumonitis.

(Reproduced by courtesy of Professor I. Doniach.)

sessile or polypoid growths with smooth surfaces growing from the wall of a large bronchus (Fig. 21.19) which present because they had caused obstructive pneumonitis following bronchial occlusion. The cut surfaces of the tumours have usually a yellowish-grey or pink colour and are often coarsely trabeculated. Myoblastomas are normally restricted to the bronchial wall though in the case described by Hebert *et al.* the tumour extended between the cartilage plates into the extrabronchial lymph glands. In the reported cases and others known to the author, the tumours have all been found in the main, segmental or beginning of subsegmental bronchi (Fig. 21.19). Bush and Plain (1968) reported a case with three separate myoblastomas arising in the right main bronchus and reviewed three previous such cases. Like other but similarly polypoidal, benign bronchial tumours they cause interference with inspiratory or expiratory airflow, leading either to obstructive pneumonitis or local emphysema respectively.

Microscopically, the tumours are composed of three types of cells, large granular and foamy cells with small dark nuclei arranged occasionally in syncytial masses, long fusiform cells with elongated nuclei (Figs. 21.20, 21.21) and other cells with an eosinophilic cytoplasm and oval dark nuclei. In appearance the tumours closely resemble "myoblastomas" found elsewhere in the body.

The age of the patients has varied between 30 and 50 years. No distant spread of any of these bronchial growths has been reported, and all the symptoms have been attributable to bronchial obstruction.

Although these growths have been described under various names in different parts of the body, they were first regarded as myoblastomas by Abrikossoff (1926). Many suggestions have been offered about their origin, and these have included derivation from myoblast cells (Abrikossoff), histiocytes (Martin, 1950), fibroblasts (Pearse, 1950), Schwann cells (Fust and Custer, 1949) and non-chromaffin paraganglion cells (Smetana and Scott, 1951). No definite conclusion has yet been reached concerning their nature but recent work strongly supports the view that they are tumours derived from nervous tissue, probably Schwann cells in the peripheral nervous system and oligodendroglial-like cells in the posterior pituitary. The discovery of similar tumours arising in the pituitary stalk contraindicates suggestions of a smooth muscle origin though the possibility that they may be related to a nerve end-organ cannot be excluded. Furthermore, like schwannomas, myoblastomas may sometimes be multiple in both the skin and the bronchial tree.

1 (e) Malignant Melanoma of the Bronchus

Very few of the previously reported cases of this rare primary bronchial neoplasm can be regarded as completely proven. Only in the case described by Salm (1963) was an alternative primary site for the tumour excluded by postmortem examination. Other likely but not completely proven surgical cases have been described by Allen and Spitz (1953) and Hsu *et al.* (1962). The case described by Rosenberg *et al.* (1965) emphasizes the importance of excluding secondary bronchial melanotic growth from a primary tumour which may have been removed surgically from the skin many years previously.

All of the tumours so far reported have occurred in the large central bronchi presenting as grey or dark brown, polypoidal or infiltrating growths which have led to bronchial obstruction. The distribution of the metastases in the fatal case described by Salm followed the same general pattern as that seen in ordinary bronchial carcinoma. *Microscopically*, the tumour is distinguished by the presence of scattered melanin-producing cells which react positively with Fontana's melanin stain (Fig. 21.22). The tumour cells are pleomorphic with many multinucleated cells and in places they are arranged in whorls. In the intact bronchial epithelium beyond the immediate site of the primary growth junctional naevus changes

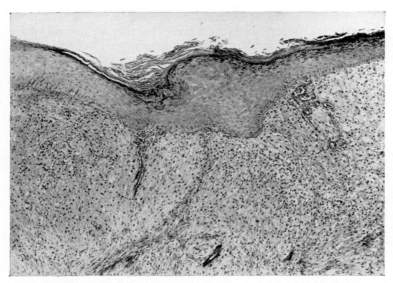

FIG. 21.20. A low-power view of a myoblastoma of the bronchus showing the uniform appearance of the tumour cells in the submucous region of the bronchus. The overlying bronchial epithelium has undergone squamous metaplasia. Approx. × 80.

(Reproduced by courtesy of Drs. P. A. Peterson, E. H. Soule and P. E. Bernatz and the Editor of *The Journal of Thoracic Surgery*.)

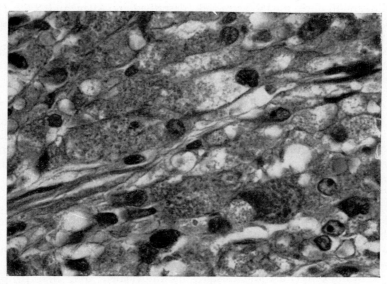

FIG. 21.21. A high-power view of the cells comprising the myoblastoma seen in Fig. 21.20 showing the granular appearance of the spindle-shaped cells. × 800.

(Reproduced by courtesy of Drs. P. A. Peterson, E. H. Soule and P. E. Bernatz and the Editor of *The Journal of Thoracic Surgery*.)

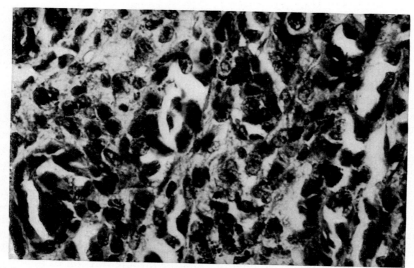

Fig. 21.22. A high-power field of a bronchial melanoma. In some of the cells fine granules of melanin pigment may be observed. × 400 H and E.

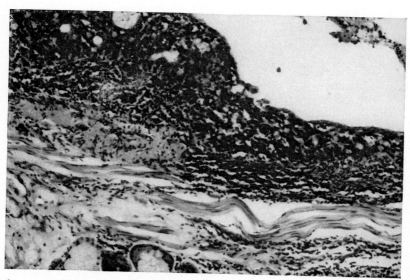

Fig. 21.23. A melanoma arising in the bronchial epithelium. The appearances are those of a junctional naevus which has undergone malignant changes. × 100 H and E.

(lentigo maligna) are found (Fig. 21.23), and these provide further evidence in favour of the origin of the main bronchial tumour from melanoblasts contained in the bronchial epithelium.

Bronchial melanomas probably arise from melanoblasts detached from the primitive oropharyngeal epithelium of the foetus by the growing laryngotracheal bud, as the migration of neuro-ectodermal melanoblasts is completed before the laryngo-tracheal bud appears. Similar tumours occasionally arise both in the larynx and the oesophagus, two other structures which both develop from the primitive oropharynx.

1 (f) Pulmonary Chemodectomas

Heppleston (1958) described an unusual lung tumour situated in the periphery of the lung of an achondroplasic dwarf which he described as a carotid body-like tumour. He considered that the tumour was related to chemodectomas found elsewhere in the body.

The tumours, currently regarded as chemodectomas in the lung, were first described by Korn *et al.* (1960). They described a series of minute tumours, the majority of which were microscopical in size. Zak and Chabes (1963) added a further small series of cases, and Barroso-Moguel and Costero (1964) were the first to demonstrate that the minute tumours possessed neural connections and that some of the tumour cells contained argentaffin granules (Figs. 21.25, 21.26). The author has also seen two cases with multiple minute tumours, but at the time of discovery was uncertain whether they represented physiological or pathological structures. Owing to the usual random method of taking lung sections from post-mortem material, and the minute fraction of the total lung which can be examined in any section, discovery of such

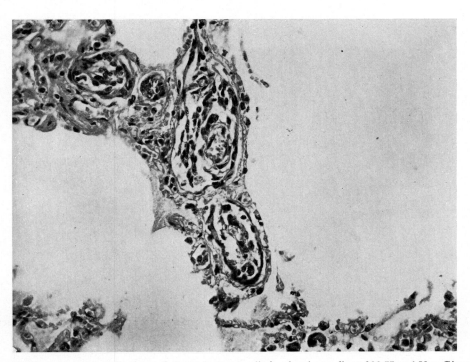

FIG. 21.24. Perivenous collections of chemodectoma cells in alveolar wall. × 300 H and Van Gieson.

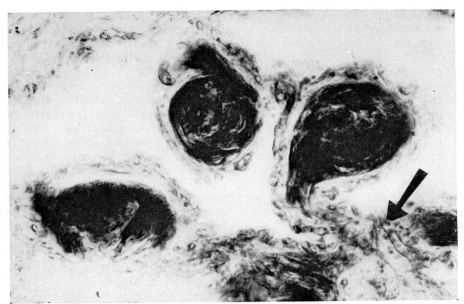

FIG. 21.25. Three chemoreceptor masses in human lung showing supplying nerve fibres (arrowed). The chemoreceptor tissue stains dark black with the silver-staining method employed. × 300 stained by Barroso-Moguel silver method.
(Reproduced by courtesy of Dr. I. Costero, Mexico City.)

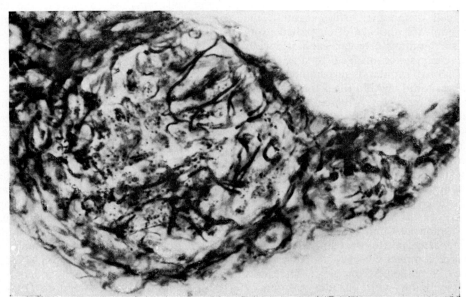

FIG. 21.26. Human chemoreceptor tumour in lung showing the argentaffin cells. × 800 stained by Rio Hortega silver method.
(Reproduced by courtesy of Dr. I. Costero, Mexico City.)

small tumours, if few in number, would indeed be fortunate. According to Korn *et al.* they are present in about 1 in 300 post-mortems though in the author's experience they are much less frequently encountered.

According to Barroso-Moguel and Costero, chemoreceptor cells are normally found in the lung in relation to pulmonary venules especially in the lungs of patients dying of chronic congestive heart failure (Fig. 21.24). They regard the minute tumours under discussion as hyperplastic collections of normal chemo-dectoma cells which they resemble in structure, staining properties and function, and they behave like non-chromaffin paraganglionic tissue found elsewhere in the body.

The patients in whom the tumours have been discovered have died from a wide variety of diseases and only very rarely were the lesions actually seen macroscopically after death as greyish-white pin-point nodules beneath the visceral pleura.

They consist of interstitially situated collec-tions, *Zellballen*, of rounded or strap-like cells with vesicular nuclei, foamy cytoplasm and ill-defined cell membranes. After fixation the *Zellballen* contract and are often pulled away from the surrounding alveolar capillary walls forming a clear space bridged by fine fibrils (Fig. 21.27). Small capillaries surrounded by fine collagen fibres ramify within the cell masses and the whole structure is closely related to branches of the pulmonary vein. The centres of some of the *Zellballen* are occupied by structureless pink material and the collections of cells are sharply delineated by argyrophilic fibres. Nerve fibres, as already stated, may be shown to supply and ramify among the tumour cells, and some of the cells possess argentaffin properties.

As Korn *et al.* stated, these structures resemble "tumourlets" but can be differentiated by their interstitial position around venules and by their more compact whorled structure. They can also be differentiated from plexiform lesions seen in association with chronic pulmonary hypertension as they bear no relationship to arteries and the other changes seen in chronic pulmonary hypertension are absent.

The large presumed chemodectomas described by Heppleston and by Fawcett and Husband (1967) were characterized by nests of small polyhedral cells reminiscent of a carotid-body tumour. The nests of tumour cells were sur-rounded by a network of reticulin fibrils and were in intimate contact with thin-walled capillary vessels. Both tumours failed to give a chromaffin reaction and neither showed argentaffinic staining, also no nerve fibres were demonstrable and the cells showed no auto-fluorescence. Such tumours need to be differen-tiated from peripheral bronchial carcinoids including the oncocytomatous variety. Larger chemodectomas may also require to be differen-tiated from a variety of bronchial "adenoma" caused by proliferation of bronchial myoe-pithelial cells.

The tumours under discussion are probably formed by a hyperplasia of normal chemo-receptor cells caused by some as yet unidentified stimulus. Both the arterial and venous supplies to these tumours are derived from and drain into the pulmonary veins, suggesting that the function of the cells may be to sample the chemical composition of the blood before it returns to the left side of the heart.

2 (a) Papilloma of the Bronchus

Papillary tumours of the bronchus comprise a group of tumours of differing degrees of malignancy. At one end of the range is the benign squamous papilloma and at the other end is the papillary squamous-celled carcinoma. This section deals only with the benign papil-lomas and the reader is referred to Chapter 20 for an account of the transitional-cell papilloma and the papillary carcinoma. Benign squamous-celled papilloma of the bronchus is among the rarest of the pulmonary tumours and an early example was described by Orton (1932). Since then further cases have been added by Davis (1939), Liebow (1952), Ashmore (1954), Stein and Volk (1959), Al-Saleem *et al.* (1968), Laubscher (1969) and Le Roux *et al.* (1969).

Bronchial papillomas should be distinguished

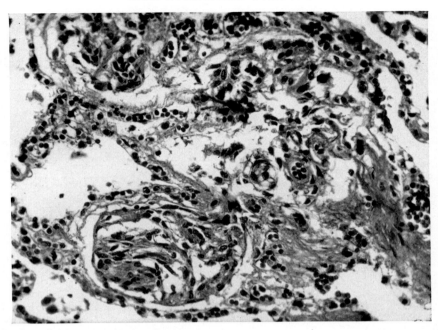

FIG. 21.27. A small chemodectoma in the periphery of the lung. The clump of tumour cells (*Zellballen*) are occupying an interstitial position in the alveolar wall, but have partly separated from the investing alveolar capillaries which lie between them and the alveolar lumen. Well-formed capillaries traversed some of the smaller clumps of cells. × 280 H and E.

from granulation tissue and other forms of bronchial polypoid tumours. They may be encountered in young persons and then the bronchial growth can form part of a generalized papillomatosis of the respiratory tract. In the case described by Hitz and Oesterlin (1932), papillomas of the larynx were associated with a bronchial papilloma and they regarded the latter as an airborne metastasis. Al-Saleem *et al.* discussed the long-term prognosis attaching to multiple respiratory papillomas confined to the major air passages in children, and found that they were often self-limiting and that permanent recovery occurred. This favourable outcome does not occur if there is broncho-alveolar extension when the mortality approaches 50 per cent (Singer *et al.*, 1966). It has been estimated that 2–3 per cent of laryngeal papillomas in children are associated with bronchial papillomas.

Bronchial papillomas in adults are often solitary growths but may be associated with papillomas elsewhere in the bronchial tree. Approximately 50 per cent of the solitary adult tumours become malignant and the long-term prognosis then becomes that attaching to lung cancer. Some of the solitary transitional-celled bronchial papillomas in adults are more correctly regarded as examples of bronchial carcinoma *in situ*.

Macroscopically, the tumour grows as a wart-like growth into the bronchial lumen (Fig. 21.28), though tumours arising from the terminal air passages may present as cystic masses.

Microscopically, they consist of a well-formed connective tissue stroma covered with either squamous metaplastic epithelium or ciliated columnar epithelium (Ashmore) (Fig. 21.29). Growths arising in the distal bronchi and

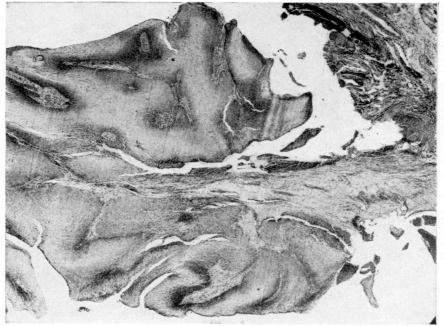

Fig. 21.29. A low-power view of a papilloma of the bronchus. × 30
H and E.

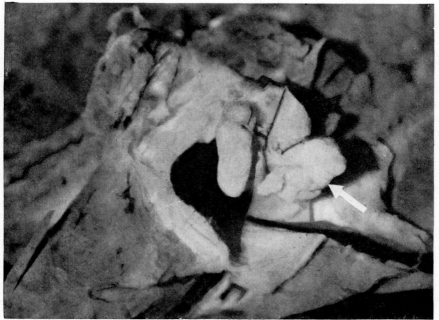

Fig. 21.28. A small polypoid papilloma of the bronchus projecting
into lumen of bronchus near the cut margin. The tumour is indicated
by an arrow. × 6.

terminal bronchioles often spread to line the adjacent alveolar spaces which are either filled with the tumour cells or distend to form epithelium-lined spaces. Careful examination of the bronchial epithelium frequently reveals multiple foci of squamous metaplasia unassociated with any gross tumour mass, suggesting that these neoplasms form part of a multicentric process. Stein and Volk consider that bronchial papillomas, like laryngeal papillomas, may be of an infective and possibly viral nature.

A wide spectrum of polypoid bronchial epithelial tumours exists varying from benign papillomas to undoubted malignant papillary types of squamous-celled carcinoma, and great difficulty may be experienced in deciding the true nature of some of these tumours. Papillary polypoid carcinomas may also arise multifocally and several benign papillomas may be found together with one that shows undoubted malignant transformation as in the case described by Kaufman and Klopstock (1964).

2 (b i) Bronchial Cystadenoma and 2 (b ii) Mucoepidermoid Adenoma

Bronchial cystadenomas are rare tumours that should be clearly differentiated from the commoner group of bronchial adenomas described previously. They are growths which undoubtedly arise from bronchial mucous glands and faithfully reproduce their structure. Unlike the cylindromatous variety of bronchial adenomas they are unassociated with any stromal mucoid changes and the tumours are confined within the cartilage layer of the bronchial wall. Tumours of this type were described by Davidson (1954) and Ramsey and Reimann (1953). Macroscopically, the growths present as rounded, polypoid tumours projecting into the bronchial lumen seldom exceeding 1·5 cm in diameter; they are usually considered pre-operatively to be bronchial cylindromatous adenomas or malignant tumours. They may cause obstructive pneumonitis.

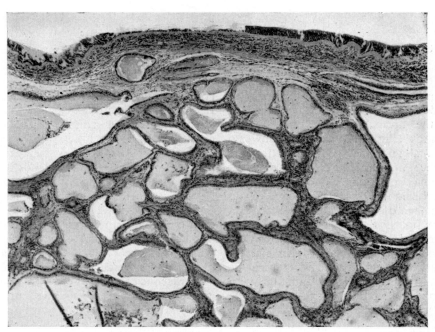

FIG. 21.30. A low-power view of a bronchial cystadenoma showing the similarity in structure to normal bronchial glands. × 40 H and E.

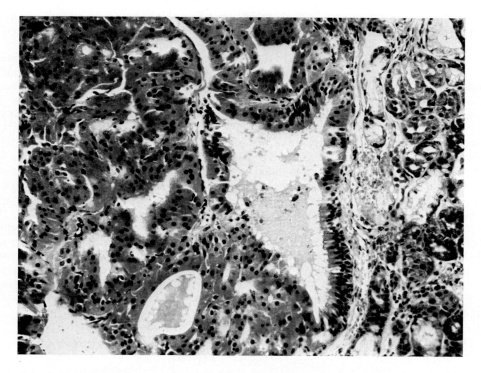

FIG. 21.31. A bronchial mucous gland oncocytoma. The tumour cells are strongly eosinophilic. × 180 approx. H and E

Microscopically, they consist of a mass of cystic, glandular spaces filled with mucus which reproduce the normal pattern of bronchial mucous glands (Fig. 21.30). They show no evidence of malignant change and the tumours are the bronchial counterparts of cystic sweat gland adenomas in the skin.

Mucoepidermoid adenomas were first recognized by Smetana quoted by Liebow (1952) and form a rare but distinctive group of mainly benign bronchial tumours derived from the mucous glands. They occur at all ages and are found equally in both sexes. They present as small mucus-coated lobulated tumours with numerous mucus-filled cysts on the cut surface (Fig. 21.32). Microscopically, they vary in appearance, in places containing solid clumps of cells with epidermoid characteristics and clearly derived from metaplastic surface bronchial epithelium or from the ducts of mucous glands. In other parts the tumour consists of a mixture of mucus-secreting acini and clumps of epidermoid cells (Figs. 21.33, 21.34).

These tumours usually behave like their counterparts in the salivary glands remaining as locally invasive neoplasms though Dowling *et al.* (1962) described an example in which malignant changes supervened. They form a distinctive group with a lesser tendency to infiltrate than the bronchial cylindromas.

Bronchial oncocytomatous adenomas are very rare tumours which arise from both the bronchial mucous glands and their ducts, and consist of collections of deeply eosinophilic cells arranged in acini and clumps (Fig. 21.31). A counterpart of this tumour occurs in the salivary glands where it is called an oxyphilic adenoma.

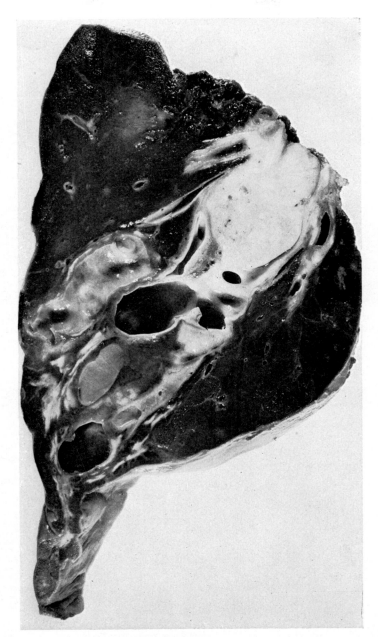

FIG. 21.32. A bronchial mucoepidermoid adenoma showing bronchocoeles filled with mucus distal to the tumour
(Reproduced by courtesy of Dr. E. Alvarez Fernandez, Madrid.)

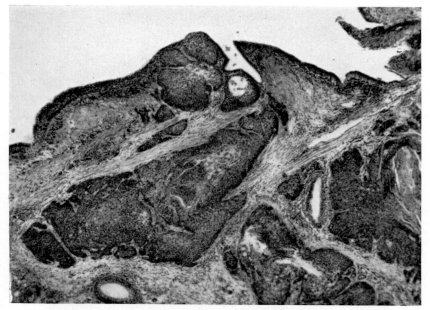

FIG. 21.33. A solid variety of mucoepidermoid adenoma found in a major bronchus of a child. Although superficially resembling a carcinoma, it was a slow-growing tumour and the child has remained free of any recurrence for many years. × 40 H and E.

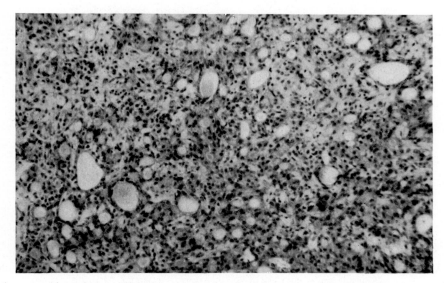

FIG. 21.34. A mucoepidermoid bronchial adenoma showing glandular spaces partly filled with mucus. × 100 H and E. (Reproduced by courtesy of Dr. E. H. Bailey.)

2 (c) Bronchial Cylindroma (Adenoid Cystic Carcinoma)

Synonyms: Bronchial mucous gland adenoma, Mixed bronchial tumours, denoid cystcar-cinoma of the bronchus

Bronchial cylindromatous adenomas (bronchial mucous gland adenomas) are uncommon tumours and make up about one-tenth of the combined group of bronchial carcinoids and bronchial cylindromatous adenomas formerly collectively referred to as bronchial adenomas. The unfortunate term cylindroma was coined by Billroth to describe an orbital growth arising from the lachrymal gland and has since been applied to several microscopically similar tumours, including a group of salivary gland adenomas and the tumours under review which arise in the tracheo-bronchial tree where mucous glands are normally found. The tumours are characterized by the growth of cells in long solid cylinders. The first cylindromatous adenoma in the bronchial tree was described by Heschl (1877). Most regard these tumours as a low-grade form of bronchial carcinoma.

Macroscopically, these tumours grow beneath the bronchial epithelium and extend directly into the subjacent bronchial wall and extrabronchial tissues. They are seldom polypoid growths but nevertheless lead to occlusion of the bronchial lumen. The cut surface is greyish-white, glary and exudes mucus and the deep margin of the tumour is often ill-defined.

Microscopically, these growths very closely resemble salivary gland adenoid cystadenomas (mixed salivary tumours). They commonly consist of interlacing cylinders of tumour cells the centres of which are frequently canalized to form tubular spaces filled with PAS positively staining epithelial mucin (Fig. 21.35). The tumour cells are stellate or pleomorphic and are darkly staining; occasionally islets of cells invest small

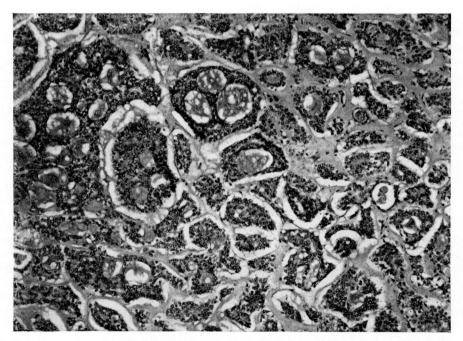

FIG. 21.35. Low-power view of a cylindromatous type of bronchial adenoma. Note the myxoid change in the stroma immediately investing the epithelial cells. × 140.

keratinized epithelial pearls. Tubular parts of the growth may be partly lined by ciliated cells (Belsey and Valentine, 1951). Many pathologists regard the bronchial cylindromatous adenoma as a low-grade carcinoma which should more correctly be called bronchial cystadenocarcinoma. Mitoses are often numerous and undoubted malignant change not uncommonly supervenes. Metastases may then be found in the liver, kidney and more rarely in other organs and retain the characteristic microscopical features of the primary tumour.

The stroma of these tumours consists of two parts. The first corresponds to the layer of lamina propria which normally invests the mucous glands and which frequently undergoes myxomatous change due to accumulation of acid mucopolysaccharide. This stains positively with alcian blue. The myxomatous change that so commonly occurs in this layer has been responsible for these tumours being mistaken for mixed tumours. The second or true collagenous stroma is formed by the connective tissue framework lying outside the lamina propria.

Histochemical studies have shown the presence of acid phosphatase in the tumour cells but no other common intracellular enzymes have been found (Willighagen et al., 1963).

3 (a) Chondroma of the Bronchus

This tumour has been confused with the local form of pulmonary hamartoma in which cartilage is an important constituent. Unlike pulmonary hamartomas, however, it grows from fully formed bronchial cartilage and is not associated with disordered growth of the other constituents of the bronchial wall. Localized pulmonary hamartomas which often consist mainly of cartilage also contain lipomatous and lymphoid tissue together with epithelium-lined spaces, but these additional tissues are absent from the true bronchial chondromas. Examples of true bronchial chondromas have been described by Blecher (1910), Davidson (1941)

and Walsh and Healy (1969). These tumours were also referred to by MacCallum (1936), Liebow (1952) and Hochberg and Schacter (1955), and the endobronchial hamartoma referred to by Young et al. (1954) was probably a further example.

Macroscopically, the tumour projects as a smooth lobulated, sessile or polypoid growth into the bronchial lumen (Fig. 21.36). It consists almost entirely of semitranslucent cartilage surrounded by a fibrous tissue wall.

Microscopically, the growths consist of hyaline cartilage containing a few elastic but no collagen fibres. As in all chondromas the cartilage cells are larger and arranged in a more disorderly fashion than are their normal counterparts. Some chondromas may show myxoid characteristics (Fig. 21.37) and others may undergo ossification and develop fatty and even red bone marrow in their centres (Figs. 21.38, 21.39).

3 (b) Lipomas of Bronchus and Lung

Lipomatous tissue is a normal constituent of the bronchial wall (Watts et al., 1946), and may also be found beneath the pleura. It is normally present in bronchi down to 1 mm in diameter (Watts et al.), being found in the submucosa between the cartilage and muscular layer surrounding the mucous glands, and occasionally it is found as a result of metaplasia inside ossified cartilage rings. Lipomas may arise from any fatty tissue within the lung and are divided into: (a) those in the bronchi and (b) those beneath the pleura.

(a) *Bronchial lipomas*. Rokitansky (1854) stated that lipomas could arise in the submucous areolar tissue of the bronchi but the first case of a bronchial lipoma was described by Kernan (1927), the tumour having been removed by bronchoscopic excision. Lipomas are found in the large bronchi where fatty tissue is normally present, and more occur in the left main and lobar bronchi than in the corresponding bronchi on the right (McCall and Harrison, 1955). The tumours are often dumb-bell-shaped with

FIG. 21.36. A chondroma of the bronchus showing its smooth lobulated surface. × 1½ natural size.

expansions filling the bronchial lumen and submucosal coat and with a narrow neck between. They usually project as smooth-walled polyps into the bronchus (Fig. 21.40) and show the microscopical structure of a fibro-lipoma (Fig. 21.41). As they grow, the normal constituents of the bronchial wall become stretched out over them including the muscle coat, which later atrophies and is replaced by connective tissue.

According to Carlisle et al. (1951a), the majority of bronchial lipomas become clinically manifest in the fifth and sixth decades. Lipomas arising in the central bronchi lead to obstructive symptoms, causing obstructive pneumonitis and chronic bronchiectasis at an early stage. Wheezing respiration may first draw attention to their presence.

Bronchial lipomas should be distinguished from intrathoracic mediastinal lipomas which may reach an enormous size and which may spread directly in the parabronchial tissues filling the lung hilum.

A rare form of bronchial lipoma may occasionally be found in which the whole circumference of a medium-sized bronchus is involved. The lipoma fills the submucous coat and extends over a considerable length of the bronchus.

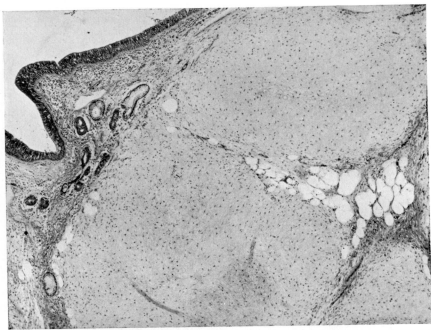

Fig. 21.37. Section of a bronchial chondroma showing its general cartilaginous structure with some tendency to myxoid change in some places. × 40 H and E.

Although it has been included as a variety of lipoma, doubt has been expressed about the neoplastic nature of this condition, some regarding it as sequel to chronic inflammation, usually chronic bronchiectasis.

(b) *Subpleural lipomas.* Peripherally situated bronchial lipomas are rare, though such a tumour was reported in the posterior apical segment of the lung near the pleural surface by Shapiro and Carter (1954). The author has seen one example of a subpleural lipoma which spread over the surface of an upper lobe.

Very rarely a subpleural lipoma may grow more rapidly and show microscopical features consistent with a liposarcoma including giant foamy lipoblasts.

3 (c) Pulmonary Fibroma and Myxoma

Like most forms of benign pulmonary connective tissue tumours, these growths are rare. They may be divided into bronchial and pulmonary fibromas, and difficulty may be experienced in differentiating them from neuro-fibromas. The distinctive features of the latter tumours are discussed elsewhere.

Pulmonary fibromas are expanding, tough, greyish-white and often fasciculated tumours which grow in the bronchial wall or in the lung substance (Fig. 21.42). When they are situated beneath the pleura they may be difficult to distinguish from pleural fibromas.

Microscopically, they are composed of spindle cells with a variable amount of collagen (Fig. 21.43), and they not infrequently undergo myxomatous change and calcification.

Pulmonary myxomas have been described by Placitelli (1953) and Littlefield and Drash (1959). The cut surfaces of the myxomatous tumours present a glistening appearance which is absent in the true fibromas.

Microscopically, the component cells are

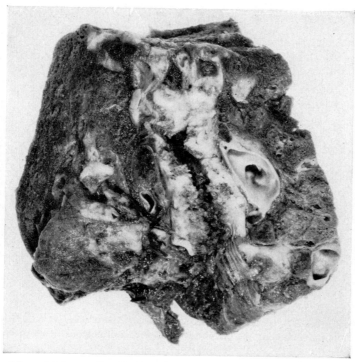

Fig. 21.38. A bronchial osteochondroma involving a length of the bronchial wall. Natural size.

stellate-shaped and lie surrounded by connective tissue mucin in which lie a few reticulin fibrils. Myxoid areas frequently occur in sarcomas and these should be distinguished from the benign tumours, though often only the ultimate clinical behaviour of the growths determines their true nature.

3 (d) Pulmonary Fibroleiomyomas

In common with pulmonary fibromas, fibroleiomyomas in the lung are divided into those which arise from the bronchial wall and those which occur in the periphery of the lung. The latter group of tumours may reach a very large size as in the case described by Williams and Daniel (1950).

Hirose and Hennigar (1955) have reviewed the literature and added a further case. In the past insufficient attention has been given to excluding a secondary metastasis from a uterine fibroleiomyosarcoma.

The term *"metastasizing fibroleiomyoma of the uterus"* was used by Steiner (1939) to describe uterine fibroids of an apparently benign and well-differentiated nature which nevertheless gave rise to secondary deposits in the lungs. The pulmonary tumours showed the features of a characteristic benign fibroleiomyoma. Two further examples were reported by Spiro and McPeak (1966). Prolonged survival usually follows removal of a solitary pulmonary metastasis and the tumours are very slowly growing. The metastases are readily confused with a primary pulmonary tumour (Pierce *et al.*, 1954). So readily may this mistake be made that it is probably unjustifiable to diagnose an intrapulmonary fibroleiomyoma in a woman in the absence of a complete examination of the uterus to exclude a malignant fibroid.

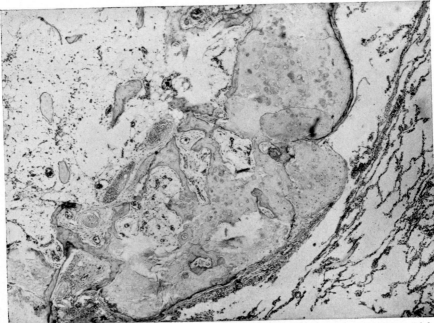

Fig. 21.39. An osteochondroma showing the peripheral zone of chondromatous tissue within which lies bone and fatty-marrow tissue. × 30 H and E.

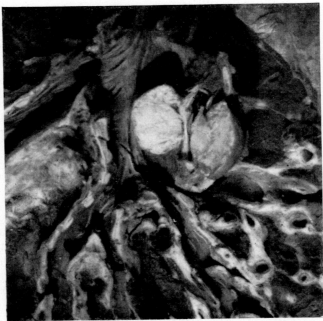

Fig. 21.40. A bronchial lipoma projecting as a smooth-walled polypus into the bronchial lumen. It was discovered by accident at post-mortem. Approx. × 1½ natural size.

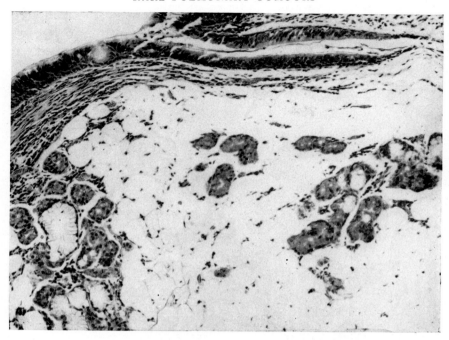

Fig. 21.41   A section of a submucous fibro-lipoma of the bronchus. The lipomatous tissue extends to just below the epithelium. × 140 H and E.

Bronchial fibroleiomyomas grow as polypoid tumours into the bronchial lumen, extending along it for a considerable distance as a tongue of soft greyish-red tumour tissue (Turkington *et al.*, 1950). They arise from the bronchial musculature (Fig. 21.44) and may undergo hyaline change and calcification (Fig. 21.45). The surface frequently ulcerates causing haemoptysis and the commonest complication is obstructive pneumonitis. Most of the peripheral tumours have only been discovered following a routine radiograph of the chest.

The peripheral tumours tend to be of a more fibrous consistency than the central bronchial growths (Figs. 21.46, 21.47) and macroscopically present a whorled appearance, consisting of plain muscle fibres and fibrous tissue together with areas which undergo hyaline degeneration and calcification. Necrosis has been described in some of the largest peripheral tumours which

nevertheless shell out readily from the surrounding lung (Randall and Blades, 1946).

3 (e) Pulmonary Sarcoma

All varieties of primary pulmonary sarcomas are rare; Passler (1896) at Breslau reported four cases of these growths in 9246 consecutive post-mortems and Rolleston and Trevor (1903) found mention of three pulmonary and two bronchial sarcomas among 3983 autopsy reports at St. George's Hospital, London.

Despite the great increase that has occurred in the incidence of lung cancer during the period that has elapsed since the above figures were quoted, no commensurate increase in the number of cases of lung sarcoma has been seen. Noehren and McKee (1954) found a reference to only one case of primary lung sarcoma among 7272 post-mortems performed at the

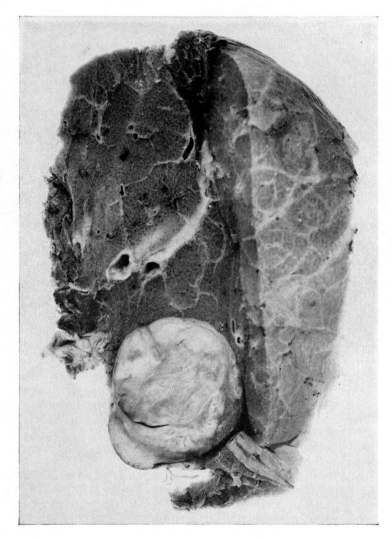

FIG. 21.42. A peripheral fibroma in lung showing its sharply circumscribed margins and coarse fasciculated appearance on the cut surface. Natural size.

University of Kansas Hospitals, and Mallory (1936) found only one case in 8000 post-mortems at the Massachusetts General Hospital. It has been estimated that for about every 500 cases of lung cancer there is one case of lung sarcoma (Iverson, 1954 and Cameron, 1975).

Primary sarcomas of the lung may be divided on clinical and pathological grounds into those originating (a) in the bronchial walls or (b) in the lung parenchyma. The bronchial variety is usually considered to be the more common (Carswell and Kraeft, 1950). Pulmonary sarcomas may arise from fibrous tissue, unstriped muscle, cartilage, nerve sheath tissue and undifferentiated primitive connective tissue cells. Neurofibrosarcomas, carcino-sarcomas of the

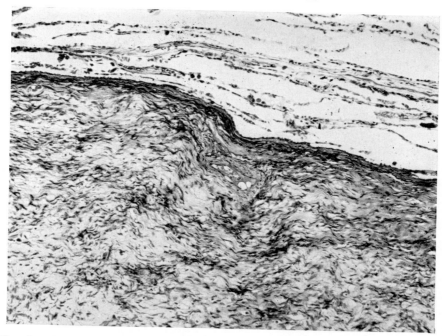

FIG. 21.43. The edge of a peripheral lung fibroma showing bundles of collagenous fibres with mature spindle-shaped fibrocytic cells. The whole growth has a sharply circumscribed margin and shows no tendency to infiltrate the lung. × 140 H and E.

bronchus and rhabdomyosarcomas are considered separately elsewhere, and the remaining groups of sarcomas are discussed in this section.

Neumann (1938) stated that myomatous growths occurred in organs where muscle fibres lie outside the vascular system, but very rarely arose from the plain muscle fibres forming the walls of vessels. In general it would appear that this statement is correct and appears to apply to the lung.

In the past much confusion arose because the undifferentiated spindle and small round-celled bronchogenic carcinomas were mistaken for both intrabronchial and intrapulmonary leiomyosarcomas. The two varieties of growth may readily be distinguished because true sarcomas produce reticulin fibres whilst carcinomatous growths fail to do so. Endobronchial sarcomas have to be distinguished from the rare polypoid squamous-celled carcinomas with sarcoma-like changes due to intracellular oedema, and from the carcino-sarcomas described in Chapter 20.

A further rare and unusual form of lung cancer, the giant-celled carcinoma, which usually arises in the periphery of the lung may also be mistaken for a rhabdomyosarcoma, though striated fibres are lacking and the behaviour and spread of these growths conform to the usual characters of lung cancer.

Despite their malignant properties all forms of pulmonary sarcoma described above are in the first instance locally invasive tumours, remaining so for a longer period than the much commoner lung cancer, and according to Black (1950) they do not give rise to distant metastases. For this reason excision of the growth, however large, by lobectomy or pneumonectomy should be considered, unless obvious distant metastases are already present. *Endobronchial fibrosarcoma and leiomyosarcoma* have been described by

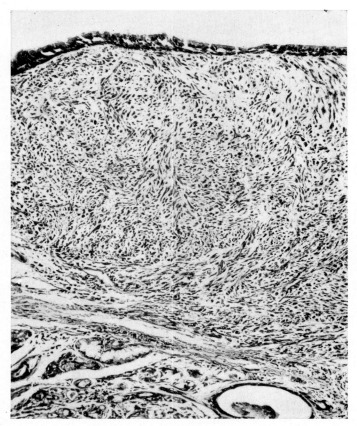

FIG. 21.44. A simple fibroleiomyoma arising in the wall of a large bronchus. × 140 H and E.

Stout (1948), Storey (1952), McEachern *et al.* (1955), Yacoubian *et al.* (1958) and Guccion and Rosen (1972); the majority arise in the large bronchi and are usually visible on bronchoscopic examination. They grow slowly, extending locally as smooth polypoid tumours within the bronchial lumen. Although the majority of growths of this type are restricted to the bronchial wall and lumen, a few invade the surrounding lung and the pulmonary veins, continuing to grow within the lumen of the latter. They are greyish-white or pink, friable growths which bleed readily and are divisible into two types, the spindle-celled and round-celled varieties. The former closely resemble benign fibroleiomyomas and great difficulty may

be experienced on histological grounds in determining whether any particular tumour is malignant or benign. The same problem arises with the peripheral pulmonary leiomyosarcomas discussed below.

Stout and Hill (1958) outlined the principal criteria by which all soft tissue sarcomas could be distinguished on histological grounds and they included the presence of bizarre giant cells, one or more mitoses in every microscopic field using ×200 magnification, and an absence of myofibrils, the general features of hyperchromatism and metastasis formation common to all forms of malignant tumours (Figs. 21.48, 21.49). Their findings are applicable to the pulmonary growths. In general the round-celled types of

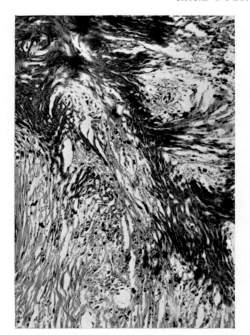

FIG. 21.45. Calcification and hyalinization of fibroleio-myoma arising in a large bronchus. × 140 H and E.

growth tend to be more malignant than the spindle-celled tumours. As with all doubtful soft tissue tumours electron microscopy should now be regarded as an essential method of examination in order to determine the cell nature of any doubtful pulmonary sarcoma.

Ulceration and inflammatory changes of the surface of these growths is common and may lead to inflammatory hyperplastic changes in the underlying tumour cells causing them to re-semble anaplastic tumours.

Diffuse Intrapulmonary Leiomyosarcomas and Fibroleiomyosarcomas

Agnos and Starkey (1958) reviewed the literature on all varieties of lung sarcoma and Guccion and Rosen in a more comprehensive and recent account analysed a series of eighty cases including thirty-two unreported examples. Individual and small numbers of cases have also

been described by Yacoubian *et al.*, Glennie *et al.* (1959), Mason and Azeem (1965) and Ramanathan (1974).

Several of these tumours have been discovered during the course of post-mortem examinations in elderly persons and the duration of their presence in the lung was unknown.

The tumours are mostly well circumscribed and tend to surround and embed the bronchi but do not invade their walls, though they cause bronchial stenosis by compression. The tumours vary in their microscopical appearances from the highly differentiated fibroleiomyosarcomas of doubtful malignancy (Merritt and Parker, 1957) to anaplastic giant-celled growths whose malignant nature is undoubted. The presence of the former type of growth in a woman should only be accepted if a metastasis from a malignant uterine fibroid is first excluded.

Some lung sarcomas can undoubtedly grow from plain muscle tissue in the interstitial plane of the alveolar walls and may continue to grow in this plane before leading finally to alveolar obliteration. This type of spread, however, can be mimicked by carcinoma of the lung which may rarely mimic a sarcomatous growth. Occasionally lung sarcomas may be multiple, and malignant change may supervene in a previously benign leiomyoma. Spread of these growths may involve the pleural surface with subsequent dissemination within the pleural cavity. Yacoubian *et al.* considered that the absence of lymph node metastases in cases of suspected pulmonary sarcomas constituted one of the more important differential diagnostic features between these tumours and lung cancer. Occasionally, however, metastasis to a hilar lymph gland occurs (Mason and Azeem, case 4) but is rare.

Endobronchial leiomyo- and fibrosarcomas metastasize late and complete surgical removal usually results in clinical cure. The prognosis attaching to intrapulmonary leiomyosarcomas is largely dependent on their size when dis-covered. Small tumours are often highly malig-nant whereas the larger tumours tend to spread more slowly though the whole group carry a less favourable prognosis than the endobronchial

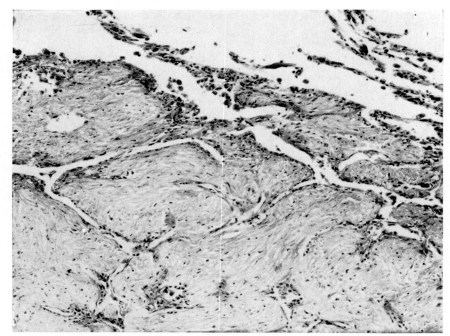

FIG. 21.46. The edge of a diffuse fibroleiomyoma arising in the periphery of the lung. × 140 H and E.

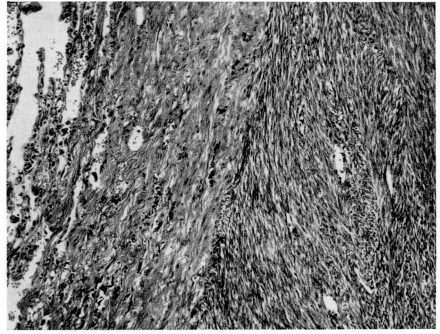

FIG. 21.47. Another part of the tumour shown in Fig. 21.46. × 140 H and E.

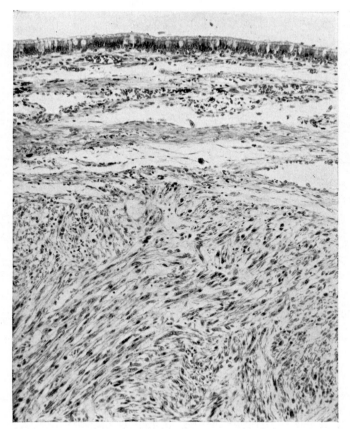

FIG. 21.48. A fibromyosarcoma arising in the wall of a large bronchus. × 140 H and E.

tumours. The large tumours may spread directly through the pleura into the chest wall.

Small intrapulmonary fibrosarcomas if near the surface of the lung may be confused with pleural tumours. As a group they carry a poor long-term prognosis and most of the tumours metastasize and prove fatal within 5 years of discovery. Large intrapulmonary fibrosarcomas are more slowly growing tumours and radical extirpation may more often be followed by cure.

Both undifferentiated pulmonary leiomyosarcoma and fibrosarcomas have been reported in foetal and early post-natal life (Killingsworth et al., 1953; Robb, 1958).

Very rarely leiomyosarcomas arise in the lung from pulmonary arteries and veins. In the case described by Wackers et al. (1969) the tumour arose from the trunk of the pulmonary artery and they reviewed twenty-one previously reported cases. These tumours cause intimal thickening in the affected vessels blocking the vessel lumen. It is possible that the tumours may arise from endothelial cells which as Bensch et al. (1964) showed possess leiomyofibrils in their cytoplasm.

Primary Chondrosarcoma and Myxosarcoma in the Lung

The first case of a primary pulmonary chondrosarcoma was presented by Wilks (1862) and

FIG. 21.49. A fibromyosarcoma of the bronchus showing malignant giant cells and large hyperchromatic spindle-shaped cells. × 140 H and E.

further cases have been described by Lowell and Tuhy (1949), Miller and Jackson (1954) and Daniels *et al.* (1967). Before any growth can be accepted as a primary lung tumour it is first essential to exclude a primary source elsewhere, usually in the skeleton, as both in their appearance and mode of spread primary and secondary chondrosarcomas in the lung are identical. Furthermore, metastatic growth from a primary skeletal tumour may not appear for at least 12 years after the primary tumour was diagnosed. In most cases a primary source of growth outside the lung can only be excluded after death by a very careful post-mortem examination. Lichtenstein and Jaffé (1943) have enumerated the features that distinguish benign chondromas from malignant bone chrondrosarcomas. The same criteria apply to pulmonary tumours, including the presence of tumour cartilage cells with large nuclei, giant cells and myxomatous change, but despite critical histological assess-

ment, considerable difficulty may be encountered. Both primary and secondary chondrosarcomas in the lung invade blood-vessels, and both may extend intravascularly as an uninterrupted tumour mass (Castren, 1931). In the primary myxosarcoma of the lung described by Miller and Jackson (Fig. 21.50), growth extended into the pulmonary veins and left atrium, and they considered that the tumour probably arose in a chondromatous type of hamartoma. A similar tumour seen by the author was probably of the same nature consisting partly of neoplastic cartilage cells and partly of undifferentiated spindle cells and was thought to have arisen from a bronchial hamartoma.

The intravascular spread of secondary chondrosarcoma in the lung may simulate a massive pulmonary embolus (Schwarz *et al.*, 1972). A subpleurally arising primary chondrosarcoma was described by Guida Filho and Pasqualucci (1963).

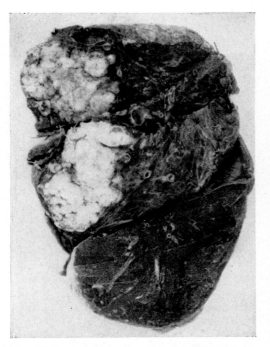

FIG. 21.50. A chondrosarcoma arising in the lung. (Reproduced by courtesy of Dr. A. A. Miller and the Editor of *J. Path. Bact.*)

3 (f) Pulmonary Rhabdomyosarcomas

Among the rare primary malignant tumours of the lung are included the rhabdomyosarcomas. Although several neoplasms have in the past been described as rhabdomyosarcomas, some including the cases described by Helbing (1898) and Zipkin (1907) would now be regarded as malignant pulmonary blastomas.

Acceptable cases of primary rhabdomyosarcomas have been described by McDonald and Heather (1939), Gordon and Boss (1955), Forbes (1955) and Drennan and McCormack (1960). Mostly the tumours have replaced one or more lobes and have shown a tendency to invade pulmonary veins and bronchi. In the case described by McDonald and Heather tumour tissue extended as far as the left atrium, whilst in the growth described by Forbes tumour tissue filled the left upper lobe bronchus. Despite their size most of the tumours have been surrounded by fibrous tissue formed of compressed lung and distant metastases have been absent.

Microscopically, they consist mainly of undifferentiated spindle and round cells together with longer strap-like cells and large giant cells, the latter containing a central mass of closely packed nuclei surrounded by a mass of vacuoles. Following the use of specialized stains cross-striations are demonstrable in some of the spindle and giant cells whilst many tumour cells are filled with glycogen. Islets of squamous epithelium formed from metaplastic bronchial mucosa may lie buried in the sarcomatous tissue.

The origin of the striated fibres has been much debated, Willis (1953) believing that they developed from undifferentiated mesoblastic cells which retained the ability to develop into rhabdomyomatous cells in post-natal life, whilst others believe that they develop from misplaced voluntary muscle tissue from the pharyngeal region. The finding of "endodermal" elements in the tumours described by Helbing and Zipkin has led to the belief that rhabdomyosarcomas were a form of intrapulmonary teratoma. It is possible that the endodermal elements formed part of a pulmonary blastoma which was mainly a polymorphic rhabdomyosarcoma. Such a tumour has been seen by the author and is illustrated in Fig. 21.52. The existence of rhabdomyomatous fibres in Wilm's tumour (nephroblastoma) of the kidney has been attributed to the plasticity of the mesoderm which normally forms both the renal cortex and abdominal parietes. Similar plasticity of the pulmonary mesoderm from which the peripheral part of the lung develops according to Waddell (1949) may be responsible for the rhabdomyosarcomatous character of some lung sarcomas. The endodermal elements reported in some of the earlier cases of pulmonary rhabdomyosarcomas were mistaken for immature bronchiolar structures which are normally found in pulmonary blastomas (see Chapter 23).

Primary rhabdomyosarcomas have also been reported to arise within the pleura, but in the case described by Duhig (1959), alternative sites of origin had not entirely been excluded.

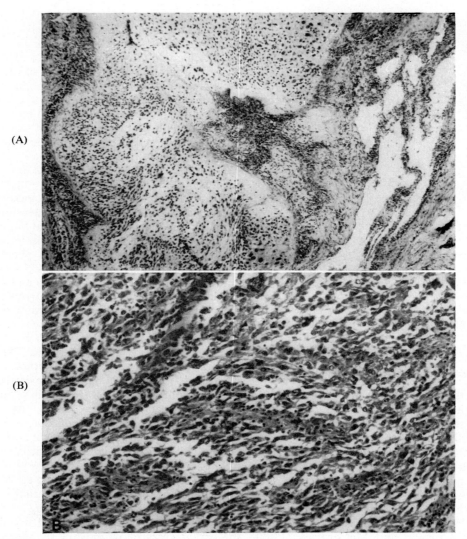

(A)

(B)

FIG. 21.51. An undifferentiated pulmonary sarcoma containing areas of chondrosarcoma. (A) shows malignant cartilage tissue and (B) undifferentiated fibrosarcoma. (A) × 40, (B) × 100. Both stained H and E.

(Reproduced by courtesy of the late Professor H. A. Magnus and Dr. B. S. Cardell.)

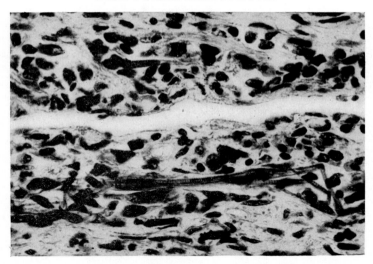

FIG. 21.52. A pulmonary blastoma containing rhabdomyomatous tissue. × 400. Mallory's PTAH stain. (Reproduced by courtesy of K. F. W. Hinson.)

3 (g) Intravascular and Sclerosing Bronchioloalveolar Tumour (IVSBAT)

This lesion of unknown aetiology and uncertain nature has only been recognized within the last 10 years. The name by which these lesions are at present known is only a provisional one and when their true nature and aetiology are better understood they will almost certainly be renamed. For the purposes of description they are included among the neoplasms and included here.

Farinacci *et al.* (1973) described a case which was probably the first to be reported though Liebow had encountered similar lesions some years previously. Dail and Liebow (1975) presented twenty cases which they had accumulated over 13 years. The author has also seen two cases. Two-thirds of the patients have been females and their ages have ranged from 14 to over 70 years though nearly half have been under the age of 30.

Many of the cases were discovered accidentally following a routine chest radiograph and for a long time the patients suffered no disability. Both lungs contain multiple small rounded opacities often mistakenly regarded as secondary tumour deposits. Subsequent follow-up radiographs at intervals show a gradual increase in the number and size of the individual opacities over a period of years until in the advanced stages, when most of the functional lung tissue has been destroyed, there is clinical evidence of the rapid onset of respiratory insufficiency. After many years the patients succumb but survival may continue for up to 12 years or more after the initial discovery of the disease.

Microscopically, the initial change is an interstitial alveolar infiltrate with lymphocytes followed by protrusion of the alveolar wall in a polypoidal manner into the alveolar lumen. The protrusion is covered with hypertrophied alveolar (type 2) epithelial cells and its core consists at first of a myxoid type of connective tissue which stains positively with the PAS and Congo-red stains. As the stroma matures it fails to stain positively with these stains and is converted into hyaline fibrous tissue in which are embedded scattered pale-staining vesicular cells (Fig. 21.53). Eventually the whole alveolus is filled and becomes totally obliterated by acellular pale-staining hyaline fibrous tissue which may undergo central calcification. The walls of partially obliterated alveoli are lined by

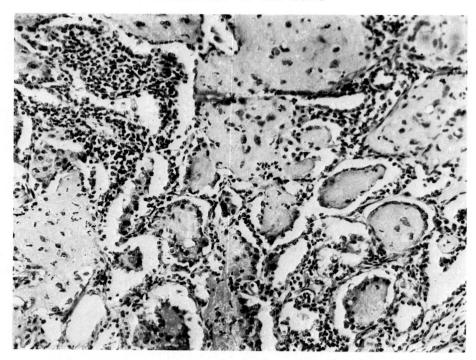

Fig. 21.53. A high-power view of an intravascular bronchiolo-alveolar tumour (IVSBAT) showing the relatively acellular protrusions into the alveolar lumens of tumour tissue. There is some interstitial lymphocytic infiltration in the alveolar walls. × 180 H and E.

swollen alveolar epithelium. The same tissue that fills and and obliterates the alveoli may spread into and along the small bronchi and bronchioles (Fig. 21.54) and also fills small branches of the pulmonary arteries and veins. This latter feature has given rise to the name by which the condition is at present known. Reticulin stains reveal a pericellular network of fibres in the intra-alveolar polyps and trichrome stain later becomes positive for collagen. The proliferative process spreads inexorably in a centrifugal manner destroying the alveolar tissue rendering it airless, avascular and fibrotic.

In some respects the changes bear a resemblance to those seen in so-called sclerosing angioma though the course and outcome are quite different.

The indolent character of the disorder and the tendency for the lesions to undergo hyaline fibrosis are very unlike the behaviour of most malignant neoplasms. No primary tumour source outside the lung has been discovered. The author considered the possibility of the disorder being primarily a metabolic disturbance rather than a neoplastic one, but so far no evidence to support this suggestion has been found.

3 (h) Pulmonary Angioma

Synonyms: Congenital arterio-venous varix, Pulmonary arterio-venous aneurysm, Pulmonary arterio-venous fistula and Pulmonary haemangioma

These tumours may be of congenital or acquired nature. The congenital lesions will be considered first.

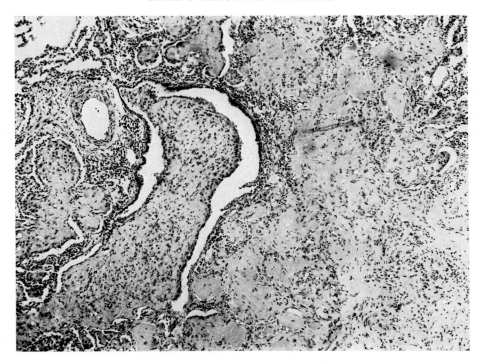

Fɪɢ. 21.54. A low-power view of an intravascular bronchiolo-alveolar tumour (IVSBAT) showing much of the lung replaced by hyaline fibrous tissue and an extension of tumour tissue occupying the lumen of a bronchus. × 75 H and E

CONGENITAL PULMONARY ANGIOMA

The term angioma or haemangioma, although commonly used for these lesions, is a misnomer, implying that they are derived from an overgrowth of angioblastic tissue. They are not formed from an overgrowth of angioblastic tissue but result from a persistence of foetal anastomotic capillaries which may transmit an ever-increasing quantity of blood from the supplying artery to the draining vein. They do not qualify for inclusion amongst the hamartomas.

The pulmonary vessels were originally developed from a network of capillaries in which free communications existed between the arterial and the venous sides of the circulation (Woollard, 1922). At first the arterial supply was derived from the dorsal aorta but later this was largely replaced by the pulmonary artery. Also initially

the veins entered the cardinal system, but later they drained through the pulmonary veins into the sinus venosus.

Persistence of short foetal capillary anastomotic channels between the arteries and veins may gradually lead to the diversion of arterial blood through such channels. As the blood-flow tends to follow the path of least resistance, an increasing volume of the arterial flow tends to be diverted through such communications which become very distended and continuously increase in size. The low diastolic pressure within the supplying artery results in its walls being thin, tortuous and very dilated owing to the larger volume of blood which it conveys. In the case of a fistulous communication between a large artery and vein, anastomotic arteries also pour their blood into the feeder artery and themselves tend to become similarly dilated.

Once established a large congenital arterio-venous anastomosis becomes a slowly expanding lesion transmitting an ever-increasing quantity of blood.

The size and effects of such arterio-venous communications depend on the sizes of the vessels involved; if the anastomoses involve small peripheral arterioles and venules, the result is a small telangiectasis which usually remains small and causes no haemodynamic changes in the pulmonary circulation. If large arteries and veins are involved in the anastomosis serious haemodynamic changes may result. The evolution of pulmonary arterio-venous anastomoses in pulmonary "angiomas" has been described in detail by Cope (1953).

Occasionally, the venous branch forming these lesions may drain into the venae cavae due to persistence of the foetal cardinal venous drainage (Grishman et al., 1949), also the arterial supply may occasionally be furnished by a branch of the bronchial artery or other systemic artery.

The first account of a pulmonary angioma is attributed to Churton (1897) who described the post-mortem findings in the lungs of a patient whom he mistakenly regarded as having died from pulmonary arterial aneurysms. His description leaves little doubt that he was really referring to the dilated, thin-walled branches of a pulmonary artery supplying a pulmonary arterio-venous aneurysm. Later, in 1917, Wilkens described a further case in a woman who had died from a massive pleural haemorrhage. Since then many cases have been added and many reviews of the subject have appeared, including those by Giampalmo (1950), Muri (1953 and 1955), Stringer et al. (1955), Le Roux (1959) and Dines et al. (1974).

Pulmonary angiomas are multiple lesions in about a third of all cases (Bosher et al., 1959; Dines et al.). In about 8 per cent of cases the lesions are bilateral and in approximately 4 per cent there is a systemic arterial supply. The individual tumours vary considerably in size (Figs. 21.55, 21.56), sometimes only two being present. In about a half of the cases with multiple angiomas the lesions are unilateral.

In 50 per cent of cases signs and symptoms attributable to them begin before the age of 15.

The multifocal nature of the condition may not be appreciated at first and may only be recognized when a second tumour appears following surgical removal of a manifest arterio-venous aneurysm in one lobe of a lung. In 60 per cent of cases pulmonary angiomas form part of the generalized disorder of hereditary telangiectiasis (Osler–Rendu–Weber disease) and may be the first manifestation of that disease.

Pulmonary angiomas may be of the cavernous or capillary varieties and each will be described separately.

CAVERNOUS ANGIOMAS OR PULMONARY
ARTERIO-VENOUS ANEURYSMS

There is no convincing evidence that pulmonary arterio-venous aneurysms bear any relationship to the arterio-venous anastomoses which are claimed to exist in the human lung by Tobin and Zariquiey (1950) and Steinberg and McClenahan (1955), but whose existence is still denied by some anatomists.

The dilated blood-filled spaces are usually supplied by more than one tortuous and dilated branch of the pulmonary artery. In the majority of pulmonary angiomas the systemic arteries (bronchial or aberrant bronchial arteries) do not communicate with the lesion. Abnormalities of the venous drainage occur and may be of surgical importance because no radical operation for the removal of a pulmonary angioma should be undertaken before the existence of an adequate normal pulmonary venous return has been confirmed.

The angiomas may be associated with congenital abnormalities of the lung such as localized bronchiolar maldevelopment (Giampalmo), and may coexist with pulmonary venous abnormalities (Grishman et al.). An example of a pulmonary angioma in a sequestrated lower lobe of the left lung supplied by an aberrant bronchial artery and draining into the hemiazygos vein was described by Bjork et al. (1963).

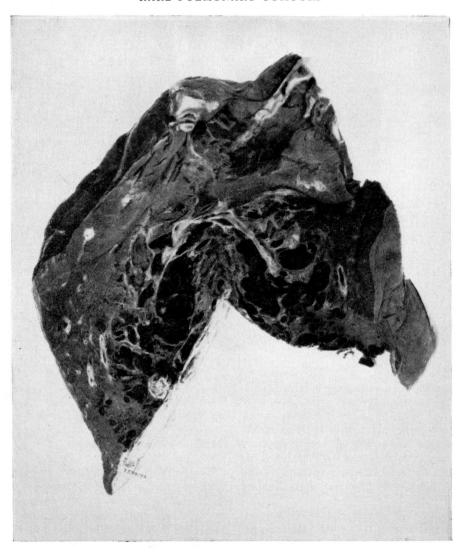

FIG. 21.55. A very extensive angioma occupying much of the lower lobe of the lung. The cavernous spaces are filled with dark blood clot. Approx. one-third natural size.

Thickening of the venous walls is not so marked as in the veins draining systemic arterio-venous anastomoses because of the lower intraluminal pressure; the media in the walls contain either few disrupted or no elastic fibres but fibro-elastoid intimal thickening and calcification of the venous walls does occur and may show in radiographs of the lungs. The veins draining the anastomoses distend and become varicose and the walls separating the varicosities atrophy, leaving large blood-filled cavities partly divided by septa which are the remains of the original walls. Disruption of the venous walls is increased by any concomitant state that raises

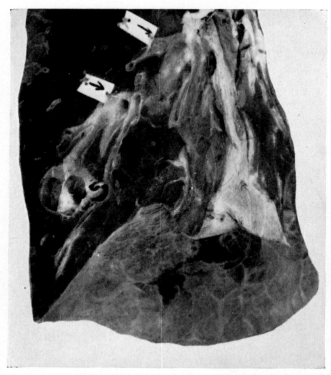

FIG. 21.56. A small cavernous angioma in the lung showing the dilated pulmonary vein draining the cavernous blood spaces. Approx. natural size.

the pulmonary venous pressure such as mitral stenosis and a chronic cough. Several pulmonary veins normally communicate with an angioma from the adjacent lung parenchyma and during atrial systole blood may flow back into the lung capillaries.

Arterio-venous aneurysms may be demonstrated in excised lung tissue by injecting the main pulmonary arteries or their lobar branches together with the main veins with different coloured latex preparations, subsequently dissolving the lung tissue with concentrated acid. In well-developed cavernous angiomas the stromal tissue investing the saccular blood-filled spaces is reduced to a minimum.

Bacterial endangeitis occasionally develops within the blood-filled spaces (Maier *et al.*, 1948) and can result in metastatic cerebral and other deep-seated abscesses.

CAPILLARY TELANGIECTASIS

This form of angioma forms an inconspicuous leash of capillaries and the lesion is often only discovered because its presence is suspected due to a large pulmonary cavernous angioma elsewhere in the lung. Like the larger cavernous lesions they tend to grow and are usually found in patients showing other evidence of Osler–Rendu–Weber disease. They lead to no haemodynamic changes in the pulmonary circulation *per se*, but they are often multiple and are sometimes associated with larger cavernous lesions (Fig. 21.58). Multiple capillary telangiectases occur occasionally in the walls of the segmental and subsegmental bronchi giving rise to repeated haemoptyses (Masson *et al.*, 1974).

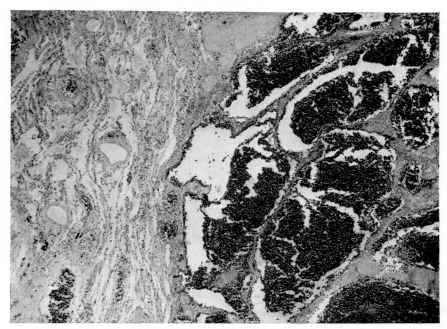

FIG. 21.57. Section of an angioma in the lung showing fibrous-walled cavernous blood spaces. × 40 H and E.

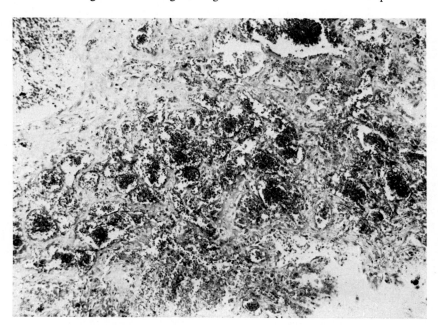

FIG. 21.58. A small capillary angioma in the lung from a case of Osler–Rendu–Weber disease (familial telangiectasia) × 100 H and E.
(Material supplied by Dr. C. Pommer, Bielefeld.)

PHYSIOLOGICAL EFFECTS AND COMPLICATIONS
OF PULMONARY ARTERIO-VENOUS
ANEURYSMS

The functional effects caused by pulmonary angiomas depend on the volume of blood they transmit, which in turn depends upon the resistance they offer to the blood-flow. Multiple small angiomas in the aggregate may cause as serious a disturbance of the pulmonary blood-flow as a single large cavernous angioma causing an arterio-venous aneurysm. The shunting of pulmonary arterial de-oxygenated blood into the systemic circulation through the arterio-venous fistula results in a varying degree of oxygen desaturation of the systemic arterial blood. If more than 30 per cent of the pulmonary blood-flow is thus shunted or alternatively if more than 5 g of reduced haemoglobin per 100 ml of blood are found in the systemic circulation cyanosis usually results, and this in turn causes a secondary polycythaemia which leads to an increase in the blood volume. Whereas arterio-venous fistulae in the systemic circulation often cause left ventricular hypertrophy due to the increased output of blood from that chamber, pulmonary arterio-venous fistulae, although they may transmit up to 66 per cent of the pulmonary blood-flow (Moyer *et al.*, 1962), rarely cause right ventricular hypertrophy because the vascular resistance normally offered to the pulmonary blood-flow is very low. The overall pulmonary vascular resistance tends to remain unchanged as the resistance offered by the smaller pulmonary arteries and arterioles in the unaffected lung tends to increase. The cause for this increase in pulmonary arterial vaso-constriction is still unknown, though in a few cases it may have been attributable to coexistent mitral stenosis. If the latter condition is responsible it leads to the establishment of a vicious circle. Increasing hypoxaemia causes increasing pulmonary arterial constriction in the unaffected lung and a greater volume of blood is then diverted through the arterio-venous fistula.

Although as previously stated right ventricular hypertrophy is uncommon it has been reported by Sloan and Cooley (1953).

Removal of a large pulmonary angioma usually leads to a reversal and correction of the pulmonary and vascular abnormalities it may have caused, but occasionally such removal may act as a stimulus to growth of other small latent pulmonary angiomas (Bosher *et al.*).

In those pulmonary arterio-venous aneurysms supplied by a normal or aberrant systemic artery a considerable shunt may develop between the systemic circulation and the pulmonary or systemic veins, thus imposing an increasing burden on the left ventricle. This type of pulmonary angioma simulates in its general circulatory effects arterio-venous fistulae occurring elsewhere in the systemic circulation.

Subpleural angiomas may burst into the pleural sac (Klinck and Hunt, 1933) (Fig. 21.59), and intrapulmonary angiomas frequently rupture causing severe haemoptysis. Intrapulmonary rupture may in turn lead to air embolism as described by Bowers (1936).

Deep-seated abscesses, especially cerebral abscesses, caused by microaerophilic streptococci conveyed in thrombotic emboli formed in the dilated veins are favoured by the cyanosis, and are similar to the abscesses complicating some forms of congenital heart disease. Also large arterio-venous aneurysms permit infected emboli from the systemic veins to pass directly into the systemic arterial circulation.

ACQUIRED PULMONARY ANGIOMA

Although this lesion has been included for convenience as a form of angioma it probably has a different aetiology from the congenital type of pulmonary angioma. Multiple pulmonary arterio-venous fistulae occur in juvenile cirrhosis (Rydell and Hoffbauer, 1956) and it has been found in a patient with metastases from carcinoma of the thyroid (Pierce *et al.*, 1959).

The association between cyanosis, chronic liver disease and distension of the pulmonary veins after death was first recognized by Fluckiger (1884, quoted by Rydell and Hoffbauer). Plastic injection studies by Rydell and Hoffbauer showed many direct vascular connections between the pulmonary arteries and

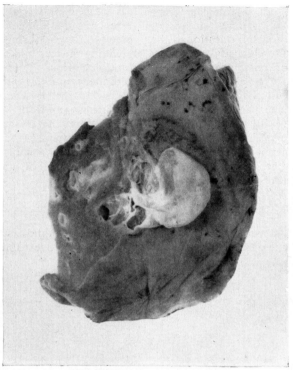

FIG. 21.59. A subpleural angioma in the lung. Rupture of such an aneurysm into the pleural cavity may result in a severe degree of haemothorax.

(Reproduced by courtesy of Dr. K. F. W. Hinson.)

veins near the hilum, in the peripheral lung tissue and beneath the pleura. In some situations the casts of the anastomotic vessels measured up to 1 mm in diameter. The factors responsible for causing the development of the abnormal vessels is not known but may be the same as those responsible for causing spider naevi in the skin in advanced liver disease.

Berthelot *et al.* (1966), using an injection technique, were able to demonstrate focal but nevertheless widespread dilatation of the smallest pulmonary muscular arteries and arterioles in thirteen cases of patients dying of advanced cirrhosis. "Spider naevi" were demonstrable on the pleural surfaces of the lung but in only one case was a pulmonary arterio-venous shunt demonstrated. The lesions were caused by excessive vasodilatation and did not provide an explanation for the arterial oxygen desaturation observed in some cases of severe cirrhosis.

Occasionally secondary deposits of growth within the lung may provoke a stromal vascular reaction in which the newly formed vessels behave as minute arterio-venous fistulae. The presence of the fistulous connections may be detected because of partial oxygen desaturation of the arterial blood following inhalation of 100 per cent oxygen.

Malignant Haemangioma of the Lung

Some doubt exists whether these tumours occur. Wollstein (1931) claimed to have described such a growth in an infant which he considered was derived from angioblasts. Tumour deposits were found in many organs at the time of death, but Lubarsch questioned the secondary nature of these deposits, regarding them as multicentric primary growths. Hall (1935) also described a tumour which he regarded as a malignant haemangioma on rather unconvincing evidence, and Powell (1958) described a further doubtful case. As in the first case difficulty was experienced in deciding on the metastatic or primary nature of the angiomatous tumours that appeared in other organs following removal of the lung neoplasm.

Pulmonary Kaposi Tumour

A very rare tumour that occasionally occurs in the lung is a Kaposi type of vascular tumour.

It is found mainly in and around the pulmonary vessels and like the cutaneous tumours tends to spread diffusely through the lungs. Normally spreading Kaposi sarcoma is only seen in very young children though a spreading form of pulmonary tumour is seen occasionally in adults. It was described by Dantzig et al. (1974) and the author has seen a further case which shows the same features that characterize the better-known cutaneous tumours (Fig. 21.60).

The nature of Kaposi's sarcoma is still in doubt and its cell of origin uncertain.

3 (i) Pulmonary Haemangiopericytoma

These rare pulmonary tumours were reviewed and three further examples added by Meade

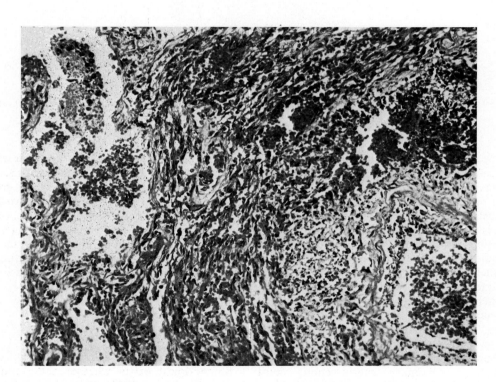

FIG. 21.60. A Kaposi sarcoma in the lung showing the typical vascular spaces filled with red blood-cells similar to those seen in the classical cutaneous Kaposi sarcoma. × 180 H and E.

et al. (1974). Haemangiopericytomas occur in many parts of the body but the lung is one of the least common sites. Originally named by Stout and Murray (1942) they are thought to arise from pericytes, cells with protoplasmic processes contained within the alveolar capillary and pulmonary venular basement membranes. Pericytes were originally thought by Roulet, their discoverer, to be contractile cells but Majno and Palade (1961) showed that they possessed phagocytic properties. It is now also thought that they may produce antibodies.

Pulmonary haemangiopericytomas occur about equally in both sexes between the ages of 10 and 73, and the average age of occurrence is 45. They are often clinically silent if small, but larger tumours cause haemoptyses, dyspnoea and chest pain, and radiologically appear as rounded shadows in the lung substance.

Macroscopically, they appear to be well encapsulated and unconnected with any bronchus (Fig. 21.61). Large tumours may extend and involve the pleura. They are greyish-white in colour, often contain areas of haemorrhage and the larger tumours often undergo some central necrosis.

Microscopically, the sheets of pale staining cells with vesicular nuclei have a perivascular arrangement and there is a network of reticulin fibrils radiating outwards from the vessel wall

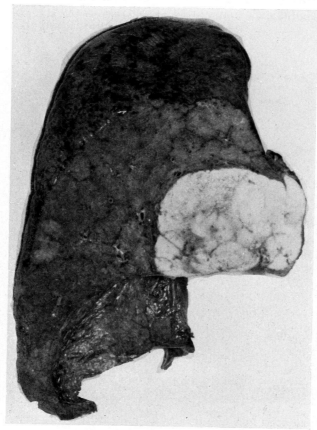

FIG. 21.61. A pulmonary haemangiopericytoma.
(Reproduced by courtesy of Dr. F. Whitwell.)

13

(Figs. 21.62, 21.63). The reticulin network intersects the tumour cells surrounding individual and small clumps of the cells. Despite the apparent vascularity of gross specimens, many of the capillaries are found to be devoid of red blood-cells presumably due to pressure from the surrounding tumour.

Electron microscopic examination of tumour cells removed from haemangiopericytomas elsewhere in the body have shown that they contain no myofibrils or attachments to the basement membrane.

In the more actively growing tumours reticulin fibres are scarcer and mitoses more numerous, and tumour cells may invade both bronchial walls and vessels. The prognosis is poor except in the case of the small tumours that can be totally removed.

Primary Pleural Tumours

In the past, even the existence of primary pleural tumours was doubted, but the present consensus of opinion now admits their existence and they have assumed a greater importance during the past few years. They exist in a benign and a malignant form and are variously known as local or generalized pleural fibromas, pleural mesotheliomas, pleural endotheliomas, or pleural fibrosarcomas. Other forms of pleural tumours of which isolated examples have been reported will be considered separately.

The first reference to such tumours was by Lieutand (1767), but Laennec (1819) considered that the earlier reported cases were carcinomas. Later Rokitansky (1842) denied the existence of primary pleural "cancer" and Robertson (1924),

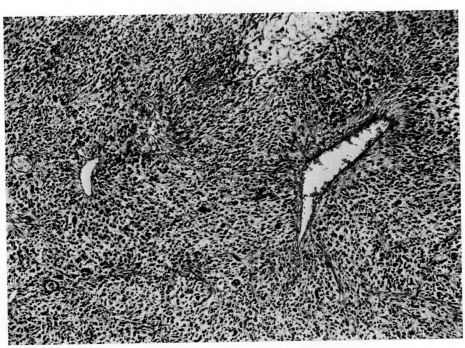

FIG. 21.62. A haemangiopericytoma from lung showing the perivascular arrangement of the tumour cells. × 180 H and E.
(Reproduced by courtesy of Dr. F. Whitwell.)

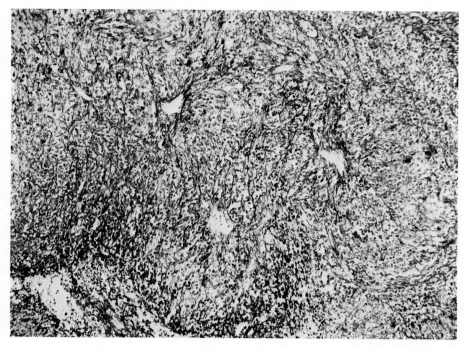

FIG. 21.63. A reticulin stain on the same tumour as shown in Fig. 21.62 showing reticulin fibrils radiating radially from the vessel walls and in places surrounding individual tumour cells. × 180 Foot's reticulin stain. (Reproduced by courtesy of Dr. F. Whitwell.)

after a long and exhaustive survey of the literature, reached the same conclusion. The latter based his findings on purely negative grounds, believing that no endothelioma or carcinoma of the pleura could be regarded as a primary tumour in the presence of a known primary growth in another organ. He also added that the primary growth elsewhere might be so small as to escape detection and that therefore no primary growth could be considered to arise from the pleura. This negative view presupposed that pathologists could never exclude the origin of a "primary" pleural growth in another organ, a doubtful precept on which to base an opinion.

In 1931 Klemperer and Rabin re-examined the problem and considered that primary pleural tumours, though rare, undoubtedly occurred, and subdivided them into (a) localized tumours and (b) generalized tumours. Since then num-

erous well-documented reports of pleural tumours have appeared, though Willis (1953) still continued to doubt their existence.

The term mesothelioma, first coined by Adami, is often applied to cells of mesoblastic origin which line coelomic cavities, including the pleura. Maximow (1927) using tissue-culture methods was able to show that the rounded surface serosal cells could develop into fibroblasts and thus demonstrated the interchangeable nature of the two cell types. Earlier Miller and Wynn (1908) considered that the lining cells of the peritoneum, a cavity of closely related structure, were able to produce tissues with both epithelial and fibroblastic characteristics. They also claimed to have shown that no true basement membrane existed between the serosal cells and the underlying stroma. Pellissier and Ouary (1952) further showed that embryonic mesenchyme could form pseudo-epithelial secretory cells which later developed into fibroblastic or phagocytic cells. In recent years the proved tumourigenic properties of asbestos both on the lung and pleura have provided an opportunity to study malignant pleural tumours in both man and animals. Also tissue culture studies have provided information on the behaviour and potentialities of mesothelial cells and have shown that such cells may develop into fibroblasts, mesothelial lining cells or phagocytes. Like other tumours of mesothelial connective tissue origin, i.e. synoviomas, primary pleural tumours may secrete hyaluronic acid, an essential constituent of much of the normal connective tissue ground substance.

Primary pleural tumours occur either as localized or generalized growths. The latter usually prove to be malignant and often completely encase the lung. Local pleural tumours though formerly more common are now outnumbered by the increasing numbers of asbestos-induced generalized growths. The local tumours are slow-growing, often attaining an enormous size, and lead to mediastinal displacement and respiratory embarrassment due to their size alone.

3 (j) Benign Local Pleural Fibroma

These tumours occur equally in both sexes; the average age of the patients when first seen is 50 years. The symptoms they cause are mainly referable to the lungs or joints and in a series of twenty-four cases described by Clagett *et al.* (1952), sixteen showed changes simulating rheumatoid arthritis in the joints of the hands, ankles, shoulders and wrists in order of frequency. This arthropathy disappears when the tumours are surgically removed but reappears if the growths recur. The incidence of osteoarthropathic joint changes in patients with pleural fibromas (66 per cent) contrasts with the low incidence of this lesion (2 per cent) in cases of carcinoma of the lung. Occasionally patients with large local pleural tumours may complain

of "something rolling about in the chest" when they alter body position. Local pleural fibromas have been described by Klemperer and Rabin (1931), Stout and Himadi (1951), Clagett *et al.* (1952), Thomas and Drew (1953) and Heaney *et al.* (1957).

Macroscopically, the tumours may reach an enormous size often being attached to the pleural surface by a pedicle and covered over most of their extent by visceral pleura (Fig. 21.64). Such pedunculated tumours are liable to undergo torsion. Pleural fibromas are smooth-walled and consist of dense, whorled fibrous tissue which may soften in places to form cysts filled with viscid fluid resembling synovial fluid. Vascular areas occur and are responsible for haemorrhages but in the main they are avascular growths which often become partly calcified. Rarely, a localized pleural tumour may consist of an overgrowth of mesothelial lining cells which form a polypoid papillary tumour containing areas of cystic softening in the interior. An example of such a local benign pleural tumour was described by Yesner and Hurwitz (1953).

The tumours remain confined to the surface of the lung unless malignant changes supervene, an event that rarely follows incomplete surgical removal. Le Roux (1962), however, regarded all pleural mesotheliomas as potentially sarcomatous tumours.

Microscopically, the tumours are composed of whorls of reticulin and collagen fibres (Fig. 21.65), interspersed among which are spindle cells resembling fibroblasts (Fig. 21.66). Haemorrhages occur in the vascular areas and calcification in the hyaline collagenous tissue. In addition to the whorled fibromatous areas, clefts lined by cubical and flattened cells resembling pleural mesothelium are found (Fig. 21.67). In the rare tumours consisting of mesothelial lining cells, the cuboidal cells line papillary processes with a fine fibrous tissue stroma in which small foci of lymphocytes occur.

The majority of local pleural fibromas are well encapsulated but a few of the more cellular growths may eventually start to invade the capsule and the underlying lung. Such cases are likely to recur following surgical removal.

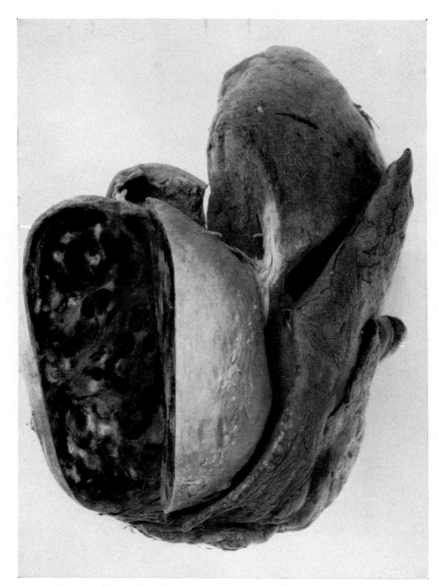

FIG. 21.64. A solitary benign pleural fibroma attached by a narrow pedicle to the pleural surface. The remainder of the lung is compressed by the large tumour mass. There are numerous haemorrhages within the tumour itself which is covered with mesothelium. Two-thirds natural size.

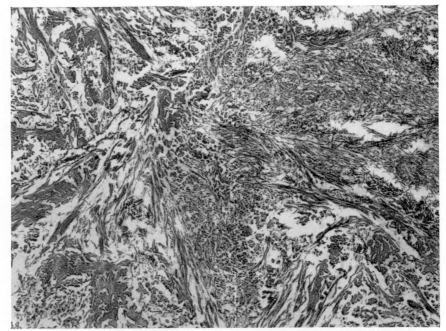

FIG. 21.65. A dense collagenous portion of a pleural fibroma. × 40 H and V.G.

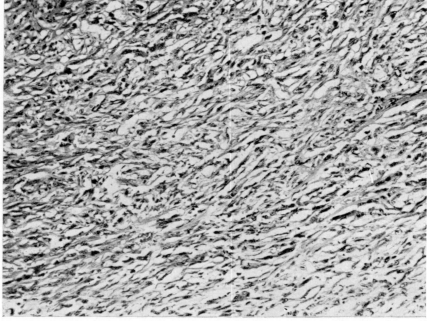

FIG. 21.66. Section of a benign pleural fibroma showing spindle-shaped fibrocytic cells and collagen fibres. × 140 H and E.

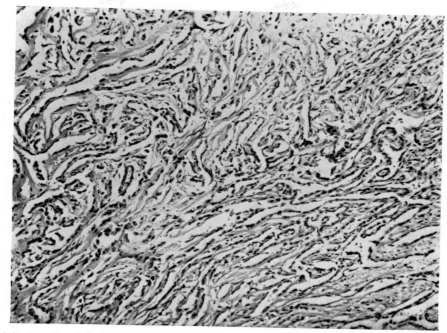

FIG. 21.67. A benign pleural fibroma from same case as shown in Fig. 21.66 showing mesothelial-lined clefts within the fibromatous tumour mass. × 140 H and E.

Although localized pleural fibromas are usually regarded as true neoplasms, Brown and Johnson (1951) described three cases in which similar tumours followed pulmonary infections. Fahr (1935) also described a pleural fibroma which followed a lung infection and it has been suggested that encysted, sterile, interlobular pleural effusions may persist after the causative infection has ceased. Such effusions may serve as culture media for mesothelial cells which develop into fibroblasts and these "cultured" cells are responsible for the fibromatous growths. In most cases of pleural fibroma there is no history of an antecedent pulmonary infection.

3 (k) Malignant Pleural Tumours

The existence of primary malignant pleural tumours was denied by Robertson (1924) and Willis (1953) who both regarded such growths as examples of secondary carcinoma.

Following the publication of cases of pleural mesotheliomas by Godwin (1957), McCaughey (1958) and the account by Wagner et al. (1960) of a series of cases that had occurred among persons who had lived and worked in the asbestos mining areas in the northern part of Cape Province in South Africa, the dogmatic statements previously denying the existence of these tumours could no longer be accepted. The ever-increasing number of pleural mesotheliomas that have been reported from most of the highly industrialized countries in Western Europe, North America and Australasia have followed the greatly increased use of asbestos for industrial purposes. In most instances the patients have given a history of having worked with or been in close contact with crocidolite asbestos or of having lived in that region of South Africa where it was mined. The close contact with this form of asbestos may have occurred only briefly 20 to 40 years previously. The epidemiology and pathology of pleural meso-theliomas have been reviewed and described by Manguikan and Prior (1963), Hourihane (1964), Newhouse and Thompson (1965), Webster (1965, 1973), Wagner (1965) and Whitwell and Raw-cliffe (1971).

Crocidolite (Cape blue asbestos), mined in the Kuruman area of Cape Province, appears to be the principal tumourigenic form of asbestos and is responsible for over 90 per cent of pleural and peritoneal mesotheliomas. Rarely amosite has been incriminated (Selikoff *et al.*, 1971) while chrysotile and anthophylite are probably not tumourigenic or minimally tumourigenic.

In Great Britain and other industrialized nations pleural mesotheliomas are encountered especially among men employed in the asbestos industry, dockyard workers engaged in various trades, those who handle and manufacture asbestos components and those who have worked in the demolition and construction industries. Women who mended or sorted the sacks in which the crude asbestos was formerly transported were also at risk. Also people of both sexes who live downwind of asbestos factories or mines or who are engaged in the transport of the mineral show a higher tumour incidence.

The relationship of asbestos to mesothelioma, though undeniable, is not an entirely direct one and this subject has been reviewed by Webster (1973). Harrington (1962) found that sedimentary asbestos mineral (crocidolite and amosite) mined in the northern part of Cape Province contained absorbed traces of 3:4-benzpyrene and other closely related polycyclic, carcinogenic hydro-carbons, but that the amount of these substances varied in different mineral samples obtained from different localities. Asbestos-induced malig-nant mesothelial tumours of the pleura take upwards of 20 years to develop following exposure to the dust. Peritoneal mesothelioma occasionally occurs in association with pleural tumours and may be presumed to be caused by the mineral fibres being swallowed and passing through the bowel wall. Unlike asbestos-induced lung cancer, cigarette smoking does not appear to exert an adjuvant carcinogenic effect. The period of exposure to asbestos varies con-siderably but in most cases extended over many years.

Experimental studies support the view that asbestos fibres display a tropism for mesothelial tissues as they tend to migrate towards such surfaces following their introduction into animals.

Asbestos-induced malignant pleural meso-theliomas may coexist with alveolar-cell carci-noma in the subjacent lung.

Macroscopically, the affected lung is en-sheathed with a thick layer of soft, gelatinous, greyish-pink tumour tissue up to 2·0 or more centimetres in thickness which may occupy most of the pleural sac. The same tumour spreads to involve the parietal pleura and extends into the interlobular fissures (Fig. 21.68). The remains of the pleural cavity are usually filled with gelati-nous, greenish-brown, slimy fluid, resembling synovial fluid, which may clot on opening the chest. The largest mass of growth is often confined to the surfaces of the lower lobe of a lung. Occasionally the pleurae on both sides of the chest may be involved.

Distant spread of the tumour occurs to the pericardial sac, diaphragm, hilar lymph nodes and through the chest wall. Invasion of the underlying lung is infrequent and may be con-fused with a coexistent alveolar-celled carcinoma. *Microscopically*, these tumours present variable features and Whitwell and Rawcliffe described four varieties: (a) tubulo-papillary, (b) sarco-matous, (c) undifferentiated polygonal-celled and (d) a mixed type. The majority of pleural meso-theliomas fall into categories (a) and (d).

(a) Tubulo-papillary mesotheliomas closely resemble adenocarcinomas. These growths con-sist of complex acinar spaces lined by flattened mesothelial cells. The latter form papillary fronds projecting into the lumen of often dilated cystic spaces. The connective tissue stroma between the acini is pale staining and there are few mitoses present (Fig. 21.69 A).

(b) In the sarcomatous tumours very few mesothelium-lined clefts are present and the neoplasm is an almost pure sarcoma consisting of a myxofibrosarcoma or densely fibrotic

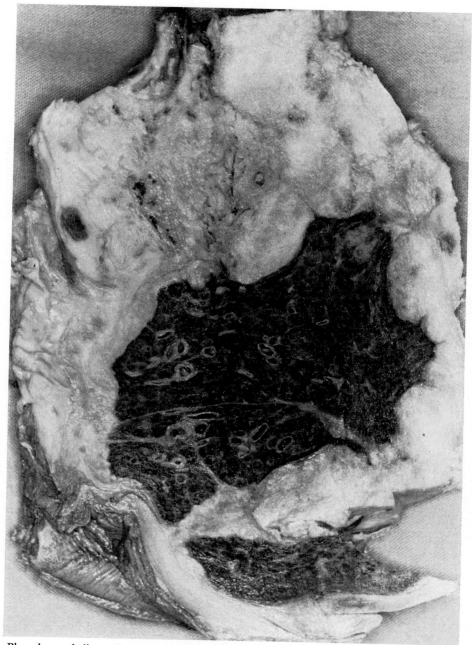

Fig. 21.68. Pleural mesothelioma showing both lobes of the lung encased in tumour which has obliterated the pleura space. Tumour tissue extends into and covers the surfaces of the interlobar fissure.
(Reproduced by courtesy of Dr. B. Goldstein, Johannesburg.)

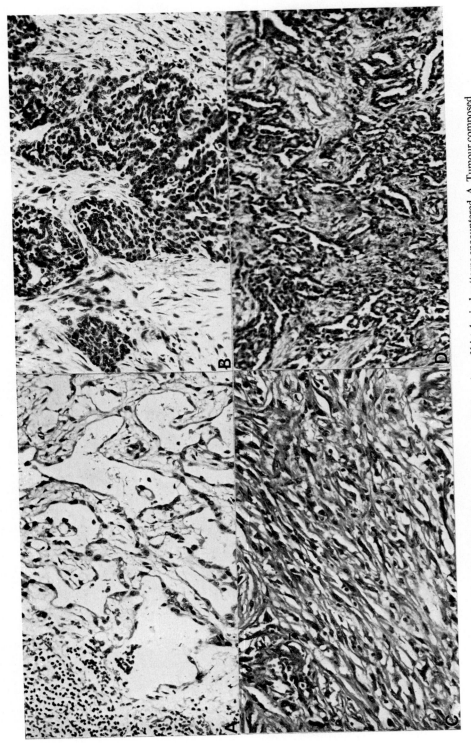

FIG. 21.69. Pleural mesotheliomas showing the wide variety of histological patterns encountered. A, Tumour composed of wide mesothelium-lined clefts. B, Solid "epithelial" type with slight spindle-celled sarcomatous stroma. C, Spindle-celled sarcomatous variety. D, Numerous mesothelium-lined clefts and more solid "epithelial" nests of cells. All × 180 H and E.

fibrosarcoma. Intermediate sarcomatous patterns occur between these two extremes (Fig. 21.69 C).

(c) The undifferentiated polygonal-cell type is in the author's experience the most difficult variety to distinguish from alveolar-cell lung cancer or secondary pleural carcinoma. In these tumours nests or sheets of polygonal epithelial-like cells are surrounded by fibrous stroma of varying cellularity but which are usually fascicular and densely collagenized (Fig. 21.69 B).

(d) The mixed variety as its name implies is a variable mixture of mainly (a) and (b) (Fig. 21.69 D).

The very variable microscopical character of these tumours was one of the principal reasons why the older pathologists were unable to accept their existence. If, however, the embryogenesis of the pleural cavities is recalled and the manner in which the embryonic mesenchyme splits to form the pleuroperitoneal canals, it will be realized that pleural mesotheliomas will bear very close resemblance to malignant synovial tumours, the joint spaces having developed in a similar fashion to the pleural cavities by splitting of mesenchyme. Furthermore, if the dual mode of embryogenesis of the lungs is correct (Waddell, 1949), both the alveolar tissue of the lung and the pleura are derived from the same mass of mesenchyme which invests the growing tips of the laryngotracheal bud and its bronchial derivatives. The electron microscopic structure of pleural mesotheliomas and bronchiolar-alveolar carcinomas are indistinguishable (Wang, 1973), lending further support to the similar nature and derivation of the two tumours.

Diffuse malignant pleural mesotheliomas secrete hyaluronic acid as shown by Wagner et al. (1962) and this can be digested by hyaluronidase. Alveolar-cell carcinomas of the lung fail to secrete this substance. Asbestos dust, though causally related to diffuse pleural mesotheliomas, is not found in the tumours themselves and asbestos fibres seldom if ever penetrate the pleural elastic laminae.

Pleural mesotheliomas have been produced experimentally in rats following the intrapleural injection of crocidolite fibres (Wagner, 1962).

Rarely a malignant pleural tumour may arise only from the surface layer of mesothelial cells.

It is often difficult to make a certain diagnosis of a primary malignant pleural tumour from biopsy material and confirmation may have to await a full post-mortem examination to exclude any other source for the growth.

Although many malignant pleural mesotheliomas do not spread outside the chest, others do spread to distant sites which include hilar lymph glands, liver, adrenals, bone and thyroid. The sarcomatous type of tumours seem more prone to behave in this manner. Whitwell and Rawcliffe have also drawn attention to the frequent slow clinical progress of patients with malignant pleural tumours unlike the usually rapid course associated with lung cancer. Whereas only about 4 per cent of their lung-cancer patients were still alive after 1 year, 43 per cent of those diagnosed as having pleural mesothelioma were still alive.

Recent work has shown that in some cases of pleural mesothelioma there may be an alteration of cellular immunity possibly related to changes in lymphocytes.

3 (l) Endometriosis of Pleura and Lung

This rare condition was first described by Bungeler and Fleury-Silveira (1939). Since then several further cases have been reported by Nicholson (1951), Lattes et al. (1956), Rodman and Jones (1962), Jelihovsky and Grant (1968) and Assor (1972). Davies (1968) reported five cases of recurrent pneumothorax occurring during menstruation and although in none of the cases was it possible to prove the presence of endometrium in the lung, the patients' histories were nevertheless very suggestive of pulmonary endometriosis.

With the exception of the single case described by Rodman and Jones in which a nodule of endometrium was found in a bronchial wall, all the endometrial deposits have been found in or just beneath the pleural surface. In the case reported by Mobbs and Pfanner (1963) a routine chest X-ray during pregnancy revealed a shadow which subsequently proved to have

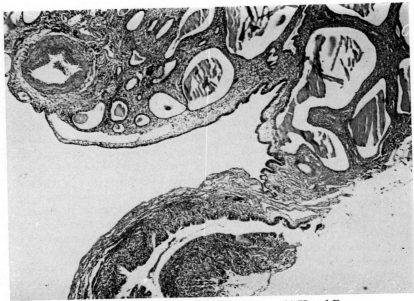

FIG. 21.70. Subpleural endometriosis. × 30 H and E.
(Reproduced by courtesy of Dr. T. Jelihovsky, Sydney, Australia.)

been caused by an endometrial deposit showing decidual changes. In some of the reported cases the patients had previously undergone pelvic surgery. In most cases, however, an undiagnosed blood-stained pleural effusion led to thoracotomy and the lung has been found to be covered with a yellow pellicle consisting of blood clot containing endometrial tissue lying just below the pleura (Fig. 21.70). The liquid blood in the pleural cavity resembled menstrual blood and failed to clot. An exception was the case reported by Rodman and Jones in which a bronchial deposit of endometrial tissue was supplied by a branch of the bronchial artery and this had led to recurrent haemoptyses.

Two explanations have been offered to account for this unusual condition. Firstly, that the endometrial cells are transported by the blood-stream to the lung from the uterus possibly aided by local trauma to the latter organ. Secondly, that it arises by a process of metaplasia of the lining coelomic mesothelium of which the pleural cavity originally formed part. Although Simpson (1927) showed that endo-

metrium could gain entry into uterine blood-vessels, a condition which he named endometriosis vascularum, the rarity of pulmonary endometriosis compared with its incidence in other sites, especially the peritoneal cavity, contraindicates blood-borne spread as the most likely cause of the lung involvement. Alternatively metaplastic change in the coelomic lining cells, including the adult pleura, provides a better explanation of most of the observed facts.

4 (a) Plasma Cell Granuloma of the Lung

Synonyms: Post-inflammatory tumour of the lung (Umiker and Iverson, 1954), Inflammatory pseudotumours, Histiocytoma of the bronchus (Bates and Hull, 1958), Histiocytoma of the lung, Vascular endothelioma (Edwards and Taylor, 1937), Plasmacytoma (Gordon and Walker, 1944) and Fibroxanthoma of the lung

The term plasma cell granuloma is used in preference to histiocytoma of the lung which was used previously to describe these lung tumours. Difficulty has been experienced in deciding whether the tumours are neoplastic or inflammatory lesions but they are now usually regarded as post-inflammatory granulomas. They are the most common tumour mass found in children under the age of 16.

From the variety of names applied to them it is evident that they may vary considerably in structure. They should be clearly differentiated from the true but very rare extramedullary pulmonary plasmacytoma described in Chapter 22. The term post-inflammatory pseudotumour has often been applied to plasma cell granulomas but the same term has also been applied to sclerosing angioma and is therefore best avoided. Bahadori and Liebow (1973) reviewed the literature and presented forty of their own collected cases. Twenty-five of these were females and the ages of the patients varied between 1 and 68 years of age. Very few patients gave any antecedent history of lung infection or complained of any symptoms. If the tumour causes symptoms, cough is the most common, and occasionally haemoptysis, chest pain and increasing dyspnoea draw attention to its presence. Extension of the granuloma to mediastinal and chest-wall structures can occur. In extremely large granulomas almost total destruction of a lung results as in a case seen by the author.

Macroscopically, they are usually spherical and sharply circumscribed tumours with well-defined capsules (Fig. 21.71). They range in size from about 2·0 cm diameter to tumours which occupy most of a lobe or even most of a lung. Some are very soft and friable, others are moderately or densely fibrotic, and they may undergo focal calcification or ossification and even cavitate. They are yellowish-white in colour but may contain areas of deep yellow colour. Rarely they grow as a tongue-like tumour into and obstruct a large bronchus or may extend through the pleural surface into the mediastinum (Sweetman *et al.*, 1958). Pleural extension can cause great difficulty in subsequent surgical removal.

In some instances the rapid development of a granuloma has been observed radiologically following a lung infection. Serial radiographs of the lung have shown the rapid development of a rounded opacity the nature of which remained uncertain until the tumour was removed at operation. When a tumour has been discovered by chance in symptomless patients, radiographs often have shown no progressive increase in its size. An uncommon but sometimes fatal complication of a plasma cell granuloma is spreading mediastinal fibrosis (sclerosing mediastinitis).

Microscopically, these tumours present a varied appearance sometimes consisting of dense collagenous fibrous tissue through which may be found interspersed plasma cells and lymphocytes. Other tumours consist mainly of a mass of mature plasma cells, Russell fuchsinophil bodies, a few lymphocytes and minimal fibrosis (Fig. 21.73). Some of the plasma cells may contain more than one nucleus but mitoses are not observed. Other plasma cells may be elongated and resemble plain muscle though they retain their characteristic nuclear structure. Plasma cell granulomas often contain spindle-shaped cells, foam cells and in some cases haemosiderin-laden phagocytes together with blood-vessels showing perivascular proliferations of cells resembling perithelial (pericytic) cells (Fig. 21.75). In the older parts of such tumours increasing amounts of collagenous fibrous tissue are formed, while elastic stains often reveal the remains of the elastic framework of the lung even though the alveolar structures are otherwise unrecognizable. Mast cells may occur in some granulomas and bronchial compression and obstruction leads to endogenous lipid pneumonia, a change responsible for some of these tumours being named fibroxanthomas. Para-amyloid may be deposited and forms part of the acellular hyaline tissue. At the edges some of the alveoli may be entrapped within the expanding tumour and appear as epithelium-lined spaces.

Although dense hyaline fibrous tissue is often the principal component of these lesions, some plasma cell granulomas consist of oedematous, poorly collagenized connective tissue cells which may be mistaken for a myxofibroma. Closer inspection, however, shows that throughout the tumour mass there is a diffuse infiltration of

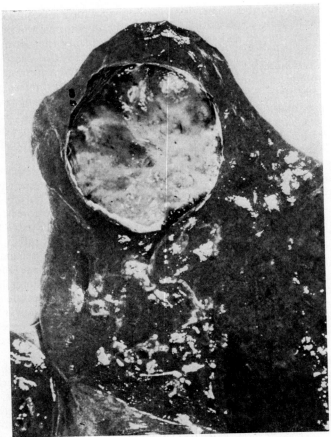

FIG. 21.71. A plasma cell granuloma showing its sharp outlines. Approx. natural size.

plasma cells and lymphocytes. This latter finding may be the only feature enabling a plasma cell granuloma to be distinguished from a true connective tissue neoplasm.

Depending upon which of the features was the most predominant, so have the tumours been named. The appearances are in many ways reminiscent of those seen in chronic villonodular synovitis in joints (Spencer and Whimster, 1950), and these tumours appear to belong to that ill-defined class of conditions which the majority consider to be granulomas but which some regard as neoplasms.

A failure to appreciate their true nature has been responsible for these tumours being regarded as extramedullary pulmonary plasmacytomas. Unlike the genuine but rare pulmonary plasmacytoma (considered in Chapter 22), plasma cell granulomas contain a variety of cells as stated above, whereas the plasmacytomas consist of uniform sheets of malignant "myeloma" cells. None of the growths so far recorded have evinced the least evidence of malignancy, and the patient described by Hill and White (1953) was alive and well 8 years after removal of the tumour. Plasma cell granulomas of the lung

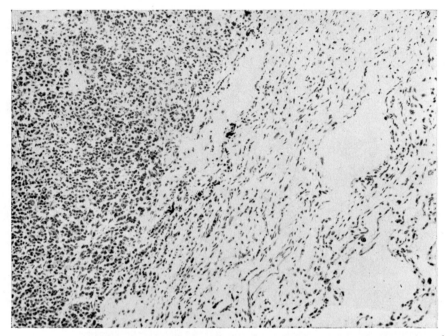

FIG. 21.72. A low-power view of the edge of a plasma cell granuloma showing the ill-defined margin and compressed and fibrous lung tissue bordering the edge of the tumour. × 140 H and E.

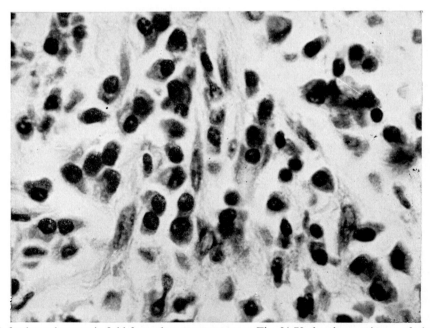

FIG. 21.73. A further microscopic field from the same tumour as Fig. 21.72 showing a mixture of plasma cells and spindle-shaped fibroblasts. × 700 H and E.

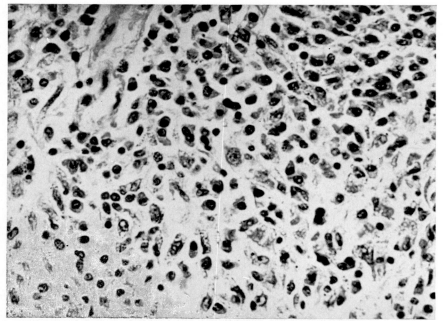

FIG. 21.74. A further section from the same histiocytoma showing plasma cells, fibroblasts and foam cells; no haemo-siderin-laden phagocytic cells are visible in this field. × 500 H and E.

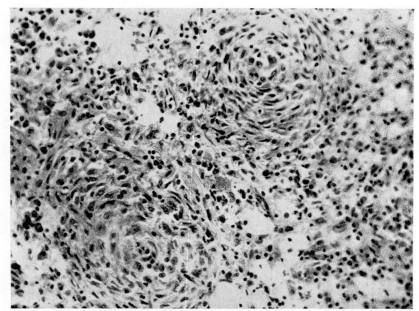

FIG. 21.75. A section of a plasma cell granuloma showing the perivascular proliferation (perithelial proliferation) and the chronic inflammatory cells lying outside this area. This appearance has been mistaken for a blood-vessel tumour. × 200 H and E.

should be clearly differentiated from sclerosing angiomas in the lung which often also contain phagocytic foam-cell collections following organization of red blood-cell extravasations.

4 (b) Sclerosing Angioma of the Lung (Sclerosing Granuloma)

Synonyms: Fibroxanthoma of lung, Vascular endothelioma of lung

This unusual condition was first described by Liebow and Hubbell (1956) though similar but unrecognized examples of these tumours were probably described under other titles. Arean and Wheat (1962) collected thirty-five cases from the literature and the author has seen several further examples.

The term sclerosing angioma has only been retained because it is the name by which these tumours are most generally known. The author does not subscribe to the view that the tumours are of vascular origin or nature, though they often contain much extravasated blood. Furthermore, they should be clearly differentiated from those fibroxanthomas which are related to

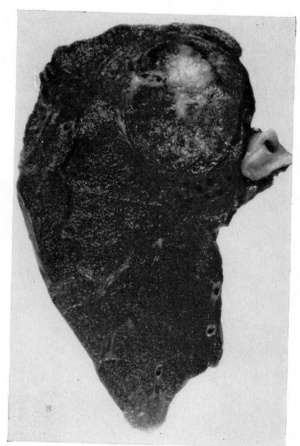

FIG. 21.76. A "sclerosing angioma" situated in an upper lobe of lung. The tumour has a well-defined margin and the cut surface is composed partly of fibrous tissue and spongy haemorrhagic tissue. One-half natural size.
(Reproduced by courtesy of the late Dr. R. D. Clay.)

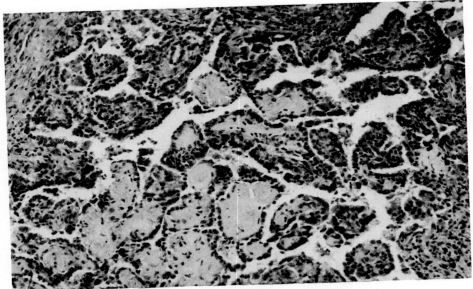

FIG. 21.77. A "sclerosing angioma" showing both the earlier cellular phase and also commencing alveolar wall fibrosis. × 140 H and E.
(Reproduced by courtesy of the late Dr. R. D. Clay.)

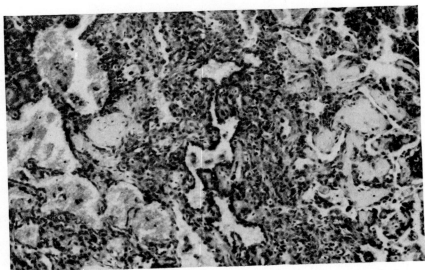

FIG. 21.78. Section of a "sclerosing angioma" showing alveolar epithelial hyperplasia. There is also commencing interstitial fibrosis and some of the remaining alveolar spaces contain extravasated blood. × 100 H and E.
(Reproduced by courtesy of the late Dr. R. D. Clay.)

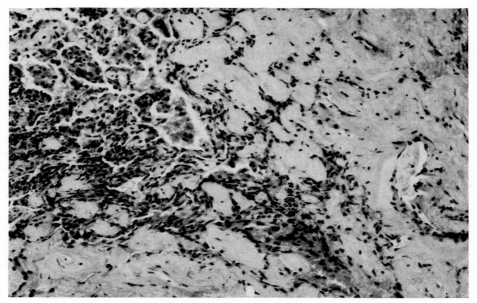

Fig. 21.79. Widespread hyaline fibrosis obliterating the early cellular phase in a "sclerosing angioma". × 140 H and E. (Reproduced by courtesy of the late Dr. R. D. Clay.)

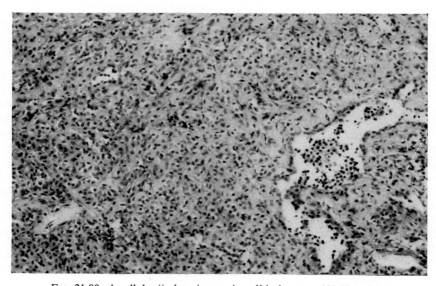

Fig. 21.80. A cellular "sclerosing angioma" in lung. × 100 H and E.

plasma cell granulomas (post-inflammatory pseudo-tumour).

The majority of sclerosing angiomas (about 65 per cent) occur in women usually below the age of 50, though occasional examples have been found in older persons. They are found anywhere in the lung but most frequently in the lower lobes.

Many of the patients suffered from previous, often large, haemoptyses extending back over a period of 30 years before the tumour was discovered. In others a history of "pneumonia" or chest pain led to a radiological examination of the chest and the discovery of a sharply circumscribed shadow in the lung. Bronchoscopic examination has usually contributed little helpful information.

Macroscopically, the tumours have a fibrous pseudocapsule and on section consist of pinkish-grey, often yellowish tissue in which are interspersed areas of old and recent haemorrhage (Fig. 21.76). Occasionally the tumours protrude into the pleural cavities covered by a layer of pleura. The rate of growth of these tumours as judged by serial radiographs is slow and in the second case described by Liebow and Hubbell little change in size was observed over a period of 12 years.

Microscopically, the earliest sequence of changes may be observed at the edge of the tumour where undifferentiated alveolar mesenchymal cells proliferate forming solid branching buds. Subsequently the surface cells covering the papillae become cuboidal and the deeper fibroblastic. In the stroma thus formed capillary blood-vessels develop into sinusoidal vessels. Increasing collagenization of the stroma produces enlargement of the papillae, throttling of the capillaries, and atrophy of the covering epithelium. Finally the collagen becomes hyaline, the alveolar epithelium disappears and adjacent processes fuse together (Figs. 21.77, 21.78, 21.79, 21.80). Haemorrhage takes place into the epithelium-lined spaces in the younger parts of the tumour and would appear to come from local capillaries and result from throttling of the veins by contraction of the surrounding older collagen. From this description it is clear that no angiomatous element is concerned in the process, and the author considers that the vascular proliferation, regarded by Liebow and Hubbell as the primary change, is secondary and follows alveolar mesenchymal proliferation.

Cholesterol crystals and numerous foamy macrophage cells appear in the remains of the alveolar spaces. The lipid may be derived both from modified intracellular lipid contained within the pneumocytes and from the breakdown of red blood-cells resulting from haemorrhages from the newly formed capillaries. Some haemosiderin-laden cells are usually present. The foam cells may disrupt and give rise to lipid-filled spaces similar to those found in a fatty liver. Finally the tumour becomes a mass of hyaline fibrous tissue, in which is interspersed clefts lined by flattened endothelium-like remnants of alveolar epithelial tissue, and in which it is often possible to see the ghost hyalinized walls of obliterated vessels. Focal calcification and even ossification may also occur. A few scattered lymphocytes and plasma cells may be seen interspersed through the tumour, but the paucity of these cells and the extent of the alveolar cell proliferation sharply differentiate it from the plasma cell granuloma (post-inflammatory tumour) of the lung.

Electron microscopic studies have yielded conflicting views of the nature of sclerosing angiomas. Haas *et al.* (1972) from their observations supported the vascular origin of the tumours and considered that endothelial cell proliferation was the initial change. A contrary view was expressed by Hill and Eggleston (1972) and Kennedy (1973) all of whom found that the initial change was a proliferation of granular (type 2) pneumocytes. Kennedy considered sclerosing angiomas might be hamartomas involving alveolar epithelium, but the nature of these lesions still remains uncertain.

CHAPTER 22

Pulmonary Reticuloses

THE pulmonary tumours to be discussed in this chapter include:

(1) Lymphoid hyperplasia.
(2) Hodgkin's disease.
(3) Pre-lymphomatous states.
(4) Lymphosarcoma and reticulum-celled sarcoma.
(5) Plasmacytoma of the lung.
(6) (a) Leukaemic changes in the lung.
     (b) Lung reactions to cytotoxic (radio-mimetic) drugs.
(7) Giant intrathoracic lymph nodes.
(8) Histiocytosis X disease (histiocytic reticuloses).

Unlike most forms of carcinoma and most sarcomas which arise from a field of cells in one organ or tissue, many of the tumours about to be discussed arise in a labile system of cells widely scattered throughout the body. Most of the tumours arise in lymphoid tissue, which in turn develops from the same widely scattered but primitive system of cells, and is the progenitor of many of the tumours in this group.

A further characteristic feature shown by some of this group of disorders is the frequency with which the tumour cell type may change, and a growth that begins as one variety may later show features more typical of another of the same group. Some of the malignant tumours in this group (malignant lymphomas) rapidly involve the entire lymphoid system throughout many regions of the body but others may start in one organ, i.e. the lung, and remain confined to it for a considerable period before general dissemination occurs. Such localized forms of the disease may offer opportunity for successful surgical removal before spread occurs to the rest of the lymphoid system.

As with other varieties of malignant disease some malignant lymphomas may be preceded by pre-malignant states of which an increasing number are being recognized. The interrelationship of these pre-lymphomatous clinical states to each other is not yet fully understood but the present state of knowledge will be discussed.

Pulmonary Lymphoid Hyperplasia

Synonym: Benign pulmonary lymphoma

The existence of subpleural lymph nodes has already been referred to in Chapter 2, and they were first described by Meinel (1869) quoted by Heller (1895). They are commonly found in the lungs of persons over the age of 25 where they appear as rounded anthracotic tumours varying in size from a few millimetres to 2–3 centimetres in diameter.

Intrapulmonary lymph nodes though less common may occasionally show up in a plain X-ray of the lung and the shadows they cause may be incorrectly diagnosed. These nodes may also be demonstrated by injecting the pulmonary lymphatics after death with a radiopaque fluid. Trapnell (1964) using pulmonary lymphangiography found that 18 per cent of adult lungs contained lymph nodes in the parenchyma: some of these possessed germinal centres and some were supplied and drained by afferent and efferent lymphatic vessels.

Generalized pulmonary lymphoid hyperplasia is a very uncommon condition and leads to hyperplasia not only of the subpleural nodes but also of lymphoid tissue lying in relation to

937

terminal and respiratory bronchioles as well as the larger air passages (Fig. 22.1). Hyperplastic lymphoid tissue is found lying adjacent to pulmonary lobular septa but the hilar lymph glands are usually unaffected. In some cases the lymph nodes contain germinal centres and well-formed sinuses. Although this condition may be discovered by chance as a result of a radiographic examination of the chest, it may occasionally lead to partial obstruction of the smaller air passages and alveolar overdistension occurs. Such a case was described by Brandes *et al.* (1943). In a more recent case seen by the author sudden chest pain of unexplained nature and slight haemoptysis were the presenting symptoms. At a subsequent exploratory operation one of the lungs was found to contain many small lymphoid nodules, but the patient has remained well for many years.

Hodgkin's Disease

When Thomas Hodgkin described the disease that now bears his name in 1832, he described not a disease but a morbid anatomical syndrome in which both the "absorbent glands" (lymphatic glands) and the spleen were enlarged.

His description of the disease was based entirely on clinical and macroscopic post-mortem changes and lacked the precision of microscopical confirmation. Even today the nature and unity of all cases regarded as Hodgkin's disease is uncertain, although it is now customary to consider it a form of neoplastic disease. The macroscopic and microscopical appearances encountered in cases of Hodgkin's disease vary enormously. At one extreme of the spectrum is the self-limiting paragranulomatous form (the lympho-reticular type) and at the other end is the highly malignant sarcomatous variety which blends imperceptibly with reticulum-celled sarcoma.

The mediastinal and pulmonary lymphoid tissue is involved in about 40 per cent of cases (Falconer and Leonard, 1936), often in conjunction with the lymphoid tissue elsewhere in the body. The presence of Hodgkin tissue deposits in the lungs usually causes little or no disability and the bronchi are seldom infiltrated or obstructed by the tumour until late in the disease.

Very occasionally Hodgkin's disease may

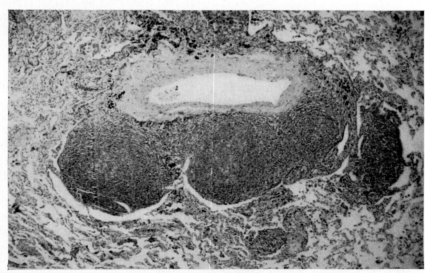

Fig. 22.1. Benign pulmonary lymphoid hyperplasia. Similar hyperplastic lymphoid nodules occurred throughout the lungs, many of which contained germinal centres. × 40 H. and E.
(Reproduced by courtesy of Dr. E. H. Bailey.)

start in the lung and although Moolten (1934) considered that 10 per cent of cases started in this organ, primary Hodgkin's disease of the lung is extremely rare but examples have been described by Hochberg and Crastnopol (1956) and Monahan (1965). Secondary involvement of the lung is much more common and most of the macroscopic forms were described by Rottino and Hoffman (1955). The lung deposits may appear as (a) solitary nodules resembling secondary carcinoma, (b) miliary nodules throughout both lungs, (c) a large confluent mass of tumour occupying most of a lobe, (d) plaques of pleural growth, and very rarely (e) deposits in the walls of the large bronchi (Figs. 22A, 22B).

Large deposits of Hodgkin's tissue may undergo central necrosis and form a cavity as described by Efskind and Wexels (1952).

Several classifications of Hodgkin's disease have been proposed only to be replaced later by new ones. The classification which is at present mainly used is that which was proposed at the Rye symposium (Lukes et al., 1966). Its main

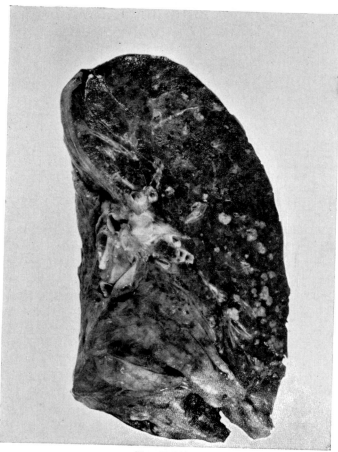

FIG. 22.2(A)

FIG. 22.2A. An example of Hodgkin's disease in the lung showing multiple small nodules throughout the lung. Approx. one-third natural size.

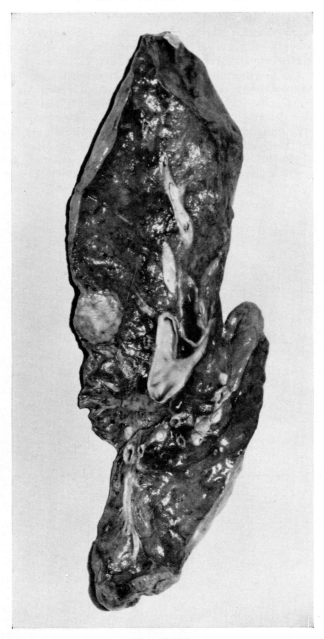

FIG. 22.2(B)

FIG. 22.2B. An example of Hodgkin's disease in the lung showing a solitary large nodule. Approx. one-third
natural size.

advantage over those used previously was its recognition of the nodular sclerosing form of the disease, a variety which carries a much better prognosis and which is the most common form encountered in the lung. The classification includes: (a) the lymphocyte predominant form, (b) the lymphocyte depleted form, (iii) mixed cellularity, (iv) the nodular sclerosing variety.

The nodular sclerosing variety accounts for about 70 per cent of all cases in which there is mediastinal or pulmonary involvement. It is characterized by destruction of the lymph gland architecture and its replacement with lymphocytes, reticulum cells, Sternberg–Reed cells, eosinophils and a variable and often considerable amount of fibrous tissue (Fig. 22.3). About 15 per cent of the patients with this form of the disease survive 15 years and increasing amounts of fibrous tissues are formed with the passage of time.

The lymphocyte depleted type of Hodgkin's disease affects mainly persons over 50 years of age and the deposits consist almost entirely of hyperchromatic and frankly malignant reticulum (histiocytic) cells together with multinucleated forms of these cells. The microscopical appearances may be indistinguishable from reticulum-celled (histiocytic) sarcoma and the survival time is short, being usually less than 2 years.

The majority of small Hodgkin deposits in the lung as already stated are asymptomatic. Later enlargement of the hilar glands and compression of the larger bronchi leads to obstructive pneumonitis causing a persistent cough. Very rarely widespread Hodgkin deposits within the lung have led to the development of cor pulmonale.

Pre-lymphomatous States including Pseudo-lymphoma, Lymphocytic Interstitial Pneumonia (LIP) and Waldenström's Macroglobulinaemia

In 1963 Saltzstein reviewed a large series of cases of malignant pulmonary lymphomas and

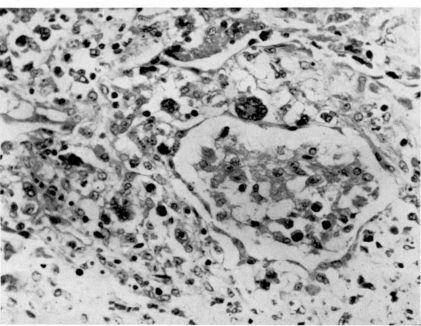

FIG. 22.3. Section of the nodular sclerosing type of Hodgkin's disease in the lung showing Sternberg–Reed giant cells, reticulum cells and eosinophil cells, the eosinophilic granules in which fail to show up. × 280 H and E.

he defined the microscopical features of a condition he named pseudo-lymphoma of the lung. He gave reasons for separating it from true primary lymphomas of the lung the chief of which was the better long-term prognosis attaching to pseudo-lymphoma.

Carrington and Liebow (1966) described what they regarded as a new disease resulting in an interstitial lymphocytic infiltration of the lung and they gave to it the name of lymphocytic interstitial pneumonia (LIP).

Since these early and first descriptions several cases and reviews of both pseudo-lymphoma of the lung and LIP have appeared. It appears that the two conditions merge and are identical, pseudo-lymphoma being the name applied when there is a localized tumour mass and LIP when it affects the lung diffusely. Intermediate forms occur showing features of both.

Pseudo-lymphoma is often clinically silent and is discovered by accident following a chest radiograph. In other cases cough, fever and pyrexia cause the patients to seek treatment. Identical clinical symptoms occur in LIP in which radiologically the changes affect mainly the lower lobes and are usually bilateral. Pseudo-lymphoma is usually localized to one lung but may involve more than one lobe and the lesion forms a localized tumour mass rather than the diffusely infiltrating lesion characteristic of LIP. The only differences between the two conditions are those based on the radiological appearances. In LIP the lower lobes of the lungs show nodular and patchy opacities and linear markings whereas in pseudo-lymphoma there is a localized tumour usually situated near the hilar region of a lung.

Both pseudo-lymphoma and LIP may be associated with dysproteinaemia. A monoclonal macroglobulinaemia (IgM) is the usual abnormality found in pseudolymphoma (Talal et al., 1967) and also occurs in LIP (Montes et al., 1968), though a polyclonal gammopathy in LIP was described by Young et al. (1969).

LIP may affect persons of all ages from infancy to 70 years of age but it occurs predominantly in persons in the fifth and sixth decades. Pseudo-lymphoma occurs almost entirely in the older age groups. The principal reason for separating pseudo-lymphoma and LIP from lymphosarcoma is the supposed different prognosis attaching to the two former conditions. It should be emphasized, however, that both pseudo-lymphoma and LIP frequently develop into malignant lymphomas though this may take several years to occur, but the disorders may prove to be non-progressive. The term pseudo-lymphoma is therefore both dangerous and misleading implying as it does little risk of malignancy developing, whereas the reverse is often true. The author agrees with McNamara et al. (1969) that the lesions should all be considered potentially or actually pre-malignant or malignant.

It has long been recognized that some lymphomatous tumours remain at first localized to their sites of origin in many tissues but ultimately after a longer or shorter period behave as a rapidly spreading systematic form of malignancy. Other lymphomas are rapidly spreading systematic forms of malignancy from the outset and soon cause death. Pseudo-lymphoma of the lung should be regarded as a similar condition to those lymphomas which remain localized to their sites of origin often for a very long period before disseminating.

Certain histological features are observed more frequently in those tumours which pursue a slow and locally infiltrative course and these are the criteria which have been regarded as diagnostic of a "pseudo-lymphoma" and which were described by Saltzstein (1963, 1969) and by Reich et al. (1974), and in LIP by Carrington and Liebow (1966), Liebow and Carrington (1969), Liebow (1968), Moran and Totten (1970) and Macfarlane and Davies (1973). The reader is also referred to the review of pseudo-lymphoma and malignant lymphoma by Greenberg et al. (1972). Because the pathologies of the two pre-lymphomatous conditions are identical they will be considered together.

Macroscopically, the ill-defined usually hilar situated mass of a pulmonary pseudo-lymphoma or the nodular infiltrates confined mainly to the lower lobes of the lungs of LIP present a similar appearance (Fig. 22.4). They consist in both

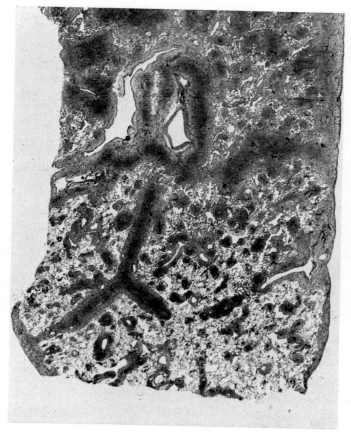

FIG. 22.4. A large section of lung from a case of diffuse lymphocytic pneumonia (LIP) showing the widespread infiltrative nature of the changes.
(Reproduced by courtesy of the Armed Forces Institute of Pathology, Washington D.C. Negative No. 59–2104.)

instances of greyish-tan-coloured rubbery tissue. In LIP the nodules tend to coalesce about branches of the pulmonary veins and in both conditions surround the bronchi.

Microscopically, the characters of the lesions are similar and therefore will be described together. The cellular infiltrate remains strictly interstitial and the predominant cell is a mature lymphocyte. These cells may form lymphoid follicles enclosing germinal centres. In addition variable numbers of plasma cells, Russell bodies and pale-staining reticulum cells including giant-cell forms of the latter are found (Figs. 22.5,

22.6). The giant-cell forms, however, do not show the atypical features of the giant cells seen in reticulum-celled sarcoma. There is a variable amount of fibrosis which increases with the passage of time and in some cases paramyloid material is present (Fig. 22.8).

The number of plasma cells are greatest in those cases showing a dysproteinaemia. In about 20 per cent of cases of LIP in addition to the diffuse interstitial alveolar and septal cellular infiltrate, small sarcoid-like collections of cells including multinucleate giant cells may be present both in the interstitium and within

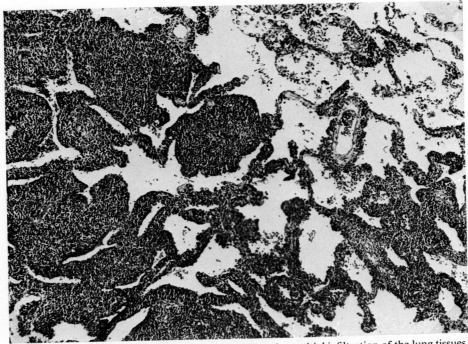

FIG. 22.5. Lymphocytic interstitial pneumonia showing the diffuse interstitial infiltration of the lung tissues which is so characteristic of this condition. × 75 H and E.

alveoli (Fig. 22.7). Some of the giant cells may be phagocytic and the condition then resembles and may be confused with giant-celled interstitial pneumonia (GIP). Although the larger bronchi remain unobstructed, the small bronchi may become compressed by the peribronchial infiltrates and changes of obstructive pneumonitis occur causing intra-alveolar collections of foamy macrophages.

Saltzstein (1963) described the principal differences between a pseudo-lymphoma and a true malignant lymphoma of the lung and these differences are as follows:

Pseudo-lymphomas (*including LIP*)	*Malignant lymphoma* (*lymphosarcoma*)
1. Mixed character of the cellular infiltrate which includes many plasma cells.	1. A more uniform cell type, plasma cells usually being absent.
2. Germinal centres often present.	2. Germinal centres absent.
3. Paramyloid sometimes present.	3. Paramyloid absent.
4. Hilar lymph nodes *not* involved.	4. Involvement of the hilar lymph nodes *indicative* of malignancy.
5. Pleura not usually infiltrated or very minimally infiltrated in its deepest layer.	5. Pleural infiltration not infrequently occurs and is infiltrated deeply.
6. Blood-vessel invasion absent.	6. Blood-vessel invasion may be present.

Of these differential features hilar lymph node involvement is the most reliable criterion of active malignancy. Both pseudo-lymphoma and LIP are at present best regarded as premalignant or dormant lymphomatous conditions, and the latter should not be regarded as a pneumonia in the usual sense of the term.

Both pseudo-lymphoma of the lung and LIP may be associated with Sjøgren's syndrome, and

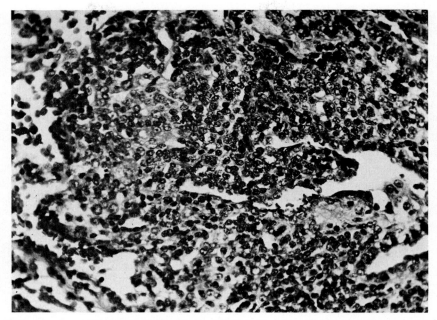

FIG. 22.6. Higher-power view of component cells of pulmonary infiltrate seen in Fig. 22.5. The cells consist of lympho-cytes, plasma cells and pale reticulum-type cells. × 280 H and E.

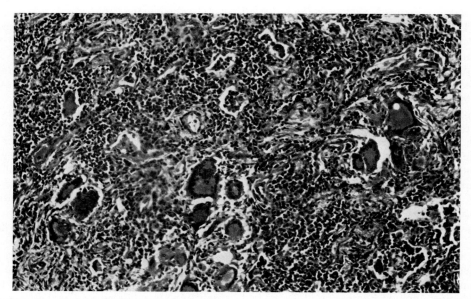

FIG. 22.7. Lymphocytic interstitial pneumonia showing numerous giant cells. The extensive interstitial lymphocytic infiltrate helps to differentiate LIP from giant-cell interstitial pneumonia. × 300 H and E.

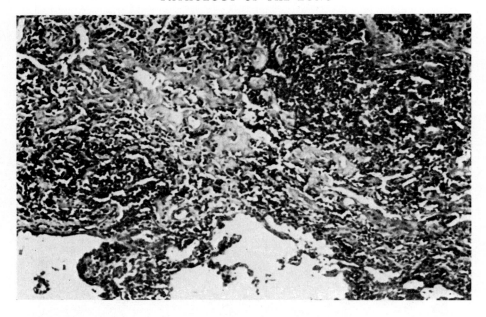

FIG. 22.8. Paramyloid formation in lymphocytic interstitial pneumonia. × 300 H and E.

Liebow (1975) claimed that one-third of all cases of LIP show this association. Eight cases of Sjøgren's syndrome were described by Talal *et al.* in which the serum IgM was greatly increased and pseudo-lymphoma-like infiltrates were present in the lungs. Sjøgren's syndrome should therefore also be included as a further pre-lymphomatous state.

A further potentially pre-lymphomatous condition is **Waldenström's macroglobulinaemia.** In this condition lung involvement may sometimes be a prominent feature. The pathological changes in the lungs show features common to plasmacytomas, to reticulum-celled (histiocytic) and lymphosarcomas. Waldenström also considered the condition he described showed some features common to lymphosarcoma. The radiological and pathological changes in the lungs were described by Öttgen and Quittman (1956) and Furgerson *et al.* (1963). The radiological changes include bilateral nodularity and occasionally enlargement of the hilar lymph nodes.

Microscopically, the lungs have shown a diffuse interstitial infiltration with lymphocytes, histiocytic cells and plasma and plasmacytoid cells. In addition the alveoli may fill with homogeneous, eosinophilic-staining, PAS-positive but lipid-free proteinaceous material, and thrombi have been found in small branches of the pulmonary arteries. Small haemorrhages may accompany these changes and have been attributed to either hyperheparinaemia or a deficiency of fibrinogen.

Localized intrapulmonary deposits of amyloid material may occur in Waldenström's macroglobulinaemia and lead to confusing radiological changes in the lungs.

Lymphomatoid granulomatosis, which has already been described in Chapter 19, though usually non-neoplastic may sometimes develop into a lymphosarcoma or may give rise to metastatic deposits in the brain and therefore qualifies to be included as a further rare potential pre-lymphomatous state.

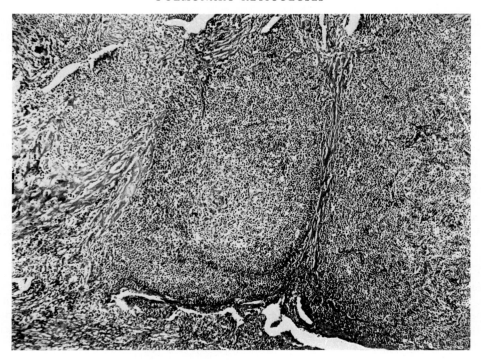

FIG. 22.9. Lymphocytic interstitial pneumonia showing an attempt at germinal centre formation. × 75 H and E.

Pulmonary Lymphomas (Lymphosarcoma and Reticulum (Histiocytic) Cell Sarcoma)

Although this group of neoplasms are responsible for less than 0·5 per cent of all primary malignant growths in the lung they are slowly increasing in importance and frequency of occurrence. According to Robbins (1953) about 7 per cent of lymphosarcomas and reticulum-celled sarcomas originate in the lung though Sugarbaker and Craver (1940) considered that only 0·5 per cent of primary lymphosarcomas started in this organ. The lung is, however, frequently involved in the spread of generalized lymphosarcoma and in about 50 per cent of such cases pulmonary deposits are found (Van Hazel and Jensik, 1956), pleural deposits being more common than intrapulmonary metastases.

Saltzstein (1963) reviewed 102 cases of primary pulmonary "lymphomas" collected from the literature and added fourteen new cases of his own. He also carried out the most extensive follow-up survey of many of these cases which has so far been attempted, and succeeded in tracing the fate of approximately 80 per cent of the patients. This investigation showed that some of the pulmonary tumours previously regarded as malignant proved to be localized tumours failing to recur following adequate surgical removal and never disseminated. This group of "benign" lymphomas have since been named pseudo-lymphomas and have already been considered together with other pre-lymphomatous states. Reticulum (histiocytic) cell sarcomas, however, more often behave as malignant tumours and disseminate widely. The most reliable criterion for determining whether a tumour is likely to prove malignant or benign is whether the hilar lymph nodes are involved at the time of removal; but even lymph node involvement when followed by efficient radio-therapy may be attended by long survival. Failure of the lymph nodes to be involved, however large or small the pulmonary tumour, is attended by a far better prognosis and frequent survival beyond 5 years.

The average age of patients with primary lymphosarcomas of the lung is 53 years and for reticulum-celled sarcomas 42 years (Saltzstein). Both groups occur more often in women than men and both tend to be located more often in the upper lobes of the lungs. The tumours may attain considerable size, even involving a whole lobe, without causing severe respiratory signs or symptoms. In several instances the presence of the tumour was first revealed in a routine chest radiograph in which it gave rise to a "coin" shadow. Symptoms may antedate by a year the surgical removal of a growth and are mainly referable to pleural and chest wall involvement and to the formation of a pleural effusion. Bronchial obstruction is usually absent except for slight narrowing due to compression by surrounding growth.

Some of the cases reported in the literature as examples of malignant pulmonary lymphoma-tous tumours would now be classified as pseudo-lymphomas of the lung. The first example was described by Pekelis (1931) though some doubt now exists whether this may have been an oat-celled lung cancer (Saltzstein). The cases reported by Churchill (1947), Maier (1948), Anlyan et al. (1950), Beck and Reganis (1951), Grimes et al. (1954) and Rose (1957) which were all reported as cases of malignant pulmonary lymphomas survived upwards of 4 years and some died after 10 or more years from unrelated causes and therefore the original tumours would now be classified as pseudo-lymphomas of the lung. Further examples were described by Hutchinson et al. (1964). Examples of pulmonary lymphosarcomas were described by Blades (1951), Baron and Whitehouse (1961, case 4), Saltzstein (1969) and Rees (1973).

Macroscopically, these tumours grow as pink-ish-white masses varying in size from 2 to 3 cm in diameter to growths which involve a whole lobe (Fig. 22.10). They possess the consistency of fish flesh and the edges are often ill-defined. They spread to the pleura and extend into the pleural

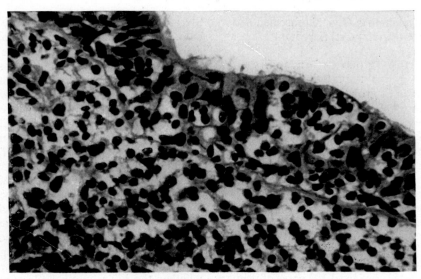

FIG. 22.10. A pseudo-lymphoma of the lung showing infiltration of the bronchial wall despite its "benign" behaviour. × 400 H and E.

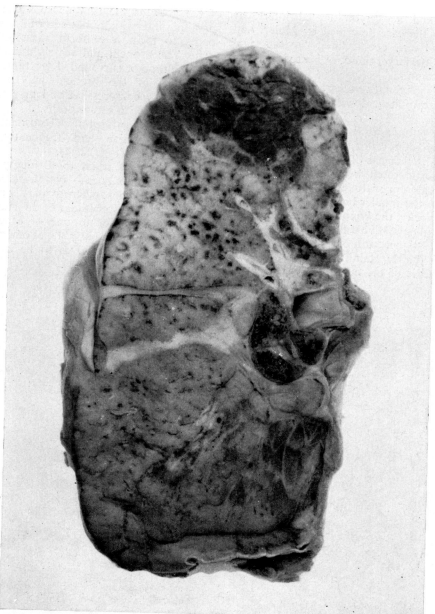

Fig. 22.11. A specimen of a primary reticulum-celled sarcoma arising in the lung. Note the almost complete absence of growth in the hilar glands despite the size of the pulmonary tumour. Approx. one-half natural size. (Reproduced by courtesy of the late Professor H. A. Magnus and Dr. B. S. Cardell.)

14

membranes and across obliterated interlobar fissures as in the case described by Maier, and central spread only occurs later. Cavitation of the centres of the larger tumours due to ischaemic necrosis is common.

In frankly malignant tumours the hilar lymph nodes become infiltrated with growth often at an early stage.

Secondary pulmonary deposits of lymphosarcoma and reticulum-celled sarcoma are common in the generalized forms of these neoplasms, the tumour deposits often being confined to the subpleural region or scattered through the lungs in relation to the smaller bronchi.

Microscopically, the lymphosarcomas, consist of small lymphocyte-like cells (Fig. 22.13) which in the very malignant tumours display mitoses, bizarre and darkly staining nuclear patterns and karyorhexis of the nuclei. At the periphery of the growths the tumour cells can be seen spreading interstitially but soon fill the alveolar spaces to form a uniform mass of tumour tissue. The bronchi and vessels usually remain patent although surrounded by tumour tissue. The walls of the bronchi may be slightly narrowed by compression and may themselves be infiltrated with tumour cells.

Reticulum (histiocytic) cell sarcomas tend to be more invasive and frequently undergo necrotic change (Fig. 22.12).

Although it is usual to divide lymphocytic pulmonary tumours into lymphosarcomas and pseudo-lymphomas, macroscopically and microscopically it may sometimes be difficult to assign a particular tumour to one or other groups, especially during the early stages of development of what eventually proves to be a malignant lymphoma.

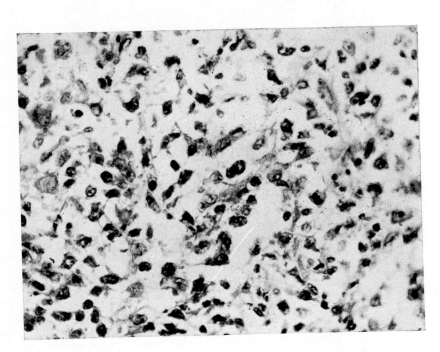

FIG. 22.12. A reticulum (histiocytic) cell sarcoma in lung. × 280 H and E.

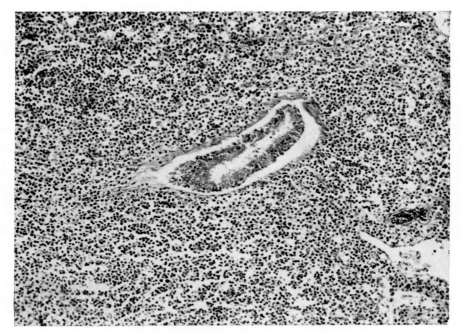

Fig. 22.13. A small round-celled type of lymphosarcoma surrounding a small bronchus. × 140 H and E.

Plasmacytoma of the Lung

These tumours are among the rarest to occur in the lung. They may arise as a primary tumour in the lung or they may be a secondary deposit from a primary growth arising from the usual primary site in the bone marrow. The majority of extramedullary plasmacytomas have occurred in the oropharyngeal region, larynx, the upper part of the trachea and the lower end of the rectum. Hellwig (1943) found no examples of primary pulmonary plasmacytomas among the sixty-five extramedullary tumours arising in the air passages which he reviewed. Romanoff and Milwidsky in 1962 claimed to have collected fourteen primary tumours arising in the lung including one of their own, and the author has seen a further example. In the latter case there was no radiological evidence that the skeleton was involved, and no tumour source other than the lung was discovered.

Most of the growths have been found in persons over the age of 50, but in the case described by Romanoff and Milwidsky the patient was only 17. The majority have been found as greyish-pink, soft tumours arising in the upper lobes of the lungs outside the bronchi. In one of the cases described by Kennedy and Kneafsey (1959), however, the growth filled the right middle lobar bronchus. Several of the reported examples of pulmonary plasmacytomas, however, would now be regarded as plasma cell granulomas and considerable confusion of these two different lesions has occurred.

Microscopically, the lung tissue is disrupted and replaced with plasma-like cells of varying degrees of maturity (Figs. 22.14, 22.15). In places compact foci of smaller cells resembling lymphoid follicles have been described which gave rise to the larger more mature myeloma cells. In the compact foci occasional very large, pale staining cells with reniform nuclei and measuring up to 30 μm in size are found. Varying numbers of Russell's fuchsinophil cells are found among the

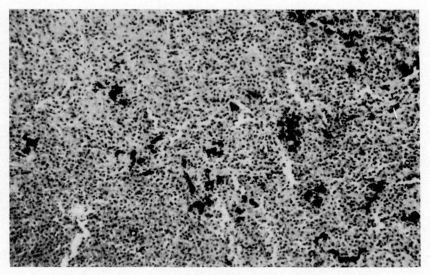

Fig. 22.14. A primary pulmonary myeloma. The lung is infiltrated with myeloma cells. × 100 H and E.
(Reproduced by courtesy of Dr. Elizabeth Davies.)

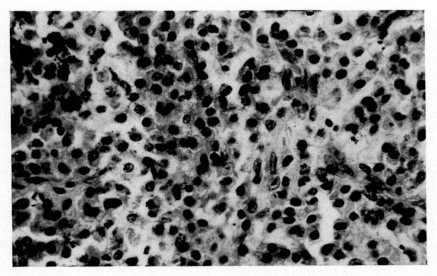

Fig. 22.15. A high-power view of myeloma cells shown in previous illustration. × 400 H and E.
(Reproduced by courtesy of Dr. Elizabeth Davies.)

mature myeloma cells. In the very rare plasma-cytoma the cells are almost uniformly plasma cells.

In some of the recorded cases (Rozsa and Frieman, 1953), the tumour had already spread to and involved adjacent ribs when discovered and death followed in a few months. In other cases surgical removal has been followed by a period of several years free from tumour recurrence. In the absence of involvement of extrapulmonary tissues radical removal may be expected to give a good result.

As in the classical form of multiple myeloma abnormal and excessive plasma globulins may be found and Bence–Jones proteose and other abnormal protein substances may be excreted in the urine.

Leukaemic Lung

Lesions in the lung directly attributable to leukaemia were first described by Fischer (1931). It is now recognized that in a quarter of all persons dying from this group of blood disorders some variety of leukaemic lesion is present in the lung, but in only 3 per cent do these changes cause disturbance of lung function (Resnick et al., 1961).

Nathad and Sanders (1955) studied a series of lungs removed post-mortem from cases of leukaemia and described a variety of changes including diffuse infiltrations and nodular lesions. Abnormal radiological changes are found more commonly in the lungs in lymphatic leukaemia

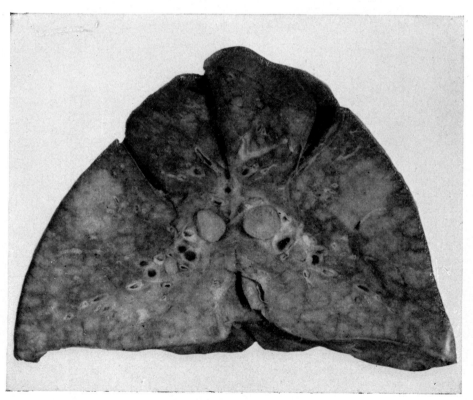

FIG. 22.16. A specimen of a child's lung showing generalized severe myeloid leukaemic infiltration of lung and hilar lymph glands. Approx. natural size.

but in both the acute and chronic stages of all three types of leukaemia, lesions may be found in the lungs.

Macroscopically, the lungs are often large, oedematous and yellowish-white in colour due to anaemia and a variable amount of intrapulmonary haemorrhage (Fig. 22.16). In the occasional case of chronic myeloid leukaemia in which terminal white blood cell counts of upwards of half a million cells per cubic mm occur, the pale blood clot filling the pulmonary vessels may be very viscous in consistency.

Microscopically, the blood-vessels in the lung, in common with those throughout the body,

often contain an excessive number of white cells which occasionally form leukaemic thrombi as described by Fiessinger and Fauvet (1941), but in the majority of cases the total white blood-cell count never reaches a sufficiently high figure to produce disturbances of blood-flow in the lung capillaries. Diffuse interstitial leukaemic infiltrations occur in the alveolar walls, around blood-vessels and bronchi, in the lobular septa and beneath the pleura (Figs. 22.17, 22.18). Usually the infiltrations are diffuse but local tumour-like deposits also occur and may undergo central cavitation following release of proteolytic enzymes from dying cells. The extent of the pulmonary leukaemic infiltration bears no

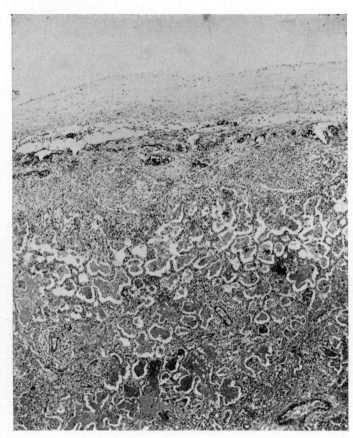

FIG. 22.17. A low-power view of a lung from a child with myeloid leukaemia showing very extensive and generalized leukaemic infiltration throughout the lung. × 40 H and E.

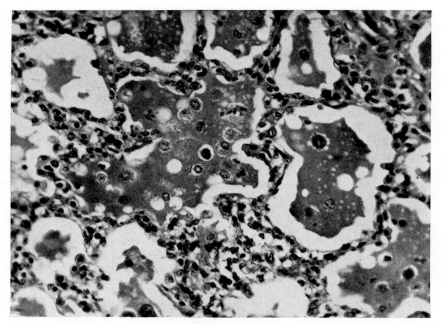

FIG. 22.18. A higher-power view of a myeloid leukaemic lung showing the diffuse alveolar wall leukaemic infiltration. × 280 H and E.

relationship to the number of circulating leukaemic cells in the blood.

Various changes in lung function occasionally result from these lesions, chief of which is an increased rate of respiration, hyperventilation and very rarely cyanosis. The interstitial leukaemic cell infiltration may lead to decreased lung compliance as was shown by Resnick *et al.* This change coupled with alveolar capillary compression and possibly some impairment of alveolar-capillary gas diffusion may occasionally result in a reduction of the arterial oxygen saturation.

Old *et al.* (1955) considered that the low blood-pressure in the pulmonary circulation, the large leukaemic cells and the increased viscosity of the blood caused stasis of the blood-flow through the lung capillaries and that this was mainly responsible for any disturbance in lung function.

In addition to the changes already described intrapulmonary haemorrhages of variable extent are often found and may add further to the terminal functional disorder.

Lung Reactions to Cytotoxic (Radiomimetic) Drugs used in Leukaemia

Synonym: Busulphan lung

Leukaemia and other forms of neoplastic disease are now being treated with a variety of cytotoxic drugs, and long-term treatment may be followed by pathological changes in the lungs and other organs.

Some of the cytotoxic drugs exert a radiomimetic effect and damage the nuclei of mitosing cells. As a result of the nuclear damage the cells may subsequently die, show bizarre nuclear changes or they may undergo proliferation depending on the severity of the nuclear damage. Although the dosage of the drugs is usually adjusted to destroy mainly the leukaemic bone marrow cells, other tissues including the lung may also suffer damage. The earliest descriptions of lung changes followed the introduction of busulphan (myleran) for the treatment of myeloid leukaemia, but similar changes occur following the use of methotrexate and chlorambucil, and can be expected to occasionally follow the use of other but similar radiomimetic drugs (Rodin et al., 1970).

Among the changes induced in the lungs are an acute fibrinous pulmonary oedema sometimes followed by the development of interstitial fibrosis, abnormal proliferations of bronchial and alveolar epithelial cells including bizarre cellular forms of these cells, and alveolar lipo-proteinosis (alveolar proteinosis) (Figs. 22.19, 22.20). Ill-defined alveolar interstitial granulomas containing giant cells and a few eosinophils were described following the use

of methotrexate by Clarysse et al. (1969).

Morgan (1959) reported the development of interstitial lung fibrosis in chronic myeloid leukaemia and Oliner et al. (1961) and Leake et al. (1963) associated the interstitial fibrosis with the use of busulphan, a drug first introduced to medicine in 1953. Similar changes can be induced in the lung by chlorambucil. The changes in the lung have been described in detail by Heard and Cooke (1968), Kirschner and Esterly (1971), Rubio (1972) and Rose (1975).

Among the first changes that may follow some weeks after the use of busulphan and chlorambucil is the appearance of a fibrinous oedema in the lung. Subsequent incorporation of the fibrin by accretion into the alveolar wall is followed by mural fibroblastic invasion and its conversion into an intramural plaque of fibrous tissue (see Chapter 7). The type 2 pneumocytes being the

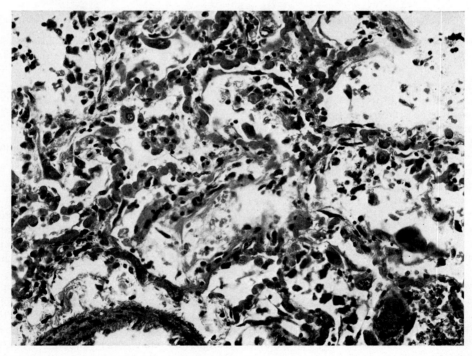

FIG. 22.19. Section of lung from a patient treated with methotrexate showing interstitial alveolar thickening with mononuclear cell infiltration and atypical, giant, strap-like alveolar epithelial cells (type 2 cells). Identical changes may follow busulphan, chlorambucil and other radiomimetic drugs. × 180 H and E.

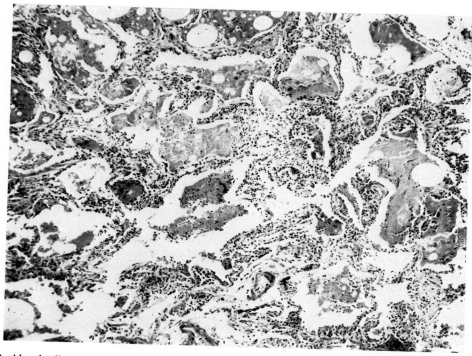

FIG. 22.20. Alveolar lipo-proteinosis following busulphan treatment for chronic myeloid leukaemia. × 75 H and E.

most rapidly dividing cells lining the alveolar wall suffer the greatest damage from these drugs and this damage is responsible for causing the pulmonary oedema. Those cells which are not destroyed may subsequently enlarge, forming rounded or strap-like cells with bizarre nuclei. Similar nuclear changes are observed in other organs containing dividing cells. The electron microscopic changes in the nuclei of these cytomegalic cells was studied by Min et al. (1974) who found that the nuclei contained tubular structures which were considered to have resulted from the action of the drug on nucleic acid metabolism and were not thought to be of viral nature.

Other later changes that may occur include proliferation of bronchiolar and alveolar epithelium similar to that which may follow some viral lung infections, i.e. influenza and measles. Such proliferative changes are self-limiting and

do not appear to undergo malignant transformation.

Busulphan has been widely used since its introduction in 1953 but few cases of lung damage have resulted. If interstitial lung fibrosis occurs it is accompanied by slight lymphocytic and plasma cell infiltration and the changes are similar to those found in idiopathic interstitial lung fibrosis (IIFL). Endothelial proliferation in the walls of the small muscular pulmonary arteries may lead to thrombosis and finally obliteration of the lumen occurs.

A further uncommon response to both busulphan and chlorambucil is the development of lipo-proteinosis (alveolar proteinosis) and the reader is referred to the account of this disorder in Chapter 17. This condition is attributed to the damage caused to the type 2 pneumocytes resulting in excessive production of surfactant or surfactant-like lipid material.

Giant Intrathoracic Lymph Nodes

Synonyms: Angiofollicular lymph node
hyperplasia, Angiomatous
lymphoid hamartoma

The presence of tumours variously described as thymomas or giant intrathoracic lymph nodes lying in relation to the structures entering the hilum of the lungs was described by Castleman (1954), Abell (1957), Mason (1959), Tung and McCormack (1967) and Anagnostou *et al.* (1972). The latter were able to trace 134 reported cases in the literature.

In the majority of cases the tumours were visualized radiographically before operation lying in the front of the chest usually in relation to the hilum of the left lung.

Clinically, the majority of cases are symptomless (90 per cent) or complain of minor symptoms due to compression of hilar structures. Ten per cent suffer from pyrexia and are found to have a hypochromic anaemia and a raised e.s.r. Myasthenia gravis which was reported in some

of the early cases was probably due to confusion with a thymic tumour and is not caused by a true giant intrathoracic lymph node.

At operation the enlarged lymphoid mass forms an ovoid, encapsulated, firm, pale, yellowish-brown mass which lies in close proximity to the hilar structures from which it is readily separated. In most instances the tumour is found in the hilar region of the upper lobe of the left lung.

Microscopically, the tumours consist of lymphoid tissue with numerous small vascular channels intersecting it but it lacks the normal sinus spaces (Fig. 22.21). The vessels are numerous and lined by swollen pale endothelial cells and their walls undergo hyaline changes so that some may be mistaken for Hassall's corpuscles. Structures resembling lymphoid follicles are a conspicuous feature. Some of these contain true germinal centres but many of the apparent germinal centres consist of pale, spindle cells arranged in whorls around a central blood-vessel. The central vessel in common with

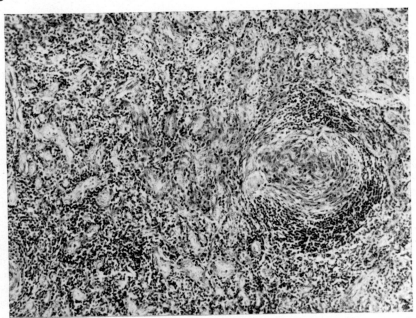

FIG. 22.21. A low-power view of a giant intrathoracic lymph node showing hyperplastic lymph nodes and prominent pale cells, perithelial cells, which have been mistaken in the past for Hassall's corpuscles. × 140 H and E.

other vessels often shows hyaline change together with adventitial cell hyperplasia which blends with the above-mentioned spindle cells (Fig. 22.22). The vessels lying in the centres of the lymphoid foci closely resemble Hassall's corpuscles (Castleman *et al.*, 1956). The whole tumour though resembling an enlarged lymph gland nevertheless lacks the architecture and structure of a normal gland. The lymphoid mass is supplied by a branch of a systemic artery which enters and supplies the capsular region and from this vessel numerous small branches supply the stroma. A few plasma cells and eosinophils are also found scattered throughout the tumour mass and foci of calcification and ossification may be present.

Tung and McCormack regard these tumours as angiomatous lymphoid hamartomas and point to their occurrence in other regions of the body than the thorax. Zettergren (1961) regarded the enlarged nodes as a benign neoplastic condition, but the failure of residual lymph gland tissue to re-grow following partial removal favours the view of Harrison and Bernatz (1963) that the glandular enlargement results from chronic venous congestion. Rarely the glandular enlargement may be associated with a microcytic iron-resistant form of anaemia, as in the case described by Lee *et al.* (1965), who considered the lymph gland enlargement was due to inflammatory changes though no cause for this was discovered.

Giant intrathoracic lymph nodes occur at all ages, having been recorded at the age of 6 weeks up to 40 years of age. Their removal is not followed by recurrence.

Histiocytosis X Disease (Histiocytic Reticuloses)

This group of disorders comprises three very closely related conditions: (a) Letterer–Siwe disease, (b) eosinophilic granuloma of lung and (c) Hand–Schüller–Christian disease.

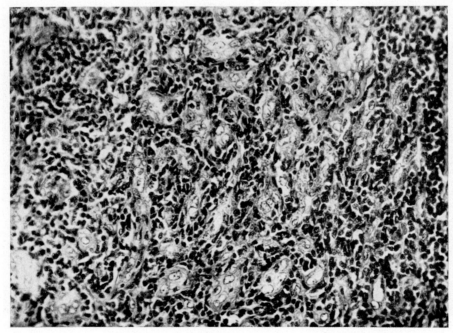

FIG. 22.22. A higher-power view of the same section showing the pale perithelial cells. × 280 H and E. (Figures 22.21 and 22.22 reproduced by kind permission of the Thoracic Society of Great Britain.)

They all have certain features in common and their natural histories show that interconversion of one form to another occurs. The close interrelationship in structure, and the evolution of one lesion into another, led Mallory (1942), Jaffé and Lichtenstein (1944) and Lichtenstein (1953) to regard the whole group as one disorder, to which the latter applied the name "Histiocytosis". The behaviour of the group shows many features consistent with a neoplastic condition but in other respects they resemble chronic granulomas. All three may affect the lung, although this organ is only one of many that may be involved in the distribution of the lesions.

LETTERER–SIWE DISEASE

Synonym: Acute histiocytic reticulosis

In 1924 Letterer described a case with characteristic clinical features in an infant aged 6 months to which he gave the name aleukaemic reticulosis. Later, Siwe (1933) described three further cases collected from the literature and added one of his own. Since these original descriptions further cases have been described and the literature reviewed by Schafer (1949), Orchard (1950) and Keats and Crane (1954), but the disorder still remains uncommon. The condition usually occurs in infants and young children and pursues a rapid course but Pruzanski and Altman (1964) described a case in an adult, and the author has known three cases in elderly people, one of whom was an octogenarian. The child becomes febrile, develops a scaly papular rash mainly on the scalp, enlargement of the spleen, liver and lymph glands and shows radiological changes in the skeleton. The upper lobes of the lungs are often involved first and the changes may later be detectable radiologically. *Macroscopically* in the later stages the lungs are voluminous, filled with cysts and the cut surfaces

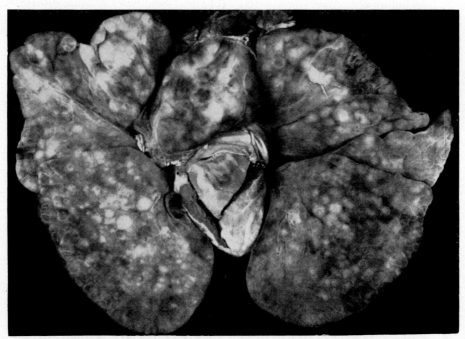

FIG. 22.23. Specimen of lungs, heart and thymus from a child with Letterer–Siwe disease showing the nodular appearance of pleural surface caused by the underlying cystic air spaces.
(Reproduced by courtesy of Professor E. F. McKeown and the Editor of *J. Path. Bact.*)

present a honeycomb appearance as seen in Figs. 22.23, 22.24. These honeycomb changes were described by Van Creveld and ter Poorten (1935), Schafer, McKeown (1954) and Keats and Crane. The cysts vary in size up to 1 cm in diameter being first found in the peripheral parts of the lung; their walls are largely formed of greyish-white tumour tissue.

Microscopically, the cell that characterizes the lesions wherever they are present is the "histiocytic cell". This cell has a large rounded or reniform nucleus with distinct nuclear membrane and contains one or two nucleoli; the cytoplasm is faintly eosinophilic. Giant cells and eosinophils are found in variable numbers, and the amount of reticulin fibres present varies

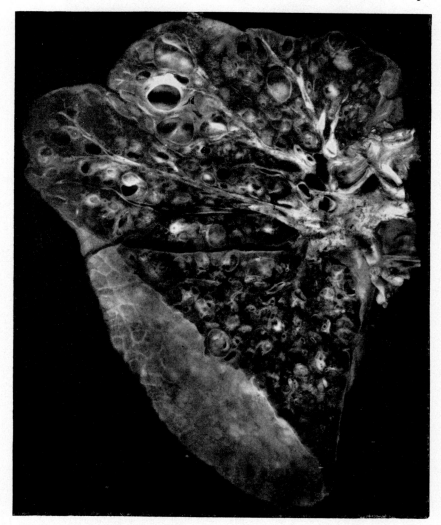

FIG. 22.24. The cut surface of the lung showing numerous air-filled cysts in the walls of which are deposits of pale tumour tissue.

(Reproduced by courtesy of Professor E. F. McKeown and the Editor of *J. Path. Bact.*)

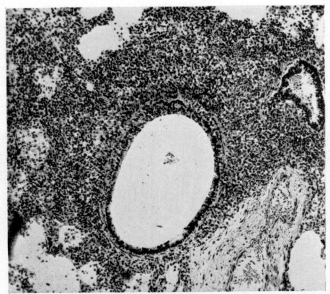

FIG. 22.25. Deposits of Letterer–Siwe tissue in the peribronchial tissues and the adjacent alevolar walls, many of which are obliterated. × 175.
(Reproduced by courtesy of Professor E. F. McKeown and the Editor of *J. Path. Bact.*)

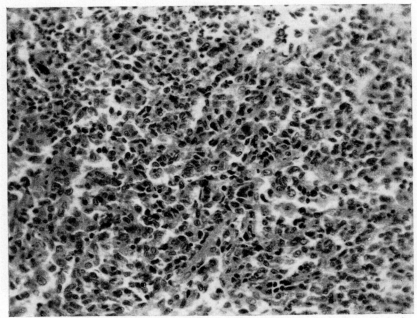

FIG. 22.26. A high-power view of the cells composing the deposits of Letterer–Siwe tissue in the lung, the so-called histiocytic cells. × 280 H and E.

considerably. In the more rapidly fatal cases few fibres are found, but in more chronic cases many are present, including even a few collagen fibres and the appearances may closely resemble an eosinophilic granuloma of the lung. The histiocytic cells often present a foamy appearance (Lichty, 1934) and may contain sudanophilic drops (Schafer). Tumour deposits are found in the bronchiolar walls, the alveolar walls, perivascular tissue, interlobular septa and beneath the pleura (Fig. 22.25). The cause of the cystic changes has been a matter of debate; some regard the cysts as evidence of weakening and dilatation of the bronchiolar walls by tumour invasion, others that bronchiolar obstruction caused by growth is the essential factor. Heppleston (1956) regards the large cysts seen in "honeycomb" lungs as bronchioles entered by oblique slit-like apertures which behave as valves, allowing inspiratory filling but preventing expiratory exhaustion distal to the valve. The author believes that bronchiolar weakening together with diffuse interalveolar infiltration of growth are both responsible; the latter by destroying alveolar tissue leads to compensatory over-distension of less-affected alveoli. Keats and Crane considered that the cysts resulted from necrotic changes in the lung and subsequent distension of these areas with air, but this is improbable as cysts occur in the complete absence of necrotic change.

EOSINOPHILIC GRANULOMA OF THE LUNG

This condition is more common than Letterer–Siwe disease. The first account of it was given by Farinacci et al. (1951) who discovered two cases from examination of biopsy material removed during an exploratory thoracotomy. Earlier Engelbreth-Holm et al. (1944) and Weinstein et al. (1947) described the condition of eosinophilic granuloma in bone and remarked upon the radiological changes present in the lungs of their cases. It is uncertain, however, whether the changes were caused by eosinophilic granuloma in the lung. Since the first case many further examples have been reported

notably by Virshup and Goldman (1956), Auld (1957), Anderson and Foraker (1959) and Williams et al. (1961).

Unlike Letterer–Siwe disease this condition is usually found in persons above the age of puberty though the author has seen a fatal case in a child of 4 years. Pulmonary eosinophilic granuloma, though sometimes a fatal disease, carries a better prognosis than either Letterer–Siwe disease or Hand–Schüller–Christian disease. Many patients have remained alive and well several years after the initial diagnosis was made and in the majority the disease appears to be completely quiescent though a few cases have developed diabetes insipidus and bone lesions.

Patients may present with a single or repeated spontaneous pneumothorax but an unexplained cough and weight loss are the more usual initial symptoms.

Macroscopically, the lesions in the lungs may take several forms and are found (a) as discrete firm greyish-white or tan masses with indistinct edges, (b) as diffuse miliary nodules, (c) as bronchial polypoid granulomatous lesions, (d) as a pleural plaque and (e) as a "honeycomb" lung. Many of these forms represent stages in the evolution of the disease which Engelbreth-Holm et al. divided into four stages: (1) a hyperplastic stage, (2) a granulomatous phase, (3) xanthomatous stage and (4) the healed fibrotic stage. The amount of fibrosis increases as the lesions age and many of the bullae and much of the puckering of the overlying pleura are produced secondarily to the fibrosis and alveolar wall destruction. The "honeycomb" cystic changes (Fig. 22.27) found in the lungs, which closely resemble those seen in Letterer–Siwe disease, are mainly the consequence of alveolar wall destruction, secondary bronchiolar dilatation following fibrosis in the lung, and to a much lesser extent air-trapping distal to partly obstructed air passages.

Microscopically, the early lesions present a granulomatous appearance and contain a variety of cells but predominantly large, irregularly shaped histiocytic cells. These are characterized by reniform or oval vesicular nuclei containing a single or two prominent nucleoli, and with a

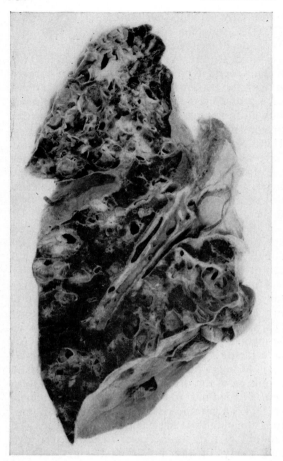

FIG. 22.27. A honeycomb lung caused by eosinophilic granuloma. Note the bullous-like spaces which may sometimes rupture causing pneumothorax. × one-half natural size.

lilac staining, often foamy cytoplasm. Binucleate forms of the cells are often present. In addition a variable and often large number of eosinophils are found, which give the lesion its name, together with lymphocytes, plasma cells and a few polymorph leucocytes (Figs. 22.29, 22.30). In the early stages there is vascular congestion and the lesions may contain areas of haemorrhage and necrosis, and some affected alveolar walls become disrupted. Perivascular deposits may infiltrate the arterial walls causing intimal

thickening and later intimal fibrosis. If the vascular lesions are sufficiently widespread pulmonary hypertension can result (Virshup and Goldman).

As the lesions age increasing amounts of reticulin and ultimately collagen fibres are formed and the whole is reduced to a mass of hyaline fibrous tissue. While these changes are progressing, collections of foam cells appear probably in response to haemorrhage and necrosis which often occur in the earlier stages.

Eosinophilic granuloma, like the other conditions composing this triad of diseases, is often a generalized condition and deposits of similar tissue may be found in the skin, lymph nodes, liver and bones. The histological changes merge with the other two forms of histiocytic reticuloses and differentiation between them rests largely on the mode of clinical presentation and the course of each disease. An unusual form of eosinophilic granuloma was described by Sivanesan (1959) in which periportal eosinophilic deposits occurred in the liver and led to distension of the intrahepatic bile ducts and periportal fibrosis, and was accompanied by "honeycomb" changes in the lung.

Progressive pulmonary lesions may lead to interference with pulmonary function. The vital capacity and total air capacity are progressively reduced and the volume of residual air increases. The changes follow destruction of the lung accompanied by increasing fibrosis. Because of the natural tendency of the disease to produce fibrosis in the lung, radiation therapy is contraindicated, but administration of steroid drugs may result in dramatic improvement provided that fibrotic changes in the lung are not already advanced.

HAND–SCHÜLLER–CHRISTIAN DISEASE

The first case of the third disorder in this group was described by Hand (1893), a further two cases were later added by Schüller (1915), and Christian (1920) recognized and described the syndrome of xanthomatous deposits in bone, defects of the membrane bones, together with

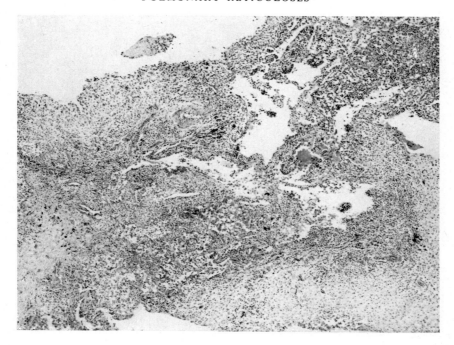

FIG. 22.28. A very low-power view of part of the lung in a case of eosinophilic granuloma showing the extensive fibrosis and destruction of the alveolar tissue which is hardly recognizable as lung tissue. × 40 H and E.

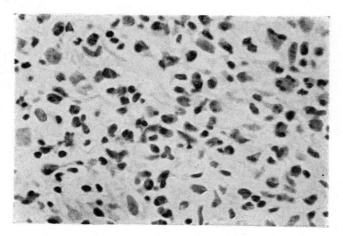

FIG. 22.29. An eosinophilic granuloma in lung showing numerous eosinophils, histiocytes and fibroblasts. × 280 H and E.

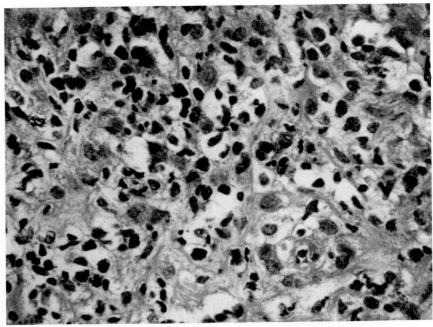

FIG. 22.30. A high-power view of the cells composing an eosinophilic granuloma of the lung showing the many varieties of cells present. These cells consist of histiocytic cells, lymphocytes and plasma cells. × 280 H and E.

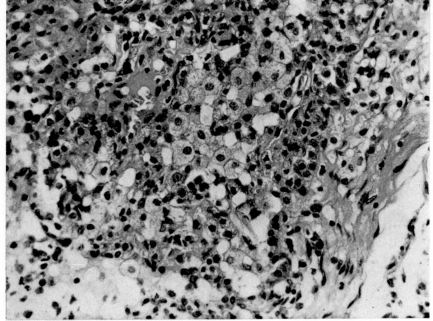

FIG. 22.31. An older lesion from an eosinophilic granuloma of the lung showing collections of foam cells in addition to the cells already seen in previous figures. × 280 H and E.

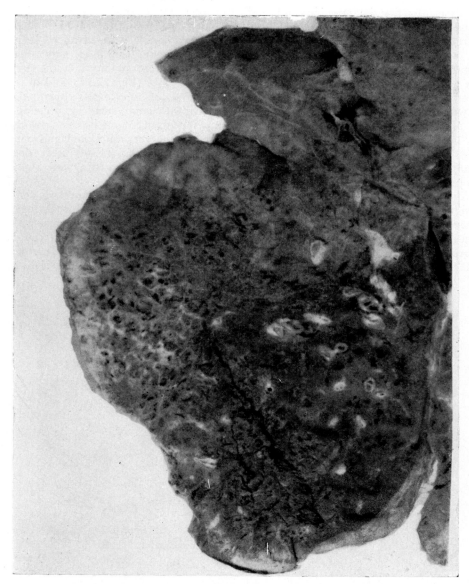

FIG. 22.32. Specimen of lung from a case of Hand–Schüller–Christian disease showing fibrosis and a honeycomb of cysts confined to the outer subpleural portion of a lower lobe.
(Reproduced by courtesy of the late Professor H. A. Magnus.)

diabetes insipidus and exophthalmos. Although the condition is often first diagnosed in childhood and adolescence, it is slowly progressive and the patients frequently reach adult life before death occurs.

Pulmonary changes were first mentioned by Henschen (1931) who remarked upon the frequent association of the bone and endocrine disorder with radiological lung changes. Cunningham and Parkinson (1950) described six cases of "honeycomb" lung and discussed the pathological changes, noting that some showed a "xanthomatous granuloma" which in retrospect was probably the result of Hand–Schüller–Christian disease. Cavanagh and Russell (1954), McKeown (1954) and McNeill and Cameron (1955) have also described the pulmonary changes in detail.

Dyspnoea and cough are common symptoms during the course of Hand–Schüller–Christian disease often preceding radiological changes in the lungs by a considerable period. Henschen also drew attention to the occasional association of cor pulmonale with advanced lung damage. *Macroscopically*, the lung may show diffuse fibrosis, most pronounced in the periphery of the lower lobes (Fig. 22.32), and in the fibrotic areas small dots of orange-yellow lipid are found together with many small cysts giving a honeycomb appearance. In some cases the interlobular septa become very thickened and conspicuous. Forsee and Blake (1954) also described a bronchial polyp consisting of xanthomatous tissue.

Microscopically, early lesions are characterized by histiocytic cells with occasional bi- and poly-nucleated forms, foamy macrophage cells and fibroblasts. The lipid material is mainly cholesterol, being doubly refractile, and is found in focal patches (Fig. 22.33). In rapidly

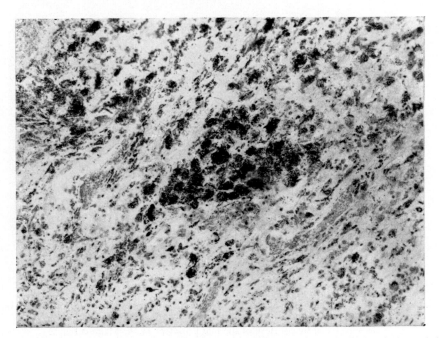

FIG. 22.33. Collections of cells laden with sudanophilic lipid material from a case of Hand–Schüller–Christian disease. The lipid-laden cells occur in focal collections. × 140 Scharlach R and haematoxylin.
(Reproduced by courtesy of the late Professor H. A. Magnus and Dr. B. S. Cardell.)

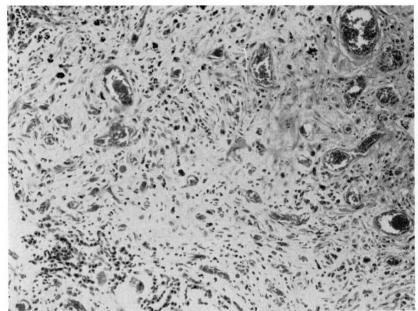

FIG. 22.34. A fibrosed area of the lung showing the variety of cells present. Note also the vascularity of the damaged lung. × 140 H and E.
(Reproduced by courtesy of the late Professor H. A. Magnus and Dr. B. S. Cardell.)

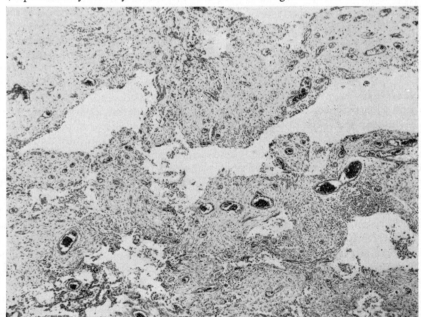

FIG. 22.35. A low-power view of the lung in a case of Hand–Schüller–Christian disease showing extensive alveolar wall fibrosis and destruction of much of the lung tissue with cystic air spaces between the fibrosed areas. × 40 H and E.
(Reproduced by courtesy of the late Professor H. A. Magnus and Dr. B. S. Cardell.)

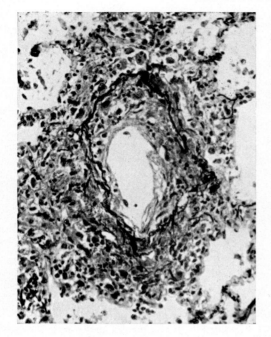

FIG. 22.36. A perivascular infiltration of the characteristic cells found in Hand–Schüller–Christian disease which is spreading into the adjacent alveolar walls. × 170 Verhoeff–Van Gieson.

(Reproduced by courtesy of Professor D. Russell and the Editor of *J. Path. Bact.*)

developing cases lipid may be entirely absent. Later, increasing fibrosis occurs and collections of lymphocytes and a few plasma cells appear (Fig. 22.34). The changes are centred mainly in the subpleural, interlobular and perivascular connective tissue but also involve bronchiolar and alveolar walls leading to the obliteration of many of the alveoli (Fig. 22.35). Perivascular lesions spread into the adventitial coat of the arteries and the vessels become narrowed by endarteritis (Figs. 22.36, 22.37). The widespread nature of the arterial lesions, together with parenchymal destruction and later alveolar fibrosis, is probably responsible for the occurrence of pulmonary hypertension that results occasionally in this condition.

Cyst formation has been variously interpreted

as being due either to bronchial obstruction by granulomatous deposits and fibrosis, or to bronchiolar mural weakening due to deposits of tumour tissue in the walls. It is probable that the honeycomb of cysts is a compensatory change and a sequel to diffuse alveolar destruction and fibrosis in those cases where active disease has ceased (McLetchie and Reynolds, 1954).

From the foregoing descriptions it is apparent from the distribution of the lesions, the microscopical appearances, and the resultant sequelae that the three diseases form a single group which merge imperceptibly one into the other; but Siwe (1949) was opposed to this view and regarded Letterer–Siwe disease as a separate entity because of its rapidly fatal nature. Hewer and Heller (1949), however, reported a case of Letterer–Siwe disease in a child aged 5, which, instead of proving rapidly fatal, gradually evolved into the clinical picture of Hand–Schüller–Christian disease. The view expressed by Lichtenstein (1953) that these disorders are the acute and chronic forms respectively of one disease process is difficult to refute.

The nature of the disease process, whether neoplastic or granulomatous, is more difficult to decide. In many respects these disorders show features akin to Hodgkin's disease inasmuch as they display some of the characters of a granuloma and some of systemic neoplasms like the reticuloses. At present this question cannot be finally settled and these three disorders have been considered as a separate group and have not been included among either the pulmonary neoplasms or chronic inflammatory conditions. Furthermore, there is no evidence to support their inclusion among the lipid storage diseases, as the lipid accumulations, which consist mainly of cholesterol and cholesterol esters, are found only in the more chronic cases of all three disorders, and probably result from obstructive pneumonitis and haemorrhages.

The origin of the histiocytic cells common to all three disorders though not fully established may be from Langerhans cells. These cells, normally found in the epidermis, are probably very much more widely distributed throughout the body. The Langerhans cells contain in their

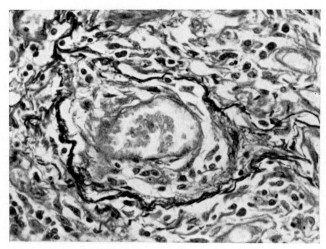

FIG. 22.37. A small branch of a pulmonary artery showing stenosis of the lumen due to a mural infiltration of cells in Hand–Schüller–Christian disease. × 280 Verhoeff–Van Gieson.
(Reproduced by courtesy of Professor D. Russell and the Editor of *J. Path. Bact.*)

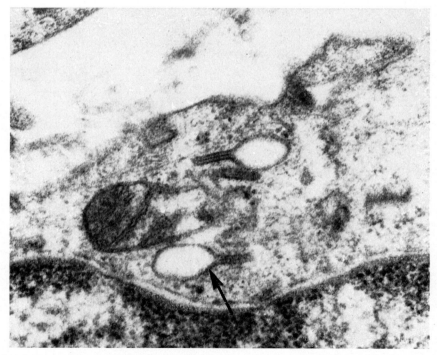

FIG. 22.38. An electron micrograph showing the characteristic cytoplasmic organelles found in the tumour cells in Letterer–Siwe disease (arrowed). In shape they resemble a tennis racket. × 50,000.

cytoplasm distinctive electron microscopic organelles resembling tennis rackets in appearance (Birbeck *et al.*, 1961). These are thought to arise either from infoldings of the plasmatic membrane or from Golgi vacuoles. Similar structures are found in the cytoplasm of the histiocytic cells seen in histiocytosis-X disease including the histiocytic cells present in the lung lesions (Basset and Turiaf, 1965) (Fig. 22.38). In all doubtful cases electron microscopic examination should be carried out to confirm the presence of these characteristic cytoplasmic organelles.

CHAPTER 23

Hamartomas, Blastoma and Teratoma of the Lung

Pulmonary Hamartoma and Blastoma (Embryoma)

The term hamartoma was coined by Albrecht (1904) (from the Greek words $\alpha\mu\alpha\varrho\iota\tau\alpha$ = error and $\hat{\omega}\mu\alpha$ = tumour) to describe certain tumours that resulted from a localized error in development of a normal constituent or constituents of an organ. Since the term was first introduced it has come to be widely applied to a great variety of tumours in many organs and because of misuse it has lost much of its original meaning.

A hamartoma is the result of unbalanced, co-ordinated, expansile but non-invasive growth confined to a limited anatomical field. The component tissues follow the general plan of development and may reach varying degrees of maturity and functional ability. Further growth of a hamartoma may continue after similar normal tissues have reached maturity.

A blastoma is a tumour developed from a pluripotent cell of a single germinal layer. It may likewise follow the normal growth pattern closely, although it is not confined to limited expansion but is capable of progressive and unlimited invasion.

Several conditions occurring in the lung may be included under the term pulmonary hamartoma and include congenital lymphangiectasis, pulmonary myomatous hyperplasia and the two tumours usually referred to as localized and generalized pulmonary hamartomas.

Pulmonary angiomas will not be described in this chapter, although possibly hamartomatous in nature, as they are customarily regarded as

benign tumours of the lung and are included with the other benign pulmonary neoplasms.

Local Pulmonary Hamartomas

These tumours are usually discovered by chance following a routine radiograph of the lungs or during a post-mortem examination on an adult.

There appear to be two distinct varieties of hamartoma: the first usually contains islands of cartilage and cleft-like spaces lined in part by ciliated epithelium but in which there is no attempt at alveolar tissue formation. The whole tumour is embedded in the lung tissue and usually lies nearer the pleural surface (Fig. 23.3).

The second and less common variety of hamartoma usually presents as small multiple tumours which are invariably found in the subpleural portions of the lungs. They consist of undifferentiated mesenchyme in which are found cystic spaces lined by peg-shaped or cuboidal epithelium from which may arise short tubules lined by an intestinal type of mucus-secreting epithelium (Fig. 23.5). This type of hamartoma often forms immature alveoli lined by cuboidal foetal-type epithelium but no ciliated epithelium is found.

Although no final opinion can as yet be expressed as to the origins of the two types of hamartoma, if the views expressed by Waddell (1949) that the lungs are developed partly from repeated branching of an entodermal bud (the bronchial tree) and partly by canalization of the mesenchyme (alveoli and respiratory bronchioles) are correct, the first type of hamartoma may represent a disorder in growth of a bronchial bud while the second and rarer variety may be derived from disordered growth of the mesenchyme. The second variety of hamartoma is readily mistaken for a well-differentiated pulmonary blastoma to which it bears a close relationship.

The synonyms for the *first or central type of hamartoma* include: adenochondroma, chondromatous hamartoma, fibro-adenoma of the lung, pulmonary lipochondroadenoma, bronchial hamartomas and mixed pulmonary tumours.

The first growth of this type was described by Lebert (1845), though Feller (1921–2) was the first to refer to these tumours as pulmonary hamartomas. Since then other cases have been described by Hickey and Simpson (1926), Möller (1933), Goldsworthy (1934), McDonald *et al.* (1945) and Carlsen and Kiaer (1950). Similar tumours occur in horses and cattle.

McDonald *et al.* found twenty hamartomas in 7972 post-mortems, an incidence of about 0·25 per cent. The majority of the tumours are single and are found nearer the pleural than the hilar regions of the lungs though occasionally multiple tumours may arise in the walls of large and small bronchi (Gudbjerg, 1961) (Fig. 23.2). The tumours are stated to be three times more common in men than in women (Stein and Poppel, 1955). They are responsible for some of the "coin lesions" seen in radiographs of the lungs.

Macroscopically, the tumours vary in size from less than 1 cm to 6 cm in diameter. Surgical exposure of a solitary pulmonary hamartoma in life may result in the ejection of the tumour which may separate cleanly from the surrounding lung. The surface is lobulated, smooth and usually formed of nodules of cartilage or dense fibrous tissue. On section a variable amount of calcified cartilage and true bone is found together with much cartilage (Fig. 23.1).

Microscopically, the tumours reproduce any of the components of bronchial tissue including areas of fibrocartilage which often calcify, metaplastic bone, lipomatous tissue, collections of lymphocytes and numerous cleft-like spaces lined in part by cuboidal epithelium and in part by ciliated epithelium (Figs. 23.3, 23.4). The ciliated epithelium occurs in patches and may be obscured by inspissated mucus contained within the epithelium-lined spaces. Bundles of plain muscle fibres are often present and surrounding the tumour is a layer of compressed alveolar tissue part of which usually adheres to the hamartoma when it is removed from the lung. A large branch of a pulmonary artery was reported traversing one of these growths by Courmount (1895) and some hamartomas may be traversed by recognizable small bronchi or bronchioles. In a few hamartomas cartilage may be entirely absent as was the case in the "fibro-adenoma" described by Scarff and Gowar (1944). Some of these latter cases may have been examples of the second type of hamartoma referred to previously. The absence of anthracotic pigment indicates that the tumours are not connected with the main bronchial pathway. In the hamartomas which arise in recognizable bronchi, all stages may be seen in the growth of the tumour cartilage which grows by a process of metaplasia from the peribronchial connective tissue. Many regard all forms of bronchial chondromas as hamartomatous in nature, but in the hamartomas referred to in this text other elements such as bundles of muscle fibres and nervous tissue occur in addition to cartilage, bone and bone marrow. Chondromas are regarded as separate entities and are discussed elsewhere (Chapter 21).

Möller considered that the formation cartilaginous tissue in these tumours was induced by the overlying epithelium, a view supported by Willis (1952). The author does not agree with this view as he finds it difficult to believe that any form of epithelium can so "differentiate" connective tissue as to produce cartilage, lymphocytes, fatty tissue and bone, a characteristic apparently possessed only by the respiratory epithelium.

Malignant change in these tumours is almost unknown, though a single large hamartoma reported by Simon and Ballon (1947) showed what the authors regarded as malignant change, one part of the tumour failing to show the usual

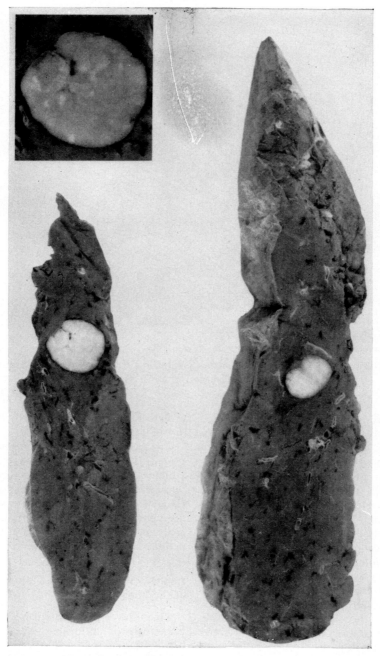

FIG. 23.1. A solitary pulmonary hamartoma showing the sharply circumscribed borders, and inset, the calcified areas in the tumour which may be recognizable in lung radiographs. Large specimen natural size. Inset × 2.

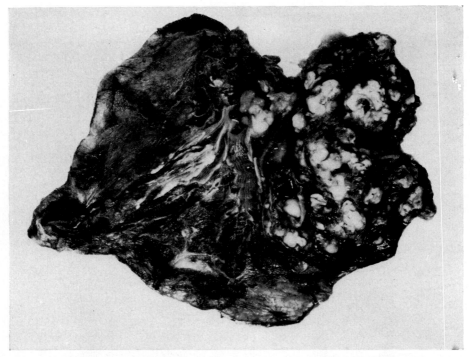

FIG. 23.2. A specimen of lung containing multiple chondromatous hamartomas. Nine-tenths natural size. (Reproduced by courtesy of Dr. E. H. Bailey.)

encapsulation by compressed and atrophied lung. Greenspan (1933) described an osteo-chondra-sarcoma arising in a tumour which contained multiple clefts lined by ciliated epithelium which was probably hamartomatous in nature. Clinically these growths are usually completely silent and are chiefly of importance in the differential diagnosis of the solitary pulmonary nodule discovered in a routine chest radiograph. Considerable importance has been attached to stippling of the tumour observed in X-rays caused by calcific deposits, an appearance more indicative of a solitary hamartoma or granulomatous condition than a lung carcinoma. These growths enlarge very slowly and probably expand in most instances over a period of many years. The growth rates of solitary pulmonary hamartomas has been studied by Jensen and Schiødt (1958) who found a gradually rising growth curve, the rate of growth being greater in the large tumours.

The main features of the *second or peripheral type of local pulmonary hamartoma* have been alluded to previously and they are closely related to pulmonary blastomas, but growth rate of the hamartomas is slower and the different components of the tumour are much more mature and show none of the features of a neoplasm such as are found in the blastomas.

Congenital Adenomatoid Malformation of the Lung

This type of malformation is usually found in premature or stillborn infants replacing either one lobe or a whole lung and it is frequently associated with other congenital pulmonary abnormalities. Occasionally the malformation

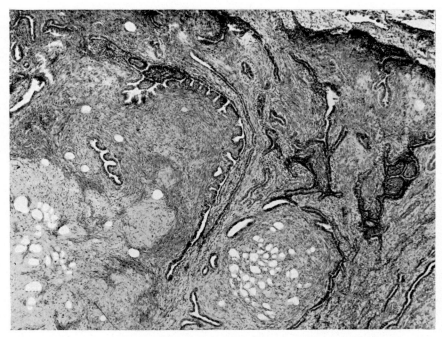

FIG. 23.3. A low-power view of a solitary pulmonary hamartoma showing the cleft-like spaces lined by columnar but non-ciliated epithelium. There is no cartilage in this specimen. × 40 H and E.

exists as a separate mass connected by an extra bronchus with the main bronchial tree, a type of malformation probably closely related to accessory lungs. In the majority of cases the pregnancy was complicated by hydramnios and premature delivery.

The first case was described by Stoerk (1897). More recent accounts have been given by Harris and Schattenberg (case 2, 1942), Ch'in and Tang (1949), Jones (1949), Goodyear and Shillitoe (1959), Spector *et al.* (1960) and Van Dijk and Wagenvoort (1973).

The majority of these malformations only involve a lobe or part of a lobe but rarely as in the case described by Goodyear and Shillitoe almost an entire lung was involved. The affected portion of lung is enlarged, firm, pale and airless but often contains small cysts (Fig. 23.6) and rarely one or two large cysts. It may lead to displacement of the other thoracic viscera including the heart and compress the inferior

vena cava. Tension pneumothorax can result from rupture of a hyperinflated cyst during attempts at resuscitation, and a distended cyst is one cause of the "expanding neonatal lung syndrome". Respiratory distress results from the compression of the normal lung tissue. Infection of the cysts and their contents commonly occurs in post-natal life.

Microscopically, the hamartomas consist of a maze of epithelium-lined tubes and spaces resembling foetal bronchioles (Fig. 23.7). Some of the tubes are lined with ciliated epithelium which is thrown into folds. Mucous glands are present and resemble those normally seen in the walls of the larger bronchi at this age. The loose stroma between the epithelial tubes may contain cartilage, as in the case described by Goodyear and Shillitoe, together with elastic and muscle fibres. In the cases described by Kwittken and Reiner (1962) some of the cysts were lined by columnar epithelium while others

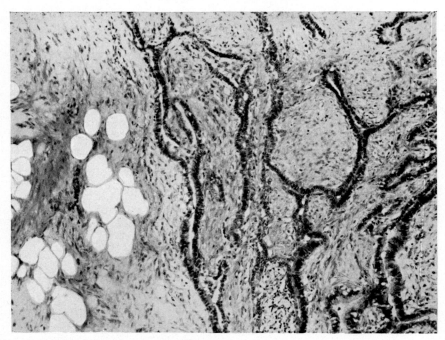

FIG. 23.4. A higher-power view of the same hamartoma seen in Fig. 23.3 showing the non-ciliated columnar epithelium lining the spaces and the surrounding fibro-fatty tissue. × 140 H and E.

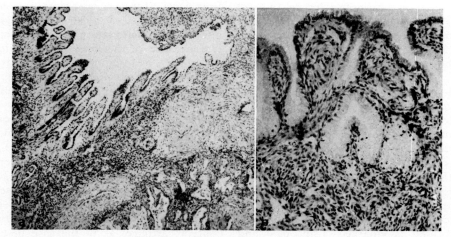

FIG. 23.5. A peripheral type of local pulmonary hamartoma. In (a) can be seen tubular structures lined by peg-shaped columnar but non-ciliated epithelium, immature alveolar tissue and undifferentiated mesenchyme. In (b) is shown a glandular structure lined by mucus-secreting cells and in part by peg-shaped epithelium with underlying undifferentiated mesenchymal tissue. Compare this lesion with the larger diffuse pulmonary hamartoma with which it is very closely related. (a) × 40, (b) × 100 stained H and E.

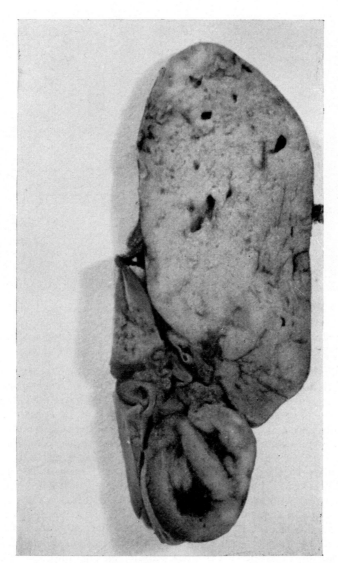

FIG. 23.6. A diffuse hamartoma occupying most of the right lung together with the opened heart. The tumour has been cut across and shows a solid mass containing several empty spaces. × 2.

(Reproduced by courtesy of Dr. J. E. Goodyear and the Editor of *J. Clin. Path.*)

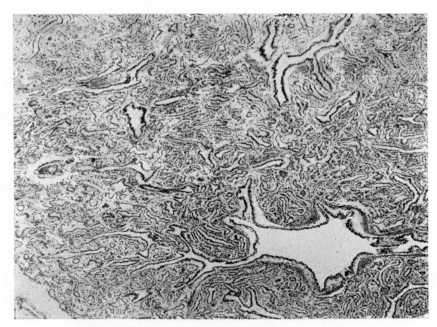

FIG. 23.7. The central part of a diffuse pulmonary hamartoma showing numerous epithelium-lined spaces closely resembling normal bronchi. × 40 H and E.

(Reproduced by courtesy of Dr. J. E. Goodyear and the Editor of *J. Clin. Path.*)

were lined by cuboidal epithelium and showed little or no elastic tissue in their walls. Each type of epithelium was continuous with the other but the transition was very abrupt. The intervening non-cystic lung parenchyma may contain alveoli lined by tall mucus-secreting epithelium and alveoli may also become more vascular and display changes similar to those described as alveolar dysplasia (MacMahon, 1948) (Fig. 23.8).

If a child born with a diffuse pulmonary hamartoma survives into adult life the lung may present an appearance somewhat similar to cystic bronchiectasis. In a case seen by the author (Fig. 23.9) multiple cysts filled with mucus replaced most of the lower lobe of a lung but no bronchial obstruction was discovered. Bounding part of the mucus-filled cystic lung, but sharply demarcated from it, was a small residual subpleural portion of totally collapsed, partly fibrosed but recognizable lung alveolar tissue. The alveoli in the cystic lung were mostly disrupted and where visible consisted of fibrosed septa lined in part by cuboidal, mucus-secreting epithelium and in other places by tall columnar, intestinal type of epithelial tissue (Fig. 23.10). The larger bronchi were distended with thick mucus and showed considerable atrophy of the bronchial cartilage and other normal mural structures. The vascular supply of the whole lung was received from normal sources.

Reiter (1936) also described a lung hamartoma involving the upper lobe of the right lung of an adult which consisted of a small firm mass composed of cartilage and intercommunicating, cystic, epithelium-lined spaces. The cyst walls contained mucous glands, nerves and large arteries and the right eparterial bronchus communicated with these cysts; the whole mass was surrounded by a cap of normal pulmonary alveolar tissue.

Compression of the adjacent unaffected lung and its supplying bronchi may cause infection,

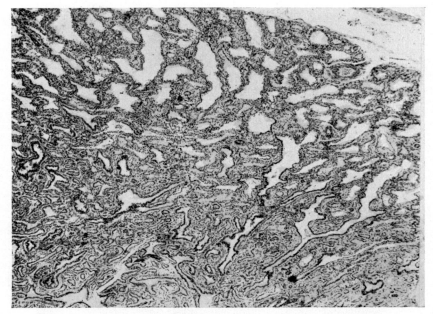

FIG. 23.8. The peripheral part of a diffuse pulmonary hamartoma showing the thickened alveolar walls which resemble those described as alveolar dysplasia. × 40 H and E.

(Reproduced by courtesy of Dr. J. E. Goodyear and the Editor of *J. Clin. Path.*)

and malignant change (myxosarcoma) has been reported in a diffuse hamartoma by Stephanopoulos and Catsaras (1963) in a young child.

The frequent existence of other congenital defects in the same patient including accessory and absent bronchi, hypoplasia of the opposite lung and accessory systemic arteries supports the view that these tumours result from developmental errors. The error was considered by Conway (1951) to be mainly a failure of the mesenchymal (respiratory) portion of the lung to canalize and join normally with the endodermal (bronchial) tree. In a diffuse pulmonary hamartoma the disorder of growth may affect both portions of the lung.

Pulmonary Fibroleiomyomatous Hamartomatous Disorders

Three varieties of pulmonary fibroleiomyomatous hamartomatous states can be recognized:

(1) a single or multiple focal muscle tissue proliferation,
(2) pulmonary lymphangioleiomyomatosis,
(3) tuberose sclerosis.

(1) A Single or Multiple Focal Overgrowth of Myomatous Tissue

A single or multiple focal overgrowth of myomatous tissue may occur in the alveolar walls, in the pleura or around the walls of the smaller bronchi. The overgrowth of muscle tissue usually involves only the alveolar walls, but may arise from muscle tissue in the respiratory bronchioles or smaller bronchi (Fig. 23.11). The blood-vessels are not involved and the focal muscle proliferations need to be differentiated in women from a metastasizing fibroleiomyoma of the uterus. Unlike the latter, however, they spread irregularly at their margins and the whole

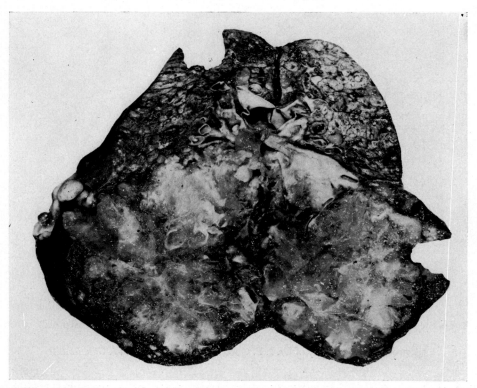

FIG. 23.9. Diffuse pulmonary hamartoma in an adult lung. The affected portion of the lung is filled with a honeycomb of spaces filled with thick mucus. A thin subpleural rind of compressed and largely destroyed lung parenchyma can still be seen. Four-fifths natural size.

lesion if macroscopically visible is often stellate and not spherical and sharply circumscribed.

Fibroleiomyomatous hamartomas begin abruptly in the subpleural zone and fade away towards the hilum but there may be a transitional band of fibrosis intervening between the hamartomatous and the normal lung. They may give rise to single or multiple coin opacities in lung radiographs. A focal fibroleiomyomatous hamartoma was described by Cruickshank and Kent-Harrison (1953), and further localized hamartomas of this type were described by Brahdy (1941) and Williams and Daniel (1950). Many minor examples are frequently found in sections of lungs removed post-mortem.

Focal fibroleiomyomatous hamartomas may be associated with chondromatous hamartomas or hamartomatous overgrowth of lymphatic channels. This condition should not be confused with leiomyomatous hyperplasia seen in chronically fibrosed and damaged lungs.

(2) Pulmonary Lymphangioleiomyomatosis

Pulmonary lymphangioleiomyomatosis is a very uncommon disorder which occurs in all races and forms part of a much more extensive hamartomatous development of plain muscle involving the lungs, lymphatics, and hilar, abdominal and lower cervical lymph nodes. The earliest cases were described by Burrell and Ross (1937) and Rosendal (1942) and were associated with honeycomb cystic changes in the lungs.

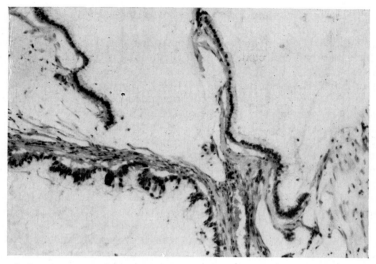

FIG. 23.10. Microscopical appearances of the mucoid zone seen in the previous figure. The walls of the "honeycomb" are lined in part by tall columnar cells which secrete mucus. × 90 H and E.

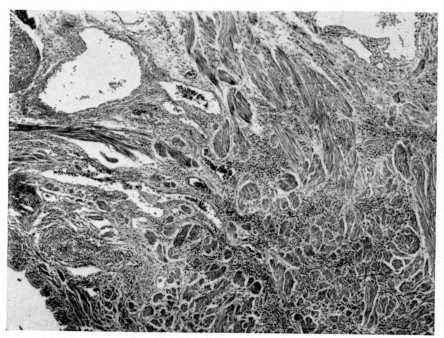

FIG. 23.11. A section of a fibroleiomyomatous hamartoma arising in peribronchial tissues. × 40 H and E.

15

Corrin *et al*. (1975) in the most extensive account to date reviewed thirty-four previously published cases and added a further twenty-three new examples which they had collected. All the patients were women. It may first present between the ages of 2 and 47 years of age.

Clinically, it usually causes the patient to complain of gradually increasing dyspnoea punctuated sometimes by a sudden severe exacerbation of the symptom caused by a spontaneous pneumothorax or the rapid onset of a chylous pleural effusion. Other modes of presentation include haemoptyses. In almost all cases death occurs within 10 years from the onset of symptoms and not infrequently within as short a period as 3 years.

Macroscopically, in a well-established case the lungs present a honeycomb appearance with thickened pleura and septa surrounding air-distended spaces. If the hilar lymph nodes are involved they become enlarged and spongy and the thoracic duct is a very dilated and sponge-like structure filled with chyle. Rupture of the larger dilated mediastinal or pulmonary lymphatic channels may result in an opalescent chylous pleural effusion. The dilated chyle-filled lymphatics extend downwards and often include those in the retroperitoneal region in the abdomen as well as those in the root of the neck. The dilated abdominal lymphatic channels may be responsible for causing a chylous peritoneal effusion.

Microscopically, pulmonary lymphangioleiomyomatosis is characterized by striking, nodular, interstitial proliferation of plain muscle tissue in the lobular septa, pleura, the walls of the alveoli, as well as in the walls of the smaller bronchi and bronchioles (Fig. 23.12). Some of the nodules consist of small round cells formed by immature myocytes. As a consequence of the bronchiolar changes air-trapping ensues and is mainly responsible for the honeycomb changes observed macroscopically. Similar mural proliferation of muscle in the walls of the pulmonary veins causes

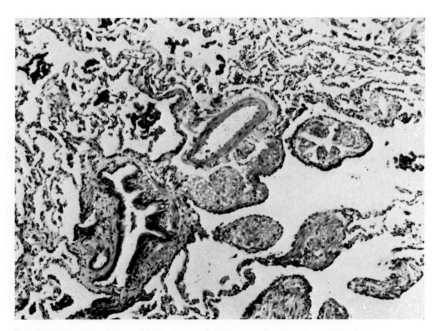

Fig. 23.12. Lymphangioleiomyomatosis showing abnormal excess muscle tissue in the walls of a perivascular lymphatic channel as well as around a small bronchus, and in the alveolar wall interstitium. × 100 H and E.

venous obstruction with capillary haemorrhages into the alveoli causing haemoptyses (Fig. 23.13). The lymphatic channels throughout the lungs are distended due to obstruction caused by mural muscle proliferation (Fig. 23.14).

As a result of the pulmonary venous obstruction siderotic changes occur in the lung including iron-pigment encrustation of alveolar elastic fibres with accompanying foreign-body giant-cell formation and accumulation of intra-alveolar siderophages. The obstructive changes in the pulmonary arteries together with increasing airway obstruction results in pulmonary hypertension and right ventricular hypertrophy. The affected lymph glands are almost entirely replaced by interlacing bundles of smooth muscle between which are a honeycomb of distended sinus spaces (Vadas *et al.*, 1967).

Pulmonary lymphangioleiomyomatosis is probably a disease *sui generis* and is not a *forme fruste* of tuberose sclerosis to which its pulmonary manifestations bear a very close resemblance. The disease occurs only in women but is not inherited, whereas males are more frequently affected in tuberose sclerosis which is inherited through a dominant gene. Patients with tuberose sclerosis who develop pulmonary changes are, however, usually women. Other stigmata of tuberose sclerosis including sebaceous adenomas, tuberose glial nodules and ocular phakomas have not been described in cases of pulmonary lymphangioleiomyomatosis though renal angiofibrolipomas do occur in both disorders.

(3) Tuberose Sclerosis

Synonyms: Epiloia, Bournville's disease

As already stated considerable doubt exists as to whether pulmonary lymphangioleiomyomatosis is not a *forme fruste* of tuberose sclerosis though reasons have been advanced previously for still regarding them as separate conditions.

Tuberose sclerosis is a rare disorder the incidence of which has been estimated at 1 in

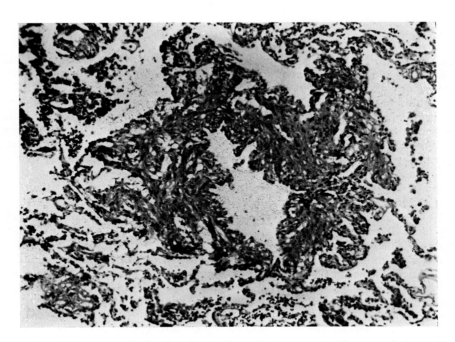

FIG. 23.13. Lymphangioleiomyomatosis showing abnormal muscle tissue surrounding a small intrapulmonary vein. × 180 H and E.

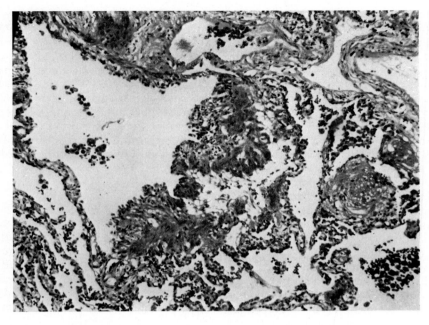

Fig. 23.14. Diffuse lymphangioleiomyomatosis showing focal patches of muscle tissue in the wall of a lymphatic. × 100 H and E.

500,000 persons. Although the condition is regarded as being primarily of neurological interest, the occurrence of sebaceous adenomas and renal tumours has long been recognized as part of this unusual disorder. Among the lesser-known changes that occur in this hereditary disease are those affecting the lung.

The earliest reference to pulmonary symptoms and pathology in tuberose sclerosis was made by Lautenbacher (1918) who described changes in the lungs of a woman aged 36 who, during the last 2 years of her life, suffered from dyspnoea and finally died from bilateral spontaneous pneumothorax. At the subsequent post-mortem, the lungs contained innumerable air-filled cystic spaces which were described as the size of "nuts" and both organs presented a spongelike appearance. In addition there were small submiliary nodules of "fibrosarcomatous" tissue in the walls of the alveoli, bronchioles and blood-vessels. Lautenbacher regarded the nodules as secondary deposits arising from the renal

tumours that were present. Berg and Vejlens (1939) described a case in Norway in which the lungs showed an overgrowth of connective tissue, involuntary muscle and blood-vessels, and the whole lung presented an emphysematous appearance. In 1946 Berg and Nordenskjöld succeeded in tracing three members of the family of the patient reported earlier in 1939, and all showed evidence of tuberose sclerosis; two showed cystic changes in their lungs post-mortem but neither had shown any clinical evidence of the lung disease until shortly before death. Further cases have since been reported by Moolten (1942) and Murphy et al. (1958).

Macroscopically, much of both lungs is replaced by a honeycomb of small cystic spaces which bear no particular relationship to any part of the lung lobules. In addition small, solid, creamy-white rounded nodules varying in size up to a centimetre in diameter are visible on the cut surface, but they are not related to the vessels or bronchi (Fig. 23.15). In advanced cases, where

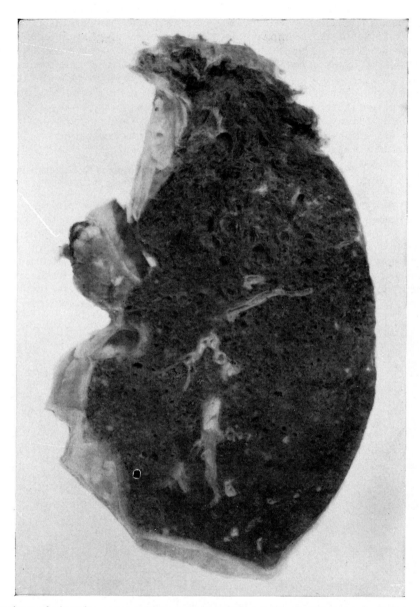

Fig. 23.15. Specimen of a lung from a case of tuberose sclerosis showing the fine honeycomb of air spaces and multiple small solid nodules of tumour tissue (hamartomas). Two-thirds natural size.
(Reproduced by courtesy of the late Professor H. A. Magnus.)

the patients have died with clinical signs of chronic lung disease, there is usually evidence of hypertensive thickening of the pulmonary arteries. Sudden dyspnoea may be caused by rupture of a subpleural cyst resulting in spontaneous pneumothorax.

Microscopically, the appearances vary enormously from one part of the lung to another. In some parts the lung has a normal appearance, in others there are emphysematous changes; elsewhere nodules of fibroleiomyomatous tissue covered with flattened or cuboidal alveolar epithelium project into the alveoli in a papillary fashion (Figs. 23.16, 23.17). In other parts of the lung dilated engorged capillaries follow obstruction of small pulmonary veins by hamartomatous tissue and lead to intra-alveolar haemorrhages and collections of siderophages. If the latter are abundant they impart to the gross specimen of the lung a brick-red colour.

In the present state of knowledge it is impossible to advance any logical explanation for the bizarre neoplastic nodules that appear in this condition in so many organs including the lungs. The presence of glial nodules in the central nervous system, sebaceous adenomas in the skin and tumours composed of involuntary muscle and fibrous tissue in the kidneys and lung shows that this is a congenital disorder of tissue organization probably involving all germinal layers. Both Moolten and Murphy *et al.* regarded the tumours as hamartomas.

Pulmonary Changes in Generalized Neurofibromatosis

In addition to the presence of neurofibromatous tumours within the lung which are considered in Chapter 21, the occurrence of cystic lung disease in association with generalized neurofibromatosis (von Recklinghausen's disease) was first described by Starck (1928). Since the first description several further examples of the association of the two changes have been

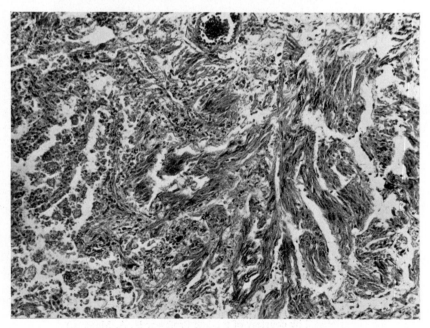

FIG. 23.16. A hamartomatous tumour mass composed mainly of plain muscle fibres from a case of tuberose sclerosis. × 140 H and E.

(Reproduced by courtesy of the late Professor H. A. Magnus and Dr. B. S. Cardell.)

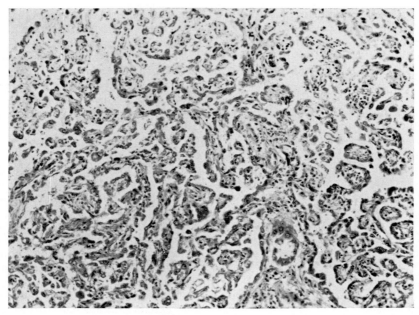

Fig. 23.17. A hamartoma from a case of tuberose sclerosis showing epithelium-lined papillary processes. × 140
H and E.
(Reproduced by courtesy of the late Professor H. A. Magnus and Dr. B. S. Cardell.)

described by Davies (1963), Massaro *et al.* (1965) and Israel-Asselain *et al.* (1965). In almost all the cases, the onset of dyspnoea in patients with neurofibromatosis led to X-ray examination of the lungs and the discovery of reticular shadows, nodular opacities and cystic changes. Subsequent angiographic examination showed a decreased blood-flow in the affected areas of the lungs which usually proved to be in the upper lobes. *Macroscopically*, the lungs show numerous bullae on the surface and on section a honeycomb appearance.

Microscopically, there is a diffuse interstitial fibrosis leading to extensive alveolar destruction and compensatory dilatation of the terminal and respiratory bronchioles and alveolar ducts. Many of the distal airways become lined by cuboidal, columnar or multilayered squamous-like epithelium and the mural muscle is frequently hyperplastic. Hyperplasia of the neurilemmal cells occurs in the intrapulmonary nerves, especially the periarterial nerve branches. Glomus-like structures occur in the small branches of the pulmonary arteries.

The functional derangements caused by these changes include a lowered arterial oxygen saturation on exercise, slight increase of the timed vital capacity, and hyperventilation leading to actual or incipient respiratory alkalosis. In the later stages pulmonary hypertension may occur.

The relationship, if any, between generalized neurofibromatosis and tuberose sclerosis remains undetermined but both conditions are familial diseases and affect the neuroectodermal tissues. In the former the neurological complications are mainly restricted to the peripheral nervous system and in the latter to the central nervous system.

Pulmonary Blastoma (Embryoma)

A blastomatous tumour as already stated is a growth originating from a pluripotent embryonic cell of one germ layer. Pulmonary blastomas are distinguished from hamartomas of single germ layer derivation because of their progressive and sometimes unlimited growth. The first case was described by Barnard (1952) who named it a pulmonary embryoma because of its close histological resemblance to foetal lung. Although the growth he described was very extensive, and both clinically and histologically resembled a malignant tumour, the patient was alive and well over 15 years later, and for this reason it was classed as a benign pulmonary blastoma.

Spencer (1961) reported a further three cases and showed that some of these tumours were malignant and he gave to them the name of pulmonary blastomas for the reasons stated below.

Additional cases have been described by Minken et al. (1968), Rao et al. (1974) and others. Although the majority of cases have occurred in adult life they have also been found in infants and children.

Macroscopically, these tumours occur beneath the pleural surface of the lung and appear as large, rounded, creamy-yellow tumours, which are soft, friable, and often contain extensive areas of haemorrhage and necrosis. Unlike a peripheral lung cancer, they have a sharp margin though the tumour is seldom completely encapsulated (Fig. 23.18). Irregular extensions of tumour tissue into the adjacent lung parenchyma seen in peripheral cancers have been absent.

Microscopically, these tumours have a very characteristic appearance, consisting of an undifferentiated, embryonic type of connective tissue in which tubes develop lined by a multilayered, vacuolated, columnar epithelium, which very closely simulate foetal bronchioles (Figs. 23.19, 23.20, 23.21). In other parts of the growth, the primitive tumour cells develop into bundles of mature involuntary and occasionally voluntary muscle fibres, cartilage and fibrous tissue (Fig. 23.22). As already stated pulmonary rhabdomyosarcomas are very closely related and are probably identical with these tumours. The pulmonary blastoma is the pulmonary counterpart of the renal nephroblastoma and its existence can only be explained by the assump-

tion that the peripheral part of the lung is derived from mesoblastic tissue (Waddell's theory, see Chapter 1). The development of tubular structures from mesoderm need occasion no surprise, as the proximal tubular system of the kidney provides a further example of the ability of mesoblastic tissues to form epithelium-lined structures in addition to the usual connective tissues.

Pulmonary blastomas have been repeatedly confused by many pathologists with carcinosarcomas of the lung. The essential difference has been summarized by Nazari et al. (1971). A carcinosarcoma is a malignant tumour arising from two separate germinal layers which in the case of a carcinosarcoma of the lung are the bronchial epithelium (endoderm) and bronchial connective tissue (mesoderm). A blastoma is a malignant tumour arising from a multipotent cell of one germinal layer which in the case of a pulmonary blastoma is the mesoderm. A bronchial carcinosarcoma must of necessity arise from a bronchus whereas a pulmonary blastoma arises from the respiratory (mesodermal) tissue of the lung and is always a peripherally situated tumour.

Very few of these unusual pulmonary tumours have so far been described, but they form a striking variety of pulmonary growth many of which ultimately metastasize by the blood-stream to the brain, and spread may also occur to the hilar lymph glands.

Intrapulmonary Teratomas

Synonym: Dermoid cysts of the lung

Mediastinal teratomas constitute the second most common group of mediastinal tumours which may require surgical intervention (Inada and Nakano, 1958), but intrapulmonary teratomas are among the rarest tumours encountered in pathology.

The first case was described by Black and Black (1918), the previously reported cases being of too doubtful nature to include.

More recent case reports and reviews of the literature have been given by Collier et al. (1959),

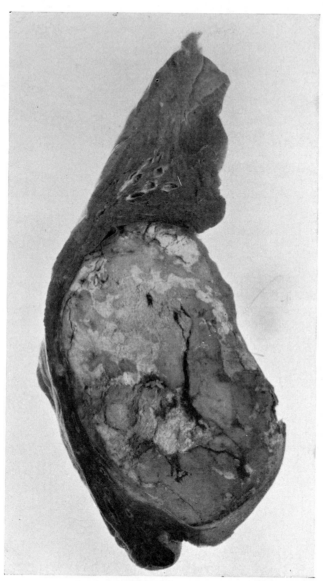

FIG. 23.18. A pulmonary blastoma showing its well-defined margins and areas of necrosis. Three-quarters natural size.

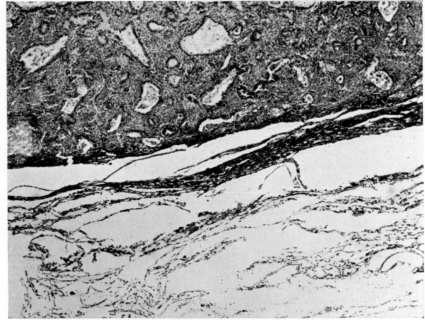

FIG. 23.19. A low-power view of a "benign" pulmonary blastoma showing its well-defined margin surrounded by compressed lung tissue. × 140 H and E.

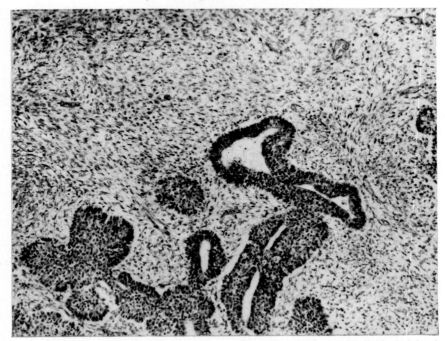

FIG. 23.20. Section of benign pulmonary blastoma showing solid and cannulated epithelial buds lying in a loose connective tissue stroma. The stroma bears a resemblance to that seen in foetal lung. × 140 H and E.

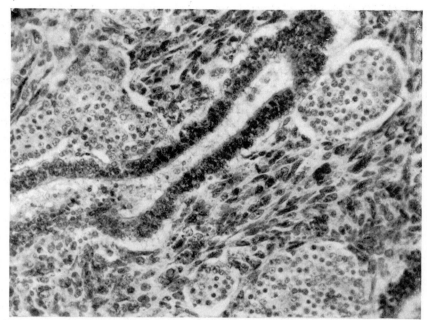

FIG. 23.21. A higher-power view of the same tumour seen in Fig. 23.19, showing the appearance of the tubes and the blending of non-cannulated epithelial masses with the surounding cellular stroma. × 300 H and E.

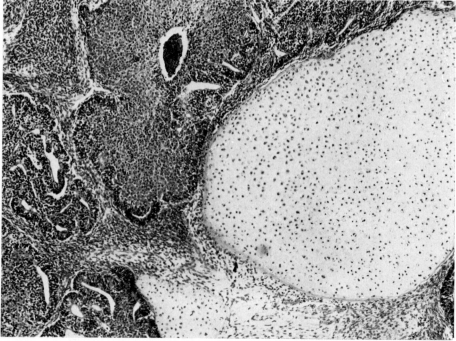

FIG. 23.22. A pulmonary blastoma showing islands of cartilage as well as the characteristic stroma in which tubular structures are present. × 75 H and E.

Ali and Wong (1964), Trivedi *et al.* (1966), Bateson *et al.* (1968), Pound and Willis (1969) and Day and Taylor (1975).

About half the reported cases have been benign tumours, the remainder malignant. With few exceptions the tumours presented in adults but the malignant teratoma described by Pound and Willis occurred in a 10-month-old infant. The great majority of the tumours have been found in the upper lobes and usually on the left side. They have been either cystic or solid tumours and have grown slowly (Figs. 23.23, 23.24). In the cases described by Collier *et al.* and Day and Taylor the cystic mass communicated with a segmental bronchus and this caused haemoptysis. Characteristically the cystic tumours have contained inspissated, brownish sebaceous material and hair, the walls being partly lined by squamous epithelium (Fig. 23.25).

The microscopic appearance of the cyst wall includes such tissues as skin and skin appendage glands, bronchial structures and elements of

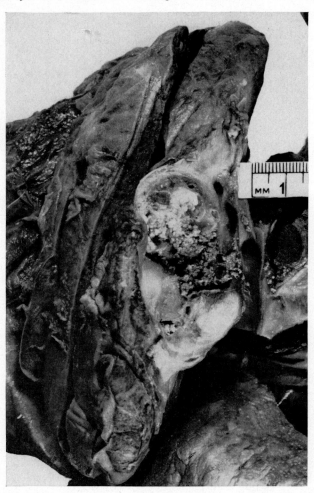

Fig. 23.23. A longitudinal section of the lung showing a well-encapsulated intrapulmonary teratoma. (Reproduced by courtesy of Drs. M. Y. Ali and P. K. Wong and the Editor of *Thorax*.)

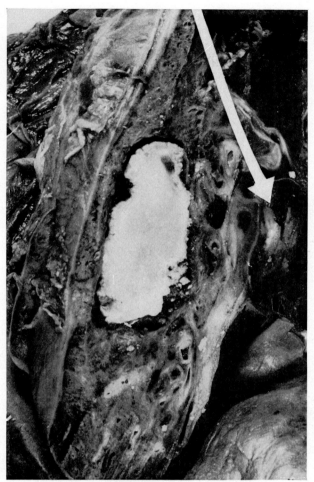

FIG. 23.24. The adipose tissue "nucleus" of the intrapulmonary teratoma surrounded by sebaceous material. The arrow indicates thrombus within an adjacent branch of the pulmonary artery.

(Reproduced by courtesy of Drs. M. Y. Ali and P. K. Wong and the Editor of *Thorax*.)

other body tissues including pancreatic tissues (Fig. 23.26). In the case reported by Pound and Willis glial tissue was found and thymic tissue was described by Day and Taylor.

The origin of pulmonary teratomas is unknown but Schlumberger (1946) considered that the mediastinal teratomas developed in relation to the thymus and its anlage, the third pharyngeal pouch. Collier *et al.* believed that the entodermal elements of the tumour were derived from the entoderm of the third pharyngeal

pouch which in turn was surrounded by mesenchyme in close relation to the ectodermal cervical sinus. Pound and Willis considered that the genesis of both mediastinal and pulmonary teratomas were closely related and that the primordial teratomatous focus in the "potential mediastinum" was "caught up and carried by the respiratory outgrowth from the foregut".

Kellett *et al.* (1962) described an unusual example of lung heteroplasia occurring in a 2-month-old negro infant in which a cyst was

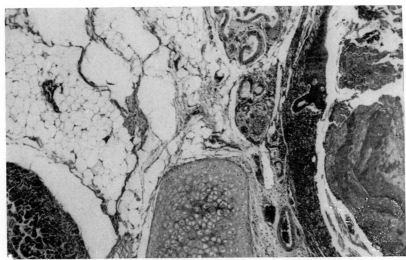

FIG. 23.25. An intrapulmonary teratoma showing an island of pancreatic tissue in the left-hand corner, cartilage and a mixture of squamous and columnar-celled epithelium lining the cyst. The cyst to the right of the figure is filled with sebaceous material. × 40 H and E.

(Reproduced by courtesy of Drs. M. Y. Ali and P. K. Wong and the Editor of *Thorax*.)

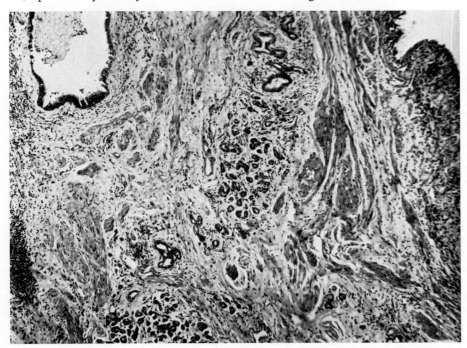

FIG. 23.26. A pulmonary teratoma showing pancreatic acinar tissue as well as columnar cell mucus-secreting epithelium in the upper left side of the illustration. × 75 H and E.

removed from the upper lobe of the right lung and was found to contain bronchial structures and heteroplastic exocrine and endocrine pancreatic tissue. No other ectopic structures were present. The development of pancreatic tissue was considered to be related to the endodermal origin of the main bronchial tree. This heteroplasia did not constitute a teratoma and examination of many of the descriptions of previously recorded "intrapulmonary teratomas" reveals an absence of all other "teratomatous" elements other than pancreas.

Some other reported cases were probably examples of persistent bronchial heteroplasia and some have been associated with other pulmonary growth defects, i.e. sequestrated lung.

CHAPTER 24

Secondary Tumours in the Lung

THE lungs in common with the liver possess an enormous capillary network but in the former the entire blood-flow has to pass through this network causing them to be ideally situated to arrest emboli of cancer cells. In addition they possess a very extensive network of lymphatic spaces continuous with the abdominal lymphatics in which embolic or direct spread of cancer cells may occur. Although the rich vascular and lymphatic supplies to the lung undoubtedly partly account for the frequency of metastatic growth in this organ, the suitability of lung tissue as a site for metastatic cell growth is also a factor of the greatest importance but one about which very little is known at present.

In common with all organs, tumour growth may also involve the lung by direct spread, and in addition two potential though less common methods of tumour metastasis exist, namely the transpleural and the intrabronchial routes.

DIRECT INFILTRATION

Although direct infiltration is not a very important method by which malignant tumours spread to the lung, carcinomas growing within the liver may occasionally extend directly through the capsule of the organ and overlying adhesions to reach the diaphragm. From here growth extends either directly into the adjacent lung or may spread transpleurally to the surface of the organ; also occasionally breast cancer may extend directly through the chest wall into the pleural cavity and involve the underlying lung, whilst direct spread of oesophageal cancer involves the main bronchi. Other rare forms of intrathoracic tumours which may spread directly into the lung include malignant tumours of the thymus, and very rarely tumours originating from the sympathetic ganglia.

LYMPHATIC-BORNE METASTASES

The pulmonary lymphatics form a closed system of spaces whereby tumour cells may spread rapidly within the lung, and they also communicate freely with the lymphatics in the abdomen and the neck. Meyer (1958) showed by injection studies that a very free anastomosis exists between the lymphatics in the lower lobes of the lungs and the lymphatic channels, including the lymph nodes, in the upper part of the abdomen especially the nodes around the coeliac axis. Direct anastomoses occur by lymphatic channels passing through the oesophageal hiatus or indirectly through the diaphragmatic lymphatic plexus. Carcinoma cells can thus reach the lungs and pleurae from tumours situated in the upper abdomen especially from primary carcinomas situated in the stomach and pancreas.

Local spread of growth through the lymphatics of a lobe or a segment of lung is a very frequent accompaniment of the spread of a primary lung cancer. Following obstruction of the normal centripetal flow through the hilar glands, lymph flows in a retrograde fashion into the intrapulmonary lymphatics often reaching and filling the subpleural plexus.

Occasionally, lymphatic-borne emboli of cancer cells from distant sites give rise to discrete deposits of growth, and these may appear in the submucosal lymphatics of large bronchi where

FIG. 24.1. Secondary carcinoma in the wall of a large bronchus simulating a primary carcinoma. Growth had reached the bronchial wall through the lymphatics. Natural size.

they closely simulate a primary growth (Fig. 24.1). Attention has already been drawn to the presence of secondary lymphatic-borne metastases in the walls of the large bronchi from a peripherally situated lung cancer which may readily be mistaken for the primary tumour. Secondary carcinoma thus invading a bronchial wall may spread and fill the lamina propria and tumour cells, then proceed to infiltrate between intact overlying bronchial epithelial cells.

Diffuse permeation of the pulmonary lymphatics is often referred to as lymphangitis carcinomatosa and gives rise to a very striking clinical and pathological picture. Bristowe (1868) is usually credited with the first description of this condition, though the clinical signs and symptoms together with the macroscopic appearances of the lung had been described earlier by Andral (1829) quoted by Wu (1936). Many different varieties of primary cancer may give rise to this form of spread within the lungs, notably carcinoma arising in the stomach, breast, lung, prostate, pancreas and occasionally the ovary.

The lungs present a characteristic appearance both macroscopically and on radiological examination. They are firm, airless, oedematous organs often partly collapsed by the associated pleural effusions, and the pleural lymphatic channels are outlined by the contained growth.

On section the lungs display greyish-white lines, standing out in relief from the cut surface and radiating outwards from the hilum to join the subpleural lymphatic network (Fig. 24.2). In the majority of fatal cases the central regional lymph glands are filled with secondary growth. On closer inspection the majority of the greyish-white lines are seen to consist of mantles of tumour tissue surrounding branches of the pulmonary arteries and bronchi whilst others consist of growth infiltrating septal tissues. Gentle pressure expresses lengths of tumour tissue from the lymphatics.

The route by which growth reaches the intrapulmonary lymphatics is still a controversial subject but the evidence will be reviewed. Girode (1889) suggested three possible routes: (a) by the lymphatic system via the thoracic duct and hilar lymph glands and thence by retrograde spread into the lung; (b) by direct extension through the diaphragm and later invasion of the pulmonary lymphatics; and (c) by the bloodstream giving rise to multiple carcinomatous emboli which lodge in the smaller pulmonary arteries and subsequently spread through the vessel walls into the perivascular lymphatics. The third method of spread will be considered at length.

Schmidt (1903) studied forty-one cases of abdominal carcinoma and showed that in almost a third it was possible to find microscopical evidence of tumour emboli in the smaller pulmonary arteries. The importance of this observation lay in the fact that it demonstrated the frequency of lung emboli of malignant cells. The same author considered the usual route of spread was by the thoracic duct and thence to the right side of the heart and the pulmonary arteries.

von Meyenburg (1919) expressed the view that it was the presence of cancer cells in the perivascular lymphatics that caused endarteritis obliterans in the smaller pulmonary arteries

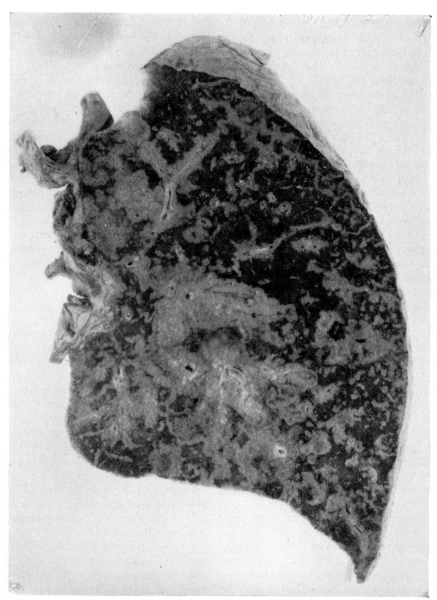

FIG. 24.2. Diffuse lymphatic infiltration by secondary carcinoma throughout the lung. The primary growth was situated in the stomach. The perivascular and peribronchial lymphatics are all clearly outlined with growth. Three-quarters natural size.

that is so often noted in lymphangitis carcinomatosa.

Schierge (1922) was the first to consider that the malignant cells might have grown inside and passed through the walls of the arteries in which they had lodged to reach the surrounding lymphatics channels.

Wu (1936) found that in a series of forty-nine cases of lymphangitis carcinomatosa eleven had right ventricular hypertrophy due to widespread obliterating endarteritis of the pulmonary vessels, and in five of these cases viable intraluminal cancer cells could be identified.

Morgan (1949) in a comprehensive review of the subject showed that it was possible to find tumour cells in the pulmonary arteries in a third of all cases of lymphangitis carcinomatosa, and he considered that blood-borne embolization was primarily responsible for this condition. According to Morgan the emboli lodge mainly in vessels between 800 and 400 μm in diameter, the cancer cells subsequently growing through the vessel walls into the accompanying lymphatics. The lymphatic infiltration commences in the periphery of the lung and spreads centripetally. Harold (1952) found that the incidence of vascular spread in a series of cases he investigated amounted to 50 per cent.

Some further supporting evidence for the blood-borne theory was provided by Costedoat and Codvelle (1932) who found that the thoracic duct was usually free of growth in cases of lymphangitis carcinomatosa. If direct and continuous lymphatic permeation were responsible this structure might be expected to share in the same change and become filled with tumour tissue. The author has also noticed that in this condition despite extensive intrapulmonary lymphatic involvement there may be little or no secondary growth present in the hilar lymph glands. Jarcho (1936) also drew attention to the frequent association of pulmonary lymphangitis carcinomatosa and diffuse blood-borne spread of small emboli of tumour cells in other organs of the body, notably the bone marrow, which often resulted in a leucoerythroblastic anaemia.

A further cogent observation was made by Hauser and Steer (1951) who noted that in one patient previously subjected to thorocoplasty, the collapsed lung escaped malignant infiltration unlike its fellow. Since recent haemodynamic studies have shown that the blood-flow is reduced in collapsed lung, the absence of cancer cell emboli might have been anticipated.

From these observations it would appear that most, but not all, cases of lymphangitis carcinomatosa result from blood-borne emboli, diffuse retrograde lymphatic permeation aided by lymphatic embolization probably accounting for the remainder. Cases of unilateral lobar or segmental lymphangitis carcinomatosa complicating a primary growth arising in the ipsilateral lung are undoubtedly due to lymphatic permeation. The usual route of secondary lymphatic invasion from the abdomen is through the coeliac lymph glands to the posterior mediastinal and paraoesophageal groups, and thence to the hilar glands from where it spreads in retrograde fashion into the lung.

BLOOD-BORNE METASTASES

Although almost all malignant tumours release malignant cells into the blood-stream (Fig. 24.3), very few succeed in establishing themselves and forming metastases, the remainder dying and leaving no trace of their existence.

The fate of tumour emboli in the small pulmonary arteries in man and animals was studied by Schmidt, whose results were mentioned above, by Iwasaki (1915–16), Warren and Gates (1936) and Saphir (1947) among many others.

The majority of malignant cells are rapidly destroyed leaving no trace, but a few adhere to the endothelium and become covered by fibrin. In a few cases the tumour cells migrate through the wall of the vessel in the wake of polymorph leucocytes and begin to multiply in the extravascular tissues; further spread then takes place along the paths of least resistance which often prove to be extravascular lymphatic channels (Fig. 24.4). Other cells grow out into the surrounding alveoli either filling them with a solid mass of malignant cells (a hilic growth) or lining their walls (a lepidic growth).

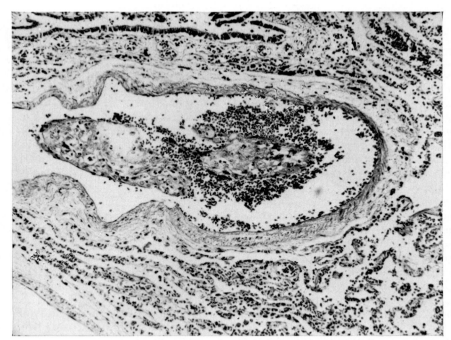

FIG. 24.3. An embolus of carcinoma cells in a branch of the pulmonary artery. The primary growth was situated in the breast. × 40 H and E.

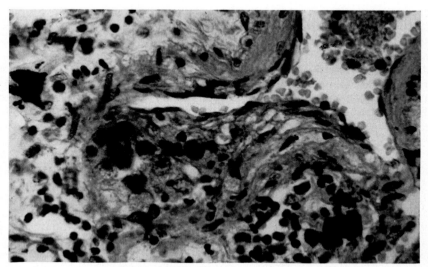

FIG. 24.4. An embolus of carcinoma cells impacted in a small branch of a pulmonary artery. The surface of the embolus has become covered with a single layer of endothelial cells and the carcinoma cells have infiltrated the subjacent arterial wall to gain the periarterial tissues. × 400 H and E.

In other cases the malignant cells attached to the endothelium ultimately become encased by fibrous tissue formed by the organization of the fibrin, finally disappearing leaving only a small intimal plaque of fibrous tissue to mark the site of the embolus.

Both Iwasaki and Warren and Gates found that following experimental injection of tumour emboli into animals, some tumour cells might fail to grow but remained viable within organizing fibrin clot over long periods of time; such tumour cells later became encased in an intimal fibrous plaque.

Zeidman and Buss (1952) showed that in animals most tumour emboli lodge at points of division in both arteries and capillaries though some individual malignant cells passed through the pulmonary vessels, appearing in the aorta.

The factors which influence the fate of tumour emboli are largely unknown but certain organs may furnish and others fail to supply the necessary metabolic, growth and environmental factors necessary for continued cell proliferation. Humoral factors at present largely unknown may inhibit the growth of some tumour cells.

Hunter (1927) and Grams (1939) drew attention to the rapid spread of breast cancer in women following deep X-ray therapy and radium implantation, and Cirio and Balestra (1930) showed that in mice irradiation of the lungs increased the likelihood of pulmonary metastases developing, and reduced the natural immunity of the animals to metastatic spread.

Considerable variation is encountered in the amount of host cellular reaction to metastatic growth. Occasionally secondary tumours in the lungs excite a very considerable plasma and lymphocytic cell reaction in which it may be difficult to identify the tumour cells. In other, often rapidly fatal cases, a diffuse form of intra-capillary metastatic spread occurs that suggests a complete breakdown of all defensive mechanisms shortly before death allowing the unhindered spread of malignant cells.

Among the more common varieties of primary tumours which metastasize readily to the lung are: carcinomas of the stomach, breast, lung, prostate, colon, liver thyroid, pancreas, liver,

(Fig. 24.5). Other growths commonly metastasizing to the lung include seminomas and teratomas of the testis, chorioncarcinoma (Fig. 24.6), bone sarcomas (Fig. 24.7), malignant melanomas, nephroblastomas of the kidney (Wilm's tumour) and neuroblastomas of the adrenals and sympathetic nervous ganglia. When considering the blood-stream spread of any tumour to the lung it is necessary not only to consider the spread from the primary site, but spread from secondary and even tertiary deposits, some of which are not infrequently sited within the liver.

Willis (1952) in a series of 500 post-mortems on cases dying of malignant disease found the incidence of metastatic growths in the lung amounted to about 30 per cent, but in the author's experience of over 1000 cases of malignant disease this figure was not above 20 per cent, despite careful routine palpation and sectioning of the lungs in all cases.

The majority of secondary tumours are usually spherical masses varying in size from the miliary nodules described by Kettle (1925) to the well-known "cannon-ball" type of growths which may measure over 5 cm in diameter. Some secondary growths may become cavitated due to the discharge of tumour tissue into a bronchus (Minor, 1950; Semple and West, 1955).

Chorioncarcinoma gives rise to extremely haemorrhagic secondary metastases (Fig. 24.6), in which tumour tissue may not be apparent macroscopically. Although this tumour almost invariably metastasizes to the lung, the invasive and erosive character of the tumour cells leads to very extensive haemorrhage; this ultimately organizes destroying the tumour cells. A chorioncarcinoma develops no host stromal vasculature, receiving its blood-supply from the blood bath created by its own invasiveness and capacity to erode host vessels. In addition as Teacher (1907–8) first showed, an inflammatory reaction occurs around the invading tumour cells before any haemorrhage takes place. This may be due to auto-immunization to placental tissue which is heterochthonous tissue. Although chorionic carcinoma is among the most malignant of all adrenal, kidney (hypernephroma) and uterus

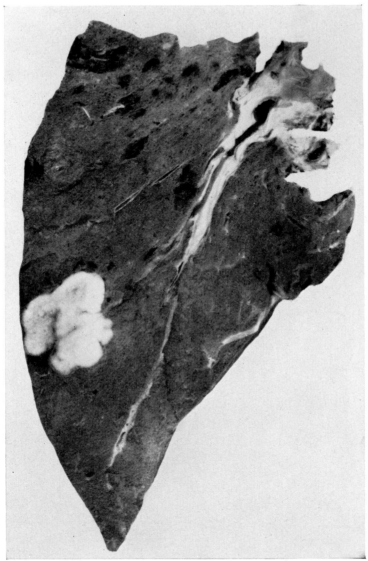

FIG. 24.5. A typical secondary blood-borne deposit of carcinoma showing well-defined margins. Two-thirds natural size.

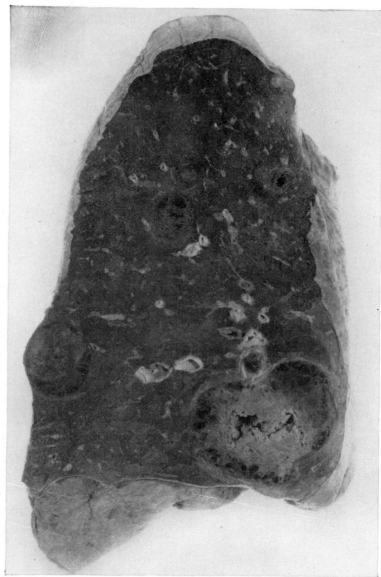

FIG. 24.6. Secondary chorioncarcinoma in the lung showing extensive haemorrhages around the deposits of growth. Two-thirds natural size.

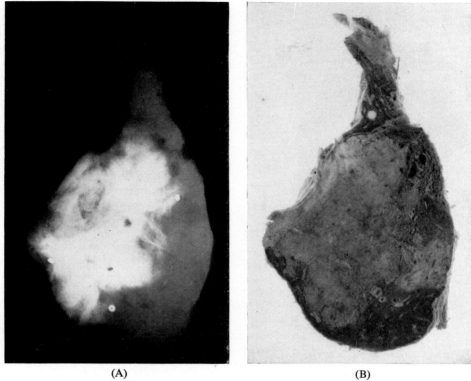

(A)   (B)

FIG. 24.7. Secondary osteogenic sarcoma in lung. In (A) a radiograph of the specimen showing the bony tumour tissue as a radiopaque (white) zone. In (B) is shown the specimen of lung containing secondary growth.

human tumours occasionally metastatic growths in the lungs may not appear until several years after the primary tumour was first discovered. In a case described by Le Brigand *et al.* (1959) an interval of 10 years elapsed between the discovery of a "hydatidiform mole" and the first accidental radiological and subsequent surgical excision of a secondary deposit of chorionic carcinoma from the lung.

Secondary metastases of osteogenic sarcoma often continue to form bone and the dense neoplastic bone requires vigorous sawing to display the interior of the growth. Metastases from secondary chondrosarcomas tend to grow intravascularly and produce distinctive greyish, translucent, cartilaginous deposits filling the pulmonary arteries (Fig. 24.8). Both in behaviour and appearance they resemble very closely the

few reported cases of primary chondrosarcomas of the lung. In the absence of a very complete examination of every accessible bone, including the vertebrae, no case of primary chondrosarcoma of the lung should be accepted as such.

Occasionally the only evidence of the existence of a sarcoma at autopsy may be the discovery of intrapulmonary arterial deposits. In such cases the walls of one or more major pulmonary arteries, including the main branch to a lower lobe of a lung, becomes thickened and the intima shows many greyish-white semi-translucent patches (Fig. 24.9). Some of the smaller branches may be found to be blocked by pale embolic tissue usually confused macroscopically with ante-mortem thrombus. Microscopic examination of the pulmonary arteries reveals metastatic tumour tissue and platelet thrombi in varying

FIG. 24.8. An intravascular deposit of secondary chondromyxosarcoma in the lung. The primary growth was situated in a cervical vertebra. The mode of spread exactly simulates that of the rare primary chondrosarcoma arising in the lung.

stages of organization attached to the intima of the affected pulmonary artery.

Massive tumour emboli leading to a fatal termination due to pulmonary arterial obstruction have been described by Storey and Goldstein (1962).

Although the majority of blood-borne metastases may reach the lung through the pulmonary arteries, a few undoubtedly are carried in the bronchial arterial flow although this may be difficult to prove.

Although probably most metastatic tumours in the walls of bronchi are lymphborne some are probably conveyed by the bronchial arteries. Among the latter are malignant melanoma, bronchial wall deposits of which may occur 25 years after the removal of a primary intraocular growth. In such cases there is often no evidence of adjacent hilar lymph gland involvement and blood-borne spread seems probable.

A further route by which blood-borne metastases reach the lung is thorugh portal–caval anastomoses, these having been shown to exist by Calabresi and Abelmann (1957) in conditions of liver fibrosis and hepatic venous obstruction. Lore *et al.* (1958) considered that some hitherto unexplained metastases from primary intra-abdominal cancers could be explained on the basis of the existence of these shunts. This finding requires further proof before it can be accepted as a recognized route for metastatic spread.

Although the majority of patients with pulmonary metastases die within a short time after they have been discovered, a few cases of long survival and spontaneous regression of lung metastases have now been reported. The most common secondary metastatic tumours to behave in this manner have been hypernephromas, melanomas and chorion carcinomas though very rarely secondary prostatic and thyroid carcinoma have regressed following oestrogen and radioactive iodine therapy (Zerman, 1943; Miller *et al.*, 1962; Sakula, 1963).

Bloom and Wallace (1964) reviewed the literature of spontaneous regression of secondary deposits of hypernephroma and found that it occurred more often in men than women and sometimes followed removal of the primary renal tumour. They also showed that testosterone given therapeutically might induce regression of secondary deposits. Following regression of the growth the site became replaced by fibrous tissue.

TRANSPLEURAL SPREAD

Tumour cells may spread within the pleural sac becoming implanted on the surface of the lung. As Willis (1952) has stated the majority of such pleural deposits are found on the surfaces of the dependent parts of the lungs. In some cases the growths spread through the diaphragmatic lymphatics from the abdomen, but in other instances, notably a tumour arising in the breast, invasion of the pleural cavity may occur either directly through the chest wall or more often from the growth-filled subpleural lymphatic spaces. The most extreme examples of this condition have been mistaken for primary pleural tumours as the whole lung becomes encased in tumour tissue.

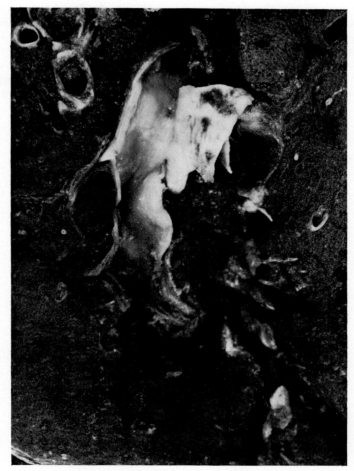

FIG. 24.9. Branches of the pulmonary artery showing mural thickening due to diffuse metastatic sarcoma. The neoplastic cells are mingled with thrombus and both are becoming incorporated into the intima. × 3 natural size.

(Reproduced by courtesy of the late Dr. S. Robinson.)

INTRABRONCHIAL SPREAD OF GROWTHS

This is an uncommon method of spread but may occur in some bronchiolar-alveolar celled carcinomas. Small portions of a parent growth become detached and are carried by respiratory movements in advance of the spreading edge of the main tumour to become attached to alveolar walls. Although such detached portions of growths are regarded as airborne metastases origin of the tumours.

Similarly the so-called airborne metastases of bronchial papillomas almost certainly represent multiple independent primary growths.

The blood-supply of metastatic growths in the lung was studied by Wright (1938) and Cudkowicz and Armstrong (1953b), both of whom showed that primary tumours were supplied by the bronchial arteries though they disagreed about the within the lung, some may represent a multifocal blood-supply to secondary growths. The former believed that secondary growths were also supplied by the bronchial arteries through the stroma of the tumour, whilst the latter were unable to show any increase in the bronchial artery blood-flow. More recent studies using an injection–corrosion technique have shown beyond all doubt that pulmonary metastatic tumours, like primary lung tumours, are supplied by branches of the bronchial arteries (Liebow, 1966). Wright also confirmed that chorioncarcinoma, a tumour which provokes no stromal formation, is entirely dependent on the blood that bathes its cells for its nutrition.

An Appendix of Technical Methods used in the Study of Lung Pathology

IN THIS short appendix a few of the staining and technical methods and apparatus used in conjunction with such methods are included. The techniques described have proved to be of considerable value and have helped to advance knowledge of lung pathology. The author wishes to express his thanks to the inventors of the various techniques described for allowing them to be included here, and also to the Editors of the Journals mentioned for allowing the reproduction of some of the illustrations.

The first technique to be described was introduced by Professor J. Gough and Mr. J. E. Wentworth of the Welsh National School of Medicine, Cardiff, Wales, and was at first used in the study of coalworker's lung and other forms of pneumoconioses. It has also proved a useful method for distinguishing the principal forms of emphysema and provides a means for obtaining a permanent record of gross whole lung pathology. Thin sections of whole lungs can be cut and mounted on paper sheets and bound together in book form. The technique has been extended to investigation of changes in other organs.

THIN SECTIONS OF ENTIRE ORGANS MOUNTED ON PAPER

by J. Gough and J. E. Wentworth

Welsh National School of Medicine, Cardiff, Wales

Descriptions of the technique and its uses have been previously published:

Recent Advances in Pathology, 7th edition, 1960. J. & A. Churchill Ltd., London.

Harvey Lectures 1957/58. Series 53, 171. Academic Press, N.Y., 1959.
J. Roy. Microscop. Soc., 1949, **69**, 231.
Proc. 9th Internat. Cong. Industr. Med., London, 1948, p. 661.
J. Fac. Radiol., 1949, **1**, 28.

The following description embodies modifications which improve the technique.

Remove the lungs from the body whole and without rupturing the pleura. If there are dense adhesions take the parietal pleura out with lung. (A few small tears do not matter except where there are large emphysematous bullae.) One or both lungs may be used. (We reserve one for bacteriological and chemical investigations.) Cut off at the hilum and fully distend by running the following solution into the major bronchi by means of a tube and cannula from a reservoir about 4 ft above the lung or by using a compressor pump producing about the same pressure.

Solution A

Liq. formaldehyde (40 per cent)	500 cc
Sodium acetate	200 g
Water	5000 cc

There is no need to tie the bronchi after expansion. Place the lung in a container of fixative large enough for it to float freely with no distortion from pressure. Cover with a cloth wet with the fixative. The amount necessary to distend the lung varies up to about 2 litres and in the containers we use there is a further 3 litres.

Fix for two days or longer and then cut a slice about $\frac{3}{4}$ in. thick. This may be in any direction but a sagittal one is most convenient.

Good results are usually obtained after a few days' fixation but, in the absence of any urgency, the slice is allowed to continue to fix for some weeks to reduce proteolytic enzymes.

Wash the slice in running water for at least 72 hours to remove the formalin. We use a syphon system and have a drip of 50 per cent liquid formaldehyde running into the washing water to give a dilution of approximately 1 in 20,000. The formalin inhibits the growth of organisms which, especially in the summer, may digest the gelatin in the next stage of the technique. Place the slice in the following solution:

Gelatin	300 g
Ethylene glycol monoethyl ether (cellosolve)	40 ml
Capryl alcohol	5 ml
Glycerin	100 ml
Water	1250 ml

With this formula the block can be cut at a lower temperature than hitherto and destructive crystal formation is less likely to occur during freezing. In very warm weather 10 cc of a 10 per cent dilution of the fixing solution A may be added as a further discouragement to the growth of contaminating organisms.

Remove the air from the slice to improve penetration by the gelatin. To do this place the slice in a dish containing the gelatin solution heated to about 60°C and place in a chamber connected to a vacuum pump. If the chamber is rectangular and constructed from perspex the sides must be at least $\frac{3}{4}$ in. thick to prevent distortion by pressure. Some of the gelatin solution can be used as a seal between the two parts of the chamber. The Speedivac Single Stage Gas Ballast Pump (Model E.S. 35) is very efficient, but where water pressure is good, a simple water pump is adequate though less efficient. If a Speedivac pump is used the exhaust should lead direct to the open air.

Place the specimen, still in the gelatin solution, in an incubator at 35°C for 48 hours in a container in which it can lie flat and be completely immersed. Cast the gelatin and specimen into a block by allowing the gelatin to set in the container. Remove the block and fix to the microtome holder by warming the latter and then put weights on top of the block. The under surface of the gelatin melts, and, as it resets, the block sticks to the holder. Put in deep freeze cabinet at minus 25°C for several hours, preferably overnight.

Cutting can usually commence as soon as the block is removed from the deep freeze cabinet. If at first the knife does not cut easily through the block or the tissue tends to split, the surface can be softened by rubbing over it a cloth damped with warm water. Put the sections into the 10 per cent formalin–sodium acetate solution A for 24–48 hours to harden the gelatin. Wash in cold water 1–2 hours to remove the formalin. For lungs the optimum thickness is 400 microns. For solid organs like liver, somewhat thinner sections, 300 microns, are usually preferable.

Mount on paper, using a fresh solution of:

Solution B

Gelatin†	75 g
Glycerin	70 cc
Cellosolve	40 cc
Water	850 cc

Trim the surplus gelatin from the edges of a section. Pour some of the warm solution over a sheet of Perspex‡ and place the section flat on the Perspex. Pour more of the solution over the section and then cover with a sheet of Whatman's No. 1 filter paper. Run a rubber roller squeegee lightly over the paper to remove surplus solution and air bubbles. Stand the Perspex sheet on end for a few seconds and then lay flat until the gelatin sets. Place in an X-ray film drying cabinet and, when thoroughly dry, strip the paper with the section attached from the Perspex. If no "dryer" is available, dry as thoroughly as possible at room temperature and when there

† In the U.S.A., gelatin of the specification 80–100 bloom should be used. In the U.K., Nelsons No. 1 or Leiners Grade B is satisfactory.

‡ Perspex (Plexiglas) is a methacrylic resin. Glass cannot be used for this purpose as the sections would then permanently adhere to the glass.

are no wet patches left, complete the drying in an incubator at 35°C or so.

The most convenient machine for cutting the large sections is the MSE "Large Section" Microtome (according to Gough and Wentworth), made by the Measuring and Scientific Equipment Ltd., 14–28 Spenser Street, London, S.W.1.

The method is applicable to liver, kidney, heart, etc.

Cellophane Stiffening Technique

Sections can be stiffened by means of water permeable cellophane grade P.T. 600, as follows. Soak the paper mounted sections and also sheets of cellophane in water for several hours. Take a sheet of Perspex and flood the whole surface with solution B. Take a sheet of cellophane and spread it over the Perspex. Flood the whole surface again with solution B and lay on the sheet of cellophane. Excess fluid and every air bubble should be squeegeed out. After thorough drying as in the main technique, the preparation can be stripped from the Perspex. If a stiffer preparation is required several sheets of cellophane each side of the paper mount may be used but in this case the sheets must be large enough to be kept taut by folding over two ends of the Perspex. The dried preparation is stripped from the Perspex after cutting the cellophane along the folded edges.

The second technique was described by Drs. A. A. Liebow, Milton R. Hales, Gustaf E. Lindskog and William E. Bloomer of Yale University School of Medicine and provides a valuable method for the preparation of rigid plastic casts of the pulmonary blood vessels and the bronchial tree. It has been used extensively for demonstrating the existence of intrapulmonary vascular anastomoses, abnormal vascular distribution in the lungs and bronchial abnormalities. The rigid casts provide a permanent record of the abnormalities and portions of the casts may be detached for closer and detailed examination.

PLASTIC DEMONSTRATIONS OF PULMONARY PATHOLOGY

by Averill A. Liebow, Milton R. Hales, Gustaf E. Lindskog and William E. Bloomer

(Modified from the original publication in the *Bulletin of the International Association of Medical Museums*, No. XXVII, 116–129, 1947)

Any material that yields an accurate cast upon injection into a hollow anatomical structure will find many uses in the study of pathology. Such casts should require no liquids nor other special care for their preservation, and should be sturdy enough to resist sensible handling over long periods of time. Vinylite, which was used as an injection medium as early as 1936 by Narat *et al.*, in considerable measure meets these specifications. Our own experience has been almost entirely with the lungs where our object was to secure casts not only of the respiratory tree, but also of the bronchial and pulmonary vessels. These structures create problems for whose solution special methods were evolved which it may be of interest to present.

Materials

In practice, a 28 per cent solution by weight of vinylite (VYHH, Specification No. 1)† is prepared in acetone. This is made in convenient lots by suspending 700 g of vinylite in 2100 cc of acetone, and shaking by hand or with the aid of an adapted machine until the solid is dissolved. The approximately saturated stock may be diluted with acetone to give a 10–15 per cent solution for the injection of finer structures such as alveoli and minute vessels.

One of the problems in the use of this material is shrinkage. This may be compensated by reinjection and by the use of "fillers". Diatomaceous earth up to 10 per cent by weight in

† Made by Union Carbon and Carbide Chemicals Corp., Bakelite Division, 427 MacCorkle Avenue, South Charleston, West Virginia.

FIG. A.1. The modified domestic pressure cooker adapted for injecting different-coloured vinylite solutions. As shown, the interior of the container has been divided up into a series of compartments each draining to the exterior through separate stopcocks.

(Reproduced by courtesy of Dr. A. A. Liebow.)

final concentration is convenient for this latter purpose.

Concentrated plastic containing the filler is the most viscous of the materials and is employed for making coarse casts or for a final filling of the largest structures. Thus, if a complete cast is desired, dilute plastic is injected first, followed in subsequent injections by the concentrated plastic, and finally by the concentrated filled plastic, as the shrinkage of the previously injected material occurs.

Multicoloured casts are easily obtained, but the colouring materials must meet three specifications: (1) The pigment must mix well with the plastic. (2) It must be unaffected by the concentrated hydrochloric acid employed in the preparation of the casts, or by the subsequent conditions of use. (3) It must not "bleed" so readily as to discolour adjacent structures. The following colouring materials have been found useful:

Green: Calco† Oil Green—0–9660 Special
Red: Calco† Oil Red—N–1700
Black: Simple Lamp Black.

Other acetone soluble dyes have been described by Narat et al. We have coloured the plastic by adding 1·5 cc of a saturated acetone solution of the dye to each 100 cc of plastic. It is often desirable to use a radio-opaque medium for injection, for then roentgenograms can be made of the injected specimen for comparison with other films. Powdered lead carbonate or bismuth oxychloride is easily suspended in the vinylite and gives the desired opacity to the X-rays. Barium sulphate has the undesirable property of rendering the casts more brittle.

A convenient chamber for storing the plastic and injecting it under pressure can be improvised from a pressure cooker (Fig. A.1) as follows:

† Made by Calco Chemical Division, American Cyanamide Co., Dyestuffs Division, Bound Brook, N.J.

A copper liner, divided into appropriate compartments, is made to fit the bottom of a 4-quart stainless steel pressure cooker provided with a pressure gauge. Each compartment is tapped by a metal tube that is brought through the wall of the cooker by a small stopcock. Pressure is applied from the laboratory compressed air line through a large inlet in the cover.

Procedure of Injection

In autopsy specimens the bronchial circulation is injected through the aorta *in situ* just above the origin of the coeliac axis. The axillary artery is ligated on both sides just distal to the origin of the costo-cervical trunk, to allow for injection of the upper intercostal vessels from which the bronchials are sometimes derived. The thyro-cervical trunk, internal mammary arteries, vertebral arteries and the common carotid are also ligated. The aorta is cannulated just above the origin of the coeliac axis. Through this cannula the aorta is flushed with water; the cardiac end of the aorta is then clamped or tied and water forced in under pressure to clear the small branches. The water is partially removed by subsequent flushing with air, and finally by flushing with acetone under pressure. The aortic cannula is then connected to the pressure chamber by means of rubber tubing and the black plastic is injected under a chamber pressure of 5–10 lb (2·5–5 kg). This is probably much more than the actual pressure in the vessels which is reduced by the friction between the plastic and the walls of the vessels. The tubing is then clamped and disconnected from the injection chamber, and after 15 min the thoracic viscera are removed en masse. By this time the plastic has hardened sufficiently in the intercostal arteries to allow their transection 5 cm from the aorta without leakage. The main pulmonary artery is cannulated preferably by means of a single large cannula that is tied in place just above the semilunar valves. The pulmonary veins can, in similar fashion, be injected through a single large cannula which is inserted into the left auricle. The latter structure, including the interatrial septum, is dissected carefully from the heart just above the mitral valve ring, avoiding any injury to the pulmonary veins. Rubber bands stretched taut are useful as ligatures and will hold where silk or cotton fails.

It is important to inject the lungs in a normally distended state. This is accomplished by suspending them in a vacuum jar (Fig. A.2) as described by Moolten (1935), who used the procedure for fixation. We have employed a cylindrical jar of pyrex glass 12 in. (5 cm) in diameter, 18 in. (7·5 cm) tall, and ½ in. (1·25 cm) thick. The top of a vacuum desiccator which comes provided with a 1½ in. (3·75 cm) central opening makes an ideal cover if the edge of the jar is ground to fit and if two additional holes are drilled to accommodate No. 11 rubber stoppers. The lumen of the trachea is made to communicate with the outside air through a cannula. We have found it wise to follow Moolten's instructions to use a short length of threaded metal pipe for cannulating the trachea, which can be conveniently connected by means of a sleeve with another short length of pipe set in the No. 8 stopper that fits the central opening in the top of the jar. Leads to the previously cannulated vessels are brought in from the pressure chamber through an additional rubber stopper. The third opening in the top of the vacuum chamber is for a stopper penetrated by a glass Y-tube, one limb of which goes to the house suction line, and the other of which leads to a mercury manometer.

The chamber is now evacuated while both the lung and the manometer are kept under constant observation. In this process it is best alternately to expand and collapse the lung, by varying the suction. This tends gently to expand whatever of the tissue is atelectatic, without over-expanding the remainder of the parenchyma. The degree of expansion is a matter of judgement and the size of the thorax should be kept in mind. A negative pressure of −10 mm of Hg is sufficient to maintain the lungs in a completely expanded state, despite small leaks, although it may be necessary to reduce the pressure momentarily to −40. It may be necessary to suspend the apices from the top of the jar and to suture adjacent lobes together in order to maintain the shape of

16

FIG. A.2. The complete apparatus showing both the pressurized container and the chamber for inflating the suspended lung under negative pressure.
(Reproduced by courtesy of Dr. A. A. Liebow.)

the lung. With the lung expanded, plastic is introduced from the chamber simultaneously into the pulmonary arteries and veins. The vessels are refilled with plastic of increasing viscosity during the next 24 hours until they will take no more. Air that is displaced rises into the tubing which joins the pressure chamber to the jar while the plastic is still soft, but does not interfere with reinjection. Filling the vessels first has the advantage of helping to maintain the shape of the lung when the respiratory passages are subsequently injected.

Immediately after the first injection of the vessels, the respiratory tree is given a first filling merely by pouring the plastic into the open end of the tracheal cannula. At this time especially, it is important to watch the chamber vacuum gauge, as there will be a drop in pressure when the trachea is obstructed by the plastic. Subsequent injections are made under the common

pressure with the vessels through an adapter made to fit the tracheal inlet cannula. The preparation is then left to harden at room temperature. When the plastic has become hard (24 to 48 hours) the tissue is ready for digestion. Blocks of tissue can first be removed for histological section and fixed in formalin. No special procedure is necessary, as the plastic dissolves in xylol. The remainder of the specimen is placed in concentrated hydrochloric acid, which dissolves the tissue in 24 to 48 hours. The casts are washed for 24 hours in running water, dried in air, and defatted with petroleum ether.

Subsequently they can be stored in air and they can be handled for close examination with only a minimum of care. Fragments of the finest structures can be broken off and mounted by means of a tiny droplet of plastic on glass slides for microscopic study or photography.

The third technique was introduced by Drs. A. A. Liebow and Milton R. Hales and was used to demonstrate the pulmonary blood vessels including the smallest branches. The method provides a convenient one for the study of the finer pulmonary vasculature and enables the vessels to be viewed stereoscopically if required. It may also be combined with the Gough and Wentworth technique enabling permanent paper mounted sections showing the pulmonary vessels to be obtained.

A MODIFICATION OF THE SCHLESINGER GELATIN MASS FOR VASCULAR INJECTION

by AVERILL A. LIEBOW and MILTON R. HALES

1. Basic Uses

A. Roentgenographic visualization of vessels.
B. Identification and visualization of vessels in larger sections, in stereoscopically viewed cleared specimens, and in routine histological sections.

2. Principle of Injection Mass

Basic ingredient is gelatin with added potassium iodide to prevent solidification at room temperature. Irreversible solidification occurs following addition of formalin just prior to injection. Barium sulphate makes mass radiopaque. Gelatin–potassium iodide base solidifies in minutes to hours following addition of formalin, depending on pH, concentration of formalin, and temperature. Useful pH range is quite narrow. Slow solidification occurs at pH below 5·5, rapid solidification at pH above 6·7. Most useful pH is 6·2 allowing mass at room temperature to solidify in about 30–60 min, a practical interval for injection purposes. In special circumstances, very slow or very fast solidification may be obtained by varying the pH. Substitution of inert pigments for the barium of Schlesinger's original mass enables the preservation of differential colours, through clearing and/or routine histological preparation.

3. Materials

1. 2-octonal (Secondary N-octyl alcohol, secondary caprylic alcohol).
2. Phenol, U.S.P., liquified.
3. Bacto-gelatin (Difco laboratories No. 0143–01).
4. Potassium iodide, U.S.P., granular.
5. Buffer salt mixture. pH 6·2 (Harleco No. 4031).
6. Barium sulphate (any U.S.P. grade marked for X-ray use).
7. "Watchung" red BW. RL-555-D.†
8. "Toluidine" yellow GW. YL-660-D.†
9. "Monastral" green BW. GP-511-D.†
10. "Monastral" blue. BP-192-D.†
11. "Sterling R" carbon black.‡

4. Preparation of Mass

A. Stock mixtures and solutions.

1. Mixture A
   2-Octanol   20 cc
   Phenol   30 cc
   Mix well, store in glass bottle or flask.

2. Mixture B
   Gelatin   230 g
   Potassium iodide   345 g
   Buffer salt   23 g
   Mix thoroughly in blender, store in suitable wide-mouthed container.

B. To prepare final mass.

1. Place 163 cc distilled water in blender.
2. Add 1·0 cc mixture A. Spin ½ min.
3. Add 61 g mixture B. Spin 1 min.
4. Add 100 g barium sulphate or 5 g of desired coloured pigment. Spin 2 min.

† The coloured pigments may be obtained from Pigments Department, E. I. Du Pont de Nemours & Co., Wilmington 98, Delaware.
‡ Carbon black is obtained from Godfrey L. Cabot Inc., 77 Franklin Street, Boston 10, Massachusetts.

5. Store in refrigerator, if mass is not to be used shortly.
6. To melt for use, warm mass to room temperature.

C. One batch (61 g mixture B + 163 cc distilled water) = 181 cc final volume. Larger or smaller volumes may be prepared with appropriate aliquots. When mixed as above, the barium sulphate mass will fill vessels of from 25 to 50 μ diameter, while the coloured pigment masses will fill capillaries. The ratio of barium sulphate or pigment to each batch may be increased if desired; this will increase the viscosity of the injection mass, resulting in a coarser injection.

D. Determination of solidification time.
1. Shake mass to give uniform solution.
2. Into each of 4 test tubes pour 4·5 cc of the mass.
3. To tube No. 1 add 0·5 cc of 50 per cent formalin (50 cc formalin + 50 cc distilled water). Repeat with the other 3 tubes, adding respectively 40, 30 and 20 per cent formalin.
4. Insert applicator stick into each tube, and stir thoroughly. Tubes are inspected after 30 min and the one which exhibits desired solidification will indicate the required formalin concentration for that particular batch. The mass with best solidification should be of a rich syrupy consistency, verging on stringiness. Some change in required formalin concentration may be noted following prolonged storage.

5. Technique of Injection

A. After cannulation of vessels, tie or clamp potential leaks. Immediately prior to injection the required volume of injection mass is mixed, rapidly but well, with the correct volume and concentration of formalin. The ratio of formalin to injection mass is 1:9.
B. Injection mass may be injected by syringe, gravity, or pressure chamber. (We use aspirator bottles for reservoirs, and Baumanometer bulbs with aneroid gauges to establish and maintain the desired injection pressures.)

C. Progress of solidification process may be observed by placing a small quantity of injection mass in a beaker. Immediately upon observing stringiness, injection apparatus should be disconnected to facilitate cleaning.

The original gelatin mass for vascular injection was described by Schlesinger M. J. (1957) *Laboratory Investigations* **6**, 1.

The fourth method to be described is used for whole lung fixation. The apparatus is a modification of that originally designed by Dr. B. E. Heard (1958). Subsequent to fixation, thick whole lung sections are lightly impregnated with barium sulphate and provide satisfactory material for the identification of the different types of emphysema. The reader is referred to the original paper for the complete technique.

AN APPARATUS FOR FIXING LUNGS TO STUDY THE PATHOLOGY OF EMPHYSEMA

by B. E. HEARD, J. R. ESTERLY and J. S. WOOTLIFF

(This account has been reproduced by courtesy of the Editor from *American Review of Respiratory Diseases* **95**, 311–312)

The apparatus described below was designed to fix by injection whole, undamaged lungs with 15 per cent aqueous formalin at a pressure of between 25 and 30 cm of fixative. Perfusion of the fixative solution is maintained for 72 hours. The whole apparatus is a modification of that originally designed by Dr. B. E. Heard (1958).

Fifteen per cent aqueous formol saline is raised by a pump (P) from container (A) to container (M) until the float (C) is lifted. This moves one of two microswitches at (D), which operate the relay (E) and stop the pump. The manifold to which the lungs are attached (G) is supplied by a tube leading from container (M). When the level of formalin in container (M) falls a short distance, the float drops and moves a second microswitch at (D) which restarts the pump. The filters (F) remove altered blood and exudates that are otherwise injected deep into the lungs, and sedimentation occurring in both containers also helps in this respect. The diagram does not show an overflow that was fitted to the original apparatus near the top of the upper container and which drains into the lower container. It is included in case the float fails to switch off the pump. The pressure in the manifold is checked with the manometer (H).

The occasional, rather than continuous, action of the pump in the new modified apparatus reduces wear in the motor and the previous continuous pumping and overflowing of formalin gave rise to excessive formalin fumes and considerable foaming. As a consequence the foam dried as a crust around the lungs and fragments of it broke away into the circulating fixative and occasionally blocked the overflow pipe. These problems do not arise in the new modified apparatus.

Some human lungs require higher initial pressures to complete expansion, especially if there is a history of severe obstructive airway disease. If the lung fails to expand, it is lowered in a separate container of formalin, but is still supplied by longer tubing from the original manifold, so that the filling pressure is increased. The lung is put back in container (A) when all parts of it are distended. Often, however, such a lung will fill very slowly but satisfactorily under the standard pressure of 25 to 30 cm of fixative, but may take 15 min to do so. Very small lungs, such as those of children and animals, require lower fixing pressures.

The only defect that has arisen during the use of the apparatus is occasional stiffness in the sleeve supporting the float (C) preventing its free movement. The filters (F) require fairly frequent cleaning. Formalin is added to replace that removed in the fixed lungs. The apparatus should be drained and thoroughly cleaned once every six months.

Although alternative methods of fixation have been described by others, including the use of

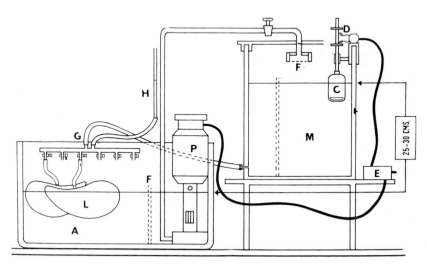

FIG. A.3. A diagram of the apparatus used for formalin fixation and distension of whole lungs. For key to lettering see the text.
(Reproduced by courtesy of Dr. Brian E. Heard and the Editor of *American Review of Respiratory Diseases*.)

formalin steam (Weibel and Vidone, 1961), with the aqueous formalin used in the present apparatus, no evidence of shrinkage can be demonstrated in measurements of lung volumes (Heard, 1960, and Laws and Heard, 1962). Also, the lung volumes closely resemble those calculated from roentgenograms of the same patients taken in life in full inspiration.

The fifth method is a staining technique for nerve fibres in human lung and was used and described by the author (Spencer and Leof, 1964).

A STAINING TECHNIQUE FOR NERVE FIBRES IN HUMAN LUNG

by H. SPENCER and D. LEOF

Like many staining methods for nerve fibres it behaves capriciously and good results are only obtained if very fresh material is used, preferably within 3 hours of death or removal.

A freshly prepared 0·01–0·05 per cent solution of vital methylene blue in normal saline is injected into the bronchi and allowed to slowly distend the lung. When filled, the surface of the lung is stained a light blue. The dye is left within the lung for 10 min and is then allowed to run out after which a stream of pure oxygen is bubbled slowly through the bronchi. To allow the stream of oxygen to penetrate into the alveoli, the pleural surface is punctured and the whole lung is kept warm and damp and warmed to body temperature. Oxygenation is continued for 1 hour after which the bronchial tree is filled with an aqueous saturated solution of ammonium molybdate at 4°C and the whole lung immersed in the same solution. After storage for 12–15 hours the molybdate solution is washed out rapidly with minimal amounts of normal saline. Frozen sections 25–50 microns thick are prepared, fixed in formalin vapour as rapidly as possible (Lane, 1976), dehydrated, cleared and mounted on glass slides in Canada balsam. Counterstaining with very dilute safranin may be used if desired. The method has given the best results in the author's hands when foetal or neonatal lung tissue is used.

References

ABELL M. K. (1957) *Arch. Path.* **64,** 584.

ABER C. P., CAMPBELL J. A. and MEECHAM J. (1963) *Brit. Heart J.* **25,** 109.

ABRAHAMSON M. L. (1959) *Lancet* **1,** 449.

ABRIKOSSOFF A. (1926) *Virchows Arch. path. Anat.* **260,** 215.

ADAMS J. M., GREEN R. G., EVANS C. A. and BEACH N. (1942) *J. Pediat.* **20,** 405.

ADAMS W. E. (1947) *Surgery* **22,** 723.

ADDISON W. (1942) *Roy. Soc. Phil. Tr.* **12,** 157.

ADKINS R. B., NAGEL C. B., COLLINS H. A., ENDE N. S. and FOSTER J. H. (1962) *Circulation* **26,** 681.

ADLER I. (1912) *Primary Malignant Growths of the Lungs and Bronchi; a Pathological and Clinical Study.* Longmans, Green & Co, New York.

ADLER R. H., MANTZ F. E. and WARE P. F. (1955) *J. thorac. Surg.* **29,** 283.

AEBY C. T. (1880) *Der Bronchialbaum der Säugetiere und des Menschen, nebst Bemerkungen über den Bronchialbaum der Vögel und Reptilien.* W. Engelmann, Leipzig.

AGEE F. O. and SHIRES D. L. (1965) *J. Amer. med. Ass.* **194,** 227.

AGNOS J. W. and STARKEY G. W. B. (1958) *New Engl. J. Med.* **258,** 12.

AHERNE W. and DAWKINS M. J. R. (1964) *Biol. Neonat.* **7,** 214.

AHMED A. (1974) *J. Path.* **112,** 1.

AHMED FATIMA S. and HARRISON C. V. (1963) *J. Path. Bact.* **85,** 357.

AITKEN W. J. and ROY K. P. (1953) *Trans. roy. Soc. trop. Med. Hyg.* **47,** 418.

AJELLO L. (1952) *Amer. J. trop. Med. Hyg.* **1,** 227.

AJELLO L. (1954) *Amer. J. trop. Med. Hyg.* **3,** 897.

AJELLO L., REED B. E., MADDY K. T., BUDURIN A. A. and MOORE J. C. (1956) *Amer. J. vet. Res.* **129,** 485.

AKAZAKI K. (1936) *Beitr. path. Anat.* **97,** 439.

AKHTAR M. and YOUNG I. (1973) *Arch. Path.* **96,** 145.

ALBERT R. E. and ARNETT L. C. (1955) *Arch. industr. Hlth.* **12,** 99.

ALBERT R. E., LIPPMAN M., SPIEGELMAN J., STREHLOW C., BRISCOE W., WOLFSON P. and NELSON N. (1967) in *Inhaled Particles and Vapours II,* Ed. C. N. Davies, p. 361, Pergamon Press, Oxford.

ALBRECHT E. (1904) *Verh. dtsch. path. Ges.* **7,** 153.

ALBRINK W. S., BROOKS S. M., BIRON R. E. and KOPEL M. (1960) *Amer. J. Path.* **36,** 457.

ALDO M. A., BENSON M. D., COMERFORD F. R. and COHEN A. S. (1970) *Arch. int. Med.* **126,** 298.

AL-DOORY Y. and KALTER S. S. (1967) *Mycopathologia (Den Haag)* **31,** 289.

ALDRIDGE W. N., BARNES J. M. and DENZ F. A. (1949) *Brit. J. exp. Path.* **30,** 375.

ALEXANDER J. K., TAKEZAWA H., ABU-NASSAR H. J. and YOW E. M. (1962) *Clin. Res.* **10,** 59.

ALI M. Y. and WONG P. K. (1964) *Thorax* **19,** 228.

ALIVISATOS G. P., PONTIKAKIS A. E. and TERZIS B. (1955) *Brit. J. industr. Med.* **12,** 43.

ALLAN G. W. and ANDERSEN D. H. (1960) *Pediatrics* **26,** 432.

ALLEN A. C. and SPITZ SOPHIE (1945) *Amer. J. Path.* **21,** 603.

ALLEN A. C. and SPITZ SOPHIE (1953) *Cancer (Philadelphia)* **6,** 1.

ALLISON A. C. (1965) Personal communication.

ALLISON A. C., HARRINGTON J. S., BIRBECK M. and NASH T. (1967) in *Inhaled Particles and Vapours II,* Edited by C. N. DAVIES, pp. 155–164, Pergamon Press, Oxford.

ALLISON A. C. and HART P. D. (1968) *Brit. J. exp. Path.* **49,** 465.

ALLISON A. C. (1971) *Arch. int. Med.* **128,** 131.

ALLISON P. R. (1947) *Thorax* **2,** 169.

AL-MALLAH Z. and QUANTOCK O. P. (1968) *Thorax* **23,** 320.

AL-SALEEM T., PEALE A. R. and MORRIS C. M. (1968) *Cancer (Philad.)* **22,** 1173.

ALTHERR F. (1936) *Virchows Arch. path. Anat.* **297,** 445.

ALTMANN H. W. and SCHÜTZ W. (1959) *Beitr. path. Anat.* **120,** 455.

ALTSCHULE M. D., GILLIGAN D. R. and ZAMCHECK N. (1942) *J. clin. Invest.* **21,** 365.

ALVIZOURI M. (1958) *Arch. Path.* **66,** 422.

AMANO S. (1933) *Trans. Soc. path. jap.* **23,** 842.

AMATRUDA T. T. Jr., MULROW P. J., GALLAGHER J. C. and SAWYER W. H. (1963) *New Engl. J. Med.* **269,** 544.

AMBERSON J. B. and SPAIN D. M. (1947) *Trans. Ass. Amer. Physns.* **60,** 92.

AMEEL D. J. (1934) *Amer. J. Hyg.* **19,** 279.

AMMICH O. (1938) *Virchows Arch. path. Anat.* **302,** 539.

AMROMIN G., SAPHIR O. and GOLDBERGER I. (1955) *Arch. Path.* **60,** 467.

ANAGNOSTOU D. and HARRISON C. V. (1972) *J. clin. Path.* **25**, 306.

ANDERSEN D. H. and HODGES R. G. (1946) *Amer. J. Dis. Child.* **72**, 62.

ANDERSON A. E. and FORAKER A. G. (1959) *Arch. Intern. Med.* **103**, 966.

ANDERSON A. E. and FORAKER A. G. (1961) *Arch. Path.* **72**, 520.

ANDERSON A. E. and FORAKER A. G. (1962) *Amer. J. Med.* **32**, 218.

ANDERSON A. E., HERNANDEZ J. A., ECKERT P. and FORAKER A. G. (1964) *Science* **144**, 1025.

ANDERSON A. E. and FORAKER A. G. (1974) in *Pathology Annual* Ed. S. C. Sommers, p. 231, Appleton-Century-Crofts, New York.

ANDERSON E. G., SIMON G. and REID L. (1973) *J. Path.* **110**, 273.

ANDERSON R. C., CHAR F. and ADAMS P. (1958) *Dis. Chest* **34**, 73.

ANDREWES C. H., LAIDLAW P. P. and SMITH W. (1934) *Lancet* **2**, 859.

ANDREWS B. E. and MCDONALD J. C. (1957) *Proc. roy. Soc. Med.* **50**, 753.

ANDREWS E. C. (1957) *Bull. Johns Hopk. Hosp.* **100**, 28.

ANDREWS H. S. and MILLER A. J. (1935) *Amer. J. Dis. Child.* **50**, 673.

ANGUS G. E. and THURLBECK W. M. (1972) *J. appl. Physiol.* **32**, 483.

ANJILVEL LILY and THURLBECK W. M. (1966) *Canad. med. ass. J.* **95**, 1179.

ANLYAN A. J., LOVINGOOD C. B. and KLASSEN K. P. (1950) *Surgery* **27**, 559.

ANSPACH W. E. and WOLMAN I. J. (1933) *Surg. Gynec. Obstet.* **59**, 635.

ANTOINE G. (1935) *Ann. Méd. lég.* **15**, 108.

ANTUNES M. L. and VIEIRA DA LUZ J. M. (1969) *Thorax* **24**, 307.

ANTWEILER H. (1960) *Gewerbepath. Gewerbehyg.* **17**, 574.

ARBLASTER P. G. (1950) *Thorax* **5**, 333.

ARCHER V. E., MAGNUSON H. J., HOLADAY D. A. and LAWRENCE P. A. (1962) *J. occup. Med.* **4**, 55.

ARCHER V. E., WAGONER J. K. and LUNDIN F. E. Jr. (1973) (a) *J. occup. Med.* **15**, 204; (b) *Hlth. Phys.* **25**, 351.

ARCHIBALD R. W. R., WELLER R. O. and MEADOW S. R. (1971) *J. Path.* **103**, 27.

AREAN V. M. and WHEAT M. W., Jr. (1962) *Amer. Rev. resp. Dis.* **85**, 261.

AREECHON W. and REID LYNNE (1963) *Brit. Med. J.* **1**, 230.

AREY L. B. (1946) *Developmental Anatomy*, 5th ed. (a textbook and laboratory manual of embryology), W. B. Saunders, Philadelphia.

ARIAS-STELLA J. and RECAVARRENS S. (1962) *Amer. J. Path.* **41**, 55.

ARIAS-STELLA J. and KRÜGER H. (1963) *Arch. Path.* **76**, 147.

ARIAS-STELLA J. and SALDANA M. (1963) *Amer. J. Path.* **43**, 30a. Also Vth Ann. Conf. on Research in Emphysema, Aspen, Colorado.

ARIAS-STELLA J. (1971) in *High Altitude Physiology. Cardiac and Respiratory Aspects*, Ciba Foundation Symposium, Ed. Ruth Porter and J. Knight, p. 31, Churchill-Livingstone, Edinburgh and London.

ARIAS-STELLA J., KRÜGER H. and RECAVARREN S. (1973) *Thorax* **28**, 701.

ARMSTRONG M. Y. K., GLEICHMANN E., GLEICHMANN H., BELDOTTI L., ANDRE-SCHWARTZ J. and SCHWARTZ R. S. (1970) *J. exp. Med.* **132**, 417.

ARNOLD H. A. and BAINBOROUGH A. R. (1957) *Canad. med. Ass. J.* **76**, 478.

ARONSON J. D., SAYLOR R. M. and PARR E. I. (1942) *Arch. Path.* **34**, 31.

ARONSON S. M. and WALLERSTEIN L. (1950) *N.Y. St. J. Med.* **50**, 2723.

ARRIGONI A. (1933) *Clin. med. ital.* **64**, 299.

ARRIGONI M. G., WOOLNER L. B. and BERNATZ P. E. (1972) *J. thorac. cardiovasc. Surg.* **64**, 413.

ARRILLAGA F. C. (1913) *Arch. Mal. Coeur* **6**, 518.

ARRILLAGA F. C. (1924) *Bull. Soc. Méd. Paris* **48**, 292.

ARVANITAKIS C., SEN S. K. and MAGNIN G. E. (1972) *Amer. Rev. resp. Dis.* **105**, 827.

ASCENZI A. and BOSMAN C. (1965) *Ann. Anat. path. (Paris)* **10**, 99.

ASCHOFF L. (1909) *Med. Klin.* **5**, 1702.

ASCHOFF L. (1910) *Verh. dtsch. path. Ges.* **14**, 125.

ASHBURN L. L. and EMMONS C. W. (1942) *Arch. Path.* **34**, 791.

ASHCROFT T. and HEPPLESTON A. G. (1973) *J. clin. Path.* **26**, 224.

ASHLEY D. J. (1970) *J. Path.* **102**, 186.

ASHLEY D. J. B., DANINO E. A. and DAVIES H. D. (1963) *Thorax* **18**, 45.

ASHMORE P. G. (1954) *J. thorac. Surg.* **27**, 293.

ASKANAZY M. (1914) *Arb. path. Anat. Bakt.* **9**, 147.

ASKANAZY M. (1919) *Korresp.-Bl. schweiz. Arz.* **49**, 465.

ASPIN J. and SHIRRAS A. F. (1951) *Brit. J. Tuberc.* **45**, 185.

ASPIN J. (1962) *Amer. Rev. resp. Dis.* **85**, 444.

ASSOR D. (1972) *Amer. J. clin. Path.* **57**, 311.

ATERMAN K. and PATEL S. (1970) *Amer. J. Anat.* **128**, 341.

ATTWOOD H. D. (1956) *J. clin. Path.* **9**, 38.

ATTWOOD H. D. (1958) *J. Path. Bact.* **76**, 211.

ATTWOOD H. D. and PARK W. W. (1961) *J. Obstet. Gynaec. Brit. Comm.* **68**, 611.

ATTWOOD H. D. (1964) *J. Path. Bact.* **88**, 285.

ATTWOOD H. D. (1972) *Amniotic Fluid Embolism*, in *Pathology Annual*, Ed. S. C. Sommers, p. 145, Appleton-Century-Crofts, New York.

ATTWOOD H. D., PRICE C. G. and RIDDELL R. J. (1972) *Thorax* **27**, 620.

AUCHINCLOSS J. H., Jr., COOK E. and RENZETTI A. D. (1955) *J. clin. Invest.* **34**, 1537.

AUERBACH O., PETRICK T. G., STOUT A. P., STATSINGER A. L., MUEHSAM G. E., FORMAN J. B. and GERE J. B. (1956) *Cancer (Philadelphia)* **9**, 76.

AUERBACH O., GERE J. B., FORMAN J. B., PETRICK T. G., SMOLIN H. J., MUEHSAM G. E., KASSOUNY D. Y. and STOUT A. P. (1957a) *New Engl. J. Med.* **256**, 97.

AUERBACH O., GERE J. B., PAWLOWSKI J. M., MUEHSAM G. E., SMOLIN H. J. and STOUT A. P. (1957b) *J. thorac. Surg.* **34**, 298.

AUERBACH O., STOUT A. P., HAMMOND E. C. and GARFINKEL L. (1962a) *New Engl. J. Med.* **267**, 111; (1962b) *ibid.* **267**, 119.

AUERBACH S. H., MIMS O. M. and GOODPASTURE E. W. (1952) *Amer. J. Path.* **28**, 69.

AULD D. (1957) *Arch. Path.* **63**, 113.

AUSTRIAN R., McCLEMENT J. H., RENZETTI A. D., DONALD K. W., RILEY R. L. and COURNAND A. (1951) *Amer. J. Med.* **11**, 667.

AUSTRIAN R. (1968) *J. clin. Path.* **21** (Suppl. 2), 93.

AVERY F. W. and BARNETT T. B. (1967) *Amer. Rev. resp. Dis.* **95**, 584.

AVERY G. B. (1963) *Pediatrics* **32**, 801.

AVERY MARY E. and MEAD J. (1959) *Amer. J. Dis. Child.* **97**, 517.

AVERY MARY E., FRANK N. R. and GRIBETZ I. (1959) *J. clin. Invest.* **38**, 456.

AVERY MARY E. (1966) in International Academy of Pathology symposium on "Pathologic Physiology and Anatomy of the Lungs", Cleveland (Williams & Wilkins Co., Baltimore).

AVERY MARY E., GATEWOOD O. B. and BRUMLEY G. (1966) *Amer. J. Dis. Child.* **111**, 380.

AXHAUSEN H. (1929) *Virchows Arch. path. Anat.* **274**, 188.

AYLWIN J. A. (1951) *Thorax* **6**, 250.

AYNAUD M. (1926) *C.R. Soc. Biol. (Paris)* **95**, 1540.

AZAR H. A. and LUNARDELLI C. (1969) *Amer. J. Path.* **57**, 81.

AZCUY A., ANDERSON A. E., BATCHELDER T. and FORAKER A. G. (1961) *Amer. Rev. resp. Dis.* **84**, 680.

AZCUY A., ANDERSON A. E. and FORAKER A. G. (1962) *Ann. int. Med.* **57**, 1.

AZMY S. and EFFAT S. (1932) *J. Egypt. med. Ass.* **15**, 87.

AZZOPARDI A. and THURLBECK W. M. (1968) *Amer. Rev. resp. Dis.* **97**, 1038.

AZZOPARDI J. G. (1959) *J. Path. Bact.* **78**, 513.

AZZOPARDI J. G. and WILLIAMS E. D. (1968) *Cancer (Philadelphia)* **22**, 274.

AZZOPARDI J. G., FREEMAN E. and POOLE G. (1970) *Brit. med. J.* **4**, 528.

BAAR H. S. and D'ABREU A. L. (1949) *Brit. J. Surg.* **37**, 220.

BAAR H. S. and BICKEL H. (1952) *Acta paediat. (Uppsala) Suppl.* **90**, 171.

BAAR H. S. (1955) *J. clin. Path.* **8**, 19.

BAAR H. S. and FERGUSON F. F. (1963) *Arch. Path.* **76**, 659.

BAARSMA P. R., DIRKEN M. N. J. and HUIZINGA E. (1948) *J. thorac. Surg.* **17**, 252.

BABUDIERI B. (1951) *Acta med. ital. Mal. infett.* **6**, 29.

BADER M. E., BADER R. A. and SELIKOFF I. J. (1961) *Amer. J. Med.* **30**, 235.

BADER M. E., BADER R. A., TIERSTEIN A. S., MILLER A. and SELIKOFF I. J. (1970) *Mt. Sinai J. Med.* **37**, 492.

BADHAM C. (1814) *An Essay on Bronchitis*, 2nd ed., Callow, London, quoted by REID L. (1958).

BAERTHLEIN K. and TOJODA A. (1913) *Zbl. Bakt.* **57**, 281 (quoted by XALABARDER C. (1961)).

BAGGENSTOSS A. H. and ROSENBERG E. F. (1943) *Arch. Path.* **35**, 503.

BAGGENSTOSS A. H. (1952) *Proc. Mayo Clin.* **27**, 412.

BAGLIO C. M., MICHEL R. D. and HUNTER W. C. (1960) *J. thorac. cardiovasc. Surg.* **39**, 659.

BAHADORI M. and LIEBOW A. A. (1973) *Cancer (Philadelphia* **31**, 191.

BAHNSON H. T., SPENCER F. C. and NEILL C. A. (1958) *J. thorac. Surg.* **36**, 777.

BAILEY R. E. (1971) *J. clin. Endocr.* **32**, 317.

BAILEY W. C., BROWN M., BUECHNER H. A., WEIL H., ICHINOSE H. and ZISKIND M. (1974) *Amer. Rev. resp. Dis.* **110**, 115.

BAINBRIDGE F. A. (1914) *J. Physiol. (Lond.)* **48**, 332.

BAKER R. D. (1942) *Amer. J. Path.* **18**, 479.

BAKER R. D. and SEVERANCE A. O. (1948) *Amer. J. Path.* **24**, 716.

BAKER R. D., WARRICK G. W. and NOOJIN R. O. (1952) *Arch. intern. Med.* **90**, 718.

BAKER R. D. (1956) *Amer. J. Path.* **32**, 287.

BAKER R. D. (1957) *J. Amer. med. Ass.* **163**, 805.

BAKER R. D. (1964) *Vth Congress Int. Acad. Path.*, London.

BALAKRISHNAN S. (1973) *Brit. med. J.* **4**, 329.

BALBONI V. G. and CASTLEMAN B. (1963) *New Engl. J. Med.* **268**, 1470.

BALCHUM O. J., JUNG R. C., TURNER A. F. and JACOBSON G. (1967) *Amer. J. Med.* **43**, 178.

BALDWIN E. De F., COURNAND A. and RICHARDS D. W. Jr. (1949) *Medicine* **28**, 210.

BALIS J. U., DELIVORIA M. and CONEN P. E. (1966) *Lab. Invest.* **15**, 530.

BALL J. H. and YOUNG D. A. (1974) *Amer. Rev. resp. Dis.* **109**, 480.

BALLANTYNE A. J., CLAGETT O. T. and McDONALD J. R. (1957) *Thorax* **12**, 294.

BALO J., JUHÁSZ E. and TEMES J. (1956) *Cancer (Philadelphia)* **9**, 918.

BALSER W. (1883) *Virchows Arch. path. Anat.* **91**, 67.

BALTISBERGER W. (1921) *Z. ges. Anat.* **61**, 283.

BAMFORTH J. (1946) *Thorax* **1**, 118.

BAN B. (1956) *J. thorac. Surg.* **32**, 254.

BANASZAK E. F., THIEDE W. H. and FINK J. N. (1970) *New Engl. J. Med.* **283**, 271.

BANISTER J. and TORRANCE R. W. (1960) *Quart. J. exper. Physiol.* **45**, 352.

BARGER J. D., CREASMAN R. W. and EDWARDS J. E. (1954) *Amer. J. clin. Path.* **24**, 441.

BARNARD P. J. (1954) *Circulation* **10**, 343.

BARNARD W. G. (1926) *J. Path. Bact.* **29**, 241.

BARNARD W. G. and DAY T. D. (1937) *J. Path. Bact.* **45**, 67.

BARNARD W. G. (1946a) *J. Path. Bact.* **58**, 625.

BARNARD W. G. (1946b) *J. Path. Bact.* **58**, 631.

BARNARD W. G. (1952) *Thorax* **7**, 299.

BARNES M. G. and BRAINERD H. (1964) *New Engl. J. Med.* **271**, 981.

BARNETT C. H. and SPENCER H. (1956) Unpublished findings.

BARNETT C. H. (1957) *Thorax* **12**, 175.

BARON M. G. and WHITEHOUSE W. M. (1961) *Amer. J. Roentgenol.* **85**, 294.

BARRACLOUGH M. A., JONES J. J. and LEE J. (1966) *Clin. Sci.* **31**, 135.

BARRATT-BOYES B. G. and WOOD E. H. (1958) *J. Lab. clin. Med.* **51**, 72.

BARRETT N. R. (1938) *J. thorac. Surg.* **8**, 169.

BARRETT N. R. (1947) *Thorax* **2**, 21.

BARRIE H. J. and HARDING H. E. (1947) *Brit. J. industr. Med.* **4**, 225.

BARRITT D. W. and JORDAN S. C. (1961) *Lancet* **1**. 729.

BARROSO-MOGUEL R. and COSTERO I. (1962) *Amer. J. Path.* **41**, 389.

BARROSO-MOGUEL R. and COSTERO I. (1964) *Amer. J. Path.* **44**, 17a.

BARSON A. J. (1971) *Arch. Dis. Childh.* **46**, 55.

BARTLETT J. G., GORBACH S. L. and FINEGOLD S. M. (1974) *Amer. J. Med.* **56**, 202.

BARTLEY T. D. and AREAN V. M. (1965) *J. thorac. cardiovasc. Surg.* **50**, 114.

BARWELL C. F. (1955) *Lancet* **2**, 1369.

BASCH S. (1915) *Med. Rec.* **87**, 539.

BASSETT FRANÇOISE and TURIAF S. (1965) *Compte rend. Acad. Sci. (Paris)* **261**, 3701.

BASSETT FRANÇOISE, POIRIER J., LECROM M. and TURIAF J. (1971) *Z. Zellforsch.* **116**, 425.

BASSET FRANÇOISE, SOLER P. and TURIAF J. (1973) *Ann. Méd. Interne (Paris)* **124**, 279.

BASSET FRANÇOISE, SOLER P., WYLLIE L., ABELANET R., LE CHARPENTIER M., KREIS B. and BREATHNACH A. S. (1974) *Virchows Arch. path. Anat.* **362**, 315.

BATAWI M. A. EL and HUSSEIN M. (1964) *Brit. J. industr. Med.* **21**, 231.

BATES D. V., VARVIS C. J., DONEVAN R. E. and CHRISTIE R. V. (1960) *J. clin. Invest.* **39**, 1401.

BATES D. V. (1972) *Amer. Rev. resp. Dis.* **105**, 1.

BATES H. R. Jr. (1965) *Amer. J. Obstet. Gynecol.* **91**, 295.

BATES M. and CRUICKSHANK G. (1957) *Thorax* **12**, 99.

BATES M. (1968) *Thorax* **23**, 311.

BATES T and HULL O. H. (1958) *Amer. J. Dis. Child.* **95**, 53.

BATESON E. M., HAYES J. A. and WOO-MING M. (1968) *Thorax* **23**, 69.

BAUM G. L. and SCHWARZ J. (1959) *Amer. J. med. Sci.* **238**, 661.

BAUM G. L., RACZ I., BUBIS J. J. and MOLHO M. (1966) *Amer. J. Med.* **40**, 578.

BAUM G. L., DONNERBERG R. L., STEWART D., MULLIGAN W. J. and PUTMAN L. R. (1969) *New Engl. J. Med.* **280**, 410.

BAUM G. L. and LERNER P. I. (1970) *Ann. int. Med.* **73**, 263.

BAYLEY E. C., LINDBERG D. O. N. and BAGGENSTOSS A. H. (1945) *Arch. Path.* **40**, 376.

BEATTIE C. P. (1957) *Trans. roy. Soc. trop. Med. Hyg.* **51**, 96.

BEATTIE J. and KNOX J. F. (1961) in *Inhaled Particles and Vapours*. Edited by C. N. DAVIES, p. 419, Pergamon Press, Oxford.

BEATTIE J. M. and HALL A. J. (1911–12) *Proc. roy. Soc. Med.* **5** (iii), 147.

BEAVER D. L. and SHAPIRO J. L. (1956) *Amer. J. Med.* **21**, 879.

BEAVER P. C. and DANARAJ T. J. (1958) *Amer. J. trop. Med. Hyg.* **7**, 100.

BECH A. O. (1961) *Thorax* **16**, 144.

BECH A. O., KIPLING M. D. and HEATHER J. C. (1962) *Brit. J. industr. Med.* **19**, 239.

BECK E. G., BRUCH J. and BROCKHAUS A. (1963) *Z. Zellforsch.* **59**, 568.

BECK W. C. and REGANIS J. C. (1951) *J. thorac. Surg.* **22**, 323.

BECKER A. E. (1966) *The Glomera in the Region of the Heart and Great Vessels*. Academisch Proefschrift, University of Amsterdam, Drukkerij Aemstelstad. Also (1966) *Path. europ.* **1**, 410.

BECKER B. J. P. (1961) *Med. Proc.* **7**, 107.

BECKER N. H. and SOIFER I. (1971) *Cancer (Philad.)* **27**, 712.

BECROFT D. M. O. (1967) *J. clin. Path.* **20**, 561.

BECROFT D. M. O. (1971) *J. clin. Path.* **24**, 72.

BEDFORD D. E., AIDAROS S. M. and GIRGIS B. (1946) *Brit. Heart J.* **8**, 87.

BEDFORD D. E. and KONSTAM G. L. S. (1946) *Brit. Heart J.* **8**, 236.

BEDSON S. P. and WESTERN G. T. (1930) *Brit. J. exp. Path.* **11**, 502.

BEDSON S. P. (1950) *Brit. med. J.* **2**, 282.

BEINTKER E. (1944) *Reichsarbeitsblatt* **3**, 37 (quoted by A. J. ROBERTSON *et al.* (1961)).

BEIRNE G. L., OCTAVIANO G. N., KOPP W. L. and BURNS R. O. (1968) *Ann. int. Med.* **69**, 1207.

BÉLAND J. E., MANKIEWICZ E. and MacINTOSH D. J. (1968) *Canad. Med. Ass. J.* **99**, 813.

BELCHER J. R. and PATTINSON J. N. (1957) *J. thorac. Surg.* **34**, 357.

BELCHER J. R. and SOMERVILLE W. (1959) *Brit. med. J.* **1**, 1280.

BELL J. H. (1880) *Lancet* **1**, 819, 871, 909.

BELL J. W., JOSEPH J. E. and LEIGHTON R. S. (1964) *J. thorac. cardiovasc. Surg.* **48**, 984.

BELLELI V. (1884–5) *Unione. med. egiz. Alessandria* Vol. 1, No. 22–23.

BELSEY R. H. R. and VALENTINE J. C. (1951) *J. Path. Bact.* **63**, 377.

BELT T. H. (1934a) *Amer. J. Path.* **10**, 129.

BELT T. H. (1934b) *Amer. J. med. Sci.* **188**, 418.

BENATAR S. R., FERGUSON A. D. and GOLDSCHMIDT G. (1972) *Quart. Med. J.* **41**, 85.

BENECKE E. (1938) *Verh. dtsch. path. Ges.* **31**, 402.

BENEKE R. (1905) *Zbl. allg. Path. path. Anat.* **16**, 812.

BENHAM R. W. (1935) *J. infect. Dis.* **57**, 255.

BENISCH B. M., FAYEMI A., GERBER M. A. and AXELROD J. (1972) *Amer. J. clin. Path.* **58**, 343.

BENOIT F. L., RULON D. B., THEIL G. B., DOOLAN P. D. and WATTEN R. H. (1964) *Amer. J. Med.* **37**, 424.

BENSCH K. G., GORDON G. B. and MILLER L. (1964) *Z. Zellforsch.* **63**, 759.

BENSCH K. G., GORDON G. B. and MILLER L. R. (1965a) *J. ultrastructure Res.* **12**, 668.

BENSCH K. G., GORDON G. B. and MILLER L. R. (1965b) *Cancer* **18**, 592.

BENSCH K., MILLER L. and GORDON G. (1965c) *Amer. J. Path.* **46**, 30a.

BENSCH K., CORRIN B., PARIENTE R. and SPENCER H. (1968) *Cancer (Philadelphia)* **22**, 1163.

BENSTEAD J. G. (1950) *Lancet* **1**, 206.

BERCOVITZ Z. (1937) *Amer. J. Trop. Med.* **17**, 101.

BERG G. and VEJLENS G. (1939) *Acta paediat. (Uppsala)* **26**, 16.

BERG G. and NORDENSKJÖLD A. (1946) *Acta med. scand.* **125**, 428.

BERG R. M., BOYDEN E. A. and SMITH F. R. (1949) *J. thorac. Surg.* **18**, 216.

BERGMAN F. and LINDER E. (1958) *J. thorac. Surg.* **35**, 628.

BERGMANN M., ACKERMAN L. V. and KEMLER R. L. (1951) *Cancer (Philadelphia)* **4**, 919.

BERGMANN M. and GRAHAM E. A. (1951) *J. thorac. Surg.* **22**, 549.

BERGOFSKY E. H., TURINO G. M. and FISHMAN A. P. (1959) *Medicine* **38**, 263.

BERKOWITZ D., GAMBESCIA J. M. and THOMSON C. M. (1952) *Gastroenterology* **20**, 653.

BERMAN E. J. (1958) *Arch. Surg. (Chicago)* **76**, 724.

BERNARD E., ISRAEL L. and DEBRIS M. M. (1962) *Amer. Rev. risp. Dis.* **85**, 22.

BERNHEIMER H., EHRINGER H., HEISTRACHER P., KRAUPP O., LACHNIT V., OBIDITSCH-MAYER I. and WENZL M. (1960) *Wien. klin. Wschr.* **72**, 867.

BERNSTEIN A. (1935) *Arch. intern. Med.* **56**, 1117.

BERNSTEIN J., NOLKE A. C. and REED J. O. (1959) *Circulation* **19**, 891.

BERSON S. D. (1975) Personal communication.

BERTALANFFY F. D. and LAU C. (1962) *Int. Rev. Cytol.* **13**, 357.

BERTALANFFY F. D. (1964) *Int. Rev. Cytol.* **17**, 213, 233.

BERTHELOT P., WALKER J. G., SHERLOCK SHEILA and REID LYNNE (1966) *New Engl. J. Med.* **274**, 291.

BERTHRONG M. and COCHRAN T. H. (1955) *Bull. Johns Hopk. Hosp.* **97**, 69.

BESKIN C. A. (1961) *J. thorac. cardiovasc. Surg.* **41**, 314.

BEST P. V. and HEATH D. (1958) *J. Path. Bact.* **75**, 281.

BEST P. V. and HEATH D. (1964) *Brit. Heart. J.* **26**, 312.

BESTERMAN E. (1961) *Brit. Heart J.* **23**, 587.

BEVERLEY J. K. A. and BEATTIE C. P. (1952) *J. clin. Path.* **5**, 350.

BHATNAGAR S. S. and SINGH K. (1935) *Indian J. med. Res.* **23**, 337.

BICKEL H. and HARRIS H. (1952) *Acta paediat. (Uppsala) Suppl.* **90**, 22.

BIDSTRUP PATRICIA L. and CASE R. A. M. (1956) *Brit. J. industr. Med.* **13**, 260.

BIGNALL J. R. and MOON A. J. (1955) *Thorax* **10**, 183.

BILS R. F. (1970) *Arch. environm. Hlth.* **20**, 468.

BIRBECK M. S., BREATHNACH A. S. and EVERALL J. D. (1961) *J. invest. Derm.* **37**, 51.

BIRD T. and THOMSON J. (1957) *Lancet* **1**, 59.

BIRESSI P. C. and CASASSA P. M. (1956) *Minerva. med. (Torino)* **47**, 930.

BISCHOFF F. and BRYSON G. (1964) in *Progress in Experimental Tumour Research*, Vol. 5, Ed. F. Homburger, S. Karger, Basel and New York.

BJÖRK V. O. and SALÉN E. F. (1950) *J. thorac. Surg.* **20**, 933.

BJÖRK V. O. (1953) *J. thorac. Surg.* **26**, 533.

BJÖRK V. O., INTONTI F., ALETRAS H. and MADSEN R. (1963) *Acta chir. scand.* **125**, 69.

BLACK H. and ACKERMAN L. V. (1952) *Ann. Surg.* **136**, 44.

BLACK H. R. and BLACK S. O. (1918) *Ann. Surg.* **67**, 73.

BLACK W. C. (1969) *Cancer (Philadelphia)* **23**, 1347.

BLACKLOCK J. W. S., KENNAWAY E. L., LEWIS G. M. and URQUHART M. E. (1954) *Brit. J. Cancer* **8**, 40.

BLACKLOCK J. W. S. (1957) *Brit. J. Cancer* **11**, 181.

BLADES B. and DUGAN D. J. (1944) *J. thorac. Surg.* **13**, 40.

BLADES B. (1951) cited by BECK W. C. and REGANIS J. C. (1951).

BLADES B. and MCCORKLE R. G. (1954) *J. thorac. Surg.* **27**, 415.

BLAKE F. G. and CECIL R. L. (1920) *J. exp. Med.* **31**, 403.

BLAKE F. G. (1931) *Ann. intern. Med.* **5**, 673.

BLAND E. F. and SWEET R. H. (1949) *J. Amer. med. Ass.* **140**, 1259.

BLECHER S. (1910) *Mitt. Grenzgeb. Med. Chir.* **21**, 837.

BLESOVSKY A. (1967) *Thorax* **22**, 351.

BLOOM H. J. and WALLACE D. M. (1964) *Brit. med. J.* **2**, 476.

BLOOMER W. E., HARRISON W., LINDSKOG G. E. and LIEBOW A. A. (1949) *Amer. J. Physiol.* **157**, 317.

BLOOMFIELD A. L. (1921) *Amer. Rev. Tuberc.* **4**, 847.

BLOT W. J. and FRAUMENI J. F. Jr. (1975) *Lancet* **2**, 142.

BLUMENFELD H. L., KILBOURNE E. D., LOURIA D. B. and ROGERS D. E. (1959) *J. clin. Invest.* **38**, 199.

BODIAN M. (1952) *Fibrocystic Disease of the Pancreas.* W. Heinemann, London.

BODIAN M. (1958) in *Fungous Diseases and their Treatment.* Edited by RIDDELL and STEWART. Butterworth, London.

BOGEDAIN W., CARPATHIOS J., KALEMKERIS K. and MCMAHON R. J. (1962) *J. Amer. med. Ass.* **182**, 247.

BOGGS T. R. and PINCOFFS M. C. (1915) *Bull. Johns Hopk. Hosp.* **26**, 407.

BÖHME A. (1934) *Ergebn. ges. Med.* **19**, 489.

BOHROD M. G. (1958) *Int. Arch. Allergy* **13**, 39.

BOLCK F. (1950) *Virchows Arch. path. Anat.* **319**, 20.

BOLLINGER O. (1870) *Virchows Arch. path. anat.* **49**, 583.

BOLLINGER O. (1877) *Zbl. med. Wiss.* **15**, 481.

BOMMER W. (1962) *Amer. J. Dis. Child.* **104**, 657.

BONANNI P. P., FRYMOYER J. W. and JACOX R. F. (1965) *Amer. J. Med.* **39**, 411.

BONHAM-CARTER R. E. (1957) *Lancet* **1**, 1292.

BONIKOS D. S., BENSCH K. G. and NORTHWAY W. H. (1975) *Amer. J. Path.* **78**, 9a.

BONNE C. (1939) *Amer. J. Cancer* **35**, 491.

BONNELL J. A. (1955) *Brit. J. industr. Med.* **12**, 181.

BONSER G. M. (1929) *J. Hyg. (Lond.)* **28**, 340.

BONSER G. M. and THOMAS G. M. (1955) *Schweiz. Z. allg. Path.* **18**, 85.

BONSER G. M., FAULDS J. S. and STEWART M. J. (1955) *Amer. J. clin. Path.* **25**, 126.

BONVENTRE P. F. and ECKERT NANCY J. (1963) *Amer. J. Path.* **43**, 201.

BOONPUCKNAVIG V., BLAMARAPRAVATI N., KAMTORN P. and SUKUMALCHANDRA Y. (1973) *Amer. J. clin. Path.* **59**, 461.

BOREN H. G. (1964) *Arch. environ. Hlth.* **8**, 119.

BORISJUK J. P. (1967) Dissertation, Univ. of Kiev, U.S.S.R. (quoted by SHABAD, L. M. (1971)).

BORRIE J. (1952) *Ann. roy. Coll. Surg. Engl.* **10**, 165.

BORSOS-NACHTNEBEL Ö. (1942) *Zbl. allg. Path. path. Anat.* **79**, 174.

BORST H. G., McGREGOR M., WHITTENBERGER J. L. and BERGLUND E. (1956) *Circulat. Res.* **4**, 393.

BOSHER L. H. Jr., BLAKE D. A. and BYRD B. R. (1959) *Surgery* **45**, 91.

BOSS J. H. and CRAIG J. M. (1962) *Pedriatrics* **29**, 890.

BOSTON COLLABORATIVE DRUG SURVEILLANCE PROGRAM (1973) *Lancet* **1**, 1399.

BOSTROEM E. (1891) *Beitr. path. Anat.* **9**, 1.

BOUHUYS A., LINDELL S. E. and LUNDIN G. (1960) *Brit. med. J.* **1**, 324.

BOURNE H. and RUSHIN W. R. (1950) *Industr. Med.* **19**, 568.

BOVORNKITTI S., PANTASUWAN P. and KANGSADAL P. (1961) *Amer. Rev. resp. Dis.* **84**, 386.

BOWDEN D. H. and WYATT J. P. (1963) *Amer. J. Path.* **43**, 31a.

BOWDEN D. H., DAVIES E. and WYATT J. P. (1968) *Arch. Path.* **86**, 667.

BOWDEN D. H. and ADAMSON I. Y. R. (1972) *Amer. J. Path.* **68**, 521.

BOWDEN K. M. (1948) *Med. J. Aust.* **2**, 311.

BOWEN D. A. L. (1959) *J. clin. Path.* **12**, 435.

BOWERS W. F. (1936) *Neb. St. med. J.* **21**, 55.

BOYCOTT A. E. and OAKLEY C. L. (1932) *J. Path. Bact.* **35**, 468.

BOYCOTT A. E. and OAKLEY C. L. (1933) *J. Path. Bact.* **36**, 205.

BOYD GLADYS (1931) *Canad. med. Ass. J.* **25**, 174.

BOYD J. D. (1937) *Contr. Embryol. Carneg. Instn.* 26, 1.

BOYD J. D. (1939–40) *London Hosp. Gaz.* **23**, clin. suppl. i, iii.

BOYD J. D. (1961) *Brit. med. Bull.* **17**, 127.

BOYD J. F. (1958) *Surg. Gynec. Obstet.* **106**, 176.

BOYD J. F. (1960) *100th Meeting of Path. Soc. of Gt. Britain and Ireland, London.*

BOYD J. T., DOLL R., FAULDS J. S. and LEIPER J. (1970) *Brit. J. industr. Med.* **27**, 97.

BOYD L. J. and McGAVACK T. H. (1939) *Amer. Heart J.* **18**, 562.

BOYDEN E. A. (1955a) *Segmental Anatomy of the Lungs,* Blakiston Division. McGraw-Hill Co., New York.

BOYDEN E. A. (1955b) *Amer. J. Surg.* **89**, 79.

BOYDEN E. A. (1958) *J. thorac. Surg.* **35**, 604.

BOYDEN E. A. and TOMPSETT D. H. (1961) *Acta Anat.* **47**, 185.

BOYDEN E. A., BILL A. H. Jr. and CREIGHTON S. A. (1962) *Surgery* **52**, 323.

BOYDEN E. A. and TOMPSETT D. H. (1962) *J. thorac. cardiovasc. Surg.* **43**, 517.

BOYDEN E. A. (1965) *Amer. J. Anat.* **116**, 413.

BOYDEN E. A. and TOMPSETT D. H. (1965) *Acta anat.* **61**, 164.

BOYDEN E. A. (1967) *Amer. J. Anat.* **121**, 749.

BOYDEN E. A. (1972) in *Development of the Human Lung,* Chap. 64, Harper & Row, Hagerstown, Md.

BOZIC C. (1963) *Pediat.* **32**, 1094.

BOZIC C. (1974) *Virchows Arch. path. Anat.* (A) **363**, 371.

BRACKEN M. N., BAILEY W. R. and THOMAS H. M. (1948) *Amer. J. Path.* **24**, 611.

BRADFORD J. K., BLALOCK J. B. and WASCOM C. M. (1961) *Amer. Rev. resp. Dis.* **84**, 582.

BRADLEY W. H. (1952) *Brit. J. Tuberc.* **46**, 196.

BRADSHAW M., MYEROWITZ R. L., SCHNEERSON R., WHISNANT J. K. and ROBBINS J. B. (1970) *Ann. int. Med.* **73**, 775.

BRAHDY L. (1941) *Amer. Rev. Tuberc.* **43**, 429.

BRAILSFORD J. F. (1938) *Brit. J. Radiol.* **11**, 393.

BRAIN R. SIR and HENSON R. A. (1958) *Lancet* **2**, 971.

BRAND M. M. and MARINKOVICH V. A. (1969) *Arch. Dis. Childh.* **44**, 536.

BRANDES W. W., COOK R. A. and OSBORNE M. P. (1943) *Arch. Path.* **36**, 465.

BRANDFONBRENER M., LANDOWNE M. and SHOCK N. W. (1955) *Circulation* **12**, 557.

BRANDT H. M. and LIEBOW A. A. (1958) *Lab. Invest.* **7**, 469.

BRANDT H. and TAMAYO R. P. (1970) *Hum. Path.* **1**, 351.

BRANTIGAN O. C. (1947) *Surg. Gynec. Obstet.* **84**, 653.

BRANTIGAN O. C. (1951) *J. thorac. Surg.* **22**, 566.

BRAS G., RICHARDS R. C., IRVINE R. A. and MILNER P. F. A. (1964) *Lancet* **2**, 1257.

BRAUN W. (1958) *Dtsch. med. Wschr.* **83**, 870 (903).

BRAUNSTEIN H. (1955) *Amer. J. Path.* **31**, 837.

BRAUNWALD E., BRAUNWALD NINA S., ROSS J. Jr. and MORROW A. G. (1965) *New Engl. J. Med.* **273**, 509.

BRECKENRIDGE R. T. and RATNOFF O. D. (1964) *New Engl. J. Med.* **270**, 298.

BRECKLER I. A. (1957) *J. thorac. Surg.* **34**, 177.

BREMER J. L. (1935) *Carnegie Inst. of Washington Publ.* 459.

BREMER J. L. (1935) *Contr. Embryol. Carneg. Instn.* **25**, 83.

BRENNER O. (1935) *Arch. intern. Med.* **56**, 211, 457, 724, 976, 1189.

BREWER D. B. and HEATH D. (1959) *J. Path. Bact.* **77**, 141.

BREWER D. B. and HUMPHREYS D. R. (1960) *Brit. Heart J.* **22**, 445.

BRIEGER H. and GROSS P. (1966) *Arch. Envir. Hlth.* **13**, 38.

BRINGHURST L. S., BYRNE R. N. and GERSHON-COHEN J. (1959) *J. Amer. med. Ass.* **171**, 15.

BRINTON H. P., FRASIER E. S. and KOVEN A. L. (1952) *Publ. Hlth. Rep. (Wash.)* **67**, 835.

BRINTON W. D. (1950) *Brit. Heart J.* **12,** 305.

BRISCOE, HOLT, MATTHEWS and SANDERSON (1936–7) *Trans. Inst. Min. Metall.* **46,** 241.

BRISTOWE J. S. (1868) *Trans. Path. Soc. Lond.* **19,** 228.

BRITISH MEDICAL JOURNAL (1970) **1,** 126.

BRITISH MEDICAL JOURNAL (1972) **2,** 672.

BRITISH MEDICAL JOURNAL (1972) **3,** 369.

BRITISH MEDICAL JOURNAL (1973) **3,** 119.

BROCH O. J. (1943) *Acta med. scand.* **113,** 311.

BROCK R. C. and BLAIR E. A (1931–2) *J. thorac. Surg.* **1,** 50.

BROCK R. C., CANN R. J. and DICKINSON J. R. (1937) *Guy's Hosp. Rep.* **87,** 295.

BROCK R. C. (1942a) *Guy's Hosp. Rep.* **91,** 111.

BROCK R. C., HODGKISS F. and JONES H. O. (1942b) *Guy's Hosp. Rep.* **91,** 131.

BROCK R. C. (1943) *Guy's Hosp. Rep.* **92,** 26, 82, 123.

BROCK R. C. (1944) *Guy's Hosp. Rep.* **93,** 90.

BROCK R. C. (1948) *Guy's Hosp. Rep.* **97,** 75.

BROCK R. C. (1950) *Thorax* **5,** 5.

BRODY H. (1942) *Arch. Path.* **33,** 221.

BRODY H. (1943) *Arch. Path.* **35,** 744.

BROMAN I. (1923) *Anat. Anz.* **57,** Suppl. 83.

BROMAN I. (1927) in *Die Entwicklung des Menschen vor der Geburt.* J. F. Bergmann, Munich.

BROTMACHER L. and CAMPBELL M. (1958) *Brit. Heart J.* **20,** 97.

BROWN C. H. and HARRISON C. V. (1966) *Lancet* **2,** 61.

BROWN C. J. O. (1958) *Postgrad. med. J.* **34,** 195.

BROWN R. K. and ROBBINS L. L. (1944) *J. thorac. Surg.* **13,** 84.

BROWN W. H. (1928) *Lancet* **2,** 1022.

BROWN W. J. and JOHNSON L. C. (1951) *Milit. Surg.* **109,** 415.

BROWNE C. R. (1955) *Brit. J. industr. Med.* **12,** 279.

BROWNE S. G. (1954) *Lancet* **1,** 393.

BRUNNER F. P., FRICK P. G. and BUHLMANN A. A. (1964) *Lancet* **1,** 1071.

BRUNNER S., ROVSING H. and WULF H. (1964) *Amer. Rev. resp. Dis.* **89,** 250.

BRYSON C. C. and SPENCER H. (1951) *Quart. J. Med.* **20** N.S., 173.

BRZOSKO I. W. and KAPUŠCIŇSKA W. (1957) *Bull. Acad. pol. Sci.* **5,** 53.

BUCHANAN W. D. (1965) *Ann. N.Y. Acad. Sci.* **132,** 507.

BUCHER U. and REID LYNNE (1961a) *Thorax* **16,** 207; (1961b) *ibid.* **16,** 219.

BUCHNER H. A., PREVATT A. L., THOMPSON J. and BLITZ O. (1958) *Amer. J. Med.* **25,** 234.

BUCHNER H. A. and ANSARI A. (1969) *Dis. Chest* **55,** 274.

BUCKINGHAM SUE, HEINEMANN H. O., SOMMERS S. C. and MCNARY W. F. (1966) *Amer. J. Path.* **48,** 1027.

BUCKINGHAM SUE, SOMMERS S. C. and SHERWIN R. P. (1967) *Amer. J. clin. Path.* **48,** 269.

BUCKLEY J. J. C. (1958) *Trans. roy. Soc. trop. Med. Hyg.* **52,** 335.

BUECHLEY R., DUNN J. E., LINDEN G. and BRESLOW L. (1957) *Cancer (Philadelphia)* **10,** 63.

BUFFMIRE D. K. and MCDONALD J. R. (1951) *Surg. Clin. N. Amer.* **31,** 1191.

BUHRMESTER C. C. (1936) *Ann. Otol. (St Louis)* **45,** 687.

BULLIVANT C. M. (1966) *Brit. Med. J.* **1,** 1272.

BULLOWA G. J. M., CHESS J. and FRIEDMAN N. B. (1937) *Arch. intern. Med.* **60,** 735.

BUNDESEN H. N., FISHBEIN W. I., DAHMS O. A. and POTTER E. L. (1937) *J. Amer. med. Ass.* **109,** 337.

BÜNGELER W. and FLEURY-SILVEIRA D. (1939) *Arch. Cirurg. clin. exp.* **3,** 169.

BURCH G. E. and ROMNEY R. B. (1954) *Amer. Heart J.* **47,** 58.

BURCHARTH F. and AXELSSON C. (1972) *Thorax* **27,** 442.

BURFORD T. H. and BURBANK B. (1945) *J. thorac. Surg.* **14,** 415.

BURGER R. A. (1947) *Amer. J. Dis. Child.* **73,** 481.

BURKE B. A. and GOOD R. A. (1973) *Medicine* **52,** 23.

BÜRKI K. (1963) *Arch. Kreisl.-Forsch.* **40,** 35.

BURMEISTER R. W., TIGERTT W. D. and OVERHOLT E. L. (1962) *Ann. int. Med.* **56,** 789.

BURNET F. M. and FREEMAN M. (1937) *Med. J. Aust.* **2,** 299.

BURR R. C., MACKAY E. N. and SELLERS A. H. (1963) *Canad. med. Ass. J.* **88,** 1181.

BURRELL L. S. and ROSS H. M. (1973) *Brit. J. Tuberc.* **31,** 38.

BURROWS H. and CLARKSON J. R. (1943) *Brit. J. Radiol.* **16** N.S., 381.

BURWELL C. S., ROBIN E. D., WHALEY R. D. and BICKELMANN A. G. (1956) *Amer. J. med.* **21,** 811.

BUSH R. W. and PLAIN G. L. (1968) *Amer. J. clin. Path.* **50,** 563.

BÜSING C. M. and BLEYL U. (1974) *Virchows Arch. path. Anat.* **363,** 113.

BUTLER H. (1952) *J. Anat. (Lond.)* **86,** 95.

BUXTON J. T. Jr., HEWITT S. C., GADSDEN R. H. and BRADHAM G. B. (1965) *J. Amer. med. Ass.* **193,** 573.

BYAM W. and ARCHIBALD R. G. (1923) *The Practice of Medicine in the Tropics.* Hodder & Stoughton, London. Vol. 3, p. 1735.

BYROM F. B. (1954) *Lancet* **2,** 201.

CADHAM F. T. (1924) *J. Amer. med. Ass.* **83,** 27.

CAFFEY J. (1940) *Amer. J. Dis. Child.* **60,** 586.

CALABRESI P. and ABELMANN W. H. (1957) *J. clin. Invest.* **36,** 1257.

CALLAHAN W. P., SUTHERLAND J. C., FULTON J. K. and KLINE J. R. (1952) *Arch. intern. Med.* **90,** 468.

CAMERON E. W. J. (1975) *Thorax* **30,** 516.

CAMERON G. R. (1948) *Brit. med. J.* **1,** 965.

CAMERON G. R. and DE S. N. (1949) *J. Path. Bact.* **61,** 375.

CAMERON G. R., DE S. N. and SHEIKH A. H. (1951) *J. Path. Bact.* **63,** 181.

CAMERON G. R. and SHEIKH A. H. (1951) *J. Path. Bact.* **63,** 609.

CAMERON G. R. (1954) *Brit. med. J.* **1,** 347.

CAMPBELL A. H. and GLOYNE S. R. (1942) *J. Path. Bact.* **54,** 75.

CAMPBELL A. H. (1961) *Aust. Ann. Med.* **10,** 129.

CAMPBELL A. H. and LEE E. J. (1963) *Brit. J. Dis. Chest* **57,** 113.

CAMPBELL C. C. (1958) in *Fungous Diseases and their Treatment*. Edited by RIDDELL and STEWART. Butterworth, London.

CAMPBELL J. A. (1955) *Amer. J. Surg.* **89**, 1009.

CAMPBELL J. A. (1958) *Thorax* **13**, 177.

CAMPBELL J. M. (1932) *Brit. med. J.* **2**, 1143.

CAMPBELL M. and HUDSON R. (1951) *Guy's Hosp. Rep.* **100**, 26.

CAMPBELL M. (1957) *Lancet* **1**, 111.

CAMPICHE M. A., GAUTIER A., HERNANDEZ E. I. and REYMOND A. (1963) *Pediat.* **32**, 976.

CAMPOS J. and IGLESIAS B. (1957) *Rev. Lat.-amer. anat. Patol.* **1**, 109.

CANADA W. J., GOODALE F. Jr. and CURRENS J. H. (1953) *New Engl. J. Med.* **248**, 309.

CANFIELD C. J., DAVIS T. E. and HERMAN R. H. (1963) *New Engl. J. Med.* **268**, 230.

CANTLIE J. (1904) *Brit. med. J.* **2**, 671.

CAPLAN A. (1953) *Thorax* **8**, 29.

CAPLAN A., COWEN E. D. H. and GOUGH J. (1958) *Thorax* **13**, 181.

CAPLAN A., PAYNE R. B. and WITHEY J. L. (1962) *Thorax* **17**, 205.

CAPLAN H. (1958) *J. Path. Bact.* **76**, 77.

CAPPELL D. F. and MCFARLANE M. N. (1947) *J. Path. Bact.* **59**, 385.

CARDELL B. S. and GURLING K. J. (1954) *J. Path. Bact.* **68**, 137.

CARDELL B. S. and PEARSON R. S. B. (1959) *Thorax* **14**, 341.

CARILLI A. D., GOHD R. S. and GORDON W. (1964) *Dis. Chest* **47**, 118.

CARINI A. (1910) *Bol. Soc. de med. cir. São Paulo* **18**, 204.

CARINI A. and MACIEL J. (1916) *Zbl. Bakt.* (orig.) **77**, 46.

CARLILE W. K., HOLLEY M. E. E. and LOGAN G. B. (1963) *J. Amer. med. Ass.* **184**, 477.

CARLISLE J. C., LEARY W. V. and MCDONALD J. R. (1951a) *Proc. Mayo Clin.* **26**, 103.

CARLISLE J. C., MCDONALD J. R. and HARRINGTON S. W. (1951b) *J. thorac. Surg.* **22**, 74.

CARLSEN C. J. and KIAER W. (1950) *Thorax* **5**, 283.

CARR D. T. and OLSEN A. M. (1954) *J. Amer. med. Ass.* **155**, 1563.

CARRÉ I. J. (1960) *Arch. Dis. Child.* **35**, 481.

CARRERA J. L. (1920) *Amer. J. Syph.* **4**, 1.

CARRINGTON C. B. and LIEBOW A. A. (1966) *Scientific Proceedings of the 63rd Meeting of the American Association of Pathologists and Bacteriologists*, p. 36a.

CARRINGTON C. B. and LIEBOW A. A. (1966) personal communication.

CARRINGTON C. B. and LIEBOW A. A. (1966) *Amer. J. Med.* **41**, 497.

CARRINGTON C. B. and LIEBOW A. A. (1966) *Amer. J. Path.* **48**, 36a.

CARRINGTON C. B. and LIEBOW A. A. (1970) *Human Path.* **1**, 322.

CARROLL D., COLIN J. E. and RILEY R. L. (1953) *J. clin. Invest.* **32**, 510.

CARROLL D. (1956) *Amer. J. Med.* **21**, 819.

CARROLL G. J. (1954) *J. Pediat.* **45**, 501.

CARSTAIRS L. S. and EDMOND R. T. D. (1963) *Proc. roy. Soc. Med.* **56**, 267.

CARSWELL J. and KRAEFT N. H. (1950) *J. thorac. Surg.* **19**, 117.

CARTER H. S. and YOUNG J. L. (1950) *J. Path. Bact.* **62**, 271.

CASSAN S. M., DIVERTIE M. B., HOLLENHORST R. W. and HARRISON E. G. (1970) *Ann. int. Med.* **72**, 687.

CASTELLANI A. (1913) *J. Ceylon. Br. Brit. med. Ass.* **10**, 20.

CASTELLANI A. and CHALMERS A. J. (1919) *Manual of Tropical Medicine*, 3rd ed., p. 1886. Baillière, Tindall and Cox, London.

CASTEX M. R. and MAZZEI E. S. (1951) *Brit. med. J.* **2**, 391.

CASTLEDEN L. I. M. and HAMILTON-PATERSON J. L. (1942) *Brit. med. J.* **2**, 478.

CASTLEMAN B. (1940) *Arch. Path.* **30**, 130.

CASTLEMAN B. and BLAND E. F. (1946) *Arch. Path.* **42**, 581.

CASTLEMAN B. (1954) *New Engl. J. Med.* **250**, 26.

CASTLEMAN B., IVERSON L. and MENENDEZ V. P. (1956) *Cancer (Philadelphia)* **9**, 822.

CASTREN H. (1931) *Acta Soc. Med. "Duodecim"* **15**, 1.

CAUMA D., TOTTEN R. S. and GROSS P. (1965) *J. Amer. med. Ass.* **192**, 371.

CAVANAGH J. B. and RUSSELL D. S. (1954) *J. Path. Bact.* **68**, 165.

CAVIN E., MASTERS J. H. and MOODY J. (1958) *J. thorac. Surg.* **35**, 816.

CAYLEY C. K., CAEZ H. J. and MERSHEIMER W. (1951) *Amer. J. Dis. Child.* **82**, 49.

CECIL R. L. (1922) *Amer. J. med. Sci.* **164**, 58.

CEELEN W. (1931) in *Handbuch der Speziellen pathologischen Anatomie und Histologie*, Eds. Henke F. and Lubarsch O. Julius Springer, Berlin, Vol. 3, part 3, p. 98.

CENTRAL TUMOR LABORATORY (MOROCCO) (1958) *Bull. Inst. nat. Hyg.* (Paris) **13**, 73.

CHAGAS C. (1909) *Mem. Inst. Osw. Cruz* **1**, 159.

CHAMBERS W. N., CRISCITIELLO M. G. and GOODALE F. (1961) *Circulation* **23**, 91.

CHAMBON L. (1955) *Ann. Inst. Pasteur.* **89**, 229.

CHANG S. C. (1957) *Cancer (Philadelphia)* **10**, 1246.

CHANNELL S., BLYTH W., LLOYD M., WEIR D., AMOS W. M. G., LITTLEWOOD A. P., RIDDLE H. F. V. and GRANT I. W. B. (1969) *Quart. J. Med.* **38** N.S., 351.

CHANOCK R. M., MUFSON M. A., BLOOM H. H., JAMES W. D., FOX H. H. and KINGSTON J. R. (1961) *J. Amer. med. Ass.* **175**, 213.

CHANOCK R. M., HAYFLICK L. and BARILE M. F. (1962) *Proc. nat. Acad. Sci. (Wash.)* **48**, 41.

CHAN-YEUNG M., CHASE W. H., TRAPP W. and GRZYBOWSKI S. (1971) *Chest* **59**, 33.

CHAPMAN B. M., SCHWARTZ H. and HAISLIP D. B. (1948) *Ann. intern. Med.* **28**, 850.

CHAPMAN J. S. (1962) *Amer. J. Med.* **33**, 471.

CHAPMAN J. S. and SPEIGHT M. (1967) in *La Sarcoïdose, Rapports de la IVe Conférence Internationale*, Paris, 1966, Eds. J. TURIAF and J. CHABOT, Masson, Paris, p. 265.
CHARMS B. L., BROFMAN B. L. and KOHN P. M. (1959) *Circulation* **20**, 850.
CHARR R. (1955) *Amer. Rev. Tuberc.* **71**, 877.
CHARR R. (1956) *Ann. intern. Med.* **44**, 806.
CHASE W. H. (1959) *Exp. Cell Res.* **18**, 15.
CHATGIDAKIS C. B. (1963) *Med. Proc.* **9**, 383.
CHATGIDAKIS C. B. (1964) *Med. Proc.* **10**, 132.
CHAUDHURI M. R. (1971) *J. thorac. cardiovasc. Surg.* **61**, 319.
CHAUDHURI R. B. and SAHA T. K. (1961) *J. Indian med. Ass.* **37**, 317.
CHEADLE W. B. (1888) *Lancet* **1**, 861.
CHENG C. L., LIN T. C. and LI C. N. (1958) *Chinese med. J.* **77**, 260.
CHIARA H. (1950) *Bull. schweiz. Akad. med Wiss* **6**, 432.
CHICKERING H. T. and PARK J. H. (1919) *J. Amer. med. Ass.* **72**, 617.
CHIKAMITSU H. (1940) *Folia Endocrin. Jap.* **16**, 85 (quoted by LUNDBERG G. D. (1963)).
CHIN D., T'ANG YEN, WEI-LIANG C., TSUNG-JEN S. and YI-HSÏN T. (1958) *Chin. med. J.* **76**, 40 (in English).
CH'IN K. Y. and TANG M. Y. (1949) *Arch. Path.* **48**, 221.
CHOFNAS I. (1963) *Amer. Rev. resp. Dis.* **87**, 280.
CHOLAK J. (1959) *Arch. industr. Hlth.* **19**, 205.
CHOWN B. (1939) *Amer. J. Dis. Child.* **57**, 489.
CHRISTELLER E. (1916–18) *Virchows Arch. path. Anat.* **223**, 40.
CHRISTIAN H. A. (1920) *Med. Clin. N. Amer.* **3**, 849.
CHRISTIE A. and PETERSON J. C. (1945) *Amer. J. publ. Hlth.* **35**, 1131.
CHRISTIE G. S. (1954) *Aust. Ann. Med.* **3**, 49.
CHRISTIE R. V. (1934) *J. clin. Invest.* **13**, 295.
CHRISTIE R. V. and MEAKINS J. C. (1934) *J. clin. Invest.* **13**, 323.
CHU C. M. and CH'EN S. H. (1955) *Chin. J. intern. Med.* **8**, 629.
CHU J., CLEMENTS J. A., COTTON E. K., KLAUS M. H., SWEET A. Y., THOMAS M. A. and TOOLEY W. H. (1965) *Pediatrics* **35**, 733.
CHUMNIJARAKIJ T. and POSHYACHINDA V. (1975) *Lancet* **1**, 1357.
CHURCH R. E. and ELLIS A. R. P. (1950) *Lancet* **1**, 392.
CHURCHILL E. D. (1947) *Surg. Clin. N. Amer.* **27**, 1113.
CHURCHILL E. D. (1949) *J. thorac. Surg.* **18**, 279.
CHURG A. and WARNOCK M. L. (1976) *Cancer (Philadelphia)* **37**, 1469.
CHURG J. and STRAUSS L. (1951) *Amer. J. Path.* **27**, 277.
CHURTON T. (1897) *Brit. med. J.* **1**, 1223.
CIBA SYMPOSIUM ON EMPHYSEMA (Report and Conclusions, 1958) (1959) *Thorax* **14**, 286.
CIBA SYMPOSIUM (1962) *Pulmonary Structure and Function*. J. and A. Churchill, London.
CIHAK R. W., ISCHIMARU T., STEER A. and YAMADA A. (1974) *Cancer (Philadelphia)* **33**, 1580.
CIRIO L. and BOLESTRA G. (1930) *Pathologica* **22**, 451.

CITRON B. P., HALPERN M., MCCARRON M., LUNDBERG G. D., MCCORMICK R., PINCUS I. J., TATTER D. and HAVERBACK B. J. (1970) *New Engl. J. Med.* **283**, 1003.
CIVIN W. H. and EDWARDS J. E. (1951) *Arch. Path. Lab. Med.* **51**, 192.
CLAGETT O. T., MCDONALD J. R. and SCHMIDT H. W. (1952) *J. thorac. Surg.* **24**, 213.
CLAIREAUX A. E. (1953) *Lancet* **2**, 749.
CLAIREAUX A. E. and FERREIRA H. P. (1958) *Arch. Dis. Childh.* **33**, 364.
CLARA M. (1937) *Z. mikr.-anat. Forsch.* **41**, 321.
CLARYSSE A. M., CATHEY W. J., CARTWRIGHT G. E. and WINTROBE M. M. (1969) *J. Amer. med. Ass.* **209**, 1861.
CLAYTON YVONNE M. (1960) University of London Ph.D. Thesis (quoted by CAMPBELL M. J. and CLAYTON YVONNE M. (1964)).
CLEMMESEN J. and NIELSEN A. (1957) *J. nat. Cancer Inst.* **19**, 989.
CLERENS J. (1953) *Arch. belges Méd. Soc.* **11**, 336.
COALSON J. J., MOHR J. A., PIRTLE S. K., DEE A. L. and RHOADES E. R. (1970) *Amer. Rev. resp. Dis.* **101**, 181.
COATES E. O. and WATSON J. H. L. (1971) *Ann. int. Med.* **75**, 709.
COCHRANE A. L. (1954) *Brit. J. Tuberc.* **48**, 274.
COCKAYNE E. A. and GLADSTONE R. F. (1917) *J. Anat. (Lond.)* **52**, 64.
COCKETT F. B. and VASS C. C. N. (1950) *Brit. J. Surg.* **38**, 97.
COCKSHOTT W. P. and LUCAS A. O. (1964) *Quart. J. Med.* **33** N.S., 223.
COETZEE T. (1970) *Thorax* **25**, 637.
COHART E. M. (1955) *Cancer (Philadelphia)* **8**, 1126.
COHEN H. I., MERIGAN T. C., KOSEK J. C. and ELDRIDGE F. (1967) *Amer. J. Med.* **43**, 785.
COHEN J. (1932) *Arch. Surg. (Chicago)* **24**, 171.
COHEN R. B., TOLL G. D. and CASTLEMAN B. (1960) *Cancer (Philadelphia)* **13**, 812.
COHEN S. M., GORDON I., RAPP F., MACAULEY J. C. and BICKLEY S. M. (1955) *Proc. Soc. exp. Biol. (N.Y.)* **90**, 118.
COKE L. R. and DUNDEE J. C. (1955) *Canad. med. Ass. J.* **72**, 12.
COLE D. B. (1951) *J. thorac. Surg.* **22**, 163.
COLEMAN P. N., EDMUNDS A. W. B. and TREGILLUS J. (1959) *Brit. Heart J.* **21**, 81.
COLLETT R. W. and EDWARDS J. E. (1949) *Surg. Clin. N. Amer.* **29**, 1245.
COLLIER F. C., ENTERLINE H. T., KYLE R. H., TRISTAN T. T. and GREENING R. (1958) *Arch. Path.* **66**, 594.
COLLIER F. C., DOWLING E. A., PLOTT D. and SCHNEIDER H. (1959) *Arch. Path.* **68**, 138.
COLLINS D. H. (1937) *J. Path. Bact.* **45**, 97.
COLLINS D. H., DARKE C. S. and DODGE O. G. (1958) *J. Path. Bact.* **76**, 531.
COLLIS E. L. (1915) *Milroy Lecture. Roy. Coll. of Physicians* quoted by HOLT P. F. (1957).

CONANT N. F. (1941) *J. Bact.* **41**, 536.

CONANT N. F. and HOWELL A. (1941) *Proc. Soc. exp. Biol.* (*N.Y.*) **46**, 426.

CONANT N. F. and HOWELL A. (1942) *J. invest. Derm.* **5**, 353.

CONANT N. F. (1950) *Amer. Rev. Tuberc.* **61**, 690.

CONANT N. F. (1954) *Manual of Clinical Mycology.* W. B. Saunders, Philadelphia.

CONDON R. E., PINKHAM R. D. and HAMES G. H. (1964) *J. thorac. cardiovasc. Surg.* **48**, 498.

CONGDON E. D. (1922) *Contr. Embryol. Carneg. Instn.* **14**, 47.

CONNAR R. G., FERGUSON T. B., SEALY W. C. and CONANT N. F. (1951) *J. thorac. Surg.* **22**, 424.

CONNING D. M. and HEPPLESTON A. G. (1966) *Brit. J. exp. Path.* **47**, 388.

CONWAY D. J. (1951) *Arch. Dis. Childh.* **26**, 263.

COOKE C. R. and HYLAND J. W. (1960) *Amer. J. Med.* **29**, 263.

COOKE F. N. and BLADES B. (1952) *J. thorac. Surg.* **23**, 546.

COOPER G. (1836) *London med. Gaz.* **18**, 600.

COOPER N. S. (1946) *Arch. Path.* **42**, 644.

COOPER R. L. and LINDSEY A. J. (1955) *Brit. J. Cancer* **9**, 304.

COORAY G. H. and LESLIE N. D. G. (1958) *Brit. J. Cancer* **12**, 1.

COPE G. C. (1953) *Brit. J. Tuberc.* **47**, 166.

COPE V. Z. (1938) *Actinomycosis.* Oxford University Press, London.

CORDERO J. (1963) *Rev. lat.-amer. anat. Path.* **1**, 25.

CORDINGLY J. L. (1972) *Thorax* **27**, 433.

CORONINE C., KOVAC W. and SALZER G. (1960) *Wien. klin. Wschr.* **72**, 531.

CORPE R. F. and STERGUS I. (1963) *Amer. Rev. resp. Dis.* **87**, 289.

CORPER H. J. and FREED H. (1922) *J. Amer. med. Ass.* **79**, 1739.

CORRIN B. and KING E. (1966) *J. Path. Bact.* **92**, p 14 supplement, *Proc. 113th Meeting Path. Soc. Gt. Britain & Ireland.*

CORRIN B. and SPENCER H. (1967) Unpublished observation.

CORRIN B. and CLARK A. E. (1968) *Histochemie* **15**, 95.

CORRIN B. and KING E. (1969) *J. Path.* **97**, 325.

CORRIN B., CLARK A. E. and SPENCER H. (1969) *J. Anat.* **104**, 65.

CORRIN B. (1970) *Thorax* **25**, 110.

CORRIN B. and KING E. (1970) *Thorax* **25**, 230.

CORRIN B. and PRICE A. B. (1972) *Thorax* **27**, 324.

CORRIN B. and VIJEYARATNAM G. S. (1972) *J. Path.* **108**, 115.

CORRIN B., ETHERTON J. E. and CONNING D. M. (1973) *J. Path.* **109**, p. v.

CORRIN B., SPENCER H., TURNER-WARWICK M., BEALES S. J. and HAMBLIN J. J. (1974) *Virchows Arch. path. Anat.* **364**, 81.

CORRIN B., LIEBOW A. A. and FRIEDMAN P. J. (1975) *Amer. J. Path.* **79**, 348.

CORRIN B. and SPENCER H. (1975) Unpublished work.

CORSSEN G. (1963) *J. Amer. med. Ass.* **183**, 314.

CORTES F. M. and WINTER W. L. (1961) *Amer. J. Med.* **33**, 223.

CORTEZ L. M. and PANKEY G. A. (1972) *Amer. Rev. resp. Dis.* **105**, 823.

CORYLLOS P. N. and BIRNBAUM G. L. (1933) *Arch. intern. Med.* **51**, 290.

COSH J. A. (1953) *Brit. Heart J.* **15**, 423.

COSTEDOAT and CODVELLE F. (1932) *Bull. Soc. Méd. Paris* **48**, 1159.

COSTELLO J. F., MORIARTY D. C., BRANTHWAITE M. A., TURNER-WARWICK M. and CORRIN B. (1975) *Thorax* **30**, 121.

COTTOM D. G. and MYERS N. A. (1957) *Brit. med. J.* **1**, 1394.

COTTON B. H., SPALDING K. and PENIDO J. R. F. (1952) *J. thorac. Surg.* **23**, 505.

COTTON R. E. (1959) *Brit. J. Dis. Chest* **53**, 142.

COTTRELL J. C., BECKER K. L., MATTHEWS M. J. and MOORE C. (1969) *Amer. J. clin. Path.* **52**, 720.

COUCH R. B., CATE T. R. and CHANOCK R. M. (1964) *J. Amer. med. Ass.* **187**, 442.

COURMONT P. (1895) *Mem. Soc. Sci. méd. Lyon.* **35** (part 2) 13 (quoted by HICKEY P. M. and SIMPSON W. M. (1926)).

COURNAND A., HIMMELSTEIN A., RILEY R. L. and LESTER C. W. (1947) *J. thor. Surg.* **16**, 30.

COURNAND A. (1950) *Circulation* **2**, 641.

COURT-BROWN W. M. and DOLL R. (1965) *Brit. med. J.* **2**, 1327.

COURTICE F. C. and PHIPPS P. J. (1946) *J. Physiol.* (*Lond.*) **105**, 186.

COURTICE F. C. (1963) *Brit. med. Bull.* **19**, 76.

COWAN and STEEL (1974) *Manual for the Identification of Medical Bacteria*, Ed. S. T. Cowan, 2nd ed. p. 111, Cambridge University Press.

COWDERY J. S. (1947) *Arch. Path.* **43**, 396.

CRAIG J. M., FENTON K. and GITLIN D. (1958) *Pediatrics* **22**, 847.

CRALLEY L. V. (1942) *J. industr. Hyg.* **24**, 193.

CRANE J. T. and GRIMES C. F. (1960) *J. thorac. cardiovasc. Surg.* **40**, 410.

CRINQUETTE J. (1962) *Rev. Tuberc.* (*Paris*) **26**, 617.

CROCKER A. C. and FARBER S. (1958) *Medicine* **37**, 1.

CROFTON J., LIVINGSTONE J., OSWALD N. and ROBERTS A. (1952) *Thorax* **7**, 1.

CROME L. and VALENTINE J. C. (1962) *J. clin. Path.* **15**, 21.

CROSS R. M. and BINFORD C. H. (1962) *Lab. Invest.* **11**, 1103.

CROWE G. G. (1954) *J. thorac. Surg.* **27**, 399.

CRUICKSHANK A. H. (1948) *J. Path. Bact.* **60**, 520.

CRUICKSHANK B. (1957) *Proc. roy. Soc. Med.* **50**, 462.

CRUICKSHANK B. (1959) *Brit. J. Dis. Chest* **53**, 226.

CRUICKSHANK D. B. and KENT-HARRISON G. (1952) *Thorax* **7**, 182.

CRUICKSHANK D. B. and KENT-HARRISON G. (1953) *Thorax* **8**, 316.

CSILLAG A. and BRANDSTEIN L. (1954) *Acta microbiol. Acad. Sci. hung.* **2,** 179.

CUBO-CAPARO A., DE LA VEGA E. and COPAIRA M. (1961) *Amer. J. vet. Res.* **22,** 673.

CUDKOWICZ L. (1952) *Brit. J. Tuberc.* **46,** 99.

CUDKOWICZ L. and ARMSTRONG J. B. (1953a) *Thorax* **8,** 46.

CUDKOWICZ L. and ARMSTRONG J. B. (1953b) *Thorax* **8,** 152.

CUDKOWICZ L. (1968) *The Human Bronchial Circulation in Health and Disease*, Williams & Wilkins Co., Baltimore.

CUEVA J. A. and LITTLE M. D. (1971) *Amer. J. trop. Med. Hyg.* **20,** 282.

CULINER M. M. (1963) *Dis. Chest* **44,** 351.

CULINER M. M. (1964) *Amer. J. Med.* **36,** 395.

CULINER M. M. and WALL A. C. (1965) *Dis. Chest* **47,** 118.

CULVER G. A., MAKEL H. P. and BEECHER H. K. (1951) *Ann. Surg.* **133,** 289.

CUMMINGS D. E. (1939) *4th Saranac Laboratory Symposium on Silicosis*, Trudeau Sch. Tuberc., Saranac Lake, N.Y.

CUMMINGS M. M. and HUDGINS P. C. (1958) *Amer. J. med. Sci.* **236,** 311.

CUMMINS S. L. and SLADDEN A. F. (1930) *J. Path. Bact.* **33,** 1095.

CUNNINGHAM G. J. and PARKINSON J. (1950) *Thorax* **5,** 43.

CUNNINGHAM G. J. and WINSTANLEY D. P. (1959) *Ann. roy. Coll. Surg. Engl.* **24,** 323.

CURETON R. J. R. and HILL I. M. (1955) *Thorax* **10,** 131.

CURRAN R. C. (1953) *J. Path. Bact.* **66,** 371.

CURRAN R. C. and ROWSELL E. V. (1958) *J. Path. Bact.* **76,** 561.

CURRAN R. C. (1960) Personal communication.

CURWEN M. P., KENNAWAY E. L. and KENNAWAY N. M. (1954) *Brit. J. Cancer.* **8,** 181.

CURZIO *Dissertations Anatomiques et Pratiques sur une Maladie de la Peau d'une Espèce Fort Rare et Fort Singulière*, No. 23–Naples, quoted by LEINWAND *et al.* (1954).

CUTTING R. A., LARSEN P. S. and LANDE A. M. (1939) *Arch. Surg. (Chicago)* **38,** 599.

D'ABRERA V. ST. E. (1958) *Ceylon med. J.* **4,** 195.

DACIE J. V. and HOYLE C. (1942) *Brit. J. Tuberc.* **36,** 158.

DAHLIN D. C. (1949) *Amer. J. Path.* **25,** 105.

DAIL D. and LIEBOW A. A. (1975) *Amer. J. Path.* **78,** 6a.

DALGAARD J. B. (1947) *Acta path. microbiol. scand.* **24,** 118.

DALLDORF F. G., PATE D. H. and LANGDELL R. D. (1968) *Arch. Path.* **85,** 149.

DALLDORF F. G., KAUFMAN A. F. and BRACHMAN P. S. (1971) *Arch. Path.* **92,** 418.

DALY C. (1954) *Brit. med. J.* **2,** 687.

DALY DE B. I., DUKE H. and WETHERALL J. (1947) *17th Internat. Physiological Cong.* Abstracts of Communications, Oxford, p. 12.

DALY DE B. I. (1958) *Quart. J. exp. Physiol.* **43,** 2.

DAMMANN J. F., BERTHRONG M. and BING R. J. (1953) *Bull. Johns Hopk. Hosp.* **92,** 128.

DAMMANN J. F., BAKER J. F. and MULLER W. H. (1957) *Surg. Gynec. Obst.* **105,** 16.

DANARAJ T. J., DA SILVA L. S. and SCHACHER J. F. (1957) *Proc. Alumni Ass. Malaya* **10,** 109.

DANARAJ T. J. (1959) *Arch. Path.* **67,** 515.

DANARAJ T. J., DA SILVA L. S. and SCHACHER J. F. (1959) *Amer. J. trop. Med. Hyg.* **8,** 151.

DANBOLT N. and FOSS M. H. (1958) *Norske Laegeforen.* **78,** 275 (291).

D'ANGELO W. A., FRIES J. F., MASI A. T. and SHULMAN L. E. (1969) *Amer. J. Med.* **46,** 428.

DANIEL D. G. (1969) *Amer. Heart J.* **78,** 720.

DANIELS A. C. (1949) *Dis. Chest* **16,** 360.

DANIELS A. C. and CHILDRESS M. E. (1956) *Calif. Med.* **85,** 369.

DANIELS A. C., CONNER G. H. and STRAUS F. H. (1967) *Arch. Path.* **84,** 615.

DANTZIG P. I., RICHARDSON D., RAYHANZADEH S., MAURO J. and SHOSS R. (1974) *Chest* **66,** 522.

DARKE C. S. and WARRACK A. J. N. (1958) *Thorax* **13,** 327.

DARLING S. T. (1906) *J. Amer. n d. Ass.* **46,** 1283.

DA ROCHA LIMA H. (1913) *Zbl. Bakt. I. Abt. Orig.* **67,** 233.

DAS J. B., DODGE O. G. and FAWCETT A. W. (1959) *Brit. J. Surg.* **46,** 582.

DASHIELL G. F. (1961) *Amer. J. trop. Med.* **10,** 37.

DASSANAYAKE W. L. P. (1948) *Brit. J. industr. Med.* **5,** 141.

D'AUNOY R. and HAAM E. VON (1934) *J. Path. Bact.* **38,** 39.

DAVIDSOHN C. (1907) *Berl. klin. Wschr.* **44,** 33.

DAVIDSON J. M. and MACLEOD W. M. (1969) *Brit. J. Dis. Chest* **63,** 13.

DAVIDSON L. and GELFAND M. (1972) *Cent. Afr. J. Med.* **18,** 129.

DAVIDSON L. R. (1944) *J. thorac. Surg.* **13,** 471.

DAVIDSON M. (1941) *Brit. J. Surg.* **28,** 571.

DAVIDSON M. (1954) *A Practical Manual of Diseases of the Chest*, 4th ed. Oxford University Press, London.

DAVIES D. (1962) *Brit. J. Dis. Chest* **56,** 171.

DAVIES D. (1972) *Quart. J. Med.* **65** N.S., 395.

DAVIES D. V. and GUNZ F. W. (1944) *J. Path. Bact.* **56,** 417.

DAVIES G. and REID L. (1970) *Thorax* **25,** 669.

DAVIES J. N. P. (1948) *E. Afr. med. J.* **25,** 117.

DAVIES P. D. B. (1963) *Thorax* **18,** 198.

DAVIES R. (1968) *Thorax* **23,** 370.

DAVIS E. W. (1939) *South. Surg.* **8,** 47.

DAVIS J. M. G. (1963) *Brit. J. exp. Path.* (a) **44,** 454; (b) *ibid.* **44,** 568.

DAVIS J. M. G., GROSS P. and DE TREVILLE R. T. P. (1970) *Arch. Path.* **89,** 364.

DAVISON K. (1958) *Brit. J. Tuberc.* **52,** 149.

DAVISON W. C., HOLM M. L. and EMMONS V. B. (1919) *Bull. Johns Hopk. Hosp.* **30,** 329.

DAVSON J. and SUSMAN W. (1937) *J. Path. Bact.* **45,** 597.

DAVSON J. (1939) *J. Path. Bact.* **49,** 483.

DAWES G. S., MOTT J. C., WIDDICOMBE J. G. and WYATT D. G. (1953) *J. Physiol.* **121,** 141.

DAWES G. S. (1958) in *"Circulation"*, *Proc. of Harvey Tercentenary Cong.*, *London.* Blackwell, Oxford.

DAWKINS SYLVIA M., EDWARDS J. M. B. and CLAYTON Y. M. (1958) in *Fungous Diseases and their Treatment.* Edited by RIDDELL and STEWART, Butterworth, London.

DAY D. W. and TAYLOR S. A. (1975) *Thorax* **30**, 582.

DAY H. F., SISSON W. R. and VOGT E. C. (1929) *Amer. J. Roentgenol.* **22**, 349.

DEAMER W. C. and ZOLLINGER H. U. (1953) *Pediatrics* **12**, 11.

DEAN G. (1959) *Brit. med. J.* **2**, 852.

DE BETTENCOURT J. M., SALDANHA A. and FRAGOSO J. C. B. (1953) *J. belge Radiol.* 36, 263.

DEINHARDT F., MAY R. D., CALHOUN H. H. and SULLIVAN H. E. (1958) *Arch. int. Med.* **102**, 816.

DELANEY L. T., SCHMIDT H. W. and STROEBEL C. F. (1956) *Proc. Mayo Clin.* **31**, 189.

DE LANGE C. (1927) *Acta paediat. (Uppsala)* **6**, 352.

DELANÖE P. and DELANÖE MME P. (1912) *C. R. Acad. Sci. (Paris)* **155**, 658.

DELARUE N. C. (1951) *J. thorac. Surg.* **21**, 535.

DELIKAT E. (1945) *Lancet* **2**, 370.

DE LORIMER A. A., TIERNEY D. F. and PARKER H. R. (1967) *Surgery* **62**, 12.

DE MARCHETTIS D. (1654) *D'anatomia Patavii.*

DE MATTEIS ANNA and ARMANI G. (1967) *J. Path. Bact.* **94**, 464.

DE MONBREUN W. A. (1934) *Amer. J. trop. Med.* **14**, 93.

DE MONBREUN W. A. (1939) *Amer. J. trop. Med.* **19**, 565.

DE NAVASQUEZ S., FORBES J. R. and HOLLING H. E. (1940) *Brit. Heart J.* **2**, 177.

DE NAVASQUEZ S. (1942) *J. Path. Bact.* **54**, 313.

DENNY J. J., ROBSON W. D. and IRWIN D. A. (1937) *Canad. med. Ass. J.* **37**, 1.

DENT J. H., NICHOLS R. L., BEAVER P. C., CARRERA G. M. and STAGGERS R. J. (1956) *Amer. J. Path.* **32**, 777.

DENTON J. (1925) *Amer. J. med. Sci.* **169**, 531.

DENTON R. (1960) *Pediatrics* **25**, 611.

DEPARTMENT of HEALTH, NEW ZEALAND (1973) *Special Report Series No.* 42, p. 17, A. R. Shearer, Wellington, New Zealand.

DE POZZIS A. (1673–4) in *Miscellanea Curiosa Medico-Physica. Academie Naturae Curiosorum Sive Ephemerigum Medico-Physicarum Germanicarum,* Bibliopolae Lipsiensis (1676) Dec. I, observatio xxx, p. 31, J. Fritzchii, Frankfurt and Leipzig (quoted by RYLAND and REID (1971)).

DEPPISCH L. M. and DONOWHO E. M. (1972) *Amer. J. clin. Path.* **58**, 489.

DERRICK E. H. (1937) *Med. J. Aust.* **2**, 281.

DERRY D. C. L., CARD W. I., WILSON R. and DUNCAN J. T. (1942) *Lancet* **1**, 224.

DESCHREIDER A. R. (1957) *Chem. and Indust.* **43**, 1412.

DETERLING R. A. and CLAGETT O. T. (1947) *Amer. Heart J.* **34**, 471.

DÉVÉ F. (1935) *Ve Cong. Ann. Fed. Sci. Med. Alg. Tunis Maroc (Oran).*

DE VILLIERS A. J. and WINDISH J. P. (1964) *Brit. J. industr. Med.* **21**, 94.

DEXTER L. (1955) *Brit. Heart J.* **18**, 209.

DEXTER L. (1962) 4th World Congress of Cardiology, Mexico City. *Lancet* **2**, 877.

DIACONIŢĂ G. (1975) *Thorax* **30**, 682.

DIAMOND S. and VAN LOON E. L. (1942) *J. Amer. med. Ass.* **118**, 771.

DI BIASI W. (1951) *Virchows Arch. path. Anat.* **319**, 505.

DICKE T. E. and NAYLOR B. (1969) *Dis. Chest* **56**, 122.

DICKEY L. B. (1950) *Dis. Chest* **17**, 151.

DICKIE HELEN A. and RANKIN J. (1958) *J. Amer. med. Ass.* **167**, 1069.

DICKSON E. C. (1937a) *Calif. west. Med.* **47**, 151.

DICKSON E. C. (1937b) *Arch. intern. Med.* **59**, 1029.

DICKSON E. C. and GIFFORD M. A. (1938) *Arch. intern. Med.* **62**, 853.

DICKSON J. A., CLAGETT O. T. and McDONALD J. R. (1946) *J. thorac. Surg.* **15**, 196.

DI MATTEI E. (1916) *Malaria e Malat. Paesi. Caldi* **7**, 225.

DIMOND E. G. and JONES T. R. (1954) *Amer. Heart J.* **47**, 105.

DINES D. E., ARMS R. A., BERNATZ P. E. and GOMES M. R. (1974) *Chest* **65**, 586.

DITTRICH J. K. and SEIFERT G. (1953) *Z. Kinderheilk.* **73**, 639.

DIVELEY W., McCRACKEN R., STONEY W., GUEST J. and McCONNELL V. (1963) *J. thorac. cardiovasc. Surg.* **45**, 101.

DOBY-DUBOIS M., CHEVREL M. L., DOBY J. M. and LOUVET M. (1964) *Bull. Soc. Path. exot.* **57**, 240.

DOENECKE F. and BELT J. H. (1931) *Frankfurt. Z. Path.* **42**, 161, 170.

DOERING P. and GOTHE H. D. (1957) *Klin. Wschr.* **35**, 1105.

DOERR W. (1953) *Virchows Arch. path. Anat.* **324**, 263.

DOIG A. T. and McLAUGHLIN A. I. G. (1936) *Lancet* **1**, 771.

DOIG A. T. (1976) *Thorax* **31**, 30.

DOLL R. (1952) *Brit. J. industr. Med.* **9**, 180.

DOLL R. and HILL A. B. (1952) *Brit. med. J.* **2**, 1271.

DOLL R. (1953) *Brit. med. J.* **2**, 521, 585.

DOLL R. and HILL A. B. (1954) *Brit. med. J.* **1**, 1451.

DOLL R. (1955) *Brit. J. industr. Med.* **12**, 81.

DOLL R. and HILL A. B. (1956) *Brit. med. J.* **2**, 1071.

DOLL R., HILL A. B. and KREYBERG L. (1957) *Brit. J. Cancer* **11**, 43.

DOLL R. (1958) *Brit. J. industr. Med.* **15**, 217.

DOLL R., HILL A. B., GRAY P. G. and PARR E. A. (1959) *Brit. med. J.* **1**, 322.

DOLL R. and HILL A. B. (1964) *Brit. med. J.* **1**, 1399, 1460.

DOLL R., FISHER R. E. W., GAMMON E. J., GUNN W., HUGHES G. O., TYRER F. H. and WILSON W. (1965) *Brit. J. industr. Med.* **22**, 1.

DOLL R., MORGAN L. G. and SPEIZER F. E. (1970) *Brit. J. Cancer* **24**, 623.

DOLLERY C. T. and HUGH-JONES P. (1963) *Brit. med. Bull.* **19**, 59.

DOLLERY C. T., GILLAM P. M. S., HUGH-JONES P. and ZORAB P. A. (1965) *Thorax* **20**, 175.

DONALD K. J., EDWARDS R. L. and MCEVOY J. D. S. (1975) *Amer. J. Med.* **59**, 642.

DONIACH I. (1947) *Amer. J. Roentgenol.* **58**, 620.

DONIACH I., MORRISON B. and STEINER R. E. (1954) *Brit. Heart J.* **16**, 101.

DONOHUE W. L., LASKI B., UCHIDA I. and MUNN J. D. (1959) *Pediatrics* **24**, 786.

DORSIT G., GIRARD R., ROUSSET H., BRUNE J., WIESENDANGER T., TOLOT F., BOURRET J. and GALY P. (1970) *Sem. Hôp. Paris* **46**, 3363.

DOWLING E. A., MILLER R. E., JOHNSON I. M. and COLLIER F. C. D. (1962) *Surgery* **52**, 600.

DOWNING S. E., VIDONE R. A., BRANDT H. M. and LIEBOW A. A. (1963) *Amer. J. Path.* **43**, 739.

DOYLE A. P., BALCERZAK S. P., WELLS C. L. and CRITTENDEN J. O. (1963) *Arch. int. Med.* **112**, 940.

DRAKE C. H. and HENRICI A. T. (1943) *Amer. Rev. Tuberc.* **48**, 184.

DRENNAN J. M. and MCCORMACK R. J. M. (1960) *J. Path. Bact.* **79**, 147.

DRESDALE D. T., SCHULTZ M. and MICHTOM R. J. (1951) *Amer. J. Med.* **11**, 686.

DRESDALE D. T., MICHTOM R. J. and SCHULTZ M. (1954) *Bull. N.Y. Acad. Med.* **30**, 195.

DRINKER C. K. and FIELD M. E. (1933) *Lymphatics, Lymph and Tissue Fluid.* Williams & Wilkins, Baltimore.

DRINKER C. K. (1942) *The Lymphatic System (Its part in regulating composition and volume of tissue fluid).* Stanford Univ. Press, Stanford.

DRINKER C. K. and WARREN M. F. (1943) *J. Amer. med. Ass.* **122**, 269.

DRINKER C. K. (1945) *Pulmonary Oedema and Inflammation.* Harvard University Press, Cambridge, Mass.

DRURY R. A. B. and STIRLAND R. M. (1959) *J. Path. Bact.* **77**, 543.

DRYMALSKI G. W., THOMPSON J. R. and SWEANY H. C. (1948) *Amer. J. Path.* **24**, 1083.

DUBOIS A., JANSSENS P. G., BRUTSAERT P. and VAN BREUSEGHEM R. (1952) *Ann. Soc. belge Med. trop.* **32**, 569.

DUBREUIL G., LACOSTE A. and RAYMOND R. (1936) *Bull. Histol. appliq. Physiol.* **13**, 235.

DUDDING B. A., WAGNER S. C., ZELLER J. A., GMELICH J. T., FRENCH G. R. and TOP F. H. Jr. (1972) *New Engl. J. Med.* **286**, 1289.

DUDGEON L. S. and WRIGLEY C. H. (1935) *J. Laryng.* **50**, 752.

DUFFY T. J. and CHOFNAS I. (1962) *Amer. J. med. Sci.* **243**, 269.

DUGGAN M. J., SOILLEUX P. J., STRONG J. C. and HOWELL D. M. (1970) *Brit. J. industr. Med.* **27**, 106.

DUGUID J. B. (1927) *Lancet* **2**, 111.

DUGUID J. B. (1959) Personal communication.

DUGUID J. B. and LAMBERT MARGARET W. (1964) *J. Path. Bact.* **88**, 389.

DUGUID J. B., YOUNG A., CAUNA D. and LAMBERT MARGARET W. (1964) *J. Path. Bact.* **88**, 405.

DUHIG J. T. (1959) *J. thorac. Surg.* **37**, 236.

DUKE H. N. (1957) *J. Physiol. (Lond.)* **135**, 45.

DUKE M. (1959) *Arch. Path.* **67**, 110.

DUMAS L. W., GREGORY R. L. and OZER F. L. (1963) *Brit. med. J.* **1**, 383.

DUNCAN D. A., DRUMMOND K. N., MICHAEL A. F. and VERNIER R. L. (1965) *Ann. int. Med.* **62**, 920.

DUNCAN J. T. (1945) *Brit. med. J.* **2**, 715.

DUNGAL N. (1946) *Amer. J. Path.* **22**, 737.

DUNNER L. (1945) *Brit. J. Radiol.* **18**, 33.

DUNNER L., HICKS M. S. and BAGNALL D. J. T. (1952) *Brit. J. Tuberc.* **46**, 43.

DUNNILL M. S. (1959) *J. Path. Bact.* **77**, 299.

DUNNILL M. S. (1960) *Brit. J. Dis. Chest* **54**, 355.

DUNNILL M. S. (1960) *J. clin. Path.* **13**, 27.

DUNNILL M. S. (1962) *Thorax* **17**, 329.

DUNNILL M. S. (1965) *Med. thorac.* **22**, 261.

DUNNILL M. S. (1968) *Brit. J. Surg.* **55**, 790.

DUNNILL M. S. (1970) *Human Path.* **1**, 265.

DUPONT V., LISSA C. and AMSTUT Z. (1962) *Revue Pratn.* **12**, 1909.

DUPREZ A., WITTEK F. and DUMOUNT A. (1956) *Thorax* **11**, 249.

DURNIN R. E., LABABIDI Z., BUTLER C., SELKE A. and FLEGE J. B. (1970) *Chest* **57**, 454.

DURNO C. and BROWN W. L. (1908) *Lancet* **1**, 1693.

DUTRA F. R. (1948) *Amer. J. Path.* **24**, 1137.

DUTZ W. (1970) in *Pathology Annual* **5**, 309.

DUTZ W., KOHOUT E. and HANKINS J. (1971) *Z. Tropenmed. Parasit.* **22**, 191.

DUTZ W., JENNINGS-KHODADAD E., POST C., KOHOUT E., NAZARIAN I. and ESMAILI H. (1974) *Z. Kinderheilk.* **117**, 241.

DVORAK ANN M. and GAVALLER B. (1966) *New Engl. J. Med.* **274**, 540.

DYKE P. C., DEMARAY M. J. and DELAVAN J. W. (1974) *Amer. J. clin. Path.* **61**, 301.

EADIE MARGARET B., STOTT E. J. and GRIST N. R. (1966) *Brit. med. J.* **2**, 671.

EARLE J. H. O., HIGHMAN J. H. and LOCKEY E. (1960) *Brit. med. J.* **1**, 607.

EAST T. and BARNARD W. G. (1938) *Lancet* **1**, 834.

EAST T. (1940) *Brit. Heart J.* **2**, 189.

EASTCOTT D. F. (1956) *Lancet* **1**, 37.

EATON M. D., MEIKLEJOHN G. and VAN HERICK W. (1944) *J. exp. Med.* **79**, 649.

ECKMANN B. H., SCHAEFER G. L. and HUPERT M. (1964) *Amer. Rev. resp. Dis.* **89**, 175.

EDGE J. R. and RICKARDS A. G. (1957) *Thorax* **12**, 352.

EDITORIAL (1959) *Circulation* **19**, 641.

EDITORIAL (1963) *Amer. J. Med.* **35**, 293.

EDWARDS A. T. and TAYLOR A. B. (1937) *Brit. J. Surg.* **25**, 487.

EDWARDS J. E., DOUGLAS J. M., BURCHELL H. B. and CHRISTENSEN N. A. (1949) *Amer. Heart J.* **38**, 205.

EDWARDS J. E. (1950) *Proc. Inst. Med. (Chicago)* **18**, 134.

EDWARDS J. E. and BURCHELL H. B. (1951) *Arch. intern. Med.* **87**, 372.

EDWARDS J. E. and CHAMBERLAIN W. B. (1951) *Circulation* **3**, 524.

EDWARDS J. E. and HELMHOLZ H. F. (1956) *Proc. Mayo Clin.* **31**, 151.

EDWARDS J. E. (1957) *Circulation* **15**, 164.

EDWARDS P. Q. (1958) in *Fungous Diseases and their Treatment.* Edited by RIDDELL and STEWART. Butterworth, London.

EFSKIND L. and WEXELS P. (1952) *J. thorac. Surg.* **23**, 377.

EHRICH J. F. and McINTOSH J. F. (1932) *Arch. Path.* **13**, 69.

EHRLICH J. C. and ROMANOFF A. (1951) *Arch. int. Med.* **87**, 259.

EIKAS J. and KIM P. K. (1960) *Acta tuberc. scand.* **39**, 140.

ELKELES A. and GLYNN L. E. (1946) *J. Path. Bact.* **58**, 517.

ELLIOT F. M. and REID LYNNE (1965) *Clin. Radiol.* **16**, 193.

ELLIS F. H., GRINDLAY J. H. and EDWARDS J. E. (1952) *Surgery* **31**, 167.

ELLMAN P. (1947) *Proc. roy. Soc. Med.* **40**, 332.

ELLMAN P. and BALL R. E. (1948) *Brit. med. J.* **2**, 816.

ELLMAN P. and CUDKOWICZ L. (1954) *Thorax* **9**, 46.

ELLMAN P., CUDKOWICZ L. and ELWOOD J. S. (1954) *J. clin. Path.* **7**, 239.

EL MALLAH S. H. and HASHEM M. (1953) *Thorax* **8**, 148.

ELO R., MÄÄTTÄ UKSILA E. and ARSTILA A. V. (1972) *Arch. Path.* **94**, 417.

ELOESSER L. (1931) *J. thorac. Surg.* **1**, 194.

ELOESSER L. (1932) *J. thorac. Surg.* **2**, 270, 373, 485.

EMANUEL B., LIEBERMAN A. D., GOLDIN M. and SANSON J. (1962) *J. Pediat.* **61**, 44.

EMANUEL D. A., LAWTON B. R. and WENZEL F. J. (1962) *New Engl. J. Med.* **266**, 333.

EMANUEL D. A. (1966) Personal communication.

EMERY J. L. (1955) *J. clin. Path.* **8**, 180.

EMERY J. L. and MITHAL A. (1960) *Arch. Dis. Childh.* **35**, 544.

EMMONS C. W. (1938) *Publ. Hlth. Rep. (Wash.)* **53**, 1967.

EMMONS C. W. (1942) *Publ. Hlth. Rep. (Wash.)* **57**, 109.

EMMONS C. W. (1944) *Mycologia* **36**, 188.

EMMONS C. W., OLSON B. J. and ELDRIDGE W. W. (1945) *Publ. Hlth. Rep. (Wash.)* **60**, 1383.

EMMONS C. W. (1949) *Publ. Hlth. Rep. (Wash.)* **64**, 892.

EMMONS C. W. (1951) *J. Bact.* **62**, 685.

EMMONS C. W. (1955) *Amer. J. Hyg.* **62**, 227.

EMMONS C. W. (1958) *Publ. Hlth. Rep. (Wash.)* **73**, 590.

EMMONS C. W. and JELLISON W. L. (1960) *Ann. N.Y. Acad. Sci.* **89**, 91.

EMMONS C. W., MURRAY I. G., LURIE H. I., KING M. H., TULLOCH J. A. and CONNOR D. H. (1964) *Sabouraudia* **3**, 306.

ENCHEV S. (1963) *Neoplasma (Bratislava)* **10**, 291.

ENDERS J. F., McCARTHY K., MITUS A. and CHEATHAM W. J. (1959) *New Engl. J. Med.* **261**, 875.

ENGEL S. (1947) *The Child's Lung.* E. Arnold, London.

ENGEL S. (1958a) *Acta anat. (Basel)* **35**, 301.

ENGEL S. (1958b) *J. clin. Path.* **11**, 302.

ENGEL S. (1959a) *Acta anat. (Basel)* **36**, 234.

ENGEL S. (1959b) *Anat. Anz.* **106**, 86.

ENGEL S. (1959c) *Anat. Anz.* **106**, 90.

ENGEL S. (1959d) Personal communication.

ENGELBRETH-HOLM J., TEILUM G. and CHRISTENSEN E. (1944) *Acta med. scand.* **118**, 292.

ENGELSTAD R. B. (1940) *Amer. J. Roentgenol.* **43**, 676.

ENRIGHT J. B. and SADLER W. W. (1954) *Proc. Soc. exp. Biol.* **85**, 466.

ENTICKNAP J. B., GALBRAITH N. S., TOMLINSON A. J. H. and ELIAS-JONES T. F. (1968) *Brit. J. industr. Med.* **25**, 72.

ENTWISTLE C. C., FENTEM P. H. and JACOBS A. (1964) *Brit. med. J.* **2**, 1504.

ENZER N. and SANDER O. A. (1938) *J. industr. Hyg.* **20**, 333.

EPPINGER H. L. (1891) *Beitr. path. Anat.* **9**, 287.

EPPINGER H. (1894) *Die Haderkrankheit, eine typische Inhalations-Milzbrandinfection beim Menschen unter besonderer Berücksichtigung ihrer pathologischen Anatomie und Pathogenesis auf Grund eigener Beobachtungen dargestellt.* G. Fischer, Jena, p. 139.

EPPINGER H. and SCHAUENSTEIN W. (1902) *Ergebn. allg. Path. Anat.* **8**, 267.

EPPINGER H. (1949) *Die Permeabilitätspathologie als Lehre vom Krankheitsbeginn.* Springer, Vienna.

EPSTEIN S., RANCHOD M. and GOLDSWAIN P. R T. (1973) *Cancer (Philadelphia)* **32**, 476.

ERASMUS L. D. (1956) *Quart. J. Med.* **25** N.S., 507.

ERB I. H. (1933) *Arch. Path.* **15**, 357.

ESTERLY J. R., OPPENHEIMER ELLA H., ROWE S. and AVERY MARY E. (1966) *J. Pediat.* **69**, 3.

ESTERLY J. R. and OPPENHEIMER E. H. (1970) *Arch. Path.* **90**, 553.

ETHERTON J. E., CONNING D. M. and CORRIN B. (1973) *Amer. J. Anat.* **138**, 11.

EURICH F. W. (1933) *Brit. med. J.* **2**, 50.

EVANS A. D. and EVANS M. (1956) *Lancet* **1**, 771.

EVANS N. (1903) *J. Amer. med. Ass.* **40**, 1772.

EVANS R. D. (1950) *Acta Un. int. Cancr.* **6**, 1229.

EVANS W., SHORT D. S. and BEDFORD D. E. (1957) *Brit. Heart J.* **19**, 93.

EVANS W. (1959) *Brit. Heart J.* **21**, 197.

EVANS W. A. and LEUCUTIA T. (1925) *Amer. J. Roentgenol.* **13**, 203.

EVERSOLE S. L. and RIENHOFF W. F. (1959) *J. thorac. Surg.* **37**, 750.

EWART W. (1889) *The Bronchi and Pulmonary Bloodvessels.* Churchill, London.

EWING J. (1940) *Neoplastic Diseases, A Treatise on Tumours,* 4th ed. W. B. Saunders, Philadelphia, 1940.

FABRONI S. M. (1935) *Med. d. Lavoro* **26**, 297.

FADELL E. J., RICHMAN ANNE D., WARD W. W. and HENDON J. R. (1962) *New Engl. J. Med.* **266**, 861.

FAHEY J. L., LEONARD E., CHURG J. and GODMAN G. (1954) *Amer. J. Med.* **17**, 168.

FAHR T. H. (1935) *Virchows Arch. path. Anat.* **295**, 502.

FALCONER E. H. and LEONARD M. E. (1936) *Amer. J. med. Sci.* **191**, 780.

FALLON J. T. (1937) *Canad. med. Ass. J.* **36**, 223.

FALOR W. H. and KYRIAKIDES A. H. (1949) *J. thorac. Surg.* **18**, 252.

FANCONI G. (1931) *Jhrb. Kinderheilk.* **133**, 257.
FARBER S. and WOLBACH S. B. (1932) *Amer. J. Path.* **8**, 123.
FARBER S. and WILSON J. L. (1933) *Amer. J. Dis. Child.* **46**, 572.
FARBER S. (1944) *Arch. Path.* **37**, 238.
FARBER S. M., MCGRATH A. K., BENIOFF M. A. and ROSENTHAL M. (1950) *J. Amer. med. Ass.* **144**, 1.
FARINACCI C. J., JEFFREY H. C. and LACKEY R. W. (1951) *U.S. Armed Forces med. J.* **2**, 1085.
FARINACCI C. J., BLUAW A. S. and JENNINGS E. M. (1973) *Amer. J. clin. Path.* **59**, 508.
FASS R. J. and SASLAW S. (1971) *Arch. int. Med.* **128**, 588.
FAULDS J. S. (1957) *J. clin. Path.* **10**, 187.
FAULDS J. S., KING E. J. and NAGELSCHMIDT G. (1959) *Brit. J. industr. Med.* **16**, 43.
FAULDS J. S. and NAGELSCHMIDT G. S. (1962) *Ann. occup. Hyg.* **4**, 255.
FAUST E. C. and HEADLEE W. H. (1936) *Amer. J. trop. Med.* **16**, 25.
FAUST E. C. and RUSSELL P. F. (1957) *Clinical Parasitology*, 6th ed. H. Kimpton, London.
FAVORITE G. O. (1934) *Amer. J. med. Sci.* **187**, 663.
FAWCETT F. J. and HUSBAND E. M. (1967) *J. clin. Path.* **20**, 260.
FAWCITT R. (1936) *Brit. J. Radiol.* **9**, 172.
FAWCITT R. (1938) *Brit. J. Radiol.* **11**, 378.
FECHNER R. E. and BENTINCK B. R. (1973) *Cancer (Philadelphia)* **31**, 1451.
FEKETY F. R. Jr., CALDWELL J., GUMP D., JOHNSON J.E., MAXSON W., MULHOLLAND J. and THOBURN R. (1971) *Amer. Rev. resp. Dis.* **104**, 499.
FELDMAN W. H., DAVIES R., MOSES H. E. and ANDBERG W. (1943) *Amer. Rev. Tuberc.* **48**, 82.
FELIX W. (1928) in *Die Chirurgie der Brustorgane.* Edited by SAUERBRUCH F. Vol. **1**, Pt. 1, 3rd ed. J. Springer, Berlin.
FELLER A. (1921–2) *Virchows Arch. path. Anat.* **236**, 470.
FELTON W. L. (1952) *Lab. Invest.* **1**, 364.
FELTON W. L. (1953) *J. thorac. Surg.* **25**, 530.
FERENCZ CHARLOTTE and DAMMANN J. F. (1957) *Circulation* **16**, 1046.
FERENCZ CHARLOTTE (1960a) *Bull. Johns Hopk. Hosp.* **106**, 81; (1960b) *ibid.* **106**, 100.
FERGUSON D. J. and VARCO R. L. (1955) *Circulation Res.* **3**, 152.
FERGUSON F. C., KOBILAK R. E. and DEITRICK J. E. (1944) *Amer. Heart J.* **28**, 445.
FERNAN-ZEGARRA L. and LAZO-TOBOADA F. (1961) *Revista Peruana de Patologia* **6**, 49.
FETTERMAN G. H. and LERNER H. (1936) *J. Lab. clin. Med.* **21**, 1157.
FETZER A. E., WERNER A. S. and HAGSTROM J. W. C. (1967) *Amer. Rev. resp. Dis.* **96**, 1121.
FEUARDENT R. (1953) M.D. Thesis presented to the University of Geneva.
FEYRTER F. (1927) *Frankfurt. Z. Path.* **35**, 213.
FEYRTER F. (1953) *Über die peripheren endokrinen (parakrinen) Drüsen des Menschen.* W. Maudrich, Wien–Düsseldorf.

FEYRTER F. (1954) *Virchows Arch. path. Anat.* **325**, 723.
FEYRTER F. (1960) *Wien. klin. Wschr.* **72**, 386.
FIALHO A. S. (1946) quoted by LACAZ C. in *Localizações pulmonares da micose de Lutz (Anat. Patol. Patogen).* Rio de Janeiro.
FIENBERG R. (1953) *Amer. J. Path.* **29**, 913.
FIENBERG R. (1955) *Amer. J. Med.* **19**, 829.
FIESSINGER N. and FAUVET J. (1941) *Presse méd.* **49**, 449.
FIGUEROA W. G., RASZKOWSKI and WEISS W. (1973) *New Engl. J. Med.* **288**, 1096.
FINDLAY D. W. (1891) *Brit. J. Derm.* **1**, 339; also *Middlesex Hosp. Rep.* (1889), p. 29.
FINDLAY-JONES J. M., PAPADIMITRIOU J. M. and BARTER R. A. (1974) *J. Path.* **112**, 117.
FINDLEY C. W. and MAIER H. C. (1951) *Surgery* **29**, 604.
FINK J. N., BANASZAK E. F., THIEDE W. H. and BARBORIAK J. J. (1971) *Ann. int. Med.* **74**, 80.
FINKE W. (1956) *Int. Rec. Med.* **169**, 61.
FINKELDEY W. (1931) *Virchows Arch. path. Anat.* **281**, 323.
FINKELSTEIN H. (1905–12) in *Lehrbuch der Säuglings-Krankheiten.* H. Kornfeld, Berlin.
FISCHER B. (1922) *Frankfurt. Z. Path.* **27**, 98.
FISCHER W. (1931) in *Handbuch der speziellen Anatomie und Histologie.* Edited by HENKE F. and LUBARSCH O. Vol. 3, p. 509. Springer, Berlin.
FISHEL C. R. (1923) *J. Amer. med. Ass.* **80**, 102.
FISHER J. H. and MACKLIN C. C. (1940) *Amer. J. Dis. Child.* **60**, 102.
FISHER J. H. and HOLLEY W. J. (1953) *Arch. Path.* **55**, 162.
FITZGERALD M. X., CARRINGTON C. B. and GAENSLER E. A. (1973) *Med. Clin. N. Amer.* **57**, 593.
FLANAGAN P. and ROECKEL I. E. (1964) *Amer. J. Med.* **36**, 214.
FLECK L. (1961) in *Inhaled Particles and Vapours.* Edited by C. N. DAVIES, p. 367. Pergamon Press, Oxford.
FLEISCHNER F. G. and REINER L. (1954) *New Engl. J. Med.* **250**, 900.
FLEMING H. A. and BAILEY S. M. (1966) *Brit. med. J.* **1**, 1322.
FLICK J. B., CLERF L. H., FINK E. H. and FARRELL J. T. (1929) *Arch. Surg. (Chicago)* **19**, 1292.
FLOYD J., CAMPBELL D. C. Jr. and DOMINY D. E. (1962) *Amer. Rev. resp. Dis.* **86**, 557.
FLUCKIGER M. (1884) *Wien. med. Wschr.* **49**, 1457 (quoted by RYDELL R. and HOFFBAUER F. W. (1956)).
FLYE M. W. and IZANT R. J. (1972) *Surgery* **71**, 744.
FOOT N. C. (1927) *Amer. J. Path.* **3**, 413.
FOOT N. C. (1945) in *Pathology in Surgery.* J. B. Lippincott Company, Philadelphia, pp. 87–88.
FORBES G. B. (1955) *J. Path. Bact.* **70**, 427.
FORBUS W. D. (1927) *Amer. Rev. Tuberc.* **16**, 599.
FORBUS W. D. and BESTEBREURTJE A. M. (1946) *Mil. Surg.* **99**, 653.
FORMICOLA P. (1932) *Morgagni* **74**, 1591 (quoted by BECH et al. (1962)).
FORSEE J. H. and BLAKE H. A. (1954) *Ann. Surg.* **139**, 76.
FOSHAY L. (1937) *Arch. intern. Med.* **60**, 22.

FOULGER M., GLAZER A. M. and FOSHAY L. (1932) *J. Amer. med. Ass.* **98**, 951.

FOWLER E. F. and BOLLINGER J. A. (1954) *Surgery* **36**, 650.

FOX B. and RISDON R. A. (1968) *J. clin. Path.* **21**, 486.

FOX H. and GUNN A. D. G. (1965) *Brit. J. Dis. Chest* **59**, 47.

FRAENKEL E. (1917) *Z. Hyg.* **84**, 369.

FRAIMOW W., CATHCART R. T., KIRSHNER J. J. and TAYLOR R. C. (1960) *Amer. J. Med.* **28**, 458.

FRANCH R. H. and GAY B. B. Jr. (1963) *Amer. J. Med.* **35**, 512.

FRANCIS T. (1940) *Science* **92**, 405.

FRANK L. (1950) *Arch. Path.* **50**, 40.

FRÄNTZEL O. (1868) *Virchows Arch. path. Anat.* **43**, 420.

FRASCA J. M., AUERBACH O., PARKS V. R. and STOECKENIUS W. (1967a) *Exp. molec. Path.* **6**, 261.

FRASCA J. M., AUERBACH O., PARKS V. R. and STOECKENIUS W. (1967b) *Exp. molec. Path.* **7**, 92.

FRAZIER A. R. and MILLER R. D. (1974) *Chest* **65**, 403.

FREDERICK W. G. and BRADLEY W. R. (1946) quoted by BECK *et al.* (1962).

FREEDMAN E. and BILLINGS J. H. (1949) *Radiology* **53**, 203.

FREIMAN D. G. (1948) *New Engl. J. Med.* **239**, 664, 709, 743.

FREIMAN D. G., SOYEMOTO J. and WESSLER S. (1965) *New Engl. J. Med.* **272**, 1278.

FREIMAN D. G. and HARDY H. L. (1970) *Human Path.* **1**, 25.

FRETHEIM B. (1952) *Thorax* **7**, 156.

FREUND J. and MCDERMOTT K. (1942) *Proc. Soc. exp. Biol. (N. Y.)* **49**, 548.

FRIBERG L. (1950) *Acta med. scand.* **138**, 1, Suppl. 240.

FRIBOURG-BLANC A. (1971) *Rev. Méd.* **8**, 421.

FRIED B. M. (1927) *Arch. intern. Med.* **40**, 340.

FRIED B. M. (1931) *Medicine (Baltimore)* **10**, 442.

FRIED B. M. (1932) *Primary Carcinoma of the Lung: Bronchiogenic Cancer, a Clinical and Pathological Study* in two parts. Williams & Wilkins Company, Baltimore.

FRIED B. M. (1938) *Acta Un. int. Cancr.* **3**, 153.

FRIED J. R. and GOLDBERG H. (1940) *Amer. J. Roentgenol.* **43**, 877.

FRIEDBERG C. K. and GROSS L. (1934) *Arch. intern. Med.* **54**, 170.

FRIEDE E. and RACHOW D. O. (1961) *Ann. int. Med.* **54**, 121.

FRIEDLÄNDER C. (1876) *Virchows Arch. path. Anat.* **68**, 325.

FRIEDLÄNDER C. (1882) *Virchows Arch. path. Anat.* **87**, 319.

FRIEDRICH G. (1939) *Virchows Arch. path. Anat.* **304**, 230.

FRIEDRICH N. (1856) *Virchows Arch. path. Anat.* **9**, 613.

FRIMODT-MÖLLER C. and BARTON R. M. (1940) *Ind. med. Gaz.* **75**, 607.

FRISSELL L. F. and KNOX L. C. (1937) *Amer. J. Cancer* **30**, 219.

FRITZ E. (1933) *Virchows Arch. path. Anat.* **289**, 264.

FRIZZERA G., MORGAN E. M. and RAPPAPORT H. (1974) *Lancet* **1**, 1070.

FRÖHLICH F. (1949) *Frankfurt. Z. Path.* **60**, 517.

FRONSTIN M. H., HOOPER G. S., BESSE B. E. and FERRERI S. (1967) *Amer. J. Dis. Child.* **114**, 330.

FROTHINGHAM L. (1902) *J. med. Res.* **9**, 31.

FÜLLEBORN F. (1920) *Arch. Schiffs- u. Tropenhyg.* **24**, 340.

FULLER C. J. (1953) *Thorax* **8**, 59.

FULLER C. J. (1958) in *Fungous Diseases and their Treatment.* Edited by RIDDELL and STEWART. Butterworth, London, p. 138.

FURCOLOW M. L. (1948) Editorial in the *Amer. J. clin. Path.* **18**, 171.

FURCOLOW M. L. and GRAYSTON J. T. (1953) *Amer. Rev. Tuberc.* **68**, 307.

FURCOLOW M. L. (1964) *Vth Cong. Int. Acad. Path. London.*

FURGERSON W. B. Jr., BACHMAN L. B. and O'TOOLE W. F. (1963) *Amer. Rev. resp. Dis.* **88**, 689.

FUST J. A. and CUSTER R. P. (1949) *Amer. J. clin. Path.* **19**, 522.

GAENSLER E. A., CARRINGTON C. B. and COUTU R. E. (1972) *Clin. Notes on Resp. Diseases* **10**, 3.

GALLUZZI S. and PAYNE P. M. (1956) *Brit. J. Cancer* **10**, 408.

GALY P., CHARCOSSET, JACOUTON, PUTHOD and THÉOCARIS (1958) *J. franç. Méd. Chir. thor.* **12**, 518.

GANS S. L. and POTTS W. J. (1951) *J. thorac. Surg.* **21**, 313.

GARDNER A. M. N. (1958) *Quart. J. Med.* **27**, N.S., 227.

GARDNER D. L., DUTHIE J. J. R., MACLEOD J. and ALLEN W. S. A. (1957) *Scot. med. J.* **2**, 183.

GARDNER L. U. (1933) *U.S. publ. Hlth. Bull. No. 208*, p. 16.

GARDNER L. U. (1934) *1st Saranac Symposium on Silicosis.* Trudeau Sch. Tuberc., Saranac Lake, N.Y.

GARDNER L. U. (1935) *Publ. Hlth. Rep. (Wash.)* **50**, 695.

GARDNER L. U. (1937) *3rd Saranac Symposium on Silicosis.* Trudeau Sch. Tuberc., Saranac Lake, N.Y.

GARDNER L. U. (1938) *Mining. Tech. Pub. No. 929*, 1–15.

GARDNER L. U. (1939) *4th Saranac Symposium on Silicosis.* Trudeau Sch. Tuberc., Saranac Lake, N.Y.

GARDNER L. U. (1940) *J. Amer. med. Ass.* **114**, 535.

GARLAND L. H. (1966) *Amer. J. Roentgenol.* **96**, 604.

GÄRTNER H. (1952) *Arch. industr. Hyg.* **6**, 339.

GATZIMOS C. D., SCHULTZ D. M. and NEWMAN R. L. (1955) *Amer. J. Path.* **31**, 791.

GAUSEWITZ P. L., JONES F. S. and WORLEY G. (1951) *Amer. J. clin. Path.* **21**, 41.

GAYLOR J. B. (1934) *Brain* **57**, 143.

GEEVER E. F. (1947) *Amer. J. med. Sci.* **214**, 292.

GEEVER E. F., NEUBUERGER K. T. and RUTLEDGE E. K. (1951) *Dis. Chest* **19**, 325.

GELL P. G. H. and COOMBS R. R. A. (1968) *Clinical Aspects of Immunology.* Blackwell, Oxford and Edinburgh.

GELLER S. A. and TOKER C. (1969) *Arch. Path.* **88**, 148.

GENDEL B. R., ENDE M. and NORMAN S. L. (1949) *Amer. J. Med.* **9**, 343.

GEORG J., MELLEMGAARD K., TYGSTRUP N. and WINKLER K. (1960) *Lancet* **1**, 852.

GERLE R. D., JARETZKI A., ASHLEY C. A. and BERNE A. S. (1968) *New Engl. J. Med.* **278**, 1413.

GERRITSEN W. B. and BUSCHMANN C. H. (1960) *Brit. J. industr. Med.* **17**, 187.

GERSTL B., TAGER M. and SZCZEPANIAK L. W. (1949) *Proc. Soc. exp. Biol. (N.Y.)* **70**, 697.

GETZOWA SOPHIA (1945) *Arch. Path.* **40**, 99.

GEUBELLE F., KARLBERG P., KOCH G., LIND J., WALLGREN G. and WEGELIUS C. (1959) *Biol. Neonat.* **1**, 169.

GHARPURE P. V. (1948) *Indian med. Gaz.* **83**, 5.

GHOREYEB A. A. and KARSNER H. T. (1913) *J. exp. Med.* **18**, 500.

GIAMPALMO A. (1950) *Acta med. scand* **39**, 1, Suppl. 248.

GIBBS F. A., GIBBS E. L., LENNOX W. G. and NIMS L. F. (1943) *J. Aviat. Med.* **14**, 250.

GIBSON J. B. (1953) *J. Path. Bact.* **65**, 239.

GIESE W. (1953) *Verh. dtsch. path. Ges.* **36**, 284.

GIESEKING R. (1958) *Verh. Deutsch. Ges Path.* **41**, 336.

GILCHRIST T. C. and STOKES W. R. (1898) *J. exper. Med.* **3**, 53.

GILLIAM A. G. (1955) *Cancer (Philadelphia)* **8**, 1130.

GILLMAN T., PENN J., BRONKS D. and ROUX M. (1955) *Arch. Path.* **59**, 733.

GILMOUR J. R. and EVANS W. (1946) *J. Path. Bact.* **58**, 687.

GILROY J. C., WILSON V. H. and MARCHAND P. (1951) *Thorax* **6**, 137.

GILSON J. C. (1961) *Amer. Rev. resp. Dis.* **83**, 407.

GIRODE J. (1889) *Arch. gén. Méd.* **1**, 50.

GIROUD P. and JADIN J. (1954) *Bull. Soc. Path. exot.* **47**, 578.

GITLIN D. and CRAIG J. M. (1956) *Pediatrics* **17**, 64.

GITLIN D. and CRAIG J. M. (1957) *Amer. J. Path.* **33**, 267.

GLANCY D. L., FRAZIER P. D. and ROBERTS W. C. (1968) *Amer. J. Med.* **45**, 198.

GLAUSER O. (1955) *Schweiz. Z. allg. Path.* **18**, 42.

GLEICHMAN T. K., LEDER M. M. and ZAHN D. W. (1949) *Amer. J. med. Sci.* **218**, 369.

GLENN W. W. L., LIEBOW A. A. and LINDSKOG G. E. (1975) *Thoracic and Cardiovascular Surgery with related Pathology*, 3rd Ed., p. 192, Appleton-Century-Crofts, New York.

GLENNER G. G., EIN D., EANES E. D., BLADEN H. A., TERRY W. and PAGE D. (1971) *Science* **174**, 712.

GLENNIE J. S., HARVEY P. and JEWSBURY P. (1959) *Thorax* **14**, 327.

GLENNIE J. S., HARVEY P. W. and SALAMA V. (1964) *J. thorac. cardiovasc. Surg.* **48**, 40.

GLOYNE S. R. (1935) *Tubercle* **17**, 5.

GLOYNE S. R., MARSHALL G. and HOYLE C. (1949) *Thorax* **4**, 31.

GLUCK L., KULOVICH M. V. and BORER R. C. (1971) *Amer. J. Obstet. Gynec.* **109**, 440.

GMELICH J. T., BENSCH K. G. and LIEBOW A. A. (1967) *Lab. Invest.* **17**, 88.

GOADBY K. (1923) *Diseases of the Gums and Oral Mucous Membrane.* Hodder and Stoughton, London, p. 112.

GODLESKI J. J. and BRAIN J. D. (1972) *J. exp. Med.* **136** 630.

GODMAN G. C. and CHURG J. (1954) *Arch. Path.* **58**, 533.

GODWIN M. C. (1957) *Cancer (Philadelphia)* **10**, 298.

GOETZ R. H. (1945) *Clin. Proc.* **4**, 337.

GOLD M. M. A. (1946) *Arch. intern. Med.* **78**, 197.

GOLDEN A. (1944) *Arch. Path.* **38**, 187.

GOLDEN A. (1948) *Amer. J. Path.* **24**, 716.

GOLDEN A. and TULLIS I. F. (1949) *Milit. Surg.* **105**, 130.

GOLDENBERG G. J. and GREENSPAN R. H. (1960) *New Engl. J. Med.* **262**, 1112.

GOLDMAN A. (1942) *Carcinoma of the Lung Duration. Medico-Surgical Tributes to Harold Brunn,* University of California Press, Beverley.

GOLDMANN E. (1907) *Proc. roy. Soc. Med.* 1 *Surg.* Sect.1.

GOLDSTEIN B. and WEBSTER I. (1972) *Ann. N.Y. Acad. Sci.* **200**, 306.

GOLDSWORTHY N. E. (1934) *J. Path. Bact.* **39**, 291.

GOODALE F. and THOMAS W. A. (1954) *Arch. Path.* **58**, 568.

GOODALE R. L., GOETZMAN B. and VISSCHER M. B. (1970) *Amer. J. Physiol.* **219**, 1226.

GOODING C. G. (1946) *Lancet* **2**, 891.

GOODPASTURE E. W. (1919) *Amer. J. med. Sci.* **158**, 863.

GOODPASTURE E. W., AUERBACH S. H., SWANSON H. S. and COTTER E. F. (1939) *Amer. J. Dis. Child.* **57**, 997.

GOODWIN J. F., STEINER R. E. and LOWE K. G. (1952) *J. Fac. Radiol.* **4**, 21.

GOODWIN J. F., STANFIELD C. A., STEINER R. E., BENTALL H. H., SAYED H. M., BLOOM V. R. and BISHOP M. B. (1962) *Thorax* **17**, 91.

GOODWIN J. F., HARRISON C. V. and WILCKEN D. E. L. (1963) *Brit. med. J.* **1**, 701, 777.

GOODYEAR J. E. and SHILLITOE A. J. (1959) *J. clin. Path.* **12**, 172.

GORALEWSKI G. (1940) *Arch. Gewerbepath. Gewerbehyg.* **9**, 676.

GORALEWSKI G. (1947) *Z. ges. inn. Med.* **2**, 665.

GORDON A. J., DONOSO E., KUHN C. L. A., RAVITCH M. M. and HIMMELSTEIN A. (1954) *New. Engl. J. Med.* **251**, 923.

GORDON H. W., MILLER R. J. and MITTMAN C. (1973) *Human Path.* **4**, 431.

GORDON J. and WALKER G. (1944) *Arch. Path.* **37**, 222.

GORDON L. Z. and BOSS H. (1955) *Cancer (Philadelphia)* **8**, 588.

GORDON R. E. and HAGAN W. A. (1936) *J. infect. Dis.* **59**, 200.

GORHAM G. W. and MERSELIS J. G. Jr. (1959) *Bull. Johns Hopk. Hosp.* **104**, 11.

GORLIN R., LEWIS B. M., HAYNES F. W., SPIEGEL R. J. and DEXTER L. (1951) *Amer. Heart J.* **41**, 834.

GOTSMAN M. S. and WHITBY J. L. (1964) *Thorax* **19**, 89.

GOTTLIEB A. J., SPIERA H., TEIRSTEIN A. S. and SILTZBACH L. E. (1965) *Amer. J. Med.* **39**, 405.

GOUGH J. (1940) *J. Path. Bact.* **51**, 277.

GOUGH J. (1947) *Occup. Med.* **4**, 86.

GOUGH J. and WENTWORTH J. E. (1949a) *J. roy. micro. Soc.* **69**, 231.

GOUGH J., JAMES W. B. L. and WENTWORTH J. E. (1949b) *J. Fac. Radiol.* **1**, 28.

GOUGH J. (1952) *Proc. roy. Soc. Med.* **45**, 576.

GOUGH J. (1955) *Lancet* **1**, 161.

GOUGH J., RIVERS D. and SEAL R. M. E. (1955) *Thorax* **10**, 9.

GOUGH J. (1959) in *Modern Trends in Pathology*. Edited by D. H. COLLINS. Butterworth, London.

GOUGH J. (1960) in *Recent Advances in Pathology*. Edited by C. V. HARRISON. J. and A. Churchill, London.

GOUGH J. (1961) *Acta allerg.* (*Kbh*) **16**, 391.

GOUGH J. (1962) in Ciba Foundation Symposium on *Pulmonary Structure and Function*, p. 259. J. and A. Churchill, London.

GOUGH J. (1965) *Bull. N.Y. Acad. Med.* **41**, 927.

GOULDEN F., KENNAWAY E. L. and URQUHART M. E. (1952) *Brit. J. Cancer* **6**, 1.

GOVAN A. D. T. (1946) *Amer. J. Surg.* **58**, 423.

GOWENLOCK A. H., PLATT D. S., CAMPBELL A. C. P. and WORMLEY K. G. (1964) *Lancet* **1**, 304.

GOWING N. F. C. and HAMLIN IRIS M. E. (1960) *J. clin. Path.* **13**, 396.

GRAEF I. (1935) *Amer. J. Path.* **11**, 862.

GRAHAM E. A., SINGER J. J. and BALLON H. C. (1935) *Surgical Diseases of the Chest*. Lea and Febiger, Philadelphia.

GRAHAM S. and LEVIN M. L. (1971) *Cancer* (*Philadelphia*) **27**, 865.

GRAINGER R. C. (1958) *Brit. J. Radiol.* **31**, 201.

GRAMLICK F. and WIETHOFF E. O. (1960) *Dtsch. med. Wschr.* **85**, 1750.

GRAMS L. R. (1939) *Arch. Path.* **28**, 865.

GRANT A. and BARWELL C. (1943) *Lancet* **1**, 199.

GRANZOW J. (1932) *Arch. Gynäk.* **151**, 612.

GRAWITZ P. (1880) *Virchows Arch. path. Anat.* **82**, 217.

GRAY S. H. and CORDONNIER J. (1929) *Arch. Surg.* (*Chicago*) **19**, 1618.

GRAYSON R. R. (1956) *Ann. intern. Med.* **45**, 393.

GREEN A. E. and SHIELD R. T. (1950) *Proc. roy. Soc.* A **202**, 407.

GREEN G. M. and KASS E. H. (1964) *J. clin. Invest.* **43**, 769.

GREEN G. M. (1970) *Amer. Rev. resp. Dis.* **102**, 691.

GREEN H. N. and STONER H. B. (1950) *Biological Actions of the Adenine Nucleotides*. H. K. Lewis and Company, London.

GREEN J. D., HARLE T. S., GREENBERG S. D., WEG J. G., NEVIN H. and JENKINS D. E. (1970) *Amer. Rev. resp. Dis.* **101**, 293.

GREENBERG E., DIVERTIE M. B. and WOOLNER L. B. (1964) *Amer. J. Med.* **36**, 106.

GREENBERG P. B., MARTIN T. J., BECK C. and BURGER H. G. (1972) *Lancet* **1**, 350.

GREENBERG S. D., JENKINS D. E., BAHAR D., HSU K. H. K., HUNSAKER M. and JONES R. J. (1963) *Texas State J. Med.* **59**, 949.

GREENBERG S. D., BOUSHY S. F. and JENKINS D. E. (1967) *Amer. Rev. resp. Dis.* **96**, 918.

GREENBERG S. D., GYÖRKEY F., WEGG J. G., JENKINS D. E. and GYÖRKEY P. (1970) *Amer. Rev. resp. Dis.* **102**, 648.

GREENBERG S. D., HEISLER J. G., GYÖRKEY F. and JENKINS D. E. (1972) *South. med. J.* **65**, 775.

GREENBERG S. D., SMITH M. N. and SPJUT H. J. (1974) *Lab. Invest.* **30**, 375.

GREENBLATT M., HEREDA R., RUBENSTEIN L. and ALPERT S. (1964) *Amer. J. clin. Path.* **41**, 188.

GREENSPAN E. B. (1933) *Amer. J. Cancer* **18**, 603.

GREENSPAN E. B. (1934) *Arch. intern. Med.* **54**, 625.

GREGORY G. A. and TOOLEY W. H. (1970) *New Engl. J. Med.* **282**, 1141.

GREGORY J. C. (1831) *Edin. med. surg. J.* **36**, 389.

GREGORY J. E., GOLDEN A. and HAYMAKER M. (1943) *Bull. Johns Hopk. Hosp.* **73**, 405.

GRIEBLE H. G., COTTON F. R. and BIRD T. J. (1970) *New Engl. J. Med.* **282**, 531.

GRIECO M. H. and RYAN S. F. (1968) *Amer. J. Med.* **45**, 811.

GRIFFIN J. W., DAESCHNER C. W., COLLINS V. P. and EATON W. L. (1954) *J. Pediat.* **45**, 13.

GRIFFITHS SYLVIA P., LEVINE O. R. and ANDERSEN D. H. (1962) *Circulation* **25**, 73.

GRILL C., SZÖGI S. and BOGREN H. (1962) *Acta Med. scand.* **171**, 329.

GRIMES O. F., WEIRICH W. and STEPHENS H. B. (1954) *J. thorac. Surg.* **27**, 378.

GRIMLEY P. M., WRIGHT L. D. Jr. and JENNINGS A. E. (1965) *Amer. J. clin. Path.* **43**, 216.

GRISHMAN A., POPPEL M. H., SIMPSON R. S. and SUSSMAN M. L. (1949) *Amer. J. Roentgenol.* **62**, 500.

GRONIOWSKI J. (1963) *Arch. Path.* **75**, 144.

GROSS P. and BENZ E. J. (1947) *Surg. Gynec. Obstet.* **85**, 315.

GROSS P., BROWN J. H. V. and HATCH T. F. (1952) *Amer. J. Path.* **28**, 211.

GROSS P. (1953) *Amer. J. clin. Path.* **23**, 116.

GROSS P. and WESTRICK M. L. (1953) *Amer. J. Path.* **29**, 576.

GROSS P. and WESTRICK M. L. (1954) *Amer. J. Path.* **30**, 195.

GROSS P., WESTRICK M. L. and MCNERNEY J. M. (1956) *Amer. J. Path.* **32**, 739.

GROSS P. (1958) *Arch. Path.* **66**, 605.

GROSS P., WESTRICK M. L. and MCNERNEY J. M. (1959) *Arch. Path.* **68**, 252.

GROSS P., WESTRICK M. L. and MCNERNEY J. M. (1960a) *Arch. Path.* **69**, 130.

GROSS P., WESTRICK M. L. and MCNERNEY J. M. (1960b) *Dis. Chest* **37**, 35.

GROSS P., WESTRICK M. L. and MCNERNEY J. M. (1961) *J. occup. Med.* **3**, 258.

GROSS P. (1962) *J. occup. Med.* **4**, 485.

GROSS P. (1962) *Amer. Rev. resp. Dis.* **85**, 828.

GROSS P., MCNERNEY J. M., WESTRICK M. L. and BABYAK MARY A. (1962) *Arch. Path.* **74**, 81.

GROSS P., McNERNEY J. M. and BABYAK MARY A. (1963) *Dis. Chest* **43**, 113.

GROSS P., RINEHART W. E. and HATCH T. (1965) *Arch. environ. Hlth.* **10**, 768.

GROSS P., PFITZER E. A. and HATCH T. F. (1966) *Amer. Rev. Resp. Dis.* **94**, 10.

GROSS P. and TOLKER ETHEL B. (1966) *Arch. Environ. Hlth.* **12**, 213.

GROSS P. and DE TREVILLE R. T. P. (1968) *Arch. Path.* **86**, 255.

GROSS P. and DE TREVILLE R. T. P. (1972) *Amer. Rev. resp. Dis.* **106**, 684.

GROSS P. (1976) *The Biologic Categorization of Inhaled Fibreglass Dust*. In the press.

GROSS R. E. and LEWIS J. E. (1945) *Surg. Gynec. Obst.* **80**, 549.

GROTH D. H., MACKAY G. R., CRABLE J. V. and COCHRAN T. H. (1972) *Arch. Path.* **94**, 171.

GRUENFELD G. E. and GRAY S. H. (1941) *Arch. Path.* **31**, 392.

GRUENWALD P. (1941) *Amer. J. Path.* **17**, 879.

GRUENWALD P. (1948) *Arch. Path.* **46**, 59.

GRUENWALD P. and JACOBI M. (1951) *J. Pediat.* **39**, 650.

GRUNOW W. A. and ESTERLY J. R. (1972) *Chest* **61**, 298.

GUCCION J. G. and ROSEN S. H. (1972) *Cancer (Philadelphia)* **30**, 836.

GUDBJERG C. E. (1961) *Amer. J. Roentgenol.* **86**, 842.

GUDJØNSSON SK. V. and BECKER K. (1936) *J. industr. Hg.* **18**, 215.

GUIDA FILHO B. and PASQUALUCCI M. (1963) *Rev. Brasil Cir.* **45**, 293.

GUILLAN R. A. and ZELMAN S. (1966) *Amer. J. clin. Path.* **46**, 427.

GUNN F. D. and NUNGESTER W. J. (1936) *Arch. Path.* **21**, 813.

GUNN F. D. (1937) *Arch. Path.* **4**, 835.

GUPTA R. K., SCHUSTER R. A. and CHRISTIAN W. D. (1972) *Arch. Path.* **93**, 42.

GURTNER H. P., GERTSCH M., SALZMANN C., SCHERRER M., STUCKI P. and WYSS F. (1968) *Schweiz. med. Wschr.* **98**, 1579, 1695.

GUTHRIE K. J. and MONTGOMERY G. L. (1947) *Lancet* **2**, 752.

GUYTON A. C. and LINDSEY A. W. (1959) *Circulation Res.* **7**, 649.

GYE W. E. and PURDY W. J. (1922) *Brit. J. exp. Path.* **3**, 75, 86.

GYLLENSWÄRD A., LODIN H., LUNDBERG A. and MÖLLER T. (1957) *Pediatrics* **19**, 399.

HAAS J. E., YUNIS E. J. and TOTTEN R. S. (1972) *Cancer (Philadelphia)* **30**, 512.

HAAS P. A. and JOHNSON H. F. (1967) *Ind. Eng. Chem.* **6**, 225 (quoted by STAUB N. C. (1974)).

HABER S. L. and BENNINGTON J. L. (1962) *J. Pediatrics* **61**. 759.

HADDEN H. N. and DIRKEN M. N. J. (1955) *J. Path. Bact.* **70**, 419.

HADFIELD G. (1938) *Lancet* **2**, 710.

HAGE E. (1972) *Acta path. microbiol. scand.* **80A**, 225.

HAGE ESTHER (1973) *Virchows Arch. path. Anat.* **361**, 121.

HAGER H. F., MIGLIACCIO A. V. and YOUNG R. M. (1949) *New Engl. J. Med.* **241**, 226.

HAGERSTAND I. and LINELL F. (1964) *Acta med. scand.* **176** (suppl. 425), 171.

HAHNE O. H. (1964) *Amer. Rev. resp. Dis.* **89**, 566.

HAIGHT C. (1942–3) *J. thorac. Surg.* **11**, 630.

HALASZ N. A., HALLORAN K. H. and LIEBOW A. A. (1956) *Circulation* **14**, 826.

HALASZ N. A., LINDSKOG G. E. and LIEBOW A. A. (1962) *Ann. Surg.* **155**, 215.

HALE L. W., GOUGH J., KING E. J. and NAGELSCHMIDT G. (1956) *Brit. J. industr. Med.* **13**, 251.

HALEY L. D. and McCABE A. (1950) *Amer. J. clin. Path.* **20**, 35.

HALL E. M. (1935) *Amer. J. Path.* **11**, 343.

HALL W. C., KOVATCH R. M. and SCHRICKER R. L. (1973) *J. Path.* **110**, 193.

HALLERMAN W. (1928) *Frankfurt. Z. Path.* **36**, 471.

HALLGREN B., KERSTELL J., RUDENSTAN C. M. and SVANBORG A. (1966) *Acta chir. scand.* **132**, 613.

HAM E. K., GREENBERG D. and REYNOLDS R. C. (1971) *Exp. molec. Path.* **14**, 362.

HAMBURGER M. and ROBERTSON O. H. (1940) *J. exper. Med.* **72**, 261, 275.

HAMER N. A. J. (1963) *Thorax* **18**, 275.

HAMILTON A. G. (1950) *Parasitology* **40**, 46.

HAMILTON B. P. M., UPTON G. V. and AMATRUDA T. T. Jr. (1972) *J. clin. Endocr.* **35**, 764.

HAMILTON J. B. and TYLER G. R. (1946) *Radiology* **47**, 149.

HAMILTON T. R. and ANGEVINE D. M. (1946) *Milit. Surg.* **99**, 450.

HAMILTON W. F., WINSLOW J. A. and HAMILTON W. F. Jr. (1950) *J. clin. Invest.* **29**, 30.

HAMMAN L. and RICH A. R. (1935) *Trans. Amer. clin. climat. Ass.* **51**, 154.

HAMMAN L. and RICH A. R. (1944) *Bull. Johns Hopk. Hosp.* **74**, 177.

HAMMOND E. C. and HORN D. (1954) *J. Amer. med. Ass.* **155**, 1316.

HAMMOND E. C. (1958) *Brit. med. J.* **2**, 649.

HAMMOND E. C. and HORN D. (1958) *J. Amer. med. Ass.* **166**, 1159, 1294.

HAMPERL H. (1931) *Virchows Arch. path. Anat.* **282**, 724.

HAMPERL H. (1937) *Virchows Arch. path. Anat.* **300**, 46.

HAMPERL H. (1956) *Amer. J. Path.* **32**, 1.

HAMPERL H. (1957a) *Virchows Arch. path. Anat.* **330**, 325.

HAMPERL H. (1957b) *J. Path. Bact.* **74**, 353.

HAMPTON A. O. and CASTLEMAN B. (1943) *Amer. J. Roentgenol.* **43**, 305.

HAMPTON A. O., PRANDONI A. G. and KING J. T. (1945) *Bull. Johns Hopk. Hosp.* **76**, 245.

HAND A. (1893) *Arch. Pediat.* **10**, 673.

HANGEN R. K. and BAKER R. D. (1954) *Amer. J. clin. Path.* **24**, 1381.

HANKINSON H. W. and EDWARDS F. R. (1959) *Thorax* **14**, 122.

HANSMAN D. H. and WEIMANN R. B. (1967) *Cancer (Philadelphia)* **20**, 1515.

HANSMAN D., GLASGOW H., STURT J., DEVITT L. and DOUGLAS R. (1971) *New Engl. J. Med.* **284**, 175.

HARBITZ F. (1918) *Arch. intern. Med.* **21**, 139.

HARDING H. E. and OLIVER G. B. (1949) *Brit. J. industr. Med.* **6**, 91.

HARDING H. E. and McLAUGHLIN A. I. G. (1955) *Brit. J. industr. Med.* **12**, 92.

HARDING H. E., McLAUGHLIN A. I. G. and DOIG A. T. (1958) *Lancet* **2**, 394.

HARDY HARRIET L. and TABERSHAW I. R. (1946) *J. industr. Hyg.* **28**, 197.

HARDY HARRIET L. (1955) *Amer. Rev. Tuberc.* **72**, 129.

HARDY HARRIET L. and STOECKLE J. D. (1959) *J. chron. Dis.* **9**, 152

HARDY HARRIET L. (1961) *Amer. J. med. Sci.* **242**, 150.

HARGREAVE F. E., PEPYS J., LONGBOTTOM J. L. and WRAITH D. G. (1966) *Lancet* **1**, 445.

HARKAVY J. (1924) *J. Path. Bact.* **27**, 366.

HARKAVY J. (1941) *Arch. intern. Med.* **67**, 709.

HAROLD J. T. (1952) *Quart. J. Med.* **21** N.S., 353.

HARPER F. R., CONDON W. B. and WIERMAN W. H. (1950) *Arch. Surg. (Chicago)* **61**, 696.

HARRINGTON J. S. (1962) *Nature* **193**, 43.

HARRIS P. (1955) *Brit. Heart J.* **17**, 85.

HARRIS P. and HEATH D. (1962) *The Human Pulmonary Circulation.* E. and S. Livingstone, Edinburgh.

HARRIS W. H. and SCHATTENBERG H. J. (1942) *Amer. J. Path.* **18**, 955.

HARRISON C. V. (1948) *J. Path. Bact.* **60**, 289.

HARRISON C. V. (1951) *J. Path. Bact.* **63**, 195.

HARRISON C. V. (1958) *Brit. J. Radiol.* **31**, 217, 226.

HARRISON C. V. (1958-9) *Lectures on the Scientific Basis of Medicine* VIII. p. 226. Athlone Press, London.

HARRISON E. G. Jr. and BERNATZ P. E. (1963) *Arch. Path.* **75**, 284.

HARRISON M. T., MONTGOMERY D. A. D., RAMSEY A. S., ROBERTSON J. H. and WELBOURN R. B. (1957) *Lancet* **1**, 23.

HARRISON W. and LIEBOW A. A. (1952) *Circulation* **5**, 824.

HÄRTING F. H. and HESSE W. (1879) *Vjschr. gerichtl. Med.* **30**, 296; **31**, 102, 313.

HARTROFT W. S. (1945) *Amer. J. Path.* **21**, 889.

HARVEY A. M. (1937) *Arch. Int. Med.* **59**, 118.

HARVEY C., BLACKET R. B. and READ J. (1957) *Aust. Ann. Med.* **6**, 16.

HARVEY J. C. (1957) *Ann. intern. Med.* **47**, 1067.

HARVEY R. M., FERRER M. I., RICHARDS D. W. and COURAND A. (1951) *Amer. J. Med.* **10**, 719.

HARVEY W. A. (1948) *Ann. intern. Med* **28**, 768.

HARVILL T. H. (1942) *J. Amer. med. Ass.* **119**, 494.

HARZ C. O. (1877) *Central-Thierarzneischule z. München.* Literarisch-artistische Anstalt. Munich.

HASAN F. M. and KAZEMI (1974) *Chest* **65**, 289.

HASELTON P. S., HEATH D. and BREWER D. B. (1968) *J. Path. Bact.* **95**, 431.

HASSAN M. A., RAHMAN E. A. and RAHMAN I. A. (1973) *Brit. med. J.* **1**, 515.

HASTINGS E. V. (1950) *Arch. Path.* **49**, 453.

HASTREITER A. R., PAUL M. H., MOLTHAN M. E. and MILLER R. A. (1962) *Circulation* **25**, 916.

HATTORI S., MATSUDA M. and WADA A. (1965) *Gann* **56**, 275.

HATTORI S., MATSUDA M., TATEISHI R., TATSUMI N. and TERAZAWA T. (1968) *Gann* **59**, 123.

HATTORI S., MATSUDA M., TATEISHI R., NISHIHARA H. and HORAI T. (1972) *Cancer (Philadelphia)* **30**, 1014.

HAUBRICH R. and SCHULER B. (1949–50) *Fortschr. Röntgenstr.* **72**, 68.

HAUCK ANNA J. (1963) *New Engl. J. Med.* **268**, 1356.

HAUSER T. E. and STEER A. (1951) *Ann. int. Med.* **34**, 881.

HAWES S. C. (1952) *Brit. J. Tuberc.* **46**, 176.

HAWORTH SHEILA G., HISLOP A. and REID L. (1974) *J. Pediat.* **84**, 783.

HAYES J. A. and SHIGA A. (1970) *J. Path.* **100**, 281.

HAYHURST R. E. and SCOTT E. (1914) *J. Amer. med. Ass.* **63**, 1570.

HAYMAN L. D. and HUNT R. E. (1952) *Dis. Chest* **21**, 691.

HAYNES F. W., KINNEY T. D., HELLEMS H. K. and DEXTER L. (1947) *Fed. Proc.* **6**, 125.

HAYWARD G. W. (1955) *Brit. med. J.* **1**, 1361.

HAYWARD J. and REID L. M. (1952) *Thorax* **7**, 89.

HEAD J. R. (1951) *Quart. Bull. Northw. Univ. med. Sch.* **25**, 210.

HEAD J. R. (1955) *Amer. J. Surg.* **89**, 1019.

HEANEY J. P., OVERTON R. C. and DE BAKEY M. E. (1957) *J. thorac. Surg.* **34**, 553.

HEARD B. E. (1958) *Thorax* **13**, 136.

HEARD B. E. (1959) *Thorax* **14**, 58.

HEARD B. E. (1960) Demonstration to the 100th meeting of the Path. Soc. of Gt. Britain and Ireland, London. Also a personal communication.

HEARD B. E. (1962) *J. Path. Bact.* **83**, 159.

HEARD B. E., HASSAN A. M. and WILSON S. M. (1962) *J. clin. Path.* **15**, 17.

HEARD B. E. and IZUKAWA T. (1963) in *Fortschritte der Staublungenforschung.* Edited by H. REPLOH and W. KLOSTERKÖTTER, p. 249. Dinslaken.

HEARD B. E. and COOKE R. A. (1968) *Thorax* **23**, 187.

HEARD K. M., POSEY E. L. and LONG J. W. (1958) *Amer. J. Med.* **24**, 157.

HEATH D. and WHITAKER W. (1955) *J. Path. Bact.* **70**, 291.

HEATH D. and WHITAKER W. (1956) *J. Path. Bact.* **72**, 531.

HEATH D., BROWN J. W. and WHITAKER W. (1956) *Brit. Heart J.* **18**, 1.

HEATH D. and WATTS G. T. (1957) *Thorax* **12**, 142.

HEATH D., COX E. V. and HARRIS-JONES J. N. (1957) *Thorax* **12**, 321.

HEATH D. and BEST P. V. (1958) *J. Path. Bact.* **76**, 165.

HEATH D., DU SHANE J. W., WOOD E. H. and EDWARDS J. E. (1958) *Thorax* **13**, 213.

HEATH D. and EDWARDS J. E. (1958) *Circulation* **18**, 533.

HEATH D., SWAN H. J. C., DU SHANE J. W. and EDWARDS J. E. (1958) *Thorax* **13**, 267.

HEATH D., DONALD D. E. and EDWARDS J. E. (1959) *Brit. Heart. J.* **21**, 187.

HEATH D., WOOD E. H., DU SHANE J. W. and EDWARDS J. E. (1959) *J. Path. Bact.* **77**, 443.

HEATH D. and HICKEN P. (1960) *Thorax* **15**, 54.

HEATH D. and THOMPSON I. McK. (1969) *Thorax* **24**, 232.

HEATH D., EDWARDS C. and HARRIS P. (1970) *Thorax* **25**, 129.

HEATH D. (1971) in *High Altitude Physiology: Cardiac and Respiratory Aspects*, Ciba Foundation Symposium, Ed. R. PORTER and J. KNIGHT, p. 52. Churchill-Livingstone, Edinburgh and London.

HEATH D., EDWARDS C., WINSON M. and SMITH P. (1973) *Thorax* **28**, 24.

HEATH D., SMITH P. and HASLETON P. S. (1973) *Thorax* **28**, 551, 559.

HEATH D. (1975) Personal communication.

HEBERT W. M., SEALE R. H. and SAMSON P. C. (1957) *J. thorac. Surg.* **34**, 409.

HECHT H. H., KUIDA H., LANGE R. L., THORNE J. L. and BROWN A. M. (1962) *Amer. J. Med.* **32**, 171.

HECHT V. (1910) *Beitr. path. Anat.* **48**, 262.

HEDBLOM C. A. (1931) *Surg. Gynec. Obstet.* **52**, 406.

HEERUP L. (1927) *Hospitalstidende* **70**, 1165.

HEILMAN R. S., TABAKIN B. S., HANSON J. S. and NAEYE R. L. (1962) *Amer. J. Med.* **32**, 298.

HEIMLICH H. J. (1953) *J. Thorac. Surg.* **26**, 29.

HELBING C. (1898) *Zbl. allg. Path.* **9**, 433.

HELLER A. (1895) *Dtsch. Arch. klin. Med.* **55**, 141.

HELLWIG C. A. (1943) *Arch. Path.* **36**, 95.

HELLY K. (1907) *Z. Heilk. (Prague)* **28**, 105.

HEMSATH F. A. and PINKERTON H. (1956) *Amer. J. clin. Path.* **26**, 36.

HENDERSON R., HISLOP A. and REID L. (1971) *Thorax* **26**, 195.

HENRICI A. T. and GARDNER E. L. (1921) *J. infect. Dis.* **28**, 232.

HENRY E. W. (1952) *Brit. Heart J.* **14**, 406.

HENSCHEN F. (1931) *Acta paediat. (Uppsala)* **12**, Suppl. 6.

HENTEL W., LONGFIELD A. N., VINCENT T. N., FILLEY G. F. and MITCHELL R. S. (1962) *Amer. Rev. resp. Dis.* **87**, 216.

HEPBURN D. (1925) *J. Anat. (Lond.)* **59**, 326.

HEPPLESTON A. G. (1947) *J. Path. Bact.* **59**, 453.

HEPPLESTON A. G. (1953) *J. Path. Bact.* **66**, 235.

HEPPLESTON A. G. (1954a) *J. Path. Bact.* **67**, 51.

HEPPLESTON A. G. (1954b) *J. Path. Bact.* **67**, 349.

HEPPLESTON A. G. (1956) *Thorax* **11**, 77.

HEPPLESTON A. G. (1958) *J. Path. Bact.* **75**, 461.

HEPPLESTON A. G. (1961) in *Inhaled Particles and Vapours.* Edited by C. N. DAVIES, p. 320. Pergamon Press, Oxford.

HEPPLESTON A. G. and LEOPOLD J. G. (1961) *Amer. J. Med.* **31**, 279.

HEPPLESTON A. G. (1962) *Amer. J. Path.* **40**, 493.

HEPPLESTON A. G. (1963) *Amer. J. Path.* **42**, 119.

HEPPLESTON A. G. and MORRIS T. G. (1965) *Amer. J. Path.* **46**, 945.

HEPPLESTON A. G. and STYLES J. A. (1967a) *Nature (Lond.)* **214**, 521.

HEPPLESTON A. G. and STYLES J. A. (1967b) *Fortschr. Staublungenforsch.* **2**, 123.

HEPPLESTON A. G. (1967c) *Nature (Lond.)* **213**, 199.

HEPPLESTON A. G. (1969) *Brit. med. Bull.* **25**, 282.

HEPPLESTON A. G., WRIGHT N. A. and STEWART J. A. (1970) *J. Path.* **101**, 293.

HEPPLESTON A. G. (1972) *Ann. N.Y. Acad. Sci.* **200**, 347.

HEPPLESTON A. G. and YOUNG A. E. (1972) *J. Path.* **107**, 107.

HEPPLESTON A. G. (1973) in *Biology of Fibroblast*, Ed. E. KULONEN and J. PIKKARAINEN, p. 529. Academic Press, London and New York.

HEPTINSTALL R. H. and SALMON M. V. (1959) *J. clin. Path.* **12**, 272.

HERBERT F. A., NAHMIAS B. B., GAENSLER E. A. and MACMAHON H. E. (1962) *Arch. int. Med.* **110**, 628.

HERBUT P. A. (1944) *Amer. J. Path.* **20**, 911.

HERBUT P. A. (1946) *Arch. Path.* **41**, 175.

HEROUT V., VORTEL V. and VONDRÁČKOVÁ A. (1966) *Amer. J. clin. Path.* **46**, 411.

HERRNHEISER G. and HINSON K. F. W. (1954) *Thorax* **9**, 198.

HERS J. F. PH. (1955) *The Histopathology of the Respiratory Tract in Human Influenza.* Verhandelingen van het Nederlands Inst. voor praeventieve Geneeskunde **26**. H. E. Stenfert Kroese N.V., Leiden.

HERS J. F. PH., MASUREL N. and MULDER J. (1958) *Lancet* **2**, 1141.

HERS J. F. PH. and MULDER J. (1961) in a report on "The International Conference on Asian Influenza", Bethesda, in the *Amer. Rev. resp. Dis.* **83**, 84.

HERXHEIMER G. (1901) *Zbl. allg. Path. path. Anat.* **12**, 529.

HESCHL R. (1877) *Wien. med. Wschr.* **27**, 385.

HESS H. (1956) *Arch. klin. Chir.* **283**, 274.

HEUTER C. (1914) *Beitr. path. Anat.* **59**, 520.

HEWER T. F. and HELLER H. (1949) *J. Path. Bact.* **61**, 499.

HICKAM J. B. and CARGILL W. H. (1948) *J. clin. Invest.* **27**, 10.

HICKEY P. M. and SIMPSON W. M. (1926) *Acta radiol. (Stockh.)* **5**, 475.

HICKIE J. B., GIMLETTE T. M. D. and BACON A. P. C. (1956) *Brit. Heart J.* **18**, 365.

HICKS J. D. (1953) *J. Path. Bact.* **65**, 333.

HIEGER I. (1949) *Brit. J. Cancer* **3**, 123.

HIGENBOTTAN T. W. and HEARD B. E. (1976) *Thorax* **31**, 226.

HIGGINS M. R., RANDALL R. E. Jr. and STILL W. J. S. (1974) *Brit. med. J.* **3**, 450.

HILD J. R. (1942) *Amer. J. Dis. Child.* **63**, 126.

HILDEEN T., KROGSGAARD A. R. and VIMTRUP B. (1958) *Lancet* **2**, 830.

HILDING A. C. (1943) *Ann. Otol.* **52**, 5.

HILDING A. C. (1949) *Acta oto-laryng. (Stockh.)* **37**, 138.

1042

REFERENCES

HILDING A. C. (1957) *New Engl. J. Med.* **256**, 634.
HILL A. B. and FANNING E. L. (1948) *Brit. J. industr. Med.* **5**, 1.
HILL G. S. and EGGLESTON J. C. (1972) *Cancer (Philadelphia)* **30**, 1092.
HILL L. D. and WHITE M. L. (1953) *J. thorac. Surg.* **25**, 187.
HIMMELWEIT F. (1943) *Lancet* **2**, 793.
HINDS J. R. and HITCHOCK G. C. (1969) *Thorax* **24**, 10.
HINES L. E. (1922) *J. Amer. med. Ass.* **79**, 720.
HINSON K. F. W., MOON A. J. and PLUMMER N. S. (1952) *Thorax* **7**, 317.
HINSON K. F. W. (1958a) in *Fungous Diseases and their Treatment.* Edited by RIDELL and STEWART. Butterworth, London.
HINSON K. F. W. (1958b) in *Carcinoma of Lung.* Edited by J. R. BIGNALL. E. and S. Livingstone, Edinburgh.
HINSON K. F. W. and NOHL H. C. (1960) *Brit. J. Dis. Chest* **54**, 54.
HINSON K. F. W. and KUPER S. W. A. (1963) *Thorax* **18**, 350.
HINSON K. F. W. (1975) Personal communication.
HIROSE F. M. and HENNIGAR G. R. (1955) *J. thorac. Surg.* **29**, 502.
HIRSCH M. J., KASS I., SCAEFER W. B. and DENST J. (1959) *Arch. intern. Med.* **103**, 814.
HISLOP A. and REID L. (1971) *Thorax* **26**, 190.
HISLOP A. and REID L. (1974) *Thorax* **29**, 90.
HITZ B. and OESTERLIN E. (1932) *Amer. J. Path.* **8**, 333.
H.M.S.O. (Her Majesty's Stationery Office) (1969) *Report on Confidential Enquiries into Maternal Deaths in England and Wales (1964–1966).* H.M.S.O., London.
HOCH W. S., PATCHEFSKY A. S. and TAKEDA M. (1974) *Cancer (Philadelphia)* **33**, 1328.
HOCHBERG L. A. and SCHACTER B. (1955) *Amer. J. Surg.* **89**, 425.
HOCHBURG L. A. and CRASTNOPOL P. (1956) *Arch. Surg.* **73**, 74.
HOCHHEIM K. (1903) *Medicinabrat*, Dr. Johannes Orth zur Feier seines 25 jährigen Professorenjubiläums gewidmet, Berlin, quoted by CLAIREAUX A. E. (1953).
HODSON C. J. (1950) *J. Fac. Radiol.* **1**, 176.
HOFFMAN E. (1962) *Rev. Lat.-amer. anat. Path.* **6**, 49.
HOFFMAN E. F. and GILLIAM A. G. (1954) *Publ. Hlth. Rep. (Wash.)* **69**, 1033.
HOLDEN W. D., SHAW B. W., CAMERON D. B., SHEA P. J. and DAVIS J. H. (1949) *Surg. Gynec. Obstet.* **88**, 23.
HOLLAND R. H., ACEVEDO A. R. and CLARK D. A. (1960) *Bit. J. Cancer* **14**, 169.
HOLLAND W. W. and REID D. D. (1965) *Lancet* **1**, 445.
HOLLEY S. W. (1946) *Milit. Surg.* **99**, 528.
HOLLING H. E. and VENNER A. (1956) *Brit. Heart J.* **18**, 103.
HOLLISTER L. E. and CULL V. L. (1956) *Amer. J. Med.* **21**, 312.
HOLM M. L. and DAVISON W. C. (1919) *Bull. Johns Hopk. Hosp.* **30**, 324.

HOLMGREN A. and SWANBORG N. (1964) *Acta med. Scand.* **176** (suppl. 425), 276.
HOLT P. F. and OSBORNE S. G. (1953) *Brit. J. industr. Med.* **10**, 152.
HOLT P. F. and KING D. T. (1955) *J. chem. Soc.* 773.
HOLT P. F. (1957) *Pneumoconiosis.* E. Arnold, London.
HOLT P. F. and WEST C. W. (1960) *Brit. J. industr. Med.* **17**, 25.
HOLT P. F., MILLS J. and YOUNG D. K. (1964) *J. Path. Bact.* **87**, 15.
HOLZEL A., PARKER L.. PATTERSON W. H., CARTMEL D., WHITE L. L. R., PURDY R., THOMPSON K. M. and TOBIN O'H. (1965) *Brit. med. J.* **1**, 614.
HOMBURGER F. (1943) *Amer. J. Path.* **19**, 797.
HONEY M. and JEPSON E. (1957) *Brit. med. J.* **2**, 1330.
HOOD W. B. Jr., SPENCER H., LASS R. W. and DALEY R. (1968) *Brit. Heart J.* **30**, 336.
HOOPER ANNE D. (1957) *Arch. Path.* **64**, 1.
HÖRA J. (1934) *Frankf. Z. Path.* **47**, 100.
HOURIHANE D. O'B. (1964) *Thorax* **19**, 268.
HOURIHANE D. O'B., LESSOF L. and RICHARDSON P. C. (1966) *Brit. med. J.* **1**, 1069.
HOUSTON C. S. (1960) *New Engl. J. Med.* **263**, 478.
HOUSTON J. C., DE NAVASQUEZ S. and TROUNCE J. R. (1953) *Thorax* **8**, 207.
HOWARD C. P. (1924) *Amer. J. Syph.* **8**, 1.
HOWARD S., KINLEN L. J., LEWINSOHN H. C., PETO J. and DOLL R. (1975) *A Mortality Study among Workers in an English Asbestos Factory.* 18th Int. Congr. on Occupational Health, Brighton, 1975.
HOWARD S. A. and WILLIAMS M. J. (1957) *Cancer (Philadelphia)* **10**, 1182.
HOYLE C., DAWSON J. and MATHER G. (1954) *Lancet* **2**, 164.
HSIUNG C.-C. (1963) *Chinese med. J.* **82**, 390.
HSU C.-W., WU S.-C., CH'EN, C.-S. (1962) *Chinese med. J.* **81**, 263.
HUCK F. F. (1947) *Occup. Med.* **3**, 411.
HÜCLEL R. (1927) *Frankfurt Z. Path.* **35**, 320.
HUDSON R. E. B. (1965) *Cardiovascular Pathology*, Vol. 2, p. 1872. E. Arnold, London.
HUEBNER R. J., ROWE W. P.. WARD T. G., PARROTT R. H. and BELL J. A. (1954) *New Engl. J. Med.* **251**, 1077.
HUEPER W. (1927) *Arch. Path.* **3**, 14.
HUEPER W. C. (1957) *Arch. Path.* **63**, 427.
HUEPER W. C. (1958) *Arch. Path.* **65**, 600.
HUEPER W. C. (1962) *Clin. Pharmacol. Ther.* **3**, 776.
HUEPER W. C., KOTIN P., TABOR E. C., PAYNE W. W., FALK H. and SAWICKI E. (1962) *Arch. Path.* **74**, 89.
HUEPER W. C. and PAYNE W. W. (1962) *Arch. environ. Hlth.* **5**, 445.
HUETER C. (1910) *Klin.-Therap. Wschr. Berlin* **17**, 1135, quoted by WARNER W. P. (1934).
HUGHES J. E. (1930) *Proc. roy. Soc. Med. (Sect. Obstet. and Gynae.)* **23**, 33.
HUGHES J. P. and STOVIN P. G. I. (1959) *Brit. J. Dis. Chest* **53**, 19.
HULTGREN H., SELZER A., PURDY A., HOLMAN E. and GERBODE F. (1953) *Circulation* **8**, 15.

HULTGREN H. N., SPICKARD W. B., HELLRIEGEL K. and HOUSTON C. S. (1961) *Medicine* **40**, 289.

HULTGREN H. N., SPICKARD W. B. and LOPEZ C. (1962) *Brit. Heart J.* **24**, 95.

HUMBERT G. (1904) *Rev. Med.* **24**, 453.

HUNT A. C. (1956) *Thorax* **11**, 287.

HUNTER D., MILTON R., PERRY K. M. A. and THOMPSON D. R. (1944) *Brit. J. industr. Med.* **1**, 159.

HUNTER D. and PERRY K. M. A. (1946) *Brit. J. industr. Med.* **3**, 64.

HUNTER J. B. (1927) *Brit. J. Surg.* **15**, 159.

HUNTER R. A., WILLCOX D. R. C. and WOOLF A. L. (1954) *Guy's Hosp. Rep.* **103**, 196.

HUNTER W. C., SNEEDEN V. D., ROBERTSON T. D. and SNYDER G. A. C. (1941) *Arch. intern. Med.* **68**, 1.

HUNTINGTON G. S. (1919) *Anat. Rec.* **17**, 165.

HUNTINGTON G. S. (1920) *Amer. J. Anat.* **27**, 99.

HURTADO A. (1937) *Rev. méd. peruana* **9**, 3; also *Aspectos fisiópatológicos y patológicos de la vida en la altura.* Imprenta Editora Rimac S.A., Lima.

HURTADO A. (1942) *J. Amer. med. Assoc.* **120**, 1278.

HURTADO A. (1960) *Ann. int. Med.* **53**, 247.

HURWITZ C. (1964) *Acta Un. int. Cancer* **20**, 648.

HURWITZ S. and STEPHENS H. B. (1937) *Amer. J. med. Sci.* **193**, 81.

HUSSON G. S. and WYATT T. C. (1959) *Pediatrics* **23**, 493.

HUSTEN K. (1959) *Arch. Gewerbepath. Gewerbehyg.* **16**, 721.

HUSTON J., WALLACH D. P. and CUNNINGHAM G. J. (1952) *Arch. Path.* **54**, 430.

HUTCHINS G. M. and OSTROW P. T. (1973) *Amer. J. Path.* **70**, 34a.

HUTCHINSON J. (1875) *Illustrations of Clinical Surgery.* J. and A. Churchill, London, p. 42.

HUTCHINSON W. B., FRIEDENBERG M. J. and SALTZSTEIN S. (1964) *Radiology* **82**, 48.

HUTCHISON H. E. (1952) *Cancer (Philadelphia)* **5**, 884.

HUTCHISON W. M., DUNACHIE J. F., SIIM J. CHR. and WORK K. (1970) *Brit. med. J.* **1**, 142.

HYATT R. W., ADELSTEIN E. R., HALAZUN J. F. and LUKENS J. N. (1972) *Amer. J. Med.* **52**, 822.

HYMES A. C., DOE R. P., SHALLCROSS RUTH, LAMUSGA, RITA and LEWIS MARY D. (1962) *Amer. J. Med.* **33**, 398.

IKEDA K. (1936) *Arch. Path.* **22**, 62.

IKEDA K. (1937a) *Amer. J. clin. Path.* **7**, 376.

IKEDA K. (1937b) *Arch. Path.* **23**, 470.

INADA K. and KISHIMOTO S. (1957) *Dis. Chest* **31**, 109.

INADA K. and NAKANO A. (1958) *Arch. Path.* **66**, 183.

INDER SINGH, KAPILA C. C., KHANNA P. K., NANDA R. B. and RAO B. D. P. (1965) *Lancet* **1**, 229.

INDER SINGH, KHANNA P. K., LAL M., HOON R. S. and RAO B. D. P. (1965) *Lancet* **2**, 146.

INKLEY S. R. and ABBOTT G. R. (1961) *Arch. int. Med.* **108**, 903.

INS A. (1878) *Virchows Arch. path. Anat.* **73**, 151 (quoted by GROSS P. and WESTRICK M. (1954)).

IRAVANI J. and VAN AS A. (1972) *J. Path.* **106**, 81.

IRVINE L. G., SIMSON F. W. and STRACHAN A. S. (1930) *International Conference on Silicosis, Johannesburg*, p. 251.

ISAAKSOHN (1871) *Virchows Arch. path. Anat.* **53**, 466.

ISENBERG J. I., GOLDSTEIN H. and KORN A. R. (1968) *New Engl. J. Med.* **279**, 1376.

ISLAND D. P., SHIMIZU N., NICHOLSON W. E., ABE K., OGATA E. and LIDDLE G. W. (1965) *J. clin. Endocr.* **25**, 975.

ISRAEL H. L., MITTERLING R. C. and FLIPPIN H. F. (1948) *New Engl. J. Med.* **238**, 205.

ISRAEL H. L. and DIAMOND P. (1962) *New Engl. J. Med.* **266**, 1024.

ISRAEL J. (1878) *Virchows Arch. path. Anat.* **74**, 15.

ISRAEL J. (1887) *Arch. klin. Chir.* **34**, 160.

ISRAEL K. S., BRASHEAR R. E., SHARMA H. M., YUM M. N. and GLOVER J. L. (1973) *Amer. Rev. resp. Dis.* **108**, 353.

ISRAEL M. S. and HARLEY B. J. S. (1956) *Thorax* **11**, 113.

ISRAEL-ASSELEIN R., CHEBAT J., SORS CH., BASSET F. and LE ROLLAND A. (1965) *Thorax* **20**, 153.

IVANOVA M. G. and OSTROVSKAJA I. S. (1950) *Gigiena* **4**, 21.

IVERSON L. (1954) *J. thorac. Surg.* **27**, 130.

IWASAKI T. (1915–16) *J. Path. Bact.* **20**, 85.

JACKSON C. and JACKSON C. L. (1933) *Ann. Surg.* **97**, 516.

JACKSON C. and JACKSON C. L. (1945) *Diseases of the Nose, Throat and Ear, including Bronchoscopy and Esophagoscopy.* W. B. Saunders, Philadelphia, p. 675.

JACKSON C. (1950) *Dis. Chest* **17**, 125.

JACKSON H. Jr. and PARKER F. (1944) *New Engl. J. Med.* **231**, 35.

JACKSON J. R., GIBBONS R. J. and MAGNER D. (1958) *Amer. J. Path.* **34**, 1051.

JACOB G. and ANSPACH M. (1965) *Ann. N.Y. Acad. Sci.* **132**, 536.

JAFFÉ F. A. (1951) *Amer. J. Path.* **27**, 909.

JAFFÉ H. L. and LICHTENSTEIN L. (1944) *Arch. Path.* **37**, 99.

JÄGER R. (1954) in *Die Staublungenerkrankungen.* Vol. 2, p. 142, Edited by JÖTTEN, KLOSTERKÖTTER and PFEFFERKORN. Steinkopff, Darmstadt.

JAKSCH-WARTENHORST R. (1923) *Wien. Arch. inn. Med.* **6**, 93.

JAMES D. G. and THOMSON A. D. (1955) *Quart. J. Med.* **24 N.S.**, 49.

JAMES D. G. and THOMSON A. D. (1959) *Lancet* **1**, 1057.

JAMES D. G. (1961) *Amer. Rev. resp. Dis.* **84**, No. 5, part 2, p. 15.

JAMES D. G. (1962) *Thorax* **17**, 284.

JAMES I. and PAGEL W. (1944) *Brit. J. Surg.* **32**, 85.

JAMES L. S. (1959) *Pediatrics* **24**, 1069.

JAMES W. R. L. (1954) *Brit. J. Tuberc.* **48**, 89.

JAMES W. R. L. (1955) *Brit. J. industr. Med.* **12**, 87.

JAMISON S. C. and HOPKINS J. (1941) *New Orleans med. surg. J.* **93**, 580.

JANKU J. (1923) *Čas. Lék. čes.* **62**, 1021.

JAQUES W. E. and MARISCAL G. C. (1951) *Bull int. Ass. med. Mus.* **32**, 63.

JARCHO S. (1936) *Arch. Path.* **22**, 674.
JARRETT W. F. H. (1954) *J. Path. Bact.* **67**, 441.
JATLOW P. and RICE J. (1964) *Amer. J. clin. Path.* **42**, 285.
JAWAHRI K. L. and SHAMMA A. (1960) *Dis. Chest* **38**, 569.
JEFFERY P. K. (1973) Ph.D. Thesis, Univ. of London, p. 100.
JELIHOVSKY TATIANA and GRANT A. F. (1968) *Thorax* **23**, 434.
JELIHOVSKY TATIANA (1972) *Pathology* **4**, 65.
JENNINGS F. L. and ARDEN A. (1962) *Arch. Path.* **74**, 351.
JENNNGS G. H. (1937) *Brit. med. J.* **2**. 963.
JENSEN K. E., MINUSE E. and ACKERMANN W. W. (1955) *J. Immunol.* **75**, 71.
JENSEN K. G. and SCHIØDT T. (1958) *Thorax* **13**, 233.
JESIONEK and KIOLEMENOGLOU (1904) *Münch. med. Wschr.* **51**, 1905.
JHA V. K. and RAVINDRAN P. (1972) *Indian J. Chest Dis.* **14**, 109.
JIMENEZ-MARTINEZ M., PERES-ALVAREZ J. J., PERES-TREVINO C., RUBIO-ALVAREZ V. and DE RUBENS J. (1965) *J. thorac. cardiovasc. Surg.* **50**, 59.
JIROVEC O. and VANÉK J. (1954) *Z. allg. Path. path. Anat.* **92**, 424.
JOHNSON H. E. and BATSON R. (1948) *Dis. Chest* **14**, 517.
JOHNSON J. (1957) *J. thorac. Surg.* **34**, 308.
JOHNSON R. T., COOK M. K., CHANOCK R. M. and BUESCHER E. L. (1960) *New Engl. J. Med.* **262**, 817.
JOHNSTON R. F. and GREEN R. A. (1965) *Amer. Rev. resp. Dis.* **91**, 35.
JOHNSTON W. H. and WAISMAN J. (1971) *Arch. Path.* **92**, 196.
JOHNSTONE H. F. (1961) in *Inhaled Particles and Vapours.* Edited by C. N. DAVIES, p. 195. Pergamon Press, Oxford.
JOHNSTONE J. M. and McCALLUM H. M. (1956) *Scot. med. J.* **1**, 360.
JOLIAT G., ABETEL G., SCHINDLER A. M. and KAPANCI Y. (1973) *Virchows Arch. path. Anat.* **358**, 215.
JONES G. P. and JULIAN D. G. (1955) *Brit. med. J.* **2**, 361.
JONES M. C. and THOMAS G. O. (1971) *Thorax* **26**, 652.
JONES W. R. (1933) *J. Hyg. (Lond.)* **33**, 307.
JONES W. R. (1934) *1st Saranac Symposium on Silicosis.* Trudeau Sch. Tuberc., Saranac Lake, N.Y.
JONES W. R. (1949) *Arch. Path.* **48**, 150.
JORESS M. H. and BUSHUEFF B. P. (1952) *Dis. Chest* **22**, 55.
JÖTTEN K. W. and EICKHOFF W. (1944) *Arch. Gewerbepath. Gewerbehyg.* **12**, 223.
JULIANELLE L. A. (1930) *J. exper. Med.* **52**, 439.
JUNGHANSS W. (1959) *Virchows Arch. path. Anat.* **332**, 538.

KAFA V. and BECO V. (1960) *Arch. Dis. Childh.* **35**, 57.
KAGEYAMA K. (1960) *Jap. J. Chest Dis.* **19**, 684.
KAHLAU G. (1947) *Frankfurt. Z. Path.* **59**, 143.
KAINDL F. (1969) *Wien. Z. inn. Med.* **50**, 451.
KAMPMEIER O. F. (1928) *Amer. Rev. Tuberc.* **18**, 360.
KANGAS S. and VIIKARI S. (1960) *Acta chir. scand.* **119**, 463.

KANJUH V. I., SELLERS R. D. and EDWARDS J. E. (1964) *Arch. Path.* **78**, 513.
KAPANCI Y., WEIBEL E. R., KAPLAN H. P. and ROBINSON F. R. (1969) *Lab. Invest.* **20**, 101.
KARLBERG P. J. E. (1957) in *Physiology of Prematurity. Trans. of 2nd Conf., Princeton.* Edited by J. T. LANMAN, New York.
KARRER H. E. (1956) *Bull. Johns Hopk. Hosp.* **98**, 65.
KARRER H. E. (1958) *J. biophys. biochem. Cytol.* **4**, 693.
KARRER H. E. (1960) *J. biophys. biochem. Cytol.* **7**, 357.
KARSHNER and KARSHNER (1920) *Ann. Med.* **1**, 371.
KARSNER H. T. and ASH J. E. (1912) *J. med. Res.* **27**, 205.
KARSNER H. T. and GHOREYEB A. A. (1913) *J. exper. Med.* **18**, 507.
KARSNER H. T. and MEYERS A. E. (1913) *Arch. intern. Med.* **11**, 534.
KARSNER H. T. (1933) *Arch. intern. Med.* **51**, 367.
KARTAGENER M. (1933) *Beitr. klin. Tuberk.* **83**, 489.
KATZ H. L. and AUERBACH O. (1951) *Dis. Chest* **20**, 366.
KATZ I., LEVINE M. and HERMAN P. (1962) *Amer. J. Roentgenol.* **88**, 1084.
KATZ R. I., BIRNBAUM H. and ECKMAN B. H. (1961) *Amer. Rev. resp. Dis.* **84**, 725.
KATZENSTEIN ANNA-LUISE, LIEBOW A. A. and FRIEDMAN P. J. (1975) *Amer. Rev. resp. Dis.* **111**, 497.
KATZNELSEN D., VAWTER G. F., FOLEY G. E. and SHWACHMAN H. (1964) *J. Pediatrics* **65**, 525.
KAUFFMAN S. L., ORES C. N. and ANDERSEN D. H. (1962) *Circulation* **25**, 376.
KAUFMAN G. and KLOPSTOCK R. (1964) *Amer. Rev. resp. Dis.* **88**, 839.
KAUFMAN N. and SPIRO R. K. (1951) *Arch. Path.* **51**, 434.
KAUFMANN E. (1929) *Pathology for Students and Practitioners*, Vol. 1, p. 352. H. K. Lewis, London.
KAWAHATA K. (1938) *Gann* **32**, 367.
KAWAI K. (1959) *Jap. J. exp. Med.* **29**, 359.
KAWAI M., MATSUYAMA T., AMAMOTO H. and NAKAMURE M. (1960) *Medical Report Yawata Iron and Steel Works, Japan* (quoted by HUEPER W. C. (1962)).
KAY J. M., HARRIS P. and HEATH D. (1967) *Thorax* **22**, 176.
KAY J. M. and EDWARDS F. R. (1973) *J. Path.* **111**, 239.
KAY J. M. and GROVER R. F. (1975) *Prog. Resp. Res.* **9**, 157.
KAYE R. L. and SONES D. A. (1964) *Ann. int. Med.* **60**, 653.
KEARNEY M. S. (1969) *J. Path.* **97**, 729.
KEATS T. E. and CRANE J. F. (1954) *Amer. J. Dis. Child.* **88**, 764.
KEELEY J. L. and SCHAIRER A. E. (1960) *Ann. Surg.* **152**, 871.
KEIBEL F. and MALL F. P. (1912) *Human Embryology*, Vol. 2. J. B. Lippincott, Philadelphia.
KELLER A. E., HILLSTROM H. T. and GASS R. S. (1932) *J. Amer. med. Ass.* **99**, 1249.
KELLETT H. S., LIPPHARD D. and WILLIS R. A. (1962) *J. Path. Bact.* **84**, 421.
KELLGREN J. H. (1952) *Brit. med. J.* **1**, 1093, 1152.
KENNAWAY E. L. and KENNAWAY N. M. (1947) *Brit. J. Cancer* **1**, 260.

KENNEDY A. (1973) *J. clin. Path.* **26,** 792.
KENNEDY J. D. and KNEAFSEY D. V. (1959) *Thorax* **14,** 353.
KENNEDY J. H. (1959) *J. thorac. Surg.* **37,** 231.
KENNEDY J. H. and WILLIAMS M. J. (1964) *Ann. Surg.* **160,** 90.
KENT G., GILBERT E. S. and MEYER H. H. (1955) *Arch. Path.* **60,** 556.
KENT T. H. and LAYTON J. M. (1962) *Amer. J. clin. Path.* **38,** 596.
KERNAN S. D. (1927) *Laryngoscope (St. Louis)* **37,** 62.
KERNEN J. A., O'NEAL R. M. and EDWARDS D. L. (1958) *Arch. Path.* **65,** 471.
KETTLE E. H. (1925) *The Pathology of Tumours*, 1st ed. H. K. Lewis, London.
KETTLE E. H. (1926) *J. industr. Hyg.* **8,** 491.
KETTLE E. H. (1932) *J. Path. Bact.* **35,** 395.
KETTLE E. H. and ARCHER H. E. (1933) *Proc. roy. Soc. Med.* **26,** 811.
KEYE J. D. and THOMAS W. A. (1955) *Amer. J. Med.* **19,** 131.
KHEIFETS S. L. (1962) *Klin. Med. (Moscow)* **40,** 39.
KIJIMA S. (1963) *Brit. med. J.* **2,** 451.
KIKUTH W. (1956) *Dtsch. med. Wschr.* **81,** 1633.
KILLINGSWORTH W. P. and HIBBS W. G. (1939) *Amer. J. Dis. Child.* **58,** 571.
KILLINGSWORTH W. P., McREYNOLDS G. S. and HARRISON A. W. (1953) *J. Pediat.* **42,** 466.
KILPATRICK G. S., HEPPLESTON A. G. and FLETCHER C. M. (1954) *Thorax* **9,** 260.
KING D. S. (1941) *Amer. J. Roentgenol.* **45,** 505.
KING E. J. (1934) *Canad. med. Ass. J.* **31,** 237.
KING E. J. (1939) *4th Saranac Symposium on Silicosis.* Trudeau Sch. Tuberc., Saranac Lake, N.Y., p. 33.
KING E. J. and NAGELSCHMIDT G. (1945) *Special Report Series M.R.C.* No. 250, H.M.S.O., London.
KING E. J. (1947) *Occup. Med.* **4,** 26.
KING E. J. and HARRISON C. V. (1948) *J. Path. Bact.* **60,** 435.
KING E. J., MOHANTY G. P., HARRISON C. V. and NAGELSCHMIDT G. (1953) *Brit. J. industr. Med.* **10,** 9.
KING E. J., HARRISON C. V., MOHANTY G. P. and NAGELSCHMIDT G. (1955) *J. Path. Bact.* **69,** 81.
KING E. J., YOGANATHAN M., HARRISON C. V. and MITCHISON D. A. (1957) *Arch. industr. Hlth.* **16,** 380.
KING L. S. (1954) *Arch. Path.* **58,** 59.
KINJO M., WATANABE T. and TANAKA K. (1974) *Chest* **65,** 458.
KINLOCH J. D., WEBB J. N., ECCLESTON D., ZEITLIN J. (1965) *Brit. med. J.* **1,** 1533.
KINNEY T. D. (1942) *Amer. J. Path.* **18,** 799.
KIPKIE G. F. and JOHNSON D. S. (1951) *Arch. Path.* **51,** 387.
KIRBY W. M. M. and McNAUGHT J. B. (1946) *Arch. intern. Med.* **78,** 578.
KIRKLIN J. W., HARP R. A., McGOON D. C. (1964) *J. thorac. cardiovasc. Surg.* **48,** 1026.
KIRSCHNER R. H. and ESTERLY J. R. (1971) *Cancer (Philadelphia)* **27,** 1074.

KISCH B. (1958) *Exp. Med. Surg.* **16,** 17.
KISS F. (1954) *Acta morph. Acad. Sci. hung.* **4,** 537.
KIVILUOTO R. and MEURMAN L. (1969) *Proc. Int. Conf. Pneumoconiosis, Johannesburg (1969).* Ed. H. A. SHAPIRO, p. 190. Oxford University Press, Cape Town.
KLASSEN K. P., MORTON D. R. and CURTIS G. M. (1951) *Surgery* **29,** 483.
KLAUS D. and ZEH E. (1961) *German med. Mth.* **6,** 226.
KLAUS M., REISS O. K., TOLLEY W. H., PIEL C. and CLEMENTS J. A. (1962) *Science* **137,** 750.
KLEIN E. (1901) *J. Hyg. (Lond.)* **1,** 78.
KLEMPERER P. and RABIN C. B. (1931) *Arch. Path.* **11.** 385.
KLINCK G. H. and HUNT H. D. (1933) *Arch. Path.* **15,** 227.
KLINGER H. (1931) *Frankfurt. Z. Path.* **42,** 455.
KLOSTERKÖTTER W. and BÜNEMAN G. (1961) in *Inhaled Particles and Vapours.* Edited by C. N. DAVIES. Pergamon Press, Oxford.
KLOSTERKÖTTER W. and EINBRODT H. J. (1967) in *Inhaled Particles and Vapours.* 2nd Ed. Edited by C. N. DAVIES. Pergamon Press, Oxford.
KLUGE R. C., WICKSMAN R. S. and WELLER T. H. (1960) *Pediatrics* **25,** 35.
KNEELAND Y. and SMETANA H. F. (1940) *Bull. Johns Hopk. Hosp.* **67,** 229.
KNEELAND Y. Jr. and PRICE KATHERINE M. (1960) *Amer. J. Med.* **29,** 967.
KNISELY W. H., WALLACE J. M., MAHALEY M. S. and SATTERWHITE W. M. (1957) *Amer. Heart J.* **54,** 483.
KNISELY W. H. (1960) *Amer. Rev. resp. Dis.* **81,** 735.
KNOWLES H. C., ZEEK P. M. and BLANKENHORN M. A. (1953) *Arch. intern. Med.* **92,** 789.
KNUDTSON K. P. (1960) *Amer. J. clin. Path.* **33,** 310.
KNYVETT A. F. (1966) *Quart. J. Med.* **35** N.S., 313.
KOBAYASHI M., STAHMANN M. A., RANKIN J. and DICKIE H. A. (1963) *Proc. Soc. exp. Biol.* **113,** 472.
KOBAYASHI S. (1921) *Jap. med. World* **1,** 14.
KÖBERLE F. (1959) *Z. Tropenmed. Parasit.* **10,** 308.
KÖBERLE F. (1960) *Verh. dtsch. Ges. Path.* **44,** 139.
KÖBERLE F. (1963) Personal communication.
KODOUSEK R., VOJTEK V., ŠERÝ Z., VORTEL V., FINGER-LAND R., HÁJEK V. and KUČERA K. (1970) *Cas. Lék. čès.* **109,** 923.
KODOUSEK R., VORTEL V., FINGERLAND A., VOJTEK V., ŠERÝ Z., HÁJEK V. and KUČERA K. (1971) *Amer J. clin. Path.* **56,** 394.
KODOUSEK R. (1974) *Adiaspiromycosis,* Acta Univ. Palackianae Olomucensis Facultatis Medicae 70, Palacky University, Olomouc, Czechoslovakia.
KOFFLER D. (1964) *Arch. Path.* **78,** 267.
KOHN P. M., TAGER M., SIEGEL M. L. and ASHE R. (1951) *New Engl. J. Med.* **245,** 640.
KOINO S. (1922) *Jap. med. World* **2,** 317.
KOLLER T. and MÜLLER P. (1957) *Schweiz. med. Wschr.* **87,** 1413.
KÖLLIKER A. (1881) *Verhandl. Phys.-Med. Ges. Würzburg* **16,** 1.
KONWALER B. E. (1950) *Amer. J. clin. Path.* **20,** 385.

KONWALER B. E. and REINGOLD I. M. (1952) *Cancer (Philadelphia)* **5**, 525.
KOONTZ A. R. (1925) *Bull. Johns Hopk. Hosp.* **37**, 340.
KORN D., BENSCH K., LIEBOW A. A. and CASTLEMAN B. (1960) *Amer. J. Path.* **37**, 641.
KORN D., GORE I., BLENKE A. and COLLINS D. P. (1962) *Amer. J. Path.* **40**, 129.
KOTIN P. and FALK H. L. (1960) *Cancer (Philadelphia)* **13**, 250.
KOTIN P., FALK H. L. and COURINGTON D. P. (1963) *Meeting of Amer. Coll. Phys. Detroit.*
KOUNTZ W. B. and ALEXANDER H. L. (1928) *Arch. Path.* **5**, 1003.
KOUNTZ W. B. and ALEXANDER H. L. (1934) *Medicine (Baltimore)* **13**, 251.
KOVÁTS F. (1965) *Die Toxomycose der Lunge* Akadémiai Kiadó, Budapest.
KOZLOWA E. W. and RODIONOW W. W. (1963) *Nowotwory* **13**, 233.
KRAHL V. E. (1960) *Bull. Sch. Med. Maryland* **45**, 36.
KRAHL V. E. (1962) *Med. Thorac.* **19**, 194 and Ciba Symposium on *Pulmonary Structure and Function* 1962, p. 53. J. & A. Churchill, London.
KRAHL V. E. (1963) *Angiology* **14**, 149.
KRAHL V. E. (1968) *Aspen Emphysema Conference* **9**, 141.
KRAKOWER C. (1936) *Arch. Path.* **22**, 113.
KRAMER R. and SOM M. L. (1936) *Arch. Otolaryng. (Chicago)* **23**, 526.
KRAMER R. (1939) *Ann. Otol. (St. Louis)* **48**, 1083.
KRAUS A. R., MELNICK P. J. and WEINBERG J. A. (1948) *J. thorac. Surg.* **17**, 382.
KRAUS O. (1957) *Med. Mschr.* **11**, 433 (434).
KREYBERG L. (1955) *Brit. J. Cancer* **9**, 495.
KREYBERG L. (1956) *Brit. J. prev. soc. Med.* **10**, 145.
KRUMWIEDE C., MCGRATH M. and OLDENBUSCH C. (1930) *Science* **71**, 268.
KUHN C., GYORKEY F., LEVINE B. E. and RAMIREZ-RIVERA J. (1966) *Lab. Invest.* **15**, 492.
KUHN C. (1972) *Cancer (Philadelphia.)* **30**, 1107.
KUHN C. and KUO T.-T. (1973) *Arch. Path.* **95**, 190.
KUIDA H., DAMMIN G. J., HAYNES F. W., RAPAPORT E. and DEXTER L. (1957) *Amer. J. Med.* **23**, 166.
KURODA S. (1937) *Industr. Med. Surg.* **6**, 304.
KUROYA M., ISHIDA N. and SHIRATORI T. (1953) *Tohoku J. exp. Med.* **58**, 62.
KWITTKEN J. and REINER L. (1962) *Pediatrics* **30**, 759.

LACAZ C. DA S. (1956) *An. Fac. Med. S. Paulo* **29**, 9.
LACAZ C. DA S. (1959) *Rev. Inst. Med. trop. S. Paulo* **1**, 150.
LACQUET L. K., FORNHOFF M., DIERICK R. and BUYSSENS N. (1971) *Thorax* **26**, 68.
LADD V. E. and SCOTT H. W. (1944) *Surgery* **16**, 815.
LADSTÄTTER L. (1954) *Klin. Wschr.* **32**, 1044.
LAENNEC R. T. H. (1819) *De l'auscultation mediate, ou traite du diagnostic des maladies des poumons et du coeur, fondé principalement sur ce nouveau moyen d'exploration*, Vol. 2. J. A. Brosson et J. S. Chaudé, Paris (French Edition).

LAENNEC R. T. H. (1829) *A Treatise on the Diseases of the Chest and on Mediate Auscultation.* Translated by J. Forbes. 1st ed. T. G. Underwood, London.
LAENNEC R. T. H. (1834) 4th edition of above English translation.
LAFFITTE H. (1937) *Mém. Acad. Chir.* **63**, 1076.
LAIPPLY T. C. and FISHER C. I. (1949) *Arch. Path.* **48**, 107.
LAIPPLY T. C., SHERRICK J. C. and CAPE W. E. (1955) *Arch. Path.* **59**, 35.
LAITINEN H., SAKSANEN S. and VIRKKUNEN M. (1964) *Ann. Chir. Gynaec. Fenn.* **53**, 4.
LAMBERT MARGARET W. (1955) *J. Path. Bact.* **70**, 311.
LAMBERTY J., HOFFMAN E. and PIZZOLATO P. (1973) *Amer. J. Path.* **70**, 34a.
LANCET (1972) **1**, 672.
LANCET (1974) **2**, 504, 706.
LANDER F. P. L. and DAVIDSON M. (1938) *Brit. J. Radiol.* **11**, 65.
LANE D. (1976) In the press.
LANE S. D. (1971) *Radiology* **101**, 291.
LANGER H. (1963) *Obstet. Gynecol.* **21**, 318.
LAPIN J. H. (1943) *Whooping Cough.* C. C. Thomas, Springfield, Ill.
LAPP H. (1951) *Frankfurt. Z. Path.* **62**, 537.
LAPTEV A. A. (1945) *Klin. Med. (Mosk.)* **23**, 75.
LARRABEE W. F., PARKER R. L. and EDWARDS J. E. (1949) *Proc. Mayo Clin.* **24**, 316.
LARSELL O. and DOW R. S. (1933) *Amer. J. Anat.* **52**, 125.
LARSEN M. C., DIAMOND H. D. and COLLINS H. S. (1959) *Arch. intern. Med.* **103**, 712.
LASNITZKI I. (1958) *Brit. J. Cancer* **12**, 547.
LASSER R. P. and LOEWE L. (1954) *Amer. Heart J.* **48**, 801.
LATTES R., SHEPARD F., TOVELL H. and WYLIE R. (1956) *Surg. Gynec. Obst.* **103**, 552.
LAUBSCHER F. A. (1969) *Amer. J. clin. Path.* **52**, 599.
LAURELL C. B. and ERIKSSON S. (1963) *Scand. J. clin. Lab. Invest.* **15**, 132.
LAURENCE K. M. (1955) *J. Path. Bact.* **70**, 325.
LAURENCE K. M. (1958) Personal communication.
LAURENCE K. M. (1959) *J. clin. Path.* **12**, 62.
LAURENZI G. A., POTTER R. T. and KASS E. H. (1961) *New Engl. J. Med.* **265**, 1273.
LAURENZI G. A., BERMAN L., FIRST M. and KASS E. H. (1964) *J. clin. Invest.* **43**, 759.
LAUTENBACHER M. (1918) *Ann. Méd.* **5**, 435.
LAUWERYNS J. M. (1966) *Science* **153**, 1275.
LAUWERYNS J. M., ST. CLAESSENS and BOUSSAUW L. (1968) *Pediatrics* **41**, 917.
LAUWERYNS J. M. and BOUSSAUW L. (1969) *Lymphology* **2**, 108.
LAUWERYNS J. M. and PEUSKENS J. C. (1969) *Life Sci.* **8**, 577.
LAUWERYNS J. M. (1970) *Human Path.* **1**, 175.
LAUWERYNS J. M., BAERT A. and BOUSSAUW L. (1970) *Amer. Rev. resp. Dis.* **101**, 448.
LAUWERYNS J. M., PEUSKENS J. C. and COKELAERE M. (1970) *Life Sci.* **9**, 1417.
LAUWERYNS J. M. and PEUSKENS J. C. (1972) *Anat. Rec.* **172**, 471.

LAUWERYNS J. M. and BAERT J. H. (1974) *Ann N.Y. Acad. Sci.* **221**, 244.

LAWS J. W. and HEARD B. E. (1962) *Brit. J. Radiol.* **35**, 750.

LAWSON R. M., RAMANATHAN L., HURLEY G., HINSON K. W. and LENNOX S. C. (1976) *Thorax* **31**, 245.

LEAKE ELEANOR, SMITH W. G. and WOODLIFF H. J. (1963) *Lancet* **2**, 432.

LEATHER H. M. (1965) *Lancet* **1**, 270.

LEBACQ E., LAUWERYNS J. and BILLIET L. (1964) *Brit. J. Dis. Chest* **58**, 31.

LEBERT H. (1845) *Physiologie Pathologique*, Vol. 2, p. 213. Baillière, Paris. Quoted by ADAMS M. J. T. (1957).

LE BRIGAND H., GRANJON A., RENAULT P., ROUSSEL A., CHRETIEN J., HOURTOULE R. and IANOTTI C. (1959) *J. franç. Méd. Chir. thor.* **13**, 511.

LE COMPTE P. M. and MEISSNER W. A. (1947) *Amer. J. Path.* **23**, 673.

LEDERER H. and TODD R. McL. (1949) *Arch. Dis. Childh.* **24**, 200.

LEE S.-C. and JOHNSON H. A. (1975) *Thorax* **30**, 178.

LEE S. L., ROSNER F., RIVERO L., FELDMAN F. and HURWITZ A. (1965) *New Engl. J. Med.* **272**, 761.

LEES A. W. and McNAUGHT W. (1959) *Lancet* **2**, 112.

LEES A. W. and McSWAN N. (1962) *Amer. Rev. resp. Dis.* **86**, 648.

LEGG M. A. (1972) *Amer. J. clin. Path.* **57**, 273.

LE GOLVAN D. P. and HEIDELBERGER K. P. (1973) *Arch. Path.* **95**, 344.

LEICHER F. (1949) *Zbl. allg. Path. path. Anat.* **85**, 49.

LEI HSÜEU-HSI, YEN CHIA-KUEI and LIU CH' ANG-MAO (1955) *Chinese J. Path.* **1**, 166.

LEINWAND I., DURYEE A. W. and RICHTER M. N. (1954) *Ann. intern. Med.* **41**, 1003.

LENDRUM A. C., SCOTT L. D. W. and PARK S. D. S. (1950) *Quart. J. Med.* **19** N.S., 249.

LENNOX B. and PRICHARD S. (1950) *Quart. J. Med.* **19** N.S., 97.

LEOPOLD J. G. and GOUGH J. (1957) *Thorax* **12**, 219.

LEOPOLD J. G. (1959) *Demonstration to the 99th Meeting of the Path. Soc. of Gt. Britain and Ireland, Cardiff.*

LEOPOLD J. G. and GOUGH J. (1963) *Thorax* **18**, 172.

LEPER M. H., KOFMAN S., BLATT N., DOWLING H. F. and JACKSON G. G. (1956) *Antibiot. Chemotherap.* **4**, 829.

LERCHE W. (1927) *Arch. Surg.* (*Chicago*) **14**, 285.

LERNER A. M., KLEIN J. O., LEVIN H. S. and FINLAND M. (1960) *New Engl. J. Med.* **263**, 1265.

LE ROUX B. T. (1959) *Quart. J. Med.* **28** N.S., 1.

LE ROUX B. T. (1962) *Thorax* **17**, 111.

LE ROUX B. T. (1969) *Thorax* **24**, 91.

LE ROUX B. T., WILLIAMS M. A. and KALLICHURUM S. (1969) *Thorax* **24**, 673.

LEROY E. P., LIEBNER E. J. and JENSIK R. J. (1965) *Amer. J. Path.* **46**, 26a.

LESSER A. (1877) *Archiv. path. Anat. Physiol. Klin. Med.* **69**, 404.

LETTERER E. (1924) *Frankfurt Z. Path.* **30**, 377.

LEVIN B. and WHITE H. (1961) *Radiology* **76**, 894.

LEVINE O. R., HARRIS R. C., BLANC W. A., WIGGER H. J. and MELLINS R. B. (1973) *J. Pediatrics* **83**, 964.

LEVINE S. B. (1973) *Arch. Path.* **96**, 183.

LEVINTHAL W. (1930) *Klin. Wschr.* **9**, 654.

LEWIS A. G., DUNBAR F. P. LASCHE E. M., BOND J. O., LERNER E. N., WHARTON D. J., HARDY A. V. and DAVIES R. (1959) *Amer. Rev. resp. Dis.* **80**, 188.

LEWIS J. F., ARNOLD C. and ALEXANDER J. (1973) *Amer. J. clin. Path.* **59**, 388.

LI Y. J. and LIU Y. M. (1958) *Nat. med. J. China* **44**, 6.

LICHTENSTEIN L. and JAFFÉ H. L. (1943) *Amer. J. Path.* **19**, 553.

LICHTENSTEIN L. and FOX L. J. (1946) *Amer. J. Path.* **22**, 665.

LICHTENSTEIN L. (1953) *Arch. Path.* **56**, 84.

LICHTY D. E. (1934) *Arch. intern. Med.* **53**, 379.

LIDDLE G. W. (1966) *Scientific Proceedings of the 63rd Meeting of the American Association of Pathologists and Bacteriologists*, also *Amer. J. Path.* **48**, 48a.

LIEBERMAN J. (1959) *New Engl. J. Med.* **260**, 619.

LIEBERMAN J. (1960) *Pediatrics* **25**, 419.

LIEBERMAN J. and KELLOGG F. (1960) *New Engl. J. Med.* **262**, 999.

LIEBOW A. A., HALES M. R. and LINDSKOG G. E. (1948) *Amer. J. Path.* **24**, 691.

LIEBOW A. A., WARREN S. and DE COURSEY E. (1949a) *Amer. J. Path.* **25**, 853.

LIEBOW A. A., HALES M. R., HARRISON W., BLOOMER W. and LINDSKOG G. E. (1949b) *Yale J. Biol. Med.* **22**, 637.

LIEBOW A. A., HALES M. R. and LINDSKOG G. E. (1949c) *Amer. J. Path.* **25**, 211.

LIEBOW A. A., HARRISON W. and HALES M. R. (1950) *Bull. int. Ass. med. Mus.* **31**, 1.

LIEBOW A. A. (1952) *Atlas of Tumor Pathology*, Sect. 5, Fasc. 17. Armed Forces Inst. Path. Washington, D.C.

LIEBOW A. A., LORING W. E. and FELTON W. L. (1953) *Amer. J. Path.* **29**, 885.

LIEBOW A. A. and HUBBELL D. S. (1956) *Cancer (Philadelphia)* **9**, 53.

LIEBOW A. A. (1959a) *Amer. Rev. resp. Dis.* **80**, 67.

LIEBOW A. A. (1959b) in *Pathology of the Heart*. Edited by GOULD S. E. 2nd ed. Blackwell Scientific Publications, Oxford.

LIEBOW A. A., STARK J. E., VOGEL J. and SCHAFFER K. E. (1959) *U.S. Armed Forces med. J.* **10**, 265.

LIEBOW A. A. (1960) *Advances int. Med.* **10**, 329.

LIEBOW A. A. (1962) 59th Scientific Proc. Amer. Assoc. Pathol. and Bacteriol. Montreal (*Amer. J. Path.* **41**, 127).

LIEBOW A. A. and CASTLEMAN B. (1963) *Amer. J. Path.* **43**, 13a.

LIEBOW A. A. (1965) *Amer. J. Anat.* **117**, 19.

LIEBOW A. A., STEER A., BILLINGSLEY J. G. (1965) *Amer. J. Med.* **39**, 369.

LIEBOW A. A. (1966) Personal communication.

LIEBOW A. A. and CARRINGTON C. B. (1966) *Trans. Stud. Coll. Phycns. Philad.* **34**, 47.

LIEBOW A. A. (1967) Personal communication.

LIEBOW A. A. (1968) in *The Lung*, Ed. A. A. LIEBOW and D. E. SMITH. Int. Acad. Path. Monograph 8, Williams & Wilkins Co., Baltimore, p. 349.

LIEBOW A. A. and CARRINGTON C. B. (1968) in *Frontiers of Pulmonary Radiology*. Ed. SIMON POCHEN and LE MAY, p. 102. Grune & Stratton, New York.

LIEBOW A. A. and CARRINGTON C. B. (1969) *Medicine* **48**, 251.

LIEBOW A. A. and CASTLEMAN B. (1971) *Yale J. Biol. Med.* **43**, 213.

LIEBOW A. A., CARRINGTON C. R. B. and FRIEDMAN P. J. (1972) *Human Path.* **3**, 457.

LIEBOW A. A. (1973) *Amer. Rev. resp. Dis.* **108**, 1.

LIEBOW A. A., MOSER K. M. and SOUTHGATE M. T. (1973) *J. Amer. med. Ass.* **233**, 1243.

LIEBOW A. A. (1975) in *Progress in Respiration Research, Alveolar Interstitium of the Lung*. Ed. F. BASSET and R. GEORGES. S. Karger, Basel, p. 1, Vol. 8.

LIEUTAND J. (1767) quoted by ROBERTSON H. E. (1924).

LIGGINS G. C. and PHILLIPS L. I. (1963) *Brit. med. J.* **1**, 711.

LIGNAC G. O. E. (1924) *Dtsch. Arch. klin. Med.* **145**, 139.

LILIENTHAL H. (1932–3) *J. thorac. Surg.* **2**, 599.

LILLEHEI C. W. (1958) *Mod. Concepts Cardiovasc. Dis.* **27**, 441.

LIND J. and WEGELIUS C. (1949) *Pediatrics* **4**, 391.

LIND J. and WEGELIUS C. (1954) *Quart. Biol.* **19**, 109 (a report on the Cold Spring Harbor Symposium).

LIND J. and WEGELIUS C. (1956) *Proc. 8th International Congress of Pediatrics* (*Copenhagen*).

LINDBERG K. (1935) *Arb. path. Inst. Univ. Helsingfors* **9**, 1.

LINDSAY M. I., HERRMANN E. C., MORROW G. W. and BROWN A. L. Jr. (1970) *J. Amer. med. Ass.* **214**, 1825.

LINDSKOG G. E. (1951) *Yale J. Biol. Med.* **23**, 310.

LINENTHAL H. and TALKOV R. (1941) *New Engl. J. Med.* **224**, 682.

LINHARTOVA A. and CHUNG W. (1963) *J. clin. Path.* **16**, 56.

LINHARTOVÁ A., ANDERSON A. E. and FORAKER A. G. (1973) *Arch. Path.* **95**, 45.

LISA J. R. and ROSENBLATT M. B. (1943) *Bronchiectasis*. Oxford University Press, New York.

LITTLEFIELD J. B. and DRASH E. C. (1959) *J. thorac. Surg.* **37**, 745.

LITTMAN M. L. and SCHNEIERSON S. S. (1954) quoted by KATZ R. I. *et al.* (1961).

LIU C. (1957) *J. exp. Med.* **106**, 455.

LIU C., EATON M. D. and HEYL J. T. (1959) *J. exp. Med.* **109**, 545.

LIVINGSTONE J. L., LEWIS J. G., REID L. and JEFFERSON K. E. (1964) *Quart. J. Med.* **33** N.S., 71.

LJUNGDAHL M. (1928) *Dtsch. Arch. klin. Med.* **160**, 1.

LOCKWOOD A. L. (1922) *Surg. Gynec. Obstet.* **35**, 461.

LOEHRY C. A. (1966) *Brit. med. J.* **1**, 1327.

LOESCHCKE H. (1921) *Beitr. path. Anat.* **68**, 213.

LOESCHCKE H. (1928) in *Handbuch der Spez. Path. Anat. u. Histol.* Edited by HENKE F. and LUBARSCH O. Vol. 3/I. Atmungswege und Lungen, Springer, Berlin.

LOESCHCKE H. (1931) *Beitr. path. Anat.* **86**, 201.

LÖFFLER W. (1932) *Beitr. klin. Tuberk.* **79**, 368.

LÖFFLER W. (1936) *Schweiz med. Wschr.* **66**, 1069.

LOKER E. F. Jr., HODGES G. R. and KELLY D. J. (1974) *Chest* **66**, 197.

LONDERO A. T. and RAMOS C. D. (1972) *Amer. J. Med.* **52**, 771.

LOOSLI C. G. (1941) *Amer. J. Path.* **17**, 454.

LOOSLI C. G. and POTTER E. L. (1951) *Anat. Rec.* **109**, 320.

LOOSLI C. G. (1968) *Yale J. Biol. Med.* **40**, 522.

LOPEZ DE FARIA J. (1954a) *Amer. J. Path.* **30**, 167.

LOPEZ DE FARIA J. (1954b) *J. Path. Bact.* **68**, 589.

LOPEZ DE FARIA J. (1956) *Amer. J. trop. Med. Hyg.* **5**, 860.

LOPEZ DE FARIA J., CZAPSKI J., LEITE M. O. R., PENNA D. DE O., FUJIOKA T. and CINTRA A. B. DE U. (1957) *Amer. Heart J.* **54**, 196.

LORE J. M., MADDEN J. L. and GEROLD F. P. (1958) *Cancer (Philadelphia)* **11**, 24.

LOURDES R., LARAYA-CUASAY, DEFOREST A., PALMER J., HUFF D. S., LISCHNER H. W. and HUANG N. N. (1974) *Amer. Rev. resp. Dis.* **109**, 703.

LOURIA D. B., BLUMENFELD H. L., ELLIS J. T., KILBOURNE E. D. and ROGERS D. E. (1959) *J. clin. Invest.* **38**, 213.

LOURIA D. B., STIFF D. P. and BENNETT B. (1962) *Medicine* **41**, 307.

LOW F. N. (1953) *Anat. Rec.* **117**, 241.

LOW F. N. and SAMPAIO M. M. (1957) *Anat. Rec.* **127**, 51.

LOW F. N. (1961) *Anat. Rec.* **139**, 105.

LOWELL L. M. and TUHY J. E. (1949) *J. thorac. Surg.* **18**, 476.

LÖWY J. (1929) *Med. Klinik.* **25**, 141.

LUBARSCH O. (1898) *Virchows Arch. path. Anat.* **151**, 546.

LUBARSCH O. and PLENGE K. (1931) in *Handbuch der spez. Path. Anat. u. Histol.* Edited by HENKE F. and LUBARSCH O., Vol. 3, p. 607. Springer, Berlin.

LUCCHESI P. F., LA BOCCETTA A. C. and PEALE A. R. (1947) *Amer. J. Dis. Child.* **73**, 44.

LUCHSINGER P. C., MOSER K. M., BÜHLMANN K. M. and ROSSIER P. H. (1957) *Amer. Heart J.* **54**, 106.

LUDIN M. and WERTHEMANN A. (1930) *Strahlentherapie* **38**, 684.

LUDLAM G. B. and BEATTIE C. P. (1963) *Lancet* **2**, 1136.

LUKES R. J., BUTLER J. J. and HICKS E. B. (1966) *Cancer* **19**, 317.

LUNA M. A., BEDROSSIAN C. W. M., LICHTIGER B. and SALEM P. A. (1972) *Amer. J. clin. Path.* **58**, 501.

LUNDBERG G. D. (1936) *J. Amer. med. Ass.* **184**, 915.

LUOTO L. (1960) *Pub. Hlth. Rep. (Wash.)* **75**, 135.

LUTZ A. (1908) *Brasil-méd.* **22**, 121.

LYNCH K. M. and McIVER F. A. (1954) *Amer. J. Path.* **30**, 1117.

LYNCH K. M. (1955) *Arch. industr. Hlth.* **11**, 185.

LYNCH M. J. G., RAPHAEL S. S. and DIXON T. P. (1959) *Arch. Path.* **67**, 68.

LYNN R. and TERRY R. D. (1964) *Amer. J. Med.* **37**, 987.

LYONS H. A., VINIJCHAIKUL K. and HENNIGAR G. R. (1961) *Arch. int. Med.* **108**, 929.

LYONS H. A., HUANG C.-T. and HENNIGAR G. R. (1964) *Ann. int. Med.* **60**, 740.

McADAMS A. J. (1955) *Amer. J. Med.* **19**, 314.

McBURNEY R. P., JAMPLIS R. W. and HEDBERG G. (1955) *J. thorac. Surg.* **29**, 271.

McCALL R. E. and HARRISON W. (1955) *J. thorac. Surg.* **29**, 317.

MacCALLUM D. K., PATEK P. R. and BERNICK S. (1966) *Arch. Path.* **81**, 509.

MacCALLUM W. G. (1920–1a) *Johns Hopk. Hosp. Rep.* **20**, 1.

MacCALLUM W. G. (1920–1b) *Johns Hopk. Hosp. Rep.* **20**, 149.

MacCALLUM W. G. (1936) *A Textbook of Pathology*, 6th ed. Saunders, Philadelphia, p. 1032.

MacCALLUM W. G. (1942) *A Textbook of Pathology*, 7th ed. Saunders, Philadelphia, p. 781.

MacCARTHY K., MITUS A., CHEATHAM W. and PEEBLES T. (1958) *Amer. J. Dis. Child.* **96**, 500.

McCARTHY K. and TOSOLINI F. A. (1975) *Lancet* **1**, 649.

McCARTNEY J. S. (1934) *Amer. J. Path.* **10**, 709.

McCARTNEY J. S. (1936) *Amer. J. Path.* **12**, 751.

McCAUGHEY W. T. E. (1958) *J. Path. Bact.* **76**, 517.

McCAUGHEY W. T. E. and THOMAS B. J. (1962) *Amer. J. clin. Path.* **38**, 577.

McCOMBS R. P. (1972) *New Engl. J. Med.* **286**, 1245.

McCONAGHIE R. J. (1962) *J. thorac. cardiovasc. Surg.* **43**, 303.

McCORDOCK H. A. and MUCKENFUSS R. S. (1933) *Amer. J. Path.* **9**, 221.

McCORDOCK H. A. and SMITH M. G. (1936) *J. exp. Med.* **63**, 303.

McCORMACK L. J., HAZARD J. B., EFFLER D. B., GROVES L. K. and BELOVICH D. (1955) *J. thorac. Surg.* **29**, 277.

McCOY G. W. and CHAPLIN C. W. (1912) *J. infect. Dis.* **10**, 61.

McCRACKEN B. H. (1948) *Thorax* **3**, 45.

McCUSKER J. J. and PARSONS D. B. (1962) *Arch. Path.* **74**, 127.

McDANIEL H. G., PITTMAN J. A., HILL S. R. and STARNES W. R. (1963) *Amer. J. Med.* **35**, 427.

MacDONALD G., PIGGOT A. P., GILDER F. W., ARNALL F. A. and RUDGE E. A. (1930) *Lancet* **2**, 846.

McDONALD J. C., McDONALD A. D., GIBBS G. W., SIEMIATYSKI J. and ROSSITER C. E. (1971) *Arch. environ. Hlth.* **22**, 677.

McDONALD J. R., HARRINGTON S. W. and CLAGETT O. T. (1945) *J. thorac. Surg.* **14**, 128.

McDONALD J. R., HARRINGTON S. W. and CLAGETT O. T. (1949) *J. thorac. Surg.* **18**, 97.

McDONALD L., EMANUEL R. and TOWERS M. (1959) *Brit. Heart J.* **21**, 279.

McDONALD S. and HEATHER J. C. (1939) *J. Path. Bact.* **48**, 533.

McDOUGAL J. S. and AZAR H. A. (1972) *Arch. Path.* **93**, 13.

McDOUGAL JENNIFER and SMITH J. F. (1975) *J. Path.* **115**, 245.

McEACHERN C. G., SULLIVAN R. E., ARATA J. E., GRIEST W. D. and SMITH R. B. (1955) *J. thorac. Surg.* **29**, 368.

McFADDEN E. R. Jr. and LUPARELLO F. (1969) *Thorax* **24**, 500.

MacFARLANE A. and DAVIES D. (1973) *Thorax* **28**, 768.

MacFARLANE P. S. and SOMMERVILLE R. G. (1957) *Lancet* **1**, 770.

MacFARLANE R. G., OAKLEY C. L. and ANDERSON C. G. (1941) *J. Path. Bact.* **52**, 99.

McGAFF C. J., ROSS R. S. and BRAUNWALD E. (1962) *Amer. J. Med.* **33**, 201.

McGARVAN M. H., BEARD C. W., BERENDT R. F. and NAKAMURA R. M. (1962) *Amer. J. Path.* **40**, 653.

McGINN S. and WHITE P. D. (1935) *J. Amer. med Ass.* **104**, 1473.

MacGREGOR AGNES R. (1936) *Arch. Dis. Childh.* **11**, 195.

MacGREGOR AGNES R. (1939a) *Arch. Dis. Childh.* **14**, 323.

MacGREGOR AGNES R. (1939b) *Arch. Dis. Childh.* **14**, 336.

MacGREGOR AGNES R. (1947) *Scottish Scientific Advisory Committee "Neonatal Deaths due to Infection, report of a subcommittee".* H.M.S.O., Edinburgh.

MacGREGOR CATHERINE S., JOHNSON R. S. and TURK K. A. D. (1960) *Thorax* **15**, 198.

McGUIRE J., WESTCOTT R. N. and FOWLER N. O. (1951) *Tr. Ass. Amer. Phys.* **64**, 404.

MACHLE W. and GREGORIUS F. (1948) *Publ. Hlth. Rep. (Wash.)* **63**, 1114.

MACIEIRA-COELHO E. and DUARTE C. S. (1967) *Amer. J. Med.* **43**, 944.

MacINTOSH D. J., SINNOTT J. C., MILNE I. G. and REID E. A. S. (1958) *Ann. intern. Med.* **49**, 1294.

MacINTYRE I. M. and RUCKLEY C. V. (1974) *Scot. Med. J.* **19**, 20.

McINTYRE MARY C. (1931) *Arch. Path.* **11**, 258.

MacKAY J. B. and CAIRNEY P. (1960) *N.Z. Med. J.* **59**, 453.

MACKAY J. M. K., NISBET D. I. and FOGGIE A. (1963) *Vet. Rec.* **75**, 550.

McKEOWN FLORENCE (1952) *Brit. Heart J.* **14**, 25.

McKEOWN FLORENCE (1954) *J. Path. Bact.* **68**, 147.

McKIM J. S. and WIGGLESWORTH F. W. (1954) *Amer. Heart J.* **47**, 845.

MACKINNON J. E., ARTAGAVEYTIA-ALLENDE R. C. and ARROYO L. (1953) *An. Fac. Med. Montevideo* **38**, 363.

MACKINNON J., WADE E. G. and VICKERS C. F. H. (1956) *Brit. Heart J.* **18**, 449.

MACKLEM P. T., FRASER R. G. and BROWN W. G. (1965) *J. clin. Invest.* **44**, 897.

MACKLIN C. C. (1936a) *Arch. Path.* **21**, 202.

MACKLIN C. C. (1936b) *J. thorac. Surg.* **6**, 82.

MACKLIN C. C. (1939) *Arch. int. Med.* **64**, 913.

MACKLIN C. C. (1951) *Lancet* **1**, 432.

MACKLIN C. C. (1954) *Lancet* **1**, 1099.

MACKLIN C. C. (1955a) *Canad. med. Ass. J.* **72**, 664.

MACKLIN C. C. (1955b) *Acta Anat.* **23**, 1.

McLAUGHLIN A. I. G., GROUT J. L. A., BARRIE H. J. and HARDING H. E. (1945) *Lancet* **1**, 337.

McLAUGHLIN A. I. G., ROGERS E. and DUNHAM K. C. (1949) *Brit. J. industr. Med.* **6**, 184.

McLAUGHLIN A. I. G. (1950) *Arch. belges Méd. soc.* **8**, 451.

McLaughlin A. I. G. and Harding H. E. (1956) *Arch. industr. Hlth.* **14**, 350.

McLaughlin A. I. G., Kazantzis G., King E., Teare D., Porter R. J. and Owen R. (1962) *Brit. J. industr. Med.* **19**, 253.

McLaughlin R. F. Jr., Tyler W. S. and Canada R. O. (1961) *J. Amer. med. Ass.* **175**, 694.

McLean K. H. (1956a) *Aust. Ann. Med.* **5**, 73.

McLean K. H. (1956b) *Aust. Ann. Med.* **5**, 254.

McLean K. H. (1957a) *Aust. Ann. Med.* **6**, 29.

McLean K. H. (1957b) *Aust. Ann. Med.* **6**, 124.

McLean K. H. (1957c) *Aust. Ann. Med.* **6**, 203.

McLean K. H. (1957d) *Aust. Ann. Med.* **6**, 282.

McLean K. H. (1958a) *Aust. Ann. Med.* **7**, 69.

McLean K. H. (1958b) *Amer. J. Med.* **25**, 62.

Macleod J. G. and Grant I. W. B. (1954) *Thorax* **9**, 71.

MacLeod M., Stalker A. L. and Ogston D. (1962) *Lancet* **1**, 191.

Macleod W. M. (1954) *Thorax* **9**, 147.

McLetchie N. G. B. and Reynolds D. P. (1954) *Canad. med. Ass. J.* **71**, 44.

MacMahon H. E. and Weiss S. (1929) *Amer. J. Path.* **5**, 623.

MacMahon H. E. (1948) *Amer. J. Path.* **24**, 919.

MacMahon H. E., Werch J. and Sorger K. (1967) *Arch. Path.* **83**, 359.

McMichael J. and Schafer E. P. S. (1944) *Quart. J. Med.* **13** N.S., 123.

McMichael J. (1947) *Advances in intern. Med.* **2**, 64.

McMichael J. (1948) *Edin. med. J.* **55**, 65.

McMillan H. A. (1949) *Arch. Path.* **48**, 377.

McMillan J. B. (1956) *Amer. J. Path.* **32**, 405.

McNally W. D. and Trostler I. S. (1941) *J. industr. Hyg.* **23**, 118.

McNamara J. J., Kingsley W. B. and Paulson D. L. (1969) *Ann. Surg.* **169**, 133.

McNeil C., MacGregor A. R. and Alexander W. A. (1929) *Arch. Dis. Childh.* **4**, 170.

McNeil R. S. and Cameron McD. H. (1955) *Thorax* **10**, 314.

McNeil R. S., Rankin J. and Forster R. E. (1958) *Clin. Sci.* **17**, 465.

McRae D. F. (1947) *Canad. med. Ass. J.* **57**, 545.

Madoff I. M., Gaensler E. A. and Strieder J. W. (1952) *New Engl. J. Med.* **247**, 149.

Magarey F. R. (1951) *J. Path. Bact.* **63**, 729.

Magidson O. and Jacobson G. (1955) *Brit. Heart J.* **17**, 207.

Magrou J. (1919) *Ann. Inst. Pasteur* **33**, 344.

Mahgoub E. S. and El Hassan A. M. (1972) *Thorax* **27**, 33.

Mahon W. E., Scott D. J., Ansell G., Manson G. L. and Fraser R. (1967) *Thorax* **22**, 13.

Maidman L. and Barnett R. N. (1957) *Arch. Path.* **64**, 104.

Maier H. C. and Fischer W. W. (1947) *J. thorac. Surg.* **16**, 392.

Maier H. C. (1948) *J. thorac. Surg.* **17**, 841.

Maier H. C., Himmelstein A., Riley R. L. and Bunin J. J. (1948) *J. thorac. Surg.* **17**, 13.

Maier H. C. (1954) *J. thorac. Surg.* **28**, 145.

Maier H. quoted by Spain D. M. and Handler B. J. (1946).

Mainwaring A. R., Williams G., Knight E. O. W. and Bassett H. F. M. (1969) *Thorax* **24**, 441.

Mainzer F. (1951) *Ergebn. inn. Med. Kinderheilk.* **2**, 388.

Majno G. and Palade G. E. (1961) *J. biophys. biochem. Cytol.* **11**, 571.

Malassez L. (1876) *Arch. Physiol. norm. path.* **3**, 353.

Mallory T. B. (1936) *New Engl. J. Med.* **215**, 837.

Mallory T. B. (1942) *New Engl. J. Med.* **227**, 955.

Mallory T. B. (1948a) *Radiology* **51**, 468

Mallory T. B. (1948b) *New Engl. J. Med.* **238**, 63.

Maltz D. L. and Nadas A. S. (1968) *Pediatrics* **42**, 175.

Mancuso T. F. and El-Attar A. A. (1969) *J. Occup. Med.* **11**, 422.

Manguikian B. and Prior J. T. (1963) *Arch. Path.* **75**, 236.

Manhoff L. J. and Howe J. S. (1949) *Arch. Path.* **48**, 155.

Mankiewicz E. and Van Walbeck M. (1962) *Arch. environ. Hlth.* **5**, 122.

Mann B. T. and Lecutier E. R. (1957) *Brit. med. J.* **2**, 921.

Manwaring J. H. (1949) *Arch. Path.* **48**, 421.

Manz A. (1954) *Beitr. Klin. Tuberk.* **111**, 598.

Marchand E. J., Marcial-Rojas R. A., Rodriguez R., Polanco G. and Díaz-Rivera R. S. (1957) *Arch. intern. Med.* **100**, 965.

Marchand P., Gilroy J. C. and Wilson V. H. (1950) *Thorax* **5**, 207.

Markowitz A. S. (1960) *Immunology* **3**, 117.

Marmion B. P. and Stoker M. G. P. (1950) *Lancet* **2**, 611.

Marmion B. P. and Stoker M. G. P. (1958) *Brit. med. J.* **2**, 809.

Marmion B. P. and Goodburn G. M. (1961) *Nature* **189**, 247.

Marsden A. T. H. (1958) *Brit. J. Cancer* **12**, 161.

Marshall A. H. E. (1946) *J. Path. Bact.* **58**, 729.

Marshall R., Sabiston D. C., Allison P. R., Bosman A. R. and Dunhill M. S. (1963) *Thorax* **18**, 1.

Martin A. E. and Bradley W. H. (1960) *Mth. Bull. Minist. Hlth. Lab. Serv.* **19**, 56.

Martin A. M. Jr. and Kurtz S. M. (1966) *Arch. Path.* **82**, 27.

Martin C. M. Kunin C. M., Gottlieb L. S. and Finland M. (1959a) *Arch. intern. Med.* **103**, 532.

Martin C. M., Kunin C. M., Gottlieb L. S., Barnes M. W., Chien Liu and Finland M. (1959b) *Arch. intern. Med.* **103**, 515.

Martin J. F., Dina M. A. and Féroldi J. (1950) *Arch. ital. Anat. Istol. pat.* **24**, 205.

Mason C. B. (1959) *J. thorac. Surg.* **37**, 251.

Mason G. A. (1949) *Lancet* **2**. 587.

Mason M. K. and Azeem P. S. (1965) *Thorax* **20**, 13.

Mason M. K. and Jordan J. W. (1969) *Thorax* **24**, 461.

Massaro D., Kayz S., Matthews M. J. and Higgins G. (1965) *Amer. J. Med.* **38**, 233.

MASSEE J. C. (1947) *Amer. J. med. Sci.* **214**, 248.

MASSON P., RIOPELLE J. L. and MARTIN P. (1937) *Ann. anat. path.* **14**, 359.

MASSON R. G., ALTOSE M. D. and MAYCOCK R. L. (1974) *Chest* **65**, 450.

MASUGI M. and MINAMI G. (1938) *Beitr. path. Anat.* **101**, 483.

MASUGI M. and YÄ-SHU (1938) *Virchows Arch. path. Anat.* **302**, 39.

MATHER G., DAWSON J. and HOYLE C. (1955) *Quart. J. Med.* **24** N.S., 331.

MATHEY J., GALEY J. J., LOGEAIS Y., SANTORO E., VANETTI A., MAUREL A. and WUERFLEIN R. (1968) *Thorax* **23**, 398.

MATS D. I. and MIZYAK L. E. (1958) *Klin. Med. (Moskva)* **36**, 36.

MATSON R. W. (1932) *Amer. Rev. Tuberc.* **25**, 419.

MATSUDA J. and THURLBECK W. M. (1972) *Amer. Rev. resp. Dis.* **105**, 908.

MATSUI S. (1924) *Mitt. med. Fak. Tokyo* **31**, 55, quoted by SPAIN D. M. and THOMAS A. G. (1950).

MATTHEW H., LOGAN A., WOODRUFF M. F. and HEARD B. (1968) *Brit. med. J.* **3**, 759.

MAVROGORDATA A. (1926) *Publ. S. African Inst. med. Res.* **3**, No. 19, 1.

MAXIMOW A. (1927) *Arch. exp. Zellforsch.* **4**, 1.

MAXIMOW A. A. and BLOOM W. (1942) *A Textbook of Histology*, 4th ed. W. B. Saunders, Philadelphia.

MAXWELL E. S., WARD T. G. and VAN METRE T. E. (1949) *J. clin. Invest.* **28**, 307.

MAXWELL J. (1934) *Quart. J. Med.* **27**, 467.

MAY J. R. (1953) *Lancet* **2**, 534.

MAYO L. E. (1942) *Virginia med. Monthly* **69**, 550.

MEAD C. H. (1932–3) *J. thorac. Surg.* **2**, 87.

MEAD J., LINDGREN I. and GAENSLER E. A. (1955) *J. clin. Invest.* **34**, 1005.

MEAD J. (1961) *Physiol. Rev.* **41**, 281.

MEADE J. B., WHITWELL F., BICKFORD B. J. and WADDINGTON J. K. B. (1974) *Thorax* **29**, 1.

MEANS J. A. and MALLORY T. B. (1931) *Ann. intern. Med.* **5**, 417.

MEBAN C. (1972) *J. Cell Biol.* **53**, 249.

MEDICAL RESEARCH COUNCIL (1942), (1943) *Special Rep. Series No. 243 and 244*, and (1945) *No. 250*.

MEDICAL RESEARCH COUNCIL (1949) Mem. No. 23.

MEDINA H. and BODZIAK C. (1948) *Arch. Biol. (S. Paulo)* **3**, 95.

MEDLAR E. M. (1947) *Amer. Rev. Tuberc.* **55**, 511.

MEHTA S. and RUBENSTONE A. I. (1967) *Amer. J. clin. Path.* **47**, 490.

MEIER G. and WURM K. (1960) *Beitr. Klin. Tuberk.* **123**, 90.

MEIRA J. A. (1942) *Arch. Cirurg. clin. exp.* **6**, 3.

MELMON K. L. and BRAUNWALD E. (1963) *New Engl. J. Med.* **269**, 770.

MELMON K. L., SJOERDSMA A. and MASON D. T. (1965) *Amer. J. Med.* **39**, 568.

MENDELSON C. L. (1946) *Amer. J. Obstet. Gynec.* **52**, 191.

MENDELSON R. W. (1921) *J. Amer. med. Ass.* **77**, 110.

MENKIN V. (1935) *Arch. Path.* **19**, 53.

MEREWETHER E. R. A. (1930) *J. industr. Hyg.* **12**, 198.

MEREWETHER E. R. A. (1933–4) *Tubercle* **15**, 109, 152.

MERRITT J. W. and PARKER K. R. (1957) *Canad. med. Ass. J.* **77**, 1031.

MEURMAN L. (1966) *Acta path. microbiol. scand. (suppl.)* **181**, 1.

MEYER E. C. and LIEBOW A. A. (1965) *Cancer (Philadelphia)* **18**, 322.

MEYER J. S. (1960) *Arch. Path.* **70**, 445.

MEYER K. F. and EDDIE B. (1947) *J. Amer. med. Ass.* **133**, 822.

MEYER K. K. (1958) *J. thorac. Surg.* **35**, 726.

MEYER R. D., YOUNG L. S., ARMSTRONG D. and YU B. (1973) *Amer. J. Med.* **54**, 6.

MEYER R. J. (1926) *Brasil méd.* **2**, 301.

MEYER W. W. and RICHTER H. (1956) *Virchows Arch. path. Anat.* **328**, 121.

MEYRICK BARBARA, STURGESS J. M. and REID L. (1969) *Thorax* **24**, 729.

MEYRICK BARBARA, MILLER J. and REID L. (1972) *Brit. J. exp. Path.* **53**, 347.

MEYRICK BARBARA, CLARKE S. W., SYMONS C., WOODGATE D. J. and REID L. (1974) *Brit. J. Dis. Chest* **68**. 11.

M'FADYEAN J. (1894) *J. comp. Path.* **7**, 31.

M'FADYEAN J. (1904) *J. comp. Path.* **17**, 295.

M'FADYEAN J. (1905) *J. comp. Path.* **18**, 23.

M'FADYEAN J. (1920) *J. comp. Path.* **33**, 1.

MICHAELS L. and LEVINE C. (1957) *J. Path. Bact.* **74**, 49.

MICHEL R. D. and MORRIS J. F. (1964) *Arch. int. Med.* **113**, 850.

MIDDLETON E. L. (1936) *Lancet* **2**, 1.

MIDDLETON E. L. (1940) *Proc. International Conference on Silicosis*. Geneva, 1938, published by the Int. Labour Office, Studies and Reports Series F (Industrial Hygiene). No. 17, pp. 25 and 134. P. S. King and Son, London, quoted by CAMPBELL A. H. and GLOYNE S. R. (1942).

MILLER A. and WYNN W. H. (1908) *J. Path. Bact.* **12**, 267.

MILLER A. A., RAMSDEN F. and GEAKE M. R. (1961) *Thorax* **16**, 388.

MILLER A., TEIRSTEIN A. S., BADER M. E., BADER R. A. and SELIKOFF I. J. (1971) *Amer. J. Med.* **50**, 395.

MILLER A., LANGER A. M., TEIRSTEIN A. S. and SELIKOFF I. J. (1975) *New Engl. J. Med.* **292**, 91.

MILLER G. J., BEADNELL H. M. S. G. and ASHCROFT M. T. (1968) *Lancet* **2**, 250.

MILLER H. C. and HAMILTON T. R. (1949) *Pediatrics* **3**, 735.

MILLER H. C., WOODRUFF M. W. and GAMBACORTA J. P. (1962) *Ann. Surg.* **156**, 852.

MILLER J. A. (1933) *J. thorac. Surg.* **3**, 246.

MILLER J. A. and JACKSON F. B. (1954) *J. Path. Bact.* **68**, 221.

MILLER J. F. (1937) *Amer. J. Dis. Child.* **53**, 1268.

MILLER K. and HARRINGTON J. S. (1972) *Brit. J. exp. Path.* **53**, 397.

MILLER W. S. (1906) *Anat. Anz.* **28**, 432.

MILLER W. S. (1919) *Amer. Rev. Tuberc.* **3**, 193.

MILLER W. S. (1947) *The Lung*, 2nd ed. C. C. Thomas, Springfield.

MIN K.-W., GYORKEY F. and GYORKEY P. (1974) *Amer. J. Path.* **74**, 107a.

MINKEN S. L., CRAVER W. L. and ADAMS J. T. (1968) *Arch. Path.* **86**, 442.

MINKOWITZ S., KOFFLER D. and ZAK F. G. (1963) *Amer. J. Med.* **34**, 252.

MINOR G. R. (1950) *J. thor. Surg.* **20**, 34.

MITCHELL D. N. and REES R. J. W. (1969) *Lancet* **2**, 81.

MITCHELL D. N. and SCADDING J. G. (1974) *Amer. Rev. resp. Dis.* **110**, 774.

MITCHELL J. (1959) *Brit. J. industr. Med.* **16**, 123.

MITTERMEYER C. and SANDRITTER W. (1971) *Thrombos. Diathes. haemorrh. (Stuttg.)* Suppl. **48**, 279.

MITUS ANNA, ENDERS J. F., CRAIG J. M. and HOLLOWAY A. (1959) *New Engl. J. Med.* **261**, 882.

MOBBS G. A. and PFANNER D. W. (1963) *Lancet* **1**, 472.

MOCHI A. and EDWARDS P. Q. (1952) *Bull. Wld. Hlth. Org.* **5**, 259.

MODY K. M. and POOLE G. (1963) *Brit. J. Dis. Chest* **57**, 200.

MOHR J. A., PATTERSON C. D., EATON B. G., RHODES E. R. and NICHOLS N. B. (1972) *Amer. Rev. resp. Dis.* **106**, 260.

MOHR R. (1909) *Beitr. path. Anat.* **45**, 333.

MÖLLER A. (1933) *Virchows Arch. path. Anat.* **291**, 478.

MOLNAR J. J., NATHENSON G. and EDBERG S. (1962) *New Engl. J. Med.* **266**, 36.

MONAHAN D. T. (1965) *J. thorac. cardiovasc. Surg.* **49**, 173.

MÖNCKEBERG J. G. (1907) *Dtsch. med. Wschr.* **32**, 1243.

MONGE M. C. (1928) *An. Fac. Med. (Lima)* **11**, 314.

MONOD O., PESLE G. and SÉGRÉTAIN G. (1951) *Presse méd.* **59**, 1557.

MONSON R. R. and PETERS J. M. (1974) *Lancet* **2**, 397.

MONTES M., ADLER R. H. and BRENNAN J. C. (1966) *Amer. Rev. resp. Dis.* **93**, 946.

MONTES M., TOMASI T. B., NOEHREN T. H. and CALVER G. J. (1968) *Amer. Rev. resp. Dis.* **98**, 277.

MONTGOMERY G. L. (1935) *J. Path. Bact.* **41**, 221.

MONTGOMERY G. L. (1942–3) *Brit. J. Surg.* **31**, 292.

MOOLTEN S. E. (1934) *Amer. J. Cancer* **21**, 253.

MOOLTEN S. E. (1935) *Arch. Path.* **20**, 77–80.

MOOLTEN S. E. (1942) *Arch. intern. Med.* **69**, 589.

MOOLTEN S. E. (1962) *Amer. J. Med.* **33**, 421.

MOORE D. B., GRAFF R. J., LANG S. and PAREIRA M. D. (1958) *Surg. Gynec. Obstet.* **107**, 615.

MOORE F. D., LYONS J. H., PIERCE E. C., MORGAN A. P., DRINKER P. A., MACARTHUR J. P. and DAMMIN G. J. (1969) *Post-traumatic Respiratory Insufficiency*, W. B. Saunders, London.

MOORE T. C. (1961) *Surgery* **50**, 886.

MOORE W. F. (1925) *Amer. J. med. Sci.* **169**, 799.

MORADOR J. L. and MORADOR S. (1959) *Antibiotics Annual* (1958–9), p. 958. Medical Encyclopedia, New York.

MORAN T. J. (1953) *Arch. Path.* **55**, 286.

MORAN T. J. and HELLSTROM H. R. (1958) *Arch. Path.* **66**, 691.

MORAN T. J. (1965) *Lecture at the Royal College of Surgeons of England.*

MORAN T. J. and TOTTEN R. S. (1970) *Amer. J. clin. Path.* **54**, 747.

MORE R. H. and MOVAT H. Z. (1959) *Arch. Path.* **67**, 679.

MORGAGNI J. B. (1769) *The Seats and Causes of Diseases investigated by Anatomy* (translated by B. Alexander, London).

MORGAN A. D. (1949) *J. Path. Bact.* **61**, 75.

MORGAN A. D., LLOYD W. E. and PRICE-THOMAS SIR C. (1952) *Thorax* **7**, 125.

MORGAN A. D. (1959) Personal communication.

MORGAN A. D. and MACKENZIE D. H. (1964) *J. Path. Bact.* **87**, 25.

MORGAN A. D. and BOGOMOLETZ W. (1968) *Thorax* **23**, 356.

MORGAN E., LANCASTER L., PEARSALL H. and LAWRENCE G. (1960) *Dis. Chest* **38**, 1.

MORGAN J. G. (1958) *Brit. J. industr. Med.* **15**, 224.

MORISON J. E. (1955) *Lancet* **2**, 941.

MORRELL M. T. and DUNNILL M. S. (1968) *Brit. J. Surg.* **55**, 347.

MORRIS J. A., SMITH R. W., BECK R. and ASSALI N. S. (1963) *Circulation* **28**, 772.

MORROW J. D., SCHROEDER H. A. and PERRY H. M. (1953) *Circulation* **8**, 829.

MORTON D. R., OSBORNE J. F. and KLASSEN K. P. (1950) *J. thorac. Surg.* **19**, 811.

MOSCHCOWITZ E. and STRAUSS L. (1963) *Arch. Path.* **75**, 582.

MOSKOWITZ R. L., LYONS H. A. and COTTLE H. R. (1964) *Amer. J. Med.* **36**, 457.

MOSSO A. (1898) *Life of Man in the High Alps.* T. F. Unwin, London.

MOTTA L. C. (1942) *An. Fac. Med. S. Paulo* **18**, 145.

MOTTURA G. (1952) *Brit. J. industr. Med.* **9**, 65.

MOUNIER-KUHN P. (1932) *Lyon. Med.* **150**, 106.

MOUNSEY J. P. D., RITZMANN L. W., SELVERSTONE N. J., BRISCOE W. A. and MCLEMORE G. A. (1952) *Brit. Heart J.* **14**, 153.

MOYER J. H., GLANTZ G. and BREST A. N. (1962) *Amer. J. Med.* **32**, 417.

MUHLEISEN J. P. (1953) *Ann. intern. Med.* **38**, 595.

MUIR C. S. (1962) *Singapore med. J.* **3**, 169.

MUIR D. C. F. and STANTON J. A. (1963) *Brit. med. J.* **1**, 1072.

MUIRHEAD E. E., MONTGOMERY P. O'B. and GORDON C. E. (1952) *Arch. int. Med.* **89**, 41.

MULDER J. and VERDONK G. J. (1949) *J. Path. Bact.* **61**, 55.

MÜLLER F. H. (1939) *Z. Krebsforsch.* **49**, 57.

MÜLLER H. (1882) *Inaugural Dissertation Univ. of Halle.* H. A. Busch, Ermsleben, quoted by RAMSEY J. N. and REIMAN D. L. (1953).

MÜLLER H. (1928) in *Handbuch der Spez. Path. Anat. u. Histol.* (Henke–Lubarsch), Vol. 3/1. J. Springer, Berlin.

MULLIGAN R. M. (1947) *Arch. Path.* **43**, 177.

MURI J. W. (1953) *Dis. Chest* **24**, 49.

MURI J. W. (1955) *Amer. J. Surg.* **89**, 265.

MURPHY E. S., FUJII Y., YASUDA A. and SASABE S. (1958) *Arch. Path.* **65**, 166.

MURPHY G. B. (1961) *Arch. environ. Hlth.* **3**, 704.

MURPHY G. H., DOCKERT M. B. and BRODERS A. C. (1949) *Amer. J. Path.* **25**, 1157.

MURPHY J. D. and BORNSTEIN S. (1950) *Ann. intern. Med.* **33**, 442.

MURRAY J. F. and BRANDT F. A. (1951) *Amer. J. Path.* **27**, 783.

MURRAY J. F., FINEGOLD S. M., FROMAN S. and WILL D. W. (1961) *Amer. Rev. resp. Dis.* **83**, 315.

MURRAY J. F. and HOWARD D. H. (1963) *Amer. Rev. resp. Dis.* **88**, 106.

MURRAY M. (1907) *Departmental Commission on Compensation for Industrial Diseases.*
Cmd. 3495, p. 14. ⎱
Cmd. 3496, p. 127. ⎰ H.M.S.O., London.

MUSCHENHEIM C. (1961) *Amer. J. med. Sci.* **241**, 279.

MUSSER J. H. (1903) *Penn. med. Bull.* **16**, 289, quoted by STOREY C. F. *et al.* (1953).

MYERSON H. C. (1922) *Laryngoscope (St. Louis)* **32**, 929.

NABARRO J. D. N. (1948) *Lancet* **1**, 982.

NAEYE R. L. (1960) *Circulation* **22**, 376.

NAEYE R. L. and LAQUEUR W. A. (1970) *Arch. Path.* **90**, 487.

NAGAYA H., ELMORE M. and FORD C. D. (1973) *Amer. Rev. resp. Dis.* **107**, 826.

NAGELSCHMIDT G., NELSON E. S., KING E. J., ATTYGALLE D. and YOGANATHAN M. (1957) *Arch. industr. Hlth.* **16**, 188.

NAGELSCHMIDT G. (1960) *Brit. J. industr. Med.* **17**, 247.

NAGELSCHMIDT G. (1960) in *Proc. Pneumoconiosis Conf. Johannesburg, 1960.* Ed. A. J. ORENSTEIN. Little Brown & Co., Boston, p. 143.

NAGELSCHMIDT G., RIVERS D. and KING E. J. (1963) *Brit. J. industr. Med.* **20**, 181.

NAIDOO P. and HIRSCH H. (1963) *Lancet* **1**, 196.

NAKAMURA T. (1955) *19th General Meeting of the Japanese Circulation Society.*

NAKAMURA T. (1958) *Presidential Lecture at 22nd Annual Meeting of the Japanese Circulation Society, Sendai.*

NANSON E. M. and WALKER R. M. (1952) *Thorax* **7**, 263.

NARAT J. K., LOEF J. A. and NARAT M. (1936) *Anat. Rec.* **64**, 155–160.

NARATH A. (1901) in *Der Bronchialbaum der Säugetiere und des Menschen*, Stuggart, quoted by KEIBEL and MALL (1912).

NASH A. D. and STOUT A. P. (1958) *Cancer (Philadelphia)* **11**, 369.

NASR A. N. M., DITCHEK T. and SCHOLTENS P. A. (1971) *J. occup. Med.* **13**, 371.

NATELSON E. A., WATTS H. D. and FRED H. L. (1970) *Chest* **57**, 333.

NATHAD D J. and SANDERS M. (1955) *New Engl. J. Med.* **252**, 797.

NAYAK N. C., ROY S. and NARAYANAN K. (1964) *Amer. J. Path.* **45**, 381.

NAZARI A., AMIR-MOKRI E., SARRAM A. and YAGHMAI I. (1971) *Chest* **60**, 187.

NEAFIE R. C. and PIGGOTT J. (1971) *Arch. Path.* **92**, 342.

NEERGARD K. (1929) *Z. Ges. exp. Med.* **66**, 373.

NEILL CATHERINE A. (1956) *Pediatrics* **18**, 880.

NEILL CATHERINE A., FERENCZ C., SABISTON D. C. and SHELDON H. (1960) *Bull. Johns Hopk. Hosp.* **107**, 1.

NEILSON D. B. (1958) *J. Path. Bact.* **76**, 419.

NELSON C. S., McMILLAN I. K. R. and BHARUCHA P. K. (1967) *Thorax* **22**, 7.

NELSON J. B. (1946) *J. exper. Med.* **84**, 7, 15.

NELSON R. L. (1932) *J. Pediat.* **1**, 233.

NELSON W. P., LUNDBERG G. D. and DICKERSON R. B. (1965) *Amer. J. Med.* **38**, 279.

NETHERCOTT S. E. and STRAWBRIDGE W. G. (1956) *Lancet* **2**, 1132.

NEUBUERGER K. T. and GEEVER E. F. (1942) *Arch. Path.* **33**, 551.

NEUBUERGER K. T., GEEVER E. F. and RUTLEDGE E. K. (1944) *Arch. Path.* **37**, 1.

NEUHOF H. and TOUROFF A. S. W. (1938) *Surg. Gynec. Obstet.* **66**, 836.

NEUMANN R. (1938) *Frankfurt. Z. Path.* **52**, 576.

NEW ENGLAND JOURNAL OF MEDICINE. Case Report 26 (1974) **291**, 35.

NEWHOUSE M. L. and THOMPSON H. (1965) *Brit. J. industr. Med.* **22**, 261.

NEWMAN W. and ADKINS P. C. (1958) *J. thorac. Surg.* **35**, 474.

NICHOLLS P. J., NICHOLLS G. R. and BOUHUYS A. (1967) in *Inhaled Particles and Vapours.* 2nd Ed. Edited by C. N. DAVIES. Pergamon Press, Oxford, pp. 93–100.

NICHOLS B. H. (1930) *Amer. J. Roentgenol.* **23**, 516.

NICHOLSON H. (1951) *Thorax* **6**, 75.

NICKERSON D. A. (1937) *Arch. Path.* **24**, 19.

NICKS R. (1957) *Thorax* **12**, 140.

NICOLAIDES A. N. (1975) Ed. *Thromboembolism, Aetiology, Advances in Prevention and Management.* Medical and Technical Publishing Co., Lancaster.

NICOLAIDES A. N. and GORDON-SMITH I. (1975) in *Thromboembolism, Aetiology, Advances in Prevention and Management*, Medical and Technical Publishing Co., Lancaster, p. 206.

NICOLLE M. M. C. and MANCEAUX L. (1909) *Arch. Inst. Pasteur Tunis* **4**, 97.

NIDEN A. H. and YAMADA E. (1966) in *Proc. VIth Internat. Congress for Electron Microscopy, Maruzen, Tokyo*, p. 599.

NIDEN A. H. (1967) *Science* **158**, 1323.

NIEWOEHNER D. E., KLEINERMAN J., and RICE D. (1974) *Amer. Rev. resp. Dis.* **109**, 725.

NIME F. A. and HUTCHINS G. M. (1973) *Johns Hopk. med. J.* **133**, 183.

NISSEN R. (1927) *Münch. med. Wschr.* **74**, 1362.

NOCARD E. (1888) *Ann. Inst. Pasteur* **2**, 293.

NOCARD E. (1890–4) *Rapport Général du Conseil d'Hygiène Publique*, p. 275. Quoted by STURDEE E. L. and SCOTT W. M. (1930).

NOEHREN T. H., McKEE F. W. (1954) *Dis. Chest* **25**, 663.

18

NOHL H. C. (1956) *Thorax* **11**, 172.

NONIDEZ J. F. (1937) *Amer. J. A\at.* **61**, 203.

NORDEN C. W., CALLERAME M. L. and BAUM J. (1970) *New Engl. J. Med.* **282**, 190.

NORDMANN M. (1938) *Z. Krebsforsch.* **47**, 288.

NORRIS G. F. and PEARD M. C. (1963) *Brit. med. J.* **1**, 378.

NORRIS R. F., KOCKENDERFER T. T. and TYSON R. M. (1941) *Amer. J. Dis. Child.* **61**, 933.

NORTHWAY W. H., ROSEN R. C. and PORTER D. Y. (1967) *New Engl. J. Med.* **276**, 357.

NOVAK E. (1952) *Gynecological and Obstetrical Pathology*, 3rd ed. W. B. Saunders Co., Philadelphia.

NOWAK J. (1966) *Acta med. Pol.* **7**, 23.

NUNGESTER W. J. and KLEPSER R. G. (1937) *Amer. J. Path.* **13**, 642.

OAKLEY CELIA, GLICK G. and McCREDIE R. M. (1963) *Amer. J. Med.* **34**, 264.

OATES J. A., MELMON K., STOERDSMA A., GILLESPIE L. and MASON D. T. (1964) *Lancet* **1**, 514.

OBERNDORFER VON S. (1930a) *Münch. med. Wschr.* **77**, 311.

OBERNDORFER VON S. (1930b) *Virchows Arch. path. Anat.* **275**, 728.

OBLATH R. W., DONATH D. H., JOHNSTONE H. G. and KERR W. J. (1951) *Ann. intern. Med.* **35**, 97.

OCHSNER A. and DE BAKEY M. (1936) *J. thorac. Surg.* **5**, 225.

OCHSNER A. and DE BAKEY M. (1941–2) *J. thorac. Surg.* **11**, 357.

ODERR C. P. (1960) *J. Amer. med. Ass.* **172**, 1991.

OEHLERT W. and DÜFFEL F. (1958) *Zbl. allg. Path. path. Anat.* **98**, 41.

OGASAWARA K. and AIDA M. (1958) *Virus (Kyoto)* **8**, 242.

OGASAWARA K. and TANAKA S. (1958) *Lancet* **2**, 589.

OGILVIE A. G. (1941) *Arch. intern. Med.* **68**, 395.

OLD J. W. and RUSSELL W. O. (1950) *Amer. J. Path.* **26**, 789.

OLD J. W., SMITH W. W. and GRAMPA G. (1955) *Amer. J. Path.* **31**, 605.

OLINER H., SCHWARTZ R., RUBIO F. Jr. and DAMESHEK W. (1961) *Amer. J. Med.* **31**, 134.

OLIVEIRA H. L. DE, LAUS-FILHO J. A. (1962) *Internat. Arch. Allergy* **20**, 298.

OLSON L. C., MILLER G. and HANSHAW J. B. (1964) *Lancet* **1**, 200.

O'NEAL R. M., AHLVIN R. C., BAUER W. C. and THOMAS W. A. (1957) *Arch. Path.* **63**, 309.

ONUIGBO W. I. B. (1957) *Brit. J. Cancer* **11**, 175.

ONUIGBO W. I. B. (1958) *Cancer (Philadelphia)* **11**, 737.

ONUIGBO W. I. B. (1963) *Amer. J. Surg.* **106**, 929.

OPDYKE D. F., DUOMARCO J., DILLON W. H., SCHREIBER H., LITTLE R. C. and SEELY R. D. (1948) *Amer. J. Physiol.* **154**, 258.

OPIE E. L. (1928) *Arch. Path.* **5**, 285.

OPPENHEIMER ELLA H. (1938) *Bull. Johns Hopk. Hosp.* **63**, 261.

OPPENHEIMER ELLA H. (1944) *Bull. Johns Hopk. Hosp.* **74**, 240.

OPPENHEIMER ELLA H. (1954) *Bull. Johns Hopk. Hosp.* **94**, 86.

OPPENHEIMER ELLA H. and GUILD H. G. (1960) *Bull. Johns Hopk. Hosp.* **88**, 101.

OPPENHEIMER ELLA H. and ESTERLY J. R. (1973) *Amer. J. Path.* **70**, 28a.

ORBISON J. A., REEVES N., LEEDHAM C. L. and BLUMBERG J. M. (1951) *Medicine (Baltimore)* **30**, 247.

ORCHARD N. P. (1950) *Arch. Dis. Childh.* **5**, 151.

ORMEROD F. C. (1941) *J. Laryng.* **56**, 277.

ORSÓS F. (1907) *Beitr. path. Anat. zur. Allg. Path. (Jena)* **41**, 95.

ORTH J. (1887) *Ätiologisches und Anatomisches über Lungenschwindsucht.* Hirschwald, Berlin.

ORTON H. B. (1932) *Ann. Otol. (St. Louis)* **41**, 933.

OSBORN G. R. (1958) *Proc. roy. Soc. Med.* **51**, 840.

OSBORN G. R. and FLETT R. L. (1962) *J. clin. Path.* **15**, 527.

OSHIMA Y., ISHIZAKI T. and MIYAMOTO T. (1964) *Amer. Rev. resp. Dis.* **90**, 572.

OSORIO J. and RUSSEK M. (1962) *Circulation Res.* **10**, 664.

OSWALD N. C. and MEDVEI V. C. (1955) *Lancet* **2**, 843.

OSWALD N. C., HINSON K. F. W., CANTI G. and MILLER A. B. (1971) *Thorax* **26**, 623.

ÖTTGEN H. F. and QUITTMAN F. (1956) *Frankfurt. Z. Path.* **67**, 599.

OWEN W. R., THOMAS W. A., CASTLEMAN B. and BLAND E. F. (1953) *New Engl. J. Med.* **249**, 919.

OYAMADA A., GASUL B. M. and HOLINGER P. H. (1953) *Amer. J. Dis. Child.* **85**, 182.

OZZELLO L. and STOUT A. P. (1961) *Cancer (Philadelphia)* **14**, 1052.

PADULA R. T. and STAYMAN J. W. (1967) *J. thorac. cardiovasc. Surg.* **54**, 272.

PAGE D. L., ISERSKY C., HARADA M. and GLENNER G. G. (1972) *Res. exp. Med.* **159**, 75.

PAI S. H., CAMERON C. T. and LEV R. (1971) *Arch. Path.* **91**, 569.

PAIGE B. H., COWEN D. and WOLF A. (1942) *Amer. J. Dis. Child.* **63**, 474.

PAINE R., BUTCHER H. R., SMITH J. R. and HOWARD F. A. (1950) *J. Lab. clin. Med.* **36**, 288.

PALMER K. N. V. (1954) *Brit. med. J.* **1**, 1473.

PALMER P. E. S. and DAYNES G. (1967) *South Afr. med. J.* **41**, 1182.

PALMOVIC V. and McCARROLL J. R. (1965) *Arch. Path.* **80**, 630.

PANCOAST H. K. (1924) *J. Amer. med. Ass.* **83**, 1407.

PANCOAST H. K. (1932) *J. Amer. med. Ass.* **99**, 1391.

PAPANICOLAOU G. N. (1942) *Science* **95**, 438.

PAPANICOLAOU G. N. and KOPROWSKA I. (1951) *Cancer (Philadelphia)* **4**, 141.

PARIENTE R., EVEN P. and BROUET G. (1967) *Presse Méd.* **75**, 183.

PARIENTE R. (1975) *Prog. Resp. Res.* **8**, 91.

PARISH D. J., CRAWFORD N. and SPENCER A. T. (1964) *Thorax* **19**, 62.

PARK W. W. (1954) *J. Path. Bact.* **67**, 563.

PARK W. W. (1957) M.D. Thesis, Univ. of Edinburgh.
PARK W. W. (1958) *J. Path. Bact.* **75**, 257.
PARKER F. and WEISS S. (1936) *Amer. J. Path.* **12**, 573.
PARKER F., JOLLIFFE L. S., BARNES M. W. and FINLAND M. (1946) *Amer. J. Path.* **22**, 797.
PARKER T. G. and SOMMERS S. C. (1956) *Arch. Surg.* **71**, 495.
PARKES W. R. (1973) *Brit. J. Dis. Chest* **67**, 261.
PARSONS R. J. and ZARAFONETIS C. J. D. (1945) *Arch. intern. Med.* **75**, 1.
PASSEY R. D. (1962) *Lancet* **2**, 107.
PASSLER (1896) *Virchows Arch. path. Anat.* **145**, 191.
PATCHEFSKY A. S., ISRAEL H. L., HOCH W. S. and GORDON G. (1973) *Thorax* **28**, 680.
PATTEN B. M. (1946) *Human Embryology.* The Blakiston Company, Philadelphia.
PATTLE R. E. (1961) in *Inhaled Particles and Vapours.* Edited by C. N. DAVIES. Pergamon Press, Oxford.
PATTLE R. E., CLAIREAUX A. E., DAVIES P. A. and CAMERON A. H. (1962) *Lancet* **2**, 469.
PATTLE R. E. (1963) *Brit. med. Bull.* **19**, 41.
PATTON M. M., MCDONALD J. R. and MOERSCH H. J. (1951) *J. thorac. Surg.* **22**, 83.
PAVILANIS V., DUVAL L., FOLEY A. R. and L'HERUEUX M. (1958) *Canad. J. pub. Hlth.* **49**, 520.
PAVLICA F. (1962) *Ann. Paediat.* **198**, 177.
PAYNE W. S., CLAGETT O. T. and HARRISON E. G. (1962) *J. thorac. cardiovasc. Surg.* **43**, 279.
PEABODY C. N. (1959) *J. thorac. Surg.* **37**, 766.
PEABODY J. W., PEABODY J. W. Jr., HAYES E. W. and HAYES E. W. Jr. (1950) *Dis. Chest* **18**, 330.
PEABODY J. W., BUECHNER H. A. and ANDERSON A. E. (1953) *Arch. intern. Med.* **92**, 806.
PEARSE A. G. E. (1950) *J. Path. Bact.* **62**, 351.
PEARSE A. G. E. (1953) *Histochemistry—Theoretical and Applied.* Churchill, London.
PEARSE A. G. E. (1966) *Nature (Lond.)* **211**, 598.
PEARSON C. M., KLINE H. M. and NEWCOMER Y. D. (1960) *New Engl. J. Med.* **263**, 51.
PEARSON R. S. R. and DE NAVASQUEZ S. (1938) *Guy's Hosp. Rep.* **88**, 1.
PECK M. E. and LEVIN S. (1952) *J. thorac. Surg.* **24**, 619.
PECK S. M. (1927) *Arch. Path.* **4**, 365.
PEKELIS E. (1931) *Pathologica* **23**, 66.
PELLER S. (1939) *Hum. Biol.* **11**, 130.
PELLISSIER A. and OUARY P. (1952) *Presse méd.* **60**, 1788.
PELTIER L. F. (1957) *Surg. Gynecol. Obstet.* **104**, 313.
PEMBERTON J. and GOLDBERG C. (1954) *Brit. med. J.* **2**, 567.
PEÑALOZA D., SIME F., BANCHERO N. and GAMBOA R. (1962) *5th Ann. Conf. on Research in Emphysema, Aspen, Colorado, U.S.A.*
PEÑALOZA D., SIME F., BANCHERO M. and GAMBOA R. (1962) *Med. Thorac.* **19**, 449.
PEÑALOZA D. and SIME F. (1971) *Amer. J. Med.* **50**, 728.
PEPYS J., RIDDELL R. W., CITRON K. M., CLAYTON Y. M. and SHORT E. I. (1959) *Amer. Rev. resp. Dis.* **80**, 167.
PEPYS J., JENKINS P. A., FESTENSTEIN G. N., GREGORY P. H., LACEY M. E. and SKINNER F. A. (1963) *Lancet* **2**, 607.

PEPYS J. and JENKINS P. A. (1965) *Thorax* **20**, 21.
PERLMAN E. and BULLOWA J. G. M. (1941) *Arch. intern. Med.* **67**, 907.
PERMAR H. H. and MACLACHLAN W. W. G. (1931) *Ann. intern. Med.* **5**, 687.
PERNIS B. and PECCHIAI L. (1954) *Med. d. Lavoro* **45**, 205.
PERRAUD R., CHAMEAUD J., LAFUMAR J. (1972) *J. Franç. Med. Chir. thorac.* **26**, 25.
PERRIN T. L. (1943) *Arch. Path.* **36**, 568.
PERRY K. M. A. and KING D. S. (1940) *Amer. Rev. Tuberc.* **41**, 531.
PETERSDORF R. G., FUSCO J. J., HARTER D. H. and ALBRINK W. S. (1959) *Arch. intern. Med.* **103**, 262.
PETERSEN A. B., HUNTER W. C. and SNEEDEN V. D. (1949) *Cancer (Philadelphia)* **2**, 991.
PETERSON E. W. and HOUGHTON J. D. (1951) *New Engl. Med.* **244**, 429.
PETERSON P. A., SOULE E. H. and BERNATZ P. E. (1957) *J. thorac. Surg.* **34**, 95.
PÉTRIAT A., CORNET L., LEGER H., CASTAING R. and TESSIER R. (1953) *Presse méd.* **61**, 1526.
PETTY T. L., MIERCORT S. R., VINCENT T., FILLEY G. F. and MITCHELL R. S. (1965) *Amer. Rev. resp. Dis.* **92**, 450.
PHILLIPS I. and SPENCER G. (1965) *Lancet* **2**, 1325.
PHILLS J. A., HARROLD A. J., WHITEMAN G. V. and PERELMUTTER L. (1972) *New Engl. J. Med.* **286**, 965.
PHILP T. (1972) *Scot. med. J.* **17**, 104.
PHILP T., SUMERLING M. D., FLEMING J. and GRAINGER R. G. (1972) *Clin. Radiol.* **23**, 153.
PIERCE J. A., REAGAN W. P. and KIMBALL R. W. (1959) *New Engl. J. Med.* **260**, 901.
PIERCE J. A. and EBERT R. V. (1965) *Thorax* **20**, 469.
PIERCE W. F., ALZNAUER R. L. and ROLLE C. (1954) *Arch. Path.* **58**, 443.
PIGGOTT J. A. and HOCHHOLZER L. (1970) *Arch. Path.* **90**, 101.
PIMENTEL J. C. (1970) *Thorax* **25**, 387.
PINCHIN S. and MORLOCK H. V. (1933) *Lancet* **1**, 1114.
PINCK R. L., BURBANK B., CUTLER S. S., SBAR S. and MANGIERI M. (1965) *Amer. Rev. resp. Dis* **91**, 909.
PINKERTON H. (1928) *Arch. Path.* **5**, 380.
PINKERTON H. and WEINMAN D. (1940) *Arch. Path.* **30**, 374.
PINKERTON H., SMILEY W. L. and ANDERSON W. A. D. (1945) *Amer. J. Path.* **21**, 1.
PINKERTON H. and CARROLL S. (1971) *Amer. J. Path.* **65**, 543.
PINKETT OLIVIA M., COWDREY C. R. and NOWELL P. C. (1966) *Amer. J. Path.* **48**, 859.
PINNER M. (1946) *Amer. Rev. Tuberc.* **54**, 582.
PIPER P. G. and KLEPPE L. W. (1958) *Arch. Path.* **65**, 131.
PIRCHAN A. and SÏKL H. (1932) *Amer. J. Cancer* **16**, 681.
PIRIE A. H. (1932) *Amer. J. Roentgenol.* **27**, 578.
PLACITELLI G. (1953) *Minerva pediat. (Torino)* **5**, 554.
PLÍHAL V., JEDLICKOVA Z., VIKLICKY J. and TOMÁNEK A. (1964) *Thorax* **19**, 104.

PLOPPER C. G., DUNGWORTH D. L. and TYLER W. S. (1973) *Amer. J. Path.* **71,** 375.

POH S. C., TJIA T. S. and SEAH H. C. (1975) *Thorax* **30,** 186.

POKORNY C. and HELLWIG C. A. (1955) *Arch. Path.* **59,** 382.

POLES F. C. and LAVERTINE J. D. O'D. (1954) *Thorax* **9,** 233.

POLICARD A. and DOUBROW S. (1929) *Presse méd.* **37,** 905.

POLICARD A. (1935) *Ann. Méd. lég.* **15,** 126.

POLICARD A. and COLLETT A. (1951) *C. R. Acad. Sci. (Paris)* **233,** 1159.

POLICARD A. (1952) *Brit. J. industr. Med.* **9,** 108.

POLICARD A., COLLETT A. and PREGERMAIN S. (1956) *Proc. of Conference on Electron Microscopy, Stockholm.*

POLICARD A. (1962) *J. clin. Path.* **15,** 394.

POLLAK ANN (1956) *Amer. J. Path.* **32,** 629.

POLLARD A., GRAINGER R. G., FLEMING O. and MEACHIM G. (1962) *Lancet* **2,** 1084.

PONCET A. and BÉRARD L. (1898) *L'Actinomycose Humaine.* Masson et Cie, Paris.

PONNAMPALAM J. T. (1964) *Brit. J. Dis. Chest* **58,** 49.

POOL P. E., VOGEL J. H. K. and BLOUNT S. G. Jr. (1962) *Amer. J. Cardiol.* **10,** 706.

POOLEY F. D. (1972) *Brit. J. industr. Med.* **29,** 146.

PORRO F. W., PATTON J. R. and HOBBS A. A. (1942) *Amer. J. Roentgenol.* **47,** 507.

PORTER C. McG., CREECH B. J. and BILLINGS F. T. Jr. (1967) *Arch. int. Med.* **120,** 224.

POSADA A. (1892) *An. Circ. med. argent. (Buenos Aires)* **15,** 481, 585.

POSKITT T. R. (1970) *Amer. J. Med.* **49,** 250.

POST G. W., JACKSON A., GARBER P. E. and VEACH G. E. (1958) *Dis. Chest* **34,** 455.

POTTER EDITH L. and YOUNG R. L. (1942) *Arch. Path.* **34,** 1009.

POTTER EDITH L. (1952) *Pathology of the Fetus and the Newborn.* The Year Book Medical Publishers Inc., Chicago.

POTTS W. J. and RIKER W. L. (1950) *Arch. Surg. (Chicago)* **61,** 684.

POTTS W. J., HOLINGER P. H. and ROSENBLUM A. H. (1954) *J. Amer. med. Ass.* **155,** 1409.

POUND A. W. and WILLIS R. A. (1969) *J. Pathol.* **98,** 111.

POWELL K. E., HAMMERMAN K. J., DAHL B. A. and TOSH F. E. (1973) *Amer. Rev. resp. Dis.* **107,** 374.

POWELL V. (1958) *Thorax* **13,** 321.

PRATT P. C. (1958) *Amer. J. Path.* **34,** 1033.

PRATT P. C. (1965) *Ann. N.Y. Acad. Sci.* **121,** 809.

PRATT P. C. (1968) *Arch. environ. Hlth.* **16,** 734.

PRATT P. C. and KILBURN K. H. (1970) *Ann. int. Med.* **73,** 134.

PRATT-THOMAS H. R. and CANNON W. M. (1946) *Amer. J. Path.* **22,** 779.

PRICE T. H. L. and SKELTON M. O. (1956) *Thorax* **11,** 234.

PRICE-THOMAS SIR C. and MORGAN A. D. (1958) *Thorax* **13,** 286.

PRICHARD M. M. L., DANIEL P. M. and ARDRAN G. M. (1954) *Brit. J. Radiol.* **27,** 93.

PRIJYANONDA B. and TANDHANAND S. (1961) *Ann. int. Med.* **54,** 795.

PRINE J. R., BROKESHOULDER S. F., McVEAN D. E. and ROBINSON F. R. (1966) *Amer. J. clin. Path.* **45,** 448.

PRINZMETAL M., ORNITZ E. M. Jr., SIMKIN B. and BERGMAN H. C. (1948) *Amer. J. Physiol.* **152,** 48.

PRIOR J. A., COLE C. R., DOCTON F. L., SASLAW S. and CHAMBERLAIN D. M. (1953) *Arch. intern. Med.* **92,** 314.

PRIOR J. T. and JONES D. B. (1952) *J. thorac. Surg.* **23,** 224.

PRIVE L., TELLEM M., MERANZE D. R. and CHODOFF R. D. (1961) *Arch. Path.* **72,** 351.

PROFFITT R. D. and WALTON B. C. (1962) *New Engl. J. Med.* **266,** 931.

PROWSE B. C. (1958) *Thorax* **13,** 308.

PRUZANSKI W. and ALTMAN R. (1964) *Arch. int. Med.* **113,** 261.

PRYCE D. M. (1946) *J. Path. Bact.* **58,** 457.

PRYCE D. M., SELLORS T. H. and BLAIR L. G. (1947–8) *Brit. J. Surg.* **35,** 18.

PRYCE D. M. (1948) *J. Path. Bact.* **60,** 259.

PRYCE D. M. and HEARD B. E. (1956) *J. Path. Bact.* **71,** 15.

PRYCE D. M. and WALTER J. B. (1960) *J. Path. Bact.* **79,** 141.

PUBLIC HEALTH LABORATORY SERVICE (1958) *Brit. med. J.* **1,** 915.

PUHR L. (1933) *Virchows Arch. path. Anat.* **290,** 156.

PUMP K. K. (1963) *Dis. Chest* **43,** 245.

PUMP K. K. (1964) *Dis. Chest* **46,** 379.

PUMP K. K. (1973) *Amer. Rev. resp. Dis.* **108,** 610.

PUMP K. K. (1974) *Chest* **65,** 431.

PUTNEY F. J. and BAKER D. C. (1938) *Dis. Chest* **4,** 20.

PUTSCH R. W., HAMILTON J. D. and WOLINSKY E. (1970) *J. Infect. Dis.* **121,** 48.

QUINN L. H. and MEYER O. O. (1929) *Arch. Otolaryng. (Chicago)* **10,** 152.

QUINTILIANI R. and HYMANS P. J. (1971) *Amer. J. Med.* **50,** 781.

RABIN C. B. and NEUHOF H. (1949) *J. thorac. Surg.* **18,** 149.

RAEBURN C. (1951) *Brit. med. J.* **2,** 517.

RAEBURN C and SPENCER H. (1953) *Thorax* **8,** 1.

RAEBURN C. and SPENCER H. (1957) *Brit. J. Tuberc.* **51,** 237.

RAFII S. and GODWIN M. C. (1961) *Arch. Path.* **72,** 424.

RAIDER L. (1971) *Chest* **60,** 504.

RAJAN V. T. and KIKKAWA Y. (1970) *Arch. Path.* **89,** 521.

RAJEWSKY B. (1939) *Radiology* **32,** 57.

RAKOV H. L. and TAYLOR J. S. (1942) *Arch. intern. Med.* **70,** 88.

RAKOWER J. (1957) *Cancer (Philadelphia)* **10,** 67.

RALL J. E., ALPERS J. B., LEWALLEN C. G., SONENBERG M., BERMAN M. and RAWSON R. W. (1957) *J. clin. Endocr.* **17,** 1263.

RAMANATHAN T. (1974) *Thorax* **29,** 482.

RAMIREZ A., GRIMES E. T. and ABELMAN W. H. (1968) *Amer. J. Med.* **45,** 975.

RAMIREZ-RIVERA J (1966) *Dis. Chest* **50,** 581.

RAMSAY B. H. and BRYON F. X. (1953) *J. thorac. Surg.* **26,** 21.

RAMSEY J. H. and REIMANN D. L. (1953) *Amer. J. Path.* **29,** 339.

RANASINHA K. W. and URAGODA C. G. (1972) *Brit. J. industr. Med.* **29,** 178.

RANDALL W. S. and BLADES B. (1946) *Arch. Path.* **42,** 543.

RAO K. M., GUPTA R. P., DAS P. B., JOHN S. and WALTER A. (1974) *Thorax* **29,** 138.

RAPPAPORT H., RAUM M. and HORRELL J. B. (1951) *Amer. J. Path.* **27,** 407.

RATNOFF O. D. and VOSBURGH G. J. (1952) *New Engl. J. Med.* **247,** 970.

RAVINES H. T. (1960) *J. thorac. cardiovasc. Surg.* **39,** 760.

RAVINES H. T. (1969) *Amer. J. clin. Path.* **52,** 767.

RAY S. C., KING E. J. and HARRISON C. V. (1951) *Brit. J. industr. Med.* **8,** 74.

RAZEMON P. and RIBET M. (1958) *Poumon* **14,** 21.

READ J. (1958) *Amer. Rev. Tuberc.* **78,** 353.

REALE F. R. and ESTERLY J. R. (1973) *Pediatrics* **51,** 91.

REATEGUI-LOPEZ L. (1969) *Rev. peru. Cardiol.* **15,** 45.

REDDY P. A., GORELICK D. F. and CHRISTIANSON C. S. (1970) *Chest* **58,** 319.

REDDY P C., CHRISTIANSON C. S., GORELICK D. F. and LARSH H. W. (1969) *Thorax* **24,** 722.

REED C. E. and BARBEE R. A. (1965) *J. Amer. med. Ass.* **193,** 261.

REES G. M. (1973) *Thorax* **28,** 429.

REES L. H., BLOOMFIELD G. A. and REES G. M. (1974) *J. clin. Endocr.* **38,** 1090.

REEVES D. L., BUTT E. M. and HAMMACK R. W. (1941) *Arch. intern. Med.* **68,** 57.

Registrar General's Decennial Suppl. England and Wales (1951) *Occupational Mortality*, Part 1. H.M.S.O., London, p. 11.

Registrar General's Statistical Review of England and Wales (1957), Part 3 Commentary. H.M.S.O., London.

REICH N. E., McCORMACK L. J. and VAN ORDSTRAND H. S. (1974) *Chest* **65,** 424.

REICHLE H. S. (1936) *Amer. J. Path.* **12,** 781.

REICHLIN S., LOVELESS M. H. and KANE E. G. (1953) *Ann. int. Med.* **38,** 113.

REID A. and HEARD B. E. (1963) *Med. thorac.* **19,** 194.

REID D. D. (1956) *Proc. roy. Soc. Med.* **49,** 767; (1964) *ibid.* **57,** 966.

RÉID D. E., WEINER A. E. and ROBY C. C. (1953) *Amer. J. Obstet. Gynec.* **66,** 465.

REID J. A. and HEARD B. E. (1963) *Thorax* **18,** 20.

REID LYNNE McA. (1950) *Thorax* **5,** 233.

REID LYNNE McA. (1954) *Lancet* **1,** 275.

REID LYNNE McA. (1955) *Thorax* **10,** 199.

REID LYNNE McA. (1958) in *Recent Trends in Chronic Bronchitis.* Lloyd-Luke, London.

REID LYNNE McA. (1958–9) *Lectures on the Scientific Basis of Medicine* **8,** p. 235. Athlone Press, London.

REID LYNNE McA. and RUBINO M. (1959a) *Thorax* **14,** 3.

REID LYNNE McA. (1959b) *Thorax* **14,** 138.

REID LYNNE McA. (1960) *Thorax* **15,** 132.

REID LYNNE McA. and LORRIMAN G. (1960) *Brit. J. Dis. Chest* **54,** 321.

REIDBORD H. E. (1974) *Arch. Path.* **98,** 122.

REIMANN H. A. (1938) *The Pneumonias*, p. 381. W. B. Saunders, Philadelphia.

REITER B. R. (1936) *Arch. Path.* **22,** 269.

REMERGER K. and HÜBNER G. (1974) *Virchows Arch. path. Anat.* **363,** 363.

RÉNON L. (1897) *Étude sur l'aspergillose chez les animaux et chez l'homme.* Masson et Cie, Paris.

RESNICK M. E., BERKOWITZ R. D. and RODMAN T. (1961) *Amer. J. Med.* **31,** 149.

RESTREPO G. and HEARD B. E. (1963a) *J. Path. Bact.* **85,** 305.

RESTREPO G. and HEARD B. E. (1963b) *Thorax* **18,** 334.

REYE R. D. K. and BALE P. M. (1973) *Arch. Path.* **96,** 427.

RHODES A. E. (1945) *Amer. J. Path.* **21,** 507.

RIBBERT H. (1900) *Dtsch. med. Wschr.* **26,** 419.

RIBBERT H. (1904) *Zbl. allg. Path. path. Anat.* **15,** 945.

RICH A. R. and GREGORY J. E. (1943) *Bull. Johns Hopk. Hosp.* **73,** 465.

RICH A. R. (1946–7) *Harvey Lect.* **42,** 106.

RICH A. R. (1948) *Bull. Johns Hopk. Hosp.* **82,** 389.

RICH A. R. (1951) *The Pathogenesis of Tuberculosis*, 2nd ed. C. C. Thomas, Springfield, Ill.

RICH A. R., COCHRAN T. H. and McGOON D. C. (1951) *Bull. Johns Hopk. Hosp.* **88,** 101.

RICHARDS D. W. (1956) *Bull. N.Y. Acad. Med.* **32,** 407.

RICHARDS M. (1956) *Trans. Brit. mycol. Soc.* **39,** 431.

RICHARDS R. L. and MILNE J. A. (1958) *Thorax* **13,** 238.

RICHARDSON H. L., HUNTER W. C., CONKLIN W. S. and PETERSEN A. B. (1949) *Amer. J. clin. Path.* **19,** 323.

RICHARDSON J. A. and PRATT-THOMAS H. R. (1951) *Amer. J. med. Sci.* **221,** 531.

RICHERT J. H. and KRAKAUR R. B. (1959) *J. Amer. med. Ass.* **169,** 1302.

RICKARDS A. G. and BARRETT G. M. (1958) *Thorax* **13,** 185.

RICKEN D. (1958) *Virchows Arch. path. Anat.* **331,** 713

RIDDELL A. R. (1948) *Occup. Med.* **5,** 716.

RIDDELL R. W. and CLAYTON Y. M. (1958) *Brit. J. Tuberc.* **52,** 34.

RIDDELL R. W. and CLAYTON Y. M. (1964) *5th Congress Int. Acad. Path., London.*

RIDGEWAY N. A., WHITCOMB F. C., ERICKSON E. E. and LAW S. W. (1962) *Amer. J. Med.* **32,** 153

RIFKIND D., CHANOCK R., KRAVETZ H., JOHNSON K. and KNIGHT V. (1962) *Amer. Rev. resp. Dis.* **85,** 479.

RIFKIND D., STARZL T. E., MARCHIORO T. L., WADDELL W. R., ROWLANDS D. T. Jr. and HILL R. B. (1964) *J. Amer. med. Ass.* **189,** 808.

RIGDON R. H. and KIRCHOFF H. (1961) *Texas Rep. Biol. Med.* **19,** 465.

RIGGS B. L. Jr. and SPRAGUE R. G. (1961) *Arch. int. Med.* **108,** 841.

RIGLER L. G., O'LAUGHLIN B. J. and TUCKER R. C. (1953) *Dis. Chest* **23,** 50.

RIGLER L. G. (1957) *J. thorac. Surg.* **34,** 283.

RIGLER L. G. (1964) *Ann. N.Y. acad. Sci.* **114,** 755.

RIGLER L. G. (1975) *Radiologic and other Biophysical Methods*, in *Tumor Diagnosis*, p. 7, Year Book Med. Publishers Inc., Chicago.

RILEY R. L., RILEY M. C. and HILL H. M. (1952) *Bull. Johns Hopk. Hosp.* **91**, 345.

RILEY R. L. (1962) in Ciba symposium on *Pulmonary Structure and Function*, p. 271. J. & A. Churchill, London.

RIMINGTON R. A. (1962) *Med. J. Australia* **1**, 50.

RINDFLEISCH E. (1872) *Zbl. med. Wiss.* **10**, 65.

RITCHIE B. C., SCHAUBERGER G. and STAUB N. C. (1969) *Circulat. Res.* **24**, 807.

RITTER J. (1879) *Dtsch. Arch. klin. Med.* **25**, 53.

RIVERS D., WISE M. E., KING E. J. and NAGELSCHMIDT G. (1960) *Brit. J. industr. Med.* **17**, 87.

RIVERS T. M. (1928) *Amer. J. Path.* **4**, 91.

RIVERS T. M., BERRY G. P. and SPRUNT D. H. (1931) *J. exp. Med.* **54**, 91.

ROBAKIEWICZ M. and GRZYBOWSKI S. (1974) *Amer. Rev. resp. Dis.* **109**, 613.

ROBB D. (1958) *Brit. J. Surg.* **46**, 173.

ROBBINS L. L. and SNIFFEN R. C. (1949) *Radiology* **53**, 187.

ROBBINS L. L. (1953) *Cancer (Philadelphia)* **6**, 80.

ROBB-SMITH A. H. T. (1941) *Lancet* **1**, 135.

ROBERG N. B. (1945) *Bull. U.S. Army med. Dept.* **4**, 97.

ROBERTS G. B. S. and BAIN A. D. (1958) *J. Path. Bact.* **76**, 111.

ROBERTSON A. J., RIVERS D., NAGELSCHMIDT G. and DUNCUMB P. (1961) *Lancet* **1**, 1089.

ROBERTSON A. J. (1964) *Lancet* **1**, 1229, 1289.

ROBERTSON H. E. (1924) *J. Cancer Res.* **8**, 317.

ROBERTSON J. L. and BRINKMAN G. L. (1961) *Amer. J. Med.* **31**, 483.

ROBERTSON O. H., COGGESHALL L. T. and TERRELL E. E. (1933) *J. clin. Invest.* **12**, 467.

ROBERTSON O. H. and UHLEY C. G. (1936) *J. clin. Invest.* **15**, 115.

ROBERTSON O. H. (1938) *J. Amer. med. Ass.* **111**, 1432.

ROBERTSON O. H. (1941) *Physiol. Rev.* **21**, 112.

ROBERTSON R. and JAMES E. S. (1951) *Pediatrics* **8**, 795.

ROBINSON G. C. (1905) *J. infect. Dis.* **2**, 498.

ROBINSON K. P., MCGRATH R. and MCGREW E. (1963) *Surgery* **53**, 630.

ROBINSON S. S. and TASKER S. (1949) *Ann. Med; west. Surg.* **3**, 55

ROBINSON W. L. (1933) *Brit. J. Surg.* **21**, 302.

ROBSON A. O. and JELLIFFE A. M. (1963) *Brit. med. J.* **2**, 207.

RODBARD S. (1950) *J. Lab. clin. Med.* **36**, 980.

RODBARD S. (1953) *Amer. J. Med.* **15**, 356.

RODGER R. C., TERRY L. L. and BINFORD C. H. (1951) *Amer. J. clin. Path.* **21**, 153.

RODIN A. E., HAGGARD M. E. and TRAVIS L. B. (1970) *Amer. J. Dis. Child.* **120**, 337.

RODMAN M. H. and JONES C. W. (1962) *New Engl. J. Med.* **266**, 805.

ROELSEN E., LUND T., SØNDERGAARD T., MØLLER B. and MYSCHETZKY A. (1959) *Acta med. scand.* **163**, 367.

ROFFO V. A. H. (1937) *Dtsch. med. Wschr.* **73**, 1200.

ROGERS E. (1946) *Lancet* **1**, 462.

ROHWEDDER J. J. and WEATHERBEE L. (1974) *Amer. Rev. resp. Dis.* **109**, 435.

ROKITANSKY VON C. (1842) *Handbuch der speciellen pathologischen Anatomie*, Vol. 2, 311. Braumüller and Seidel, Vienna.

ROKITANSKY VON C. (1854) *A Manual of Pathological Anatomy*, English translation for the Sydenham Society, London.

ROKITANSKY VON C. (1855) *Lehrbuch der Path. Anat.* Vol. 2. Braumüller, Vienna.

ROKOS J., VAEUSORN O., NACHMAN R. and AVERY M. E. (1968) *Pediatrics* **42**, 205.

ROKSETH R. and SORSTEIN O. (1960) *Acta med. scand.* **167**, 23.

ROLLESTON H. and TREVOR R. (1903) *Brit. med. J.* **1**, 361.

ROMANO N., ROMANO F. and CAROLLO F. (1971) *New Engl. J. Med.* **285**, 950.

ROMANOFF H. and MILWIDSKY H. (1962) *Brit. J. Chest* **56**, 139.

ROQUE F. T., LUDWICK R. W. and BELL J. C. (1953) *Ann. intern. Med.* **38**, 1206.

ROSE A. H. (1957) *J. thorac. Surg.* **33**, 254.

ROSE G. A. and SPENCER H. (1957) *Quart. J. Med.* **26** N.S., 43.

ROSE M. S. (1975) *Brit. med. J.* **1**, 123.

ROSEN L., BOWDEN D. H. and UCHIDA I. (1957) *Arch. Path.* **63**, 316.

ROSEN S. H., CASTLEMAN B. and LIEBOW A. A. (1958) *New Engl. J. Med.* **258**, 1123.

ROSENBERG B. F., SPJUT H. J. and GEDNEY M. M. (1959) *New Engl. J. Med.* **261**, 226.

ROSENBERG L. M. (1965) *J. Amer. med. Ass.* **192**, 717.

ROSENDAL T. (1942) *Acta radiol. (Stockholm)* **23**, 138.

ROSENOW E. C., DE REMEE R. A. and DINES D. E. (1968) *New Engl. J. Med.* **279**, 1258.

ROSENTHAL S. R. (1930–1) *J. Lab. clin. Med.* **16**, 107.

ROSS E. J. (1963) *Quart. J. Med.* **32** N.S., 297.

ROSS E. J. (1966) *Proc. roy. Soc. Med.* **59**, 335.

ROSS JOAN M. (1957) *J. Path. Bact.* **73**, 485.

ROSS P. J., SEATON A., FOREMAN H. M. and MORRIS EVAN W. H. (1974) *Thorax* **29**, 659.

RÖSSLE R. (1923) *Verh. dtsch. path. Ges.* **19**, 18.

RÖSSLE R. (1928) *Verh. dtsch. path. Ges.* **23**, 289.

RÖSSLE R. (1933) *Klin. Wschr.* **12**, 574.

RÖSSLE R. (1937) *Virchows Arch. path. Anat.* **300**, 180.

RÖSSLE R. (1943) *Schweiz. med. Wschr.* **73**, 1200.

ROSTOSKI O., SAUPE E. and SCHMORL G. (1926) *Z. Krebsforsch.* **23**, 360.

ROTH F. (1960) *Z. allg. Path.* **9**, 529.

ROTTA A. (1947) *Amer. Heart J.* **33**, 669.

ROTTINO A. and HOFFMAN G. (1955) *Amer. J. Surg.* **89**, 550.

ROUGHTON F. J. W. (1945) *Amer. J. Physiol.* **143**, 621.

ROUVIÈRE H. (1939) *Anatomy of the Human Lymphatic System*. Edwards, Ann Arbor.

ROYES K. (1938) *Brit. med. J.* **2**, 659.

ROYSTON G. R. (1949) *Brit. med. J.* **1**, 1030.

ROZSA S. and FRIEMAN H. (1953) *Amer. J. Roentgenol.* **70**, 982.

RUBIN E. and ZAK F. G. (1960) *New Engl. J. Med.* **262**, 1315.

RUBIN E. and STRAUSS L. (1961) *Amer. J. Path.* **39**, 145.

RUBIN E. H. and ARONSON W. (1940) *Amer. Rev. Tuberc.* **41**, 801.

RUBIN E. H. (1955) *Amer. J. Med.* **19**, 569.

RUBIN E. H. and LUBLINER R. (1957) *Medicine (Baltimore)* **36**, 397.

RUBIO F. A. (1972) *New Engl. J. Med.* **287**, 1150.

RUDOLPH A. M., DRORBAUGH J. E., AULD P. A. M., RUDOLPH A. J., NADAS A. S., SMITH C. A. and HUBBELL J. P. (1961) *Pediatrics* **27**, 551.

RUNYON E. H. (1959a) *Med. Clin. N. America* **43**, 273.

RUNYON E. H. (1959b) *Amer. Rev. resp. Dis.* **80**, 277.

RUSBY N. L. and WILSON C. (1960) *Quart. J. Med.* **29** N.S., 501.

RÜTTNER J. R., WILLY W. and BAUMANN A. (1954) *Schweiz Z. allg. Path.* **17**, 352.

RÜTTNER J. R. and STOFER A. R. (1954) *Schweiz. med. Wschr.* **84**, 1433.

RÜTTNER J. R., SPYCHER M. A. and ENGELER M.-L. (1968) *Path. et Microbiol.* **32**, 1.

RÜTTNER J. R. (1969) *Proc. Int. Conf. Pneumoconiosis, Johannesburg, 1969,* Ed. H. A. Shapiro, p. 448. Oxford Univ. Press, Cape Town.

RUYSCH F. (1696) *Epistola Anatomica Probl. Sexto.* Amsterdam (quoted by CUDKOWICZ L., 1968).

RYAN R. F. and MCDONALD J. R. (1956) *Mayo Clin. Proc.* **31**, 478.

RYAN S. F. (1963) *Amer. J. Path.* **43**, 767.

RYAN S. F., CIANNELLA A. and DUMAIS C. (1969) *Anat. Rec.* **165**, 467.

RYDELL R. and HOFFBAUER F. W. (1956) *Amer. J. Med.* **21**, 450.

RYDER R. C., DUNNILL M. S. and ANDERSON J. A. (1971) *J. Path.* **104**, 59.

RYLAND D. and REID L. (1971) *Thorax* **26**, 602.

SABAR I. R. (1963) *Beitr. klin. Tuberk.* **126**, 145.

SABIN A. B. and FELDMAN H. A. (1948) *Science* **108**, 660.

SABISTON D. C. (1958) *J. thorac. Surg.* **36**, 653.

SACCOMANNO G., ARCHER V. E. and SAUNDERS R. P. (1964) *Hlth. Phys.* **10**, 1195.

SACKNER M. A., AKGUN N., KIMBEL P. and LEWIS D. H. (1964) *Ann. Int. Med.* **60**, 611.

SADUN E. H. and VAJRASTHIRA S. (1952) *J. Parasitol.* **34**, 22 (Section 2).

SAITA G. and ARRIGONI-MARTELLI E. (1956) *Med. d. Lavoro* **47**, 367.

SAKABE H. (1964) *Proc. roy. Soc. Med.* **57**, 1005.

SAKNYN A. V. and SHABYNINA N. K. (1973) *Gig. Tr. prof. Zabol.* **9**, 25 (in Russian).

SAKULA A. (1963) *Brit. J. Dis. Chest* **57**, 147.

SALE L. and WOOD W. B. (1947) *J. exp. Med.* **86**, 239.

SALÉK J., PAZDERKA S. and ZÁK F. (1958) *J. thorac. Surg.* **35**, 807.

SALFELDER K. and SCHWARZ J. (1967) *Mykosen* **10**, 337.

SALFELDER K., DOEHNERT G. and DOEHNERT H-R. (1969) *Virchows Arch. path. Anat.* **348**, 51.

SALFELDER K., FINGERLAND A., DE MENDÉLOVICI M. and ZAMBRANO Z. P. (1973) *Beitr. Path.* **148**, 94.

SALM R. (1963) *J. Path. Bact.* **85**, 121.

SALTZSTEIN S. L. (1963) *Cancer* **16**, 928.

SALTZSTEIN S. L. (1969) in *Pathology Annual,* Ed. S. C. SOMMERS. Appleton-Century-Crofts, New York, p. 160.

SALVAGGIO J. E., BUECHNER H. A., SEABURY J. H. and ARQUEMBOUR P. (1966) *Ann. int. Med.* **64**, 748.

SALYER J. M. and HARRISON H. N. (1958) *J. thorac. Surg.* **36**, 818.

SAMÁNEK M., TŮMA S., BENEŠOVÁ D., POVÝŠILOVÁ V., PRAŽSKÝ F. and CÁPOVÁ E. (1974) *Thorax* **29**, 446.

SANBORN E. B. and PURCELL E. M. (1957) *J. thorac. Surg.* **34**, 85.

SANDERUD K. (1958) *Acta path. microbiol. scand.* (a) **42**, 247; (b) **43**, 47.

SANDISON A. T. (1955) *Arch. Dis. Childh.* **30**, 475–7.

SANDLER M., SCHEVER P. J. and WATT P. J. (1961) *Lancet* **2**, 1067.

SANDOZ E. (1907) *Ziegler's Beitr.* **41**, 495.

SANERKIN N. G., SEAL R. M. E. and LEOPOLD J. G. (1966) *Ann. Allergy* **24**, 586.

SANFELICE (1894) quoted by CARTER and YOUNG (1950).

SANGER P. W., ROBICZEK F., TAYLOR F. H., MAGISTRO R. and FOTI E. (1959) *J. thorac. Surg.* **37**, 774.

SAPHIR O. (1947) *Amer. J. Path.* **23**, 245.

SAPHRA I. and WINTER J. W. (1957) *New Engl. J. Med.* **256**, 1128.

SAPPINGTON S. W., DAVIE J. H. and HORNEFF J. A. (1942) *J. Lab. clin. Med.* **27**, 882.

SARNOFF S. J. (1951) *Fed. Proc.* **10**, 118.

SARNOFF S. J. and KAUFMAN H. E. (1951) *Proc. Soc. exp. Biol. (N.Y.)* **78**, 829.

SARNOFF S. J. and BERGLUND E. (1952a) *Amer. J. Physiol.* **170**, 588.

SARNOFF S. J. and BERGLUND E. (1952b) *Amer. J. Physiol.* **171**, 238.

SAROSI G. A., HAMMERMAN K. J., TOSH F. E. and KRONENBERG R. S. (1974) *New Engl. J. Med.* **290**, 540.

SAUERBRUCH F. (1934) *Arch. klin. Chir.* **180**, 312.

SAUTTER R. D., FLETCHER F. W., EMANUEL D. A., LAWTON B. R. and OLSEN T. G. (1964) *J. Amer. med. Ass.* **189**, 948.

SAVAGE M. B. (1951) *Amer. J. Obstet. Gynec.* **62**, 346.

SAWYER K. C., SAWYER R. B., LUBCHENCO A. E., MCKINNON D. A. and HILL K. A. (1967) *Cancer (Philadelphia)* **20**, 451.

SAWYERS J. L., SESSIONS R. T., KILLEN D. A. and FOSTER J. H. (1964) *J. thorac. cardiovasc. Surg.* **48**, 661.

SAYERS R. R. (1937) *3rd Saranac Symposium on Silicosis.* Trudeau Sch. Tuberc., Saranac Lake, N.Y.

SCADDING J. G. (1937) *Brit. med. J.* **2**, 956.

SCADDING J. G. (1950) *Brit. med. J.* **1**, 745.

SCADDING J. G. and HINSON K. F. W. (1967) *Thorax* **22**, 291.

SCADDING J. G. (1971) *Proc. roy. Soc. Med.* **64**, 381.

SCHAEFER K. E., AVERY M. E. and BENSCH K. G. (1964) *J. clin. Invest.* **43**, 2080.

SCHAFTER E. L. (1949) *Amer. J. Path.* **25**, 49.

SCHARFF R. W. and GOWAR F. J. S. (1944) *J. Path. Bact.* **56**, 257.

SCHAUMANN J. (1917) *Ann. Derm. Syph. (Paris)*, 5th Series, **6**, 357.

SCHAUMANN J. (1941) *Acta med. scand.* **106**, 239.

SCHEEL L. D., FLEISHER E. and KLEMPERER F. W. (1953) *Arch. Industr. Hyg.* **8**, 564.

SCHEEL L. D., SMITH B., VAN RIPER J. and FLEISHER E. (1954) *Arch. industr. Hyg.* **9**, 29.

SCHEID K. F. (1932) *Beitr. path. Anat.* **88**, 224.

SCHEIDEGGER S. (1936) *Frankfurt. Z. Path.* **49**, 362.

SCHENKEN J. R. and COLEMAN F. C. (1943) *Amer. J. Surg.* **61**, 126.

SCHEPERS G. W. H. and DURKAN T. M. (1955) *Arch. industr. Hlth.* **12**, 182, 317.

SCHEPERS G. W. H., SMART R. H., SMITH C. R., DWORSKI M. and DELAHANT A. B. (1958) *Industr. Med. Surg.* **27**, 27.

SCHEPERS G. W. H. (1961) *Exper. Tum. Res.* **2**, 203 (quoted by HUEPER W. C. (1962)).

SCHERRER M. (1970) *Schweiz med. Wschr.* **100**, 2251.

SCHIERGE M. (1922) *Virchows Arch. path. Anat.* **237**, 129.

SCHILDKNECHT O. (1932) *Virchows Arch. path. Anat.* **285**, 466.

SCHILLING R. S. F. (1956) *Lancet* **2**, 261, 319.

SCHLAEPFER K. (1924) *Arch. Surg. (Chicago)* **9**, 25.

SCHLUMBERGER H. G. (1946) *Arch. Path.* **41**, 398.

SCHMIDT M. B. (1903) *Die Verbreitungswege der Karzinome und die Beziehung generalisierter Sarkome zu den leukämischen Neubildungen.* G. Fischer, Jena.

SCHMIT H. (1893) *Virchows Arch. path. Anat.* **134**, 25.

SCHMITZ-MOORMANN P., HÖRLEIN H. and HANEFELD F. (1964) *Beitr. Silikose-Forsch.* **80**, 1.

SCHMORL G. (1888) *Dtsch. Arch. klin. Med.* **42**, 409.

SCHMORL G. (1893) *Pathologisch-anatomische Untersuchungen über Puerperal-Eklampsie.* F. C. Vogel, Leipzig.

SCHMORL G. (1905) *Zbl. Gynäk.* **29**, 129.

SCHMORL G. (1926) *Verh. dtsch. path. Ges.* **21**, 420.

SCHNEIDER C. L. (1951) *Surg. Gynec. Obstet.* **92**, 27.

SCHNEIDER C. L. (1952) *Amer. J. Obstet. Gynec.* **63**, 1078.

SCHNEIDER H. (1957) *Münch. med. Wschr.* **99**, 1858.

SCHNEIDER P. (1912a part 2) in *Die Morphologie der Mißbildungen des Menschen und der Tiere*, Vol. 3, p. 772. Ed. by SCHWALBE E. G. Fischer, Jena.

SCHNEIDER P. (1912b) in *Die Morphologie der Mißbildungen des Menschen und der Tiere. Ein Hand- und Lehrbuch für Morphologen, Physiologen, praktische Ärzte und Studierende*, Vol. 3, p. 763. Ed. by SCHWALBE E. G. Fischer, Jena.

SCHNOOR E. E. and CONNOLLY J. E. (1958) *Amer. J. Surg.* **96**, 107.

SCHRAG P. E. and GULLETT A. D. (1970) *Amer. Rev. resp. Dis.* **101**, 497.

SCHÜLLER A. (1915) *Fortschr. Röntgenstr.* **23**, 12.

SCHULTE H. W. (1957) *Tuberk.-Arzt* **11**, 751.

SCHULTZ H. (1962) Ciba Foundation Symposium on *Pulmonary Structure and Function*, p. 205. J. and A. Churchill, London.

SCHULZ D. M. (1950) *Arch. Path.* **50**, 457.

SCHULZ D. M. (1954) *Amer. J. clin. Path.* **24**, 11.

SCHULZ H. (1959) *The Submicroscopic Anatomy and Pathology of the Lung.* Springer-Verlag, Berlin.

SCHWARTZ W. B., BENNETT W., CURELOP S. and BARTTER F. C. (1957) *Amer. J. Med.* **23**, 529.

SCHWARZ J. and BAUM G. L. (1951) *Amer. J. clin. Path.* **21**, 999.

SCHWARZ J. and BAUM G. L. (1961) *Amer. Rev. resp. Dis.* **84**, 114.

SCHWARZ J. and BAUM G. L. (1963) *Arch. Path.* **75**, 475.

SCHWARZ J. (1971) in *Handb. der Spez. Path. Anat. u. Histol.* III/5b, 67. Springer, Berlin–Heidelberg–New York.

SCHWARZ M. I., GOLDMAN A. L., ROYCROFT D. W. and HUNT K. K. (1972) *Amer. Rev. resp. Dis.* **106**, 109.

SCHWARZ P. (1958) *Presse méd.* **66**, 1369.

SCOFIELD G. F. and BEAIRD J. B. (1957) *Amer. J. clin. Path.* **28**, 400.

SCOTT R. F., THOMAS W. A. and KISSANE J. M. (1959) *J. Pediat.* **54**, 60.

SEARS M. R., CHANG A. R. and TAYLOR A. J. (1971) *Thorax* **26**, 704.

SEILER E. (1960) *Schweiz. Z. Tuberk.* **17**, 205.

SELIKOFF I. J., HAMMOND E. C. and CHURG J. (1968) *J. Amer. med. Ass.* **204**, 106.

SELIKOFF I. J., HAMMOND, E. C. and CHURG J. (1970) in *Proc. Int. Pneumoconiosis Conf., Johannesburg, 1969.* Ed. H. A. SHAPIRO, Oxford Univ. Press, Cape Town, p. 183.

SELIKOFF I. J., HAMMOND E. C. and CHURG J. (1971) in *Proc. 4th Internat. Pneumoconiosis Conference, Bucharest.*

SELL S. H. W. (1960) *J. Dis. Child.* **100**, 7.

SEMPLE J. and WEST L. R. (1955) *Thorax* **10**, 287.

SEMSROTH K. H. (1939) *Arch. Path.* **28**, 386.

SEPKE G. (1957) *Zbl. Arbeitsmed.* **7**, 114.

SEREDA M. M., SHERMAN L. and SMITH E. M. G. (1961) *Canadian med. Ass. J.* **84**, 1136.

SEVITT S. and GALLAGHER N. G. (1959) *Lancet* **2**, 981.

SEVITT S. and GALLAGHER N. G. (1961) *Brit. J. Surg.* **48**, 475.

SEVITT S. (1962) *Fat Embolism.* Butterworths, London.

SEVITT S. (1966) *Lancet* **2**, 1203.

SEVITT S. (1973) in *Recent Advances in Thrombosis.* Ed. L. POLLER, p. 17. Churchill-Livingstone, Edinburgh and London.

SEVITT S. (1974) *J. clin. Path.* **27**, 21.

SEYFARTH C. (1924) *Dtsch. med. Wschr.* **1**, 1497.

SHABAD L. M. (1971) *Cancer (Philadelphia)* **27**, 51.

SHAFTER H. A. and BLISS H. A. (1959) *Amer. J. Med.* **26**, 517.

SHAMES J. M. DHURANDHAR N. R. and BLACKARD W. G. (1968) *Amer. J. Med.* **44**, 632.

SHANER R. F. (1961) *Anat. Rec.* **140**, 159.

SHAPIRO M., NICHOLSON W. E. and ORTH D. N. (1971) *J. clin. Endocr.* **33**, 377.

SHAPIRO R. and CARTER M. G. (1954) *Amer. Rev. Tuberc.* **69**, 1042.

SHARP MARY E. and DANINO E. A. (1953) *J. Path. Bact.* **65**, 389.

SHARPEY W. (1836) in *The Cyclopaedia of Anatomy and Physiology*, Vol. 1, p. 606. Ed. R. B. TODD. Sherwood, Gilbert and Piper Publications, London.

SHATALOV M. N. (1956) *Bjull. eksp. Biol. Med.* **41**, 74.

SHAVER C. G. (1948) *Occup. Med.* **5**, 718.

SHAW A. F. B. and GHAREEB A. A. (1938) *J. Path. Bact.* **46**, 401.

SHAW R. R. (1949) *Surg. Gynec. Obstet.* **88**, 753

SHAW R. R. (1951) *J. thorac. Surg.* **22**, 149.

SHEFFIELD INVESTIGATION INTO CHRONIC BRONCHITIS (1957) *Brit. med. J.* **1**, 261.

SHELDON W. H. (1959) *Amer. J. Dis. Child.* **97**, 287

SHELLEY W. M. and CURTIS E. M. (1958) *Bull. Johns Hopk. Hosp.* **103**, 8.

SHENNAN T. and MILLER J. (1910) *J. Path. Bact.* **14**, 556.

SHEPARD F. M., JOHNSTON R. B. Jr., KLATTE E. C., BURKO H. and STAHLMAN M. (1968) *New Engl. J. Med.* **279**, 1063.

SHEPHERD J. T., SEMLER H. J., HELMHOLTZ H. F. and WOOD E. H. (1959) *Circulation* **20**, 381.

SHERMAN F. E. and RUCKLE G. (1958) *Arch Path.* **65**, 587.

SHONE J. D., AMPLATZ K., ANDERSON R. C., ADAMS P. Jr. and EDWARDS J. E. (1962) *Circulation* 26, 274.

SHORT D. S. (1956) *J. Fac. Radiol.* **8**, 118.

SHORT R. H. D. (1942) *Porton Report No. 2349*, quoted by CAMERON G. R. (1948)

SHORT R. H. D. (1950) *Phil. Trans. B*, **235**, 35.

SHORTER R. G., TITUS J. L. and DIVERTIE M. B. (1964) *Dis. Chest* 46, 138.

SHORTER R. G., TITUS J. L. and DIVERTIE M. B. (1966) *Thorax* 21, 32.

SHREWSBURY J. F. D. (1936) *Quart. J. Med.* **29**, 375.

SHRINIWAS and BHATIA V. N. (1973) *Amer. Rev. resp. Dis.* **108**, 378.

SIASSI R., EMMANOUILIDES G. C., CLEVELAND R. J. and HIROSE F. (1969) *J. Pediat.* **74**, 11.

SICARD J. A. and FORESTIER J. (1922) *Bull. Soc. méd. Hôp. Paris* 46, 463.

SIKL H. (1950) *Acta Un. int. Canc.* **6**, 1366.

SILTZBACH L. E. (1962) *Thorax* **17**, 284.

SILTZBACH L. E. (1964) *Acta med. Scand.* **176** (suppl. 425), 178.

SILTZBACH L. E. (1965) *Amer. J. Med.* **39**, 361.

SILVERMAN F. N., SCHWARZ J., LAHEY M. E. and CARSON R. P. (1955) *Amer. J. Med.* **19**, 410.

SIMMONS D. H. and HEMINGWAY A. (1959) *Circulation Res.* **7**, 93.

SIMON G. and REID L. M. (1963) *Brit. J. Dis. Chest* **57**, 126.

SIMON M. A. and BALLON H. C. (1947) *J. thorac. Surg.* **16**, 379.

SIMONS E. J. (1937) *Primary Carcinoma of the Lung.* Year Book Publishing, Chicago.

SIMOPOULOS A. P., ROSENBLUM D. J., MAZUMDAR H. and KIELY B. (1959) *Amer. J. Dis. Child.* **97**, 796.

SIMPSON J. A. (1927) *Amer. J. Path.* **3**, 93.

SIMPSON S. L. (1929) *Quart. J. Med.* **22**, 413.

SIMSON F. W. (1929) *Ann. Rep. South African Inst. Med. Res.*, p. 64.

SIMSON F. W. and STRACHAN A. S. (1931) *J. Path. Bact.* **34**, 1.

SIMSON F. W. (1935) *J. Path. Bact.* **40**, 37.

SIMSON I. W. and HEINZ H. J. (1960) *Leech (Johannesburg)* **30**, 140.

SINCLAIR R. J. G. and CRUICKSHANK B. (1956) *Quart. J. Med.* 25 N.S., 313.

SINCLAIR-SMITH C. C., EMERY J. L., GADSON D., DINSDALE F. and BADDELEY J. (1976) *Thorax* 31, 40.

SINGER D. B., GREENBERG S. D. and HARRISON G. M. (1966) *Amer. Rev. resp. Dis.* **94**, 777.

SIR G. (1962) *Zbl. allg. Path.* **103**, 129.

SIVANESAN S. (1959) *Arch. Dis. Childh.* **34**, 426.

SIWE S. A. (1933) *Kinderheilk.* **55**, 212.

SIWE S. A. (1949) *Advances in Pediat.* **4**, 117.

SIXTH SARANAC SYMPOSIUM (1950) *L. U. Gardner Memorial Vol.* Paul B. Hoeber, New York.

SJØGREN H. (1933) *Acta ophthal (Kbh.)*, suppl. No. 2.

SJØGREN H. (1951) *Acta ophthal. (Kbh.)* **29**, 33.

SKIKNE M. I., PRINSLOO I. and WEBSTER I. (1972) *J. Path.* **106**, 119.

SKINNER D. B. (1963) *New Engl. J. Med.* **268**, 1324.

SKRAMOVSKY V. (1963) *Neoplasma (Bratislava)* **10**, 413.

SLOAN H. (1953) *J. thorac. Surg.* **26**, 1.

SLOAN R. D. and COOLEY R. N. (1953) *Amer. J. Roentgenol.* **70**, 183.

SLUIS-CREMER G. K. and DU TOIT R. S. J. (1968) *Brit. J. industr. Med.* **25**, 63.

SLUTZKER B., KNOLL H. C., ELLIS F. E. and SILVERSTONE I. A. (1961) *Arch. int. Med.* **107**, 264.

SLUYTER F. T. (1847) Inaugural Dissertation, Berlin, cited by RENON L. (1897).

SMALL M. J. (1959) *Arch. intern. Med.* **104**, 730.

SMART J. (1946) *Quart. J. Med.* **15** N.S., 125.

SMETANA H. F. and SCOTT W. F. (1951) *Mil. Surg.* **109**, 330.

SMILLIE W. G. and LEEDER F. S. (1920) *Amer. J. publ. Hlth.* **21**, 129.

SMITH A. G. and GILLOTTE J. P. (1960) *Amer. J. clin. Path.* **34**, 477.

SMITH F. R. and BOYDEN E. A. (1948) *J. thorac. Surg.* **18**, 195.

SMITH H. (1948) *Amer. J. Path.* **24**, 223.

SMITH J. C. (1963) *Amer. Rev. resp. Dis.* **87**, 647.

SMITH J. F. (1948) *J. Path. Bact.* **60**, 489.

SMITH J. F., DEXTER D. (1963) *Thorax* **18**, 340.

SMITH J. G., HARRIS J. S., CONANT N. F. and SMITH D. T. (1955) *J. Amer. med. Ass.* **158**, 641.

SMITH M. G. and VELLIOS F. (1950) *Arch. Path.* **50**, 862.

SMITH M. J. and NAYLOR B. (1972) *Amer. J. clin. Path.* **58**, 250.

SMITH P., HEATH D. and MOOSAVI H. (1974) *Thorax* **29**, 147.

SMITH P. and HEATH D. (1974) *Thorax* **29**, 643.

SMITH R. A. (1954) *Brit. J. Tuberc.* **48**, 311.

SMITH R. A. (1956) *Thorax* **11**, 10.

SMITH R. A. (1962) *Surg. Gynec. Obst.* **114**, 57.

SMITH R. C., BURCHELL H. B. and EDWARDS J. E. (1954) *Circulation* **10**, 801.

SMITH R. W., MORRIS J. A., MANSON W., BECK R. and ASSALI N. S. (1963) *Circulation* **28**, 808.

SMITH W., ANDREWES C. H. and LAIDLAW P. P. (1933) *Lancet* **2**, 66.

SMITH W. G. (1958) *Thorax* **13**, 194.

SNAPPER I. and POMPEN A. W. N. (1938) *Pseudo-tuberculosis in Man*, Parts I and II. Bohn, Haarlem.

SNEDDON I. B. (1955) *Brit. med. J.* **1**, 1448.

SNIDER G. L., HAYES J. A., KORTHY A. L. and LEWIS G. P. (1973) *Amer. Rev. resp. Dis.* **108**, 40.

SNIJDER J. (1965) *J. Path. Bact.* **90**, 65.

SOBOTTA R. H J. (1930) *Textbook of Human Histology and Microscopic Anatomy*, translated by W. H. PIERSOL, 4th ed. p. 77. G. E. Stechert, New York.

SODEMAN W. A. and PULLEN R. L. (1944) *Arch. intern. Med.* **73**, 365.

SOERGEL K. H. (1957) *Pediatrics* **19**, 1101.

SOERGEL K. H. and SOMMERS S. C. (1962a) *Amer. J. Med.* **32**, 399.

SOERGEL K. H. and SOMMERS S. C. (1962b) *Amer. Rev. resp. Dis.* **85**, 540.

SOLLIDAY N. H., WILLIAMS J. A., GAENSLER E. A., COUTU R. E. and CARRINGTON C. B. (1973) *Amer. Rev. resp. Dis.* **108**, 193.

SOLOMON S. (1940) *J. Amer. med. Ass.* **115**, 1527.

SOLWAY L. J., KOHAN M. and PRITZKER H. G. (1939) *Canad. med. Ass. J.* **41**, 331.

SOMMERS S. C. and McMANUS R. G. (1953) *Cancer (Philadelphia)* **6**, 347.

SOMMERS S. C. (1958) *Arch. Path.* **65**, 104.

SONTAG L. W. and ALLEN J. E. (1947) *J. Pediat.* **30**, 657.

SOREL R., LASSERRE J. and SALVADOR R. (1943) *Bull. Soc. méd. Hôp. Paris* **59**, 267.

SORENSEN S. C. and SEVERINGHAUS J. W. (1968) *J. appl. Physiol.* **25**, 217.

SOROKIN S., PADYKYULA H. A. and HERMAN E. (1959) *Develop. Biol.* **1**, 125.

SOROKIN S. P. (1970) *Arch. int. Med.* **126**, 450.

SOROUR M. F. (1932) *C.R. Congrès int. Méd. trop. Hyg. Cairo*, **4**, 321.

SOSMAN M. C., DODD G. D., JONES W. D. and PILLMORE G. U. (1957) *Amer. J. Roentgenol.* **77**, 947.

SOYSA E. and JAYWARDENA M. D. S. (1945) *Brit. med. J.* **1**, 1.

SPAIN D. M. and HANDLER B. J. (1946) *Arch. intern. Med.* **77**, 37.

SPAIN D. M. and MOSES J. B. (1946) *Amer. J. med. Sci.* **212**, 707.

SPAIN D. M. (1948) *Amer. J. Path.* **24**, 727.

SPAIN D. M. and PARSONNET V. (1951) *Cancer (Philadelphia)* **4**, 277.

SPAIN D. M. and KAUFMAN G. (1953) *Amer. Rev. Tuberc.* **68**, 24.

SPAIN D. M. (1954) *Dis. Chest* **25**, 550.

SPAIN D. M. (1957) *Amer. J. Path.* **33**, 582.

SPANNAGEL H. (1953) *Lungenkrebs und andere Organschäden durch Chromverbindungen (Arbeitsmedizin*, Heft 28). Barth, Leipzig.

SPEAR H. C., DAUGHTRY DE W. C., CHESNEY J. G., GENTSCH T. O. and LARSEN P. B. (1968) *New Engl. J. Med.* **278**, 832.

SPECTOR R. G., CLAIREAUX A. E. and ROHAN WILLIAMS E. (1960) *Arch. Dis. Childh.* **35**, 475.

SPECTOR W. G., WALTERS M. N.-I. and WILLOUGHBY D. A. (1965a) *J. Path. Bact.* **90**, 181.

SPECTOR W. G. and COOTE E. (1965b) *J. Path. Bact.* **90**, 589.

SPENCER H. (1950) *J. Path. Bact.* **62**, 75.

SPENCER H. and WHIMSTER I. W. (1950) *J. Path. Bact.* **62**, 411.

SPENCER H. and RAEBURN C. (1954) *J. Path. Bact.* **67**, 187.

SPENCER H. (1954) Ph.D. thesis, University of London.

SPENCER H. and RAEBURN C. (1956) *J. Path. Bact.* **71**, 145.

SPENCER H. (1957) *Brit. J. Tuberc.* **51**, 123.

SPENCER H. (1961) *J. Path. Bact.* **82**, 161.

SPENCER H. and SHORTER R. G. (1962) *Nature* **194**, 880.

SPENCER H. and LEOF D. (1964) *J. Anat.* **98**, 599.

SPENCER H. (1967) in *Annual Review of Medicine*, Vol. 18. Ed. A. C. DE GRAFF and W. P. CREGER. Ann. Reviews Inc., Palo Alto, p. 423.

SPENCER H. (1968) in *The Lung*. Ed. A. A. LIEBOW and D. E. SMITH. Williams & Wilkins Co., Baltimore, p. 134.

SPENCER H. (1973) in *Spezielle pathologische Anatomie*, Vol. 8, *Tropical Pathology*. Ed. H. SPENCER. Springer-Verlag, Berlin–Heidelberg–New York, p. 731.

SPENCER H. (1975) *Progress in Resp. Research* **8**, 34.

SPIRO R. H. and McPEAK C. J. (1966) *Cancer (Philadelphia)* **19**, 544.

SPJUT H. J., FIER D. J. and ACKERMAN L. V. (1955) *J. thorac. Surg.* **3**, 90.

SPLENDORE A. (1908) *Rev. Soc. Sci. S. Paulo* **3**, 109.

SPLENDORE A. (1912) in *Volume in onore del Prof. Angelo Celli nel 25° anno di insegnamento*. Rome, Tip. Naz. di G. Bertero E.C., quoted by LACAZ C. DA S.

SPRUNT D. H. and McBRYDE A. (1936) *Arch. Path.* **21**, 217.

STANKIEWICZ R., PRZESMYCKI F., KOLAGO J. and ZMIGRYDER L. (1932) *Rev. franç. Pédiat.* **7**, 324.

STANTON A. T. and FLETCHER W. (1921) *Trans. 4th Cong. Far East. Ass. trop. Med.* **2**, 196.

STANTON M. F. and WRENCH C. (1972) *J. nat. Cancer Inst.* **48**, 797.

STANTON R. E. and FYLER D. C. (1961) *Pediatrics* **27**, 621.

STARCK H. (1928) *Dtsch. Arch. klin. Med.* **162**, 68.

STARRS R. A. and KLOTZ M. O. (1948) *Arch. intern. Med.* **82**, 1, 29.

STAUB N. C. and SCHULTZ E. L. (1968) *Resp. Physiol.* **5**, 371.

STAUB N. C. (1974) *Physiol. Rev.* **54**, 678.
STEHBENS W. E. (1959) *J. Path. Bact.* **78**, 179.
STEIN A. A. and VOLK B. M. (1959) *Arch. Path.* **68**, 468.
STEIN J. and POPPEL M. H. (1955) *Amer. J. Surg.* **89**, 439.
STEINBERG B. and MUNDY C. S. (1936) *Arch. Path.* **22**, 529.
STEINBERG I. and MCCLENAHAN J. (1955) *Amer. J. Med.* **19**, 549.
STEINBERG I. (1958) *Amer. J. Med.* **24**, 559.
STEINER B. (1954) *Arch. Dis. Childh.* **29**, 391.
STEINER P. (1939) *Amer. J. Path.* **15**, 89.
STEINER P. E. and LUSHBAUGH C. C. (1941) *J. Amer. med. Ass.* **117**, 1245, 1340.
STEINER P. E. (1953) *Symposium on Cancer of the Lung held by the Unionis Internat. contra Cancrum*, p. 24. Louvain.
STENHOUSE A. C. (1967) *Brit. med. J.* **3**, 461.
STEPHANOPOULOS C. and CATSARAS H. (1963) *Thorax* **18**, 144.
STEPHENS G. A. (1920) *J. Industr. Hyg.* **2**, 129.
STEPHENS G. A. (1933) *Med. Press Circ.* **187**, 194, 216, 283.
STERN L. (1936) *J. thorac. Surg.* **6**, 2, 202.
STEVENSON J. G. and REID J. M. (1957) *Thorax* **12**, 300.
STEWART F. H. (1916) *Brit. med. J.* **2**, 5.
STEWART M. J. and FAULDS J. S. (1934) *J. Path. Bact.* **39**, 233.
STEWART M. J. and ALLISON P. R. (1943) *J. Path. Bact.* **55**, 105.
STEWART M. J., WRAY S. and HALL M. (1954) *J. Path. Bact.* **67**, 423.
STITT E. R., CLOUGH P. W. and BRANHAM S. E. (1948) *Practical Bacteriology, Hematology and Parasitology*, 10th ed. The Blakiston Company, New York.
STOCKS P. (1947) General Register Office, *Studies on Medical and Population Subjects* No. 1. H.M.S.O., London.
STOCKS P. and CAMPBELL J. M. (1955) *Brit. med. J.* **2**, 923.
STOCKS P. (1959) *Brit. med. J.* **1**, 74.
STOEBNER P., CUSSAC Y., PORTE A. and LE GAL Y. (1967) *Cancer (Philadelphia)* **20**, 286.
STOERK O. (1897) *Wien. klin. Wschr.* **10**, 25.
STOKER M. G. P. (1950) *Brit. med. J.* **2**, 282.
STOKES W. (1882) *A Treatise on the Diagnosis and Treatment of Diseases of the Chest*, Part 1, p. 129. Edited by A. HUDSON. The New Sydenham Society, London.
STOREY C. F. (1952) *J. thorac. Surg.* **24**, 16.
STOREY C. F., KNUDTSON K. P. and LAURENCE B. J. (1953) *J. thorac. Surg.* **26**, 331.
STOREY C. F. and MARRANGONI A. G. (1954) *J. thorac. Surg.* **28**, 536.
STOREY P. B. and GOLDSTEIN W. (1962) *Arch. int. Med.* **110**, 262.
STORRS R. P., MCDONALD J. R. and GOOD C. A. (1949) *J. thorac. Surg.* **18**, 561.
STORSTEIN O. (1951) *Circulation* **4**, 913.
STOUT A. P. (1935) *Amer. J. Cancer* **25**, 1.
STOUT A. P. and MURRAY M. R. (1942) *Ann. Surg.* **116**, 26.
STOUT A. P. (1943) *Arch. Path.* **35**, 803.
STOUT A. P. (1948) *Cancer (Philadelphia)* **1**, 30.
STOUT A. P. and HIMADI G. M. (1951) *Ann. Surg.* **133**, 50.
STOUT A. P. and HILL W. T. (1958) *Cancer (Philadelphia)* **11**, 844.
STOVIN P. G. I. and MITCHINSON M. J. (1965) *Thorax* **20**, 106.
STRACHAN A. S. (1934) *J. Path. Bact.* **39**, 209.
STRACHAN A. S. (1947) *Proc. of Conference on Silicosis, Pneumoconiosis and Dust Suppression in Mines.* Institution of Mining Engineers and the Institution of Mining and Metallurgy, London.
STRANG L. B. (1973) in *Foetal and Neonatal Physiology.* Ed. K. W. CROSS. Cambridge Univ. Press, Cambridge, p. 186.
STRATTON T. M. L. (1838) *Edin. med. surg. J.* **49**, 490.
STRAUB M. (1936) *Ned. T. Geneesk.* **80**, 1468.
STRAUB M. and MULDER J. (1948) *J. Path. Bact.* **60**, 429.
STRAUS I. and GAMALEYA (1891) *Arch. Méd. exp.* **3**, 457.
STRAUS R. (1942) *Arch. Path.* **33**, 69.
STRAUSS J. M., GROVES M. G., MARIAPPAN M. and ELLISON D. W. (1969) *Amer. J. trop. Med. Hyg.* **18**, 698.
STRAWBRIDGE H. T. G. (1960) *Amer. J. Path.* **37**, 391.
STREETER G. L. (1945) *Contr. Embryol. Carneg. Instn.* **31**, 29.
STRELINGER A. (1941–2) *J. lab. clin. Med.* **27**, 1510.
STRINGER C. J., STANLEY A. L., BATES R. C. and SUMMERS J. E. (1955) *Amer. J. Surg.* **89**, 1054.
STRONG L. C. (1936) *J. Hered.* **27**, 21.
STRONG R. P. (1944) *Stitt's Diagnosis, Prevention and Treatment of Tropical Diseases*, 7th ed. H. K. Lewis, London.
STUART B. M. and PULLEN R. L. (1945) *Amer. J. med. Sci.* **210**, 223.
STUART-HARRIS C. H. (1950) *Brit. med. J.* **2**, 282.
STUART-HARRIS C. H. (1954) *Brit. J. Tuberc.* **48**, 169.
STURDEE E. L. and SCOTT W. M. (1930) *Ministry of Health Reports on Public Health and Medical Subjects* No. 61, "A Disease of Parrots communicable to Man (Psittacosis)". H.M.S.O., London.
STURM A. (1946) *Dtsch. med. Wschr.* **71**, 201.
SUGARBAKER E. D. and CRAVER L. F. (1940) *J. Amer. med. Ass.* **115**, 17.
SULLIVAN M. X. (1926) *Publ. Hlth. Rep. (Wash.)* **41**, 1030.
SUMNER J., LICHTER A. I. and NASSAU E. (1951) *Thorax* **6**, 193.
SUNDERMANN F. W., DONNELLY A. J., WEST B. and KINCAID J. F. (1959) *Arch. industr. Hlth.* **20**, 36.
SUSMAN M. P. (1953) *J. thorac. Surg.* **26**, 111.
SUTLIFF W. D. and FRIEDEMANN T. E. (1938) *J. Immunol.* **34**, 455.
SUZUKI Y., CHURG J. and ONO T. (1972) *Amer. J. Path.* **69**, 373.
SWAN L. L. (1949) *Arch. Path.* **47**, 517.
SWEANY H. C., GORELICK D., COLLER F. C. and JONES J. L. (1958) *Dis. Chest* **34**, 119.
SWEET R. H. (1953) *J. thorac. Surg.* **26**, 18.
SWEETMAN W. R., HARTLEY L. J., BAUER A. J. and SALYER J. M. (1958) *J. thorac. Surg.* **35**, 802.

SWINBURN P. (1962) *N.Z. med. J.* **61**, 481.
SWYER P. R. and JAMES G. C. W. (1953) *Thorax* **8**, 133.
SYMMERS D. and HOFFMAN A. M. (1923) *J. Amer. med. Ass.* **81**, 297.
SYMMERS W. StC. (1952) *J. clin. Path.* **5**, 36.
SYMMERS W. StC. (1953) *Lancet* **2**, 1068.
SYMMERS W. StC. (1954) quoted by ROSE G. A. and SPENCER H. (1957).
SYMMERS W. StC. (1959) Personal communication.
SYMMERS W. StC. (1960) *J. clin. Path.* **13**, 1.
SYMMERS W. StC. (1964) *Proc. roy. Soc. Med.* **57**, 405.
SYMMERS W. StC. (1965) *Proc. roy. Soc. Med.* **58**, 341.
SYMPOSIUM No. 6 on the "Medical and Epidemiological Aspects of Air Pollution" (1964) *Proc. roy. Soc. Med.* **57**, 965.
SZUR L. and BROMLEY L. L. (1956) *Brit. med. J.* **2**, 1273.

TAFT E. B. and MALLORY G. K. (1946) *Arch. Path.* **42**, 630.
TAKARO T. and BOND W. M. (1958) *Int. Abstr. Surg.* **107**, 209.
TAKIZAWA T. (1956) *Tohuku Igaku Zasshi* **53**, suppl. No. 6, 97.
TALA E. (1967) *Acta radiol.* (suppl.) **268**, 1.
TALAL N., SOKOLOFF L. and BARTH W. F. (1967) *Amer. J. Med.* **43**, 50.
TALALAK P. (1960) *Arch. Dis. Child.* **35**, 57.
TAMMELING G. J., DE VRIES K., SLUITER H. J., ORIE N. G. M., TEN HAVE H., WITROP J. and ZUIDERWEG A. (1964) *Acta med. scand.* **176** (suppl. 425), 275.
TANNER F. W., LAMPERT E. N. and LAMPERT M. (1927) *Zbl. Bakt. I. Abt.* **103**, 94.
TARSHIS M. S. (1958) *Amer. Rev. Tuberc.* **78**, 921.
TAUB R. N. and SILTZBACH L. E. (1974) in *Proc. VIth Internat. Conference on Sarcoidosis*. Ed. K. IWAI and Y. HOSODA, p. 20. Univ. of Tokyo Press, Tokyo.
TAUSSIG HELEN B. (1947) *Congenital Malformations of the Heart*. Commonwealth Fund, London.
TAYLOR F. B. Jr. and ABRAMS M. E. (1966) *Amer. J. Med.* **40**, 346.
TAYLOR R. M. (1949) *Amer. J. Pub. Hlth.* **39**, 171.
TEABEAUT J. R. (1952) *Amer. J. Path.* **28**, 51.
TEACHER J. H. (1907–8) *J. Path. Bact.* **12**, 487.
TEILUM G. (1946) *Acta med. scand.* **123**, 126.
TEILUM G. (1964) *Acta med. scand.* **176** (suppl. 425), 14.
TEIR H., KOIVUNIEMI A., KYLLÖNEN K. E. J., TASKINEN E. and SAKAI Y. (1963) *Ann. Chir. Gynaec. Fenn.* **52**, 502.
TELISCHI M. and RUBENSTONE A. I. (1961) *Arch. Path.* **72**, 234.
TENNANT R., JOHNSTON H. J. and WELLS J. B. (1961) *Conn. Med.* **25**, 106.
TENNEY S. M. and REMMERS J. E. (1963) *Nature* **197**, 54.
TEPLITZ C. (1965) *Arch. Path.* **80**, 297.
TERRY R. D., SPERRY W. M. and BRODOFF B. (1954) *Amer. J. Path.* **30**, 263.
TESLER U. F., BALSARA R. H. and NIGUIDULA F. N. (1974) *Chest* **66**, 402.

THACKRAH C. T. (1831) "The Effects of the principal arts, trades, and professions, and of civic states and habits of living on the health and longevity, with particular reference to the trades and manufactures of Leeds and suggestions for the removal of many of the agents which produce disease and shorten the duration of life". Longmans, London.
THADANI U., BURROW C., WHITAKER W. and HEATH D. (1975) *Quart. J. Med.* **44**, 133.
THEILEN E., GREGG D. and ROTTA A. (1955) *Circulation* **12**, 383.
THEILER A. (1918) *7th and 8th Reports of the Director of Veterinary Research, Dept. of Agriculture, Union of South Africa*, p. 59.
THEMEL K. G. and LÜDERS C. J. (1955) *Dtsch. med. Wschr.* **80**, 1360 (1376).
THIMM A. (1931) *Med. Klin.* **27**, 1069.
THOMA R. (1893) *Untersuchungen über die Histogenese und Histomechanik des Gefäßsystems*. Enke, Stuttgart.
THOMAS SIR C. P. and DREW C. E. (1953) *Thorax* **8**, 180.
THOMAS SIR C. P. and MORGAN A. D. (1958) *Thorax* **13**, 286.
THOMAS G. C., WHITELAW D. M. and TAYLOR H. E. (1955) *Arch. Path.* **60**, 99.
THOMAS L., MIRRICK G. S., CURNEN E. C., ZIEGLER J. E. and HORSFALL F. L. (1945) *J. clin. Invest.* **24**, 227.
THOMAS L. B. and BOYDEN E. A. (1952) *Surgery* **31**, 429.
THOMAS M. R. (1949) *J. Path. Bact.* **61**, 599.
THOMAS T. V. (1971) *Postgrad. Med.* **49**, 65.
THOMAS W. A., LEE K. T., RABIN E. R. and O'NEAL R. M. (1956) *Arch. Path.* **62**, 257.
THOMSON J. G. (1945) *J. Path. Bact.* **57**, 213.
THOMSON W. B. (1959) *Thorax* **14**, 76.
THOREL C. (1896) *Beitr. path. Anat.* **20**, 85.
THORN N. A. and TRANSBØL I. (1963) *Amer. J. Med.* **35**, 257.
THORNTON N., OTKEN L. B. and SELMONOSKY C. A. (1970) *Amer. J. Dis. Child.* **119**, 454.
THURLBECK W. M. (1963a) *Thorax* **18**, 59.
THURLBECK W. M. (1963b) *Amer. J. med. Sci.* **246**, 332.
THURLBECK W. M. (1963c) *Amer. Rev. resp. Dis.* **87**, 206.
THURLBECK W. M. and ANGUS G. E. (1964) *Thorax* **19**, 436.
THURLBECK W. M., DUNNILL M. S., HARTUNG W., HEARD B. E., HEPPLESTON A. G. and RYDER R. C. (1970) *Human Path.* **1**, 215.
THURLBECK W. M. (1973) *Med. Clin. N. Amer.* **57**, 651.
THURLBECK W. M., PUN R., TOTH J. and FRAZER R. G. (1974) *Amer. Rev. resp. Dis.* **109**, 73.
TIERNEY R. B. H. (1952) *J. clin. Path.* **5**, 63.
TILSON M. D. and TOULOUKIAN R. J. (1972) *Ann. Surg.* **176**, 669.
TIMBRELL V., POOLEY F. and WAGNER J. C. (1970) in *Proc. Internat. Conf. Pneumoconiosis, Johannesburg, 1969*. Ed. H. A. SHAPIRO, p. 120. Oxford Univ. Press, Cape Town.
TIMPLE A. and RUNYON E. H. (1954) *J. Lab. clin. Med.* **44**, 202.

Tobacco Research Council (Gt. Britain) Report (1972).

Tobin C. E. and Zariquiey M. O. (1950) *Proc. Soc. exp. Biol.* (*N.Y.*) **75**, 827.

Tobin C. E. (1952) *Surg. Gynec. Obst.* **95**, 741.

Tobin C. E. (1960) *Surg. Gynec. Obst.* **111**, 297.

Toigo A., Imarisio J. J., Murmall H. and Lepper M. N. (1963) *Amer. Rev. resp. Dis.* **87**, 487.

Tomasi T. B. Jr., Fudenberg H. H. and Finby N. (1962) *Amer. J. Med.* **33**, 243.

Toole H., Propatoridis J. and Pangalos N. (1953) *Thorax* **8**, 274.

Toomey F. and Felson B. (1960) *Amer. J. Roentgenol.* **83**, 709.

Topilow A. A., Rothenberg S. P. and Cottrell T. S. (1973) *Amer. Rev. resp. Dis.* **108**, 114.

Topley and Wilson's *Principles of Bacteriology, Virology and Immunity*, 6th Ed. Eds. G. S. Wilson and A. Miles. Edward Arnold, London.

Torack R. M. (1958) *Arch. Path.* **65**, 574.

Totten R. S., Reid D. H. S., Davis H. D. and Moran T. J. (1958) *Amer. J. Med.* **25**, 803.

Totten R. S. and Moran T. J. (1961) *Amer. J. Path.* **38**, 575.

Touraine R. G. (1958) *Poumon* **14**, 881.

Touroff A. S. W. and Moolten S. E. (1934) *J. thorac. Surg.* **4**, 558.

Towey J. W., Sweany H. C. and Huron W. H. (1932) *J. Amer. med. Ass.* **99**, 453.

Trapnell D. H. (1964) *Thorax* **19**, 44.

Trapnell D. H. (1970) in *Modern Trends in Diagnostic Radiology*. Ed. J. W. McLaren, p. 39. Butterworth, London.

Traube L. (1846) *Ges. Beitr. Path. Physiol.* **1**, 65.

Tredall S. M., Carter J. B. and Edwards J. E. (1974) *Arch. Path.* **97**, 183.

Trivedi S. A., Mehta K. N. and Nanavaty J. M. (1966) *Brit. J. Dis. Chest* **60**, 156.

Tuller M. A. (1957) *Amer. J. Obstet. Gynec.* **73**, 273.

Tuller M. A., Lehr D. E. and Fishman A. P. (1961) *Fed. Proc.* **20**, 106.

Tung H. L. and Liebow A. A. (1952) *Lab. Invest.* **1**, 382.

Tung K. S. K. and McCormack L. J. (1967) *Cancer* (*Philadelphia*) **20**, 525.

Turkington S. I., Scott G. A. and Smiley T. B. (1950) *Thorax* **5**, 138.

Turnbull H. M. (1914–15) *Quart. J. Med.* **8**, 201.

Turner H. M. and Grace H. G. (1938) *J. Hyg.* (*London*) **38**, 90.

Turner-Warwick Margaret and Doniach D. (1965) *Brit. med. J.* **1**, 886.

Turner-Warwick Margaret (1963) *Thorax* **18**, 225.

Turner-Warwick Margaret (1966) *Meeting of the Thoracic Society, London. Thorax* **21**, 290.

Turner-Warwick Margaret (1968) *Quart. J. Med.* **37**, 133.

Turner-Warwick Margaret, Haslam P. and Weeks S. (1971) *Clin. Allergy* **1**, 209.

Tuynman P. E. and Gardner L. W. (1952) *Arch. Path.* **54**, 306.

Tyler W. S. and Pearse A. G. E. (1965) *Thorax* **20**, 149.

Uehlinger E. (1964) *Acta med. scand.* **176** (suppl. 425), 7.

Ugon V. A. and Tomalino D. (1958) *Tórax* **7**, 188.

U.I.C.C. (Union Internationale Contre Cancer) *Cancer Incidence in Five Continents*, Vol. II. Ed. Doll R., Muir C. and Waterhouse J. Springer-Verlag, Berlin.

U.I.C.C. (Union Internationale Contre Cancer) (1976) Technical Report Series No. 25, Geneva.

Umiker W. and Storey C. (1952) *Cancer* (*Philadelphia*) **5**, 369.

Umiker W. and Iverson L. (1954) *J. thorac. Surg.* **28**, 55.

Umiker W. (1960) *Dis. Chest* **37**, 82.

Umiker W. (1961) *Dis. Chest* **40**, 154.

Unger R. H., Lochner Jan De V. and Eisentraut A. M. (1964) *J. clin. Endocr.* **24**, 823.

Unterman D. H. and Reingold I. M. (1972) *Amer. J. clin. Path.* **57**, 297.

Ustvedt H. J. (1948) *Tubercle* **29**, 107.

Utidjian H. M. D. and De Treville R. T. P. (1970) *Fibrous Glass Manufacturing and Health.* Report of an Epidemiological Study, 35th Industrial Health Foundation Annual Meeting, October 1970, p. 1.

Utidjian H. M. D., Gross P. and De Treville R. T. P. (1968) *Arch. environm. Hlth.* **17**, 327.

Vadas G., Pare J. A. P. and Thurlbeck W. M. (1967) *Canad. med. Ass. J.* **96**, 420.

Valdes-Dapena M. A. and Arey J. B. (1967) *Arch. Path.* **84**, 643.

Valdivia E., Lalich J. J., Hayashi Y. and Sonnad J. (1967) *Arch. Path.* **84**, 64.

Valentine E. H. (1957) *Cancer* (*Philadelphia*) **10**, 272.

Valenzuela C. T., Toriello J. and Thomas W. A. (1954) *Arch. Path.* **57**, 51.

Valle A. R. and Graham E. A. (1944) *J. thorac. Surg.* **13**, 345.

Van Allen C. M. and Jung T. S. (1931) *J. thorac. Surg.* **1**, 3.

Van Allen C. M. (1932) *Surg. Gynec. Obstet.* **55**, 303.

Vance B. M. (1945) *Arch. Path.* **40**, 395.

Van Creveld S. and ter Poorten F. H. (1935) *Arch. Dis. Childh.* **10**, 125.

Van der Meer G. and Brug S. L. (1942) *Ann. Soc. belge Méd. trop.* **22**, 301.

Van Dijk C. and Wagenvoort C. A. (1973) *J. Path.* **110**, 131.

Van Duuren B. L., Goldschmidt B. M. and Katz B. S. (1968) *Arch. environm. Hlth.* **16**, 472.

Vaněk J. (1951) *Čas. Lék. čes.* **90**, 1121.

Vaněk J. and Jírovec O. (1952) *Zbl. Bakt. I. Abt.* **158**, 120.

Vaněk J., Jírovec O. and Lukes J. (1953) *Ann. Paediat.* **1**, 1.

Vaněk J. (1968) *Beitr. path. Anat.* **136**, 303.

Vaněk J. and Schwartz J. (1970) *Amer. Rev. resp. Dis.* **101**, 395.

VAN HAZEL W. and JENSIK R. J. (1956) *J. thorac. Surg.* **31**, 19.

VAN TOORN D. W. (1970) *Thorax* **25**, 399.

VAN WIJK A. M. and PATTERSON N. S. (1940) *J. industr. Hyg. Toxicol.* **22**, 31.

VEEZE P. (1968) *Rationale and Methods of Early Detection in Lung Cancer.* Van Gorcum Co., Groningen.

VERBRYCKE J. R. (1924) *J. Amer. med. Ass.* **82**, 1577.

VERITY M. A., LARSON W. M. and MADDEN S. C. (1963) *Amer. J. Path.* **42**, 251.

VERLOOP M. C. (1948) *Acta Anat.* **5**, 171.

VESSEY M. P., DOLL R., FAIRBAIRN A. S. and GLOBER G. (1970) *Brit. med. J.* **3**, 123.

VIERORDT O. (1883) *Berl. klin. Wschr.* **20**, 437.

VIERSMA H. J. (1955) *Ned. T. Geneesk.* **99**, 3593 (quoted by HEARD B. E. (1962)).

VIGLIANI E. C., BOSELLI A. and PECCHIAL L. (1950) *Med. d. Lavoro* **41**, 33.

VIGLIANI E. C. and PERNIS B. (1958) *Brit. J. industr. Med.* **15**, 8.

VIJEYARATNAM G. S. and CORRIN B. (1971) *J. Path.* **103**, 123.

VIJEYARATNAM G. S. and CORRIN B. (1972) *J. Path.* **108**, 115.

VIJEYARATNAM G. S. and CORRIN B. (1973) *Virchows Arch. path. Anat.* **358**, 1.

VILLAR T. G., CORTEZ PIMENTEL J. and COSTA M. F. E. (1962) *Thorax* **17**, 22.

VIRCHOW R. (1846) *Beitr. exp. Path. Physiol.* **2**, 1; quoted by CAMERON G. R. *et al.* (1951).

VIRCHOW R. (1847) *Virchows Arch. path. Anat.* **1**, 1.

VIRCHOW R. (1851) *Virchows Arch. path. Anat.* **3**, 427.

VIRCHOW R. (1855) *Virchows Arch. path. Anat.* **8**, 103.

VIRCHOW R. (1856) *Gesammelte Abhandlungen zur wissenschaftlichen Medicin*, p. 982. Meidinger Sohn u. Co., Frankfurt.

VIRCHOW R. (1858) *Virchows Arch. path. Anat.* **15**, 310.

VIRCHOW R. (1863–7) *Die krankhaften Geschwülste.* Vol. 2, p. 182. A Hirschwald, Berlin.

VIRSHUP M. and GOLDMAN A. (1956) *J. thorac. Surg.* **31**, 226.

VISWANATHAN R. (1947) *Indian med. Gaz.* **82**, 49.

VISWANATHAN R. (1949) *Dis. Chest* **17**, 460.

VISWANATHAN R., CHAKRAUARTY S. C., RANDHAWA H. S. and DE MONTE A. J. H. (1960) *Brit. med. J.* **1**, 399.

VISWANATHAN R., SEN GUPTA and KRISHNA IYER P. V. (1962) *Thorax* **17**, 73.

VISWANATHAN R. (1964) *Lancet* **1**, 1138.

VOEGT H. (1938) *Virchows Arch. path. Anat.* **302**, 468.

VOGEL M. D., KEATING F. R. Jr. and BAHN R. C. (1961) *Proc. Mayo Clin.* **36**, 387.

VOLINI F., COLTON R. and LESTER W. (1965) *Amer. J. clin. Path.* **43**, 39.

VOLK B. W., NATHANSON L., LOSNER S., SLADE W. R. and JACOBI M. (1951) *Amer. J. Med.* **10**, 316.

VOLLAND W. (1941) *Virchows Arch. path. Anat.* **307**, 85.

VON EULER U. S. and LILJESTRAND G. (1946) *Acta physiol. scand.* **12**, 301.

VON GLAHN W. C. and PAPPENHEIMER A. M. (1925) *Amer. J. Path.* **1**, 445.

VON GLAHN W. C. and PAPPENHEIMER A. M. (1926) *Amer. J. Path.* **2**, 235.

VON GLAHN W. C. and HALL J. W. (1949) *Amer. J. Path.* **25**, 575.

VON GLAHN W. C., HALL J. W. and SUN S-C. (1954) *Amer. J. Path.* **30**, 1129.

VON HAYEK J. W. (1940) *Z. Anat. Entwickl.-Gesch.* **110**, 412.

VON HAYEK H. (1940a) *Anat. Anz.* **89**, 216.

VON HAYEK H. (1942) *Anat. Anz.* **93**, 155.

VON HAYEK H. (1953) *Die menschliche Lunge*, 2nd Ed. Springer, Heidelberg.

VON HAYEK H. (1960) *The Human Lung.* Translated by KRAHL V. E. Hafner Publishing Co. Inc., New York.

VON MEYENBURG H. (1919) *Korresp.-Bl. Schweiz. Ärz.* **49**, 1668.

VON MEYENBURG H. (1936) *Schweiz. med. Wschr.* **17**, 1239.

VON MEYENBURG H. (1942) *Virchows Arch. path. Anat.* **309**, 258.

VON MEYENBURG H. (1942a) *Schweiz. med. Wschr.* **72**, 809.

VON MEYENBURG H. (1942b) *Virchows Arch. path. Anat.* **309**, 258.

VORWALD A. J. (1940) *Amer. J. Path.* **16**, 653.

VORWALD A. J. (1941) *Amer. J. Path.* **17**, 709.

VORWALD A. J. (1950) *6th Saranac Symposium. L. U. Gardner Memorial Volume.* Paul B. Hoeber, New York.

VORWALD A. J., DURKAN T. M. and PRATT P. C. (1951) *Arch. industr. Hyg* **3**, 1.

WACHS E. (1953) *Arch. klin. Chir.* **275**, 567.

WACK J. P., DUBUQUE T. and WYATT J. P. (1958) *Arch. Path.* **65**, 675.

WACKER W. E. C. and SNODGRASS P. J. (1960) *J. Amer. med. Ass.* **174**, 2142.

WACKERS F. J. TH., VAN DER SCHOOT J. B. and HAMPE J. F. (1969) *Cancer (Philadelphia)* **23**, 339.

WADDELL W. R. (1949) *Arch. Path.* **47**, 227.

WADDELL W. R., SNIFFEN R. C. and SWEET R. H. (1949) *J. thorac. Surg.* **18**, 707.

WADDELL W. R., SNIFFEN R. C. and WHYTEHEAD L. L. (1954) *Amer. J. Path.* **30**, 757.

WADDLE N. (1950) *Aust. N.Z. J. Surg.* **19**, 273.

WADE G. and BALL J. (1957) *Quart. J. Med.* **26** N.S., 83.

WADE H. W. (1916) *J. infect. Dis.* **18**, 618.

WADSWORTH A. B. (1918–19) *J. med. Res.* **39**, 147.

WAGENVOORT C. A., NEUFELD H. N., DuSHANE J. W., and EDWARDS J. E. (1961a) *Circulation* **23**, 740; (1961b) *ibid.* **23**, 733.

WAGENVOORT C. A., NEUFELD H. N. and EDWARDS J. E. (1961c) *Lab. Invest.* **10**, 751.

WAGENVOORT C. A., NEUFELD H. N., BIRGE R. F., CAFFREY J. A. and EDWARDS J. E. (1961d) *Circulation* **23**, 84.

WAGENVOORT C. A. (1962) *Med. thorac.* **19**, 354.

WAGENVOORT C. A., HEATH D. and EDWARDS J. E. (1964) *The Pathology of the Pulmonary Vasculature.* C. C. Thomas, Springfield, Ill.

WAGENVOORT C. A. and EDWARDS J. E. (1965) *Lab. Invest.* **10**, 924.

WAGENVOORT C. A. and WAGENVOORT N. (1965) *Arch. Path.* **79**, 524.

WAGENVOORT C. A., WAGENVOORT N. and VOGEL J. H. K. (1969) *J. comp. Path.* **79**, 517.

WAGENVOORT C. A. (1970) *Human Path.* **1**, 205.

WAGENVOORT C. A., LOSEKOOT G. and MULDER E. (1971) *Thorax* **26**, 429.

WAGENVOORT C. A. (1973) *Chest* **64**, 503.

WAGENVOORT C. A. and WAGENVOORT N. (1974) *Virchows Arch. path. Anat.* **364**, 69.

WAGENVOORT C. A., WAGENVOORT N. and DIJK H. J. (1974) *Thorax* **29**, 522.

WAGNER J. C., ADLER D. I. and FULLER D. N. (1955) *Thorax* **10**, 157.

WAGNER J. C., SLEGGS C. A. and MARCHAND P. (1960) *Brit. J. industr. Med.* **17**, 260.

WAGNER J. C. (1962) *Nature (Lond.)* **196**, 180.

WAGNER J. C., MUNDAY D. E. and HARINGTON J. S. (1962) *J. Path. Bact.* **84**, 73.

WAGNER J. C. (1965) *Ann. N.Y. Acad. Sci.* **132**, 575.

WAGNER J. C. (1969) *Proc. Int. Pneumoconiosis Conference Johannesburg, 1969.* Ed. H. A. SHAPIRO. Oxford Univ. Press, Cape Town, p. 306.

WAGNER J. C., WUSTEMAN F. S., EDWARDS J. H. and HILL R. J. (1975) *Thorax* **30**, 382.

WAGNER J. C. (1976) Personal communication.

WAGONER J. K., MILLER R. W., LUNDIN F. E. Jr., FRAUMENI J. F. Jr. and HAIJ M. E. (1963) *New Engl. J. Med.* **269**, 284.

WAGONER J. K., ARCHER V. E., CARROL B. E., HOLADAY D. A. and LAWRENCE P. A. (1964) *J. nat. Cancer Inst.* **32**, 787.

WAGONER J. K., ARCHER V. E., LUNDIN F. E., HOLADAY D. A. and LLOYD J. W. (1965) *New Engl. J. Med.* **273**, 181.

WAGONER J. K., ARCHER V. E. and GILLAN J. D. (1975) in *Proc. XIth Int. Cancer Cong., Florence, 1974,* Vol. 3. Eds. BUCALOSSI P., VERONESI U. and CASCINELLI N. Excerpta Medica, Amsterdam, p. 102.

WALCH H. A., PRIBNOW J. F., WYBORNEY V. J. and WALCH R. K. (1961) *Amer. Rev. resp. Dis.* **84**, 359.

WALCOTT G., BURCHELL H. B. and BROWN A. L. Jr. (1970) *Amer. J. Med.* **49**, 70.

WALFORD R. L. and KAPLAN L. (1957) *Arch. Path.* **63**, 75.

WALKER J. W. and MONTGOMERY F. H. (1902) *J. Amer. med. Ass.* **38**, 867.

WALSH E. G. and WHITTERIDGE D. (1945) *J. Physiol. (London)* **103**, 370.

WALSH T. J. and HEALY T. M. (1969) *Thorax* **24**, 327.

WALTER J. B. and PRYCE D. M. (1955a) *Thorax* **10**, 107.

WALTER J. B. and PRYCE D. M. (1955b) *Thorax* **10**, 117.

WALTON E. W. and LEGGATT P. O. (1956) *J. clin. Path.* **9**, 31.

WALTON E. W. (1958) *Brit. med. J.* **2**, 265.

WALTON K. W. (1954) *J. Path. Bact.* **68**, 565.

WANEBO C. K., JOHNSON K. G., SATO K. and THORSLUND T. W. (1968) *Amer. Rev. resp. Dis.* **98**, 778.

WANG N-S., HUANG S. N. and THURLBECK W. M. (1970) *Arch. Path.* **90**, 529.

WANG N-S. and THURLBECK W. M. (1970) *Human Path.* **1**, 227.

WANG N-S. (1973) *Cancer (Philadelphia)* **31**, 1046

WANG N-S., TAENSCH H. W. Jr., THURLBECK W. M. and AVERY M. E. (1973) *Amer. J. Path.* **73**, 365.

WANG N-S., SEEMAYER T. A., AHMED M. N. and KNAACK J. (1976) *Human Path.* **7**, 3.

WANSTROM R. C. (1933) *Amer. J. Path.* **9**, 623.

WARD I. M. and KRAHL J. B. (1942) *Amer. J. Dis. Child.* **63**, 924.

WARDLE E. N. (1974) *J. roy. Coll. Phycns. Lond.* **8**, 251.

WARING J. I. (1933) *Amer. J. med. Sci.* **185**, 325.

WARING J. J., NEUBUERGER K. and GEEVER E. F. (1942) *Arch. int. Med.* **69**, 384.

WARNER A. L., PALLADINO N. W., SCHWARTZ W. and SCHUSTER A. (1955) *J. Pediat.* **46**, 200.

WARNER F. S., McGRAW C. T., PETERSON H. G. Jr., CLELAND R. S. and MEYER B. W. (1961) *Amer. J. Dis. Child.* **101**, 514.

WARNER R. R. P. and SOUTHERN A. L. (1958) *Amer. J. Med.* **24**, 903.

WARNER R. R. P., KIRSCHNER P. A. and WARNER G. M. (1961) *J. Amer. med. Ass.* **178**, 1175.

WARNER W. P. (1935) *J. Amer. med. Ass.* **105**, 1666.

WARRELL D. A., HARRISON B. D. W., FAWCETT I. W., MOHAMMED Y., MOHAMMED W. S., POPE H. M. and WATKINS B. J. (1975) *Thorax* **30**, 389.

WARREN M. F. and DRINKER C. K. (1942) *Amer. J. Physiol.* **136**, 207.

WARREN S. and GATES O. (1936) *Amer. J. Cancer* **27**, 485.

WARREN S. and GATES O. (1940) *Arch. Path.* **30**, 440.

WARREN S. and SPENCER J. (1940) *Amer. J. Roentgenol.* **43**, 682.

WARREN S. (1942) *Arch. Path.* **34**, 917.

WARREN S. (1946) *Amer. J. Path.* **22**, 69.

WARREN S. and DIXON F. J. (1948) *Amer. J. med. Sci.* **216**, 136.

WARRING F. C. and HOWLETT K. S. (1952) *Amer. Rev. Tuberc.* **65**, 235.

WARTHIN A. S. (1917) *Amer. J. Syph.* **1**, 693.

WARTHIN A. S. (1931) *Arch. Path.* **11**, 864.

WARWICK MARGARET (1933) *Amer. J. Path.* **9**, 961.

WATERMAN D. H., DOMM S. E. and RODGERS W. K. (1955) *Amer. J. Surg.* **55**, 995.

WATERS A. T. H. (1862) *Emphysema of the Lungs.* Churchill, London.

WATKINS E. and HERING A. C. (1958) *J. thorac. Surg.* **36**, 642.

WATSON A. J. (1955) *J. Path. Bact.* **69**, 207.

WATSON A. J., BLACK J., DOIG A. T. and NAGELSCHMIDT G. (1959) *Brit. J. industr. Med.* **16**, 274.

WATSON W. L., CROMWELL H., CRAVER L. and PAPANI-
COLAOU G. N. (1949) *J. thorac. Surg.* **18**, 113.

WATTS C. F., CLAGETT O. T. and McDONALD J. R.
(1946) *J. thorac. Surg.* **15**, 132.

WATTS C. F. and McDONALD J. R. (1948) *Arch. Path.* **45**,
742.

WATTS J. C., CALLAWAY C. S., CHANDLER F. W. and
KAPLAN W. (1975) *Arch. Path.* **99**, 11.

WAYL P. and RAKOWER J. (1954) *Thorax* **9**, 216.

WEARN J. T., BARR J. S. and GERMAN W. J. (1926) *Proc.
Soc. exp. Biol* (*N.Y.*) **24**, 114.

WEARN J. T., ERNSTENE A. C., BROMER A. W., BARR J. S.,
BERMAN W. J and ZSCHIESCHE L. J. (1934) *Amer. J.
Physiol.* **109**, 236.

WEBB A. C. (1946) *Arch. Path.* **42**, 427.

WEBB J. K. G., JOB C. K. and GAULT E. W. (1960)
Lancet **1**, 835.

WEBB W. R. (1969) *J. Trauma* **9**, 700.

WEBER H. H. and ENGELHARDT W. E. (1933) *Zlb.
Gewerbehyg.* **10**, 41 (quoted by NORRIS G. F. and
PEARD M. C. (1963)).

WEBER J. H. (1957) *J. Sci. Food Agric.* **8**, 490.

WEBSTER B. H. (1956) *Amer. Rev. Tuberc.* **73**, 485.

WEBSTER B. H. (1960) *Amer. Rev. resp. Dis.* **81**,
683.

WEBSTER I. (1965) *Ann. N.Y. Acad. Sci.* **132**, 623.

WEBSTER I. (1969) *Proc. Int. Conf. Pneumoconiosis,
Johannesburg.* Ed. H. A. SHAPIRO. Oxford Univ. Press
Cape Town, p. 572.

WEBSTER I. (1973) *S. Afr. med. J.* **47**, 165.

WEDLER H. W. (1943) *Dtsch. med. Wschr.* **69**, 575
(quoted by GLUCKMAN J. and HURWITZ M. in *Proc.
Int. Conf. Pneumoconiosis, Johannesburg, 1969*, p. 143.
Oxford Univ. Press, Cape Town).

WEED L. A., SLOSS P. T. and CLAGETT O. T. (1956) *J.
Amer. med. Ass.* **161**, 1044.

WEGENER F. (1936) *Verhandl dtsch. Ges. Path.* **29**, 202.

WEGENER F. (1939) *Beitr. path. Anat.* **102**, 36.

WEIBEL E. R. (1959) *Z. Zellforsch.* **50**, 653.

WEIBEL E. (1960) *Circulation Res.* **8**, 353.

WEIBEL E. R. and VIDONE R. A. (1961) *Amer. Rev. resp.
Dis.* **84**, 856.

WEIBEL E. R. (1962) *Z. Zellforsch.* **57**, 648.

WEIBEL E. R. (1963) *Morphometry of the Human Lung.*
Springer-Verlag, Berlin.

WEIBEL E. R. and GIL J. (1968) *Resp. Physiol.* **4**, 42.

WEIBEL E. R. (1971) *Acta anat.* **78**, 425.

WEIBEL E. R. (1974) *Microvasc. Res.* **8**, 218.

WEIBEL E. R. (1974) *Chest* **65**, 5S.

WEICHSELBAUM A. (1886) *Med. Jahrb.* **1**, 483.

WEIL C. S., SMYTH H. F. and NALE T. W. (1952) *Arch.
industr. Hyg.* **5**, 535.

WEINBERG H. B. and TILLINGHAST A. J. (1946) *Amer. J.
trop. Med.* **26**, 801.

WEINER A. E., REID D. E. and ROBY C. C. (1949) *Science*
110, 190.

WEINGARTEN R. J. (1943) *Lancet* **1**, 103.

WEINSTEIN A., FRANCIS H. C. and SPROFKIN B. F. (1947)
Arch. intern. Med. **79**, 176.

WEINTRAUB B. D. and ROSEN S. W. (1971) *J. clin. Endocr.*
32, 94.

WEISMANN R. E., CLAGETT O. T. and McDONALD J. R.
(1947) *J. thorac. Surg.* **16**, 269.

WELCH C. C., TOMBRIDGE T. L., BAKER W. J. and KINNEY
R. J. (1961) *Amer. J. med. Sci.* **242**, 157.

WELCH K. J. and KINNEY T. D. (1948) *Amer. J. Path.*
24, 729.

WELCH W. H. (1878) *Virchows Arch. path. Anat.* **72**, 375.

WELLS A. L. (1954) *J. Path. Bact.* **68**, 573.

WELLS H. G. and DUNLAP C. E. (1943) *Arch. Path.* **35**,
420.

WELSCH K. (1928) *Frankfurt. Z. Path.* **36**, 192.

WENDT V. E., PURO H. E., SHAPIRO J., MATHEWS W. and
WOLF P. L. (1964) *J. Amer. med. Ass.* **188**, 755.

WENYON C. M. (1926) *Protozoology*, Vol. **2**. Ballière,
Tindall and Cox, London.

WERNER S. B., PAPPAGIANIS D., HEINDL I. and MICKEL
A. (1972) *New Engl. J. Med.* **286**, 507.

WERNICKE R. (1892) *Zbl. Bakt. I. Abt. Orig.* **12**, 89.

WESSEL W. and RICKEN D. (1958) *Virchows Arch. path.
Anat.* **331**, 545.

WEST J. B. (1963) *Brit. med. Bull.* **19**, 53.

WEST J. B., DOLLERY C. T. and HEARD B. E. (1964)
Lancet **2**, 181.

WEST J. B., DOLLERY C. T. and HEARD B. E. (1965)
Circulat. Res. **17**, 191.

WEST J. B. and DOLLERY C. T. (1965) *J. appl. Physiol.* **20**,
175.

WEST J. R., BALDWIN E. DE F., COURNAND A. and
RICHARDS D. W. Jr. (1951) *Amer. J. Med.* **10**, 481.

WESTERHEIDE R. L. (1964) *J. thorac. cardiovasc. Surg.*
47, 389.

WEXELS P. (1951) *Thorax* **6**, 171.

WHIMSTER W. F. (1970) *Thorax* **25**, 141.

WHITAKER W., HEATH D. and BROWN J. W. (1955) *Brit.
Heart J.* **17**, 121.

WHITELEY H. J. (1954) *J. Path. Bact.* **67**, 521.

WHITFIELD A. G. W., BOND W. H. and KUNKLER P. B.
(1963) *Thorax* **18**, 371.

WHITMORE A. and KRISHNASWAMI C. S. (1912) *Indian
med. Gaz.* **47**, 262.

WHITMORE A. (1913) *J. Hyg.* (*London*) **13**, 1.

WHITTICK J. W. (1950) *Brit. med. J.* **1**, 979.

WHITWELL F. (1952) *Thorax* **7**, 213.

WHITWELL F. (1953) *Thorax* **8**, 309.

WHITWELL F. (1955) *J. Path. Bact.* **70**, 529.

WHITWELL F. and RAWCLIFFE R. M. (1971) *Thorax* **26**, 6.

WHITWELL F., NEWHOUSE M. and BENNETT D. R. (1974)
Brit. J. indust. Med. **31**, 298.

WILD C. (1886) *Beitr. path. Anat.* **1**, 175.

WILHELMSEN L., SELANDER S., SÖDERHOLM B., PAULIN S.,
VARNAUSKAS E. and WERKÖ L. (1963) *Medicine* **42**, 335.

WILKINS G. D. (1917) *Beitr. Klin. Tuberk.* **38**, 1.

WILKS S. (1857) *Trans. path. Soc. Lond.* **8**, 88.

WILKS S. (1862) *Trans. path. Soc. Lond.* **13**, 27.

WILL D. W., MURRAY J. F., FINEGOLD S. M., SUTTER
V. L. and FISHKIN B F. (1961) *Amer. Rev. resp. Dis.*
84, 114.

WILLACH P. (1888) *Beiträge zur Entwicklung der Lunge bei Säugetieren.* A. W. Zickfeldt, Osterwieck/Harz.

WILLERSON J. T., BAGGETT A. E. Jr., THOMAS J. W. and GOLDBLATT A. (1971) *New Engl. J. Med.* **285,** 157.

WILLIAMS A. W., DUNNINGTON W. G. and BERTIE S. J. (1961) *Ann. int. Med.* **54,** 30.

WILLIAMS E. D. and AZZOPARDI J. G. (1960) *Thorax* **15,** 54.

WILLIAMS E. D. and CELESTIN L. R. (1962) *Thorax* **17,** 120.

WILLIAMS H. and CAMPBELL P. (1960) *Arch. Dis. Child.* **35,** 182.

WILLIAMS R. B. and DANIEL R. A. (1950) *J. thorac. Surg.* **19,** 806.

WILLIAMS R. T. (1963) *Brit. med. J.* **1,** 233.

WILLIAMS T. C. G. (1965) M.D. Thesis, University of London.

WILLIAMS W. J. (1958) *Brit. J. industr. Med.* **15,** 84.

WILLIAMS W. J. (1958) *Brit. J. industr. Med.* **15,** 235.

WILLIAMS W. J. (1960) *J. clin. Path.* **13,** 273.

WILLIGHAGEN R. G., VAN DER HEUL R. O. and VAN RIJSSEL TH. G. (1963) *J. Path. Bact.* **85,** 279.

WILLIS R. A. (1948) *J. Path. Bact.* **47,** 35.

WILLIS R. A. (1952) *The Spread of Tumours in the Human Body.* 1st ed. 1933, 2nd ed. 1952. Butterworth, London.

WILLIS R. A. (1953) *Pathology of Tumours,* 2nd ed. Butterworth and Company, London, 1st ed. 1948.

WILLSON H. G. (1928) *Amer. J. Anat.* **41,** 97.

WILSON J. V. and SALISBURY C. V. (1943–4) *Brit. J. Surg.* **31,** 384.

WILSON M. G. and MIKITY V. G. (1960) *Amer. J. Dis. Child.* **99,** 489.

WILSON R. J., RODNAN G. P. and ROBIN E. D. (1964) *Amer. J. Med.* **36,** 361.

WINN W. A. (1941) *Arch. intern. Med.* **68,** 1179.

WINN W. A. (1951) *Arch. intern. Med.* **87,** 541.

WINSLOW D. J. and CHAMBLIN S. A. (1960) *Amer. J. clin. Path.* **33,** 43.

WINSSER J. and ALTIERI R. H. (1942) *Amer. J. Med. Sci.* **247,** 269.

WINTERNITZ M. C., WASON I. M. and McNAMARA F. P. (1920) *The Pathology of Influenza.* Yale University Press, New Haven.

WINTROBE M. M. (1961) *Clinical Haematology,* 5th Ed., p. 673. Lea & Febiger, Philadelphia.

WOLBACH S. B. (1904) *J. med. Res.* **13,** 53.

WOLBACH S. B., TODD J. L. and PALFREY F. W. (1922) *The Etiology and Pathology of Typhus,* p. 165. The League of Red Cross Societies. Harvard University Press, Cambridge, Mass.

WOLF K. (1895) *Fortschr. Med.* **13,** 725.

WOLFGARTEN M. and MAGAREY F. R. (1959) *J. Path. Bact.* **77,** 597.

WOLINS W. (1948) *Amer. J. med. Sci.* **216,** 551.

WOLLSTEIN M. (1931) *Arch. Path.* **12,** 562.

WOLMAN I. J. (1930) *Bull. Ayer. clin. Lab.* **12,** 49.

WOLMAN I. J. (1941) *Amer. J. Dis. Child.* **61,** 1263.

WOLMAN M., IZAK G., FREUND E. and SHAMIR Z. (1952) *Amer. J. Dis. Child.* **83,** 573.

WOMACK N. A. and GRAHAM E. A. (1941) *Amer. J. Path.* **17,** 645.

WONG T-W. and WARNER N. E. (1962) *Arch. Path.* **74,** 403.

WOOD D. A., CREVER J. W. and MILLER M. (1937) *J. tech. Meth.* **17,** 78.

WOOD D. A. and MILLER M. (1938) *J. thorac. Surg.* **7,** 649.

WOOD P. (1956) *Diseases of the Heart and Circulation,* 2nd ed. Eyre and Spottiswoode, Ltd., London.

WOOD P., BESTERMAN E. M., TOWERS M. K. and McILROY M. B. (1957) *Brit. Heart J.* **19,** 279.

WOOD W. B. and GLOYNE S. R. (1931) *Lancet* **2,** 954.

WOODRUFF C. E. and MOERKE A. G. (1940) *Amer. J. Path.* **16,** 652.

WOOLLARD H. H. (1922) *Contr. Embryol. Carneg. Instn.* **14,** 139.

WOOLNER L. B., ANDERSEN H. A. and BERNATZ P. E. (1960) *Dis. Chest* **37,** 278.

WRIGHT G. H. and KLEINERMAN J. (1963) *Amer. Rev. resp. Dis.* **88,** 605.

WRIGHT G. W. (1968) *Arch. environm. Hlth.* **16,** 175.

WRIGHT H. T. Jr., BECKWITH J. B. and GWINN J. L (1964) *J. Pediat.* **64,** 528.

WRIGHT R. D. (1938) *J. Path. Bact.* **47,** 489.

WRIGHT R. R. (1960) *Amer. J. Path.* **37,** 63.

WRIGHT R. R. (1961) *Amer. J. Path.* **39,** 355.

WRIGHT R. R. and STUART C. M. (1965) *Med. thorac.* **22,** 210.

WU LIEN-TEH and WOODHEAD G. S. (1914–15) *J. Path. Bact.* **19,** 1.

WU LIEN-TEH (1926) *A Treatise on Pneumonic Plague.* Publications of the League of Nations, III Health. No. 13.

WU LIEN-TEH, CHUN J. W. H., POLLITZER R. and WU C. Y. (1936) *Plague, a Manual for Medical and Public Health Workers.* Mercury Press, Shanghai.

WU T. T. (1936) *J. Path. Bact.* **43,** 61.

WURM H. and RÜGER H. (1942) *Beitr. Klin. Tuberk.* **98,** 396.

WYATT J. P. and GOLDENBERG H. (1948) *Arch. Path.* **45,** 366.

WYATT J. P. and RIDDELL A. C. R. (1949) *Amer. J. Path.* **25,** 447.

WYATT J. P., SAXTON J., LEE R. S. and PINKERTON H. (1950) *J. Pediat.* **36,** 271.

WYATT J. P., HEMSATH F. A. and SOASH M. D. (1951) *Amer. J. clin. Path.* **21,** 50.

WYATT J. P., SIMON T., TRUMBULL M. L. and EVANS M. (1953) *Amer. J. clin. Path.* **23,** 353.

WYATT J. P. and SWEET H. (1961) *Amer. Rev. resp. Dis.* **83,** 426.

WYATT J. P., FISCHER V. W. and SWEET H. C. (1962) *Dis. Chest* **41,** 239.

WYATT J. P., FISCHER V. W. and SWEET H. C. (1964) *Amer. Rev. resp. Dis.* **89,** 533.

WYLIE W. G., SHELDON W., BODIAN M. and BARLOW A. (1948) *Quart. J. Med.* **17** N.S., 25.

WYNDER E. L. and GRAHAM E. A. (1950) *J. Amer. med. Ass.* **143**, 329.

WYNDER E. L. and GRAHAM E. A. and CRONINGER A. B. (1953) *Cancer Res.* **13**, 855.

WYNDER E. L., COVEY L. S. and MABUCHI K. (1973) *J. nat. Cancer Inst.* **51**, 391.

WYNN-WILLIAMS N. and YOUNG R. D. (1956) *Thorax* **11**, 101.

XALABARDER C. (1961) *Amer. Rev. resp. Dis.* **83**, 1.

YACOUBIAN H., CONNOLLY J. E. and WYLIER R. H. (1958) *Ann. Surg.* **147**, 116.

YAMADA A. (1963) *Acta. path. jap.* **13**, 131.

YAMADA S. (1933) *Z. ges. exp. Med.* **99**, 342.

YAMAGUCHI B. T. Jr., ADRIANO S. and BRAUNSTEIN H. (1963) *Amer. J. Path.* **43**, 713.

YATES W. M. and HANSMANN G. H. (1936) *Amer. J. med. Sci.* **191**, 474.

YÉPEZ C. G., PUIGBÓ J. J., HIRSCHAUT E., CARBONELL L., PABLO BLANCO, SUÁREZ H. and SUÁREZ J. A. (1962) *Amer. J. Cardiol.* **10**, 30.

YESNER R. and HURWITZ A. (1953) *J. thorac. Surg.* **26**, 325.

YESNER R. (1956) *Amer. J. Path.* **32**, 611.

YESNER R. (1961) *Amer. Rev. resp. Dis.* **84**, 130.

YOUNG J. M., JONES E., HUGHES F. A., FOLEY F. E. and FOX J. R. (1954) *J. thorac. Surg.* **27**, 300.

YOUNG J. M. (1956) *Amer. J. Path.* **32**, 253.

YOUNG J. S. and GRIFFITH H. D. (1950) *J. Path. Bact.* **62**, 293.

YOUNG R. C. Jr., TILLMAN R. L., BURTON A. F. and SAMPSON C. C. (1969) *J. nat. med. Ass.* **61**, 310.

YOUNG R. D. (1955) *Lancet* **2**, 750.

YOUNT F. (1948) *Ariz. Med.* **5**, 48.

YU P. N., LOVEJOY F. W., JOOS H. A., NYE R. E. and BEATTY D. C. (1954) *Amer. J. Dis. Child.* **88**, 636.

ZAIDI S. H., HARRISON C. V., KING E. J. and MITCHISON D. A. (1955) *Brit. J. exp. Path.* **36**, 553.

ZAK F. G. and CHABES A. (1963) *J. Amer. med. Ass.* **183**, 887.

ZAK F. (1964) *Acta med. scand.* **146** (suppl. 425), 21.

ZAKY H. A., EL-HENEIDY A. R., TAWFIK I., GEMEI Y. and KHADR A. (1959) *Dis. Chest* **36**, 164.

ZAKY H. A., EL-HENEIDY A. R. and FODA M. T. (1962) *Brit. med. J.* **1**, 367.

ZAKY H. A., EL-HENEIDY A. R. and KHALIL M. (1964) *Brit. med. J.* **1**, 1021.

ZAKY H. A., EL-HENEIDY M., FODA M., KHALIL M. and TARABEIH A. A. (1967) in *Int. Acad. Path.* Special Monograph "Bilharziasis", p. 30. Springer-Verlag, Berlin.

ZATUCHNI J., CAMPBELL W. N. and ZARAFONETIS C. J. D. (1953) *Cancer (Philadelphia)* **6**, 1147.

ZEEK PEARL M., SMITH C. C. and WEETER J. C. (1948) *Amer. J. Path.* **24**, 889.

ZEEK PEARL M. (1952) *Amer. J. clin. Path.* **22**, 777.

ZEIDMAN I. and BUSS J. M. (1952) *Amer. J. Path.* **28**, 535.

ZELLOS S. (1962) *Thorax* **17**, 61.

ZENKER F. A. (1861) quoted by REEVES D. L., BUTT E. M. and HAMMOCK R. W. (1941).

ZENKER F. A. (1862) in *Beiträge zur normalen und pathologischen Anatomie der Lunge.* J. Braunsdorf, Dresden.

ZERMAN P. (1943) *Med. J. Aust.* **2**, 315.

ZETTERGREN L. (1961) *Acta path. microbiol scand.* **51**, 113.

ZIENTARA M. and MOORE S. (1970) *Human Path.* **1**, 324.

ZIETLHOFER J. and REIFFENSTUHL G. (1952) *Wien. klin. Wschr.* **64**, 446.

ZINSERLING A. (1972) *Virchows Arch. path. Anat.* **356**, 259.

ZIPKIN R. (1907) *Virchows Arch. path. Anat.* **187**, 244.

ZIVERLING H. B. and PALMER C. E. (1946) *Radiology* **47**, 59.

ZOLLINGER H. U. and HEGGLIN R. (1959) *Schweiz. med. Wschr.* **88**, 439.

ZUCKERKANDL E. (1881) *S.-B. Akad. Wiss. Wien, math.-nat. Kl.*, Abt. 3, **84**, 110.

ZUCKERKANDL E. (1883) *S.-B. Akad. Wiss. Wien, math.-nat. Kl.*, Abt. 3, **87**, 171.

ZUELZER W. W. (1944) *Arch. Path.* **38**, 1.

ZUELZER W. W. and STULBERG C. S. (1952) *Amer. J. Dis. Child.* **83**, 421.

ZUNDEL W. E. and PRIOR A. P. (1971) *Thorax* **26**, 357.

Index

Page references in **bold figures** indicate page of commencement of item